Paleoseismology

This is Volume 95 in the
INTERNATIONAL GEOPHYSICS SERIES
A series of monographs and textbooks
Edited by RENATA DMOWSKA, DENNIS HARTMANN and H. THOMAS ROSSBY

A complete list of books in this series appears at the end of this volume.

Paleoseismology

2nd edition

Edited by

James P. McCalpin

GEO-HAZ CONSULTING, INC.
ESTES PARK, COLORADO

AMSTERDAM • BOSTON • HEIDELBERG • LONDON
NEW YORK • OXFORD • PARIS • SAN DIEGO
SAN FRANCISCO • SINGAPORE • SYDNEY • TOKYO
Academic Press is an imprint of Elsevier

ELSEVIER

Academic Press is an imprint of Elsevier
30 Corporate Drive, Suite 400, Burlington, MA 01803, USA
525 B Street, Suite 1900, San Diego, California 92101-4495, USA
84 Theobald's Road, London WC1X 8RR, UK

Library of Congress Cataloging-in-Publication Data
Paleoseismology / edited by James P. McCalpin. – 2nd ed.
    p. cm.
  Includes bibliographical references and index.
  ISBN 978-0-12-373576-8
1. Paleoseismology. I. McCalpin, James.
  QE539.2.P34P35 2008
  551.22–dc22   2009013122

British Library Cataloguing-in-Publication Data
A catalogue record for this book is available from the British Library.

ISBN: 978-0-12-373576-8

For information on all Academic Press publications
visit our Web site at www.elsevierdirect.com

Printed and bound by CPI Group (UK) Ltd, Croydon, CR0 4YY

Transferred to Digital Print 2011

Contents

Supplemental materials, including a chapter on *Applications of Paleoseismic Data to Seismic Hazard Assessment* and the book's Appendices and References, can be found on the companion website at http://www.elsevierdirect.com/companions/9780123735768

Contributors

Gary A. Carver   Carver Geologic, Kodiak, Alaska, USA

Chris Goldfinger   College of Oceanic and Atmospheric Sciences, Oregon State University, Corvallis, Oregon 97331 5503, USA

William R. Hackett   Consulting Geologist, 2007 Cherokee Circle, Ogden, Utah 84403, USA

Randall W. Jibson   U.S. Geological Survey, Golden, Colorado 80401, USA

James P. McCalpin   GEO-HAZ Consulting, Inc., Crestone, Colorado 81131, USA

Alan R. Nelson   U.S. Geological Survey Golden, Colorado 80402, USA

Stephen F. Obermeier   Emeritus, U.S. Geological Survey, EqLiq Consulting, Rockport, Indiana 47635, USA

Suzette J. Payne   Idaho National Laboratory, Idaho Falls, Idaho 83415-2203, USA

Thomas K. Rockwell   Department of Geological Sciences, San Diego State University, San Diego, California 92182-1020, USA

Richard P. Smith   Consulting Geologist, Nathrop, Colorado 81236, USA

Ray J. Weldon II   Department of Geological Sciences, University of Oregon, Eugene, Oregon 97403-1272, USA

Preface to the Second Edition

This edition represents a major revision and expansion of the 1996 edition, in order to respond to the explosive growth of paleoseismic research in the past 13 years. The team of 1st-Edition contributors has updated their respective chapters and, with 13 more years of experience, provide a somewhat more integrated, global, and (hopefully) mature view of their subjects than in the 1st Edition. Unfortunately, Chapter 9 and the all-chapter Reference List have been omitted from the hard-copy book and are available only as on-line Website content from Elsevier.com; this decision was solely by the publisher.

The period 1996–2009 included numerous deadly earthquakes worldwide on faults that had not experienced significant historic seismicity, and which could only have been recognized and characterized with the tools of paleoseismology. Almost all of these deadly earthquakes were triggered by reverse faults, many of them blind thrusts. Therefore, we added new material to this 2nd Edition on how to identify and characterize coseismic folding and blind thrusting, using paleoseismic methods. Another recent need has arisen, due to a worldwide energy crisis, for paleoseismic investigations as part of Seismic Hazard Assessments (SHA) of all types for energy projects such as power plants (including nuclear), dams, pipelines, and waste disposal sites (also nuclear).

This edition acknowledges the "Digital Revolution" that covers imaging ("geoinformatics"; Sinha, 2005), monitoring of earth surface processes (Vita-Finzi, 2002), data processing, and modeling. Paleoseismologists can now sit in their offices and bring up spatial datasets on desktop GIS (geographic information) systems from all over the world, and integrate them their own field observations via a common georeferencing scheme. They can then subject the data layers to 2D and 3D visualization and computer modeling, in ways undreamed of 13 years ago. Although digital methods are used in all fields of paleoseismology, one field particularly affected by the digital revolution is subaqueous paleoseismology, which is now described in its own chapter (Chapter 2B) by new contributor Chris Goldfinger (Oregon State University).

Throughout this edition we have increased the emphasis on how paleoseismic data (displacement, magnitude, recurrence, slip rate) are used in SHA, both deterministic and probabilistic (Chapter 9), as required by Federal, State, and local laws. This includes: (1) new sections on surface-faulting hazard assessment (e.g., Alquist-Priolo studies in California) and mitigation via setbacks, for various fault types, and (2) extended descriptions of how paleoseismic data are input into logic trees for deterministic and probabilistic SHA. Item (2) includes a new focus on differentiating seismogenic faults from nonseismogenic and nontectonic faults, a topic now discussed in Chapters 3, 5, and 6.

For this edition we created a more standard format for Chapters 3, 5, and 6 (the core chapters on faults) to emphasize the geometric and behavioral settings of faulting; the earthquake deformation cycle in each seismotectonic regime; and statistical properties of data on displacement and recurrence, among other things. We also added an on-line electronic data supplement at Elsevier.com, which includes additional text and graphical data such as color and grayscale versions of all the figures shown in the book, as well as oversized color graphics such as trench logs published in large format sizes.

James P. McCalpin
Crestone, Colorado
May 2009

CHAPTER 1

Introduction to Paleoseismology

James P. McCalpin* and Alan R. Nelson[†]

*GEO-HAZ Consulting, Inc., Crestone, Colorado 81131, USA, mccalpin@geohaz.com
[†]U.S. Geological Survey Golden, Colorado 80402, USA, anelson@usgs.gov

1.1 The Scope of Paleoseismology

1.1.1 Definition and Objectives

Paleoseismology is the study of prehistoric earthquakes, especially their location, timing, and size. Whereas seismologists work with data recorded by instruments during earthquakes, paleoseismologists interpret geologic evidence created during individual *paleoearthquakes*. Paleoseismology differs from more general geologic studies of slow to rapid crustal movements during the late Cenozoic (e.g., neotectonics) in its focus on the almost instantaneous deformation of landforms and sediments during earthquakes. This focus permits study of the distribution of individual paleoearthquakes in space and over time periods of thousands or tens of thousands of years. Such long paleoseismic histories, in turn, help us understand many aspects of neotectonics, such as regional patterns of seismicity and tectonic deformation as well as the seismogenic behavior of specific faults. Paleoseismology also is part of the broader field of earthquake geology, which includes aspects of modern instrumental studies of earthquakes (seismology), tectonics and structural geology, historical surface deformation (geodesy), and the geomorphology of tectonic landscapes (tectonic geomorphology). Books by Yeats *et al.* (1997), Burbank and Anderson (2001), and Keller and Pinter (2002) give different perspectives on the field of paleoseismology.

The driving force behind most paleoseismic studies is society's need to assess the probability and severity of future earthquakes (Reiter, 1995; Gurpinar, 2005). In the decade since the first edition of this book was published in 1996, deadly earthquakes occurred in 1999 in Turkey (17,118 dead), in 2001 in India (20,023 dead), in 2003 in Iran (31,000 dead), in 2004 in Indonesia (>250,000 dead from tsunamis accompanying the earthquake), in 2005 in Pakistan (80,361 dead), and in 2008 in Sichuan province, China (69,000 dead; U.S. National Earthquake Information Center). The 2004 Sumatra–Andaman Islands earthquake in Indonesia was the fourth most deadly earthquake in human history, and strained worldwide relief capacity. With the exception of the Turkish event, these earthquakes occurred on faults that had not generated a surface-rupturing earthquake in historical times or been studied by paleoseismologists. However, even where paleoseismologists have pointed out potentially dangerous faults, local governments have often not used that information to increase public awareness of seismic hazards or mitigate the effects of future earthquakes (e.g., Bilham and Hough, 2006).

International Geophysics, Volume 95

ISSN 0074-6142, DOI: 10.1016/S0074-6142(09)95001-X

Before 1980, the assessment of earthquake hazard in industrialized countries such as the United States, Japan, and the USSR was based almost solely on the historical earthquake record. Although many geologists (e.g., Allen, 1975; Research Group for Active Faults, 1980) pointed out the danger in this approach, early maps of predicted strong ground motion were derived solely from historical data. In the twenty-first century most countries with seismically active faults consider paleoseismic data in both regional (e.g., Stuchi, 2004, http://zonesismiche.mi.ingv.it/; The Headquarters for Earthquake Research Promotion, 2005, http://www.jishin.go.jp/main/index-e.html; Petersen *et al.*, 2008, http://pubs.usgs.gov/of/2008/1128/) and site-specific seismic hazard analyses (Gurpinar, 2005). In the USA, the Quaternary Fault and Fold Database of the United States (U.S. Geological Survey, 2006) contains a summary of available paleoseismic data for most known active faults.

Paleoseismology supplements historical and instrumental records of seismicity by characterizing and dating large prehistoric earthquakes. In many countries such as the USA, useful seismicity records extend back only a few centuries (Gutenberg and Richter, 1954; Stover and Coffman, 1993) and many active fault zones have no historical record of large earthquakes. For example, studies of prehistoric faulting along the Wasatch fault (Utah) show that the average recurrence interval between magnitude-7 earthquakes is probably three times longer than the 145-year period of historical settlement (McCalpin and Nishenko, 1996). In Europe, catalogs of historical large earthquakes are often considered complete back four to five centuries; for example, the Italian catalog (Stuchi *et al.*, 2004) is considered complete for Intensity 8 and above back to the year 1600. Even in parts of China and the Middle East where earthquake catalogs extend back thousands of years (Ambraseys, 1982; Gu *et al.*, 1989), historical observations are insufficient to identify all seismogenic faults. On a fault that has slipped episodically (Figure 1.1) for many hundreds of thousands of years, even a 3000-year earthquake history such as China's covers only a tiny fraction of the history of the fault. Much of the seismic history of most major faults is accessible only through the techniques of paleoseismology.

For the most part, the paleoseismic record is a record of large (moment magnitude, $M_w > 6.5$) or great ($M_w > 7.8$) earthquakes because geologic evidence of small and moderate-sized earthquakes is rarely created or preserved near the surface. Evidence of past earthquakes can range from local deformation of the ground surface along a crustal fault (fault scarps, sag ponds, laterally offset stream valleys, monoclinally folded marine terraces, scarp-dammed lakes), to indicators of the sudden uplift or subsidence of large regions above a plate-boundary fault (warped river terraces, uplifted or subsided shorelines, drowned tidal marshes), to stratigraphic or geomorphic effects of strong ground shaking or tsunamis far from the seismogenic fault (landslides, rockfalls, liquefaction features, tsunami deposits). A characteristic of most such features is that they formed instantaneously (from a geologic perspective) during or immediately after an earthquake.

Features (deposits or landforms) formed during an earthquake are described as *coseismic* and are commonly contrasted with *nonseismic* features formed by processes of erosion, deposition, and deformation unrelated to large earthquakes. For example, seismogenic faults may creep between earthquakes or slip small amounts during small to moderate earthquakes that leave no signs of sudden slip. For this reason, *nontectonic* and *nonseismic* are better adjectives than *aseismic* (no detectable seismicity) for features unrelated to fault slip or strong earthquake shaking. The term *aseismic* should be restricted to seismology.

Paleoseismologists can only study earthquakes that produce recognizable deformation (in the form of deformed stratigraphic units, displaced landforms, or earthquake-induced sedimentation). Vittori *et al.* (1991) and many succeeding European authors propose that the term *seismites* be used to describe all geologic structures and sediments genetically related to earthquakes. However, such usage poses two

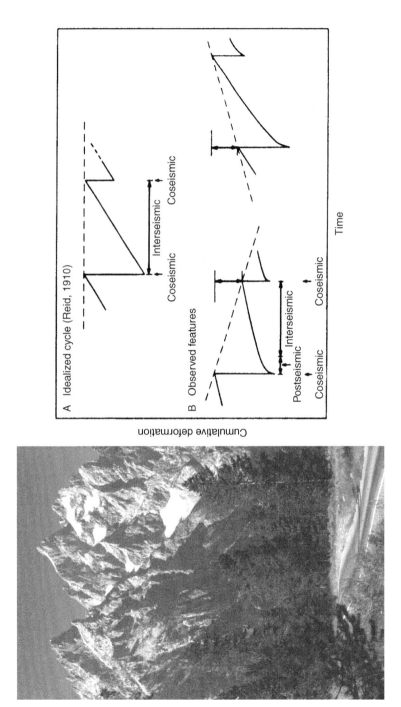

Figure 1.1: (A) The Teton Range, Wyoming (background) has been uplifted 2000 m above the floor of Jackson Hole (foreground) by hundreds of Neogene surface-rupturing earthquakes on the Teton normal fault; (B) Simplified forms of the earthquake deformation cycle. Cumulative deformation (e.g., strain, tilt, displacement) is plotted as a function of time. Step offsets correspond to the occurrence times of major earthquakes. Dashed lines show level at which failure occurs; the level varies with the effects of long-term inelastic deformation. [Part B from Thatcher (1986b). Reprinted with permission from Active Tectonics. Copyright 1986 © by the National Academy of Sciences. Courtesy of the National Academy Press, Washington, D.C.]

problems: (1) the original definition of seismites referred only to stratigraphic units containing sedimentary structures produced by shaking (Seilacher, 1969; AGI, 2007) and (2) there is considerable uncertainty in relating various geologic structures and strata to earthquakes (Wheeler, 2002). Thus, we suggest that the term "seismites" be restricted to its original definition.

Another ambiguous term used in paleoseismology is "event" (used without a modifier) or "event horizon." Many paleoseismic publications use the term *event* too freely as a synonym for *earthquake*. Erosional, depositional, and deformational "events" are only inferred responses to earthquakes; commonly it is unclear whether an earthquake or some other type of "event" is being discussed. Including a modifier with "event" avoids some of this ambiguity (e.g., Scharer *et al.*, 2007). For example, "fracturing events" are extensively documented in fault zone trenches near Yucca Mountain, Nevada, in the western U.S. (Keefer *et al.*, 2004), and the addition of "fracturing" to "event" expresses the uncertainty about whether the fracturing resulted from coseismic slip on the exposed fault, slip triggered by movement on an adjacent coseismic fault, shaking-induced compaction, or nonseismic compaction. In a similar example, to avoid implying a tsunami origin for all anomalous sandy beds in a 7000-year sequence of lake sediment on the Oregon coast, Kelsey *et al.* (2005) termed the processes that deposited the beds "disturbance events." Even greater ambiguity may occur with the widely used term "event horizon," which combines the ambiguity of "event" with the uncertainty of "horizon," a term that may mean a former surface in some paleoseismologic contexts, a bed with finite thickness in others, and is easily confused with the "horizons" of soils that are central to many paleoseismologic interpretations. Among the specific terms that paleoseismologists might use instead of "event horizon" include "unconformity" and "disconformity," terms widely used in geology for more than a century (AGI, 2007). "Earthquake horizon" (e.g., Scharer *et al.*, 2007) avoids ambiguity as long as the meaning of "horizon" is clear.

Paleoearthquakes are *prehistoric* by definition, but does "prehistoric" mean the time before oral records, or the time before contemporaneous written accounts, or the time before written accounts with some quantitative observations of earthquakes? Most paleoseismologists follow the latter definition for "prehistoric." This broad definition reduces problems caused by uncertainty in the times of transition from oral to written history around the globe; for example, from 1831 BC in parts of China (Gu *et al.*, 1989), from 550 BC in the eastern Mediterranean (Ambraseys and White, 1997), and from later than AD 1700 in New Zealand and northwestern North America (Stover and Coffman, 1993). Archaeology has contributed much to the understanding of the history of large earthquakes in some regions (Vita-Finzi, 1986); much of what we know of the seismic history of the Middle East and Mediterranean regions before the Christian era has come from *archaeoseismic* investigations (e.g., Stiros and Jones, 1996; see Chapter 2A). The older boundary on the time interval encompassed by paleoseismology studies is commonly the middle (4–6 ka, that is, 4000–6000 years ago) to early (7–10 ka) Holocene, but records of individual earthquakes may extend back into the late Pleistocene in regions of long recurrence (e.g., Crone *et al.*, 2003; Keefer *et al.*, 2004) or unusually well-preserved evidence (e.g., Ota *et al.*, 1993; Marco and Agnon, 2005).

Paleoseismology also enhances our understanding of some large historical earthquakes. As in studies of prehistoric earthquakes, such *historical paleoseismology* studies (Yeats, 1994) concentrate on measuring the amount or lateral extent of surface displacement on faults or describing the size and distribution of landforms or deposits produced by earthquake shaking. Historical paleoseismology includes studies of relatively recent earthquakes (e.g., AD 1886, Charleston, South Carolina, USA, Obermeier *et al.*, 1990; AD 1857 Ft. Tejon, San Andreas fault, USA: Sieh, 1978a; Harris and Arrowsmith, 2006; 1811–1812, New Madrid, Missouri: Tuttle *et al.*, 2002; AD 1739, Yinchuan, China, Zhang *et al.*, 1986; AD 1703, central Italy earthquakes: Blumetti, 1995), to earthquakes that occurred many centuries ago (AD 1638 and 1783, Calabria, Italy: Galli and Bosi (2002, 2003); AD 1510, Kondayama, Japan: Sangawa (1986); the Early

Byzantine tectonic paroxysm (series of large earthquakes during the Early Byzantine period of the mid-fourth to mid-sixth centuries) between the mid-fourth and mid-sixth centuries A.D: Pirazzoli *et al.* (1996); 31 BC, Jericho, Israel: Reches and Hoexter (1981)). Historical paleoseismology has even shed light on some mythical events, for example the Oracle of Delphi (Piccardi, 2000; de Boer *et al.*, 2001), the collapse of Mycenaean civilization in ancient Greece (French, 1996), and the destruction of Jericho in the Middle East in 1550 BC (Nur, 1991).

1.1.2   Organization and Scope of This Book

Paleoseismology is a field-oriented discipline and this book reflects that focus. The book chapters are organized by tectonic environment and by whether paleoseismic evidence is primary or secondary (after the distinction of Richter, 1958). As explained in Section 1.2, primary evidence reflects seismic surface faulting or folding, whereas secondary evidence is usually a response to strong ground shaking. Primary and secondary evidence are further grouped into on-fault and off-fault evidence, and as to whether the evidence consists of stratigraphic or geomorphic features. In this chapter, we explain the scope of paleoseismology, its relationship to other fields of seismology and geology, and briefly outline its early development. In Chapter 2A we describe and comment on paleoseismic field methods applicable to the study of paleoseismic evidence on land, including various surface mapping, geophysical, and subsurface exploration techniques. Chapter 2B describes unique methods and concepts applicable to offshore paleoseismic studies. Chapters 3–6 cover the application of common paleoseismic methods to the development of earthquake histories in extensional, volcanic, compressional, and strike-slip tectonic environments, respectively. Techniques for studying secondary features, such as landslides and liquefaction features, are similar in all on-land tectonic environments, including the intraplate interiors of continents where tectonic processes proceed slowly and easily studied surface fault traces are rare or absent. Chapter 7 deals primarily with liquefaction features and their interpretation, and includes discussion of the problematic field of soft-sediment deformation features. Chapter 8 describes landslide evidence for prehistoric earthquakes, including movement on "sackungen" scarps in mountainous terrain. In Chapter 9 (See Book's companion web site) we show how paleoseismic data contribute to the understanding of regional tectonic frameworks, through a brief discussion of seismogenic models, empirical sets of worldwide data, and case studies. Only when paleoseismic data are viewed within such frameworks can they form the basis for comprehensive seismic hazard assessments.

1.1.3   The Relation of Paleoseismology to Other Neotectonic Studies

Paleoseismology is a subdiscipline within the much broader fields of *neotectonics, active tectonics, and earthquake geology* (Figure 1.2 and Table 1.1). Paleoseismology adapts many concepts from seismology, structural geology, and tectonics, but its field methods and techniques are derived primarily from Quaternary geology and related disciplines, such as geomorphology, soil mechanics, sedimentology, archaeology, paleoecology, photogrammetry, age dating, or pedology (soil science). Most paleoseismic field studies require extensive training or experience in Quaternary geology, itself a highly interdisciplinary field. Thus, many advances in paleoseismology have been made by geomorphologists and other Quaternary specialists, working at the interface between tectonics and seismology. Paleoseismology is a particularly successful example of applied Quaternary geology (Wallace, 1986).

Another subfield of neotectonics, closely aligned with paleoseismology, is *tectonic geomorphology* (European usage, *morphotectonics*). Tectonic landforms have long interested geomorphologists

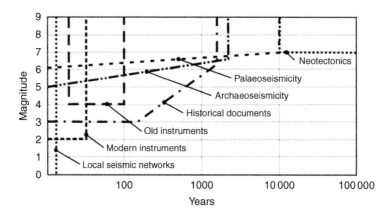

Figure 1.2: The range of earthquake magnitudes detected and time spans covered by different disciplines applied to the study of past earthquakes, From Caputo and Helly (2008), adapted from Levret (2002).

(e.g., Merritts and Ellis, 1994) and by the 1980s tectonic geomorphology was the subject of several conferences and monographs (Ollier, 1981; Yoshikawa *et al.*, 1981; Morisawa and Hack, 1985; Embleton, 1987). Burbank and Anderson (2001) provide a recent comprehensive overview of tectonic geomorphology with excellent graphics. Geomorphology has traditionally focused on surface form and modern surface processes rather than chronology and earth history, and so normally does not yield data on coseismic fold growth, fault displacement, or earthquake recurrence, which are the objective of paleoseismic studies. Weathering and erosion prevent many tectonic landforms, particularly those formed by multiple surface-faulting events over long periods of time, from preserving evidence of individual paleoearthquakes. The development of tectonic geomorphology, including its integration with paleoseismology, tectonics, and seismology, is discussed by Merritts and Ellis (1994). More recently, tectonic geomorphology forms the basis for Michetti *et al.*'s (2005) concept of the "seismic landscape," an assemblage of landforms that uniquely reflects the slip sense and slip rate of local faults (Figure 1.3).

Theoretical models that describe the seismogenic behavior of faults, developed from a wide variety of studies in seismology, geodesy, rock mechanics, and structural geology (Scholz, 2002), provide another type of framework for interpreting paleoseismic histories for comprehensive seismic hazard analyses. Perhaps the two most important models, covered more fully in Chapter 9 (See Book's companion web site) describe the *segmentation* of faults and the *earthquake deformation cycle* (Figures 1.1 and 1.4). In the segmentation model, large earthquakes, commonly of a *characteristic* size, repeatedly rupture the same part or segment of a fault, less commonly extending into adjacent segments (Schwartz and Coppersmith, 1984). In many plate-boundary settings, segments of faults have been modeled with a *time-predictable* cycle of tectonic strain accumulation followed by strain release during characteristic earthquakes (Shimizaki and Nakata, 1980; Thatcher, 1986a; Stein *et al.*, 1988). Other studies have emphasized the differing characteristics of successive earthquakes, and long-term patterns of earthquake clustering have been proposed (Sieh *et al.*, 1989; Thatcher, 1990; Weaver and Dolan, 2000; Weldon *et al.*, 2004; Kelsey *et al.*, 2005; Mazzotti and Adams, 2005; Zhang *et al.*, 2005; Dolan *et al.*, 2007; Chapter 9, See Book's companion web site).

Table 1.1: Comparison of definitions for paleoseismology, and related fields

Paleoseismology (British spelling, *palaeoseismology*)	The study of prehistoric earthquakes, especially their location, timing, and size (this book)
	The study of the timing, location, and size of prehistoric earthquakes. Paleoseismology differs from other aspects of earthquake geology in its focuses on the almost instantaneous deformation of landforms and sediments during individual earthquakes[a]
	The subdiscipline of geology that employs features of the geological record to deduce the fault displacement and age of individual, prehistoric earthquakes (NRC, 2003)
	The geological investigation of individual earthquakes decades, centuries, or millennia after their occurrence (Yeats and Prentice, 1994; also Aki, 2003)
	The study of ground effects from past earthquakes as preserved in the geologic and geomorphic record[b]
Historical paleoseismology	The establishment of lengths and displacements of ruptures on specific active faults due to historical earthquakes (Yeats, 1994)
Paleoseismicity	Prehistoric earthquakes resulting in slip on faults (Engelder, 1974)[c]
Earthquake geology	In the broad sense, is the study of the history, effects, and mechanics of earthquakes within and on the Earth's crust. Most often, earthquake geology is synonymous with active tectonics, a term used to describe the study of tectonic movements that are expected to occur within a future time span of concern to society. Important aspects of earthquake geology include the study of tectonic landforms on the Earth's surface and folds and faults within its crust produced by many earthquakes over thousands to millions of years[a]
	… commonly regarded as synonymous with neotectonics (Yeats *et al.*, 1996, p.4)
Neotectonics	The study of the post-Miocene structures and structural history of the Earth's crust[c]
Active tectonics	Tectonic movements that are expected to occur within a future time span of concern to society (Wallace, 1986). See book of the same title by Keller and Pinter (2002)
Morphotectonics	See tectonic geomorphology[c]
Seismotectonics	Study of the role of seismic activity in tectonics; includes examinations of the processes precursory to and accompanying earthquakes, the regionally significant geologic structures generated by earthquakes, and the temporal or spatial variations in processes or structures[c]

(Continued)

Table 1.1: Comparison of definitions for paleoseismology, and related fields (Cont'd)

Tectonic geomorphology	The tectonic interpretation of the morphological or topographic features of the Earth's surface; it deals with their tectonic or structural relations and origins, rather than their origins by surficial processes of erosion and sedimentation. Cf: orogeny. Obsolescent and less preferred synonym: *morphotectonics*[c]
	The unrelenting competition between tectonic processes that tend to build topography and surface processes that tend to tear them down[d]

[a] USGS web site http://earthquake.usgs.gov/research/topics.php?areaID = 10.
[b] Michetti *et al.* (2005).
[c] Glossary of Geology, on-line version, http://glossary.agiweb.org.
[d] Burbank and Anderson (2001).

1.2   Identifying Prehistoric Earthquakes from Primary and Secondary Evidence

1.2.1   Classification of Paleoseismic Evidence

We modify earlier classifications from Russia (Solonenko, 1970, 1973; Nikonov, 1988a, 1995a,b), USA (Richter, 1958, pp. 80–84) and Japan (Hagiwara, 1982), to create three hierarchical levels of evidence based on genesis, location, and timing (Table 1.2). At the highest level of classification, evidence is either *primary* or *secondary*. Primary paleoseismic evidence is produced by tectonic deformation resulting from coseismic slip along a fault plane (including growth of fault-related folds) and is equivalent to "seismotectonic deformation" in the Russian system (Nikonov, 1995a). Secondary paleoseismic evidence is produced by earthquake shaking, or by erosional and depositional responses to shaking and coseismic elevation changes. At the second level of classification, paleoseismic features are further distinguished as being on or above a fault trace (*on-fault* or near-field features), or away from or far above a fault trace (*off-fault* or far-field features). A third level of classification distinguishes *instantaneous* features formed at the time of the earthquake (coseismic) from *delayed response* (postseismic) features formed by geological processes after coseismic deformation and seismic shaking cease. Finally, paleoseismic evidence is preserved as either landforms (*geomorphic evidence*) or as deposits and structures (*stratigraphic evidence*), and this distinction often determines how we approach paleoseismic field investigations.

This combined genetic and descriptive classification scheme yields 16 categories of paleoseismic evidence (Table 1.2). The geomorphic approach commonly involves inferring the amount of fault displacement during paleoearthquakes from measurements of landform deformation. A strength of geomorphic studies is that many measurements can be made over a large area, but geomorphic evidence alone rarely provides precise ages for paleoearthquakes. The stratigraphic approach focuses on inferring fault displacement and earthquake recurrence by measuring and dating deformed strata in exposures. However, because detailed data are typically obtained from only a few sites, it is commonly unclear to what extent site data represent deformation throughout the length and width of a fault zone. Both approaches are subject to ambiguity in interpretation because both seismic and nonseismic processes can create landforms and deformation

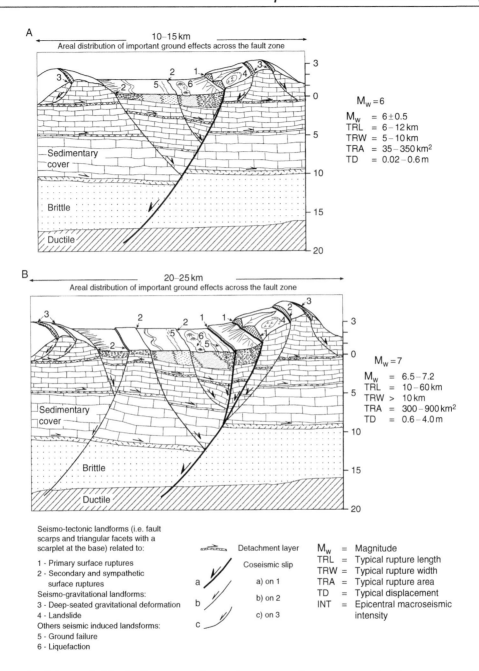

Figure 1.3: Schematic block diagrams of two Quaternary intermontane basins associated with characteristic earthquakes of $M_w = 6$ (A, upper) and $M_w = 7$ (B, lower). Diagrams illustrate the typical seismo-tectonic and seismo-gravitational landforms and underlying structures created by repetitive surface-rupturing earthquakes. On the right side are surface faulting parameters. This ensemble of primary and secondary paleoseismic evidence contributes to a unique "seismic landscape" (from Dramis and Blumetti, 2005).

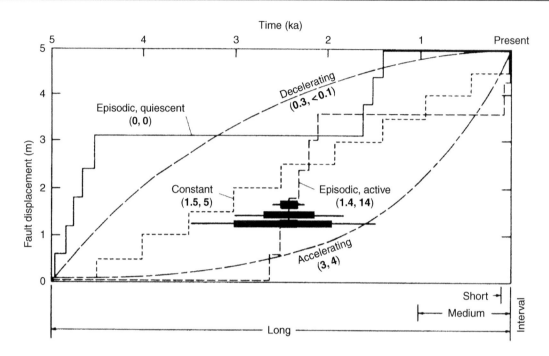

Figure 1.4: Five patterns of fault displacement versus time. Slip rates for all five patterns are the same over the long time interval (bottom), but vary when viewed over medium or short time intervals. Numbers in parentheses under each pattern label are medium-term and short-term rates for that pattern. In the three patterns with a "staircase" shape (episodic and quiescent patterns), vertical steps represent coseismic displacements like those shown in Figure 1.1. Filled squares with age error bars demonstrate the increasing difficulty of distinguishing among the different patterns of fault slip with increasingly larger errors on ages for fault displacements. Modified from Pierce (1986).

structures that may appear similar. These two complimentary approaches to field investigations (geomorphic versus stratigraphic) form the basis for the subdivisions of Chapters 3–6.

Primary paleoseismic evidence is commonly easy to associate with a particular causative fault—a fresh fault scarp along the base of a mountain front, for example, clearly indicates surface rupture of at least the section of the fault marked by the scarp. Fault scarps, fissures, and folds along the trace of a fault are typical examples of primary, on-fault, instantaneous (geomorphic) evidence (Figure 1.5).

In outcrops or trenches, corresponding stratigraphic evidence consists of displaced or folded strata, zones of sheared sediment, and fissures (Chapters 3, 5, and 6). Primary, off-fault, instantaneous evidence commonly takes the form of uplifted, subsided, or tilted surfaces some distance from the fault (as long as this deformation is directly produced by seismic slip on a fault at depth). Zones of regional uplift or subsidence during the largest subduction-zone earthquakes, measuring many tens of kilometers wide and hundreds of kilometers long, are unmistakable primary evidence of slip along this type of plate-boundary fault (Plafker, 1972; Chieh *et al.*, 2007; Chapter 5). More localized surface deformation caused by slip on shallower crustal faults that do not break the surface (*blind faults*) may be more difficult to associate with a particular fault (Yeats, 1986b; Chapters 5 and 6). In contrast, primary, delayed response evidence occurs

Table 1.2: Hierarchical classification of paleoseismic evidence, with examples of features[a]

Level 1: Genesis	Primary (Chapters 3–6) (created by tectonic deformation)			
Level 2: Location	On-fault		Off-fault	
Level 3: Timing	Instantaneous (coseismic)	Delayed response (postseismic)	Instantaneous (coseismic)	Delayed response (postseismic)[b]
Geomorphic expression	1. • Fault scarps • Fissures • Folds • Moletracks • Pressure ridges	2. • Afterslip contributions to features at left • Colluvial aprons	3. • Tilted surfaces • Uplifted shorelines • Subsided shorelines	4. • Tectonic alluvial terraces • Afterslip contributions to features at left
Stratigraphic expression	5. • Faulted strata • Folded strata • Unconformities or disconformities	6. • Scarp-derived colluvial wedges • Fissure fills	7. • Tsunami deposits and erosional unconformities caused by tsunamis	8. • Erosional unconformities and deposits induced by uplift, subsidence, and tilting
Abundance of similar nonseismic features	Few	Few	Some	Common
Level 1: Genesis	Secondary (Chapters 7–8) (Created by seismic shaking)			
Level 2: Location	On-fault		Off-fault	
Level 3: Timing	Instantaneous (coseismic)	Delayed response (postseismic)	Instantaneous (coseismic)	Delayed response (postseismic)
Geomorphic expression	9. • Sand blows • Landslides and lateral spreads in the fault zone • Disturbed trees and tree-throw craters	10. • Retrogressive landslides originating in the fault zone	11. • Sand blows • Landslides and lateral spreads beyond the fault zone • Disturbed trees and tree-throw craters • Fissures and Sackungen • Subsidence from sediment compaction	12. • Retrogressive landslides beyond the fault zone

(Continued)

Table 1.2: Hierarchical classification of paleoseismic evidence, with examples of features[a] (Cont'd)

	13.	14.	15.	16.
Stratigraphic expression	• Sand dikes and sills • Soft-sediment deformation • Landslide toe thrusts	• Sediments deposited from retrogressive landslides	• Sand dikes • Filled craters • Soft-sediment deformation structures • Turbidites	• Erosion or deposition (change in sedimentation rates) in response to retrogressive landslides or surface features such as fissures, lateral spreads, or sand blows, or other forms of landscape disturbance
Abundance of similar nonseismic features	Some	Very common	Some	Very common

[a] This classification scheme yields 16 types of paleoseismic features, as numbered consecutively in the categories "Geomorphic expression" and "Stratigraphic expression."

[b] Does not include delayed response movement on other faults due to stress changes induced by initial faulting.

Figure 1.5: Photograph of three paleoseismologists scaling the fault scarp produced during the 1957 Gobi Altai earthquake (from top to bottom, Tom Rockwell/Kelvin Berryman/Shmulik Marco). This photo exemplifies how modern paleoseismologists, like most scientists, "stand on the shoulders of giants." Photo courtesy of Carol Prentice (1995).

days to hundreds of years after the earthquake, and results from erosion or deposition induced by uplift, subsidence, lateral displacement, or tilting. Although tsunami deposits may be found thousands of kilometers from the earthquake source, tsunamis are produced by primary fault displacement of the seafloor, so tsunami deposits are considered primary off-fault evidence (Chapter 5). Whether recorded by geomorphic or stratigraphic evidence, on-fault evidence commonly indicates greater deformation than off-fault evidence and, therefore, off-fault evidence is more easily confused with features produced by nonseismic processes (Table 1.2). For example, anomalously rapid stream aggradation has been observed after earthquakes, due to either tectonic flattening of the stream's gradient (Harden and Fox, 1994) or overloading the stream with earthquake-induced landslide debris (Keefer, 1999; Almond *et al.*, 2000; Dadson *et al.*, 2004). However, the normal cause of stream aggradation is changes in the sediment:water ratio unrelated to earthquakes.

Primary evidence, especially along surface faults, was the major focus of most early paleoseismic investigations. But a major weakness of primary evidence, when used in seismic hazard assessment, is that it provides no direct evidence of the strength of shaking or the spatial extent of ground failures induced by paleoearthquakes. Using primary evidence alone, these critical components of hazard assessment can only be estimated *indirectly* by applying empirical relations between rupture dimensions, earthquake magnitude, and ground motion and ground failure to off-fault sites (Chapter 9, See Book's companion web site).

Secondary paleoseismic evidence consists of diverse phenomena, many caused by earthquake shaking. Geomorphic examples include sand blows, rockfalls, landslides, water level changes, and damaged trees; stratigraphic examples include sand dikes, and other liquefaction features, load structures in soft sediment, beds that record anomalous siltation events in lakes, and turbidity current deposits (Lucchi, 1995; Ettensohn *et al.*, 2002; Chapters 7 and 8). The distinction between instantaneous and delayed response evidence is more gradational for secondary evidence than for primary evidence because some features (e.g., sand blows, beds of rapidly deposited silt in lakes) may originate during shaking and continue to form for many hours after an earthquake. In other cases, the response of local geomorphic systems does not even begin until earthquake shaking ceases, but the length of the delay varies widely, ranging from minutes (large perennial streams incising into fault scarps) to hours (tsunamis depositing sand in coastal marshes) to years or decades (e.g., development of erosional marine or alluvial terraces, headward erosion above the headscarp of a coseismic landslide, depositional changes on alluvial fans [Keefer, 1999], or rapid building of dunes on beaches adjacent to streams choked with earthquake dislodged sediment). For the same reasons, secondary paleoseismic features are more difficult to distinguish from features produced by nonseismic processes than are primary features. The greatest ambiguity arises with secondary, off-fault, delayed response landforms and deposits (Table 1.2), as described more fully in Chapters 2–6. As we incorporate more secondary, delayed response evidence in constructing paleoseismic histories, there should be correspondingly more detailed studies of similar nonseismic features in differing tectonic, climatic, and geomorphic environments (e.g., Bull, 1991; Wheeler, 2002).

The distinction between primary and secondary evidence is not always clear (Richter, 1958, p. 83). For example, slip on shallow faults may be induced by strong ground motions rather than by slip on seismic faults (or faults connected to them), or may be caused by folding above a seismic fault. The surface features thus look like primary effects of seismic fault rupture, but are actually secondary effects (Yeats *et al.*, 1997; Chapter 5).

A limitation of much secondary evidence is that the seismogenic fault responsible for secondary deformation cannot be identified because a given strength of prehistoric ground shaking could have been generated by a nearby, small earthquake or a distant, larger one. However, secondary evidence can also commonly be used more directly than primary evidence to estimate the severity and spatial distribution of paleoearthquake ground motions. With some types of secondary evidence, such as sand blows or lateral spread landslides, one may be able to calculate quantitative measures of ground motion (Chapters 7 and 8) or the recurrence time of strong ground shaking at the site.

A recent trend in regions with low slip-rate faults or where most evidence is off-fault or secondary are "multi-archive" paleoseismic studies, which use many types of off-fault paleoseismic evidence, in addition to any on-fault evidence. As the number of types or sites of evidence for an earthquake increase, investigators' confidence that the evidence records an earthquake increases (e.g., Nelson *et al.*, 1996; Becker *et al.*, 2005; Scharer *et al.*, 2007).

The study of *modern analog* features, produced as a result of historical earthquakes, is a cornerstone of paleoseismology. Most types of geomorphic and stratigraphic evidence used to identify paleoearthquakes were first observed during or following large historical earthquakes. Because paleoseismic interpretations are often more rooted in analogy than theory, the more familiar paleoseismologists are with the effects of large, historical earthquakes, the better prepared they will be to recognize and correctly interpret the wide variety of features encountered in paleoseismic studies. Examples of well-studied historical earthquakes are cited in Chapters 3–8. Paleoseismologists should be familiar with the features produced by large historical earthquakes in their regions of study or in analogous regions, and they should never pass up an opportunity to visit the epicentral area of a recent large earthquake, where primary and secondary features may be observed firsthand.

1.2.2 The Incompleteness of the Paleoseismic Record

The geologic record of paleoearthquakes is incomplete because (1) many earthquakes are too small to produce observable primary or secondary evidence, (2) special conditions found at few sites are needed to form the most unambiguous types of evidence, and (3) much evidence created in response to even large earthquakes is quickly modified, obscured, or removed by common surficial processes. These three factors mean that only earthquakes of a given type and magnitude (size; see Appendix 1 of this book's companion web site for an explanation of magnitude scales) in a particular tectonic setting will result in identifiable primary or secondary evidence at a particular site. In addition to geophysical factors such as the depth of an earthquake or the attenuation characteristics of its seismic waves, the lower magnitude limit for earthquakes that produce identifiable paleoseismic evidence depends on two types of thresholds—creation thresholds and preservation thresholds—which vary greatly from site to site (e.g., Nelson *et al.*, 2006). To exceed creation thresholds, earthquake evidence must be distinct from similar evidence that might be produced by nonseismic processes in the same tectonic and geomorphic setting. Relatively flat surfaces in arid areas unobscured by vegetation make even small fault scarps easily mappable, whereas similar scarps go undetected in rugged, heavily forested terrain (Chapter 3). Sand blows and dikes require a capping layer of low permeability sediment into which they can intrude (Chapter 7). Tsunamis rarely deposit identifiable beds unless abundant sand lies in their paths (Chapter 5). Lateral slip along strike-slip faults may be difficult to recognize unless stream valleys, shorelines, or other linear landforms cross the fault at a high angle. In a similar manner, the absence of earthquake-induced ground failures at a site may reflect a lack of material susceptible to failure, rather than an absence of strong ground shaking (Chapter 8). For example, fluctuations in the groundwater table may make sediment susceptible to liquefaction for only part of the year. However, where surface materials can be reasonably inferred to have been susceptible to shaking-induced failure for long periods of time and have not failed, this negative evidence is a strong indication that ground shaking has not exceeded certain limits (see Chapters 7 and 8). Compilations of historical data indicate that earthquakes smaller than M_w 5 have not ruptured the surface (Wells and Coppersmith, 1994; Stirling *et al.*, 2002) and their ground motions are rarely strong enough to produce noticeable geologic effects such as landslides or liquefaction (Jibson and Keefer, 1993). Thus, even for shallow earthquakes, M_w 5 is probably a lower limit for detecting paleoearthquakes from surface evidence (Chapter 9, See Book's companion web site).

Many secondary types of paleoseismic evidence have potentially lower creation thresholds than does primary evidence (Figure 1.6). Even at long durations of shaking, ground accelerations on the order of 0.1 g are necessary for liquefaction (Chapter 7). For a site located directly over the hypocenter of an earthquake, these accelerations can be produced by earthquakes as small as M_w 5 (Joyner and Boore, 1988), which means that the creation threshold for producing liquefaction at the best paleoliquefaction sites is similar to the lower magnitude limit for producing significant earthquake damage. Keefer (1984) suggests that even smaller earthquakes may trigger various types of landslides (Chapter 8).

Once paleoearthquake evidence is created, it must also be preserved long enough and well enough for paleoseismologists to identify it. The threshold of preservation for particular evidence at a particular site is the point at which the balance among erosion, deposition, and other processes (such as bioturbation or soil

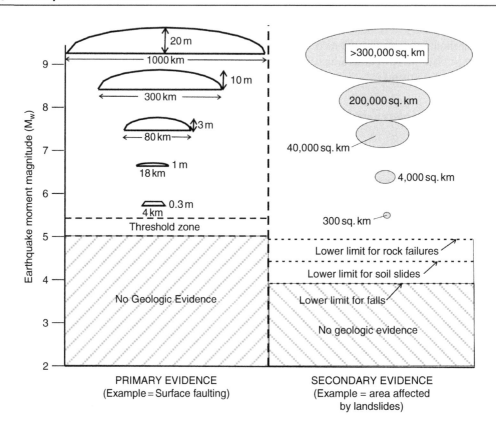

Figure 1.6: Schematic diagram showing the increase in size and distribution of primary and secondary paleoseismic evidence with increasing earthquake moment magnitude (M_w), based on measurements following historical earthquakes. The left side of the diagram shows the dimensions of surface faulting (primary evidence) observed in historical earthquakes of various magnitudes. Shaded areas schematically represent the dimensions of surface deformation but are not to scale. Values for surface rupture lengths (beneath shaded areas) and maximum displacement (to right of shaded areas) are from Wells and Coppersmith (1994). The lower magnitude limit of surface faulting earthquakes follows Bonilla (1988). The right side of the diagram shows areas affected by coseismic landsliding (an example of secondary evidence) from Keefer (1984); areas are not to scale. The largest area ($>$300,000 km^2) is for the M_w 9.2 1964 Alaska earthquake.

development) at a site begin to favor preservation of identifiable evidence (Figure 1.6). Along coasts that suddenly subside during great subduction-zone earthquakes, the preservation of sand sheets spread by tsunamis accompanying great earthquakes is ensured by quick burial with estuarine mud. But on nonsubsiding coasts impacted by tsunamis, such sheets are commonly removed by during the highest tides or made unrecognizable by root stirring (Chapter 5). Tree throw and other forms of soil bioturbation also commonly destroy evidence of surface faulting and liquefaction in humid climates (Chapters 3, 5, and 7). Due to these types of active surface processes, the most recent prehistoric earthquakes are the most easily studied because evidence of them is most abundant and best preserved. To a large degree, the success of paleoseismic studies depends on the ability of investigators to locate sites where surficial processes quickly bury paleoseismic evidence, or where those processes proceed at rates that at least do not significantly erode or disturb the evidence. The spectacular history of paleoearthquakes on the San Andreas fault developed at Pallet Creek in southern California (Sieh, 1978) is primarily due to the unusual preservation of an exceptionally detailed stratigraphic record.

In many cases, the preservation of primary or secondary evidence is determined by the relative rates of erosion and deposition versus deformation (e.g., Schumm, 1986; Wallace, 1986; Bull, 1991). Where deformation rates exceed the rates of geomorphic processes, paleoseismic landforms are created; where the reverse is true, if paleoseismic evidence is preserved, it is likely to be stratigraphic (Figure 1.7). For example, small fault scarps on steep slopes or in humid climates (where erosion rates are high) may be eroded or buried in a few hundred years, whereas scarps on flat surfaces or in arid climates (where erosion rates are low) may survive for tens of thousands of years. At present, the number of published studies of paleoearthquake evidence in arid and semi-arid climates dwarfs the number of studies of evidence in humid and polar climates. This unbalanced climatic emphasis, a greater reflection of where paleoseismologists have worked than of where paleoearthquake evidence is preserved, results in an unavoidable emphasis in this book on features formed in dry climates where paleoseismic landforms are easily identified.

1.2.3 *Underrepresentation Versus Overrepresentation of the Paleoseismic Record*

A critical issue in many seismic hazard assessments is the degree to which a particular paleoseismic record underrepresents or overrepresents the number of paleoearthquakes during a given period of time. The difficulty of finding well-preserved paleoseismic features has made *underrepresentation* an accepted limitation of paleoseismology since its beginnings (Sieh, 1981), that is, the paleoseismologist underestimates the number of paleoearthquakes of a given magnitude that have occurred at a site. In contrast, the potential for *overrepresentation* (overestimate of the number of paleoearthquakes) is less widely discussed. The potential for *both* underrepresentation and overrepresentation should always be addressed in reports of paleoseismology investigations.

A number of factors result in the record of large earthquakes in most climatic and tectonic settings being incomplete. Both creation and preservation thresholds for paleoseismic evidence vary—sometimes dramatically—from site to site. In addition, deformation during large recent earthquakes may obscure evidence of smaller, older earthquakes. This phenomenon affects geomorphic evidence, as shown by two examples. Coseismic uplift of several meters can raise a marine abrasion platform well above the zone of wave erosion, but uplift of 0.3 m during a much smaller earthquake may produce only a small marine notch, which will be quickly destroyed by erosion, removing evidence of the later earthquake. In a similar way, a small, steep fault scarp produced by a recent earthquake may be recognizable on a large, gentle scarp produced by an earlier earthquake. But if a large recent scarp forms on an earlier small scarp, erosion may quickly obscure the earlier scarp. Stratigraphic evidence of paleoseismicity can be removed by erosion, which leads to unconformities in the stratigraphic section. In such cases, all the evidence created by paleoearthquakes in the interval of "missing time" has been removed. The initial incompleteness of paleoseismic records caused by variable creation thresholds and increasingly higher preservation thresholds over time means that the longer the span of time considered, the more incomplete records are likely to be (Ager, 1993).

Paleoseismologists may overestimate the number of paleoearthquakes by incorrectly interpreting nonseismic features as evidence of paleoearthquakes (see Chapter 2A, Sec. 2A.4). The number of paleoearthquakes may also be overestimated if earthquakes on a structure are clustered within a time period of well-preserved paleoseismic evidence. Closely spaced earthquakes recorded by evidence during

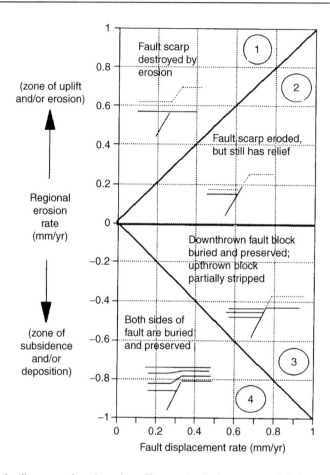

Figure 1.7: Schematic diagram showing the effects of relative rates of deformation versus geomorphic process on the preservation of a fault scarp (an example of primary, on-fault evidence). Many other types of paleoseismic features are subject to the same effects. In quadrant 1 (circled number), the regional erosion rate exceeds the fault displacement rate and the scarp is rapidly destroyed. In quadrant 2, the fault displacement rate is greater than the regional erosion rate, so the scarp is partially eroded yet retains some relief. In quadrant 3, the fault outcrops on a landscape undergoing slow subsidence and deposition, but the scarp is still partially preserved because the fault displacement rate is greater than the regional deposition rate. In quadrant 4, both sides of the fault are buried by sediments deposited at a more rapid rate than the rate of fault displacement. No surface scarp is formed under these conditions, but evidence of past earthquakes is preserved as onlapping strata in the subsurface. In depositional settings, the closer the fault can be traced upward toward the ground surface, the more recent its displacement has been; in erosional settings this is not true, and even old, inactive faults can be traced near or to the ground surface.

the short or medium-term window of Figure 1.4, if extrapolated to longer periods of time, might lead to an overestimate of the number of earthquakes that have occurred.

The issues of the overrepresentation and underrepresentation of the paleoseismic record are frequently complicated by attempts to include qualitative assessments of seismic risk in paleoseismic interpretations.

For example, in instances where the origin of some features in an exposure is ambiguous, all features may be assumed to represent past large earthquakes in order to produce a conservative seismic hazard assessment. Such procedures incorrectly mix paleoseismic interpretations with engineering, societal, and political concerns (Reiter, 1990). The most useful assessments are those where all assumptions are clearly stated and the reasoning behind them fully explained.

Problems arising from overrepresentation and underrepresentation can be reduced by applying large doses of caution when interpreting paleoseismic evidence. Tests of hypotheses should employ several different types of data (e.g., trenching, shallow geophysics, paleontology) with constant attention to possible nonseismic causes for field relations. Multiple, independent lines of evidence pointing to the same hypothesis instill confidence in interpretations. We limit our remaining advice on identifying and interpreting paleoseismic features to the following four quotations:

The great investigator is primarily and preeminently the man who is rich in hypotheses. In the plentitude of his wealth he can spare the weaklings without regret; and having many from which to select, his mind maintains a judicial attitude. The man who can produce but one, cherishes and champions that one as his own, and is blind to its faults. (Gilbert, 1886, p. 287)

I cannot give any scientist of any age better advice than this: the intensity of the conviction that a hypothesis is true has no bearing on whether it is true or not. (Medawar, 1979, p. 39)

Perhaps some kinds of field evidence are destined to remain ambiguous. (Vita-Finzi, 1986, p. 23)

Progress in paleoseismology has been exciting and gratifying; new advances seem to be just ahead. I fear, however, that at times we have exploited some of our new ideas too enthusiastically. In our enthusiasm, our interpretations can be, and some may have been, carried too far, too soon. I believe that some statements that I, and possibly others, have made in public have been too far-reaching, too positive, and have lacked appropriate caveats of uncertainty. (Wallace, 1987, p. 12)

1.3 Prehistoric Earthquake Dating and Recurrence

Dating a paleoearthquake is similar to dating any other geologic event in the Quaternary. That is, the paleoseismologist attempts to bracket the age of paleoseismic evidence with minimum and maximum ages. For stratigraphic evidence, maximum ages are obtained on material dating from before a paleoearthquake, typically in sediment deposited shortly before the earthquake or derived from sediment of that age. Minimum ages are obtained on material younger than the earthquake, ideally on material newly created, or whose age properties were reset, shortly after an earthquake. The approach is the same where landforms are used to date paleoearthquakes; although methods used to date landforms are typically less precise than radiometric methods used to date material in sediment. Landforms displaced vertically or offset laterally by faults, or tilted by folding, are older than an earthquake, whereas landforms created in response to an earthquake are clearly younger. At unusual sites, minimum and maximum ages may closely limit the time of an earthquake. Much more frequently, only maximum ages are available or available limiting ages span a considerable amount of time.

A widening array of Quaternary dating methods is potentially applicable to solving problems in paleoseismology. In the past decade new methods have been developed and other methods have become far more precise, accurate, and applicable to new types of sample materials. Papers about dating methods of potential use to paleoseismologists are published every month. We urge, however, all paleoseismologists to keep abreast of recent developments in radiocarbon dating (Fairbanks *et al.*, 2005), the most widely applicable dating method in paleoseismology. Few other methods are as accurate as ^{14}C dating in the time span of most interest to paleoseismologists (the Holocene). Overviews of methods applicable to dating paleoearthquakes include Noller (2000), Walker (2005), and the dating chapter in Elias' (2007) Encyclopedia of Quaternary Sciences. Detailed reviews of individual dating methods appear in the journals Quaternary International, Quaternary Science Reviews, and Quaternary Geochronology (Elsevier), and more specialized journals, such as Radiocarbon.

Following Colman and Pierce (2000), Quaternary dating methods can be grouped (in order of increasing precision) into relative age, correlated age, numerical age, and calibrated age methods based on the type of result they produce (Table 1.3). *Relative age* methods provide only a relative ranking of ages on an ordinal scale. *Correlated age* methods are not really dating methods; they rely on a comparison to a standardized series of measurements (Rutter *et al.*, 1989). Although they do not yield ages with easily quantified errors, relative age and correlated age methods are of fundamental importance in providing crosschecks on numerical ages and in allowing numerical ages to be applied to other sites that lack numerical age control. *Calibrated ages* are based on systematic changes that depend on environmental variables such as temperature and must be calibrated using numerical ages (this use of the term calibrated differs from that used in radiocarbon dating (Trumbore, 2000; Jull, 2007), which is a numerical-age method). *Numerical age* methods yield ages with stated errors derived from analytical standards. Table 1.3 provides an initial guide to available methods; examples of particular methods appear in the dating sections of later chapters.

When paleoseismologists suspect that a particular dating method may be useful in solving a problem they should contact a specialist in that dating method. Specialists who have current experience with a particular method can provide advice on optimum sampling strategies, sample requirements, costs, and time required for analysis—factors that frequently differ from year to year and from laboratory to laboratory. Samples collected without the advice of a geochronology specialist may not be worth analyzing. Specialists are also aware of other related dating methods, particularly those still in the developmental stage, and they can suggest other specialists or laboratories which may be able to help with dating problems. Keep in mind, however, that experimental dating methods or those in the early stage of development are unlikely to provide paleoseismologists with useful ages unless a substantial budget for a large number of analyses is available. Paleoseismologists should thoroughly discuss all sampling issues with potential dating laboratories *before* spending valuable field time collecting samples.

Table 1.3: Classification of Quaternary Dating Methods Applicable to Paleoseismology[a]

Type of result					
Numerical age		**Calibrated age**		**Relative age**	**Correlated age**
Type of method					
Calendar-year	Isotopic	Radiogenic	Chemical and biological	Geomorphic	Correlation
Historical records	^{14}C	*Luminescence*	Amino acid raceimization	Soil profile development	*Lithostratigraphy*
Dendrochronology	K-Ar and ^{39}Ar-^{40}Ar	Electron spin resonance	Obsidian and tephra hydration	Rock and mineral weathering	*Tephrochronology*
Varve chronology	*Uranium series*		*Lichenometry*	*Progressive landform modification*	Paleomagnetism
	Cosmogenic isotopes other than ^{14}C (^{210}pb, ^{36}Cl)		Soil chemistry	Rate of deposition	*Fossils*
			Rock varnish chemistry	*Relative geomorphic position*	*Artifacts*
					Stable isotopes

[a] Modified from Colman *et al.* (1987). Methods in italics are particularly applicable to dating paleoearthquakes in the Holocene. Thick line indicates the type of result most commonly produced by the methods below it; thin line indicates the type of result less commonly produced by the methods below it.

1.3.1 *Dating Accuracy and Precision and Their Relation to Recurrence*

Few paleoseismic studies adequately discuss the relations among the accuracy and precision of the methods used to date landforms or sediments associated with paleoearthquakes and the times the earthquakes occurred, although the situation has improved since the first edition of this book. Particularly at plate boundaries where earthquakes occur frequently, determining recurrence patterns is commonly limited less by the applicable types of dating methods than by the accuracy and precision of the methods used. A few ages of high accuracy and precision may provide as much information about the history of a fault as many ages of low precision (Figure 1.4; Trumbore, 2000).

Accuracy is a measure of how close an age is to the time of formation or death of the dated material; *precision* is a measure of the analytical reproducibility of a method, commonly expressed as two standard deviations about a mean. But neither accuracy nor precision measures what is frequently the largest source of error in paleoearthquake dating—*sample context error*, or the error involved in inferring the time of an earthquake from the age of an accurately dated sample (Table 1.4). In many paleoseismic studies, errors

Table 1.4: Sources of uncertainty in relating the age of a dated geologic sample, to the age of a paleoearthquake (from Noller et al., 2000)

Type of uncertainty	Description
Analytical error	For numerical and radiogenic ages, the standard deviation of the computed sample age, based on the counting statistics of the method
Natural variability in sample quality or suitability	Variability that arises when the dated stratum contains a heterogeneous mixture of materials, each of which may yield a different age because: (1) they violate the assumption of the dating method to different degrees or (2) they are different-age materials within the same stratum
Geologic context error	The uncertainty of relating the age of the dated sample, to the age of the surrounding stratigraphy
Calibration error	The error induced when converting a relative-age result to a numerical-age result, via use of a calibration curve
Violation of assumption	Error induced when the physical properties or history of the sample (burial history, exhumation history) violate the assumptions of a dating method, as to initial conditions, or later conditions

involving the stratigraphic or geomorphic context of samples are frequently inadequately discussed or not even acknowledged because such errors are difficult to estimate. For many samples the latter type of errors is commonly a substantial percentage of the age (Colman and Pierce, 2000). By ignoring context errors and applying ages with only their analytical errors as if they were accurate estimates of the times of past earthquakes, many paleoearthquake histories have been presented with an unjustified degree of accuracy. Such errors arise from several different causes and are discussed more fully in Chapter 2A.

Our final comment about dating is that it is difficult to overemphasize the value of obtaining (1) multiple ages on separate samples associated with the same event using the same methods and (2) ages from different types of dating methods used on different types of sample materials associated with the same event. We know of no field situations where a single analysis on a single sample by any dating method provides a definitive age for a paleoearthquake. The advantage of measuring multiple ages associated with the same paleoearthquake is that one can average multiple ages of the same type on the same type of sample to reduce errors. Preferably, one can sample above and below the horizon of interest to "bracket" the age of the paleoearthquake (Chapter 6).

The advantage of determining the age of a single paleoearthquake via different dating methods is that it permits one to assess anomalous age estimates that may result when using only a single dating technique. For example, when dating multiple samples from a stratigraphic section with a single dating technique, it is common to have anomalous age "outliers" and age reversals in the stratigraphic section. The investigator is then faced with a dilemma of which ages to believe, and which to disregard. The decision to disregard one or more samples is made easier if one has samples from the same stratigraphic units dated by a second or third method, methods that are not subject to the same type of inferred contamination process as is the suspect ages. Different dating methods also allow dating of different intervals of time, which has an important bearing on understanding the complete recent history of faults (Figure 1.1).

1.3.2 Patterns in Recurrence

The *recurrence* (timing) of paleoearthquakes is a critical characteristic in assessing the hazard from large earthquakes. In active tectonic regions the recurrence period of moderate to large earthquakes may be so short that they pose a greater hazard than the largest earthquakes, which occur much less frequently. In contrast, in regions where fault slip rates are low, the recurrence intervals for large earthquakes may be so long that the earthquakes are not a significant hazard for facilities with design lives of less than 100 years.

The assumption that earthquakes of roughly the same size occur with regular recurrence has been attractive and powerful in the sense that it allows calculation of average recurrence intervals from short- or long-term deformation (slip) rates. But this assumption may not apply to the long-term history of many faults; historical records of large earthquakes show a great deal of variability in their spatial and temporal patterns of recurrence. In the plate-boundary regions of the Pacific, such variability appears to be the rule rather than the exception (Thatcher, 1990). Not surprisingly, the results of some of the most detailed paleoseismology studies mirror those of the historical record: as the number and precision of ages for past earthquakes increase recurrence patterns become more complex (e.g., McCalpin and Nishenko, 1996; Biasi *et al.*, 2002). The degree of *earthquake clustering* and the extent to which earthquakes on one fault (or segment) trigger earthquakes on adjacent faults or segments (termed *contagion* by Perkins, 1987; and earthquake *stress triggering* by Stein *et al.*, 1997) are of increasing concern in paleoseismology studies (Chapter 9, See Book's companion web site).

The relations among fault slip rates, recurrence patterns, and the time interval of observations are nicely illustrated by Pierce (1986; Figure 1.4). Depending on the length and age of the *window of observation*, slip rates (and hence inferred earthquake recurrence and perceived hazard) may vary by an order of magnitude. Figure 1.4 also clearly illustrates the importance of attempting to date individual earthquakes when assessing earthquake hazards. Without multiple ages that span much of the recent history of a fault there is little hope of accurately forecasting its future activity.

Earthquake recurrence may also have a significant impact on the types of evidence that can be used to infer paleoearthquakes. Where large earthquakes occur frequently, episodic shaking is commonly the dominant process in producing some types of slope or ground failures (Chapter 8). In such highly active areas these features can be reasonably inferred to have been caused by earthquakes. For example, recent work in New Zealand shows strong correlations between the times of major historical earthquakes on the Alpine fault and ages for rockfall events derived from lichen diameter measurements (Bull *et al.*, 1994; Bull and Brandon, 1998). But where earthquakes are smaller or much less frequent than along major fault zones with high slip rates, nontectonic processes may produce most slope and ground failures. In another example, if great earthquakes occur every few hundred years off the coast of central western North America, then most abruptly buried tidal-marsh soils and continental-shelf turbidites in the region were probably produced during such earthquakes. However, if great earthquake recurrence was a thousand or more years, then nonseismic processes that produce similar features would be much more difficult to distinguish from evidence of great earthquakes (Nelson *et al.*, 1996; Goldfinger *et al.*, 2008; Chapter 5).

1.4 Estimating the Magnitude of Prehistoric Earthquakes

Both primary and secondary evidence are used to estimate the magnitude of paleoearthquakes. Paleomagnitude estimates involve many assumptions, and large errors are often associated with the many parameters that need to be measured in order to estimate magnitudes. The best approach is to use several methods to estimate magnitudes. For a more detailed discussion of this topic see (Chapter 9, See Book's companion web site).

Because primary evidence is produced directly by fault rupture, the geographic distribution of primary evidence is related to the area of fault rupture and, hence, to earthquake magnitude. The length of fault surface rupture is the parameter most commonly used to estimate magnitude (Figure 1.6). Estimates of the *surface rupture length* during past earthquakes are compared to compilations of worldwide data on surface rupture lengths during historical earthquakes to estimate ranges of earthquake magnitudes for particular paleoearthquakes (Wells and Coppersmith, 1994; Stirling *et al.*, 2002). Variations on this procedure involve estimating the area of the fault plane that slipped during past ruptures or the seismic moment, a measure of the total energy released during the earthquake.

Primary evidence that reflects the relative amount of slip on faults or the folding of surfaces above faults may also be used to estimate magnitude. As with fault length, these methods rely on empirical comparisons with amounts of deformation recorded during historical earthquakes of different magnitudes (Figure 1.6). Examples include use of the amount of lateral offset of young stream channels, the thickness of colluvial wedges in fault exposures, or the amount of uplift of former shorelines to distinguish M_w 6 earthquakes from earthquakes of M_w 7–8 (Chapters 3–6). Hemphill-Haley and Weldon (1999) show how such displacement measurements can be statistically related to paleoearthquake magnitude. In the past decade, the complex relations among shallow surface-rupturing faults, variable geomorphic site parameters and processes, and slip on deeper master faults, especially from one earthquake to the next, make even empirical estimates of paleoearthquake magnitude from measurements of surface rupture or fold growth uncertain (e.g., Kelsey *et al.*, 2008).

Secondary evidence is less commonly used to infer paleomagnitudes than primary evidence, but in some cases it may provide more accurate estimates of magnitude than primary evidence. More importantly, secondary evidence is the only evidence available for those earthquakes in which the seismogenic fault does not rupture or fold the surface. Mapping the distribution of secondary evidence, such as liquefaction features or earthquake-induced landslides, over a large area may reveal a pattern of variable ground motion intensity. From the areal extent of the features and their relative size, paleoearthquake magnitude may be inferred using empirical methods based on historical observations (Figure 1.6; Chapters 7 and 8). In addition, engineering-based static and dynamic analyses (e.g., a pseudostatic limit equilibrium analysis of a landslide) of the failed material can yield estimates of shaking strength that are independent of historical–empirical correlations.

The accuracy of estimating magnitudes from both primary and secondary features is dependent on the assumption that correlated features are the product of the same paleoearthquake. But several moderate-magnitude paleoearthquakes closely spaced in time may have created features that, after several hundred or thousand years, appear to be the product of one large paleoearthquake. The Rainbow Mountain–Fairview Peak–Dixie Valley (Nevada) earthquake sequence of 1954 (M_w 6.2, 6.5; M_w 7.2; M_w 6.7) created three zones of fault scarps in 6 months; the latter two surface ruptures were only 4 min apart (Doser and Smith, 1989; Hodgkinson *et al.*, 1996; Caskey *et al.*, 2004). If the entire 123 km-long zone of Fairview–Dixie scarps were attributed to a single earthquake in some future paleoseismic study, empirical relations would indicate a single M_w 7.8 earthquake. Multiple great earthquakes ($M_w > 7.8$) have also uplifted or subsided many tens of kilometers of coast as little as a few minutes to a few months apart (e.g., Ando, 1975; Briggs *et al.*, 2007). Thus, even with significant improvements in dating precision and accuracy, paleoseismological methods will rarely show whether widespread primary or secondary evidence was produced by one very large earthquake or a series of smaller earthquakes occurring less than a few decades apart.

1.5   The Early Development of Paleoseismology, 1890–1980

Not surprisingly, paleoseismology developed first and has grown fastest in countries where rates of tectonic processes are high and where geologic investigations are supported by a sophisticated scientific infrastructure. Our brief outline of some early developments of paleoseismology focuses on some of the significant advances in countries such as the United States, Japan, Russia, and New Zealand.

The notion that the dramatic topography of some regions was created little by little during repeated earthquakes has a long history, extending through most of the current millennium (Vita-Finzi, 1986, p. 9; Bonilla, 1991). The main impetus to understand the history of recent faults arose in the late 1800s and early 1900s from detailed accounts of historical ruptures along active fault zones (Richter, 1958). For example, McKay (1886) recognized that scarps produced by earthquakes in New Zealand in 1848 and 1855 were identical to larger fault-zone features with similar origins, and ruptures produced by the 1891 Nobi earthquake in central Japan led Koto (1893) to similar conclusions. Other earthquakes of this period that stimulated much initial paleoseismic research include the 1906 San Francisco event in the United States (Lawson *et al.*, 1908), the Assam, India, earthquake of 1897 (Oldham, 1899), and the 1923 Kanto earthquake in central Japan (Kaizuka, 1976).

North American investigators frequently look back to the work of Gilbert (1890, pp. 340–362; 1928) who used a number of modern paleoseismic concepts to interpret normal fault scarps along the Wasatch fault in Utah (Wallace, 1980b; Machette and Scott, 1988). To Gilbert, variable fault scarp heights on different alluvial surfaces suggested multiple surface faulting events. By measuring fault scarp heights, he also estimated fault displacement per event and deduced that faulting events were no more frequent during the late Pleistocene high stand of pluvial Lake Bonneville than during its Holocene low stand. His letter to the inhabitants of Salt Lake City, Utah (Gilbert, 1884), on the inevitability of future large earthquakes is one of the first earthquake forecasts based on prehistoric, rather than historical, seismicity. Wallace and Scott (1996) consider Gilbert to be the "father of paleoseismology" in the USA (although that same appellation was applied to Wallace himself by many, including us in the first edition of this book, Figure 1.8).

Studies of the late nineteenth and early twentieth centuries made little attempt to relate earthquake features, which had been carefully described by early geologists, to a specific number of, or magnitude of, paleoearthquakes (Wallace, 1980b). Seismology was still a developing science and could not provide much theoretical support for descriptive fault studies.

In the period during and shortly after the two world wars most paleoseismic studies remained largely descriptive and focused on large landforms, such as fault scarps and raised alluvial and marine terraces. Surface faulting during the 1929 Murchison earthquake (Henderson, 1937) prodded New Zealand geologists into a concerted search for evidence of prehistoric faulting (e.g., Speight, 1938; Cotton, 1950) that culminated in several papers summarizing the geomorphic evidence of paleoseismicity throughout New Zealand (Wellman, 1952, 1953, 1955; Bowen, 1954; Lensen, 1958). In a similar way, Kuno (1936) recognized that the 1-km offset of Pleistocene features on the Tanna fault in Japan was of a similar nature to the 2- to 3-m offset created during the 1930 Idu earthquake. The height and distribution of marine terraces in Japan were used to infer several scales of active folding in Japan during this period (Otuka, 1932), and interseismic crustal movements were distinguished from coseismic movements (Yamasaki and Tada, 1928). Coastal studies of uplift of shorelines and tilting of marine terraces, which began following the 1923 Kanto earthquake (Kaizuka, 1993), progressed rapidly in the decades after the subduction-zone earthquakes of 1946 along the Nankai trough (e.g., Sugimura and Naruse, 1954; Yoshikawa *et al.*, 1964). In Russia, Florensov and Solonenko (1963, 1965) used their observations on the rupture trace of the

Figure 1.8: Photograph of a subsidiary fault scarp of the 16 December 1954 Fairview Peak, Nevada, earthquake. Photo was taken approximately 100 m north of U.S. Highway 50 in October 1984, nearly 30 years after formation. At right is Robert E. Wallace (1916–2007), widely regarded in the United States as the "father of paleoseismology" (see Wallace and Scott, 1996).

1957 Gobi-Altai earthquake (M_w 8.1) with their mapping of late Quaternary fault scarps in Russia and Mongolia to infer that some fault-zone landforms record identifiable paleoearthquakes. In a similar study in Nevada, USA, Slemmons (1957) mapped normal fault scarps along the trace of the Dixie Valley–Fairview Peak earthquakes of 1954. In 1958 Richter briefly summarized much of this early work in his book *Elementary Seismology*.

Paleoseismology emerged as a distinct discipline during the late 1960s and early 1970s. Florensov (1960) and Solonenko (1962, 1970, 1973) were among the first to propose a *paleoseismogeological method*, with an emphasis on the traces of prehistoric surface faulting or "paleoseismodislocations" and seismo-gravitational failures. The word *paleoseismicity* first appeared in the title of an English-language article about fault plane microgrooves (Engelder, 1974) and the term was being used in Japanese publications (e.g., Huzita and Ota, 1977, p. 135) shortly after. This was the beginning of the "modern" period of paleoseismology in the sense that stratigraphic and geomorphic evidence were used to interpret the characteristics of individual prehistoric earthquakes. For example, slip rates were calculated from the cumulative displacements of landforms of approximately known age, in settings ranging from strike-slip offset of river terraces (Lensen, 1968) to uplift of marine terraces during subduction-zone earthquakes (Yoshikawa *et al.*, 1964). The displacements and rupture lengths observed in historical earthquakes were used to infer the magnitudes of paleoearthquakes, based on the heights and lengths of faults scarps (Slemmons, 1977; Bonilla *et al.*, 1984). Studies of subduction-zone features during this period in Japan (Ota, 1975; Matsuda *et al.*, 1978; Nakata *et al.*, 1979) and documentation of the effects of the great earthquakes of 1960 (Plafker and Savage, 1970; Kaizuka *et al.*, 1973) and 1964 (Plafker, 1965, 1969b, 1972; Plafker and Rubin, 1978) in Alaska and Chile led to advances in the understanding of subducting plate boundaries that have strongly influenced paleoseismic studies since that time. In the continental United States in the 1960s, early work on the San Andreas fault by Wallace (1970) and others, the 1971 San Fernando earthquake (M_w 6.6) in southern California and trenches across the surface ruptures (Heath and Leighton, 1973), and the discovery of young faults in excavations during construction of nuclear power plants and other large facilities in California spurred many investigations into the stratigraphic expression of Quaternary faulting.

In his Presidential Address to the Geological Society of America in 1975, Clarence Allen provided the first overview to a wide audience of paleoseismic methods and assumptions (Allen, 1975: ". . . the most important single contribution to gaining a better understanding of long-term seismicity, which is critical to the siting and design of safe structures and to the establishment of realistic building codes, is to learn more, region by region, of the late Quaternary history of deformation, and particularly that of the Holocene epoch. More specifically, I see a special need for geomorphological studies in these regions, better and more radiometric dates, and accurate detailed geological field mapping that utilizes trenches and boreholes as well as surface exposures. Studies of these kinds, in my opinion, offer the best hope of inferring what has happened during earthquakes within the very recent geologic past, and therefore what is likely to happen again in the near future." Allen and Scott (2002) provide the historical background for this address.

During the late 1970s the development of modern paleoseismology in the United States was strongly influenced by Sieh's (1978a,b) detailed geomorphic and stratigraphic work on the San Andreas fault in California. Although Sieh was not the first to use trenches to expose a fault zone (e.g., Converse, Davis and Associates, 1968; Clark *et al.*, 1972), he was the first to demonstrate to a wide audience that earthquake chronologies comparable to some of those based on historical records could be reconstructed through the meticulous mapping and dating of trench-wall sediments. Meehan (1984) and Bonilla (1991) describe these early days of paleoseismology in the United States, particularly the extensive efforts made to derive paleoearthquake magnitude, displacement, and recurrence data from fault-zone exposures.

Acknowledgments

Comments by Silvio Pezzopane, Steve Obermeier, and Bob Bucknam improved this chapter in the first edition.

Field Techniques in Paleoseismology— Terrestrial Environments

James P. McCalpin

GEO-HAZ Consulting, Inc., Crestone, Colorado 81131, USA, mccalpin@geohaz.com

2A.1   Introduction

2A.1.1   Scope of the Chapter

This chapter describes common techniques used on-land to collect field data in most paleoseismic investigations, regardless of the tectonic setting (offshore methods are described in Chapter 2B). The methods fall into two broad categories, depending on whether paleoseismic features are *landforms* (Section 2A.2) or have *stratigraphic expression* (Section 2A.3). Basic geomorphic techniques include locating paleoseismic features with remotely sensed imagery and making detailed topographic maps of paleoseismic landforms. Stratigraphic techniques emphasize the finding of buried faults with geophysical methods and mapping of paleoseismic deformation in subsurface exposures. The section on trenching (Section 2A.3.2) is particularly detailed in view of the importance of trench studies in identifying and dating paleoearthquakes.

The techniques described in this chapter are mostly basic methods of geologic investigation as typically applied to unconsolidated sediments by Quaternary geologists (cf. Goudie, 1981). We describe those methods that have been most widely used in previous paleoseismic investigations. Overall, there is a slight emphasis in this chapter on primary paleoseismic evidence, because the majority of published reports have focused on primary evidence, particularly along continental fault traces. Studies of secondary evidence (e.g., paleoliquefaction research; Chapter 7) use similar techniques, whereas others (subduction-zone coastal studies of Nelson and Personius, 1996) use the more classical techniques of Quaternary stratigraphy. Other uncommon methods used in some Quaternary studies (e.g., palynology) may be the most useful in both types of paleoseismic studies, especially in those climatic or geomorphic settings where paleoseismologists have performed little work to date.

Many of the generic techniques described in this chapter, such as fault scarp profiling and trench logging, were developed specifically for paleoseismic studies. Other techniques, such as geomorphic mapping (Section 2A.2), shallow coring, and shallow geophysics (Section 2A.3), were developed for other types of investigations and have been adapted for use in paleoseismology. In this chapter we describe only the application of these latter techniques to paleoseismic studies; readers should consult the cited literature for further details on theory, methods, and equipment.

International Geophysics, Volume 95

ISSN 0074-6142, DOI: 10.1016/S0074-6142(09)95002-1

2A.1.2 Preferred Sequence of Investigations

An ideal sequence of paleoseismic investigations would progress from the *regional scale* (thousands of square kilometers), to the *local scale* (a few square kilometers), to the *site scale* (1 ha to a few square meters). In many areas data on the regional neotectonic setting are already available, so the emphasis in most paleoseismic studies is on the local and especially the site scales. The sequence of topics in this chapter likewise proceeds from the regional scale (e.g., remote sensing), to the local scale (geomorphic mapping), to the site scale (geophysics, fault-zone trenching) (Figure 2A.1).

With the success and growing popularity of *fault trenching,* a misconception may have arisen that regional- and local-scale Quaternary geologic mapping is no longer necessary in paleoseismic studies, because trenches provide all needed data on paleoearthquake magnitude and recurrence. Nothing could be farther from the truth. Without careful geomorphic mapping along a fault trace, and establishment of a Quaternary stratigraphic framework, trenching is unlikely to be productive. Every time the author has sited trenches without the benefit of local geomorphic maps, the trenches have yielded useless or ambiguous data, and a second round of trenching has been required based on subsequent detailed local geomorphic mapping. The importance of preliminary Quaternary geologic mapping before subsurface investigation cannot be overemphasized.

2A.2 Mapping Paleoseismic Landforms

The first, and often most fruitful, approach to paleoseismic reconstruction should be careful mapping of Quaternary landforms and deposits in the zone of deformation. *Surficial geologic mapping* helps identify deformed geomorphic surfaces and reveals trends in deformation styles and rates across landforms of different ages. Based on the estimated ages of faulted geomorphic surfaces and/or deposits (Section 2A.2.5), and their measured displacements, the following variables can be estimated without recourse to stratigraphic investigation: (1) fault slip or surface deformation rates, (2) displacements or tilting per event, and (3) bracketing ages for deformation events. Geomorphic mapping helps indicate where fault traces are the result of single versus multiple events, the approximate size and timing of paleoearthquakes, and whether paleoearthquakes are within the range of certain dating techniques. As stated earlier, we feel that *trenches should never be sited until the geomorphic relations in the area of investigation are thoroughly understood.*

2A.2.1 Locating Surface Deformation

Surface deformation from paleoearthquakes can be located by database query, remote sensing, aerial reconnaissance, or field inspection. The two latter approaches are based on the traditional (manual) analysis of analog data, whereas the two former approaches are mainly digital. These latter approaches have undergone a "digital revolution" in the past 15 years, giving rise to the field of "geoinformatics" (Sinha, 2006) which has been fruitfully applied to paleoseismology (Arrowsmith, 2006). Geoinformatics covers the integral use of digital data sources such as remote sensing images, digital elevation models (DEMs), GPS measurements, orthophotographs, and raster or vector maps of geology, soils, and so on; analysis software such as geographic information systems (GISs), 2D and 3D visualization software, and spatial decision support systems; and computational methods such as map algebra. It is now possible to merge multiple datasets on a desktop GIS and to create sophisticated 2D and pseudo-3D visualizations, in a way not possible even 10 years ago.

The first step in collecting primary paleoseismic evidence is to identify and map a zone or area of crustal deformation. Active fault traces are often expressed at the surface as linear belts of anomalous landforms

Generic Flow Chart for Paleoseismic Trenching Studies

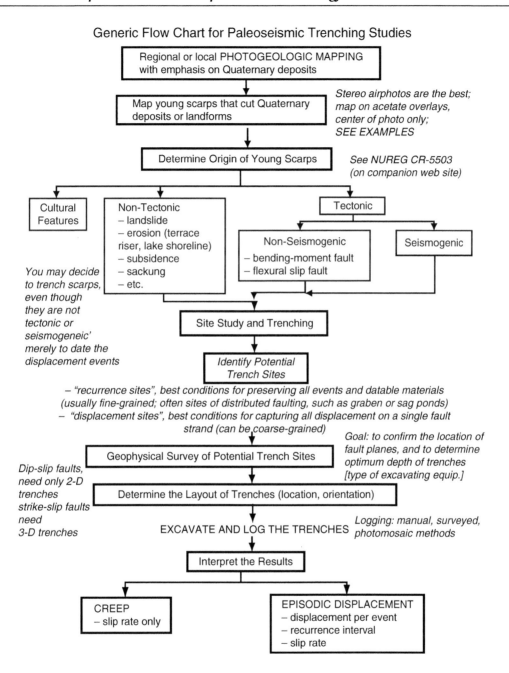

Figure 2A.1: **A generic sequence of paleoseismic investigations that ends in trenching a young fault scarp. NUREG CR-5503 refers to Hanson *et al*. (1999).**

across which vertical relief is evident (i.e., dip-slip faults; see diagrams in Chapters 3 and 5) or where terrain elements are shifted laterally (i.e., strike-slip faults; see diagrams in Chapter 6). Active folds are identified by more indirect perturbations of topography and drainage networks. For an excellent survey of fault-related landforms, see Burbank and Anderson (2001).

2A.2.1.1   Database Query

Quaternary fault databases have been compiled for many countries as part of the World Map of Major Active faults (Trifonov and Machette, 1993). In the USA the U.S. Geological Survey maintains a digital version of the Quaternary Fault and Fold Database of the United States (http://earthquake.usgs.gov/regional/qfaults). This database is available as GIS files and as Google Earth.kmz files (http://earthquake.usgs.gov/regional/qfaults/google.php). Some State geological surveys maintain their own fault databases (e.g., Colorado, Widmann *et al.*, 2002; California, Bryant, 2005); the Colorado database includes all faults with late Cenozoic movement. For consultants performing applied paleoseismic studies required by regulations, these fault databases are among the first data acquired in a new project (along with digital geologic maps).

2A.2.1.2   Remote Sensing and Aerial Mapping

Remote sensing imagery can include aerial photographs (Figure 2A.2), satellite images, radar images, or maps made from DEMs. The choices above include both hard copy or digital files, and of the latter, georeferenced or nongeoreferenced images. However, the best imagery for *identifying* neotectonic features may not always constitute the best base map on which to *map them*, as explained below. With the widespread use of GISs today, it is often desirable that the base map be a digital, georeferenced map, rather than a nongeoreferenced image such as an aerial photograph. Computer graphic versions of a map

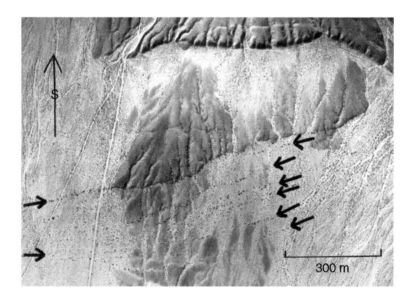

Figure 2A.2: Vertical aerial photograph showing east–west-trending normal fault scarps (in shadow at top, and in center) and parallel lineaments (between arrows) that overlie buried faults. Ephemeral streams flow from top to bottom. Note dissection of scarp at top of photo and secondary alluvial deposition on the downthrown fault block at upper center. Lone Mountain fault zone, Nevada (photograph courtesy of J. C. Yount).

can always be produced from a georeferenced digital map, but the reverse is not always true, as explained in following sections.

Aerial Photography: Prior to the late-1990s the preferred method for identifying faults or other neotectonic deformation, in almost all areas except densely forested terrain, was stereoscopic examination of overlapping aerial photographs. Practicing paleoseismologists were required to be proficient in photogeologic interpretation from stereoscopic photographs, following the classic texts from the "golden age" of photogeology in the 1960s (Ray, 1960; Miller, 1961; Lattman and Ray, 1965). This expertise could also be drawn upon throughout one's career for every geological investigation, not merely the paleoseismic ones.

In the USA, stereoscopic aerial photographs are readily available for purchase at scales ranging from 1:60,000 through 1:15,000, so they can be utilized at all scales of investigation (regional, local, and site scale). Cluff and Slemmons (1971) proposed that *low-sun-angle photography* (LSAP) was best for mapping fault zones, because it emphasized the subtle fault-related topography (Figure 2A.3). Unfortunately, this type of specialized photography is rarely available for purchase, because a general-purpose airphoto must minimize shadowing by taking the photographs at the highest possible sun angles. However, with the advent of DEMs it is possible to nearly duplicate LSAP images by creating hillshaded DEMs (discussed later).

Almost all commercially available aerial photography in the USA is unrectified vertical photography (Figure 2A.2). The geometric properties of such unrectified photographs are well known, and *photogrammetric measurements* of true horizontal and vertical distances can be made from stereo pairs, using mechanical instruments such as parallax bars or optical instruments such as analytical stereoplotters. For example, Nelson and Personius (1993) and Nelson *et al.* (2006) made 298 measurements of vertical surface offset on fault scarps along the Wasatch fault zone from 35-year-old black-and-white prints mounted in a analytical stereoplotter (Pillmore, 1989) in a fraction of the time it would have taken to measure 298 scarp profiles in the field.

A disadvantage of unrectified airphotos for mapping is that the horizontal scale varies within the image, depending on relief of the imaged terrain and distance from the photo center. Thus, a map made by tracing contacts directly on such airphotos (or on an attached transparent overlay) has variable horizontal scale and must be spatially corrected (orthorectified) to make the horizontal scale constant, before it can be georegistered. One method used to identify faults stereoscopically and still produce a georegistered fault map, is to identify the neotectonic features by stereoscopic examination of overlapping airphotos, but to map the relevant contacts on orthorectified versions (*orthophotos*) of those same photographs. If using a hard copy orthophoto, the mapped contacts visible in stereo can be transferred manually to the orthophoto (or on an overlay) and will have constant horizontal scale. It can then be digitized on a digitizing table or by scanned and finally georegistered. Alternatively, if a digital version of the orthophoto is available, it can be displayed on the computer monitor, and contacts seen in the stereoscope can be transferred by eye to the screen via "heads-up" digitizing.

Satellite Imagery: Satellite imagery has advantages and disadvantages compared to airphotos, for identifying and mapping neotectonic features. The main disadvantage of satellite imagery is the inability to view them stereoscopically. Advantages of satellite imagery for identifying neotectonic features include that (1) the image covers a much larger area than an airphoto, so large features can be recognized which might be overlooked on separate airphotos; (2) images are digital, so can be manipulated and enhanced to accentuate slight tonal/color anomalies associated with neotectonic features. Advantages of satellite imagery for mapping neotectonic features include that (1) images are relatively scale-invariant

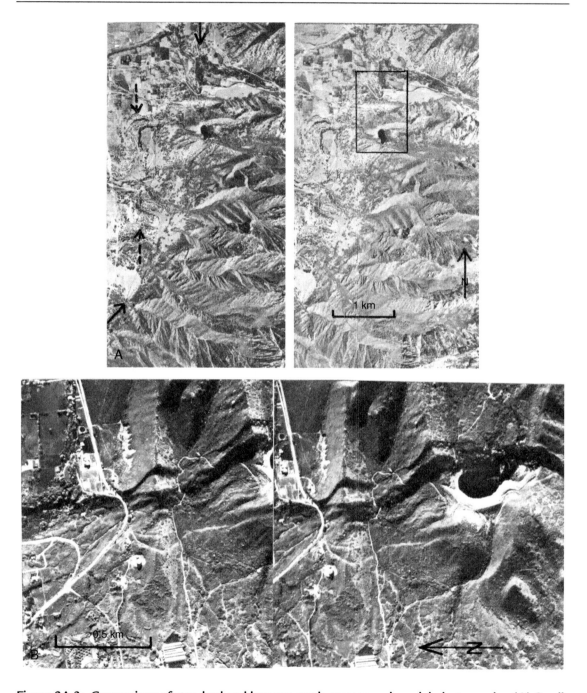

Figure 2A.3: Comparison of standard and low-sun-angle stereoscopic aerial photographs. (A) Small-scale (original scale 1:50,000) vertical aerial photograph in standard lighting conditions. Fault scarps of the Wasatch fault zone, Utah (between solid arrows) offset Pleistocene lacustrine deposits (lower left and center) and moraines (upper center) at the base of a steep, faceted range front. Box at upper

and do not require orthorectification for mapping; (2) most satellite images are provided with accompanying georegistration files, so are automatically georegistered when imported into a GIS.

The majority of technical papers describing satellite images used for mapping faults and lineaments in the USA, arose from work done in the 1960s–1970s for regional-scale seismic hazard studies for nuclear power plants (NPPs). Examples include satellite scanner images (Glass and Slemmons, 1978; Slemmons, 1981); satellite photography (Muehlberger *et al.,* 1985), side-looking-airborne radar (Wing, 1970), and thermal infrared imagery (Sabins, 1967, 1969; Wallace and Moxham, 1967). However, after about 1980 no new NPPs were licensed in the USA and interest in the topic waned. There was a slight resurgence of interest in this topic in the 1990s, focused on identifying secondary paleoseismic evidence (Wiley *et al.,* 1991) and primary evidence in countries where paleoseismic investigations were just beginning (e.g., Astaras and Soulakellis, 1991). The most recent application of satellite imagery to neotectonic mapping is the Spaceborne Imaging Radar (SIR) data acquired by NASA from the early 1990s to present, including the Shuttle Radar Topography Mission. These data are now widely used to create topographic models (Burgmann *et al.,* 2000), as well as interferometric studies of contemporary vertical movements of the crust.

Digital Elevation Models: A DEM is a digital grid file with elevation data posted at every point (pixel) on a regularly spaced grid. Two different types of DEMs are used in fault mapping: (1) DEMs made by scanning preexisting topographic maps and (2) DEMs made by altimetry surveys.

Included in the first type are "off-the-shelf" DEMs such as the 7.5′ DEMs produced and sold by the U.S. Geological Survey, which are interpolated from scanned hard copy topographic maps at 1:24,000 scale, with contour intervals ranging from 10 to 40 ft. The normal grid spacing in this series is 30 m, but many DEMs have been further interpolated to a 10 m spacing. DEMs of this type contain any inaccuracies (vertical or horizontal) that existed in the published, hard copy map. It is generally impossible to remove these inaccuracies, because in most cases the contour lines that form the basis of the grid interpolation were drawn visually (i.e., subjectively) by a human operator viewing stereo airphotos on an analytical stereoplotter. For example, in forested regions the operator was estimating where the ground surface might be, based on some subjective data. Since the advent of GPS surveying, it is common in detailed mapping to find minor conflicts between the GPS coordinates of a feature on the ground and the coordinates of that same point derived from a scanned topographic map; many times these discrepancies are traced to inaccuracies in the topographic map, particularly at very large scales.

DEMs can also be created by airborne or spaceborne altimetry surveys using LiDAR or synthetic-aperture radar (SAR). In their initial, "unfiltered" version, these surveys record the elevations of the tops of all features on the earth's surface such as bare ground, tops of vegetation (trees, shrubs), tops of buildings, roads, and so on (Figure 2A.4B). In sparsely vegetated areas these unfiltered DEMs can be used for mapping and visualization. In heavily vegetated areas, a second step of "filtering" removes all the signal returns from vegetation tops to yield a "bare-earth" DEM (Figure 2A.4C). Removal is achieved by manual editing and by interpolation of elevations from bare ground areas in the original LiDAR or radar data; the resulting "bare-earth" DEMs are superior for mapping subtle fault traces, particularly in forested areas

center shows area of figure B. (B) Large-scale (original scale 1:6000) low-sun-angle aerial photograph of faulted moraines on the Wasatch fault zone. Multiple synthetic fault scarps (shadowed) and antithetic fault scarps form a graben. Reservoir at far right was placed in a depression behind a terminal moraine, further deepened by graben formation (photographs courtesy of Brigham Young University and the Utah Geological Survey).

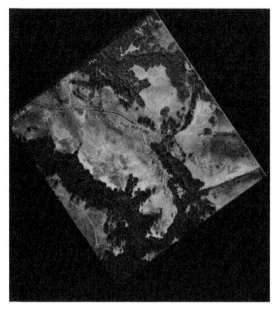

A. Traditional aerial photograph of the southeastern part of the study area.

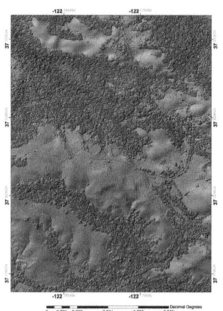

B. Unfiltered LiDAR return data show all surfaces imaged including vegetation. LiDAR images prepared by the National Center for Airborne Laser Mapping (NCALM)

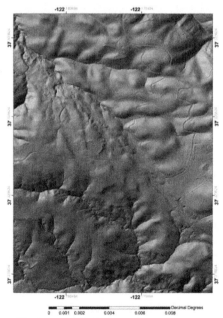

C. Filtered LiDAR data reveal 'bare Earth' surface beneath vegetation

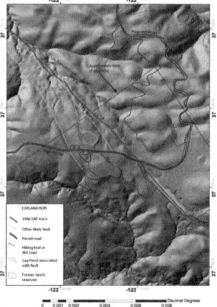

D. Fault trace and other features identified and interpreted from LiDAR imagery. Preliminary interpretation and air photo provided by Dr. Carol Prentice, USGS.

(Figure 2A.4D). LiDAR-derived DEMs have been used to identify and characterize faults in densely forested areas (e.g., Nelson *et al.*, 2003; Cunnigham *et al.*, 2006); in heavily urbanized areas (Kondo *et al.*, 2008); in lightly urbanized areas of subtle, creeping faults (Engelkemeir and Khan, 2008); and in deserts (Frankel *et al.*, 2007). The future of open-source LiDAR DEMs can be seen at the Open Topography Portal on the World Wide Web (http://opentopography.org), which contains a Google Earth.kmz file that links the Google Earth browser to all the 1 m, bare-earth LiDAR-DEM hillshade images created in northern California along major active faults. Using this kmz file, Google Earth users can zoom from a global view to the 1 m DEMs along the faults. The detail on the hillshade images is incredible and shows abundant landslide terrain as well as tectonic landforms along the fault traces.

The use of DEMs in paleoseismic studies has greatly increased in the past decade and is gradually replacing the use of conventional aerial photography in fault mapping studies, for several reasons. First, an increasing number of paleoseismic studies are being performed on low-slip-rate faults, on subtle folding over blind thrusts, and on poorly expressed secondary/off-fault evidence. These features can best be recognized by rather subtle topographic and geomorphic anomalies, of the type that can be enhanced on DEMs by draping satellite images or airphotos over DEMs and exaggerating the topographic relief (Figure 2A.5), or by making derivative maps of slope angle or of slope curvature, or by subenvelope maps (showing departures of local topography from a predefined smooth surface). Second, 30 m and 10 m DEMs (interpolated from published topographic maps) are now readily available and inexpensive in the USA. A 90-m DEM is available for the entire world based on SAR acquired during the space shuttle radar topography mission (SRTM) of 2000, available for download (in GeoTIFF or ArcInfo ASCII format) by the CGIAR consortium for spatial information (http://srtm.csi.cgiar.org).

Third, most paleoseismic studies now use GIS software which can easily georegister and then manipulate DEMs, making them easier to use than hard copy topographic maps when integrating map data of various sources and scales. Finally, recent use of LiDAR-generated and interferometric synthetic-aperture radar (IFSAR)-generated DEMs in forested regions has proved that neotectonic features can be detected which were previously undetectable on either topographic maps, DEMs made from topographic maps, or aerial photographs (e.g., Sherrod *et al.*, 2004; Cunningham *et al.*, 2006). A comparison of these methods is shown in Table 2A.1.

Interferometry from Synthetic-Aperture Radar (InSAR): Radar interferometry permits one to measure elevation changes of the earth's surface between two successive passes of a radar satellite. INSAR analyses have become commonplace in studies of active crustal deformation, but being limited to detecting contemporary elevation changes, one might think InSAR has no applicability to paleoseismic studies. This is generally true, but InSAR has two indirect applications for identifying active faults not shared

---

Figure 2A.4: Comparison of fault visualization and mapping on aerial photographs and LiDAR images (A) Vertical aerial photograph of the San Andreas fault near southern end of the Mill Creek, California study area. Note heavy forest obscures details of fault geomorphology. **(B)** Hillshade image produced from the unfiltered DEM derived from first returns of LiDAR data showing tree canopy tops. **(C)** Hillshade produced from "bare-earth" DEM derived from filtered returns of LiDAR data showing ground surface. Note prominent fault features that are almost completely obscured by forest in aerial photographs. **(D)** Preliminary fault mapping on "bare-earth" LiDAR image. Mapping by Carol Prentice, U.S. Geological Survey, Menlo Park, CA; Rich Koehler and John Baldwin, William Lettis & Associates, Walnut Creek, CA. Source: U.S. Geological Survey, Menlo Park, CA. **(See Color Insert.)**

Figure 2A.5: Three-dimensional perspective view of Karakax valley (northwestern Tibet). (A) from a DEM created by interferometry from two radar images (center at 36.1°N latitude, 79.2°E longitude). Scale varies in this perspective view, but the area is about 20 km (12 miles) wide in the middle of the image, and there is no vertical exaggeration. Elevations range from 4000 m (13,100 ft.) in the valley to over 6000 m (19,700 ft.) at the peaks of the glaciated Kun Lun mountains running from the front

by other remote sensing techniques. First, InSAR may be used to identify the slight surface folding accompanying blind thrust faulting. It was InSAR that identified the causative blind thrust fault responsible for the Bam, Iran earthquake (M_w 6.5) of 23-Dec-2003 (e.g., Stramondo *et al.*, 2005), which did not produce a surface rupture.

Second, InSAR can identify structures that are currently creeping without creating earthquakes. Such a distinction might be helpful in determining, for example, whether a short prehistoric scarp of unknown origin was an old coseismic fault scarp, or perhaps some type of slowly evolving nontectonic scarp (sackung, landslide, subsidence, etc.). In such a case, if InSAR shows there is contemporary creep occurring across the scarp, the scarp is more likely to be the result of an ongoing nontectonic, nonseismogenic process.

2A.2.1.3 Aerial Reconnaissance

Prior to the advent of free satellite imagery viewing on the Internet (e.g., Google Earth), the only way to visually "zoom in" on suspected paleoseismic features on the earth's surface was to hire an airplane or helicopter and fly over the area. This technique is obviously still available, albeit somewhat expensive. If the flight is scheduled when sun angles are low, one can take low-sun-angle photographs using a hand-held camera (Figure 2A.6). The scale and detail of photographs and visual observations are only limited by the quality of the camera (or eyes), and the minimum altitude that the pilot will fly.

2A.2.1.4 Field Inspection

In two situations, it may ultimately be necessary to make your reconnaissance map of neotectonic features in the field, because: (1) the ground surface is obscured by dense vegetation or (2) neither hard copy base maps nor airphotos are unavailable, or if available, are too small in scale to be of use. For example, in densely forested regions the small landforms created by dip-slip or strike-slip faulting may not be visible on any remote sensing imagery. Recent cases of this include the discovery of the Seattle fault (Washington State, USA) in a dense boreal forest (Nelson *et al.*, 2003). Such reconnaissance mapping can be made from direct field measurements and plotting of the horizontal locations of neotectonic features, by either using a distance-and-bearing technique to relate points to each other (e.g., the old "Brunton-and-pace" method), or by surveying points with a handheld geographic positioning systems (GPS). This latter method has become popular since May 2000, when the deliberate degradation of GPS satellite signals was eliminated. Since that time, even inexpensive handheld GPS receivers can achieve point accuracies of <5 m. Figure 2A.7 shows a reconnaissance map of surface fault ruptures associated with the 26 January 2001 Bhuj, India earthquake, in a remote area (Rann of Kachchh) where the only topographic base maps available were much too small scale (1:50,000; contour interval 100 ft.) to be of use in portraying the fault traces.

---

right toward the back. The active strand of the Altyn Tagh fault is visible as a sharp break in slope running diagonally up the valley side (between arrows). The original two radar images were acquired with spaceborne imaging radar C/X band (SIR-C/X-SAR) synthetic-aperture radar, aboard the space shuttle endeavor in October 1994. In this drape of C-band radar over the DEM, the L-band amplitude is assigned to red, L- and C-band (24 and 6 cm wavelengths) average to green, and C-band to blue (from NASA, http://southport.jpl.nasa.gov/pio/srl2/sirc/krkx.html); (B) from Google Earth; fault trace is between arrows. (See Color Insert.)

Table 2A.1: Comparison of various remote sensing methods used for fault mapping

Method	Advantages	Disadvantages
Standard stereo aerial photography	1—Widely available and inexpensive 2—True stereoscopic viewing is possible 3—Terrain is imaged in natural colors/tones (visible or infrared)	1—Often low contrast, due to high sun angles 2—Not orthorectified, so cannot map directly on them without either: (1) removing the relief displacement with photogrammetric hardware or software or (2) hand-transferring map contacts to an orthophotograph
Low-sun-angle aerial photography (LSAP, stereo)	1—Landforms in fault zone are accentuated by shadowing 2—Outside of shadowed areas, terrain is in natural colors/tones	1—With extreme shadowing, some image information is lost 2—Not orthorectified, so cannot map directly on them without either (1) removing the relief displacement with photogrammetric hardware or software or (2) hand-transferring map contacts to an orthophotograph 3—Not widely available; expensive to acquire
Free satellite images in an Web-based viewer (e.g., Google Earth; see Figure 2A.5B)	1—Available worldwide, and free 2—Views can be vertical, or oblique with image draped over a DEM	1—Resolution of the images varies; best in cities in North America, where resolution can be ca. 1 m
DEM-low resolution (>50 m pixels) (Ex. GTOPO 90 m, 3 arc s)	1—Widely vailable and inexpensive	1—DEMs contain no information about the true colors/tones of image area (i.e., not a photographic image)
DEM-low resolution (5–50 m pixels) (Ex. GTOPO 30 m, 1 arc s; Shuttle Imaging Radar DEMs (Figure 2A.5A); USGS 10 m)	2—Hillshade models can be made in GIS with any illumination azimuth, inclination, and vertical exaggeration 2—Orthorectified, so can be mapped on, and integrated with other georeferenced layers in a GIS	2—True stereo viewing is not possible; however, an oblique, orthometric view of the DEM can be made in most GIS software, including "draping" orthorectified digital airphotos or satellite images over the DEM
DEM-low resolution (<5 m pixels) (LiDAR; IFSAR)	1—LiDAR-generated DEMs can be processed to show "bare-earth" ground surface; good if vegetation obscures ground	1—If vegetation is too dense, "bare-earth" image cannot be generated 2—Not widely available; expensive to acquire (US $1000–2000 per mi^2 for 2–3 m postings)

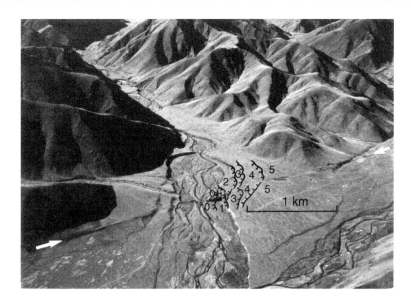

Figure 2A.6: Low-oblique aerial photograph taken from a small airplane showing the trace of the dextral Awatere fault (between arrows) at the Saxton River, South Island, New Zealand. Six alluvial terraces (0–5), dated by weathering rind thicknesses, are dextrally offset from 8 ± 2 m (terrace 0/1 riser) to 66 ± 5 m (terrace 4/5 riser). Terrace risers are accentuated with black lines, hachures toward lower terrace. Offsets were used by Knuepfer (1992) and McCalpin (1996b) to calculate slip rates (3.8–7.3 mm/yr) and displacements per event (6–8 m).

2A.2.1.5 Mapping Conventions

The suite of map symbols used to depict neotectonic features such as faults depends on the scale of the map. On small-scale (say, ≥1:100,000) and medium-scale (say, 1:50,000) maps, neotectonic features are generally too small to be portrayed at their real size and so are depicted somewhat schematically by convention, using point, line, and polygon symbols. A good example of such fault mapping is the series of maps along the Wasatch fault zone, Utah (scale 1:50,000; Machette, 1992), in which only the three "traditional" geologic line symbols are used for faults. Faults are mapped either as solid lines (i.e., the map unit is displaced, and the fault can be precisely located), dashed lines (i.e., the map unit is displaced, but the fault can only be approximately located), or dotted lines (i.e., the map unit is not displaced, but buries and overlies the fault, which is concealed). These different line symbols express very different interpretations of the location of and age of faulting, and should not be treated in a cavalier manner by either the mapmaker or the reader.

On large-scale maps (say, <1:25,000), it may be possible to portray neotectonic features at their true scale. In this case, additional line symbols, borrowed from the field of *geomorphological mapping*, may be useful in emphasizing the style of surface fault expression (Section 2A.2.2). At large scales more detailed subdivision of tectonic landforms is also desirable.

Surface zones of diffuse faulting, folding, and tilting are more difficult to identify on aerial imagery. *Broad zones* of warping, such as that created by blind thrusting or subduction megathrust earthquakes, must be mapped as areas (polygons) rather than as linear features like faults or narrow folds. These broad zones can often be mapped only after considerable field work has defined the areal limits of deformation,

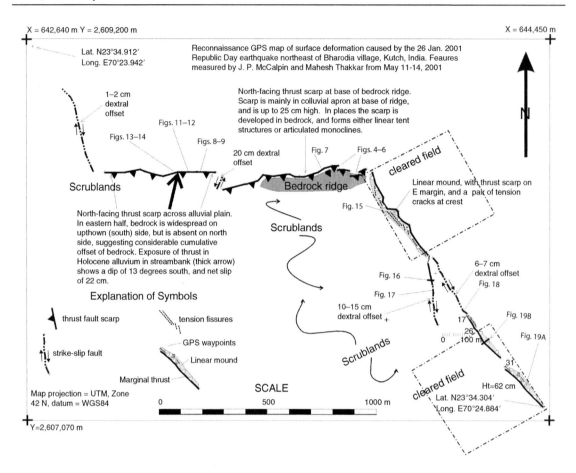

Figure 2A.7: Reconnaissance neotectonic map made by a handheld GPS receiver, showing surface rupture features associated with the 26 January 2001, Bhuj (India) earthquake on the featureless plain south of the Rann of Kutch. Labels "Figs." refer to photographs of surface ruptures in McCalpin and Thakkar (2003).

in contrast to well-defined fault zones where photogeologic mapping typically precedes field work. Secondary paleoseismic evidence, such as sand blows or landslides, is often accentuated by variable tones and textures on aerial photographs, rather than by topographic relief (see Figure 7.13).

2A.2.2   Mapping Deposits Versus Landforms in Seismic Areas

Paleoseismic features can be mapped either as *structures* deforming geologic deposits (the geologic approach) or as *landforms* (the geomorphic approach). Because Quaternary geologic mapping and geomorphology in North America are grounded in a geologic tradition, we tend to portray paleoseismic features on geologic maps. For example, geologic maps along fault zones (termed *strip maps*) usually include a detailed subdivision of Quaternary deposits, across which fault traces are portrayed by the line symbols previously described. The paleoseismic history of the fault is then interpreted from the interaction of the fault with Quaternary deposits, often aided by displacement measurements plotted on the map at key localities along strike. Examples of such strip maps are a series along the San Andreas fault zone, California (see multiple citations in Wallace, 1990, pp. 19–21), the Wasatch fault zone, Utah (see citations in Machette *et al.*, 1992a), and the Tectonic Map Series of the Geological Survey of Japan (e.g., Tsukuda *et al.*, 1993).

In a *deposit-oriented mapping* system, Quaternary deposits are typically differentiated based on inferred genesis (alluvial, glacial, eolian, lacustrine, mass movement), as interpreted from geomorphology and lithology (Flint, 1971; Catt, 1988). Such broad genetic classes are further subdivided into *second-order map units* based on the particular landform- or process-defined geomorphic environments. For example, within the alluvial class, alluvial fans, stream terraces, and floodplains might be differentiated. Each second-order map unit can be further subdivided by age to form *third-order map units*. In general, a detailed differentiation of Quaternary deposits increases the usefulness of the map for paleoseismic interpretation. The strip maps in the Wasatch fault zone series (scale 1:50,000) with up to 44 Quaternary geologic map units (e.g., Machette, 1992) contain more useful paleoseismic data than, for example, the strip map of the San Andreas fault by Davis and Dubendorfer (1987), which contains only 12 Quaternary map units.

Paleoseismic features can also be portrayed using a landform-oriented mapping system which emphasizes morphology rather than deposits. Geomorphological mapping has a long history in Europe (e.g., Klimaszewski, 1963, 1968; Tricart, 1965; Crofts, 1981), with several sets of mapping symbols proposed (Verstappen and van Zuidam, 1968; Demek, 1972; Demek and Embleton, 1978). Only recently has geomorphological mapping been applied to neotectonics (e.g., Goy *et al.*, 1988; Zuchiewicz, 1989). Most geomorphological maps published to date have scales between 1:25,000 and 1:100,000 and at such scales landforms created by single paleoearthquakes usually cannot be distinguished. Therefore, these maps are best suited for regional neotectonic studies and the reconnaissance phases of paleoseismic investigations; although they may be helpful in locating optimal sites for trenches. We suggest the reader consult Goy *et al.* (1991) for an excellent system of geomorphological map units adapted for neotectonic studies (Figure 2A.8).

2A.2.3 Detailed Topographic Mapping

Many landforms produced by paleoseismic deformation are too small (<1–5 m high) to be well characterized on existing topographic maps (typical contour interval of 3–15 m). Documenting the critical geomorphic details of faulting, from which net displacement or recurrent movement might be deduced, can be done in two ways. Landforms may be mapped directly using commonly used sets of *symbols* for landforms (Figure 2A.8). Although geomorphic maps portray landform shape and spatial relationships well, they are subjective and do not usually contain the actual topographic data by which the landforms were defined. An alternative objective approach is to create *very large-scale topographic maps* of paleoseismic landforms from which readers may make their own measurement or conclusions. Such detailed maps are very useful for measuring horizontal offsets along strike-slip or oblique-slip faults (e.g., McGill and Sieh, 1991; Grant and Sieh, 1994). A topographic map with landform mapping superposed would combine the subjective and objective approaches described here, but may become too cluttered if too many lines are drawn. We recommend that critical locations along a fault be documented by large-scale geomorphic maps with contour intervals small enough (<0.5 m) to permit precise map measurements of offsets.

Topographic maps with scales larger than 1:1000 were traditionally constructed by geologists using a plane table, alidade, and stadia rod (e.g., Sieh, 1978b). This time-consuming process has been replaced by

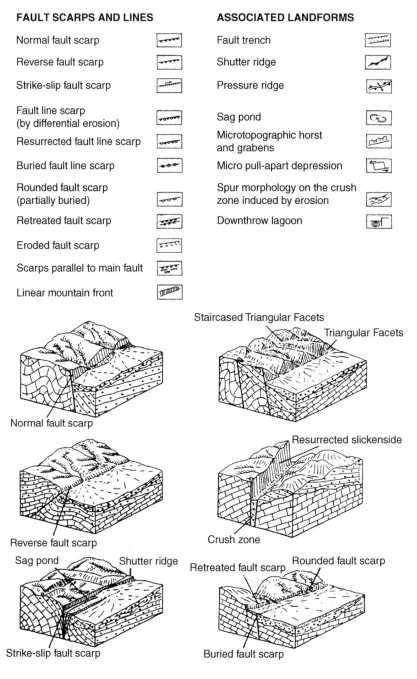

FAULT SCARPS AND LINES

Normal fault scarp

Reverse fault scarp

Strike-slip fault scarp

Fault line scarp
(by differential erosion)

Resurrected fault line scarp

Buried fault line scarp

Rounded fault scarp
(partially buried)

Retreated fault scarp

Eroded fault scarp

Scarps parallel to main fault

Linear mountain front

ASSOCIATED LANDFORMS

Fault trench

Shutter ridge

Pressure ridge

Sag pond

Microtopographic horst
and grabens

Micro pull-apart depression

Spur morphology on the crush
zone induced by erosion

Downthrow lagoon

Staircased Triangular Facets

Triangular Facets

Normal fault scarp

Reverse fault scarp

Resurrected slickenside

Crush zone

Sag pond   Shutter ridge

Strike-slip fault scarp

Retreated fault scarp   Rounded fault scarp

Buried fault scarp

Figure 2A.8: Geomorphological mapping symbols and corresponding block diagrams of neotectonic features. Top: block diagrams illustrating fault scarps and fault lines. Bottom: geomorphic mapping symbols related to fault scarps and fault lines. For the complete set of block diagrams and an additional 62 geomorphic mapping symbols, see Goy *et al.* (1991) (reprinted with permission from INQUA Neotectonics Commission).

two modern field surveying techniques: (1) total station surveying and (2) GPS surveying. In total station surveying, the elevation and horizontal position of survey points are measured by a survey total station instrument (*theodolite* with integral *electronic distance meter*), by trigonometry from the vertical and horizontal angles between the base station and the measurement points and the slope distance. This method was pioneered by Sieh in California, and excellent examples of maps thus derived are shown in McGill and Sieh (1991) and Grant and Sieh (1994) (Figure 2A.9).

Small paleoseismic landforms can also be mapped at ultra-large scale by GPS surveys. Handheld 12-channel GPS receivers used in stand-alone mode have limited horizontal (ca. 3–5 m) and vertical (ca. 9–15 m) accuracy, so cannot produce topographic maps such as shown in Figure 2A.9. Additional accuracy can be gained by adding a stationary GPS base station to the 12-channel GPS survey, or for vertical accuracies in the decimeter and centimeter range, employing a survey-grade GPS system with postprocessing software to correct for variable atmospheric conditions.

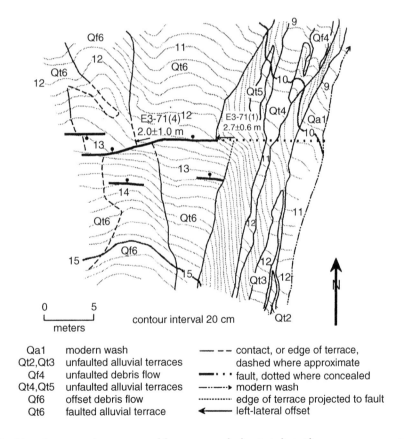

Qa1	modern wash	— – – contact, or edge of terrace,
Qt2,Qt3	unfaulted alluvial terraces	dashed where approximate
Qf4	unfaulted debris flow	▬▬···· fault, dotted where concealed
Qt4,Qt5	unfaulted alluvial terraces	—·—··→ modern wash
Qf6	offset debris flow	············· edge of terrace projected to fault
Qt6	faulted alluvial terrace	←——— left-lateral offset

Figure 2A.9: Very large-scale topographic map made by total station measurements, showing fine-scale topography in a zone of 2-m sinistral offsets on the Garlock fault, California. Surficial mapping and surveying of the 620 control points took 3 days. Critical geological contacts for measuring lateral offsets have been superimposed on the base contours. From McGill and Sieh (1991); reprinted with permission of the American Geophysical Union.

The recent development of terrestrial laser surveying systems (laser scanning) has opened up a faster way to collect tens of thousands of terrain survey points for making ultra-large-scale maps via DEM generation and contouring. To date, ground-based laser scanning has been applied to detailed landslide mapping (e.g., Rowlands *et al.*, 2003) and neotectonic mapping of well-preserved landforms where displaced by faulting (e.g., Oldow and Singleton, 2008).

2A.2.4   Topographic Profiling

Topographic profiling is often the easiest method of documenting the vertical component of paleoseismic faulting, folding, or tilt. Profiles measured at right angles to fault scarps provide a measurement of vertical surface offset, which is related to fault displacement by geometric relationships explained (for normal faults) in Section 3.3.1, Chapter 3. Other geomorphic nickpoints, such as shoreline angles or alluvial terrace flat/riser junctions, can be profiled along their strike to detect paleoseismic warping or uplift.

2A.2.4.1   Fault Scarp Profiling

Fault scarp profiles are the main source of data on vertical displacement and age of faulting in reconnaissance paleoseismic investigations of dip-slip faults (Wallace, 1977; Bucknam and Anderson, 1979; Haller, 1988). The interpretation of normal fault scarp profiles is well advanced (Chapter 3), while reverse fault scarp profiles have received less attention (Chapter 5).

The most common profiling strategy in paleoseismic studies of fault scarps is to measure one or more perpendicular topographic profiles across the fault scarp, extending the ends of the profile past the zone of deformation. From these profiles both displacement and age of faulting can be deduced. Vertical fault displacement is related to fault scarp height by various trigonometric formulas described in Chapter 3. If one systematically measures many (>1 per km) scarp profiles along the strike of the fault scarp, variations of displacement in space and time will be revealed, and it will be possible to calculate the maximum and average displacements per event that are needed in seismic hazard analysis (Chapter 9, See Book's companion web site).

Most fault scarp profiles are oriented perpendicular to the fault scarp, so that the surface offset measured from the profile can be graphically related to the vertical component of fault displacement (details in Chapter 3). However, if profiles are intended primarily for dating the scarp using erosion-based methods (Chapter 3) rather than for measuring displacement, the profile line should follow the line of steepest slope, so that it parallels the transport vector of material across the scarp. Where fault scarps trend perpendicular to the local slope of terrain (the typical case), a profile perpendicular to fault strike will also parallel the local fall line, and such profiles can be used for both displacement and age estimates. Where fault scarps trend diagonally across local slopes, the fall line runs diagonally across the fault scarp, so scarp profiles laid out to estimate age (as described earlier) will exaggerate displacement. If surface faulting had an oblique component, measurements of vertical surface offset based on fault scarp profiles will underestimate true net slip. This shortcoming cannot be overcome unless well-defined *piercing points* can be identified above and below the fault scarp (see Chapter 6).

If profiles are to be dated with the empirical regression technique (Section 3.4.1), one should measure multiple scarp profiles with as wide a range in height and maximum slope angle as possible. The *maximum scarp slope angle,* also critical for scarp dating, should be averaged from four to eight measurements made close to each profile site (Machette, 1989), rather than from the single value measured on the scarp profile.

The best sites for profiling, especially if *diffusion dating* is to be performed, are those where creep and rain splash have been the main scarp-modifying agents, because the strength of these processes is not dependent on slope position. Such sites are often found on the interfluves between gullies that dissect the scarp. Areas affected by sheetwash, rillwash, gullying, slumping, spring sapping, animal tracks, or human disturbances should be avoided. Likewise, areas of eolian or alluvial fan deposition against the scarp should be avoided. If diffusion dating will be performed on profiles, steep faulted surfaces should be avoided (see Section 3.4.1).

The preferred method of fault scarp profiling is somewhat dependent on the degree of surface vegetation cover. On sparsely vegetated fault scarps in the western United States, Wallace (1977) and Bucknam and Anderson (1979) laid a telescoping *stadia rod* directly on the ground surface and measured its inclination with an *Abney level*. The rod is moved sequentially along the profile with the length and inclination of every segment recorded. Profile segments at the scarp base and crest (see definitions in Table 3.3) must be short enough to portray *scarp curvature* accurately, because curvature is a critical factor in diffusion dating (Section 3.4.1). If vegetation density precludes laying a rod directly on the surface, scarp profiles can be made by more traditional leveling methods using the rod in a vertical position. Profiles over long distances (across broad scarps or folds) have been made via trigonometric leveling using instruments such as total stations, and most recently, by handheld laser rangefinders. The more advanced handheld models, such as the laser guns manufactured by Laser Atlanta, have a built-in compass, and a range of 600 m reflecting off of natural terrain (9 km with prisms).

Errors in slope angles arise from natural undulations in the ground surface, presence of low vegetation and roots beneath the rod, inaccuracy of the Abney level, and unquantifiable "operator error." Mayer (1984, pp. 305) reports slope angle errors (1 standard deviation) of ± 1 to $2°$ using the Wallace method. Slope segment lengths may contain error from incorrect positioning or misalignment of the rod in each segment, as well as errors in reading the rod length. Scarp heights measured from profiles are affected by uncertainties in both angles and segment lengths. Mayer (1984) reported 1-sigma errors of ± 6–15% in scarp height when making repeated profiles at the same location with three different individuals and two types of rods. This degree of error has significant implications for direct methods of fault scarp dating (Table 3.8). Error might be diminished by staking or stringing the profile line (to reduce rod misalignments), removing surface vegetation, and using more accurate rods and inclinometers, but no rigorous field tests of these corrective practices have been made.

Recent studies have measured fault scarp profiles with survey-grade GPS receivers, with horizontal and vertical accuracies measured in centimeters. GPS surveys should have smaller measurement errors than the older manual methods described above. However, even GPS fault scarp profiles will vary along the strike of a fault scarp, due to natural variations in the scarp height and slope angle (intrinsic variability of the measured phenomenon).

2A.2.4.2 Topographic Riser Profiling

Topographic risers are created by fluvial or coastal erosion (e.g., wave-cut cliffs) and define *linear geomorphic datums* that can record paleoseismic deformation. The most commonly used topographic riser in paleoseismology is the wave-cut cliff at the landward edge of a marine terrace. The *nickpoint* at the junction of the wave-cut cliff and marine platform (termed the *shoreline angle*) defines an originally horizontal datum. Similar nickpoints at the base of fluvial terrace risers define a datum that slopes downstream at the stream gradient existing at the time of riser formation. The locus of points defined by these nickpoints forms a *paleodatum* that should record deformation from paleoearthquakes.

In most cases nickpoints have been covered with colluvium, so their elevation must be reconstructed by drilling, geophysics, or profile projection. Bradley and Griggs (1976) describe the use of all three methods to measure the present elevation of shoreline angles on uplifted and tilted marine terraces in California. Most elevations are estimated as the graphical *intersection point* of the projected shoreline platform and the projected wave-cut cliff, using average angles for both projections, and assuming that the original position of the wave-cut cliff was in the center of the present degraded cliff (e.g., McCalpin, 1994, for deformation of the Pleistocene shorelines of Lake Bonneville, Utah). LaJoie (1986) provides an extensive summary of measurements made on deformed marine terrace nickpoints.

2A.2.5 Dating Methods for Late Quaternary Landforms

At the reconnaissance stage, the most commonly used methods for dating displaced Quaternary landforms belong to the geomorphic and correlation categories (Table 1.3); primarily geomorphic position, soil profile development, and climatic correlation. These techniques are often sufficient to establish the local Quaternary stratigraphic framework and to establish whether more detailed studies are warranted. For detailed Quaternary geologic mapping in a seismic area, however, the ages of the displaced/undisplaced landforms must be known with more precision (e.g., to determine preliminary fault source parameters such as recurrence intervals and slip rates). These more precise ages are determined by either: (1) directly dating the surface of the displaced/undisplaced landform, (2) dating the soil profile that has developed subsequent to stabilization, (3) dating the uppermost deposit that underlies the landform, or (4) directly dating the deformation feature that displaces the surface (e.g., a fault scarp). In this section we only discuss items 1–3; item 4 is discussed separately in Chapters 3 (normal fault scarps) and 5 (reverse fault scarps). It must be remembered that deformation features that displace the surface of a landform (item 4, above) may be younger, to much younger, than the landform stabilization age.

Directly Dating the Landform: Prior to the 1990s, dating the stabilization age of a Quaternary landform was accomplished by chemical/biological methods (obsidian hydration, rock varnish cation ration, dendrochronology, lichenometry), and geomorphic methods (rock and mineral weathering, rock varnish development). All of these methods except dendrochronology predict surface age by measuring a degree of surface alteration, and dividing that by an assumed rate of alteration (i.e., they are calibrated ages, in the terminology of Table 1.3). The methods all suffer from a common weakness, in that the rate of alteration is dependent on variables other than the age of the surface (e.g., rock composition, mean annual precipitation, and temperature, etc.). In most paleoseismic investigations there is not enough time or money to conduct dating experiments to establish the effects of these other "extra-temporal" variables and thus correct the calibrated ages. This drawback then limits the precision of the age estimates.

However, in the late 1980s it was discovered that cosmogenically produced isotopes within surface rocks (^{36}Cl, ^{10}Be, ^{26}Al, ^{14}C, ^{3}He) can be used to numerically date the stabilization age of a geomorphic surface (see review papers by Zreda and Phillips, 2000; Gosse and Phillips, 2001). Since that time, numerous studies have used the method to establish a precise Quaternary geomorphic framework for their study area (Brown *et al.*, 1998 [Tien Shan, ^{36}Cl]; van der Woerd *et al.*, 1998 [Kunlun Fault]; Ritz *et al.*, 2003 [Gobi-Altai, ^{10}Be]; van der Woerd *et al.*, 2006 [San Andreas fault, ^{10}Be-^{26}Al]; Frankel *et al.*, 2007a,b). Even the so-called precarious rocks (untoppled by past earthquakes) have been cosmogenically dated by Bell *et al.* (1998).

Dating the Soil Profile that Underlies the Surface: Once a geomorphic surface stabilizes (ceases to be actively eroded or deposited), weathering processes create soil horizons (pedogenic horizons) in the deposit (parent material of the soil) underlying the surface. With increasing time soil horizons become thicker and better developed, a process that has led to the concept of the *soil chronosequence* in pedology (see Birkeland, 1999). Prior to the development of cosmogenic dating, almost all geomorphic surfaces were initially "dated" based on their degree of soil profile development, particularly in arid and semiarid areas where datable carbon was lacking. Soil profiles were described qualitatively, or characterized semi-quantitatively by computing a profile development index (PDI) or soil development index (SDI; see Harden, 1982, Harden and Taylor, 1983). PDI and SDI quantify several time-dependent soil horizon properties discernable in the field. These include horizon color (compared to the unaltered parent material), horizon texture, dry consistence, soil structure (development of peds), and morphology of secondary calcium carbonate. The age of the geomorphic surface would then be estimated by reference to calibrated curves of PDI as a function of surface age (as independently dated by some other means, typically radiocarbon). In most cases PDI-based age estimates could not be independently checked, due to the lack of appropriate material for any other dating method. However, Zehfuss *et al.* (2001, GSA data repository) compared PDI values from soil pits in Owens Valley, California, to the cosmogenic ages from nearby boulders to create a calibration study.

Dating the Deposit that Underlies the Landform: Dating the uppermost part of the deposit (parent material) that underlies a geomorphic surface yields a maximum age for stabilization of the overlying surface. An example is the optically stimulated luminescence (OSL) dating by Zuchiewicz *et al.* (2004) on terraces displaced by the Dien Bien Phu fault zone, NW Vietnam. The one danger in this method is inadvertently sampling material for dating that has been physically intruded or introduced into shallow deposits after (perhaps long after) the surface was stabilized. There are many processes that can intrude younger material into older deposits in the soil-process zone, including plowing and digging by humans, burrowing by animals (bioturbation), evacuation and back-filling of tree roots, frost churning in periglacial environments (cryoturbation),and so on. One can avoid this by sampling beneath the soil-process zone, but the farther below the geomorphic surface one samples, the less closely the sample age will constrain the age of the overlying geomorphic surface.

2A.3   Mapping Paleoseismic Stratigraphy

The stratigraphic expression of paleoearthquakes can range from displaced strata and angular unconformities (primary evidence) to clastic dikes and soft-sediment deformation (secondary evidence; Table 1.1). Such stratigraphic evidence may or may not be accompanied by geomorphic evidence of paleoseismicity. Early paleoseismic investigations concentrated on geomorphic evidence because natural vertical exposures in fault zones are rare. In the absence of such exposures, however, early paleoseismic investigators were faced with the choice of (1) relying only on geomorphic data for the paleoseismic analysis, (2) drilling boreholes or collecting geophysical data, or (3) excavating artificial trench exposures across deformation zones. The development of modern paleoseismology owes much to developments in the latter two of these fields. In the following sections we contrast geophysics, trenching, and drilling as paleoseismic techniques and describe methods used successfully in previous paleoseismic investigations.

The ultimate goal of studying stratigraphy in earthquake areas is to identify "event indicators" in the stratigraphic sequence, and to then determine if the "events" that created these indicators are paleoearthquakes, or some other type of deformational, erosional, or depositional event. If the characteristics of an event indicator unambiguously indicate an earthquake origin, then the event indicator can be further interpreted as an *earthquake horizon* (formerly termed an *event horizon* in paleoseismic literature). This 2-stage terminology was proposed by Scharer *et al.* (2007) and permits paleoseismologists to differentiate between observational evidence (event indicators) and interpretations (earthquake horizons).

Event indicators are "morphologic and sedimentologic evidence of ground deformation documented in trench exposures. Event indicators are the basic evidence from which we build the case for each interpreted earthquake horizon. . ." In their system, "elevating an event indicator or a set of event indicators to the status of a *paleoearthquake horizon* is based on a combination of quality and frequency of indicators at a particular horizon. . . .. The highest quality event indicator has only one interpretation; that the morphologic and sedimentologic features could only have been caused by earthquake-induced ground deformation. In contrast, low-quality event indicators may be interpreted in more than one way" (see Table 1.2, row labeled "Abundance of similar nonseismic features"). By ranking the quality and abundance of event indicators, it is easier to identify unambiguous earthquake horizons from other unconformities in the stratigraphic section created by nonearthquake processes. Examples of event indicators and earthquake horizons are described in Chapters 3–6 for various tectonic environments.

One of the most important advances in paleoseismology since the first edition was published is the increased use of shallow geophysical surveys after the mapping phase of study, but prior to the trenching phase (Figure 2A.1). This practice has increased the success level of trenching studies, because investigators can now know where the major paleoseismic deformation structures are prior to trenching. They can compare several possible trenching sites to determine how well the structures are expressed; whether the deformation zone is narrow or broad, shallow or deeply buried by postfaulting sediments; whether the groundwater level will likely interfere with trenching; and whether hard bedrock will be encountered on the upthrown side of the fault. Knowing these practical facts permits them to choose the best trenching site among several, to plan in advance for any necessary dewatering, to plan the length, width, and depth of the trench, and to hire the appropriate excavating machinery. In the past, many of these topics were simply left to chance, and the answers to the above questions unfolded, for better or worse, as the trench was dug. But most investigators today would agree that it is worthwhile investing some time and expense in a campaign of geophysical surveys, rather than trenching based only on visible surface evidence.

2A.3.1   Geophysical Techniques in Paleoseismology

Geophysical methods can be useful in terrestrial paleoseismology in three ways (1) as a reconnaissance technique to define shallow stratigraphy and structure in a fault zone (or liquefaction area), in order to locate the optimum sites for further trenching or drilling; (2) for tracing faults to depths greater than can be reached by trenching (>5–6 m) or drilling (typically 15–25 m); and (3) for detecting buried faults that have no surface expression. (Offshore geophysical methods are described in Chapter 2B). The following discussion is based partly on the excellent collection of recent case histories in Stephenson and McBride (2003).

The required depth of penetration depends on whether geophysics is being used (1) to merely locate sites for future trenching (in which case, ≤5 m), (2) as a substitute for trenching (in which case, 5–10 m), or (3) if geophysics is being used to trace faults downward in crustal-scale cross sections (tens to thousands of meters). To be useful in paleoseismic investigations, geophysical surveys of the first two types need to have high spatial resolution and the ability to distinguish between unconsolidated deposits with (often) very similar material properties. The field of engineering geophysics includes shallow site exploration, and texts in that field provide the background for the methods described next. The historic survey of geophysics applied to fault assessment (Krinitsky, 1974) has now been updated by Stephenson and McBride (2003) to include new methods such as ground-penetrating radar and seismic refraction tomography. A special application of geophysics in volcanic-extensional terranes is locating dike swarms beneath volcanic rift zones (Section 4.3.3).

Most of the successful investigations described below relied on only a single geophysical technique. However, a few recent investigations have used multiple techniques; e.g., Wise *et al.* (2003) used gravity, vertical electrical soundings, 2D resistivity, seismic reflection, and seismic refraction; Kurcer *et al.* (2008) used electrical resistivity tomography (ERT) and P-wave seismic tomography.

2A.3.1.1 Seismic Methods

Seismic reflection and *seismic refraction* methods can contribute to paleoseismology in two ways (1) by detecting faults and (2) by characterizing subsurface strata that have been offset, folded, or tilted by faulting. However, the simple detection of a fault plane may not, in itself, provide any data on the timing and magnitude of individual paleoearthquakes. Because seismic reflection methods image horizontal and dipping reflectors, the method is best suited to detecting vertical separations associated with a dip-slip component of faulting. In contrast, seismic refraction usually does not possess the resolution to detect small paleoseismic features. For example, Pelton *et al.* (1985) could only distinguish the largest faulted units along the 1983 Borah Peak, Idaho, surface rupture trace with seismic refraction.

In the late 1990s 2D seismic refraction tomography was collected across several fault scarps, to see if the structural detail observed in paleoseismic trenches could be also interpreted from seismic tomograms. Early studies were performed on normal faults in the US Basin and Range Province (Utah), at the sites of previous paleoseismic trenches (Morey and Schuster, 1999, Oquirrh fault; Mattson, 2004, Mercur fault; Sheley *et al.*, 2003, Oquirrh fault). Geophysicists used a 3.6 kg sledgehammer source and 40–48 Hz vertical motion geophones to collect the data, with geophones spaced at either 0.5 or 0.66 m intervals, and a hammer shot point occupying every geophone position (Figure 2A.10). They used the multigrid SIRT algorithm of Morey and Schuster (1999) to reduce the data.

The authors coined the term "seismic trenching" for their use of 2D refraction tomograms, a term which, in hindsight, has proved a bit premature. In the studies cited above (all of which were of fault scarps on gravelly late Pleistocene alluvial fans) the authors determined that the lowest velocity regions beneath the fault scarp corresponded to the scarp-derived colluvial wedge. Although this interpretation was valid for the gravelly alluvial fans in their area, it was hardly universal, as they discovered when they tried to export that paradigm to fault scarps in other depositional environments. For example, on paleoshoreline deposits with interbedded loess along the Wasatch fault (McCalpin, 2002), they misinterpreted the subsurface deposits (Sheley *et al.*, 2003), because loess and shoreline sands there had lower P-wave velocities than the gravelly, scarp-derived colluvial wedges (Figure 2A.11). The same problem was encountered by Buddensiek *et al.* (2008) on the 9-m-deep Mapleton "megatrench" on the Wasatch fault. Although several identified "low-velocity zones" corresponded to colluvial wedges mapped in the trench, the largest wedge in the trench could not be identified in the tomograms, and there were many other small low-velocity zones in the tomogram that did not correspond to any colluvial wedge.

In another case of "mistaken identity," Kurcer *et al.* (2008) used P-wave refraction tomography across a scarp in Turkey and identified a 3 m vertical displacement of the underlying seismic horizons. However, when the scarp was trenched later, strata beneath it extended continuously beneath the scarp and were not faulted; the authors then concluded it was an erosional scarp. Despite these failures, the refraction tomography method could be further refined for use in different depositional environments, via calibration studies at existing trench sites so that the true subsurface conditions could be observed, at least in the upper 4–8 m.

Figure 2A.10: Photograph of a seismic refraction tomography line extending from the downthrown block (foreground) up the scarp face (middle ground), from Sheley *et al.* (2003). White flags indicate geophone locations, spaced 0.5 m apart.

High-resolution *seismic reflection* techniques have been successfully used to map subsurface fault structure in zones of Quaternary dip-slip faulting. An early reconnaissance study along the Wasatch fault zone, USA, used the *Mini-Sosie system* (Crone and Harding, 1984a). The system was optimized for detecting normal faults that displace late Quaternary lacustrine and alluvial gravels. In a similar study, Crone and Harding (1984b) used Mini-Sosie to detect buried normal faults between surface fault scarps in a swarm of normal faults. The Mini-Sosie system has also been used to characterize the shallow geometry of the 1983 Borah Peak rupture (Miller and Steeples, 1986; Treadway *et al.*, 1988) and the Quaternary oblique-slip Meers fault, Oklahoma, USA (Myers *et al.*, 1987; Miller *et al.*, 1990). More recent studies of the Wasatch fault zone, USA, have refined the high-resolution method (Benson and Mustoe, 1991; Stephenson *et al.*, 1993). At the Kaysville trench site (Swan *et al.*, 1980; McCalpin *et al.*, 1994), Stephenson *et al.* (1993) were successful in detecting not only fault planes in a complex fault scarp-graben system, but also imaged several subsurface contacts tilted toward the fault. Comparison of detailed seismic reflection records with the trench log (Figure 2A.12) revealed that most details of trench macrostratigraphy could be imaged. However, seismic methods could not differentiate individual colluvial wedges 1 to 2 m thick against the main fault plane, and thus could not independently determine the number of or size of prehistoric displacements. High-resolution seismic reflection seems to work best in well-stratified, thinly bedded deposits such as found in sag ponds or playas (Figure 2A.13), where even relatively small displacements can be seen and used to map fault traces (Zilberman *et al.*, 2005). In such

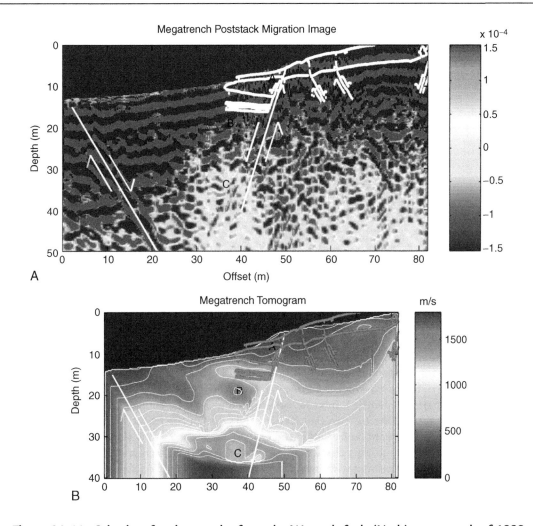

Figure 2A.11: Seismic refraction results from the Wasatch fault (Utah) megatrench of 1999 (McCalpin, 2002). (A) Poststack migration image and (B) refraction tomogram. White lines show faults interpreted by Sheley *et al.* (2003), black lines (red in color version) are their rendition of McCalpin's faults. The ground surface of McCalpin's trench does not exactly match the ground surface from the seismic profile because the survey line was about 20 m to the north and subparallel to the trench. From Sheley *et al.* (2003). (See Color Insert.)

environments P-wave refraction tomography would not work so well, because most of the sediments share a similar P-wave velocity.

In areas of shallow groundwater, seismic reflection surveys based on S-waves are superior to those based on P-waves, because S-waves are not affected by the "masking effect of water in very low-velocity sediment" (Woolery, 2005). For example, looking for dip-slip faults buried beneath saturated late Quaternary sediments, Woolery (2005) claimed subsurface resolutions of 1.5–2.2 m using S-wave seismic reflection (spread of 48 horizontally polarized geophones, spaced 6–20 m apart, sledgehammer source).

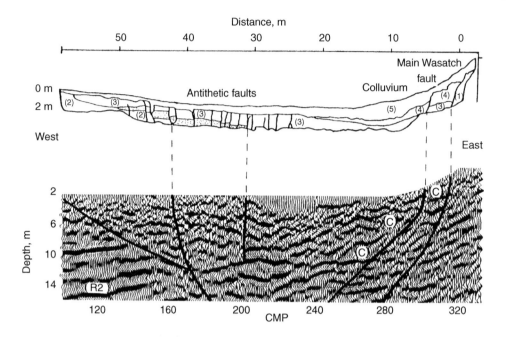

Figure 2A.12: Comparison of high-resolution seismic reflection data (lower part) with a trench log (upper part) across a Quaternary normal fault scarp and graben, Wasatch fault zone, Utah. Map units in upper part: 1, late Pleistocene lacustrine sand and silt; 2, Holocene alluvial fan; 3, older colluvium and graben sediments; 4, intermediate-age colluvium; 5, younger colluvium and graben sediments. Figure 2A.16 shows the two faults at the right end of the trench log. Seismic data cannot image all the small-displacement faults within the graben (between 25 and 40 m on upper scale), but do indicate a decrease in fault dips that could not be inferred from trench data. From Stephenson *et al.* (1993); reprinted with permission of the American Geophysical Union.

2A.3.1.2 Ground-Penetrating Radar

Ground-penetrating radar (GPR) produces subsurface images that superficially resemble seismic reflection results. The similarities arise from the common use of transmitted waves that are reflected and then detected by a receiver on the surface. In contrast to the compressional elastic waves utilized in seismic reflection (frequency = 100 Hz), radar uses transmitted electromagnetic radiation with frequencies from 80 to 300 MHz. Materials with high electrical conductivity such as clay or fluids with high dissolved solids will rapidly decrease the depth of penetration. Subsurface contacts with higher contrasts in dielectric properties return stronger reflections. Studies that have tested GPR for shallow exploration for engineering purposes include Hammond *et al.* (1986) and Kuo and Stangland (1989).

When our first edition was published in 1996, few published studies had applied GPR to zones of active faulting (Bilham, 1985; Bilham and Seeber, 1985; Smith and Jol, 1995). They found that coarse colluvium and fan alluvium at faulted range fronts did not possess dielectric layering, and so returned few reflections on GPR images. Likewise, massive, moist saline soils were opaque to GPR. However, prehistoric colluvial wedges made of dry gravels were detected at the Borah Peak, Idaho, fault scarp. The colluvial wedges there were imaged by detecting the underlying, clay-bearing buried soils that formed during interseismic periods of landscape stability. Wide zones of faulting were imaged by GPR on the Wasatch

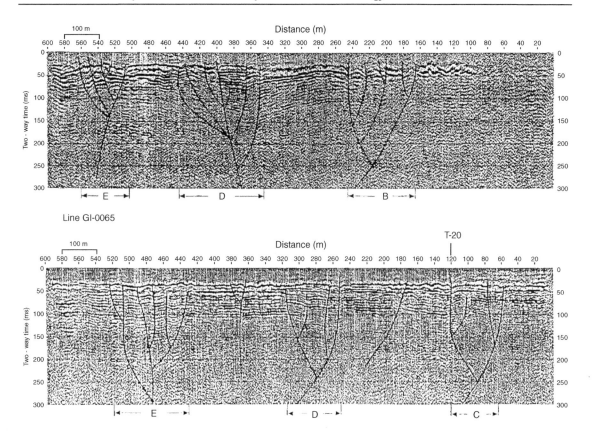

Figure 2A.13: Profile of a high-resolution seismic refraction line across the Avrona playa shear zone, Israel. This is section GI-0065 of Zilberman *et al.* (2005) across the Elat fault. The Elat fault is primarily a sinistral strike-slip fault, and the groups of faults seen in the section (B–D) are each positive flower structures. Note the fine stratification in the playa sediments. From Zilberman *et al.* (2005).

fault zone (Smith and Jol, 1995) and the San Andreas fault zone. Thus, GPR promised some potential in environments with sediments that were neither too clayey nor too saline.

Studies between 1996 and 2008 have refined the method somewhat, expanding from dry environments to progressively wetter environments. The GPR surveys were mainly run prior to trenching, to identify the location of major structures (normal faults, Meghraoui *et al.*, 2000; reverse faults, Anderson *et al.*, 2003; strike-slip faults, Gross *et al.*, 2002, 2004; Green *et al.*, 2003; Ferry *et al.*, 2004). Anderson *et al.* (2003) discovered that GPR could be useful for locating faults in very coarse alluvial fans at range fronts, even given the bouldery texture and coarse bedding in the fans (Figure 2A.14).

McCalpin and Harrison (2000) found that normal faults displacing buried paleosols developed on eolian and fine-fluvial deposits were easily visible on GPR profiles, as a disruption and displacement of the subhorizontal radar horizons (Figure 2A.15). A similar conclusion was made by Meghraoui *et al.* (2000). However, the radar horizons are not always sufficiently unique to permit a bed-by-bed correlation across the fault. As shown in Figure 2A.15, the contacts correlated across the fault by the radar geophysicist were discovered to be different paleosols once the trench was excavated, such that the GPR correlation underestimated the total vertical displacement. This underestimation is probably a common feature on

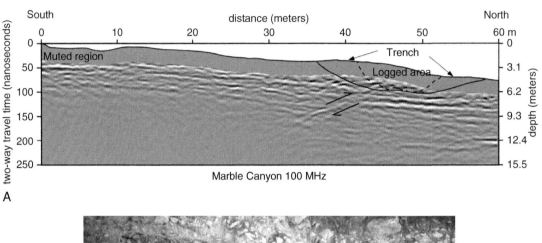

Figure 2A.14: GPR defines a thrust fault target on an alluvial fan. (A) GPR profile across the fault scarp on the alluvial fan; (B) Photograph of the thrust fault in the trench wall. The outline of the trench is shown. The thrust fault is well defined by a reflector dipping to the west (left), which is well defined below the depth of the trench, but poorly defined in the trench; From Anderson *et al.* (2003).

dip-slip faults and was ultimately caused by inter-event erosion on the upthrown fault block but not on the downthrown block. This resulted in a truncated stratigraphic section on the upthrown block, as opposed to a complete section on the downthrown block. Given this difference in the stratigraphic sections, the second or third reflector below the surface on the upthrown block will not necessarily be the same unit as the second or third reflector below the surface on the downthrown block.

2A.3.1.3 Electrical Methods

In the past 10 years there has been a marked increase in the use of 2D electrical resistivity surveys across fault traces or fault scarps, particularly in Europe, in order to locate structures prior to trenching. Caputo *et al.* (2003) is often cited as the definitive reference for this technique, of electrical resistivity

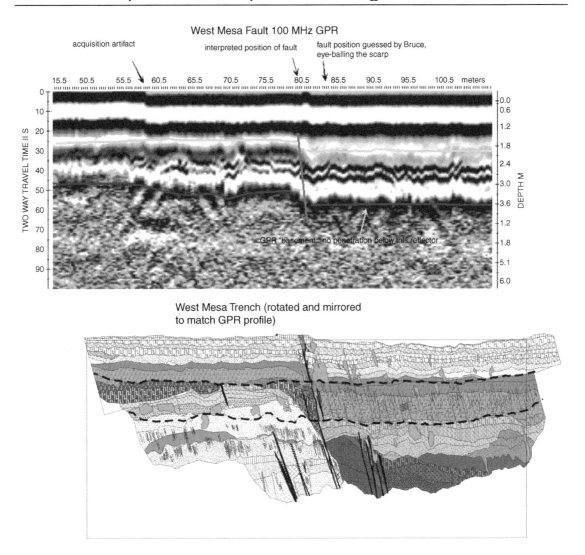

Figure 2A.15: 100 MHz GPR profile (top) and trench log (bottom) of the Calabacillas normal fault, Rio Grande rift, New Mexico, USA. Scale is in meters, total penetration depth is about 3.6 m, limited by a clay-rich paleosol. The fault position interpreted from GPR (green line at top) corresponds to the fault exposed in the trench walls. However, the stratigraphic contacts interpreted from GPR (yellow and blue lines on profile) did not represent the same stratigraphic unit on the footwall and hanging wall. Thus, the true vertical displacement was much greater than inferred from GPR, as can be appreciated from the trench log. From McCalpin and Harrison (2000). (See Color Insert.)

tomography (ERT). ERT has several advantages over other methods. First, its depth penetration is not limited by moisture, salinity, or clays as is GPR. Instead, penetration depths of 25–30 m are common (Figure 2A.16; see also Similox-Tohon *et al.*, 2005). Second, resistivity values are not as ambiguous about the effect of the water table (depth of saturation) as is P-wave seismic tomography, in which a velocity of ca. 1500 m/s can be interpreted as either saturated soft material or dry harder material. A disadvantage is that metal objects in the ground and electrical wires can induce false signals, particularly in urban areas.

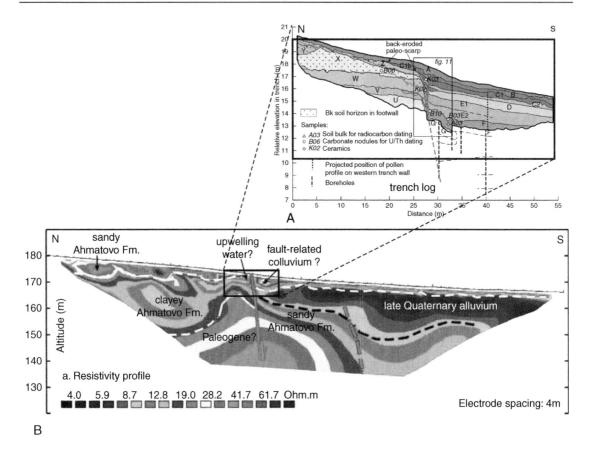

Figure 2A.16: ERT profile across the Chirpan normal fault, Bulgaria. This fault experienced 0.45 m displacement in the 1928 Chirpan earthquake. The trench log (top, ca. 4 m deep) shows three earlier colluvial wedges of middle to early Holocene age. The ERT profile (bottom) penetrates to about 30 m and shows that the major vertical displacement of pre-Quaternary deposits occurs on a buried fault farther south than the Holocene fault, something that is not obvious from the modern topography. From Vanneste *et al.* (2006). (See Color Insert.)

Caputo *et al.* (2007) used ERT to trace a normal fault in southern Italy buried under a 1-km-wide late Pleistocene alluvial plain; the fault was buried by up to 10 m of alluvium in places, yet was clearly imaged on the ERT. Despite the success of this technique in Europe, it has not been widely adopted in North America, for reasons unknown; the ERT method is not mentioned in the collection of papers edited by Stephenson and McBride (2003).

2A.3.1.4   *Electromagnetic Methods*

Electromagnetic induction surveys have been used in conjunction with standard Wenner-array electrical resistivity profiling in the USA to detect liquefaction-induced deformation (Wolf *et al.*, 1998, 2006). Whereas resistivity was more successful in locating long sand dikes (typically related to lateral spreading), Wolf *et al.* (2006) concluded that "The electromagnetic induction method (EM-31), while less sensitive to dike locations, was useful in characterizing depositional facies changes by their differences in electrical conductivity. Data from the study site support the interpretation that the earthquake-induced liquefaction features occurred near the boundary of a facies change, which may have constituted a zone of weakness along which excess pore fluids and sand escaped." Thus, EM methods may be helpful in defining the stratigraphic setting for paleoearthquake deformation.

2A.3.1.5 Magnetic Methods

Aeromagnetic surveys are a standard exploration tool for detecting large-scale faulting, particularly for dip-slip faults as in the Basin and Range Province, USA (early paper by Smith, 1967). Grauch (2001, 2002) and Grauch *et al.* (2001) refined this technique in the Basin and Range province, USA, to locate poorly expressed intrabasin normal faults using low-elevation acquisition flights (100–150 m terrain clearance) and closely spaced flight lines (100–150 m). The resulting linear magnetic anomalies (Figure 2A.17) result from the abrupt thickening of syntectonic sediments (coarser, more magnetic) on the downthrown side of the normal fault. Steep aeromagnetic gradients are assumed to overlie near-vertical contacts (faults) between deposits with different magnetic properties. To accentuate these steep aeromagnetic gradients, Grauch (2002) used a "gradient window method" which computes the aeromagnetic horizontal gradient by moving a 1 km × 1 km window over the gridded data. The gradient map is treated as a DEM and, when illuminated, depicts steep gravity gradients as if they were surface ridges and troughs (Figure 2A.17A). Interestingly, aeromagnetic surveys in these extensional basins reveal many more faults than intersect the ground surface (Figure 2A.17B). These "hidden" or "blind" normal faults are not all deeply buried; most of them extend to within 100 m of the present ground surface. Grauch's (2001, 2002) results pose somewhat of a dilemma for seismic hazard analysis in extensional basins, because it is unknown whether all these blind faults are seismogenic.

Fewer studies have used on-ground measurements to detect faults on the scale of meters to tens of meters. Bailey (1974) detected fault gouge with a magnetometer in California. Salyards *et al.* (1992) used the paleomagnetic signature of sediments within 50 m of the San Andreas fault to identify rotations resulting from plastic deformation in a wide dextral shear zone. This plastic deformation accommodated as much displacement as brittle faulting at the main fault, but had not previously been recognized. Recognition of the plastic deformation resulted in doubling the estimate of the Holocene slip rate for this fault location.

Deep imaging (down to ca. 2 km) can be achieved by magnetotelluric (MT) profiling using long profile lines. For example, Park *et al.* (2003) measured a 5.4-km-long MT profile across a range-front thrust fault in the Kyrgyz Tien Shan, central Asia. The profile revealed a 1- to 2-km-thick section of Neogene strata overthrust by crystalline rocks along a gently dipping decollement. A cross section based on the MT data suggests >2.5–7.5 km of total shortening across the basin margin, of which at least 2.5–4.8 km clearly involves the decollement in late Neogene time. That information allowed them to calculate a long-term Neogene slip rate to compare with late Quaternary slip rates based on fault scarp heights.

2A.3.1.6 Gravity Methods

The structural relief resulting from displacement on dip-slip faults can be well expressed as differences in gravitational attraction across the fault. Near-surface faults in unconsolidated material, however, can only be interpreted from gravity data if stations are closely spaced and very accurately surveyed. Benson and Baer (1987) used a Worden gravimeter, with stations on 7.6- to 12.2-m spacings, surveyed to a 3 cm vertical accuracy, to detect buried normal faults in alluvium on the Wasatch fault zone, USA. Residual Bouger gravity anomalies of ±0.2 milligals correlated reasonably well to mapped normal faults, faults discovered in creek banks during their study, and newly inferred faults with no surface expression.

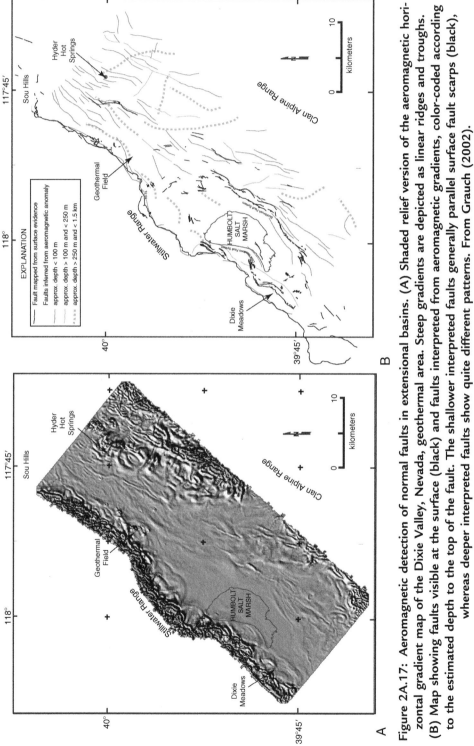

Figure 2A.17: Aeromagnetic detection of normal faults in extensional basins. (A) Shaded relief version of the aeromagnetic horizontal gradient map of the Dixie Valley, Nevada, geothermal area. Steep gradients are depicted as linear ridges and troughs. (B) Map showing faults visible at the surface (black) and faults interpreted from aeromagnetic gradients, color-coded according to the estimated depth to the top of the fault. The shallower interpreted faults generally parallel surface fault scarps (black), whereas deeper interpreted faults show quite different patterns. From Grauch (2002).

However, this application occurred in an optimum setting for gravity contrasts, where low-density alluvium overlying high-density bedrock became abruptly thicker across each fault.

2A.3.2 Trenching

Excavation of trenches in deformation zones has become a major element of paleoseismic studies in most countries. Following the early use of trenches in NPP investigations in the United States (Hatheway and Leighton, 1979; Hatheway, 1982), trenching techniques have expanded to address problems of paleoearthquake faulting, folding, ground failure, and faulting-induced sedimentation. In the following sections we describe in considerable detail the mechanics of excavating and logging trenches. In contrast, siting and interpreting of trenches depend heavily on the types of features being investigated, so those topics are addressed as appropriate in Chapters 3–8.

Much delicate paleoseismic evidence may also be destroyed by trenching and backfilling. In this regard paleoseismic trenching is similar to archeologic excavations that so disturb a site that future interpretation must rely on the initial excavation. *Conservation archeology* (Schiffer and Guterman, 1977) is the concept that some critical sites are better left undisturbed and saved for future excavation using improved techniques, rather than being hastily excavated at present. This concept has not been widely applied to paleoseismology, but it may be appropriate in the future.

2A.3.2.1 Location, Orientation, and Pattern of Trenches

The best location, orientation, and pattern for trenches are highly site dependent, so only the most general guidelines are presented here. However, past work has shown that trench placement is such a critical element in paleoseismic investigations that success or failure often depends on it. Stated another way, once a trench is sited the stratigraphy and structure to be exposed is somewhat predetermined. If the investigation is to be successful, the trench must yield the type and quality of data anticipated by the investigator, a result which is not always accomplished.

Trenches across faults are typically sited to optimize data on either *paleoearthquake displacement* or *paleoearthquake recurrence* (Sieh, 1981). The best sites for measuring displacement are where all displacement is concentrated on a single, narrow fault strand, and subsidiary faulting and folding are negligible. The best sites for measuring recurrence (i.e., for dating individual paleoearthquakes) are local fault-zone depressions filled with fine-grained and/or organic interfaulting sediments. Such depressions are often associated with distributed faulting and/or folding, where subsidiary faults created by successive paleoearthquakes are separated vertically and horizontally and can be distinguished. On dip-slip faults, the best trench locations for measuring displacement and recurrence often coincide, such as single-trace, high fault scarps fronted by sediment-filled graben (Figure 2A.18) or back-tilted areas (e.g., Swan *et al.*, 1980; McCalpin *et al.*, 1994). On strike-slip faults good sites for measuring displacement are usually poor for measuring recurrence, and vice versa (Chapter 6).

Trench location is also dictated by the number of paleoearthquakes the investigator wishes to observe. Trenches across faults on very young Quaternary surfaces may expose only one or two paleoearthquake displacements, so trench structure and stratigraphy may be relatively simple. Trenches on progressively older surfaces are likely to expose the cumulative deformation from many paleoearthquakes, where the effects of earlier displacements are obscured by those of later displacements. In narrow zones of deformation, multiple displacements often result in complex shearing, cross-cutting, and interfaulting sedimentation and weathering that are difficult to reconstruct; in such situations individual paleoearthquakes, especially older

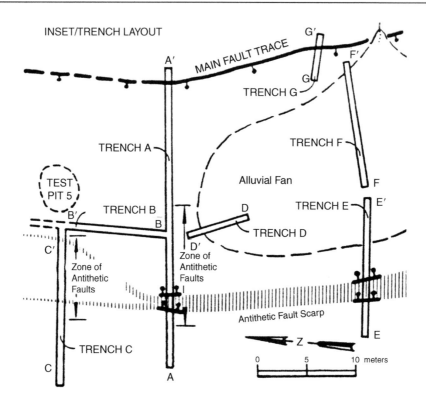

Figure 2A.18: Map of multiple trench locations and orientations placed in a graben at the foot of a large normal fault scarp. Fault-perpendicular trenches (A, C, E, G) expose fault planes and colluvium from which the paleoseismic interpretation is made. Trench F was dug to find datable material in the unfaulted alluvial fan. Fault-parallel trenches B and D trace strata deposited in the graben by fault-parallel streams. From Swan *et al.* (1981).

ones, may be overlooked. For example, few investigations have recognized more than four paleoearthquakes in a single trench unless faulting was widely distributed (e.g., Sieh, 1978a; Fumal *et al.*, 1993). Beyond a certain height, dip-slip fault scarps may be too large to be completely trenched by available commercial excavating equipment. For example, on the Wasatch fault, Utah, trenches have generally been limited to scarps 10 m high or smaller, and these scarps typically represent only three or four surface-faulting events. Early attempts to trench larger scarps (such as the 23 m-high scarp on the Wasatch fault zone at Mapleton, Utah; Swan *et al.* 1980) with standard size construction backhoes resulted in incomplete penetration through the colluvial wedge sequence at the scarp base. However, more recently benched "megatrenches" have achieved deeper penetration (Figure 2A.19).

Based on Quaternary geologic and geomorphic mapping and knowledge of local slip rates, the paleoseismologist can usually estimate the number of paleoearthquakes that might be exposed in trenches on various landforms. Trench location is then partly dictated by the goals of the investigation, for example, to characterize only the most recent paleoearthquakes (by trenching young deposits) or to compile a long history of deformation (by trenching older deposits). In addition, trench locations are often restricted by nongeological considerations, such as road access, land ownership, and previous ground surface disturbance. Whenever possible, trenching sites should have undergone minimal prior surface

Figure 2A.19: A large benched trench across a 23 m-high normal fault scarp on the Wasatch fault zone at Mapleton, Utah. Note the deep inner slot which is shored, and small vertical boards on the first bench, to assist in climbing up the trench. Preliminary results are described by Olig *et al.* (2005). (See Color Insert.)

disturbance (grading, filling) which might destroy the critical relationships of faults to shallow deposits, especially the modern soil. However, trenching has been successful where the fault zone was buried beneath artificial fill (Section 6.3.3.2).

Trench orientation is dictated by the inferred sense of fault displacement, with trenches being aligned roughly parallel to the sense of movement (perpendicular to fault strike for dip-slip faults, Figure 2A.20; parallel to fault strike for strike-slip faults). *Fault-perpendicular trenches* are often used to locate and define the width of strike-slip fault zones, with parallel trenching following to define offsets of piercing points (Chapter 6). Dip-slip paleoearthquakes can often be adequately characterized by a single trench at each site along the fault, especially where displacement is concentrated beneath a single fault scarp. For oblique-slip and strike-slip displacements, multiple trenches are needed to capture the three-dimensional components of slip.

Trenches have also been excavated to study folds (Chapter 5) and various off-fault paleoseismic features such as sand blows (Chapter 7). These trenches are used mainly to expose stratigraphy that was deformed by seismic shaking or deposited in response to earthquake deformation. Trench orientation and placement are not as critical in these cases as long as the trench intersects the features of interest.

A final category of trenches is trenches that are excavated across a development site to prove or disprove the existence of young faulting not visible at the surface (e.g., Hatheway and Leighton, 1979). The location of such trenches is determined by construction site dimensions rather than by any geological considerations. These types of trenches are hopefully oriented perpendicular to local structural trends, which maximizes their probability of intersecting a fault. It would be coincidental if trenches so sited were optimal for measuring either paleoearthquake displacement or recurrence, so paleoseismic interpretation of the site may often be supplemented by data from better sited trenches beyond the construction site. More details on development-related trenching are given in Chapter 9 (See Book's companion web site).

Figure 2A.20: Photograph showing the orientation and placement of a trench (between arrows) across an 8 m-high normal fault scarp (in shadow). The trench completely traverses the scarp, from well onto the upthrown block, to well onto the downthrown block, thus exposing prefaulting strata on both the footwall and (if the trench is deep enough) on the hanging wall. A zone 10 m wide at the foot of the scarp has been back-tilted toward the scarp; the hinge line of tilting is shown by a dashed line. Grey's River fault, Wyoming (Jones, 1995).

2A.3.2.2   Excavating (or Reexcavating) the Trench

The choices of excavating equipment, trench cross-sectional shape, and shoring strategy are all interrelated and depend on what kind of material is being trenched, the topography at the trench site, how deep the trench is, how stable trench walls are, and whether the wall needs to be photographed. As pointed out by Hatheway and Leighton (1979, p. 178) "the method of excavation that proves to be least disturbing to the host soil and/or rock will also likely prove to be the least expensive." Trenches in unconsolidated deposits are usually excavated by hand if fault scarps are very small (less than 2 m high); although house-sized trenches have been excavated by hand in China (Li and Zheng, 1992).

Larger trenches require powered excavating machines. There are five basic types used in paleoseismic trenching (Table 2A.2): (1) rubber-tired backhoe loaders, (2) tracked hydraulic excavators ("trackhoes"), (3) rubber-tired (wheel) loaders, (4) scrapers, and (5) track-type tractors ("bulldozers).

Early trenches for regional paleoseismic studies in the United States were narrow, deep single slots (*California-style trenches*; Figures 2A.21, B1 and 2A.22) dug by backhoes and small trackhoes across moderate-to-large scarps. This trench shape involves a minimum of material excavated and time consumed, and it can be easily shored. However, due to narrowness only small portions of the trench wall can be seen from any one location, and photographing the wall is difficult (Figure 2A.23). Large numbers of people cannot view the trench at once, as is occasionally required during regulatory review. One advantage of such trenches, however, is their minimal disturbance to already-developed areas (Figure 2A.24).

Table 2A.2: Excavating capabilities of various types of machines

Type of machine	Photographs of typical machines made by Caterpillar company	Digging depth	Width of digging bucket	Capacity of digging bucket	Advantages or disadvantages for paleoseismology
Rubber-tired backhoe loader	Cat 420	Typically 4–5 m	0.3–0.9 m	Up to 0.6 m^3	Inexpensive; widely available; good for single-slot trenches up to ca. 4 m; can backfill trenches as well as dig them
Tracked hydraulic excavator ("trackhoe")	Cat 325	Up to 11 m	Up to 2.4 m	Up to 5.8 m^3	Widely available; digs over twice as deep as a backhoe, and moves material much faster; more maneuverable on steep terrain; can dig single-slots deeper than 4 m, OR move larger volumes of material needed in 1- or 2-bench trenches; can dig deep slots within in wider trenches by walking onto their floors (if wide enough)
Rubber-tired (wheel) loader	Cat 994	unlimited	Typically 3.2 m	Up to 36 m^3, but typically 3–5 m^3	Good for multibenched trenches, if material is unconsolidated and soft, such as eolian deposits; has the most flexibility in placing the spoil dirt

(Continued)

Table 2A.2: Excavating capabilities of various types of machines—Cont'd

Type of machine	Photographs of typical machines made by Caterpillar company	Digging depth	Width of digging bucket	Capacity of digging bucket	Advantages or disadvantages for paleoseismology
Scraper (Cat 621)		unlimited	Up to 3.7 m	Up to 34 m^3	Expensive; not widely available; best for very large benched trenches, due to its large capacity (see Figure 2A.28)
Track-type tractor ("bulldozer") Cat D8		unlimited	Up to 3 m	Not applicable	Widely available, but expensive in larger sizes; can rip hard rocks, unlike scrapers and loaders; best for deep trenches in semi-consolidated deposits or bedrock; to distribute soil dirt from trench, may need to work in combination with a loader

The largest machines of each type are generally made for the mining industry, rather than for the construction industry.

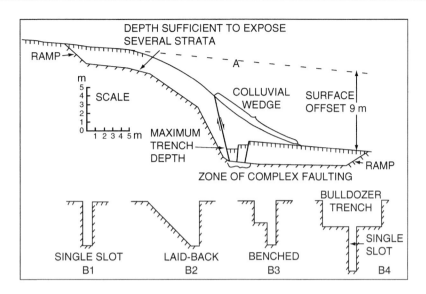

Figure 2A.21: (A) Typical longitudinal section of a trench across a fault scarp; compare to Figure 2A.8. (B) Cross sections of various types of trenches. B1, single slot (California-style); B2, laid-back; B3, single-benched trench (double-benched trench not shown); B4, backhoe trench within a bulldozer trench. From McCalpin (1989b); reprinted with permission of (A).
(A) Balkema Publishers.

Figure 2A.22: Single-slot trench excavated into the face and toe of an 8 m-high normal fault scarp, eastern Bear Lake fault zone, Utah. Note the tops of 2.1 m-high hydraulic aluminum shores visible in the trench.

Figure 2A.23: Photograph of the interior of a 0.9 m-wide, 4.3 m-deep single slot trench (Kaysville trench, Wasatch fault zone, Utah; see McCalpin *et al.*, 1994). Hydraulic aluminum shoring supports, each 2.15 m high, are stacked in two tiers. Sediments exposed in walls are fine-grained graben-fill sediments; red markers show faults, yellow and blue markers show stratigraphic units.

By the mid-1990s in the USA most workers had switched to wider *double-benched trenches* because North American paleoseismologists prefer to map on vertical trench walls. There were three reasons for this: (1) the shape allows a better view of trench wall stratigraphy and fault relations and can be much more easily photographed, (2) deeper trenches were needed to expose more faulting events, and (3) shoring, which is expensive and sometimes difficult to obtain, is not necessary. According to safety regulations in the USA (OSHA, 1989), the vertical walls in a benched trench cannot exceed 1.5 m high (in strong, type A soils) or 1.2 m (in weak, type B and C soils), with intervening benches 1.5 m wide. Thus, for every additional depth increment of 1.2–1.5 m, the trench must contain another 1.5 m-wide horizontal bench on each side, increasing the width of the trench by 3 m. A minor drawback to such trenches is the apparent shift of dipping planar features (faults, contacts, unconformities) on the trench log when passing across a bench, if the feature does not strike perpendicular to the wall (Figure 2A.25). However, as long as the strike and dip of the feature is marked on the log, the reader will not become confused.

The excavation volumes required for deep benched trenches soon exceed the capabilities of backhoes and trackhoes, and other machines that move larger dirt volumes (bulldozers, wheeled loaders, and scrapers) become more economic. Early deep benched trenches were excavated by bulldozers (Figure 2A.26), but a

Figure 2A.24: An example of "clean trenching" in a developed area (golf course). The turf at this site was cut out and rolled (lower right) and all material excavated by a small backhoe loader was placed directly in a dump truck and hauled off-site. The trench was then fenced and covered with plywood sheets (at right) when loggers were not present. The trench was logged simultaneously with golf course use, and neither activity interfered with the other. However, golfers who hit balls into the trench were assessed a two-stroke penalty.

drawback is that the dozer has difficulty moving the excavated dirt once it reaches the end of the trench, resulting in large piles of dirt at each trench end. These rising piles ultimately limit the depth of the trench, because they become too steep for the dozer to push more excavated dirt up them.

Wheeled loaders and scarpers do not have this limitation, and the spoil dirt can be taken out of the excavation and piled anywhere on the site. However, scrapers and especially wheeled loaders can only excavate into relatively soft geologic deposits. McCalpin and Harrison (2000) excavated two long trenches with wheeled loaders and then had a trackhoe cut a deep slot in the floor of the trench (Figure 2A.27), which was shored with aluminum hydraulic shores. By digging a 3 m-deep slot into the floor a benched trench, the paleoseismologist gets another 3 m of depth, without having to excavate another pair of benches, which saves a large amount of excavation.

Cotton *et al.* (1988) used *scrapers* to excavate a 14 m-deep trench across the San Gabriel fault, California; McCalpin and Shlemon (1996) show an even larger 20 m-deep scraper trench (Figure 2A.28).

An alternative style of trenching is the *open-pit excavation,* in which all trench walls are laid back to slopes ranging from 45° (Figure 2A.29 common in Japan) to a 75° (termed Venezuelan-style by Audemard, 2005, but used worldwide; Figure 2A.30). In flat areas such trenches are often nearly square in plan view, becoming more elongated if the trench crosses a scarp with significant vertical relief (e.g., Okumura *et al.*, 1994). Open-pit trenches have several advantages: (1) They can be very deep (note the 13 m depth in Figure 2A.29B, cross section), (2) they do not require shoring, (3) the walls can be logged without the need for elaborate scaffolding, (4) the trench walls are easy to view and photograph

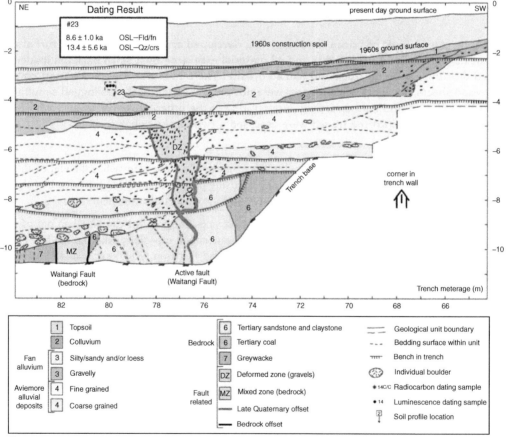

from outside of the trench, and (5) the excavation is stable enough to form a quasi-permanent exposure, such as the Neo-Dani Fault Museum in central Japan (Okada *et al.*, 1992; Sato *et al.*, 1992). The main disadvantage of the open-pit trench style is the size of the excavation and the volume of material that must be removed. A minor disadvantage is the need to *rectify the trench log* drawn on nonvertical trench walls to a true vertical cross section. However, if the trench is logged by electronic or photogrammetric techniques (Section 2A.3.2.6) projection of points to a vertical plane is simplified.

The trench types described above lie on flat ground or ascend up a relatively smooth fault scarp. However, sometimes the best vertical exposure of a fault is where it ascends an erosional escarpment that is not parallel to the strike of the fault. In that case, the fault can be exposed in a series of vertical cuts, or "half-trenches," that follow the fault's outcrop on the erosional escarpment. An example is shown in Figure 2A.31, where a normal fault descends from the top of a faulted pediment surface, down an irregular, ca. 20 m-high erosional escarpment. The fault scarp on the pediment was too broad (ca. 800 m) to trench from end-to-end and the exact location of the fault plane under the scarp was unclear. In addition, to dig even a 10 m-deep trench on the broad scarp would have required a large, double-benched trench. But by digging a series of half-trenches that stepped down the erosional escarpment, 17 vertical meters of the fault plane and colluvial wedge section were exposed, with only the removal of a small volume of dirt. As long as the top strata exposed in one trench overlapped the bottom strata exposed in the next trench upslope, the entire paleoseismic record will be uncovered.

Trenches in consolidated bedrock rarely preserve evidence of individual paleoseismic events, but at critical facility sites where bedrock is at the surface, trenches must be excavated into competent rock merely to prove or disprove the existence of faults. Much of the early trenching done for NPPs in the United States was performed in rock, by drilling and blasting. Hatheway and Leighton (1979) provide a good summary of the excavation and logging techniques utilized in bedrock excavations.

Trenches may be excavated completely before logging begins, or dug incrementally as logging proceeds ahead of backfilling. On dip-slip faults complete excavation is advantageous because the entire trench can be viewed at once and critical areas reexamined. However, due to the instability of trench walls or lack of shores some dip-slip fault trenches may have to be dug and logged in increments. For strike-slip faults incremental excavation and logging of the vertical walls of fault-parallel and fault-perpendicular trenches is often required to measure three-dimensional deformation (e.g., Wesnousky *et al.*, 1991; also see Chapter 6). *Incremental trenching* refers to the progressive excavation of closely spaced, parallel trench walls. After the initial trench is excavated and logged, the logged trench wall is excavated back 20–50 cm, and the new exposure is logged. That wall is then cut back 20–50 cm parallel to the previous wall and is again logged. As the trench wall is progressively cut back, many successive wall positions are mapped. The resulting closely spaced, parallel trench logs can then be used to create a three-dimensional diagram of structural and stratigraphic relations in the deformation zone. At present, this technique has mainly been used on strike-slip fault traces, where a 3D representation is needed to calculate displacement

Figure 2A.25: Trench and trench log across the Waitangi fault, New Zealand. (A) Photograph of the trench, which has 6 wall levels at right. At 180 m long and 10 m deep, this was the largest trench excavated in New Zealand as of 2004. (B) Log of the left (south) wall of the trench, showing the currently active fault trace (thick black lines) and an older fault trace not active in the late Quaternary. Note how the active fault appears to shift to the left on the trench log as it crosses the benches, because the fault trends more easterly than a perpendicular to the trench walls. From Barrell *et al.* (2005). (See Color Insert.)

Figure 2A.26: Photographs of an early double-benched trench excavated by bulldozer, at the site of the proposed Point Conception (California) liquefied natural gas terminal. Note person at far left for scale. The project was proposed in 1972, but by the time site suitability studies were completed in 1980 (including seismic hazard studies; Rice *et al.*, 1981), oil prices had fallen so much as to make the project uneconomic, and it was abandoned (photograph courtesy of T. K. Rockwell).

vectors. This method of trenching is totally destructive, however, because excavation consumes the entire feature being mapped. The floors of trenches have also been logged to make *isopach maps* showing horizontal displacement and folding (Sieh, 1978a).

During excavation senior personnel must decide when the trench is deep enough to expose the desired stratigraphy. This decision must often be made while the excavating machinery is still positioned over that portion of the trench in question, since it may be difficult to get excavating equipment back into position to deepen a narrow trench once it has been completed. It is usually worthwhile to have the backhoe/trackhoe excavate a trench to its full depth reach, if such depth does not initiate caving. This method has the added advantage of often exposing small faults near the bottom of the trench which cannot be seen closer to the surface, either due to masking by surface disturbances or by upward truncations within the stratigraphic package. Because the backhoe bucket or bulldozer blade smears a thin film of cohesive soil onto trench walls, it often is difficult to see stratigraphy in the walls as they are cut. Accordingly, geologists are often tempted to enter the unshored, just-excavated part of the trench to scrape off the walls, inspect the stratigraphy, and determine if key strata have been exposed or whether the trench needs to be deepened.

Reexcavating a trench. In some cases reexcavating an earlier trench may be necessary. An example is where: (1) the interpretation of the earlier trench seems suspect in light of more recent data, (2) larger excavating equipment can dig a deeper trench, or (3) when new sample collecting or dating techniques have become available that were not developed when the first trench was excavated. An example of all three reasons is the 1988 reexcavation (McCalpin *et al.*, 1994) of the 1978 Kaysville trench on the

Figure 2A.27: Photograph of a trench dug by wheeled loaders, with a central slot dug by a trackhoe. The slot had previously been shored with hydraulic shoring, but those were removed by the time this photo was taken, just prior to backfilling. West Trench, Calabacillas fault, Albuquerque, New Mexico (McCalpin and Harrison, 2000). Photograph taken by J. P. McCalpin on June 1, 1999.

Wasatch fault, Utah (Swan *et al.*, 1980). The Swan *et al.* chronology was based on a single radiocarbon date that, based on work performed after 1978 on this and adjacent segments, appeared suspiciously too young. Meanwhile, AMS-radiocarbon dating had been applied successfully to low-organic-content buried soils and fissure fills, which were abundant in the 1978 trench (but not dated), and luminescence dating had been developed and applied to fine-grained sag pond deposits, which were also abundant. The 1988 trench was about 50% deeper in the graben than the 1978 trench, so that it exposed critical faulted strata not exposed by the 1978 trench. The end result of the 1988 reexcavation was a paleoseismic history constrained by four radiocarbon and nine luminescence ages. Ages for the latest three paleoearthquakes were established, and the resulting recurrence interval was four times longer than assumed in 1978; it was also irregular, as were the displacements in each event. Thus, the interpretation arising from the reexcavation was very different than that from the original trench.

2A.3.2.3 Dewatering the Trench

Dewatering trenches increases their safety and makes the trench a more pleasant place to work. There are two main approaches to dewatering a trench, passive and active. On sloping ground, it may be possible to perform passive dewatering by digging a shallow ditch downslope from the toe of the trench, such that groundwater entering the trench will simply continue flowing down the trench axis and out the lower end. Audemard (2005) describes this method of draining sag ponds. On flat ground there are two options, both active dewatering. First, one can periodically pump the water out of the trench and release it some distance away (Figure 2A.30). However, this means that groundwater will continue to intersect the trench walls and seep out, that the bottom of the trench will always have some water in it, and the part of the wall submerged will have to be re-cleaned before logging. A preferable approach is to dig shallow pits or drill relief wells around the trench and to lower the groundwater level by pumping, until it falls below the

Figure 2A.28: Multibench trench excavated by scrapers. The trench sidewalls are composed of 13 vertical walls each 1.5 m high, separated by 12 horizontal benches 1.5 m wide; total depth is ca. 20 m, total volume ca. 23,000 m³. Total excavation cost was about US $100,000 (1995 dollars). The trench was excavated to determine the origin of a 20-cm-wide ground fissure in an area of proposed commercial development at Lakeview Hot Springs, southwestern Riverside County, California. The fissure proved to be the surface expression of an old hot springs vent and was unrelated to previously suspected faults or to local differential settlement. From McCalpin and Shlemon (1996).

trench floor. While this method keeps the trench walls stable and the trench floor dry, it is more expensive than simply pumping the water out of the trench floor.

2A.3.2.4 Trench Safety

Numerous fatalities have occurred when vertical walls have collapsed on geologists who were crouched at the base of the trench looking for key contacts. It is probably true that more geologists have been killed in trenches than in earthquakes. Thompson and Tannenbaum (1977) report that about 100 workers are killed each year in construction trench collapses in the United States. Trench walls typically slump in several geologic or topographic situations: (1) at the fault plane, where crushing has reduced material cohesion and possibly created open voids; (2) in the low-density colluvium immediately downslope of the fault; (3) where groundwater outflow is strong, and (4) where cohesionless material (often saturated) ravels out of the freshly cut trench walls and undermines overlying cohesive units—this leads to massive slumping and caving, often along preexisting vertical cracks. Most slumping occurs within minutes after the face has been excavated, due to vibrations from the machinery during continued excavation, but vertical walls may continue to collapse for days after excavation, especially in wet weather. Personnel can be lowered into the trench while standing on the backhoe bucket, but this will not prevent caving.

In the United States, federal regulations require that trenches deeper than 1.5 m be stabilized with shores, or excavated sufficiently wide (or with sloping walls) that personnel can avoid collapsing wall sections (U.S. Occupational Safety and Health Administration (OSHA), 1989). Single-slot trenches are usually shored with 7-ft-long *hydraulic aluminum shores* which expand from 0.7 to 1.4 m wide (Figure 2A.23).

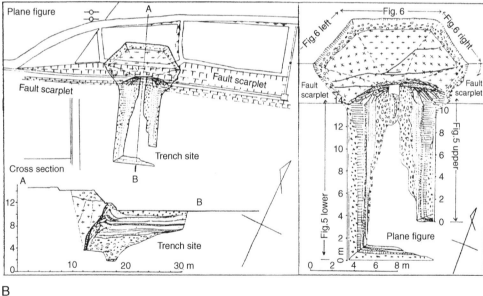

Figure 2A.29: (A) Photograph of the 13 m-deep "open-pit" style trench across the Atotsugawa fault at Nokubi, central Japan; (B) Plan views and cross section of the trench shown in (A). The fault scarplet transected by the deepest part of the trench is 5 m high. Patterns shown on the lower trench walls indicate various unconsolidated deposits, primarily terrace gravels (open circles) and sands.

The "X" pattern at the upper end of the trench indicates granite that has been thrust over the alluvium. From Okada *et al.* (1992); reprinted with permission of the *Journal of Geography* (Japan).

Figure 2A.30: "Venezuelan-style" trench dug by Audemard (2005). Trench is 90 m long, 8 m deep, 8 m wide at top, 4 m wide at bottom; excavated volume is ca. 3600 m³. Note the steep sidewalls, lack of benches, and string reference grid on right wall. To make contacts more visible the walls were wetted by the water truck at right; at left a vacuum truck is pumping out the collected water in the trench bottom. Photo courtesy of Franck Audemard.

Cass and Wall (1989) provide an extensive description of shoring options with these units. Trenches too wide for the use of any hydraulic or *screw-type shores* are usually shored with lumber or heavy timbers cut onsite. The trench should be surrounded by a fence immediately after excavation, with "no trespassing" signs prominently displayed. Trench shoring standards used in California are quite comprehensive and may be used as conservative guidelines in the absence of local regulations (California Department of Transportation, 1977; Cass and Wall, 1989).

2A.3.2.5   Preparing for Logging

Before a trench can be logged the walls must be cleaned well enough to expose the structures, stratigraphy, and soil horizons. In most trench investigations in the United States across dip-slip faults only one wall is cleaned and mapped (logged). This wall is usually chosen to be the shaded one, because it is difficult to trace contacts on a wall which is partly lit by direct sunlight and partly in shadow (e.g., in an east–west-oriented trench, only the southern wall would be logged). Because of the extra labor involved in cleaning the opposite trench wall, this is usually only done in places to confirm features seen in the logged wall (e.g., the trend of faults and paleochannels). However, we note that the custom in Japan is to log all walls of a trench, and this practice certainly facilitates 3D reconstructions of slip.

Trench walls are scraped off with various tools to remove soil smeared on walls during excavation (Figure 2A.32); typically 2–5 cm will be removed. The goal of cleaning is to best expose mappable contacts, and this dictates the use of different tools in different materials. In coarse gravels, rough smoothing can be performed with any coarse tool, to remove major undulations in the wall. Further cleaning can be accomplished with coarse brushes or blasts of compressed air. In loose to dense sands, rough wall undulations can be carved away with a large smoothing tool such as a stirrup hoe; harder sands

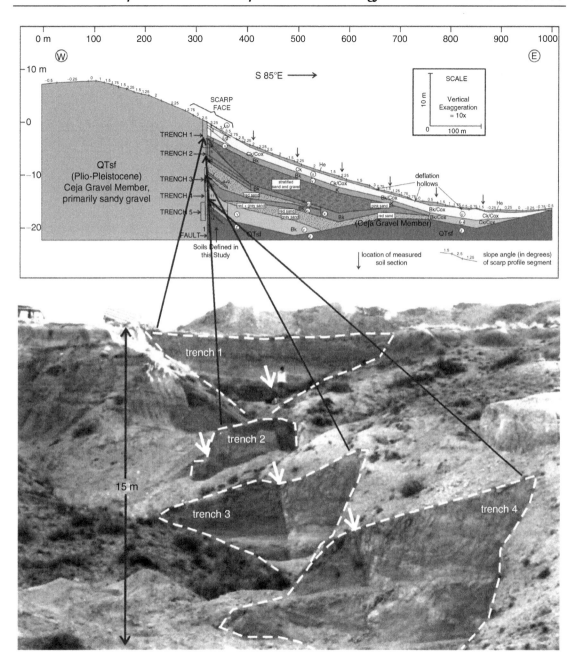

Figure 2A.31: A series of "half-trenches" (vertical cuts) used to expose a fault as it ascends up an erosional escarpment that trends roughly perpendicular to fault strike. (Top) Schematic cross section of the fault zone and colluvial wedge, showing the location of the half-trenches (rectangles along the fault plane). Note the high vertical exaggeration of this section; (Bottom) Telephoto view of the upper parts of trenches 1–4, outlined with white dashed lines. From McCalpin *et al.* (2006). Trench logs from this site are available on on-line Content. (See Color Insert.)

Figure 2A.32: Cleaning tools commonly used in North America: (a) masonry trowels; (b) claw hammer, claws are used for plucking; (c) geologist's hammer; (d) lightweight gardening mattock; (e) heavier gardening mattock; (f) stirrup hoe; (g) weeding hoe; (h) broom; (j) drywall brush; (k) scrub brush; (l, m) paint brushes; (n) portable water sprayer. Inset: a Nishiri gama hoe from Japan, favored by many professionals.

will require a smaller and thicker flat tool, such as the flat blade of a mattock or hoe. Fine smoothing can be done with trowels. Cleaning walls of massive silt is similar to that for sand. Before cleaning, sharpen the cutting edge with a flat file, and keep it scarp; this will reduce join lines between scraped areas.

Very cohesive soils (clays, sandy clays, silty clays; common in the textural B horizons of paleosols) pose a special challenge for cleaning, immediately after trench excavation the trench walls may be moist enough for limited scraping with bladed tools, but once the walls dry out scraping will merely smear dried clay over the wall and obscure any contacts. In fact, the clay smeared onto the walls by the excavating machine must be removed before the walls dry out, or even the rough cleaning of the walls will be very difficult. Rough and fine cleaning of clay trench walls must be done by hacking out (plucking out) blocks of hard cohesive sediment from the wall. One should practice restraint when plucking, using the desiccation cracks as a starting point, to pry out angular "peds" from the wall, but trying to remove only enough thickness to reveal a mappable contact. Stratigraphic and pedologic contacts can be seen on plucked walls, by sighting down the length of the trench walls. However, to map structures in clayey deposits, sometimes it is necessary to scrape the wall, since structures are not typically visible on a plucked wall. When faced with a trench wall of clay, I typically pluck the wall first, map the stratigraphic or soils contacts, and then scrape selected parts of the wall below any vertical anomalies in the stratigraphic contacts, looking for fault planes there.

If a source of water is available, another option for cleaning hard, clay-rich trench walls is a 2000–4000 psi pressure washer. These machines will literally blow the dried surface clay right off the trench wall, but if not moved quickly, the water jet will gouge large pits in the trench wall. Care is advised.

Some coarse grained and massive deposits change appearance when they dry out, such that bedding and structures visible when the wall is moist are very difficult to see when the wall dries out. In this case, the wall can be artificially wetted just prior to logging, to accentuate the visibility of contacts. In the western USA we often use a 12-l spray can that is pressurized by hand-pumping (Figure 2A.32n).

2A.3.2.6 The Reference Grid

After the walls between the shores are cleaned, a *reference grid* for mapping must be constructed if the trench is to be logged using the manual or photomosaic method. Typically the grid is composed of horizontal lines of low-stretch nylon string, spaced 1 m apart, attached to the trench wall by large (5 mm × 10–15 cm) nails. Flagging or tape attached to these lines at 1-m intervals provides the horizontal control or, alternatively, vertical string lines can be placed on 1 m centers (vertical lines are mandatory for the photomosaic method). It is often difficult to attach nails to trench walls composed of noncohesive gravelly or friable material. In such cases the level line may be attached directly to a hydraulic shore or fence postdriven into the trench floor. However, it is best to keep the grid lines entirely separate from the shoring system, because a shift in shoring units (due to loss of hydraulic pressure, weight of people climbing on them during logging, etc.) will distort the line grid.

After the first horizontal grid line is set, others are set parallel to it at 1 m vertical intervals. The horizontality of each successive line can be checked by the line (bubble) level, and by measuring the vertical distance between successive horizontal string lines. If the vertical distance varies by more than 2% (i.e., ±2 cm over 1 m), the line should be releveled. Once all horizontal lines have been strung, flagging or tape is attached to the first level line at 0.5 m intervals. Horizontal distance marks are transferred between the parallel string lines of different vertical heights by means of a plumb line. These plumb line measurements should be checked on each horizontal line by measuring between horizontal tape marks to ensure 2% precision.

If the trench wall is vertical and smooth (a fortunate but rare occurrence), the string line grid will lie tightly against the wall and parallax problems will be minimal during manual or photomosaic logging. However, if the trench wall to be logged is not vertical, or is very irregular with cavities, the vertical string line may diverge significantly from the trench wall. Two approaches to logging are possible in this case. In the first approach, the level lines are kept tightly up against the wall, even though the wall is not a smooth vertical plane; this is the practice in open-pit trenches. To avoid scale changes on the log, actual measurements in the plane of the (nonvertical) trench wall must be *trigonometrically corrected* later to project onto a vertical plane. Such reduction in the field can be tedious, considering that it may involve hundreds or thousands of measurements. The second approach is to string the grid lines within a vertical plane, using the shoring system as support, with the result that the grid lines may be several decimeters (or even meters) from the logged wall. In this approach contacts on the wall are "sighted" in relation to the grid system, but *parallax problems* may arise if the lines are far from the wall. Such parallax errors may amount to several decimeters and will make construction of the final log difficult.

In most fault-perpendicular trenches on dip-slip faults, the trench reference grid should be tied to a georeferenced datum by measuring the GPS coordinates of the ground surface near the ends of the trench, on some even meter interval on the trench grid system. Fault-parallel and fault-perpendicular trenches across a strike-slip fault, if not contiguous, must be tied together by a *common surveyed datum* if oblique displacements are to be measured accurately (e.g., Grant and Sieh, 1994). This common datum is typically artificial and local and is not tied to any larger geodetic survey or coordinate system. Trenches excavated at construction sites are also tied to the *site survey grid* to ensure that the trench is properly located with respect to planned construction (e.g., Lund and Euge, 1984).

2A.3.2.7   Identifying and Marking Contacts

Lithologic units are differentiated as discrete sedimentary deposits characterized by a consistent texture, sorting, bedding, fabric, or color (previous section). Soil units, in contrast, are weathering zones or profiles which may be developed on a single unit, or may be developed across multiple lithologic units. Identifying lithologic and soil units on trench walls is facilitated if lithologic contrasts are emphasized by use of some wall-treatment technique. For example, slight differences in deposit cohesion may be accentuated if the trench wall is left to "weather" for several days or weeks. Wind and rain can then etch out differential relief between different strata and even reveal subtle structures such as cross-bedding in loose sands. Similar relief can sometimes be created by repeated brushing of the face with brooms or paintbrushes. Conversely, some contacts appear sharper when moist, so walls can be sprayed or misted with a portable water sprayer immediately before logging.

Vague stratigraphic contacts with little textural or color contrasts can often be seen better in diffuse lighting conditions than in direct sunlight. Midday sunlight falls on trench walls at low angles and accentuates (via shadowing) minor irregularities and tool marks at the expense of subtle textural or color variations. Frost *et al.* (1991) suggest that logging at night with artificial illumination, the direction and intensity of which can be controlled, can accentuate subtle stratigraphic and structural features. A corollary use is that of ultraviolet illumination of trench walls at night, which accentuates the fine structure of soils containing calcium carbonate (Kim Thorup, personal communication, 1995). In daylight, vague contacts can often be located by sighting down the length of the trench, nearly parallel to the trench wall.

Contacts identified visually are usually accentuated by scribing a line on the trench wall with a knife or edge of a trowel (in cohesive sediments), or placing nails with attached colored flagging along the contact (see Figure 2A.23). In the corresponding trench log, tectonic features (faults, tension cracks, liquefaction features) are rendered with the thickest lines, lithologic contacts with thinner lines, and soil horizon boundaries or facies boundaries within major (genetic) depositional units with very thin or dashed lines.

The critical features in a paleoseismic trench are the deformation features which must be depicted to emphasize their relations with stratigraphic units. The expression of faults and folds in near-surface unconsolidated materials is often more subtle than for faults in bedrock. At times fault traces near the surface are not visible, even though they are known to have ruptured to the surface (Bonilla and Lienkaemper, 1991). Fault traces with a vertical component of displacement are easy to see if multiple stratigraphic units or soils are faulted (Figure 2A.33). This type of stratigraphic offset is primary evidence for dip-slip faulting and also commonly occurs on strike-slip faults. In the absence of stratigraphic offset, faults in unconsolidated deposits are identified from changes in material texture, hardness, or clast fabric along the fault trace.

Fault gouge, created by mechanical crushing of rock and smearing along the fault plane, is rare along faults in unconsolidated deposits because confining pressure near the surface is too low. However, thin (1–5 cm), tabular bodies of *translocated sediment* are often found along fault planes. In many cases those bodies are composed of soft-sediments (silt, clay, marl) dragged along the fault plane from displaced strata. Such zones of smeared cohesive material often contain granular material or blocks of adjacent strata, often termed "mixture of adjoining materials," "mixed rock," or "tectonic mixing" on trench logs (Bonilla and Lienkaemper, 1991).

Fault zones in clast-rich deposits are usually identified by a consistent *clast fabric* different from that observed in adjacent strata. Shear on the fault may rotate clast long axes parallel to the fault plane,

Figure 2A.33: Photographs contrasting easily visible faults (thick arrows) in well-stratified lacustrine sand and silt with an obscure fault strand (thin arrows) in massive sandy debris-facies colluvium. The main fault (center) places laminated lacustrine sands and silts (left) in fault contact in fault contact with the scarp-derived colluvium (right). Horizontal string lines at center and right are 1 m apart. Kaysville trench on the Wasatch fault zone, Utah. See McCalpin *et al.* (1994).

resulting in what paleoseismologists often term *shear fabric* (Figure 2A.34). Yount *et al.* (1987) reasoned that on steeply dipping faults, dip-slip fault shear would twirl clasts so their long axes became horizontal (aligned with strike), whereas strike-slip movement would twirl clasts into near-vertical orientations (long axes aligned with dip direction). This speculation is not supported by most exposures along normal faults, in which long axes of fault-zone pebbles typically parallel the dip of the fault (Figure 2A.34). The author is unaware of any laboratory experiments that have reproduced the aligned elongate pebbles seen in many trenches, but such experiments would permit a more confident interpretation of shear fabric.

Faults in unconsolidated deposits are often accompanied by *fissures, open voids,* and *fault-related rubble,* particularly with extensional faulting. The rubble consists of blocks of adjacent stratigraphy that may have fallen downward into open fissures, been dragged upward along the fault, or both. Fissures and fissure-filling materials (see Chapter 3) are most common along normal faults (Bonilla and Lienkaemper, 1991).

2A.3.2.8   Mapping Soil Horizons in Trenches

Soil horizons are important markers in trench exposures because they indicate the location of past ground surfaces in the stratigraphic sequence, and their degree of development may indicate the length of time that surface was stabilized. The techniques for recognizing and delineating soil horizon contacts are beyond the scope of this book; see Birkeland (1999) for an excellent summary. Birkeland *et al.* (1991) also describe some applications of pedology to neotectonics. In fact, many of the advances in paleoseismology in the USA during the 1970s–1990s resulted from applying Birkeland's teachings about soil stratigraphy to fault exposures (Figure 2A.35).

Figure 2A.34: Close-up photograph of two parallel normal fault strands (between arrows) in gravelly late Pleistocene deltaic deposits, East Bear Lake fault, Utah. Numbers on rod are 10 cm apart. Note anomalous parallel clast fabric and slight discoloration of gravel along the fault strands.

The concept of mapping soil horizons separately from lithologic units often eludes geologists with no training in pedology (e.g., geologists often map soil horizons as lithologic units and become confused when a soil horizon crosses from one lithologic unit into another). However, the interaction of soil profiles with lithologic units is often critical to understanding the sequence of depositional versus tectonic events and their relative timing (Shlemon, 1985). For example, if a soil is developed on tectonically displaced strata and is truncated at the fault scarp along with those strata, the period of soil formation entirely predates faulting. In contrast, if strata are displaced but the soil horizons extend across the fault plane and are developed on other faulted strata, some (perhaps all) of soil formation postdates the faulting.

In most cases soil horizon boundaries will parallel stratigraphic units in the trench, such as buried A (organic) horizons. In this case soils can be mapped as if they were stratigraphic units. An example is shown in Figure 2A.36, where the homogenous prefaulting deposits (parent materials) were so strongly overprinted by calcareous soil formation that the mappable units are soil horizons rather than lithologic units. The right 1/3 of the trench log shows three buried soils on the fault footwall; from oldest to youngest, buried soil 5 (horizon Btb5), buried soil 4 (horizons Btb4/Bkb4), and buried soil 3 (horizons Btkb3/Kb3). The central 1/3 of the log shows an intermediate, fault-bounded structural block composed of forward-tilted, prefaulting stratigraphy of buried soils 1 and 2 (horizons Bkb1/Bkb2/Cub2), overlain by

Figure 2A.35: Dr. Peter W. Birkeland taught two generations of students how to combine pedogenic science and geomorphology, which turned out to be a critical topic during the development of paleoseismology in the USA (1970s–1990s). He was considered a "Superman" by his students. Photo by J. McCalpin, May 1975, eastern base of the Sierra Nevada, USA.

blocks of footwall units that broke off the free face and fell into the zone (Bkb2/Btkb3/Kb3/Btb4/Bkb4). The left 1/3 of the log shows the youngest prefaulting paleosol (buried soil 1, composed of horizons Btkb1/Kb1/Bkb1). Overlying all the prefaulting soils is postfaulting colluvium (A + Bt/Bk/Cu), mainly retransported eolian sand. When restoring the mapped units to their prefaulting geometry (retrodeformation), these soil horizons can be treated just the same as stratigraphic units.

Where soils are composed of multiple horizons there are two options for trench mapping. The first option is to map the entire soil profile as a single unit, without representing any individual soil horizons. As long as the constituent horizons within the soil profile maintain relatively constant properties and thickness laterally, a single soil description can be used for the entire soil profile. A second option must be used if soil horizons within a soil profile pinch out laterally, change profile properties, or cross from one lithologic unit (parent material) to another. Such complications are common where soils are developed across fault scarps. Because soil contacts and lithologic contacts cross each other, a unique line symbol should be given to soil contacts. Figure 2A.37 shows an example of the complex relations created by soil formation that began before the initial faulting and continued during and after faulting, and also fault-induced erosion and colluvial deposition.

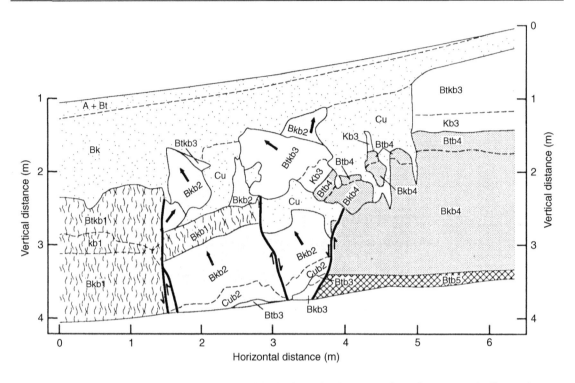

Figure 2A.36: Example of soils parallel to lithologic units in a trench wall. Arrows indicate the up direction in each of the rotated soil blocks. Horizon abbreviations A, B, and C denote master horizons; k, calcium carbonate; t, textural B horizon; b, buried horizon; u, material unaffected by pedogenesis. The fallen and rotated blocks of soil in the center fell from buried soils 1, 2, and 3 which must have been exposed on the fault free face, even though buried soils 1 and 2 are no longer present directly upslope of the main fault (they were eroded away). From Birkeland *et al.* (1991, p. 48); reprinted with permission of the Utah Geological Survey.

The complex interactions between soils and lithologic units are not merely of academic interest. In the absence of material suitable for numerical dating, the degree of soil development can yield an estimate of the relative lengths of time portrayed by soils in the different stratigraphic positions. In Figure 2A.37, the entire 130-ka time span since the formation of the faulted surface is represented on the downthrown block by the development of soils 1, 2, 3, 4, and the buried soil. The ratio of development between these soils reflects the amounts of time between successive faulting events. Quantitative estimates of these relative time spans can be made based on the relative amounts of *pedogenic clay* or *calcium carbonate* in each soil; examples are given by Machette (1978), Nelson and Weisser (1985), Birkeland *et al.* (1991), McCalpin (1994), and McCalpin and Berry (1996).

2A.3.2.9   Defining and Labeling Map Units

For a trench log to communicate information to a wide audience, the map units must be defined in a way that emphasizes the sequence of deformation, sedimentation, and weathering exposed at a site. Hatheway and Leighton (1979) differentiate the subjective versus objective approaches to trench logging.

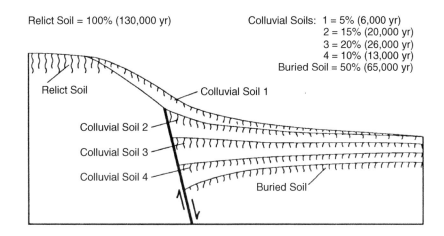

Figure 2A.37: Complex soil relations on a fault scarp. On the upthrown fault block, all 130,000 years of soil formation is contained within a single soil profile (Relict Soil), whereas on the downthrown block that same length of soil development is spread out among five soil profiles. The relict soil above the scarp has been continuously developing since stabilization of the geomorphic surface (130,000 years BP). In contrast, the "Buried soil" on the downthrown block developed only in the time span between the stabilization of the geomorphic surface, and the deposition of the earliest scarp-derived colluvium (130,000–65,000 years BP). In apportioning geologic time to soil development, the degree of development of the relict soil (as measured by Profile Development Index, clay or carbonate accumulation) is defined as 100%. The sum of similar soil development indices for the four buried soils and one surface soil on the downthrown block should also equal 100%, neglecting catena variations (see Section 3.4.2). The length of time represented by each soil on the downthrown block can be estimated as the product of (1) the ratio of their development to the relict soil (percentage values in upper right), multiplied by (2) the age of the relict soil, resulting in the age spans in parentheses. From Birkeland *et al.* (1991, p. 45); reprinted with permission of the Utah Geological Survey.]

In *subjective logging,* the logger first observes the trench wall and makes a geologic interpretation of the structural and stratigraphic relations exposed in the wall. The correctly scaled log is then made to illustrate the primary geologic features. The rock or soil matrix is added in secondary importance; small features that do not bear on the major interpreted structures or strata may not be logged at all. The log is thus schematic (Figure 2A.38) but planimetrically accurate. The subjective approach to trench logging developed during NPP investigations (Hatheway and Leighton, 1979), where the log was meant to answer specific regulatory questions, such as "Is a fault present?" and, if so, "Is the age of faulting older than some predefined regulatory criterion?" The advantages of a subjective log are that it can be made rapidly and is easy to interpret with respect to regulatory criteria, because all extraneous features that do not bear on the major interpretation have been omitted. The disadvantage of this type of log is that it is difficult to advance alternative interpretations of the log, because the interpretation was integral to drafting the log, and many details have been omitted. The initial enthusiasm for this type of trench log in nuclear safety investigations soon waned when field reviews showed that alternative interpretations of structural relations might be valid, requiring complete re-logging of the trench walls.

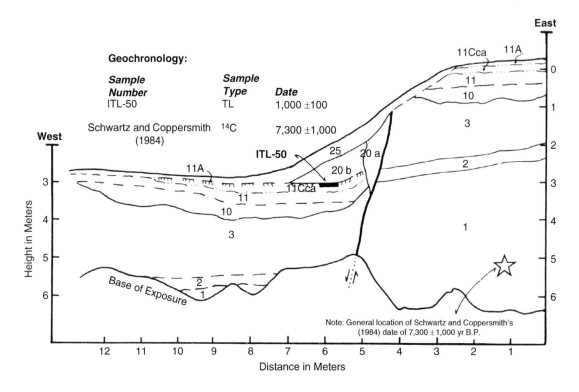

Figure 2A.38: Example of a subjective trench log. Stratigraphic units and the fault are outlined, but no details of deposit sedimentology or sedimentary structures are portrayed. Unit numbers were changed for this book to represent time gaps associated with erosional unconformities (between units 3 and 10) and soil horizons (units 11A/11Cca). The number gap between units 11 and 20 represents the time required to develop the soil horizons. From Jackson (1991), Plate 1; reprinted with permission of the Utah Geological Survey.

In contrast, *objective logging* "attempts to portray equally all physical features of the trench face, larger than a threshold of resolution, in an impartial manner and without regard to relative importance. Both obvious and subtle features are shown with equal resolution, and little subjective interpretation is made during the recording process" (Hatheway and Leighton, 1979, p. 173). This approach seeks only to document what the trench wall looks like (Figure 2A.39). The most extreme example of an objective log would be an unannotated photograph of the trench wall, which showed no interpreted features. The advantage of an objective trench log is that several interpretations can be proposed and tested against the stratigraphic relations portrayed on the log. The log also acts as an archival record of how the trench wall appeared, which may be recognized in the future as containing some newly discovered phenomenon. The disadvantage of this approach is that objective logs may not be readily interpretable as drawn; even an expert may have to study the relations in a log for a time before deducing an interpretation.

In practice, a judicious combination of the subjective and objective approaches produces trench logs that are sufficiently detailed to act as archival records, yet have sufficient interpretational emphasis that the major elements can be separated at a glance from the minor elements. One easy way to make an objective/subjective log is to make a photomosaic of the trench wall and add annotations to it (Sections 2A.3.2.11 and 12).

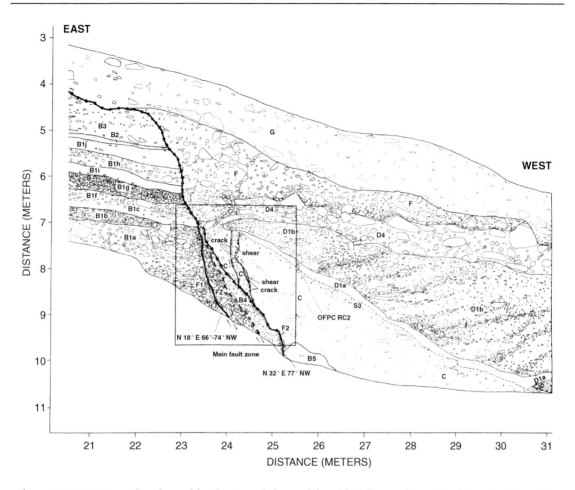

Figure 2A.39: Example of an objective trench log, with added linework emphasizing stratigraphic and structural features. Note how the accurate representation of clasts defines sedimentary structures within mappable units, and shear fabric in the main fault zone (in box). From Olig *et al.* (1994); reprinted with permission of Elsevier Publishing Company B.V.

The *mappable units* defined in a trench exposure are based on the experience of the trench logger. As a general rule, units are distinguished as discrete deposits that are composed of consistent lithology and weathering characteristics (e.g., Miall, 1990). A typical lithologic description of a unit will include the following:

1. Color (Munsell system)
2. Dominant grain size class (gravel, sand, silt, clay), with appropriate modifiers
3. Volume percentage of clasts >2 mm in diameter (gravel)
4. Clast diameter (average and maximum)
5. Clast shape
6. Clast sorting
7. Matrix grain size (often estimated from rheological properties)
8. Matrix compactness (hard versus soft)

9. Bedding thickness
10. Sedimentary structures
11. Weathering or soil formation
12. Fossils
13. Nature of bounding contacts
14. Deformation structures
15. Genetic interpretation

A complementary scheme for defining trench units is *lithofacies codes*, such as that proposed by Nelson (1992b) for deposits found in normal fault zones in semiarid climates. Lithofacies are differentiated based on the main grain size, modifying grain size, internal structure, an interpretive modifier, and soil horizon development (for examples, see Section 3.3).

The best labeling/numbering scheme for trench map units is one that tells the reader at a glance the age and origin of each map unit. The simplest system is to number the units in order of their inferred age, based on stratigraphic position; I prefer to number the oldest unit as 1 and have numbers increase with decreasing age. Lower-case letters can be used to distinguish beds of different grain size or facies within a series of strata of similar origin (e.g., fluvial beds of coarse gravel, fine gravel, and sand). Hiatuses can be reflected by creating a gap in the unit numbering sequence. For example, in a trench across a dip-slip fault, there is a time gap between the prefaulting and postfaulting deposits; if a soil profile developed between them, the implication is that the time gap was long enough for soil formation. Another time gap can exist if the amount of vertical fault displacement is greater than the depth of the trench. In that case, all the beds exposed on the upthrown block will be older than even the oldest bed on the downthrown block. There may also be erosional unconformities or paleosols in the hanging wall or footwall sequences, and those too represent time gaps. The larger the displacement, the larger this time gap will be. Time gaps can be "represented" by a gap in the unit numbers, as shown in Figure 2A.38. Based on the stratigraphy shown in that figure, there are three time gaps: (1) an erosional unconformity between units 3 and 10, on both the upthrown and downthrown blocks; (2) a paleosol developed on unit 11, prior to faulting (horizons 11A and 11Cca developed on parent material 11); and (3) a time gap between the prefaulting soil and the deposition of postfaulting colluvium. The latter is a very small time gap close to the fault, where soil horizon 11A is quickly buried by debris-facies scarp-derived colluvium (unit 20), but becomes a longer time gap farther away from the fault, where 11A is buried by wash-facies colluvium (unit 25). Thus, that time gap is time-transgressive. The gaps in the unit numbering sequence reflect the existence of these time gaps. Lower-case letters distinguish the near-contemporaneous facies of early postfaulting fissure-fill colluvium (20a) from slightly later debris-facies colluvium (20b). If the trench logger selects the unit numbers to "tell the story" of the events displayed in the trench walls, it will assist the reader to understand the time sequence of events.

Such simple numbering systems are sufficient for a single trench, but become insufficient when numbering units from a cluster of trenches (as needed in many strike-slip fault investigations), because a temporal hiatus (erosion, soil formation) in Trench A may be correspond to periods of deposition in Trench B (or C, D, etc.). Thus, there has to be enough "room" in the unit numbering scheme to insert new unit numbers to represent the beds in Trench B. This common occurrence in California trenches has given rise to three-digit numbering systems, where major stratigraphic packages are numbered in the 100s, 200s, 300s, and so on, and individual beds in those sequences given discrete numbers within those ranges. A unique feature of many California trenches is that they are in sag pond environments where deposition is nearly continuous, and earthquake recurrence intervals are short (a few hundred years). This combination means that pedogenic soils (weathering profiles) are poorly developed or nonexistent.

The minor hiatuses present are represented by thin A horizons or peat layers, and these are treated and numbered as stratigraphic units (see Chapter 6). This situation contrasts with trenches across dip-slip faults and faults with long recurrence intervals (several kyr to tens of kyrs), where most of the time represented in the trench is "soil formation time" rather than "depositional time" (e.g., Figure 2A.38). Ultimately, the best choice for a trench unit numbering scheme is one that "tells the story," however complex, to the reader.

2A.3.2.10   The Problem of Fault Nonvisibility

In most applications of structural geology, strata that overlie a fault and are not visibly faulted are assumed to postdate faulting. In unconsolidated deposits, however, faults may lose *visibility* even though the host deposits can be proved to have been displaced. Bonilla and Lienkaemper (1991) provide a comprehensive discussion of this problem, which we briefly summarize next. Terms used in the discussion are defined in Figure 2A.40 and Table 2A.3.

Nonvisibility of fault strands means they are hard to see in the trench wall. It can be caused by two general mechanisms, *concealment* or *termination of displacement.* Concealment means that the fault displacement exists, but cannot easily be seen. For example, where a fault is nonvisible under a surface rupture trace, or where it is visible both below and above a stratum but not within the stratum, the cause must be concealment of a fault trace that does displace the deposits. Fault strands that are nonvisible immediately after surface faulting might be concealed due to intergranular movements, bending of the affected stratum, or many small-displacement distributed ruptures. Fault strands that were visible immediately following a surface rupture are often progressively obscured through time by soil formation, bioturbation, freeze thaw, shrink-swell, plastic flow of clay, rearrangement of grains in granular material, or human activities such as plowing. Nonvisible fault segments are far less common in normal faults (5–10% of fault strands) than in

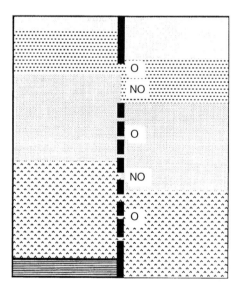

Figure 2A.40: Schematic diagram of a simple fault exposed on a trench wall showing designation of obscure segments (O) and not obscure (visible) segments (NO), as defined by Bonilla and Lienkaemper (1991; Figure 2A.1).

Table 2A.3: Definitions of terms describing the visibility of fault strands in vertical exposures

Term	Definition
Strand	A part of a fault trace exposed in a trench wall; the term is preferable to "segment" as used by Bonilla and Leinkaemper, which has planimetric connotations (Chapter 9, See Book's companion web site)
Principal strand	The fault strand that has the largest real or apparent displacement in a given exposure
Obscure segment	Part of a known fault strand where the fault is not clearly visible in the trench wall, but is visible both higher and lower on the trench wall, or was known to have displaced the ground surface at time of faulting; usually caused by concealment
Dieout up	Where a fault strand ends, or seems to end, upward; caused by concealment or by termination of displacement
Dieout down	Where a fault strand ends, or seems to end, downward; caused by concealment or by termination of displacement
Nonvisibility	A general term that encompasses obscure segments, dieout up, and dieout down
Depth of dieout up	Vertical distance from the ground surface at the time of faulting to the top of the visible part of the fault strand

After Bonilla and Lienkaemper (1991).

strike-slip faults (60–70% of fault strands) and reverse faults (30–60% of fault strands) (Bonilla and Lienkaemper, 1991, p. 18). *Obscure segments* are most common in sand (due to intergranular adjustments) and soil horizons (due to bioturbation and pedoturbation), less common in silt and clay (which are often well stratified), and least common in gravel (where rotated pebbles show the fault trace).

A second reason for nonvisibility of a fault strand is actual termination of displacement. "Reasoning indicates that all faults must actually end somewhere, and observational evidence supporting this conclusion is provided by experimental fault studies and mine mapping showing fault strands that die out upward, downward, or both" (Bonilla and Lienkaemper, 1991, p. 29). For those faults where the position of the ground surface at time of faulting is known, an astounding 73% of strike-slip fault strands, and 75% of reverse fault strands exposed in trenches, show *dieout up* (Table 2A.3), whereas only 15% of normal faults dieout up. At most paleoseismic trench sites it is not possible to distinguish whether dieout up results from concealment, distributed deformation, or termination of displacement, because the position of the ground surface at the time of faulting is not generally known. However, the material properties responsible for concealment of true faults should be similar for strike-slip, reverse, and normal faults, which implies that the minimum of 15% of obscure segments found in normal faults probably represents concealment, while the additional 60% obscure segments in reverse and strike-slip faults may result from termination of displacement.

Repeated faulting alternating with deposition of strata will result in increasing displacement on the fault with depth (Figure 2A.41A). Such *differential displacements* may result from (1) episodic faulting or (2) *attenuation of displacement* in the vertical direction. The proportion of cases caused by these two factors is unknown, but Collins (1990) argues that attenuation is more prevalent. To prove a recurrent faulting origin, the displacements must abruptly increase with depth only at unconformable contacts of each depositional unit; within each unit displacement must be constant. If this condition is *not* met, the increase of displacement with depth may be due to vertical attenuation of displacement in near-surface materials. Such a situation might result from a single faulting event, as shown in Figure 2A.41B. In this example the fault died out upward before it reached the ground surface existing at the time of rupture. This discussion assumes that the faults can be easily observed on the trench wall and are not concealed due to one of the recognition factors discussed earlier.

2A.3.2.11 Imaging the Trench Wall

In the early days of trench logging in the USA, when fault trenches were typically deep single-slots less than 1 m wide, it was difficult to photograph the trench walls and so it was rarely done. The trench logs were drawn manually on graph paper, and perhaps a few photographs were taken of key features, but the entire trench wall was seldom photographed. However, with the advent of digital photography, computer graphics software, and double-benched/laid-back trenches in the 1990s, that practice has changed. Today, imaging the trench wall has become an integral part of the trench logging process, so much so that the single-slot trench geometry has fallen out of use except in reconnaissance investigations or in densely developed areas.

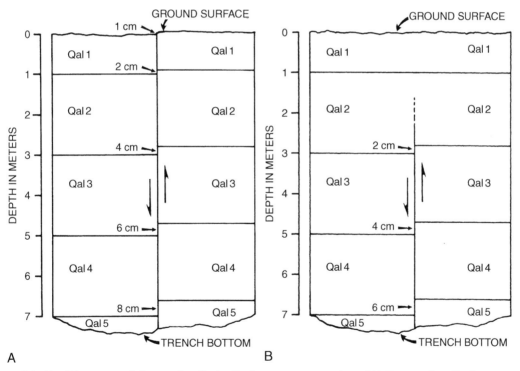

Figure 2A.41: Diagrams of decreasing fault displacement upsection. (A) Decreasing displacement of this type may be due to recurrent faulting or to a vertical displacement gradient. (B) Example of a fault trace that dies out before it reaches the surface. This geometry can be caused by (1) latest faulting midway during deposition of Qal2, or dieout up in a faulting event that ruptured the modern ground surface at other locations. From Collins (1990); reprinted with permission of the Association of Engineering Geologists.

Imaging the trench walls can be accomplished by photography or by more exotic means such as scanning (push-broom scanner). Frost *et al.* (1991) showed that digital photographs can be subjected to computer image enhancement techniques for identifying vague structures on trench walls. Suggested enhancements include tone stretching, edge enhancement, band ratioing, and image differencing applied visible-spectrum images to detect features hard to see with the naked eye. Carrying this idea further, Ragona *et al.* (2006) used a portable push-broom AISA hyperspectral scanner on trench exposures to collect 500 narrow contiguous spectral bands in the visible near infrared and short-wave infrared. There were three advantages to this hyperspectral imaging: (1) the infrared images showed features not visible to either the naked eye or on digital photography; (2) the reflectance spectra of each pixel can be processed in ways that digital photographs cannot; and (3) the fuller-spectrum reflectance data provide an unbiased archive that can be processed at later dates, even long after the trench has been backfilled, as processing techniques improve. These techniques for "seeing the unseen," while considered promising even in 1996, have not yet widely replaced our traditional (albeit subjective) visual methods of identifying strata and structures in trenches.

2A.3.2.12   Logging the Trench

Once all contacts have been marked on the trench wall, the task of trench logging is mainly mechanical and can be accomplished by a variety of manual or instrumental techniques.

Trench wall photography: Once all the relevant geologic contacts have been marked on the trench wall, the entire trench wall is photographed. If the trench is to be logged in the manual or electronic methods, these photographs will merely serve as a backup archive to supplement the field notes about stratigraphic units and structures in the trench walls. In that case, the photographs need only contain enough spatial reference information (reference grid) that their location on the trench wall can be determined; they do not have to be perfectly aligned with the grid, nor all at the same scale. Such photography creates an archival record of the trench wall for future reference. Photography can also be used in any trench to document wall relations (Goodman, 1976); see also McClay (1985) for the use of clear overlays, and Frost *et al.* (1991), for description of video and audio documentation.

Manual trench logging method: In the *manual method,* features on the trench wall are measured in relation to the reference grid with a tape measure. While one person measures the horizontal and vertical distances from trench features to the nearest grid line, a second person plots the position of these *control points* on the trench log (Figure 2A.42). Contacts are then drawn on the log by connecting the control points with lines that mimic the natural, irregular nature of contacts on the trench wall. Some artistry is required (Figures 2A.35, 2A.37, and 2A.38 were drawn manually). The trench log thus drawn is completed before leaving the trench site, which facilitates plotting sample locations and making preliminary interpretations.

The manual logging method has many advantages: (1) it is inexpensive and requires no equipment more sophisticated than a tape measure and some gridded paper; (2) one person can do the logging if necessary although two should be used for safety; (3) the detail of the log is limited only by the time and artistic ability of the logger; (4) it is easy to make revisions by erasing and redrawing; and (5) the log is drawn as the trench is traversed and is in nearly final form when the end of the trench is reached. Disadvantages of the method are: (1) the log does not contain an actual image of the trench wall, but is an interpretive (subjective) drawing; (2) the hard copy log has to be digitized for publication or distribution; (3) if the reference grid lines lie at some distance from the logged wall, there are possible parallax errors in the drawing; and (4) considerable time is consumed by measuring thousands of points by hand. Still, any trench can be logged satisfactorily via the manual method. In case of equipment failures (digital camera breaks, total station batteries go dead), all paleoseismologists should be able to log a trench manually.

Figure 2A.42: The author learning to log in the manual method, under the guidance of Dave Plaskett, University of Alaska. Note the hundreds of toothpicks in the wall at left, used as control points. Dry Creek archeologic site, central Alaska, June 1976.

Electronic trench logging method: The *electronic trench logging* method involves surveying the positions of stratigraphic contacts and structures on the trench wall with electronic surveying instruments, such as the total station (e.g., Lund and Euge, 1984). In this method no string line reference grid need be constructed in the trench. Instead, the total station is mounted at one end of the trench, or outside the trench if the entire wall is visible from a single point. One person holds the EDM reflector against the trench wall at unit contacts, while the instrument operator surveys the angle and distance to that point (however, many new EDMs can bounce a signal back off the trench wall without the use of a reflector). This process is repeated for as many measurement points as desired. The total station operator can enter codes that distinguish points surveyed on fault planes, contacts of individual stratigraphic units, and soils.

To create a trench log, the digitized *x*, *y*, and *z* coordinates of trench contacts must be plotted. Most plotting programs connect control points with straight line segments, so many more control points must be measured on curving or irregular contacts than is necessary in the manual method if lines are to be realistic. The advantages of the total station method of logging are increased speed and planimetric accuracy compared to the manual method, the ability to plot trench logs at any scale, and the ability to project points onto a vertical plane from irregular or nonvertical trench walls. The main disadvantages are cost of the equipment, and the need to have a printer or plotter in the field so that a hard copy log is available for on-site use.

Photomosaic trench logging method: This method has now become standard in research-grade trench investigations. The trench walls are digitally photographed and the photographs are combined into a digital mosaic covering the entire walls. To ease mosaicking, the photographs should be taken perpendicular to the trench wall with frame edges aligned parallel to the reference grid; this will minimize scale differences and parallax within each individual photograph. Each photo will have to overlap adjacent photos to avoid gaps in coverage. The author prefers to take a small number of photographs that cover relatively large areas of the trench wall. For example, in a 3-m-deep trench gridded with a 1 m × 1 m string grid, each photograph in portrait orientation covers 3 m high by 2 m wide. Each photo must contain the vertical string lines 2 m apart, because they will be used to trim the photograph and mosaic it to the adjacent photographs. Using an 8 megapixel digital camera yields sufficient detail in each photograph for mosaic logging. The only drawback to this "large-wall" method is the difficulty in taking the photograph exactly perpendicular to the trench wall, so most images contain parallax, which has to be corrected in the image-processing software.

Other authors prefer to take many more photographs covering smaller areas (1 m × 1 m, 0.5 m × 0.5 m; see Figure 2A.43) to reduce scale changes and parallax within each image. Such photos can be taken "freehand" with respect to the reference grid stringlines, or with the assistance of a rigid frame containing the camera. An example of the latter is the "Trench-O-Matic" invented by Tom Fumal (U.S. Geol. Survey), which uses a 20 mm rectilinear lens on a 35 mm camera, mounted to a PVC pipe frame about 1 m square, with the camera held >1 m from wall. This "small-wall" method requires very many photos to cover the entire trench wall, but each photo requires little rectification before mosaicking. The choice is ultimately a matter of personal preference.

Once the photomosaic has been created, it can be turned into a trench log in two ways. First, if all the contacts and structures on the trench wall were marked in a visible way (colored flagging, spray paint) before the photos were taken, then the marks will be visible also on the photomosaic. In this case, the trench log can be created in computer graphics software by simply drawing all the contacts and structures as overlay vector lines/polygons onto the photomosaic as a raster backdrop. In this technique, the essential work of trench logging was finished when the trench walls were marked, and annotations added to the photomosaic simply make the contacts easier to see. This is the preferred method.

Alternatively, one can take the photographs for the mosaic after the trench walls are cleaned and gridded, but before the contacts are marked on the wall. In this technique, one prints a hard copy of the photomosaic, mounts it on a rigid board, takes it into the trench, and draws the trench wall contacts on the photomosaic (or on an overlay) while observing the trench wall. Thus, the trench log linework is drawn while the logger is in the trench, rather than back in the office. The lines can be scanned later and digitized, resulting in a final product that is indistinguishable from one produced by the first method.

Photogrammetric method: *Photogrammetric trench logging* is an application of terrestrial photogrammetry to trench logging and may be required for very large, irregular fault exposures that cannot be accessed everywhere. The method is described by Fairer *et al.* (1989) and Coe *et al.* (1991) as follows. First, all lithologic, soil, and fault contacts must be marked on the wall clearly enough to be seen on a photograph. Second, at least four surveyed control points (with x, y, and z coordinates) must lie within each stereo pair of photographs that will be taken of the trench wall, and these too must be visible on the photos. Third, geologic notes, Polaroid photographs, and sketches are made relating geologic features to the surveyed control points, to supply critical data to the plotter operator if uncertainties arise. Fourth, the trench wall is photographed from a constant distance (ca. 2–3 m) such that adjacent photographs overlap by 60% (Figure 2A.44). Fairer *et al.* (1989) used a 70-mm-format camera mounted on a dual-tripod system, but any small-format camera will work. Photographs taken at night with strobe lighting are not plagued by daytime shadows.

The overlapping photographs are analyzed in an analytical stereoplotter, which is used in photogrammetric analyses of vertical aerial photographs. Fairer *et al.* (1989) claim that the photogrammetric method requires only about one-quarter to one-sixth the time of the manual measurement method to produce a final trench log. The main advantage of the photogrammetric method is its accuracy (±6 mm; Coe *et al.*, 1991) and its sophisticated analysis of planimetric measurement errors, which may be required by stringent

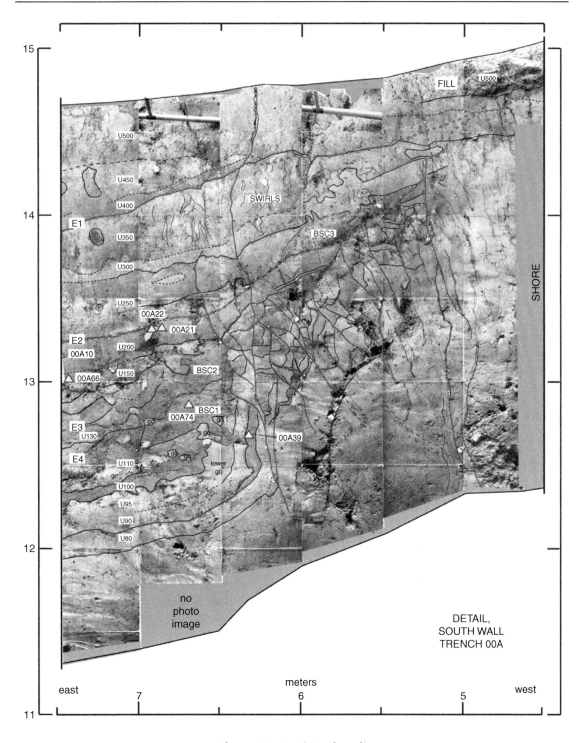

Figure 2A.43: (Continued)

Figure 2A.44: Stereo photograph pair used in the photogrammetric method of trench wall logging. The main fault zone (between arrows) stands out in relief because it has been cemented by calcium carbonate and silica. The irregular, sloping walls of this natural exposure would be difficult to log accurately with the manual or photomosaic methods. Shovels at upper right (circled) are 1.5 m long. Paintbrush Canyon fault at Busted Butte, Nevada Test Site, USA (photographs courtesy of J. A. Coe).

quality assurance programs. However, millimeter-scale accuracy is rarely necessary for an adequate interpretation of either paleoearthquake displacement or age. The drawbacks to the method are the cost of buying (>US $100,000) or renting an analytical stereoplotter, and finding a competent photogrammetrist.

The final trench log: The final trench log has two functions: (1) it is an archival graphic record of the excavation and (2) it "tells the story" of paleoearthquake history, as interpreted by the trench logger. To perform these functions, the *scale* of the trench log must allow accurate delineation of features as small as 1 cm, but must not be so large as to make the log unwieldy. A scale of 1:20 is typically used in the United States; at this scale a 5-cm-high feature is 2.5 mm high on the log, about the practical limit of legibility. Depending on desired detail, logs can be mapped at scales of 1:10 to 1:50, but should rarely be smaller scale than 1:60, or the log will merely be a "cartoon" of actual wall relations. Some states in the United States have guidelines or regulations requiring appropriate methods and scale used in trench logging (e.g., Christenson *et al.*, 2003).

Figure 2A.43: Example of a photomosaic trench log. Individual photographs in the mosaic cover a 0.5 m by 0.5 m area. Trench units (labeled) are highlighted by thin lines, and faults (such as at right center) are highlighted by slightly thicker lines. Selected units can be further emphasized by semi-transparent color fills (see color version on the companion web site, Chapter 2A). Log is from the Hayward fault, a dextral fault east of the San Andreas fault in northern California. From Lienkaemper *et al.* (2002). (See Color Inset.)

2A.3.2.13 Manipulating and Storing Digital Trench Data

Digital trench data can include raster files of various sorts (e.g., individual trench photos, the trench wall photomosaic, infrared scan images, GPR surveys, ERT and seismic tomograms) and vector line and polygon layers (e.g., stratigraphic contacts, soil horizon contacts, faults, fold axes, fractures/fissures, unit labels, sample locations and labels, etc.). These data can be saved as separate layers in the computer graphics software, allowing them to be toggled between visible and invisible to create different visualizations (Table 2A.4)

A more important need is for visualization of the trench's relationship to other trenches, and to larger spatial features (landforms, boreholes, etc.) beyond the trench. The easiest way to make such visualizations is in a GIS, where all trenches and features beyond trenches are georegistered to real-world coordinates. To date, few trench logs have been digitized into a GIS, but see the proposal for a prototype "trench-specific geologic information system (TSGIS)" by the PALEOSEIS Project (Swiss Seismological Service at ETH-Zurich). Storing trench information in a GIS has a number of advantages. First, the GIS software permits storage of features in separate layers (or coverages in ArcInfo terminology). GIS systems have two advantages over computer graphics programs for trench logging. First, GIS software can link each graphic object on the trench log with a table of attributes, so text and quantitative information can be linked with any graphical feature, such as strike and dip information for faults, sample information for sampling locations, etc. Second, the 2D or 3D locations of the features can be referenced to a worldwide or local datum and projection. Such georeferencing permits one to: (1) accurately compare the location of features among multiple trenches at a single site, to determine net slip vectors and (2) permits future workers to exactly relocate your trench exposures.

2A.3.3 Drilling, Coring, Slicing, and Peeling

Trenches have several inherent limitations for paleoseismology: (1) their depth is limited, (2) once they penetrate the groundwater table, they fill with water, and (3) the trench cannot be "taken back to the laboratory" to work on after field season ends and the trench is backfilled. Although geophysical surveys partially solve the limited depth problem, there are often ambiguities in geophysical interpretation. The traditional way to collect geologic information deeper than the floor of a trench is by drilling and coring, as described below. We also describe some recent techniques for "taking the trench back to the laboratory" that go beyond mere imaging.

2A.3.3.1 Drilling

Although it is not generally appreciated, *boreholes* and *shallow cores* have several advantages over trenches for collecting paleoseismic data in certain situations; they can be placed in areas of standing water, shallow groundwater, and intense urbanization with buried pipelines. They are relatively inexpensive, nondisruptive, safe, and can be placed as closely as needed to maintain interhole correlation. Disadvantages are the disruption of delicate structural and stratigraphic features during sampling (depending on the sampling method), and lack of the continuous two- or three-dimensional view that is often essential for interpreting paleoseismic features.

Drilling as described herein includes standard powered well-drilling methods (cabletool, auger, rotary, or percussion) in which the hole is logged by: (1) sending a geologist down the hole to log the walls (Johnson and Cole, 2001), or (2) logging either chips brought up by bailers or in the circulating medium, or by drive samples taken at irregular intervals. Human downhole logging of 1-m-diameter "bucket auger" holes is

Table 2A.4: Various options for storing digital trench log data, as a function of the number of layers and georeferencing

Graphic layers in drawing	Georeferencing of graphic layers		
	None	**Informal datum**	**Formal datum**
Single	2D; Example: trench log stored as a single-layer drawing in a vector graphics program; spatial units are inches (cm) on the drawing pane; log created by the manual or photomosaic method; *most common storage method to date*	2D; same as at left, but units in the drawing pane are replaced by real-world dimensions, keyed to a local survey datum; common in electronic trench logging, engineering projects and AutoCad drawings	2D; same as at left, but units in drawing are real-world coordinates tied to a formal map projection and datum (e.g., UTM, NAD27); can be assigned by obtaining an *xyz* GPS coordinate from somewhere in the trench reference grid
Multiple	2D; Example: multiple raster images stored in separate layers (e.g., photomosaic of trench wall; GPR survey; ERT or seismic tomogram); multiple vector line sets stored in separate layers (stratigraphic contacts, soil horizon contacts, unit labels, faults, dating sample locations, etc.)	2D; same as at left, but units in all drawings or images are replaced by real-world dimensions, keyed to a local survey datum; common in electronic trench logging, engineering projects and AutoCad drawings	2D; same as at left, but units in drawing are real-world coordinates tied to a formal map projection and datum (e.g., UTM, NAD27); images are saved as GeoTiff; georegistration can be assigned by obtaining an *xyz* GPS coordinate from somewhere in the trench reference grid
	3D; *Example*: same as above, but with addition of 3D points collected during electronic or photogrammetric logging; 3D points stored separately from all 2D projections to a single plan	3D; same as at left, but spatial units are replaced by real-world dimensions, keyed to a local survey datum	3D; same as at left, but spatial units are real-world coordinates tied to a formal map projection and datum (e.g., UTM, NAD27); can be assigned by obtaining an *xyz* GPS coordinate from somewhere in the trench reference grid

used extensively in landslide investigations in California, but rarely in paleoseismic investigations. Various *thin-wall sampling tubes* are available that will preserve sedimentary structures in all but gravelly materials (Soiltest, 1977). A common approach in urban paleoseismic drilling programs is hollow-stem augering with split-spoon sampling done every 1.5 m in conjunction with the Standard Penetration test (cf. Carter, 1982; Hunt, 1984). A truck-mounted auger can drill up to 15 m in most unconsolidated deposits, which is the depth limit in many geotechnical investigations.

The advantages of drilling are its great depth of penetration and the ability to penetrate gravelly, compacted, or cemented Quaternary deposits. Disadvantages of drilling are that it is expensive, obtaining core samples is tedious, and thin stratigraphic units and soils are mixed together by drilling and seldom recognized from chips and cuttings. The main use of drilling in paleoseismology is to *locate correlative strata* across a fault zone so estimates of net displacement can be made. Drilling may also uncover faults that are not expressed at the surface or, conversely, may prove that no fault displacement exists within a certain area. Robison and Burr (1991) drilled 19 borings across a strand of the Wasatch fault zone in an urbanized area in downtown Salt Lake City, Utah, where trenches were impractical (Figure 2A.45). From the drilling data they not only located the fault trace and measured its minimum vertical separation, but the *structure-contour map* based on the 19 borings proved that the fault projection on published geologic maps was incorrect. The destructive nature and limited spatial data from drilling typically precludes recognition of individual paleoearthquake displacements.

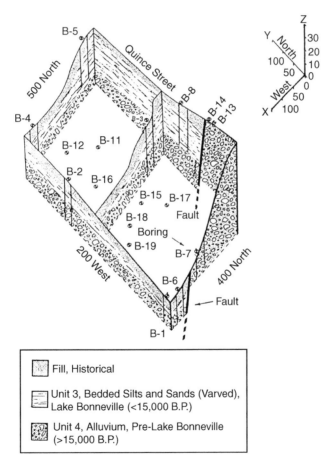

Figure 2A.45: Fence diagram showing the generalized subsurface geology in the Wasatch fault zone, downtown Salt Lake City, Utah, based on 19 borings and three shallow trenches. Boring logs within the central portion of the site were omitted for clarity; although boring locations are indicated by numbers. Scale at upper right is in feet. Reprinted with permission of the Symposium on Engineering Geology and Geotechnical Engineering.

2A.3.3.2   Coring

Coring can refer to either shallow hand-coring or continuous coring with powered equipment. Hand-coring has several advantages: It is cheap, portable, and results in continuous core with good preservation of sedimentary structures. Disadvantages are the limited depth penetration (usually <3 m) and inability to penetrate hard sediments. The main use of hand-coring in paleoseismology has been to correlate strata in areas of Holocene tectonically induced sedimentation. For example, Nelson (1992a) used a 1 m-long, 2.5 cm-diameter half-cylinder *gouge corer* to sample and correlate peat and mud marsh sediments in subsided areas of the Oregon coast, USA. Clague and Bobrowsky (1994b) used a *sonic drill* to obtain cores up to 11 m long in marsh sediments. Coring in lakes utilizes *Kuhlenberg piston corers* or the self-contained, gas-operated *Mackereth piston-coring system*, which returns lake-sediment cores up to 10 cm in diameter (Ager and Sims, 1981; Rymer and Sims, 1982; Perkins and Sims, 1983). Cores can then be X-rayed to detect sediment deformation from seismic shaking (see Chapters 7 and 8).

Geotechnical-type coring for paleoseismology is becoming a common substitute for trenching in urban areas where large trenches are impractical. For example, Dolan *et al.* (1997) drilled 30 continuous 9-cm-diameter cores across the poorly preserved Hollywood fault in Los Angeles, which is an oblique reverse fault heavily modified by urbanization. They not only located the buried fault plane and measured its displacement from boreholes (Figure 2A.46), but dated the most recent movement between 4 and 20 ka (closer to the younger age). Coring is also helpful to extend the stratigraphic record deeper than trench depths and to estimate vertical displacements on older stratigraphic datums.

2A.3.3.3   Slicing

The "Geoslicer" was invented by Nakata and Shimazaki (1997) to retrieve meter-size intact "slices" of subsurface materials in fault zones. A pair of steel plates is driven vertically into the ground and then pulled out, bringing with it a relatively undisturbed "slice" of geologic deposits (Nakata and Shimazaki, 2000). Three models of geoslicer are manufactured by Fukkon Corporation, Japan (http://www.geoslicer.com). The "Wide Geoslicer" comes in two sizes, 1.2 m wide × 2 m long, and 1.5 m wide × 4 m long (shown in Figure 2A.47); each retrieves a 10–15-cm-thick slice. The "Long Geoslicer" is custom made in sizes up to 12 m wide and 10–11 m long. The personal-size "Handy Geoslicer" retrieves slices 10 cm wide, 3 cm thick, and 1.5 m long.

These geoslicers have generally been used in clast-free sediments where the groundwater level is very shallow, and trenching would not be feasible. That includes sag ponds, tidal flats, low-lying coastal plains, and alluvial and marine terraces (Takada and Atwater, 2004; Komatsubara *et al.*, 2008), where the target may be faults, liquefaction features, or tsunami deposits. The geoslicer can also be driven into the floor of a fault trench to sample deeper stratigraphic levels. Geoslicers have also been used to study the folded strata above blind thrust faults (Kaneda *et al.*, 2008a). However, to date there have been few detailed quantitative studies on the sedimentology or macro- or microfabrics of retrieved sediment slices, as have been performed on peels (see Section 2A.3.3.4).

2A.3.3.4   Peeling

The collection of "soil peels" or "soil monoliths" from vertical surfaces has a long history in soil science (e.g., Smith and Moodie, 1947), but is a relatively new addition to paleoseismology. The traditional methods are to spray a liquid onto the vertical wall that penetrates into the wall materials for some

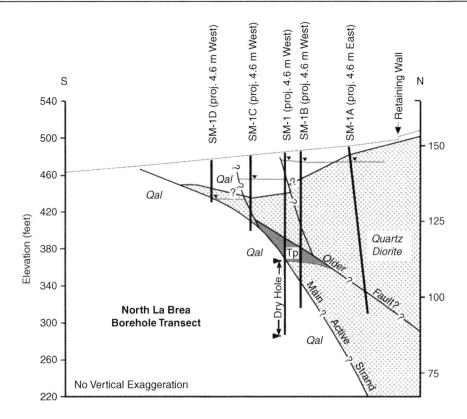

Figure 2A.46: Cross section of the North La Brea Avenue borehole transect, showing that the Hollywood fault dips steeply at depth but flattens near the surface. Thick vertical lines denote continuous cores. Small triangles and gray lines denote groundwater levels in boreholes. From Dolan *et al.* (1997). Reprinted with permission of the Geological Society of America.

distance and then dries and hardens. Successive coats of the liquid are added, interspersed with flexible backing grids or cloths, until the whole mass is strong enough to be removed. Many liquids have been used over the years, including latexes, lacquers, and epoxies. The soil peel method is most suitable for geologic deposits that do not contain a large proportion of clasts.

McCalpin *et al.* (1993) collected soil monoliths from fragile, clast-rich scarp-derived colluvium by using expanding urethane foam, which is widely available in small pressurized cans for sealing cracks in houses. Beads of the urethane were applied to a 60 cm × 60 cm sheet of 3/8″ plywood, which was then pressed against a vertical trench wall and left to dry and harden overnight. Once the urethane had hardened, an 8–10-cm-thick monolith was cut away from the host deposit parallel to the plywood and removed along with the plywood (Figure 2A.48). Most of the clasts in the deposits were tightly bonded to the plywood via the hardened urethane, and those clasts held the looser interstitial matrix in place long enough for transport to the laboratory. The monoliths were then slowly disassembled in the lab while measuring clast fabrics. Although their samples were of colluvial wedge sediments, the same technique could be used to collect gravelly deposits in fault zones, to preserve the shear fabric shown by clasts (e.g., Figure 2A.34).

Figure 2A.47: Photograph of a wide geoslicer mounted on a trackhoe arm with a hammering attachment. Photo courtesy of K. Okumura.

Figure 2A.48: Soil monolith collected from the post-1983 colluvial wedge shed from the Borah Peak, Idaho, fault scarp. Shown from left to right, the plywood board, the urethane foam (yellow), and the monolith. The monolith soil is darker because it was sprayed with an aerosol clear fixative. Subsequent to this photograph, the standing monolith was secured by wrapping it in rubber carpet padding. The base was then cut away and the monolith was detached and laid down for transport. Photo by J. P. McCalpin, 1985.

2A.3.4 Dating Methods for Late Quaternary Deposits

Late Quaternary deposits can be dated by a wide array of correlated age, relative age, numerical age, and calibrated age methods, as outlined in Chapter 1 (Section 1.3). These methods are not unique to paleoseismology, and their application to dating paleoearthquakes requires a certain approach to sampling and date interpretation (as discussed in detail in later chapters). However, the paleoseismologist should have an appreciation of some generic issues in dating Quaternary deposits to date paleoearthquakes. One important issue is the nature of earthquake horizons as unconformities, and general approaches for dating an unconformity. Because unconformities represent "missing" geologic time removed by erosion, there is the chance that the actual time of the paleoearthquake will fall into the missing time interval.

An example is shown in Figure 2A.49. The site that was faulted had been a site of deposition from 20 to 16 ka (Stage 1), but then deposition ceased at 16 ka and soil profile development dominated until 10 ka (Stage 2). At 10 ka the ground surface was faulted (Stage 3), and then the area was again buried by deposition from 10 to 7 ka (Stage 4). The preservation of the paleoearthquake horizon depends on what happens next. If there is no subsequent erosion, then the paleoearthquake horizon is preserved as the unconformity between the top of the buried paleosol, and the bottom of the scarp-derived colluvial wedge (labeled 9.5). Its age could be relatively tightly constrained by age samples below (10 ka) and above (9.5 ka) the unconformity.

But what if later erosion removes part or all of the stratigraphic section? In Stage 5A, erosion at 3 ka removes the deposits laid down between 9 and 7 ka, and replaces them with deposits from 3 to 1 ka. The erosional unconformity now places 3 ka sediments atop sediments of various ages, some of which still preserve the 10 ka prefaulting paleosol, but some of which do not. In the footwall, for example, the unconformity can only be dated as occurring between 17 and 3 ka. Fortunately, erosion was incomplete, so there is still part of the unconformity where 9.5 ka colluvium lies atop the 10 ka paleosol. But in scenario Stage 5B, erosion has even removed the colluvial wedge. In this scenario the unconformity can only be dated as forming between 17 and 3 ka.

So despite the accuracy and precision of the dating method(s) employed, the paleoseismologist must appreciate how ages of deposits (however obtained) relate to the age of unconformities, and thus to the age of the paleoearthquake. The above example falls under the topic of "sample context error," and it is only one part of that topic.

A related concept is judging the age of a paleoearthquake fault rupture based on how close to the modern ground surface the fault plane can be traced. In depositional settings, the closer the fault plane can be traced to the surface, the younger the faulting is. But in erosional settings this principal does not hold, and even faults that have not moved for millions of years can be traced to the surface, because erosional processes are continually lowering the landscape and exposing older structures. In such settings the interaction of the fault plane and the weathering profile give nearly the only clues to the youngest age of movement (see Section 9.4.2 in Chapter 9, See Book's companion web site).

One technique used in trying to date paleoearthquakes in bedrock-erosional terrain, where Quaternary deposits are absent, is electron spin resonance (ESR) of fault gouge. The theory is that mechanical crushing and perhaps brief frictional heating during fault movement would reset the ESR age to zero, which would then provide a basis for dating the latest movement. Calibration studies on faults with known historic ruptures, however, indicate that a single surface rupture only partly resets the ESR signal. Fukuchi (2000) showed that the 1995 Kobe earthquake did not reset fault gouge ESR ages to zero on the Nojima fault. Instead, gouge within 0–3 mm of the fault plane dated at 0.15–0.29 Ma, and ages became older

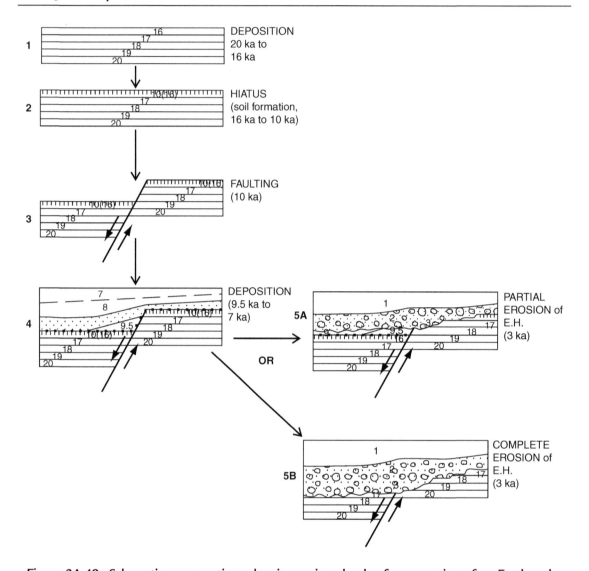

Figure 2A.49: Schematic cross sections showing various levels of preservation of an Earthquake Horizon (stages 4, 5A, 5B). The earliest stage (1, at top) shows four conformable, tabular, fluvial strata deposited between 20 and 16 ka (numbers indicate age of deposition). From 16 to 10 ka the landscape stabilized and a soil formed (stage 2, Hiatus). Faulting occurred at 10 ka (stage 3), and was rapidly followed by deposition of colluvium (stage 4, 9.5 ka) and more alluvium (9–7 ka), burying the soil (buried paleosol). The Earthquake Horizon is the ground surface that existed at time of faulting (top of buried paleosol), as shown by the thick dots in stage 4. Stage 4 shows an Earthquake Horizon fully preserved via deposition, and easily dated between 10 and 9.5 ka. Alternative degrees of preservation include partial erosion of the E.H. (stage 5A), where the E.H. has been eroded from upthrown block but preserved on downthrown block; in this case, displacement can still be constrained between 10 and 9.5 ka. However, in the case of complete erosion of the E.H. and the buried paleosol (stage 5B), displacement can only be constrained as having occurred between 17 and 3 ka.

farther from the fault plane (at 35–50 mm away, ESR ages were 0.39–0.73 ma). They concluded that ESR ages usually dated either the beginning of fault movement (as found by Ulusoy, 2004), or the last period of accelerated movement that was followed by a period of lower slip rate/longer recurrence (as found by Yao *et al.*, 2008).

2A.4 Distinguishing Paleoseismic Features from Nonseismic or Nontectonic Features

Not all faults that break the earth's surface are tectonic, and not all tectonic faults are seismogenic. Because paleoseismology is the study of prehistoric *earthquakes*, we want to consciously avoid mistaking geologic features created by nontectonic (or nonseismic) processes, for geologic features created by paleoearthquakes. Such a determination is not as easy as it sounds. Determining the true origin of a geologic feature can be complicated by the principle of *geomorphic convergence* or *equifinality* (Chorley *et al.*, 1984; Schumm, 1991), which states that similar-appearing landforms can be produced by different, unrelated geomorphic processes. For example, Solonenko (1977a, pp. 41–42) defined seven types of *pseudotectonic* features that mimicked surface fault rupture in central Asia. They include (1) fissures associated with volcanic doming (and collapse) and diapirs from salt tectonics or fractures at fold axes, (2) linear grooves and troughs related to glaciation and ice-marginal drainage, (3) linear canyons excavated along zones of weak rock, (4) gravitational spreading features in high-relief, glaciated areas, (5) selective erosion along fault-line scarps, (6) permafrost features such as frost churning and thermokarst, and (7) man-made features such as old irrigation systems and roads. In the realm of secondary evidence, widespread landslides or rapid sedimentation events in rivers, lakes, and coastal environments caused by infrequent, but severe storms can produce evidence similar to that induced by earthquake shaking (Chapter 5). In addition, a wide variety of nonseismic soft-sediment deformation features mimic similar structures formed by earthquake-induced liquefaction (Chapter 7). Thus, not all deformation of near-surface earth materials results from earthquakes. In fact, fault-like features are created by many nonseismic processes, and in the field it is often difficult to distinguish such "faults" from those produced during paleoearthquakes. A similar ambiguity exists for off-fault features and deformed sediments, as described in Table 1.2.

A rigorous approach to distinguishing seismic from nonseismic deformation features was published in a lengthy monograph by Hanson *et al.* (1999), from which we draw the following discussion. The monograph was written to be a practical guideline for the staff of the U.S. Nuclear Regulatory Commission, for applied purposes in seismic hazard assessment such as those discussed in Chapter 9 (See Book's companion web site). However, we discuss the monograph's fundamental premises here, because they dictate how paleoseismologists evaluate field evidence as being seismic or nonseismic in origin. Obviously, no paleoseismologist wishes to perform a lengthy "paleoseismic study" on one or more deformation features, complete with calculations of paleoearthquake magnitude and recurrence, only to discover later that the feature was formed by processes unrelated to any earthquake!

Figure 2A.50 shows Hanson *et al.*'s (1999) distinction between faults that can produce surface deformation (right circle), and faults that can produce vibratory (earthquake) ground motion (left circle). Paleoearthquakes are caused by sudden slip on the faults common to both circles, that is, "capable tectonic sources." These faults are both tectonic AND seismogenic, according to the following definitions:

> *Tectonic fault*: is produced by deep-seated crustal-scale processes acting at or below seismogenic depths

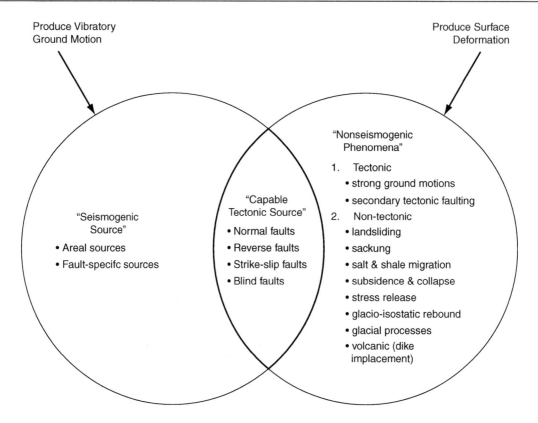

Figure 2A.50: Hanson *et al.'s* (1999) distinction between faults that can produce surface deformation (right circle), and faults that can produce vibratory (earthquake) ground motion (left circle). Any of the 14 types of faults listed in the right circle could produce a fault scarp at the surface, or a displacement of Quaternary deposits. But only four of those fault types can generate $M_w > 5$ earthquakes.

Nontectonic fault: a feature produced by shallow crustal or surficial processes acting above seismogenic depths (a "rootless" structure)

Seismogenic fault: a fault capable of producing a moderate to large earthquake ($M_w > 5$)

Nonseismogenic fault: a fault incapable of producing $M_w > 5$ earthquakes

In paleoseismic studies, we study primary and secondary evidence created during paleoearthquakes, from sources in the left-hand circle ("seismogenic sources" and "capable tectonic sources"). We may also encounter prehistoric surface deformation features produced by "Nonseismic phenomena," and we must be able to distinguish those from the effects of paleoearthquakes. To make this distinction, Hanson *et al.* (1999) differentiate "tectonic faults" from "nontectonic faults" (two mutually exclusive categories), and "seismogenic faults" from "nonseismogenic faults" (two mutually exclusive categories, and independent of tectonic status) (Figure 2A.51).

According to Hansen *et al.* (1999), these distinctions are important in seismic hazard analysis; "Because nonseismogenic features may sometimes be interpreted as seismogenic features (and vice

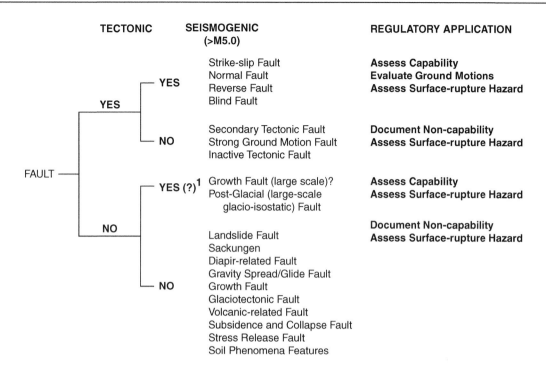

Figure 2A.51: Classification of faults as tectonic/nontectonic and seismogenic/nonseismogenic, with examples (central column), and types of hazard assessment required by the U.S. Nuclear Regulatory Commission ("Regulatory Application"). From Hanson *et al.* (1999).

versa), a primary objective… is to provide criteria to help differentiate seismogenic–tectonic features from nonseismogenic features." Faults produced by tectonic processes may or may not be seismogenic. Tectonic faults include both primary faults capable of producing earthquakes and secondary faults that are produced by earthquakes but are not themselves capable of generating an earthquake. Primary tectonic faults typically are classified into one of four categories based on sense of slip and fault geometry: strike-slip (transcurrent or transform) faults, normal faults, reverse faults, and blind or buried thrust faults; although the transitions between each category are gradational, and individual earthquakes on a specific fault may exhibit attributes of more than one category. Examples of secondary tectonic faults include hanging-wall deformations above a blind thrust fault and various types of strong ground motion phenomena (e.g., ridge-crest shattering, basin-margin fracturing). Examples of nontectonic faults include those produced by gravitational processes (e.g., landslide features, sackungen), dissolution phenomena (e.g., karst collapse features), evaporate migration (e.g., salt domes and salt flowage structures), sediment compaction (e.g., growth faults, subsidence structures), and isostatic adjustments (e.g., glacial rebound structures).

There is a transition class of faults associated with crustal-scale landsliding that can generate $M_w > 5$ earthquakes yet do not extend deeper than 8–10 km in the crust. Examples include the Hilina fault system on the island of Hawaii, which generated an M_w 7.2 earthquake in 1975 and an M_w 7.9 earthquake in 1868, yet soles into a subhorizontal detachment failure plane at 8–10 km (Cannon and Burgmann, 2001; Cannon *et al.*, 2001). These faults are discussed further in Chapter 4.

The paleoseismologist thus should be aware that all faults that displace landforms and near-surface deposits (e.g., as exposed in a trench wall), and all deformed and liquefied sediment layers may not represent paleoearthquakes. To assess whether exposed faults are likely to be tectonic or seismogenic, one should consult the detailed criteria of Hansen *et al.* (1999. In Chapters 3–6 we describe, for each tectonic environment, the geomorphic and stratigraphic features that carry the greatest ambiguity as to seismic versus nonseismic origin.

2A.4.1   Special Case: Stable Continental Interiors

Intraplate cratonic environments, sometimes called "stable continental regions" (SCRs) by seismologists or "stable cratonic cores" (SCCs; Fenton *et al.*, 2006), are not areas where one would expect large earthquakes or abundant paleoearthquake evidence. However, large damaging earthquakes have occurred historically in SCRs, and in the past few decades more geologic evidence of large prehistoric ruptures has been identified in such areas. We devote a brief section to SCR paleoseismology herein, for two reasons: (1) the types of paleoseismic evidence, and thus optimum paleoseismic techniques, differ from those in plate boundary regions and (2) much of the world's population, and many of the world's critical facilities (e.g., NPPs), lie in SCRs.

Fenton *et al.* (2006) define SCC regions as an even more stable subset of SCRs, as described below. Beginning with the SCRs defined by Johnston (1994) (i.e., continents minus their actively deforming regions), Fenton *et al.* (2006) excluded all regions of non-Precambrian crust indicated by Johnston (1994). They then further reduced the regions of Precambrian crust by excluding passive margins together with a zone of approximately 200 km inland from passive margin coasts, and by excluding a similar 200 km zone around regions of Phanerozoic deformation. The remaining regions (Figure 2A.52), with a total area of 94.51×10^6 km^2, constitute the most SCRs on the planet. Four SCCs (Africa, North America, South America, and Australia) comprise 80% of the total SCC area.

These SCC regions are typically underlain by Precambrian bedrock, generally medium-to high-grade metamorphic assemblages of silicic continental and mafic oceanic rocks. With the exception of glacio-isostatic rebound following late Pleistocene glaciation of high latitude areas, SCC regions have undergone little or no significant tectonism during the Phanerozoic (since 500 Ma). SCCs are far removed from plate boundaries, including passive margins, and contain no embedded rift structures (by definition). From experience in eastern Canada (Adams and Basham, 1991), and around the world (Johnston, 1994), it is clear that regions of rifted crust embedded within the continents are significantly more seismically active than adjoining unrifted regions. Rifts of Precambrian age, however, show seismic activity no higher than unrifted crust and are assumed to be "healed."

2A.4.1.1   Unglaciated Continental Interiors

The dominant characteristic of unglaciated SCR/SCC regions is the long recurrence interval between characteristic (or morphogenic) earthquakes. For example, Clark and McCue (2003) describe Australia as follows: "Although the data are still few, there appears to be a pattern emerging where surface rupturing earthquakes in Australia occur preferentially on a few preexisting and often ancient faults. These faults seem to be quiescent for long periods of time between surface rupturing events, in the order of tens of thousands to a hundred thousand years or more."

Because of such long recurrence, the dominant geomorphic processes in SCRs are exogenetic (erosion, deposition) rather than endogenetic (uplift, subsidence). Thus the landscape is not a "seismic landscape"

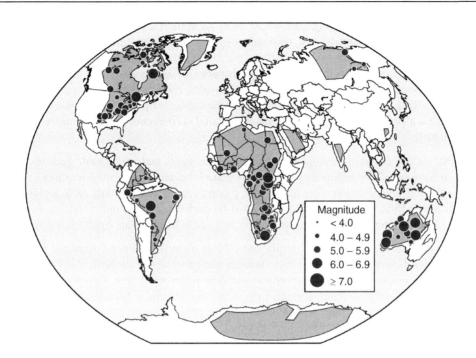

Figure 2A.52: Distribution of the world's stable cratonic cores (SCCs, shaded gray). Historic earthquakes magnitudes in SCCs are shown by black circles. From Fenton *et al.* (2006).

in the terms of Michetti *et al.* (2005), but is mainly an "aseismic landscape" in which the traces of active faults are not prominent. The typical geomorphic signatures of rapid uplift (steep mountain fronts), subsidence (deep sediment-filled troughs), or lateral offset (linear "rift" valleys) are absent, and instead the surface trace of the seismogenic fault may be very obscure or completely obscured (buried). In such environments, faults that do break the surface may be so heavily modified by erosion that they have been assumed by geologists to be (1) erosional features, (2) ancient (pre-Quaternary) faults, or (3) fault-line scarps (e.g., the Meers fault, Oklahoma, USA: Crone and Wheeler, 2000; Wheeler and Crone, 2001; Wheeler, 2005). Such interpretations are understandable, especially if the fault crops out only in "bedrock" terrain where Quaternary deposits are absent.

Guccione (2005) lists four types of paleoseismic evidence commonly used in the depositional intraplate region of central USA, and three are secondary, off-fault evidence: (1) paleoliquefaction features, (2) sediment that is involved, that is penecontemporaneous, and that postdates fault-related folding and faulting, (3) changes in geomorphology and sediment patterns where subtle deformation causes changes in surface hydrology, and (4) biotic responses to a seismic event, such as trauma and related hydrologic changes.

A unique fault type on margins of SCCs is the active growth fault, a normal fault that forms in response to progressive differential compaction of thick sequences of Quaternary coastal plain and deltaic sediments. These faults creep, as shown by deformation of modern infrastructure, and have created normal fault scarps tens of km long, for example, those in the Houston, Texas area (Shah and Lanning-Rush, 2005; Engelkemeir and Khan, 2008). They are a good example of a tectonic, nonseismogenic fault.

Primary, on-fault evidence: Identifying fault traces in SCRs is more difficult than elsewhere and may require recognizing previously overlooked, discontinuous geomorphic or structural anomalies that suggest Quaternary faulting. This approach falls under the topic of tectonic geomorphology (not a subject of this book, but see Burbank and Anderson, 2001). In the absence of continuous scarps, faults may be recognized by alignments of: linear stream valleys and swales, escarpments, drainages, springs, and bedrock notches. Blind faults have been recognized by the subtle warping of Quaternary geomorphic surfaces (due to fault-propagation folding) that may only total <10 m over large regions (e.g., Mihills and Van Arsdale, 1999).

To confirm that such geomorphic anomalies actually overlie faults and are tectonic features (something typically taken for granted for continuous scarps in a plate boundary area), often requires extensive programs of subsurface verification. Due to the high groundwater levels in many SCR regions, trenching is often infeasible and the "tectonic verification" must be performed via geophysical surveys (e.g., Van Arsdale *et al.*, 1995) or via shallow coring programs (e.g., Mihills and Van Arsdale, 1999).

Primary, off-fault evidence: Distributed faulting and folding away from the seismogenic fault can create subtle, second-order deformation of Quaternary deposits and geomorphic surfaces, that is often difficult to distinguish from the original Quaternary erosional and deposition relief. For example, Mihills and Van Arsdale (1999) constructed a structure-contour map of the unconformity between Tertiary strata and the overlying Quaternary alluvium in the New Madrid seismic zone, in order to measure regional paleoseismic deformation. This effort required 253 shallow cores, and the resulting map reveals relief that mirrors the (much smaller and subtler) surface deformation. They interpreted the structure-contour map as a more robust representation of the latest Pleistocene to present strain field of the New Madrid seismic zone.

SCR paleoearthquakes may also cause minor tilting of fault-bounded blocks, giving rise to geomorphic anomalies for which there is no likely erosional or depositional explanation. For example, crustal tilting can perturb the channels of alluvial rivers in predictable ways (changes in sinuosity or gradient, or deflections; Schumm *et al.*, 2000). Crustal tilt can also create anomalies in smaller streams and their depositional and erosional patterns (Guccione *et al.*, 2002; Guccione, 2005). The drainage network itself may display an asymmetry unexplainable by local stratigraphy or structure, created by tilt. For example, Garrote *et al.* (2006) interpreted structural domains (blocks) with different tilt in the Central USA SCR by use of a "transverse topographical drainage-basin asymmetry index."

The above examples indicate that paleoseismic studies in SCRs, more so than in plate boundary areas, require a more sophisticated portrayal of local and regional topography than may be available from standard topographic maps in order to identify geomorphic anomalies. This portrayal may take the form of multiple types of aerial/satellite imagery, on which image enhancement processing is performed such as edge enhancement algorithms to detect lineaments (e.g., Moore and Waltz, 1983; Tripathi *et al.*, 2000). DEMs made from topographic maps, synthetic-aperture radar (SAR), or LiDAR are also useful, and hillshade renditions of DEMs can be illuminated from multiple directions to emphasize subtle topographic anomalies.

Secondary evidence: Much of the paleoseismic history of unglaciated SCCs has been reconstructed from secondary evidence, primarily liquefaction. Chapter 7 goes into considerable detail about paleoseismic evidence from the central and eastern USA, so that detail is not repeated here.

2A.4.1.2   Formerly Glaciated Continental Interiors

SCCs glaciated in the late Quaternary constitute, for paleoseismology, a "special case of a special case." In northern latitudes, such as Canada, the United Kingdom, and Fennoscandia, many if not most geomorphic features associated with prior (>10,000 year old) surface faulting did not survive the erosive effects of late Pleistocene glaciations, further hampering the recognition of potentially seismogenic structures through paleoseismic studies (Adams, 1996). In addition, many phenomena active in postglacial or periglacial environments produce features that are similar to fault-related surface deformation (e.g., Adams *et al.*, 1993; Olesen *et al.*, 2004). Thus, as well as trying to identify potentially seismogenic structures, care must be taken to discard scarp-like structures produced by nonseismogenic processes from consideration in seismic source characterization studies. A good overview of neotectonics of formerly glaciated areas is given by a Special Issue of Quaternary Science Reviews (2000, vol. 19, issues 4–5).

For example, Morner (2003, 2005) lists nine types of paleoseismic evidence found in Sweden, seven of which are secondary: (1) primary fault structures (scarps), (2) secondary bedrock fracturing, (3) earth slides and rock falls, (4) paleoliquefaction, (5) shaking structures (soft-sediment deformation), (6) turbidites, (7) tsunami deposits, (8) boulder trails, and (9) methane gas venting. The list includes subaerial and subaqueous features, because much of Sweden was submerged for a brief time after deglaciation. All features in the list are also formed by nontectonic processes, so it is important to employ formal hypothesis testing to distinguish tectonic and pseudotectonic features (geomorphic and stratigraphic). In the absence of rigorous tests, paleoseismic studies will be controversial and lead to unanswered questions (e.g., Godin *et al.*, 2002).

Primary, on-fault evidence: Fault scarps displacing glaciated bedrock surfaces have been described in Scandinavia (Olesen, 1988; Dehls *et al.*, 2000), Scotland (Fenton, 1992; Firth and Stewart, 2000) and Canada (Fenton, 1994). In Scandinavia, many of the "active faults" inferred by early studies in mountainous regions, were later determined to be sackungen (e.g., Blikra *et al.*, 2002). In addition to these bedrock faults, many exposures of till, drumlins, and eskers expose one or more faults, some of which can be demonstrated to be glaciotectonic structures resulting from glacier ice movement and melting (e.g., Aber and Ber, 2007). Fenton (1999) and Firth and Stewart (2000) propose criteria to distinguish postglacial tectonic, seismogenic fault scarps from other types of scarps (Table 2A.5). Using criteria such as those above, Stewart *et al.* (2001) concluded that some scarps in Scotland previously identified as postglacial fault scarps were not.

Secondary evidence: Soft-sediment deformation and liquefaction features are common in sediments deposited by glaciers, both modern and ancient. As mentioned previously, liquefaction is the most important type of paleoseismic evidence in many intraplate, SCR regions. Thus, when working in intraplate areas formerly glaciated, it is critical to distinguish glaciotectonic liquefaction from seismic shaking liquefaction. Aber and Ber (2007) describe this distinction from the glaciologist's viewpoint; so far, no paleoseismologist had tackled this question from the paleoseismology viewpoint. This remains a fruitful field for further research (see also the discussion of soft-sediment deformation in Chapter 7).

2A.5   Specialized Subfields of Paleoseismology

Paleoseismology is related to at least three subfields that have been defined previously in published papers, *archeoseismology* (Galadini *et al.*, 2006); *dendroseismology* (Jacoby, 1997); and *speleoseismology* (Becker *et al.*, 2006). These three subfields are defined by the use of a specific type of paleoseismic evidence, typically secondary evidence, that is used to identify and characterize a paleoearthquake. Each of these fields began primarily as an offshoot of nongeological disciplines, wherein practitioners of archeology, dendrochronology,

Table 2A.5: Criteria for distinguishing postglacial faults from other fault types

	Postglacial faults	Reverse tectonic faults (Tectonic, Seismic)	Glaciotectonic deformation (Nontectonic, Nonseismic)
Length	10 m to 100s km	>10 km	M to km (<3 km)
Continuity	Generally continuous	Continuous to discontinuous	Discontinuous
No. of scarps	Single	Single to multiple	Generally multiple
Sense/style	Predominantly reverse	All	Reverse and normal
Plan pattern	Linear, angular	Linear, arcuate	Irregular
Scarp height	mm to 10s of m	Up to km	Up to several m
Displacement history	Single event	Repeated events	Continuous deformation
Secondary deformation	Minor faulting	Faulting and folding	Faulting and folding
Relationship to ice cover	Within area of former ice cover	No relation	Margins of former ice cover
Timing	Postglacial	No constraint	Synglacial

From Fenton (1999).

and speleology (respectively) began to investigate anomalous relationships they observed in the field. Subsequently, the practitioners developed a suite of specialized techniques for data collection and interpretation that lay outside the normal scope of their own disciplines, as well as outside of geology or seismology. In the past decade archeoseismology in particular has evolved to the point where investigations are now so multidisciplinary and complex, that it may be considered as a companion field to paleoseismology rather than a subfield (Galadini *et al.*, 2006, p. 411). A detailed treatment of these three research fields is beyond the scope this book, but in the following brief sections we offer a summary of principles and key literature of the first two fields; speleoseismology is described in Chapter 8.

2A.5.1 Archeoseismology

The term archeoseismology refers to "investigations related to the seismic effects on ancient structures, uncovered by means of archeological excavations or pertaining to the monumental heritage" (Galadini *et al.*, 2006). According to Stiros and Jones (1996, p. 1), archeoseismology "focuses on individual seismic events occurring at precise moments over relatively recent time (the last few millennia), whose action affected precise locations—human constructions and their environment—which in turn can be studied in detail through the archaeological record." A lesser-used but related term is "seismic archeology," which according to Guidoboni (1996) is defined as "understanding the effects of seismic activity on historic buildings, ancient cities or archeologic sites by the use of archeological methods." Interestingly, the first North American paper to mention the term "archeoseismology" (Herrmann *et al.*, 1978) used it in the title but never defined it or mentioned it in the text, and was not an archeoseismic study at all by definition. The brief discussion below provides only the barest introduction to the subject; interested readers should consult Stiros and Jones (1996), McGuire *et al.* (2000), and especially Galadini *et al.* (2006) for details. New results will be arising from the 2008 to 2012 IGCP Project 567, "Earthquake Archaeology; Archaeoseismology along the Alpine–Himalayan seismic zone."

Archeoseismic damage falls into four general categories (Table 2A.6). First, ancient structures may be deformed and displaced by surface fault rupture, or affected by rapid geodetic changes. Second, buildings, walls and columns may collapse or topple from high ground accelerations during seismic ground shaking. Third, sites can be damaged by secondary coseismic geologic processes such as landsliding or tsunamis. Fourth, there may be evidence of human response to earthquake damage. Readers interested in case histories are referred to the large compilation in Stiros and Jones (1996), and a smaller but more current collection in the Special Issue of Journal of Seismology, Archeoseismology at the beginning of the twenty-first century (Galadini *et al.*, 2006). To date, few archeoseismic studies have been performed in the Western Hemisphere, but the effects of earthquakes on ancient cities in Central and South America is discussed briefly by Kovach (2004). Recently, Nur and Burgess (2008) published a book on archeoseismology written for nonscientists.

2A.5.1.1 Phenomena Related to Faulting and Ground Rupture

These phenomena are generally the least ambiguous phenomenon in archeoseismology, and include a Holocene fault displacing an archeologic structure, rather than a natural landform or deposit. Instances of displacement of archeologic sites by Holocene faulting have become increasingly recognized in the past few decades. Where the displacement is large (>1 m), the tectonic cause of the deformation is usually obvious, because the active fault trace (scarp) can be easily traced on either side of the deformed archeologic site. For example, Trifonov (1978) measured dextral offsets of the Middle Age Chungundor fortress (central Asia) of up to 2.5 m along the main Kopet-Dagh fault, and as much as 9 m offset on older irrigation ditches. Zhang *et al.* (1986) measured 3 m of normal oblique fault offset of the Great Wall of China, which occurred during the earthquake of 1739 AD. Korjenkov *et al.* (2006) describe how a blind thrust fold destroyed the Kamenka medieval fortress, Kyrgyzstan in the 12th century AD. In contrast, Koukouvelas *et al.* (2005) were performing a paleoseismic trenching study of the Helike fault scarp in Greece, when they unexpectedly encountered archeologic sites in their trenches; these remains predated the latest faulting, and provided additional age control on paleoearthquakes.

Where fault displacements are <ca. 2 m, the evidence becomes more ambiguous, for two reasons. First, the fault trace may be obscured (eroded, buried) beyond the archeologic site, so its projection into the site may go unrecognized. Second, the fault may not form a sharp, distinct surface rupture. Instead, the ground surface above the fault may deform in a plastic or distributed manner, creating similar broad deformation in the archeologic remains that might be mistaken for another type of damage. Such a case was described by Reches and Hoexter (1981) at the eighth century Hisham's Palace (Israel), where the rectangular floor plan was distorted by a broad (50 m wide) zone of sinistral horizontal shear into a rhomboid shape. Earlier studies had noticed the anomalous shape, but ascribed it to poor construction. Similarly, Marco *et al.* (1997) and Ellenblum *et al.* (1998) described an offset Crusader castle (ca. 1179 AD) in Israel. "The offset, fully expressed in the southern and northern defense walls, reaches 2.1 m in sinistral displacement with <5 cm of vertical slip.... Displacement is distributed over about a 10-m-wide zone, and the deformation is accommodated primarily by small offsets and rotations of the carved limestone blocks. All the displacements on the southern wall are purely horizontal (all the blocks retain their original level), and all the rotations are about vertical axes."

Table 2A.6: Classification of types of archeoseismological evidence

Origin of damage	Mechanics of phenomenon	Examples
Phenomena related to faulting and ground rupture	Faulting	Offset or deformation of artifact-bearing strata
		Offset of deformation of formerly continuous walls, irrigation channels, "roads," etc.
		Transported but unbroken relics adjacent to fault plane
	Fissuring	Sediment-filled fissures cutting artifact-bearing strata
		Dilated walls, irrigation channels, etc.
	Geodetic changes[a]	Destruction or abandonment due to rapid vertical tectonic movements, particularly at coastal sites (emergence or submergence[a])
Damage related to ground shaking	Indicators that are mainly diagnostic of earthquake shaking	Rotated or translated drums of dry masonry columns
		Some fractures initiated by axial loading of columns or column drums
		Some "conjugate" diagonal fractures in walls, especially where nucleated on openings
		Widespread and coeval destruction layers
	Less certain indicators of earthquake shaking	Toppled columns
		Collapsed, tilted, toppled walls, etc.
		Victims and/or artifacts buried beneath toppled walls, etc.
Secondary natural phenomena	Mainly diagnostic	Seismic tsunami damage at a coastal site
	Less certain indicators	Hot-springs travertines that have been deposited unconformably on relics
		Earthquake-related landslip phenomena disturbing site
		Aligned hot springs at a site
Secondary anthropogenic phenomena		Archeological evidence of antiseismic construction
		Inscriptions recording earthquake deaths or damages

Footnote "a" indicates additions to original published table.

In contrast, normal faulting tends to be recognizable even at the centimeter levels of displacement, due to its lack of plastic deformation, as described in a graben that disrupts the ancient city of Hierapolis (Turkey) (Hancock *et al.*, 2000). In general, it may be difficult to distinguish submeter displacements caused by surface faulting, from nontectonic causes such as local ground settlement, differential compaction, and local sliding (Karcz and Kafri, 1978).

Destruction/abandonment of archeologic sites from rapid coseismic vertical tectonic movements: Both Nikonov (1996) and Spondilis (1996) describe well preserved but submerged ruins that they infer resulted from coseismic subsidence. Spondilis (1996, p. 124) explains the applicable criteria as follows, "The submergence of the building remains at Methoni cannot be regarded as the result of eustatic change like the simple subsidence or erosion of the beach, because in that case everything should have vanished... For the same reason a tectonic subsidence of around one or two meters per millennium... which may be rapid in geological terms but is too slow for the archeological data, is hardly likely to have been the cause. The degree of preservation of these remains can only be explained by a very rapid vertical movement..." Similar reasoning was made by Stiros and Papageorgiou (2001), who concluded that the ancient harbor of Phalasarna (Crete) was uplifted about 6.5 m during the earthquake of AD 365. On the west coast of North America, abandonment of prehistoric coastal villages has also been ascribed subduction-zone paleoearthquakes (Hutchinson and McMillan, 1997).

2A.5.1.2 Damage Related to Ground Shaking

This type of archeoseismic evidence is inherently ambiguous, because there are many other causes of prehistoric destruction of structures beside earthquakes. For example, tilted and cracked walls can result from many possible causes, as pointed out by Karcz and Kafri (1978), so these are some of the most ambiguous evidence. A less ambiguous feature is laterally offset drums of columns (Figure 2A.53A), which indicate a horizontal force. Although offset drums are commonly interpreted as evidence of earthquake shaking, in rare instances the force has been proved to have been caused by horizontal impact forces from other parts of a collapsing building. Parallel fallen columns ("oriented destruction") are typically taken as proof of a paleoearthquake (Figure 2A.53B), but the absence of orientation does not disprove an earthquake origin for collapse, because building "failure under a seismic load is an extremely complicated process" (Stiros, 1996, p.141).

More confidence can be placed in interpreting multiple lines of archeoseismic evidence for a paleoearthquake. For example, the presence of many buried precious products and artifacts in collapse debris, yet no sign of fire (which would be expected from war and looting), plus the existence of quickly built buttresses and timber beams suggesting postearthquake repairs. Or, site wide, total destruction may occur for which earthquakes are the only reasonable explanation. In such destruction floors and roofs have collapsed, precious artifacts (potential loot) have been indiscriminately destroyed, and human bodies remain trapped in debris, in contradiction to local burial customs. Rapp (1982) describes this type of evidence at Troy.

Secondary natural phenomena: *Tsunamis*; At some coastal sites that display destruction layers, the pervasive ruins are overlain by "exotic" marine sand deposits interpreted as tsunami (?) deposits, in Greece (Dakoronia, 1996) and in Crete (Vallianou, 1996; Stiros and Papageorgiou, 2001).

Criteria for interpreting an earthquake origin for damage. Since the field of archeoseismology was established, there has been heated debate about the criteria for distinguishing coseismic from nonseismic damage in ancient structures. Karcz and Kafri (1978) adopted the conservative view that a paleoseismic origin can only be accepted if all other causes, such as poor construction, soil settlement, or human destruction, could be ruled out. They noted that "the critical examination of field evidence often cited in

Figure 2A.53: Typical archaeoseismic damage. (A) Laterally offset column drums at the Heraion Temple, Samos Island, Greece; (B) fallen columns at the Temple of Zeus, Olympia, Greece. Photographs courtesy of Stathis Stiros.

support of ancient seismicity has shown that the individual features are difficult to distinguish from the features of damage due to poor construction and adverse geotechnical effects." To assess the probability of nonseismic causes, Rapp (1986, p. 56) suggests the following parameters must be analyzed (1) the mechanical properties of the building materials; (2) the nature and quality of construction; (3) special characteristics of the regolith (overburden), including topography, earth and soil materials, and hydrology; (4) the regional earthquake regime; and (5) archeological evidence for destructive human forces. Similar methodologies for evaluating archeologic damage are given by Karcz and Kafri (1981), Nikonov (1988b), Stiros (1988a,b), Stiros (1996), and Galadini *et al.* (2006).

According to Galadini *et al.* (2006), a typical archeoseismic site investigation involves six general tasks of data collection from the site itself:

1. Reconstruction of the local archeological stratigraphy aimed at defining the correct position and chronology of a destruction layer, presumably related to an earthquake
2. Analysis of the deformations potentially due to seismic shaking or secondary earthquake effects, detectable on walls
3. Analysis of the depositional characteristics of the collapsed material
4. Investigations of the local geology and geomorphology to define possible natural cause(s) of the destruction
5. Investigations of the local factors affecting the ground motion amplifications
6. Estimation of the dynamic excitation, which affected the site under investigation

The six tasks above can lead to testable hypotheses on the possible origin of a particular deformation feature. In most published studies to date, the formulation and testing of hypotheses was approached in an ad hoc manner. In contrast, Sintubin and Stewart (2008) propose a more formal approach to rank the geologic setting and archeoseismic damage as to its likelihood of being seismic. They refer to this as a "logic tree" approach.

Archeoseismology, as broadly defined, also includes using the presence of artifacts in faulted/unfaulted sediments as a correlation tool or as a Quaternary dating method (cf., Table 1.2). For example, loose archeologic artifacts in unconsolidated deposits can be used as stratigraphic markers to correlate strata across faults, and thus to measure and date displacement simultaneously. This application of archeoseismology is more common in North America, where there are few prehistoric stone buildings. An example is Noller *et al.* (1994), who measured horizontal displacements across the San Andreas fault, California, based on correlation of artifact-bearing colluviums on opposite sides of the fault. Lafferty (1996) described a similar use of artifact-bearing strata in dating paleoearthquakes in the New Madrid fault zone, central USA.

2A.5.2 *Dendroseismology*

Dendroseismology is the study and dating of prehistoric earthquakes based on their effects on trees (Jacoby, 1997). This subfield may involve merely using dendrochronology as a method to date geomorphic features or deposits deformed by paleoearthquakes. However, the trees themselves may record growth disturbances due to paleoearthquakes, in which case tree rings contain both the evidence for the event and its age.

Tree rings can record evidence of seismic events in three ways (Jacoby, 1997). *Primary dendroseismologic evidence* is created by surface rupture, such as tree roots sheared off by surface faulting, or trees tilted by differential shearing. Trees can also be damaged by coseismic subsidence and drowning, as occurs in subduction-zone megathrust earthquakes (Atwater and Yamaguchi, 1991). *Secondary evidence* is caused by seismic shaking or geomorphic processes induced by the earthquake. Shaking can cause the tree crown or large branches to snap off during high ground accelerations, which decreases the photosynthetic surface and leads to tree death or slower growth. For example, Jacoby *et al.* (1988) report that numerous traumatized trees, some of which had lost their crowns, occur within 10 m of the active (1857) trace of the San Andreas fault; similar trees are not found away from the fault trace. Very narrow or missing growth rings occur beginning in 1813 and 1857, which they interpret to date earthquakes. Trees may also be tilted from coseismic liquefaction or landsliding. Delayed geomorphic responses to earthquakes, such as gully creation, can undermine trees and lead to tilting.

Despite the promise of dendroseismology for very precise dating of paleoearthquakes, the method is limited by the distribution and age of trees, as well as by other factors. Trees can suffer trauma from nonseismic causes such as wind storms, lightning, fire, and being impacted by other falling trees. Less drastic changes in growth rate can be caused by many environmental factors such as moisture availability, disease, and nonseismic geomorphic disturbances such as floods and landslides. To eliminate these nonseismic features as causes of growth changes, dendrochronologists study many undisturbed trees near fault zones. The growth rings of these trees constitute a *control set* that reflects the growth patterns of undisturbed trees of the same species in a similar growth environment. G. C. Jacoby (personal communication) suggests three requirements for a successful application of dendroseismology: (1) The tree damage must be tied to a geologic rupture feature, (2) multiple lines of evidence must exist for the paleoearthquake, and (3) a *master chronology* must be assembled from trees unaffected by the paleoearthquake. For details on methodology, readers are referred to general treatments (Jacoby, 1997) and the following case histories, arranged by tectonic environment: (1) strike-slip faults, where damage is caused by fault rupture and shaking (Page, 1970; LaMarche and Wallace, 1972; Wallace and LaMarche, 1979; Meisling and Sieh, 1980; Jacoby *et al.*, 1988; Yadav and Kulieshius, 1992); (2) normal faults and rift zones (Ruzhich *et al.*, 1982; Stahle *et al.*, 1992; Sheppard and White, 1995); and (3) subduction zones, where damage is caused by subsidence and drowning (Jacoby and Ulan, 1983; Sheppard and Jacoby, 1989; Atwater and Yamaguchi, 1991; Jacoby *et al.*, 1995, 1997).

Sub-Aqueous Paleoseismology

Chris Goldfinger

College of Oceanic and Atmospheric Sciences, Oregon State University, Corvallis, Oregon 97331 5503, USA

2B.1   Introduction

2B.1.1   Scope of the Chapter

Many of the largest earthquakes are fundamentally marine events, generated by submarine subduction zone or other plate boundary earthquakes, as well as volcano-tectonic explosions. A large proportion of the world's population lives near coastlines, thus a high proportion of hazard from active tectonics comes from submarine fault systems and volcanic and landslide generators of tsunami. During and shortly after large earthquakes, in the coastal and marine environment, a spectrum of evidence is left behind. Onshore, land levels change with elastic unflexing of the formerly coupled plates, resulting in coastal subsidence, uplift or lateral shift, and the generation of familiar onshore paleoseismic evidence such as fault scarps, colluvial wedges, damaged trees, landslides, and offset features. If the seafloor is shaken or displaced, another suite of events may result in further geologic and geodetic evidence of the event, including turbidity currents, submarine landslides, tsunami (which may be recorded both onshore and offshore), soft-sediment deformation, as well as virtually all of the evidence normally associated with onshore faults, including coseismic and post-seismic displacement.

Offshore and lacustrine records offer the potential of good preservation, good spatial coverage, and long temporal span. Marine deposits also offer opportunities for stratigraphic correlation along the source zone, something typically difficult with land paleoseismology. Stratigraphic correlation methods have potential to address source zone spatial extent, segmentation, and because of the longer time intervals available, can be used to examine recurrence models, fault interactions, clustering and other phenomenon commonly limited by short temporal records. Offshore deposits can be investigated geologically and geophysically to define their extent, stratigraphic relationships, and timing. Detailed investigations of marine deposits at the millimeter scale is now routine, and high-resolution geophysical techniques allow subsurface mapping and correlation with core samples to delineate mass transport deposits and turbidites. In some cases, direct evidence of earthquake slip is available and can be imaged using geophysical techniques. Many deposits, however, do not have a direct physical link to their causative sources and must be distinguished from other deposits through either regional correlation, dating, or sedimentological character. Submarine deposits may include a wide range of features and structures which overlap with those of onshore deposits. This chapter discusses mostly submarine deposits of transported nature and direct fault observations. Sub-aqueous soft sediment deformation is included in Chapter 7.

International Geophysics, Volume 95

ISSN 0074-6142, DOI: 10.1016/S0074-6142(09)95050-1

2B.2   Mapping and Dating Paleoseismic Landforms Offshore

2B.2.1   Submarine Mapping and Imaging Methods

Due to the inherent difficulties of working in the submarine environment, only a few examples exist of successful determinations of slip rates and paleoseismic histories through direct observation and imaging of active faults. Much more common are rough estimates of recent activity based on morphological evidence of scarp freshness, presence of associated landslides, mud volcanoes, fluid venting, tsunami deposits, and other secondary evidence. Methods of investigation in the submarine environment include some commonality with paleoseismic and tectonic investigations onshore, particularly topographic analysis and seismic reflection profiling, while other methods are relatively unique to the marine environment such as sidescan sonar imaging and acoustic geodetics.

2B.2.1.1   Seafloor Mapping Techniques

The earliest seabed mapping tool was a lead line, a simple lead weight on the end of a line, heaved by a sailor to manually sound the depth, and also collect a bottom sample with the lead which was "armed" with tallow or beeswax to bring back some of the bottom sediment. Modern mapping technologies now mostly use acoustic techniques. Because light and electromagnetic transmission in water is generally poor, modern photographic, radar, laser, and other such techniques common to terrestrial mapping are largely ineffective in the marine environment except for very short range applications. The exceptions include laser line-scan imaging, which can be done from surface vessels or towed vehicles at a low altitude above the seabed in clear conditions; towed, lowered, or vehicle mounted cameras; and bathymetric LiDAR, which can map the seabed from aircraft in very clear and shallow water conditions (Figure 2B.1). Most seafloor mapping is performed with acoustic technologies using sound waves reflected from the seafloor to establish both elevations (bathymetry) and backscatter strength of seafloor materials, which is a function of hardness, grain size, and material properties. Early mapping began with single-beam echosounders and recording fathometers, which emit an acoustic "ping" toward the seafloor that is then reflected back to the surface. The travel time of the ping is measured and converted to water depth using the sound velocity of water ($\sim$1500 m/s). The transducer typically emits a cone-shaped beam pattern spanning 5–40° of arc, thus the beam pattern on the seafloor becomes larger with water depth, a limiting resolution factor in most sonar mapping tools, a problem that can be overcome through mounting mapping instruments on deep-towed vehicles and other submersibles.

Multibeam Bathymetric Sonars   Modern seafloor mapping is performed with two primary tools: multibeam bathymetric sonars and sidescan sonars. Multibeam sonars operate on the same principle as the earlier single-beam sonars, but with multiple elements arranged in an array to establish a fan-shaped pattern of multiple beams that maximize across-track width (Figure 2B.2). In this way, a broad swath of seafloor can be mapped in a single pass. Typical swath widths are three to seven times the water depth, with individual beam widths of 1–2°. Ship motion (roll, pitch, yaw and heave, the vertical motion from waves) is recorded during data acquisition and removed during processing. More sophisticated systems employ "beam steering" to "form" the individual beams electronically in real time as the ship moves, keeping the fan of beams pointed downward as the ship rolls. The data are also corrected for tides and water velocity (further processing details are discussed in Blondel and Murton, 1997). The result is a bathymetric surface along the ship track, and a large area is surveyed by arranging the ships course in a series of linear tracks arranged to best image the seafloor in the prevailing weather conditions at the time

LiDAR bathymetry

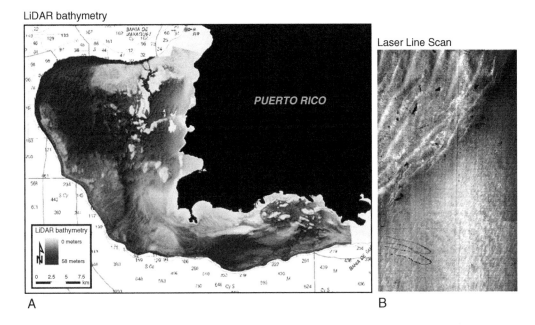

A   B

Figure 2B.1: (A) Example of LiDAR bathymetry. Images of the southwest coast of Puerto Rico were taken using a LiDAR ADS Mk II Airborne System. The 900 Hertz (1065 nm) A Nd: Yag laser acquired 4 × 4 m spot spacing and 200% seabed coverage. Image source: NOAA biogeography products: http://ccma.nos.noaa.gov/images/biogeo/LiDAR_pr.jpg. (B) Laser Line-Scan Survey (LLS) image showing sharp boundary between sand waves (top left corner) and rippled seafloor. Dark objects in the area of sand waves are pieces of drift kelp and a Salp chain shadow is seen at bottom. Altitude of LLS system is approximately 6 m off seafloor, and swath width is approximately 8 m. Image in "B" from Yoklavich *et al.* (2003), reprinted by permission of the Marine Technology Society.

of survey. In addition to bathymetric data, the strength of the returning beam can be quantified, providing a map of reflection intensity from the seafloor. When corrected for the effect of grazing angle at the seafloor, these data provide a map of relative reflectivity, which is commonly interpreted as patterns of outcropping rock, sand and mud a the seafloor. Multibeam sonars typically use 16–200 beams arrayed as a fan with a span of ~90°–150°. Frequency of these systems ranges from 12 to 400 kHz, with lower frequency systems used for deeper water applications to overcome the attenuation of high-frequency signals in the water column.

Sidescan Sonars Sidescan sonar similarly maps the seafloor in a wide swath, but instead of forming discrete beams, the entire returning signal is digitized, yielding a higher across-track resolution of returning signal strength. Because there is no angular control of the signal, these data are processed as a signal strength image, rather than bathymetry. Sidescan imagery typically consists of 2048 across-track samples per ping, yielding much higher spatial resolution (Figure 2B.3). By using two frequencies, interfering beam patterns can be processed to extract bathymetric data as well; although typically these data are not as high in quality as multibeam bathymetric sonars (Johnson and Helferty, 1990; Blondel and Murton, 1997). Sidescan sonar offers detailed imagery that is generally superior for mapping of tectonic features, and when combined with multibeam bathymetry, yields a 3D surface map that images both

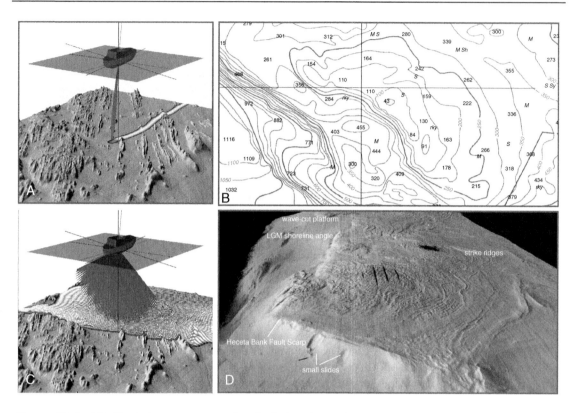

Figure 2B.2: (A) Single-beam sonar and typical ping foot print. (B) Contour plot from single-beam sounding data. (C) Typical multibeam sonar and swath footprint. Swath width varies by system, but ranges between two and seven times water depth. (D) Example of swath bathymetric survey of Heceta Bank, Oregon. Pixel size in this image is 10 m. Fault scarps, strike ridges, landslide scars, and a submerged last glacial maximum (LGM) submerged shoreline angle, and associated wave-cut platform are clearly visible. (A) and (C) courtesy John Hughes Clarke, University of New Brunswick. Image by the author, (D) Data courtesy R. W. Embley, NOAA Pacific Marine Environmental Laboratory, Newport Oregon. (See Color Insert.)

morphology and material properties in the area of interest. Such information is difficult to duplicate on land except in desert environments. Sidescan sonars generally operate in the frequency range ~6 to 400 kHz, with the lower frequencies used for deep water applications, and the higher frequencies used for shallow water or deep towed applications due to the attenuation of higher frequencies in the water column. Both types of systems may be hull-mounted, mounted on retractable or temporary overside poles, towed vehicles, or on autonomous underwater vehicles (AUVs), remotely operated vehicles (ROVs) and manned submersibles.

Sidescan imagery and multibeam bathymetry can be combined to produce a 3D image showing topographic details, as well as and backscatter strength, analogous to radar imagery draped over subaerial topography.

Seismic Reflection Profiling   Subsurface structures are imaged in the marine environment much as they are on land, using primarily seismic reflection profiling, with ground truth from drill holes and core

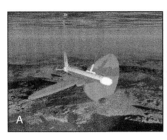
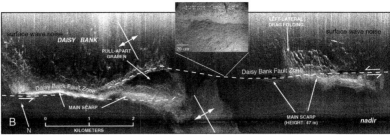

Figure 2B.3: (A) Sidescan towfish, typical of side scan sonar systems. Two transducer arrays transmit beams to port and starboard. This figure illustrates the port beam only. The beam is narrow in the horizontal plane and broad in the vertical plane, shown to the right in the figure. In a typical single-beam 100 kHz system this would be ~1° in the horizontal and 40° in the vertical and for a 500 kHz system, ~0.2° in the horizontal and 40° in the vertical. Radiation occurs also out the rear of the transducer, shown here as the pattern extending horizontally to the left of the figure (starboard for the towfish). Image courtesy L-3 Klein Associates, reprinted by permission. (B) SeaMARC 1A starboard swath image of the Daisy Bank Fault Zone, Cascadia margin. A left stepover in the left-lateral fault, associated pull-apart basin and anticline offset are shown. Offset anticlinal axis is highly reflective rubble. Drag folding and other fault details are shown. Inset shows active Holocene scarp from DELTA submersible photograph. Modified after Goldfinger *et al.* (1996). (See Color Insert.)

samples when possible. Seismic profiling again uses sound waves, but at a much lower frequencies to penetrate the seafloor substrate. Marine seismic profiling is a sophisticated process developed largely for the petrochemical industry which requires both deep penetration in deep water, and high resolution. Spatial resolution is improved through the use of large arrays of tuned sound sources, and long arrays of towed receivers which receive each sound pulse at many different receive points, allowing "stacking" of weak signals, as well as calculation of velocities of subsurface units. Paleoseismology generally makes use of simpler shallower penetration systems that focus on the upper subsurface section using higher frequencies to identify disruption of young strata, fault details, fault terminations, colluvial wedges and in general the same types of features observed in terrestrial trenches, though not quite at the same resolution (e.g., Seitz and Kent, 2005; Ridente *et al.*, 2008; Brothers *et al.*, 2009). Seismic profiling can also image the spatial continuity of turbidites and subsurface mass wasting deposits, as well as key reflectors that can be used to establish temporal control of structural movement (Figure 2B.3). Such systems tend to use operating frequencies from 3 to 300 kHz, commonly from 3.5 to 20 kHz, and may sweep through a range of frequencies in their outgoing pulses known as "CHIRP" (Compressed High Intensity Radar Pulse, Figure 2B.4). This has the effect of increasing resolution by including high frequencies, but still maintaining enough energy in a longer outgoing pulse to achieve penetration of the water column and subsurface (Verbeeka and McGee, 1995). CHIRP sonars may also be used to invert the reflection amplitude and phase data to obtain material properties such as density, porosity, and sound speed (Turgut *et al.*, 2002; Schock, 2004).

2B.2.1.2 Sampling Methods

Coring Tools Among the most common sampling tools in the marine environment are coring tools, shown in Figure 2B.5. Cores are commonly taken with a device lowered from a vessel on a wire or synthetic rope to the seabed. Common tools are the gravity corer, piston corer, box corer, Kasten corer,

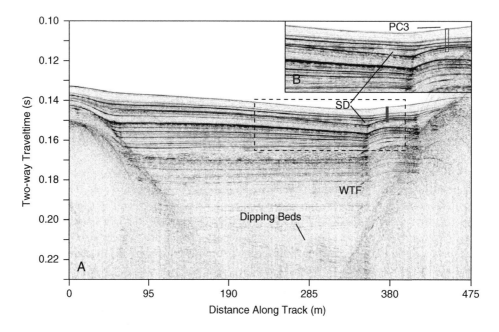

Figure 2B.4: (A) Northeast oriented, fault-perpendicular CHIRP seismic profile across the Fallen Leaf Segment of the West Tahoe Fault, Lake Tahoe California. Slip during the most recent event (MRE) was up to ~3 m in the upper 30 m of sediments. The increase in offset with depth is likely due to changes in sediment compaction rather than multiple events. A slide deposit (SD) mantles the event horizon and infills the accommodation created during the MRE. Piston core 3 sampled a sandy turbidite layer that is interpreted to be the distal reaches of the same slide sampled in piston core 1. Radiocarbon dating and the presence of the 7600–8000-year-old Tsoyowata Ash place the MRE at ~4000–5000 years BP and constrain the sedimentation rate at ~1 mm/yr over the last ~8 ka. Down-section, faintly imaged strata along the hanging wall abruptly change from horizontal to a ~4° dip toward the fault. The change in dip is interpreted to have formed during older events along the WTF. (B) Enlarged section showing deformation associated with the MRE and the projected location of PC3 onto the profile (the two are offset by <30 m). Figure and interpretation from Brothers *et al.* (2009), used with permission from the seismological society of America.

and also include the vibracorer and freeze corer. Gravity corers are the simplest of devices, consisting of a barrel, a cutter at the nose, and a weight at the top end with a valve to allow expulsion of water. A simple one way "catcher" at the nose cone prevents the sample from sliding back out the core tube. Small push corers are routinely deployed from submersibles and ROVs to collect short soft sediment samples.

The Kasten corer is a large diameter, usually square, variant of the gravity corer used when large sample volume is desired. An improvement to this scheme is the piston corer, which uses a piston inside the core liner that is attached directly to the lowering wire, and a trigger corer that touches down first, triggering a freefall of the main corer. As the piston corer penetrates, the piston remains near the level of the seafloor, and generates a vacuum that helps pull the sample into the tube. This method helps overcome the friction of penetration and allows much longer samples that also have reduced internal compaction of the sample. Very long piston coring systems are in use that are capable of collecting 50–70 m core samples from very

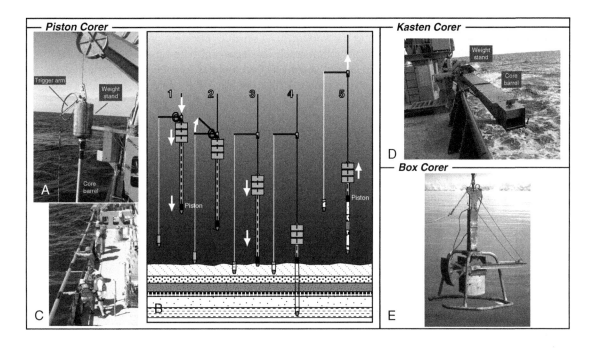

Figure 2B.5: Commonly used coring devices include the piston corer, box corer and Kasten corer. A piston corer is comprised of a piston positioned at the cutting head of a weighted, round core barrel of variable length (A). The piston corer is released to free-fall into the sediment when the trigger weight (or trigger core) hits the seafloor. The piston stays at the sediment surface as the core barrel punctures the sediment. The piston prevents the loss of sediment as the core is recovered (see sequence in (B)). A piston core ready for deployment is shown in (C). A Kasten corer is comprised of a weight stand on a rectangular core barrel (D), and does not require a trigger weight. Kasten cores are limited in length and are used where more volume of sediment is required. A box corer (E) is a weighted box with an open bottom, with a spring-loaded shovel that close over the bottom to prevent loss during recovery. Box cores penetrate a short distance into the seafloor and are typically used to capture a large amount of surface sediment. Photos (A), (C), and (D), OSU. Panel (B) by J. Patton and A. Morey. E. Courtesy Hannes Grobe. (See Color Insert.)

soft materials without requiring a drilling vessel. These systems use lightweight synthetic fiber ropes to overcome the limitations of heavy wire in deep water applications.

The box corer is a broad box device that captures the uppermost ∼0.5–1 m of sediment, and the multicorer collects and undisturbed set of six to eight samples of the seafloor surface, including the overlying water. Two important variants include vibracoring and the freeze core. A vibracorer consists of a core barrel with a motor mounted at the head to generate vibrations in the core barrel. The vibration causes sediments in contact with the outer barrel to liquefy, making possible the penetration of sandy lithologies that would be problematic for other techniques. The freeze corer consists of a simple tube and weight, or multiple tubes filled with dry ice, liquid nitrogen, or other cold source. The tube is lowered into the sediment, which freezes to the outside of the metal tube (Hill, 1999; Kondolf and Piégay, 2003). This device can collect delicate samples from very soft sediments difficult to recover with other methods.

Drilling   Drilling from platform rigs and surface ships has for decades provided age control as well as lithologic and geophysical ground truth as a compliment to seismic reflection profiling. Commercial and academic drilling vessels can extend the sampling reach far deeper into the substrate than passive sampling techniques and can employ downhole geophysical techniques to collect geophysical and physical property data from the walls of the boreholes. Both of the current Integrated Ocean Drilling Program (IODP) platforms, the JOIDES Resolution and the Chikyu have investigations of earthquakes (and thus paleoseismology) as a priority mission. The JOIDES Resolution is a conventional nonriser drilling vessel, meaning that it cannot compensate for formations under pressure at depth. The Chikyu, the first academic riser vessel, uses heavy drilling "mud" weight to balance formation pressure, allowing deeper drilling into less stable formations. Maximum drilling depth depends on the water depth, but can reach up to 8 km, though 1–3 km is more typical for the D/V Resolution. The D/V Chikyu was designed in part to reach the seismogenic plate boundary in subduction zones. These drilling vessels typically recover core, which is logged, analyzed and sampled much as other types of cores. In addition, they have the capability to install instruments in the boreholes and to plug the borehole with a pressure instrument to record time series of formation pressure to monitor and thus reveal patterns of stress change related to earthquakes and other phenomenon (Figure 2B.6).

Other drilling devices such as the portable remotely operated drill (PROD and it's derivatives; Kelleher and Randolph, 2005; Freudenthal and Wefer, 2007) and the benthic marine sampler (BM; Petters and Asakawa, 1997) are devices that fill the gap between coring and ship-based drilling, and are lowered to the seafloor on cables, from where they can drill and core up to ∼100 m depth from ships of opportunity.

2B.2.2 Dating Submarine Structures, Landforms, and Deposits Using Paleoseismic Stratigraphy

2B.2.2.1 Radiocarbon Dating

To date submarine events, the most common technique is to date calcareous microfossils, commonly planktonic foraminifera using ^{14}C. This technique can be used to date turbidites, submarine landslides, and other marine disturbance events by sampling the youngest material below, or the oldest material above the event. In dating turbidites, samples are commonly taken below each turbidite because the boundary between the top of the turbidite tail and the hemipelagic sediment is difficult to identify reliably and bioturbation is concentrated at this boundary (Goldfinger *et al.*, 2008, 2009, Figure 2B.7). Sediment samples are taken to avoid visible or undetected deformation and friction drag along the core walls. Further processing details are given in Goldfinger *et al.* (2009).

Foraminiferal samples are dated using accelerator mass spectrometry (AMS) methods which can make use of as little as a ∼1 mg carbon sample. Sensitivity tests for species-specific biases and other techniques are presented in Goldfinger *et al.* (2007a).

All radiocarbon ages must be calibrated to account for variability of carbon isotopes in the atmosphere. This variability has been captured as discussed in Chapter 2; however, marine ages have several additional complications, most importantly the reservoir correction. This value, representing the age of the seawater populated by microfossils used to date marine events, is a published spatially varying value specific to the locality of interest (e.g., Reimer *et al.*, 2004). The published value is commonly derived from paired shell/wood dates that establish the age of the water in which the marine animal lived relative to stratigraphically correlated terrestrial material. The published values are almost exclusively from the twentieth century; although it is known that these values change through time (i.e., Kovanen and Esterbrook, 2002). Time

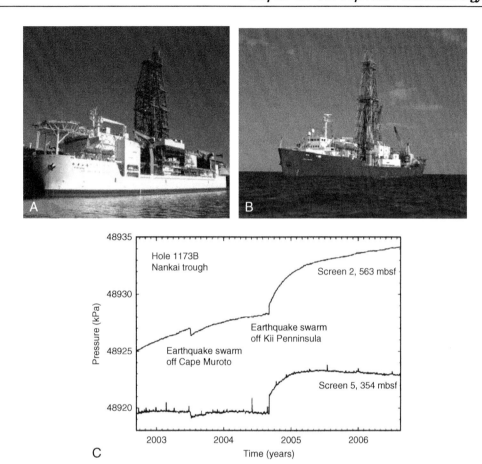

Figure 2B.6: (A) Conventional IODP drilling vessel Joides Resolution. (B) Riser capable IODP drilling vessel Chikyu. (C) Example of *in situ* pressure data applied to earthquake investigations. Pressures recorded at two levels in Hole 1173 drilled into the sediments of the subducting Philippine Sea plate, just seaward of the Nankai Trough. Screen 5 is situated only about 10 m below the top of the fine-grained lower Shikoku Basin lithologic unit and appears not to capture the secular signal seen at deeper screens (e.g., screen 2 shown). Shorter-term transients associated with elastic and viscoelastic strain are seen at all levels. Panel (C) Courtesy Earl Davis.

variation of the reservoir age is usually ignored because little data on the time history are available (Stuvier *et al.*, 1998). Development of time and space variant reservoir models is underway, and will help refine marine radiocarbon dating (see Goldfinger *et al.*, 2009).

Because the sedimentations rates in the deep sea are relatively stable over periods of interest to paleoseismology, radiocarbon ages can be corrected for such factors as basal erosion and sample thickness. To correct ages for the thickness of the radiocarbon sample, it is necessary to subtract the time representing half the sample thickness from the ^{14}C age. This correction attempts to bring the age as close as possible to the age of the deposition of the turbidite (barring basal erosion). Sedimentation rate curves can be constructed for each core using the pelagic or hemipelagic interval thicknesses and radiocarbon data, and this simple correction calculated from the curves.

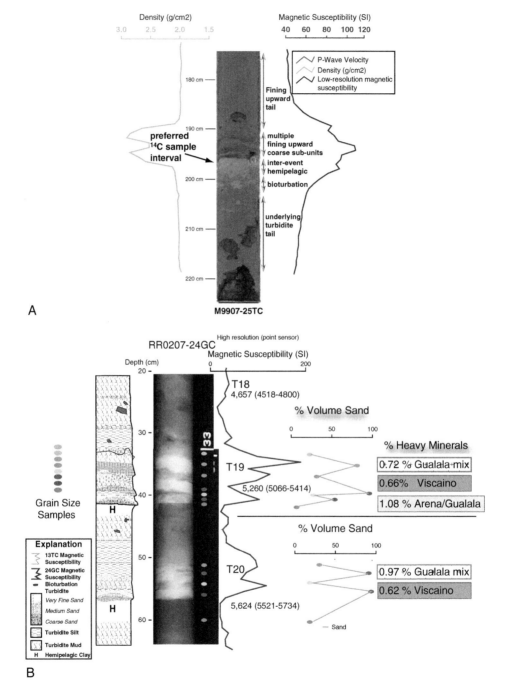

Figure 2B.7: (A) Detailed image of a Cascadia paleoseismic turbidite, its subunits and preferred ^{14}C sampling site. In this example, the turbidite tail/hemipelagic boundary is distinct visually, and variably disturbed by bioturbation. While turbidite bases can be erosive, dating is commonly done

Basal erosion is a primary concern when dating landslides or turbidites below their bases. In the case of landslide deposits, it may be advantageous to date material directly above the deposit as it is both easier to sample and lacks the erosion concern. One can mitigate the basal erosion problem to some extent by estimating the degree of basal erosion. Examination based on the morphology of the turbidite bases for the degree of roughness and obvious truncations is of value, though such analyses are imperfect. Another technique is to examine the underlying hemipelagic sediment for thickness variations among a local group of cores with the assumption that erosion is likely the primary cause of such variability. Missing section can be estimated from the difference between the thickest and the other intervals (Goldfinger *et al.*, 2008, 2009; Gutierrez-Pastor, in press). This method obviously underestimates erosion in the case where all samples are eroded.

Radiocarbon ages can be reported in a variety of ways, and these vary considerably. These are sometimes reported as simply a 1σ or 2σ range, a conservative approach. However, the probability distribution generated during the calibration process contains more information, including a probability peak that may also be significant. The probability peak can be considered the most likely age of the event of interest (Blaauw *et al.*, 2005); though of course many other external factors are involved in the correspondence between the ^{14}C age and the event age. Goldfinger *et al.* (2007) show the strong tendency for the probability peaks to lie along the age model sedimentation rate, despite smoothing of the rate curve. Goldfinger *et al.* (2009) suggest the prudent use of probability peaks and ranges as shown by Goslar and others (2005).

To develop an age model for sediment cores, it is necessary to determine hemipelagic thickness between events in the case of multiple turbidites or landslides. To establish this thickness, the boundary between the gradational turbidite tail and the overlying hemipelagic sediment must be determined as precisely as possible, though it is uniquely difficult. The reason is that the differences between the very fine-grained turbidite tail and the overlying hemipelagic may be nearly nonexistent. In the case of Cascadia cores, many tend to have obvious boundaries that are clearly visible to the eye (Goldfinger *et al.*, 2008, 2009). For many other regions, the problem is more difficult. Many attempts have been made to find universal

from planktonic foraminifera in the upper part of the underlying hemipelagic interval as the least problematic option. Typical sample location shown, with small "gap" above the sample. (B) Detail from Core RR0507-25TC event T4 along the Northern San Andreas margin. Example grain size analysis, magnetic susceptibility/density signatures and X-radiography in turbidites T19 and T20 in core 24GC below the Gualala–Noyo–Viscaino channel confluence. Light tones in the X-radiograph represent dense sand/silt intervals; darker gray tones represent clay/mud. Oval dots are grain size samples. Heavy trace is the magnetic susceptibility signature. Right plot is percent sand (obtained with Coulter laser counter method). The good correspondence between grain size, density, and magnetic susceptibility for the lithologies in both Cascadia and NSAF cores is established with selected analyses and permits the use of density and magnetics as mass/grain size proxies that show much greater resolution than possible with grain size analysis. These typical turbidites are composed of 1–3 fining upward sequences, each truncated by the overlying pulse. No hemipelagic clay exists between pulses, indicating the three pulses were deposited in a short time interval. Only the last pulse has a fine tail, indicating final waning of the turbidity current. We interpret these signatures as resulting from a single multipulse turbidity current. Number of coarse pulses commonly remains constant in multiple channel systems for a given event. Source provenance affinity for each sand pulse is shown to the right. Mineralogically distinct sandy units stacked vertically in order of arrival at a confluence near the core site. See Goldfinger *et al.* (2007) for further details and core locations. Modified after Goldfinger *et al.* (2007, 2008). (See Color Insert.)

methods for defining this boundary including clay fabric orientation (O'Brien *et al.*, 1980; Azmon, 1981), color (Rogerson *et al.*, 2006), hydraulic sorting of microfossils (e.g., Brunner and Ledbetter, 1987), XRF and XRD (e.g., Bernd *et al.*, 2002), and grain size (the most common method, i.e., Brunner and Ledbetter, 1987; Joseph *et al.*, 1998; St.-Onge *et al.*, 2004), resistivity, and other methods. Once an acceptable criterion for the boundary is determined, sedimentation rates and an age model can be constructed.

2B.2.2.2   OxCal Analysis

OxCal is radiocarbon calibration software that also includes multiple methods to allow the use of external age constraints, multiple ^{14}C ages and geological constraints such as sedimentation rates to constrain radiocarbon ages. The technique uses Bayesian statistics to combine multiple probability distributions and trim probability density functions (PDFs) (Ramsey, 1995, 2001). The external constraints may include (1) the time represented by sediment deposited between events, (2) historical information, (3) stratigraphic ordering, and (4) other external stratigraphic constraints such as dated ashes, pollen, or other biostratigraphic markers (Biasi *et al.*, 2002; Goldfinger *et al.*, 2007). Where age data are missing, sedimentation rates alone can be used to model event ages. Since calculated sedimentation rates are also dependent on the radiocarbon ages, and on basal erosion, there is some unavoidable circularity in this process unless varves or other independent rates are available (e.g., Kelsey *et al.*, 2005). Figure 2B.8 shows an example of this using hemipelagic sedimentation and historical constraints for the AD 1700 Cascadia earthquake and the NSAF 1906 and penultimate NSAF earthquakes. Further details of constrained age models are give in Goldfinger *et al.* (2007, 2008).

2B.2.2.3   Sedimentation Rate Ages

The age model of a marine core with good age control is a powerful tool. Using the sedimentation rates and hemipelagic thicknesses, one can calculate the age of an undated event based on a dated turbidite below or above (or both when possible) using sedimentation rates alone. Goldfinger *et al.* (2009) present calculations for undated Cascadia events, while, Gutierrez-Pastor *et al.* (in press) present additional analysis of hemipelagic age calculations from the Cascadia margin with hemipelagic intervals treated as a semi-independent time line. A similar analysis is presented in Kelsey *et al.* (2005) for a tsunami record in Bradley Lake, Cascadia margin.

2B.2.2.4   Event Ages and Potential Biases

The question of how well the radiocarbon ages from marine deposits represent earthquake ages is complex. In land paleoseismology, ages commonly represent maximum or minimum ages when dated using sample material below or above the event, respectively. Typically the best one can do is to collect material from as close below and as close above an event, and refer to these ages as "close maximum" and "close minimum" ages, respectively (i.e., Nelson *et al.*, 2006, 2008). The sample materials are commonly detrital, and thus certain to be of different age than the earthquake (see Chapter 2). These are commonly reported in the literature and usually indicated on space–time diagrams with arrows pointing upward or downward for maximum and minimum ages, respectively. Marine ages may, however, include reasonable attempts to correct known biases based on continuous marine sedimentation. The sedimentation rate corrections, erosion analyses, and OxCal analyses using hemipelagic intervals discussed previously are designed to approach event ages by attempting to remove these biases. These tools are unavailable in most land settings due to the absence of continuous sedimentation, though are used when other constraints are available (see Kelsey *et al.*, 2005). Goldfinger *et al.* (2009) report several examples of testing these methods against events of known age, with good results (see also Figure 2B.8).

Sequence NSAF 1906 test

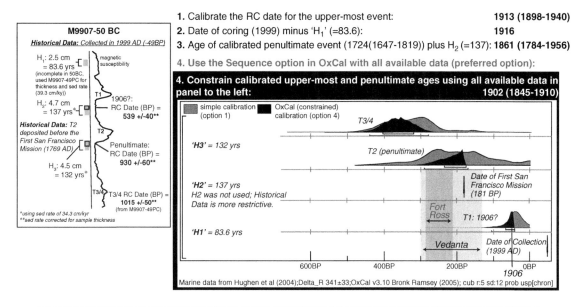

M9907-50 BC

Historical Data: Collected in 1999 AD (-49BP)

H₁: 2.5 cm = 83.6 yrs
(incomplete in 50BC, used M9907-49PC for thickness and sed rate (39.3 cm/ky))

H₂: 4.7 cm = 137 yrs*

Historical Data: T2 deposited before the First San Francisco Mission (1769 AD)

H₃: 4.5 cm = 132 yrs*

*using sed rate of 34.3 cm/kyr
**sed rate corrected for sample thickness

magnetic susceptibility

1906?:
RC Date (BP) = **539 +/-40****

Penultimate:
RC Date (BP) = **930 +/-60****

T3/4 RC Date (BP) = **1015 +/-50****
(from M9907-49PC)

1. Calibrate the RC date for the upper-most event: **1913 (1898-1940)**
2. Date of coring (1999) minus 'H₁' (=83.6): **1916**
3. Age of calibrated penultimate event (1724(1647-1819)) plus H₂ (=137): **1861 (1784-1956)**
4. Use the Sequence option in OxCal with all available data (preferred option):

4. Constrain calibrated upper-most and penultimate ages using all available data in panel to the left: **1902 (1845-1910)**

simple calibration (option 1)   OxCal (constrained) calibration (option 4)

T3/4

'H3' = 132 yrs

T2 (penultimate)

'H2' = 137 yrs
H2 was not used; Historical Data is more restrictive.

Date of First San Francisco Mission (181 BP)

Fort Ross   T1: 1906?

'H1' = 83.6 yrs

Vedanta   Date of Collection (1999 AD)

600BP   400BP   200BP   0BP
1906

Marine data from Hughen et al (2004);Delta_R 341±33;OxCal v3.10 Bronk Ramsey (2005); cub r:5 sd:12 prob usp[chron]

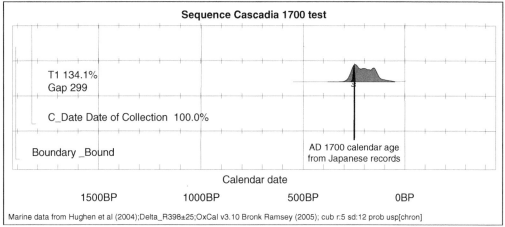

Sequence Cascadia 1700 test

T1 134.1%
Gap 299

C_Date Date of Collection 100.0%

Boundary _Bound

AD 1700 calendar age from Japanese records

Calendar date

1500BP   1000BP   500BP   0BP

Marine data from Hughen et al (2004);Delta_R398±25;OxCal v3.10 Bronk Ramsey (2005); cub r:5 sd:12 prob usp[chron]

Figure 2B.8: OxCal methods example using the well constrained 1906 San Andreas earthquake and the AD 1700 Cascadia event. The left panel shows the hemipelagic (H) data determined from visual observation, physical property data, smear slide mineralogy and X-radiography. H data are then input to OxCal with raw ¹⁴C ages converted to time via sedimentation rate curves developed for each site. Right panel shows four ways to calculate the age of the 1906 earthquake, with the preferred method being the use of underlying and overlying hemipelagic intervals, historical data (no written record of an earthquake between the date of the first San Francisco Mission built in 1769, and the 1838 earthquake). Lower panel shows result of similar application of constraining hemipelagic data to the AD 1700 Cascadia earthquake, the age of which is known independently (Satake *et al.*, 2003). This method commonly resolves the ambiguities inherent in radiocarbon dating where probability density functions (PDFs) have multiple peaks or broad distributions due to the slope or complexity

2B.2.2.5 ^{210}Pb and ^{137}Cs Activity

The ^{210}Pb method can be used to constrain the ages of young events and also to determine whether or not piston cores captured the youngest material at the seafloor. The free-falling piston corer sometimes does not sample the interface, which is blown away by the force of the falling core barrel, whereas the slowly lowered trigger core almost always includes the seafloor. ^{210}Pb activity rates can be used to either determine the age of the uppermost sediment, or determine that the uppermost material was older than the maximum typical age when ^{210}Pb reached background levels ($\sim$150 years; Robbins and Edgington, 1975).

Numerous observations from multicore samples and submersibles show that there is a very low-density material near the nepheloid layer at the seafloor. This layer is usually not recovered in piston and gravity cores, but is easily observed in multicore samples. The logarithmic decay of ^{210}Pb begins below the mixed layer, and most ^{210}Pb analyses assume that the mixed layer is entrained in any turbid flow and completely removed from the record we observe in sediment cores. This apparently presents no significant problem, as there is no time lost by removing the mixed layer as its ^{210}Pb age is constant and near zero on the seafloor prior to the rapid deposition of turbidites and landslides (Nittrouer, 1978).

^{137}Cs is a similar technique for even younger materials. The half life of ^{137}Cs is 30.3 years. Its presence is due to the atmospheric testing of nuclear devices during the 1950s and early 1960s (Schuller et al., 1993, 2002). Since that time, there has been no ^{137}Cs released to the atmosphere following the termination of atmospheric nuclear testing. Very recent small releases include the Chernobyl accident and testing by India and Pakistan. When using ^{137}Cs activity in sediments, most investigators assume that the maximum value is associated with the high fluxes of ^{137}Cs between 1962 and 1965, with the peak value commonly assumed to be 1963 $\pm$ 2 years, giving a sharp peak that can be used to constrain sedimentation rates and calculate ages of postbomb strata. In some settings including marsh peat and organic materials, ^{137}Cs has been shown to be mobile and not useful for dating. The technique works well in clay rich systems.

2B.2.2.6   Bioturbation and its Effect on Radiocarbon Dating of Interseismic Hemipelagic Sediments

A number of attempts to test the dependence of vertical rates of mixing during bioturbation on various parameters have been made. When considering single species, it has been shown that rates can be dependant on temperature, particle size, and particle shape (Wheatcroft et al., 1992 and references therein). Abyssal plain temperatures are relatively constant, and therefore unlikely to contribute to variable bioturbation. Thus we would like to know what effect particle size has on vertical mixing rates to evaluate radiocarbon age results in terms of foram size within samples, as well as any effect such rate changes may have toward biasing ^{14}C ages in either direction.

The process of bioturbation is highly complex, and equally complex to unravel. Experimental results from several settings suggest, however, that bioturbation in the deep sea is dominated by deposit feeders, and that deposit feeders in turn preferentially ingest and retain fine particles (Thomson et al., 1988, 1994; Wheatcroft, 1992). The impact of this in the context of dating marine deposits is that relatively large particles such as foraminifers used for dating are not selected by deposit feeders for retention, and

---

of the calibration curve. In this example, the overlying 300 years of hemipelagic in Cascadia sediment restricts the PDF to the earlier of two peaks. Such constraints are typically not as strong for events deeper in the core section because the present day upper boundary layer is absolute in these examples. Modified after Goldfinger et al. (2009).

apparently not vertically mixed as much as the finer fractions of material. For dating of turbidites, these are important results, and may help explain the surprising consistency we see in dating correlative turbidites when other variables such as reservoir age, basal erosion, and contamination are minimized.

2B.2.2.7 Stratigraphic Datum Ages

In Cascadia, the widespread deposition of ash sourced from the eruption of Mount Mazama (now Crater Lake) provides a clear datum throughout most of the Cascadia Basin system and provides independent age control. The age of the cataclysmic Mazama eruption is well constrained by recent work, yielding an age of 7630 ± 150 cal BP (Zdanowicz *et al.*, 1999) from Greenland ice cores, and 7600 ± 30 in British Columbia lake sediments (calibrated from Hallett *et al.*, 1997). Throughout Cascadia Basin, the first turbidite containing the Mazama ash is easily identified and has been dated in five localities, with an average age of 7130 ± 120, ~500 years after the Mazama eruption. Earlier work identified this Mazama ash-bearing turbidite as the thirteenth event down from the surface in many Cascadia basin cores (Adams, 1990; Goldfinger *et al.*, 2003a,b). Subsequent work demonstrated that Rogue Channel events could be well correlated locally, as well as correlated to other Cascadia Basin sites (Goldfinger *et al.*, 2008; see below) and that the Mazama ash first appearance was in the fourteenth margin-wide turbidite down from the surface, not the thirteenth as in other systems. Similar stratigraphic control can be found in many other settings, such as the Sumatran margin where multiple tephras with good elemental fingerprints constrain the turbidite sequence there (M. Salisbury, Oregon State University, personal communication 2008). Similarly, turbidite ages have been constrained by tephra markers in Lake Biwa, Japan (Inouchi *et al.*, 1996). Twenty turbidites were correlated to the historical record of earthquakes, and their ages calculated using sedimentation rates above the known tephra markers. Similarly, Noda *et al.* (2008) used a tephra marker to help support the dating and correlation of turbidite stratigraphy on a fan setting in the Kurile Trench.

2B.3 Locating Primary Evidence: Active Faulting and Structures

2B.3.1 Direct Fault Investigations

2B.3.1.1 Wecoma Fault, Cascadia Subduction Zone

An early example of slip rate determination based on submarine imaging and limited sampling is a group of unusual active strike-slip faults cutting both plates of the Cascadia subduction margin (Appelgate *et al.*, 1992; Goldfinger *et al.*, 1992, 1996, 1996b, 1997). Investigation of one of these, the Wecoma Fault revealed that a youthful submarine channel was clearly offset in high-resolution deep-towed sidescan sonar imagery (Figure 2B.9; Appelgate *et al.*, 1992; Goldfinger *et al.*, 1992). Cores showed that the channel had been abandoned at the start of the deglacial, setting a timeframe for the measured offset and establishing a minimum slip rate of 8.5 ± 4 mm/yr. No information on individual slip events could be determined from these data.

Matching of subsurface sedimentary packages imaged in a detailed seismic reflection grid allowed estimates of total fault slip since inception of the Wecoma Fault, and linkage of a key reflector to DSDP site 174 allowed calculation of the average slip rate since fault inception of 7–11 mm/yr, similar to the late Quaternary rate. Four of the nine mapped faults were investigated with the Alvin and SeaCliff deep diving submersibles, and two by the shallow Delta submersible. These faults (the Wecoma, Thompson Ridge Daisy Bank, and North Nitinat Faults) all exhibited evidence of recent surface rupture

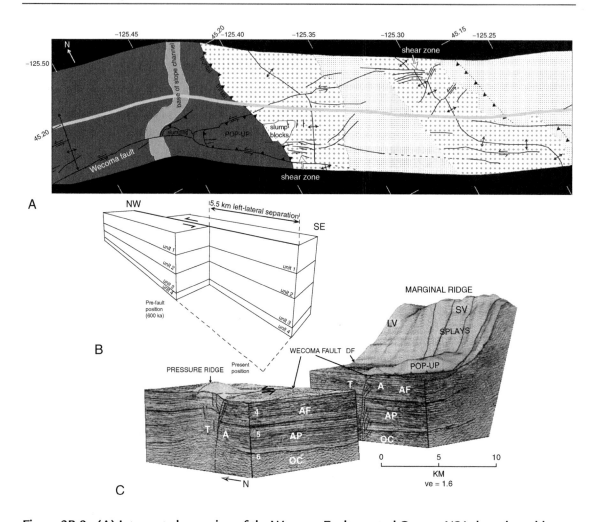

Figure 2B.9: (A) Interpreted map view of the Wecoma Fault, central Oregon USA, based on sidescan imagery, deep reflection profile, shallow sub-bottom profiles, and ALVIN dives. The main fault bifurcates as in intersects the megathrust, labeled "frontal thrust" creating a triangular pop-up. Two thrust ridges cut and offset by the basement fault shown, with sigmoidally deformed anticlinal axes. (B) Schematic view of retrodeformed trench-fill wedges used to establish net slip since fault inception. (C) Composite 3D block diagram of the intersection of the Wecoma Fault with the Cascadia megathrust, viewed toward the northeast. Migrated seismic sections scaled in two-way travel time (seconds). AP, abyssal plain section; AF, Astoria Fan; A, motion away; T, motion toward; SV, seaward vergent thrust, LV, landward vergent thrust; OC, Oceanic Crust. Modified after Goldfinger *et al.* (1992, 1997).

(Goldfinger *et al.*, 1997). At the intersection of the Wecoma Fault, with the frontal thrust ridge of the accretionary wedge, samples and video showing strongly developed sub-horizontal slickensides were collected (Goldfinger *et al.*, 1997). Samples revealed carbonate chemistry elevated in ^{3}He, indicative of rupture of the Juan de Fuca plate slab, also indicated in reflection profiles (Sample *et al.*, 1993).

The North Nitinat Fault also offset a seafloor channel and localized an elongate mud volcano aligned along the fault (Goldfinger *et al.*, 1997). Shallow water observations of the Daisy Bank Fault on the Oregon shelf revealed a fresh scarp of ~0.5 m in height, offsetting the Holocene–Pleistocene conformable horizon, and breaking carbonate pavements formed during a previous interseismic period, indicating high accelerations (Figure 2B.3; Goldfinger *et al.*, 1996).

2B.3.1.2 Lake Tahoe, California

In cases where faults traverse the shoreline, both onshore and offshore techniques can be used to best advantage. Offshore, geophysical imaging is efficient, can be of very high resolution, and is commonly relatively free of cultural noise that can plague onshore high-resolution geophysics. Fault morphology is commonly well expressed in the marine environment where erosion is minimized and scarps preserved longer due to the reduced gravitation. This makes bathymetric, sidescan sonar, and reflection imaging of submarine faults unparalleled except perhaps for faults in desert environments. Many of the same cross cutting relationships such as landsliding, offset channels, offset structures, and other features are available in the submarine environment. On the other hand, sampling and dating individual events is more difficult and costly offshore, but relatively inexpensive onshore with trenching and coring. An excellent example of the use of both techniques has been developed for faults in the Lake Tahoe Basin.

Kent *et al.* (2005) report evidence of deformation across three major fault strands within the Lake Tahoe Basin based on combination of high-resolution CHIRP seismic, a bathymetric grid combining airborne laser and multibeam sonar, and sediment cores (Figures 2B.4 and 2B.10). These faults offset submerged erosional terraces of late Pleistocene age (19.2 6 1.8 ka) and record 10–15 m of vertical deformation. A major submarine landslide, the McKinney Bay slide, has spread blocks across much of the central lake floor. This deposit is also offset vertically across the Stateline Fault by ~21–25 m.

Age constraints on the landslide deposit, and thus on the fault slip rates are uncertain, and reported variously as 60 ka (Kent *et al.*, 2005), 15–17 ka (Moore *et al.*, 2006) 300 ka (Gardner *et al.*, 2000), to Holocene (Schweickert *et al.*, 2000). Kent *et al.* (2005) calculate a deformation across several marker beds, and ^{14}C and optically stimulated luminescence (OSL) age control suggest an extension rate across the Lake Tahoe basin that of 0.4–0.5 mm/yr, assuming the 60 ka age of the McKinney Bay Slide. Moore *et al.* (2006) have suggested this slide generated a large tsunami, which may have had a role in the deposition or modification of linear boulder lines on the western shelf of the lake. Seitz and Kent (2005) investigated one of the Lake Tahoe faults, the Incline Village Fault using a grid of high-resolution CHIRP seismic profiles offshore. They imaged the fault zone in considerable detail, and the data suggest multiple events offsetting the mostly glacial stratigraphy and multiple colluvial wedges. Onshore, the fault scarp was trenched with a 7.5 m deep trench across the 5-m-high scarp, yielding event ages of 500 years, ~36.7 ka, and an older as yet undated deposit, demonstrating the utility of the integrated onshore–offshore technique (Seitz and Kent, 2005) (Further details of the most recent earthquake on one of the Tahoe faults are given in Brothers *et al.*, (2009).

2B.3.1.3 Palos Verdes Fault

Offshore Los Angeles, the Palos Verdes Fault, represents one of many proximal seismic hazards to the Los Angeles area represented by the San Andreas–parallel Peninsular Range faults. The Palos Verdes Fault is on the shallow shelf for the most part and partially located onshore on the Palos Verdes Peninsula (Figure 2B.11). Because the onshore part of the fault is located in a heavily populated and industrial area, onshore investigation is all but impossible. Offshore, McNeilan *et al.* (1996) used high-resolution seismic-reflection profiles and borehole data from the Los Angeles Outer Harbor to estimate the slip rate for the San Pedro segment of the Palos Verdes Fault based on subsurface piercing point offsets. Seismic profiling

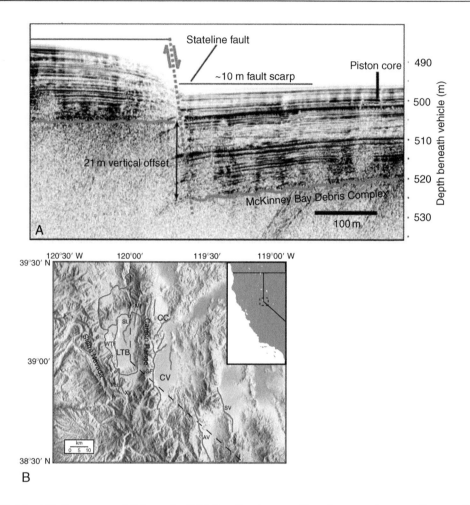

Figure 2B.10: (A) Deep-towed Edgetech CHIRP reflection profile of across the Stateline Fault (SLF), Lake Tahoe California. The top of the McKinney Bay slide complex is offset 21 m, with possible small colluvial wedges visible against the fault plane. Location map shown in B. From Kent *et al.* (2005). Reprinted with permission from the Geological Society of America. (See Color Insert.)

revealed two paleochannels offset by the fault, as well as mismatching structure contours. Using numerous ^{14}C ages, they then estimated the slip rate to be between 2.7 and 3.0 mm/yr for the past 7.8–8.0 ky based on reconstruction of offset paleochannels and subtle structure contours (Figure 2B.11). The slip rate obtained in this way is in good agreement with estimates based on uplift of marine terraces onshore (Ward and Valensise, 1994) using the combined vertical and horizontal components of separation.

This study is a classic application of land paleoseismic techniques adapted and applied to a submarine fault. Seismic profiles serve as geophysical trenches, and can be collected in a much larger spatial array and longer extent. While this technique lacks the spatial resolution to identify individual event details as is common in a trench wall, the slip rate is probably more robust due to the larger extent of sampling of the fault trace. McNeilan *et al.* (1996) used their reconstructed slip rate and observed segmentation to calculate that the fault is capable of generating an $M_w = 7.0$–7.2 earthquake every 400–900 years.

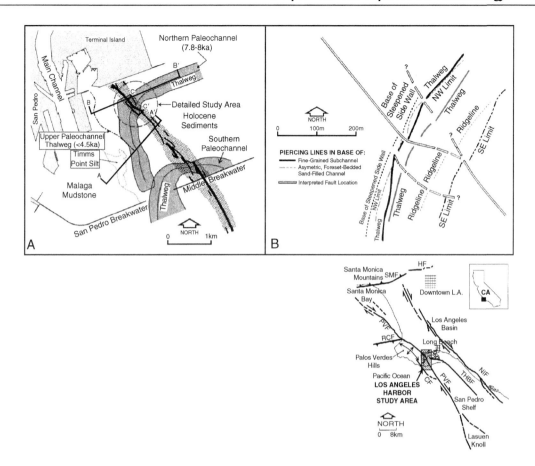

Figure 2B.11: Example of subsurface mapping of piercing point offsets in the subsurface along the Palos Verdes Fault, southern California. (A) Map of the Los Angeles Outer Harbor showing geologic exposures, Holocene paleochannels, Holocene fault strands, and zone of Pleistocene fault deformation (hachured). For the northern paleochannel only, the thalweg and limits of the sub-channel are shown. (B) Map of piercing lines of northern 7.8–8 ka paleochannel across the Palos Verdes Fault; solid lines represent thalwegs and dashed lines represent other piercing lines. From McNeilan *et al.* (1996). Reprinted with permission from the American Geophysical Union.

Marlow *et al.* (2000) used high-resolution multibeam sonar to map the surface trace of the fault extending to the southwest of the McNeilan study, and showed strong evidence of surface faulting, associated active anticlines, and interaction with the sub-parallel Avalon Knoll Fault farther offshore. Fisher *et al.* (2004) and Bohannon *et al.* (2004) used deeper seismic reflection profiles to characterize the geometry of the Palos Verdes Fault, and its along-strike variability in dip to link the structure to growth of anticlines such as Lasuen Knoll along the fault trace (Figure 2B.12).

2B.3.1.4 San Clemente Fault, California

The southern California Borderland province hosts many other active marine structures that have been the targets of paleoseismic and active fault studies. The San Clemente Fault is one of the most significant of

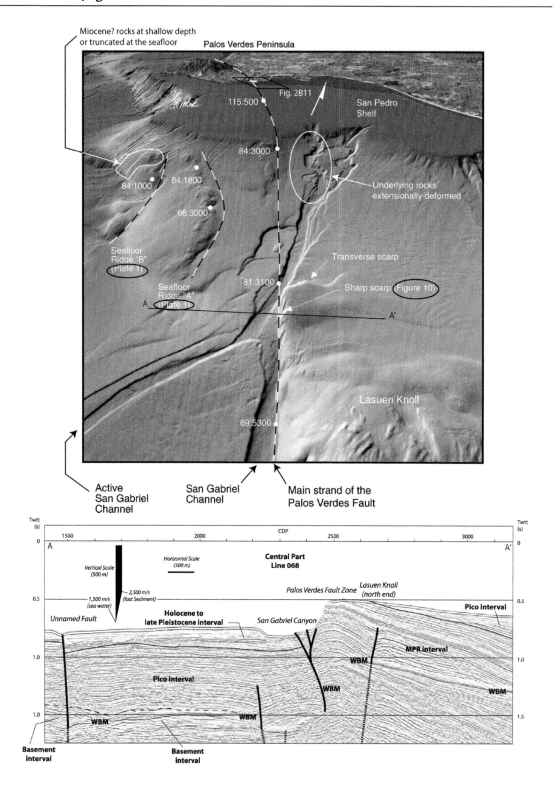

Miocene? rocks at shallow depth
or truncated at the seafloor

Palos Verdes Peninsula

Fig. 2B11

115:500

San Pedro
Shelf

84:3000

Underlying rocks
extensionally deformed

84:1000

84:1800

66:3000

Seafloor
Ridge "B"
(Plate 1)

Seafloor
Ridge "A"
(Plate 1)

Transverse scarp

81:3100

Sharp scarp (Figure 10)

A   A'

Lasuen Knoll

69:5300

Active
San Gabriel
Channel

San Gabriel
Channel

Main strand of the
Palos Verdes Fault

Twtt
(s)   Twtt
(s)

1500   2000   CDP   2500   3000

0   0
A   A'

Vertical Scale
(500 m)

Horizontal Scale
(500 m)   Central Part
Line 068

2,500 m/s
(fast Sediment)

Palos Verdes Fault Zone   Lasuen Knoll
(north end)

0.5   0.5
1,500 m/s
(sea water)

Unnamed Fault   Holocene to
late Pleistocene interval   San Gabriel Canyon   Pico interval

WBM   MPR interval

1.0   1.0
WBM

Pico interval   WBM

WBM

1.0   1.5
WBM   WBM

Basement
interval   Basement
interval

the Peninsular Range Faults, and traverses the borderland from south of the US–Mexico border to its intersection with the limit of the province at the Channel Islands Thrust (CIT), the southern limit of the Western Transverse Ranges (Figure 2B.13).

The San Clemente Fault is an active structure with instrumental seismicity and occasional moderate dextral earthquakes along its mapped trace. The San Clemente Fault has apparently been a strike-slip fault for all of its history, whereas many of the peninsular range faults have an earlier extensional history derived from the failed rifting of the Borderland, followed by a contractional episode. The San Clemente Fault clearly offsets large scale features in the borderland that can be retrodeformed in a relatively straightforward way. Construction of a hybrid multibeam/singlebeam bathymetric grid for the borderland revealed both broad and fine scale tectonic geomorphic relationships along the San Clemente, San Diego Trough and other fault systems. Slip sense for the San Clemente Fault is revealed by offset drainages, basement highs, and the numerous restraining and releasing bends that control the vertical tectonics on both local and regional scales. Retrodeformation of regional piercing points along the San Clement Fault, including a dextral separation of San Clemente Island from Fortymile Bank to the southeast established the regional net slip and average slip rate since inception of the fault of ~50–62 km and ~7 mm/yr. On a smaller scale, numerous restraining and releasing bends control the development of related folds along the San Clemente fault, also indicated by shifting channels and Holocene–Pleistocene growth strata (Goldfinger *et al.*, 2000; Legg *et al.*, 2007). Superimposed on this broad uplift are four smaller restraining-releasing bend pairs, mirroring the larger uplift that results from a left bend in the main fault trace near the Descanso Basin. ALVIN observations of the San Clemente fault on the northern flank of Navy Fan (Legg *et al.*, 2007) reveal a recent Holocene scarp 0.3–1.5 m in height with apparent horizontal slickensides exposed. The scarp is interpreted as a single event scarp, indicated by the lack of multiple slope breaks, and uniform "weathering" and bioturbation of the exposed Holocene and late Pleistocene strata. Scarp height suggests a Holocene event greater than $M_w = 6$ (Goldfinger *et al.*, 2000).

2B.3.1.5 Marmara Sea

The Sea of Marmara in Turkey represents a series of linked pull-apart basins along the North Anatolian Fault (NAF). The 1999 Izmit earthquake, coupled with the likelihood that the next NAF earthquake may strike Istanbul, has spurred research efforts in the Sea of Marmara, which contains the submarine segment of the NAF and the likely site of the next significant earthquake. Le Pichon *et al.* (2001, 2003) and Armijo *et al.* (2002) present results from high-resolution multibeam bathymetric surveys with backscatter data, sidescan sonar, and seismic reflection profiles newly acquired in the Marmara basin. These data revealed a series of en echelon pull-apart basins within the larger Marmara basin, interpreted by Armijo *et al.* (2002) as the apparent lack of a throughgoing dextral NAF. LePichon *et al.* (2003) argue for a main Marmara fault that ruptures most of the system, and a Çinarcik basin segment to the east with extension expected there based on GPS data and kinematic reconstruction. They compare this proposed rupture mode to

Figure 2B.12 (A) Shaded relief view of high-resolution multibeam bathymetry along the Palos Verdes Fault showing tectonic features of this strike-slip fault. Location of Figure 2B.11 shown on shelf at upper portion of image. Other figure callouts refer to the original publication. From Fisher *et al.* (2004). Reprinted with permission from the Geological Society of America. (B) Migrated multichannel reflection profile across the Palos Verdes Fault and Lasuen Knoll, a small restraining bend uplift along the fault (Legg *et al.*, 2007). Line of section A–A' shown in A above. Figure from Bohannon *et al.* (2004). (See Color Insert.)

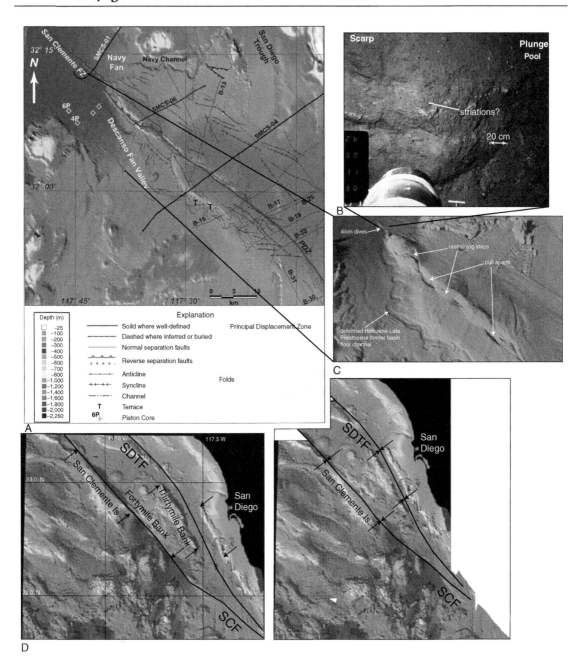

Figure 2B.13: (A) Bathymetric compilation map of the southern California Borderland showing the San Clemente Fault and the San Diego Trough Fault (SDTF). (B) Photograph from the DSV Alvin of fault scarp along the San Clemente Fault crossing the Navy fan (location in B). Sub-horizontal lineations may be slickensides consistent with strike slip motion. The scarp is composed of mud and layers of shells associated with ancient benthic communities at former cold seeps. Holocene scarp 0.3–1.5 m in height appears to be a single event scarp, indicated by the lack of multiple slope breaks

historical earthquakes to support a correspondence between onshore damage and these two primary segments. Rangin *et al.* (2004) further propose that the throughgoing fault is a late development of the most recent 100–200 ka, and that this recent propagation of the throughgoing main Marmara fault deactivated the pull-apart basins, as evidenced by undisturbed sedimentary overlap of the bounding transfer faults, inversion of the pull-apart basins, and the apparent crosscutting of the main Marmara fault across the corners of the en echelon pull aparts (Figure 2B.14). In this model, the preexisting pull-apart basin system accumulated about 30 km of dextral slip and probably appeared in late Pliocene or early Pleistocene time. Subsequently, the Sea of Marmara has been the site of progressive localization of strain. This model bears similarities to experimental physical model results of evolving strike-slip systems. Thus in a relatively short time since the 1999 earthquake, collection of marine geophysical data has resulted in the rapid evolution of a series of tectonic models for the area that was previously quite poorly known. Discussion of paleoseismic stratigraphy in the Marmara Sea in relation to historical earthquakes is included in a subsequent section.

2B.3.2 Off-Fault Investigation

2B.3.2.1 Vertical Tectonics in a Strike-Slip Setting: Channel Islands Thrust and the Catalina Ridge–San Clemente Fault Zone

Bathymetric data can be used in a number of ways and are particularly amenable to strike-slip environments as previously discussed. Another way to use these types of data is to establish strain markers such as previously level low-stand shorelines as strain markers. Chaytor *et al.* (2008) used submerged last glacial maximum (LGM) and younger paleoshorelines preserved around the Northern Channel Islands submarine banks atop the Santa Cruz–Catalina Ridge to determine the vertical strain history at the intersection of the Santa Cruz–Catalina Fault and the southern Transverse Ranges, marked by the CIT. They used high-resolution multibeam mapping combined with submersible observations to establish the nature of the shorelines (Figure 2B.15). The morphology of the slope breaks and surficial sediments revealed clear evidence of the former wave-cut shoreline angles, with a sharp slope break, wave-cut undercuts, coarse sediment on the former shoreface, decreasing in grain size downslope of the gently sloping planar platform. Intertidal mussels and barnacles were collected manually from the submersible along the submerged shorelines, presently at a depth of 100–130 m. Radiocarbon results bracket the LGM (LGM $\sim$−120 m water depth) at $\sim$19 ka. On the eastern Northern Channel Islands platform, as much as

---

and uniform "weathering" and bioturbation. The lightly bioturbated fresh scarp offsets Holocene and late Pleistocene strata, indicating a Holocene event that likely had a magnitude greater than 6 (Goldfinger *et al.*, 2000). Photo by C. Goldfinger, from Legg *et al.* (2007). (C) View of a restraining bend along the San Clemente Fault south of Navy Fan showing multiple pull apart and restraining bend features, superimposed on the larger uplift which itself is a restraining bend uplift due to a 5° strike change in the San Clemente Fault visible in (A). Location shown in (A). Channel at left is presently on the flank of the uplift, reflecting recent growth of this feature. (D) Retrodeformed San Clemente and SDTFs using morphologic and geologic piercing lines (Goldfinger *et al.*, 2000). San Clemente Fault (SCF) has a minimum horizontal separation of 50 km based on four piercing points (two are shown). The SDTF horizontal separation is 32 km, with 15 km extension as well (extension retrodeformation partially shown here to illustrate fit of offset features. (A and C) Reprinted with permission of the Royal Society of London. (See Color Insert.)

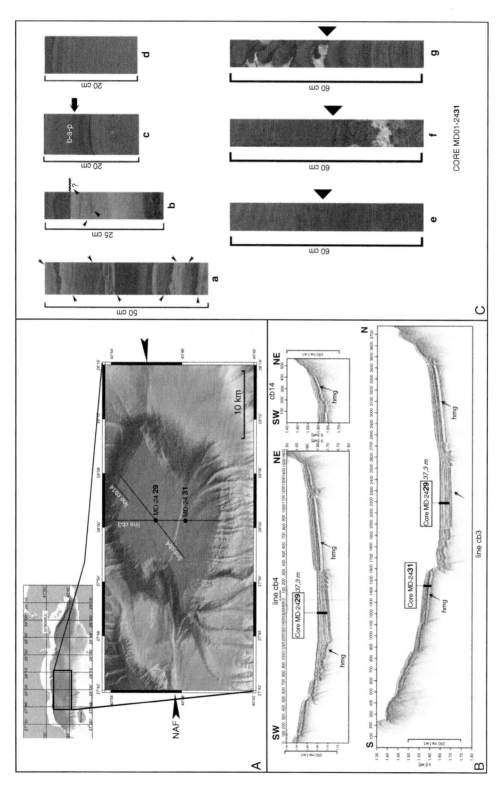

Figure 2B.14: (A) Detailed morpho-bathymetry of the Marmara Sea's Central Basin, location of giant piston cores and seismic profiles. (B) 3.5 kHz profiles across the Central Basin, showing active faults and evidence for a "homogenite." (C) Close-ups of selected portions of core MD01-2431. (a: conjugate microfractures; b: microfracturing with possible sealing by coeval turbidite arrival; c: possible *in situ* liquefaction, evidenced by ball-and-pillow –b-a-p– structure; d–g: details of the pre-Late Glacial event. In the continuous core section from 10.50 to 12.00 m X-ray scanning shows a constant orientation of microfractures). From Beck *et al.* (2007). Reprinted with permission from Elsevier. (See Color Insert.)

1.50 ± 0.59 mm/yr of late Pleistocene to Holocene uplift of the islands above the blind CIT was observed based on the uplifted shoreline. This result is higher than onshore terrace uplift estimates over a period of 125 ka since stage 5e terrace formation (Pinter *et al.*, 1998).

South of the intersection of this fault with the CIT, similar shorelines were observed rimming the submarine Pilgrim Banks atop the Santa Cruz–Catalina Ridge. Shorelines there show no net vertical tectonic motion, but are instead tilted to the north, possibly reflecting flexural bending or limited underthrusting of this block beneath the CIT.

Collectively, the submerged shorelines revealed significant uplift from underthrusting along the CIT, and northward tilt of the underthrust block. It also appears that a significant amount of differential motion at the intersection of the Peninsular Range faults and the southern Transverse Range thrusts offshore may be distributed into upper-crustal deformation both at and south of the intersection, along the length of the major dextral fault systems. Detailed bathymetry and reflection profiles suggest some of this deformation is partitioned into splay fault terminations and thrusting on the western side of the Santa Cruz–Catalina Ridge, and more subdued deformation in the basin to the east. This may be an effect of limited "subduction" of the dextral strike-slip fault beneath the CIT, and the resulting change in slip rates on the CIT across the intersection point with the strike-slip fault. Similar kinematics were reported for subduction of strike-slip faults along the Cascadia margin, discussed previously (Goldfinger *et al.*, 1997).

Other notable localities in which detailed submarine faulting studies of individual faults have been carried out, though not to the level of detail in these case histories, include the Gulf of Corinth, the Carboneras Fault off southern Spain, offshore Lebanon, and many subduction zones including Cascadia, Nankai, Hikurangi, Sumatra Costa Rica, Chile, and others.

2B.4 Locating Secondary Evidence: Landslides, Turbidites, Submarine Tsunami Deposits

The classic paper by Heezen and Ewing (1952) demonstrated that large offshore earthquakes can trigger turbidity currents having regional extent. They described the Grand Banks turbidity current, which was triggered in the epicentral area of a magnitude 7.2 earthquake on 28 November 1929. This event involved detachment and downslope movement of submarine sediment along 240 km of the continental shelf; after traveling 650 km from its source, the turbidity current still was moving faster than 20 km/h and therefore probably continued for hundreds of kilometers. Heezen and Ewing (1952) postulated that the earthquake triggered submarine slumps along an extensive length of the continental shelf corresponding to the epicentral zone of the earthquake and that these slumps transformed into turbidity currents that moved as rapidly as 100 km/h down slopes averaging only about 1.5°.

The triggering of turbidity currents and landslides from submarine canyons, shelf edges, and seamount edifices are becoming reasonably well known. In particular, turbidity currents triggered along the Cascadia margin, the northern San Andreas margin, Chile, the Japan Trench and other localities are under investigation and yielding coherent earthquake records. Many of these events have been linked temporally to onshore tsunami deposits and are becoming recognized as viable event pairs that document the occurrence of earthquake-triggered tsunami. These events can be dated and correlated in the marine environment, providing long continuous records that also provide good evidence for spatial continuity.

The use of secondary evidence such as landslides and turbidites adds some complexity to this aspect of paleoseismology. The techniques do not use fault outcrops because the faults are inaccessible and must

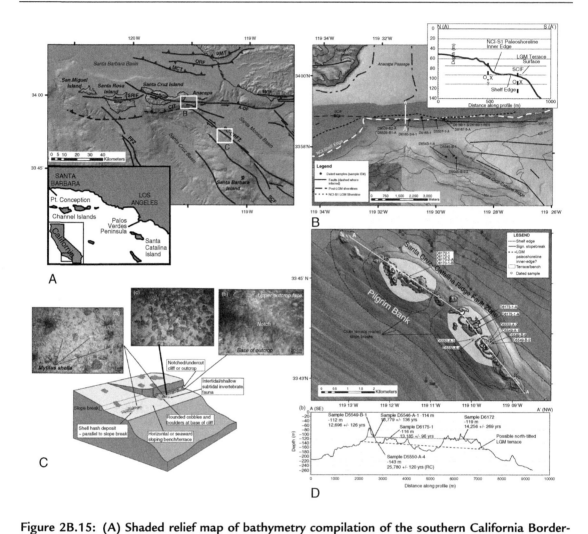

Figure 2B.15: (A) Shaded relief map of bathymetry compilation of the southern California Border-lands showing two study sites at Santa Cruz Island and Pilgrim Bank. (B) View of the bathymetry, NCI-S1 paleoshoreline, and terrace features on the southern Northern Channel Islands platform shelf edge between Santa Cruz and Anacapa Islands. The trace of the Santa Cruz Island fault (SCIF) is indicated as it crosses onto the platform, where it is likely related to destabilization of the slope. Inset: profile A–A', showing the morphology of the LGM terrace, with approximate locations of the SCIF and an additional fault indicated (fault movement indicated by X-away, O-toward).
(C) Composite schematic diagram of paleoshoreline features observed during submersible dives on the Northern Channel Islands platform and Pilgrim Banks. Examples of these features can be seen in photographs taken from the submersible: (A) Large, well-preserved Mytilus Califorianus shells on bench, Pilgrim Banks. (B) Notched, undercut rock outcrop on south side of Santa Cruz Island. (C) Rounded cobbles-boulders on probable paleoshoreline between Santa Cruz and Anacapa Islands, southern Northern Channel Islands platform. (D) Map of paleoshorelines and benches-terraces on Pilgrim Banks, based on analysis of the bathymetry and submersible observations. Terrace-bench areas are shown in white. Possible pre-LGM inner edges–terraces are indicated. Line of profile A–A' is

demonstrate that the events they are investigating are uniquely generated by earthquakes and not some other natural phenomenon. Nevertheless, these problems can be overcome, and the techniques can be powerful tools for deciphering the earthquake history along an active continental margin. These methods are complementary: the onshore record can provide temporal precision for the most recent events via radiocarbon dating, coral chronology and dendrochronology (tree-ring dating), while the marine sedimentary record generally extends further back in time, more than enough to encompass many earthquake cycles. In recent years, turbidite paleoseismology has been attempted in Cascadia (Adams, 1990; Goldfinger and Nelson, 1999; Blais-Stevens and Clague, 2001; Goldfinger *et al.*, 2003a,b, 2007, 2008, in review), Puget Sound (Karlin and Abella, 1992; Karlin *et al.*, 2004), Japan (Inouchi *et al.*, 1996), the Mediterranean (Kastens, 1984; Anastasakis and Piper, 1991; Nelson *et al.*, 1995), the Dead Sea (Niemi and Ben-Avraham, 1994), northern California (Field *et al.*, 1982; Field, 1984; Garfield *et al.*, 1994; Goldfinger *et al.*, 2007, 2008) Lake Lucerne (Schnellmann *et al.*, 2002), Taiwan (Huh *et al.*, 2006), the southwest Iberian margin (Viscaino *et al.*, submitted), the Chile margin (Blumberg *et al.*, 2008; Völker *et al.*, 2008), the Marmara Sea (McHugh *et al.*, 2006), the Sunda margin (Patton *et al.*, 2007) and even the Arctic ocean (Grantz *et al.*, 1996). Results from these studies suggest the turbidite paleoseismologic technique is evolving as a useful tool for seismotectonics.

2B.4.1 Distinguishing Earthquake and Nonearthquake Triggering Mechanisms

In off-fault paleoseismology, considerable effort must go toward distinguishing earthquake and nonearthquake sources. In the following sections, we consider this issue in some detail.

Common stratigraphic evidence of earthquakes includes submarine landslides and turbidity currents. Triggering events for these deposits may include (1) earthquakes; (2) volcanic explosions; (3) tsunami; (4) subaerial landslides into the marine environment; (5) storm wave loading, and (6) hyperpycnal flow. These primary triggers are distinguished from factors that may destabilize slope through longer term processes, such as sediment self-loading, gas hydrate thermal destabilization, sea-level change, shelf edges destabilized by groundwater input, volcanic seamount or island edifice destabilization, tectonic folding/tilting, and other factors. Triggering mechanisms have been discussed by Adams (1990), Nakajima (2000), Goldfinger *et al.* (2003a,b, 2008, 2009). Factors such as gas hydrate destabilization, sea-level change, tectonic steepening, and so on are factors that reduce seafloor stability, but do not generally trigger submarine mass movements. For example, the Storegga slide generated a large tsunami and occurred as a result of the massive deposition of glacial sediments and associated gas hydrate disassociation that destabilized the region, likely multiple times (e.g., Solheim *et al.*, 2005). The slide itself though was most likely triggered by an earthquake (Bryna *et al.*, 2005). Factors reducing slope stability may eventually lead to failure without other triggers, however, such failures are random, and are unlikely to be regional. As all of the triggering mechanisms may trigger turbidity currents and are inherently difficult to distinguish, how can earthquake-triggered turbidites be distinguished from other turbidites? An equally important question is whether environments can be found that favor preservation of earthquake deposits, while disfavoring others. Essentially two methods can be used to differentiate

indicated. Bathymetric contour interval is 100 m. Lower panel shows southeast–northwest profile (A–A'), showing the morphology of Pilgrim Banks in relation to several of the dated shell samples. The increase in depth of the samples of approximately the same age (short-dashed line) and the tilt of the possible LGM terrace surface may indicate a north-directed tilt of the Santa Cruz–Catalina Ridge. RC, radiocarbon. Reprinted with permission from the Geological Society of America.

earthquake-generated turbidites from those originating from other processes: (1) Sedimentological examination; and (2) Tests for synchronous triggering of multiple turbidite systems that can eliminate non-earthquake origins. Both of these methods may be augmented by historical earthquake records and land paleoseismic data if available.

In the following sections, we discuss these two methods and their global application, followed by specific applications to Cascadia, the Iberian margin, Japan, Sumatra, and other localities.

2B.4.1.1   Sedimentological and Mineralogical Characteristics

Japanese investigators have attempted to distinguish seismically generated turbidites (seismo-turbidites) from storm, tsunami, and other deposits. Nakajima and Kanai (2000) Nakajima (2000) and Shiki *et al.* (1996, 2000a,b) argue that seismo-turbidites may in some cases be distinguished sedimentologically. Shiki *et al.* (2000b) carefully examined known seismo-turbidites in Lake Biwa, Japan, including the 1185 AD Lake Biwa/Kyoto earthquake ($\sim M_w = 7.4$; Inouchi *et al.*, 1996). These deposits are characterized by wide areal extent, multiple coarse-fraction pulses, variable mineralogical provenance (from multiple or line sources), greater organic content, greater depositional mass and coarser deposits than the barely visible storm-generated events (Figure 2B.16). They also concluded that defining the triggering mechanism of even known earthquake-related deposits was problematic, and that further study was needed. Nakajima and Kanai (2000) observe that a known seismo-turbidite from the 1983 Japan Sea earthquake caused multiple slump events in many tributaries of a canyon system, resulting in multiple coarse sediment pulses. The stacked multipulsed turbidite subunits had distinct mineralogies and were found deposited in order of travel time to their lithologic sources, demonstrating synchronous triggering of multiple parts of the canyon system (Nakajima and Kanai, 2000). Goldfinger *et al.* (2007) found a similar relationship with vertical stacking of separate mineralogic sources along the Northern San Andreas Fault. Gorsline *et al.* (2000) find that complexity, thickness, and areal extent also serve to distinguish Holocene seismo-turbidites in the Santa Monica and Alfonso Basins of the California borderland and Gulf of California, respectively. In the Santa Monica Basin, both flood generated and earthquake-generated turbidites are present. The flood turbidites are one-tenth to one-fifth the volume of the earthquake-generated events, which are more widespread. Similarly, turbidites in the Alfonso Basin were also found to be thicker and greater in aerial extent when earthquake generated. Gorsline *et al.* (2000) argued that reasonable estimates of discharge, sediment input, and source area can be used to constrain the sediment budget for flooding episodes to define upper bounds for what sediment volumes could be available for nonseismic turbidites.

2B.4.1.2   Distinguishing Hyperpycnal Underflows

Hyperpycnal flow is the density driven underflow from storm flood discharge of rivers into marine or lacustrine systems, and proposed as a link to turbidity currents in a variety of settings. Hyperpycnites are commonly reported to have reverse-then-normal grading stemming from the waxing then waning nature of flood events (Figure 2B.17). The literature includes several reported cases and compares them to normally graded failure deposits such as those in the Var River system (Mulder *et al.*, 2001), Lake Biwa (Shiki *et al.*, 2000a), and the Toyama deep sea fan (Nakajima, 2006). The dynamics of longitudinal and temporal variability and their effects have been discussed in detail by Kneller and McCaffrey (2003), and Mulder *et al.* (2003).

In some proximal settings such as large lakes, shelf basins, and fjords, records of both earthquakes and flood deposits have been found. In one of the best comparisons, St.-Onge *et al.* (2004) show that details of both seismic and hyperpycnal deposition in the Saguenay Fjord in eastern Canada are diagnostic, and argue that hyperpycnal deposits are distinguished by reverse grading at the base, followed by normal

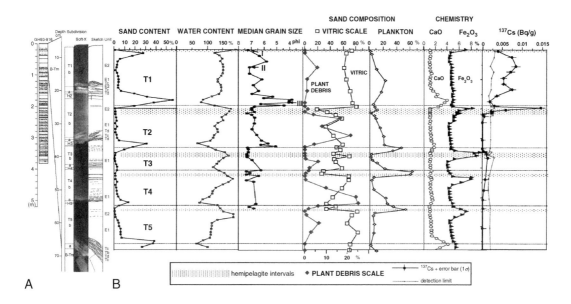

Figure 2B.16: (A) Soft X-radiograph (negative) showing sedimentary structures of upper 70 cm of core GH93-816 with a log of the core. Subdivisions T1–T5 represent turbidites while H1–H5 represent hemipelagites. Subdivisions (A) and (B) represent amalgamated beds within turbidites. Tb: parallel laminated sand; Tc: cross laminated sand/silt; Td: parallel laminated silt; E1: laminated mud; E2: graded mud. The lower part of the T5 bed and the B-Tm tephra layer has been disturbed by a coring effect. (B) Description of upper 70 cm of core GH93-816 showing sand content, water content, median grain size, sand composition, chemical composition and ^{137}Cs concentration. Median grain size in phi. (I)–(III) in median grain size column represent sampling points for grain size distributions shown in Figure 2B.9. From Nakajima and Kanai (2000). Reprinted with permission from Elsevier.

grading. The diagnostic reverse-then-normal grading for hyperpycnal deposits has been widely reported and is attributed to waxing, then waning flow associated with the storm, although the waxing portion may later be eroded during later peak flows (Guyard *et al.*, 2007). In the Saguenay Fjord, six events have normal grading alone and are inferred to be earthquake generated. Four others have similar basal units, but are topped by a reverse graded unit, and then a normally graded unit, with no evidence of hemipelagic sediment between the multiple units. These events are interpreted as an earthquake, followed by a hyperpycnite that resulted from the breaching of a landslide dam caused by the original earthquake. Dam breaching is a variant of the more common hyperpycnal scenario involving waxing and waning depletive flow (Kneller, 1995), but would likely result from a similar flow hydrograph (St.-Onge *et al.*, 2004).

Documentation of hyperpycnal flows into lakes and shelf basins is abundant; however, evidence of such flows entering canyons systems and moving into deep water is relatively sparse. Most, if not all examples involve short distances between the river mouth and canyon head, either during Pleistocene low-stand conditions or in systems that have very narrow shelves during high-stand conditions. Hyperpycnal flows extend further from river mouths with high discharge (Alexander and Mulder, 2002), but documentation is sparse. Wright *et al.* (2001) observe that hyperpycnal flow is strongly affected by ambient currents and generally delivers sediment to the slope only upon relaxation of longshore currents. Most investigators

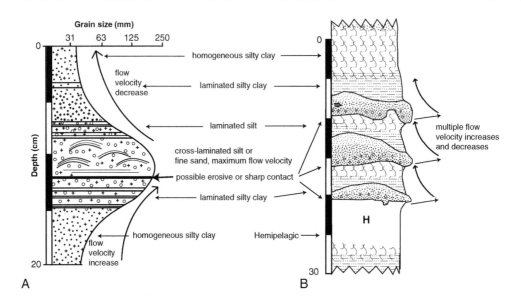

Figure 2B.17: (A) Idealized stratigraphy resulting from hyperpycnal flow, characterized by a coarsening upward sequence followed by a fining upward sequence attributed to a waxing then waning hydrographic profile during a storm event. Other events with a similar hydrographs, such as a gradual dam breaching may produce similar stratigraphy (after Mulder, 2001). (B) Typical stratigraphic sequence from a turbidite with multiple fining upward pulses from core M9907-12PC in Juan de Fuca Channel, Cascadia margin. This turbidite and many others in the Holocene Cascadia Basin turbidite sequence exhibit multipulsed stratigraphy, with no waxing phase. Multiple fining upward sequences are capped by a fine mud tail signaling the final waning of the turbidity current (From Goldfinger *et al.*, 2009).

cite Pleistocene examples or examples with little or no shelf width when referring to flows reaching the abyssal plain or lower fan reaches (e.g., Normark *et al.*, 1998; Piper *et al.*, 1999; Mulder *et al.*, 2003; Normark and Reid, 2003). This is an expected result of sea-level change, or the near direct connection between a river and a canyon in the case of narrow shelves. Under low-stand conditions, rivers and canyons are more directly connected, and such flows are expected to dominate sediment delivery to the deep sea.

Thus the deep water deposition of hyperpycnites is closely coupled to sea-level control, or alternatively to climate shifts. An example of high-stand hyperpycnal flow has been reported for the Var River, in which the canyon and river mouth are less than 1 km apart (Mulder *et al.*, 1998; Klaucke *et al.*, 2000). Many large river systems deposit most of their load in river mouth bars, with lesser quantities making it past such bars in to a delta front slope (e.g., Yellow River, Li *et al.*, 1998). Many canyon systems on continental margins were largely incised during Pleistocene sea-level low-stands (e.g., McNeill *et al.*, 2000; Curray *et al.*, 2002; Evans *et al.*, 2005; LeRoux *et al.*, 2005).

A good example is the 1969 El Nino flood, which input ~25 million tons of sediment (5× the present yearly Columbia River sediment load; Sherwood *et al.*, 1990) to the Santa Ana River in southern California over a 24 h period, in close proximity to nearby canyon heads (Drake *et al.*, 1972). Sediment from this extreme flood did not continue down canyons as hyperpycnal flow, but deposited as a distinct yellow unit on the shelf and upper slope. Over the next 10 years, the flood sediment moved downslope as turbid layer transport caused by storm wave resuspension, and deposited as yellow layers between varves of the Santa Barbara Basin (Drake *et al.*, 1972).

Hyperpycnites are also commonly organic rich as compared to seismic turbidites, having their sources in floods rather than in resuspension of older canyon wall material as in earthquake triggering (Shiki *et al.*, 1996, 2000b; Nakajima and Kanai, 2000; Mulder *et al.*, 2001). It has been suggested that this distinction may be used as a basis for distinguishing earthquake and storm deposits using OSL dating (Shirai *et al.*, 2004). However, we suspect that this generalization may easily be violated as in the case of floods in very arid regions, or earthquakes in heavily vegetated areas. For example, because the river drainage basins feeding Cascadia Basin are heavily vegetated, the Holocene turbidites, linked to earthquake origins through a variety of methods (Goldfinger *et al.*, 2008, 2009) have tails characterized by significant quantities of plant fragments (Nelson, 1976). Similarly, west Sumatran turbidites near the offshore forearc islands are very organic rich, whereas others are not.

Whether hyperpycnal flows can reach deep water via canyon systems incised during the sea-level low stands appears to be a function of shelf width, steepness, river peak storm discharge, high-stand aggradation, and the wave and current climate during peak storm discharge. However, the requirements for and evidence of hyperpycnal flows to the deep ocean under high-stand conditions (excepting very narrow shelves) remain poorly known at best (Mulder *et al.*, 2001). A well-documented example for the Toyama Channel and fan is given by Nakajima (2006) in which long-traveled and long-lived pulsed flows traveled 700 km to a deep sea fan. As with other examples, no shelf width buffers the river source from the canyon channel system in the Toyama system. In cases of narrow shelves, a turbidite record in an offshore basin or abyssal plain may well contain a mixture of hyperpycnal, sediment failure, and earthquake-generated turbidites. For systems that minimize these effects, those with wide continental shelves, or topographic barriers isolating the slope and abyssal plain the turbidite record is more likely to contain a dominantly earthquake record (Nakajima and Kanai, 2000; Abdeldayem *et al.*, 2004; Goldfinger *et al.*, 2008, 2009). The implication is that caution must be exercised to examine the river systems, their relationship to sea level during periods of interest, and the physiographic conditions of shelf width, forearc basins, and other barriers to hyperpycnal flow when evaluating a particular setting for turbidite paleoseismology.

2B.4.1.3 Synchronous Triggering

While there are few definitive sedimentological studies linking earthquakes directly with turbidites on the basis of the deposits themselves, most studies have focused on aspects of earthquake processes that are unique, and therefore eliminate most or all of the turbidite triggering mechanisms other than earthquakes. The primary characteristic that can easily be distinguished in sediment cores is spatial extent. When turbidite deposits can be correlated among widely spaced sites, synchronous deposition can be established or inferred, and if the spatial extent exceeds that reasonable for other mechanisms, then earthquake triggering is likely. Virtually all studies that make the linkage between earthquake triggering and turbidites invoke this test in some fashion, including those cited previously under sedimentological examination (Adams, 1990; Nakajima and Kanai, 2000; Gorsline, 2000; Goldfinger *et al.*, 2003b, 2007, 2008).

2B.4.1.4 Numerical Coincidence and Relative Dating Tests

In his synthesis of Cascadia Basin turbidite events, Adams (1990) observed that in several canyons feeding into a confluence, cores contained 13–14 turbidites above a regional tephra, the Mazama ash. Below the confluence, cores in the main Cascadia channel also contained 13 turbidites (Figure 2B.18). He reasoned that these events must have been synchronously triggered because if they had been independently triggered with more than a few hours separation in time, cores taken below the confluence

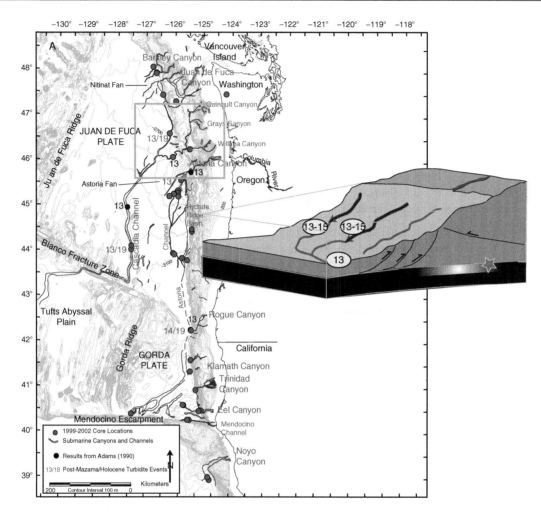

Figure 2B.18: Synchroneity test at a channel confluence as applied where Washington channels merge into the Cascadia Deep Sea Channel, indicated by box. The number of events downstream should be the sum of events in the tributaries, unless the turbidity currents were triggered simultaneously. Remarkable similarity of records in northern Cascadia supports the initial conclusion of Adams (1990) that these events are likely of earthquake origin. Modified after Goldfinger *et al.* (2009). (See Color Insert.)

should contain from 26 to 28 turbidites, not 13 as observed. The only alternative is that 13 turbidites also dropped out of the sequence due to the more distal position of the downstream core, an unlikely coincidence. The importance of this simple observation is that it demonstrates synchronous triggering of turbidity currents in tributaries the headwaters of which are separated by 50–150 km. The synchroneity demonstrated by this "confluence test" is also supported by the similar numbers of events alone, without the existence of the confluence, suggesting either synchronous triggering, or a regionally coherent coincidence. Off the California margin, Goldfinger *et al.* (2007) demonstrate that turbidites adjacent to the Northern San Andreas Fault also converge at a number of channel confluences and follow a similar pattern to that observed in Cascadia, remaining constant in number above and below the confluences.

2B.4.1.5 Stratigraphic Correlation

The lithostratigraphic correlation of turbidite stratigraphy offers a straightforward method to test for event synchroneity. The detailed geophysical "fingerprinting" of turbidites through their grain size distributions and other physical properties has direct implications for synchronous origins of the deposits. Geophysical signatures, commonly in the form of density, magnetic susceptibility, velocity, XRF composition, and other parameters can serve to establish a stratigraphic fingerprint (Figure 2B.19). Goldfinger *et al.* (2007, 2008, 2009) found that these "fingerprints" can be persistent among sites and over considerable distances. Stratigraphic fingerprints sometimes retain a remarkable similarity at sites along strike, but also commonly evolve somewhat along strike and downchannel in subtle ways that can be traced from one site to another. That such grain-size "fingerprints" exist suggest that triggering mechanisms that produced them, or the hydrodynamics of the separate canyon systems must have some commonality, as producing matching grain size patterns by coincidence is unlikely. Goldfinger *et al.* (2008, 2009) observe that in Cascadia the individual stratigraphic signatures can be traced across multiple canyon/channel systems, and at least one slope basin. Some of these sites have no physical connection, and the basin site is isolated from all other sites and sources of fluvial input.

2B.4.2 Turbidite Paleoseismology

2B.4.2.1 Cascadia

Goldfinger *et al.* (2003, 2008, 2009) investigated turbidite systems located on the continental margin of Cascadia Basin from Vancouver Island, Canada to Cape Mendocino California, USA. Cascadia Basin contains a variety of types and scales of turbidite systems including multiple canyon sources on the Washington margin that funnel turbidites into Cascadia Channel (1000 km length); Astoria Canyon on the northern Oregon margin that feeds Astoria submarine fan (300 km diameter) containing channel splays with depositional lobes; Rogue Canyon on the southern Oregon margin that feeds a small (<5 km) base-of-slope apron, and Trinidad, Eel, and Mendocino canyons (30–100 km length) on the northern California margin that feed into plunge pools, sediment wave fields, and channels. Detailed swath bathymetric data and core sampling procedures verify that key turbidite channel pathways of Cascadia Basin are open and provide a good turbidite event record. Proximal canyon mouth and inner fan channel areas have erratic turbidite event records because of extensive cut and fill episodes; however, even in these difficult locations, complete records can be found in some point bars, terraces and canyon walls that are slightly elevated above the channel thalweg. The most consistent turbidite event records occur in distal locations of continuous deep-sea channel systems such as Cascadia Channel (Figure 2B.18).

Multiple tributary channels with 50–150 km spacing and a wide variety of turbidite systems with different sedimentary sources contain 13 post-Mazama ash and 19 Holocene turbidites in Cascadia Channel, Juan de Fuca Channel off Washington, Hydrate Ridge slope basin, and Astoria Fan off northern and central Oregon. All of these events are also recorded on Rogue Apron of southern Oregon, with the addition of smaller local events recorded as silt or mud turbidites. Nineteen Holocene turbidites are found along the northern and central margin and are recorded in southern cores with 22 interspersed smaller events.

Goldfinger *et al.* (2008, 2009) used ^{14}C ages, the previously described "confluence test," and stratigraphic correlation of turbidites to determine whether turbidites deposited in separate channel systems were correlative and pass tests of synchronous deposition to test for earthquake origin (Figure 2B.20). The confluence test shows that a coherent record of 19 Holocene turbidites pass this test along the northern margin. This record represents the entire record of Holocene turbidites along the northern Cascadia margin, leaving no deposits from other sources.

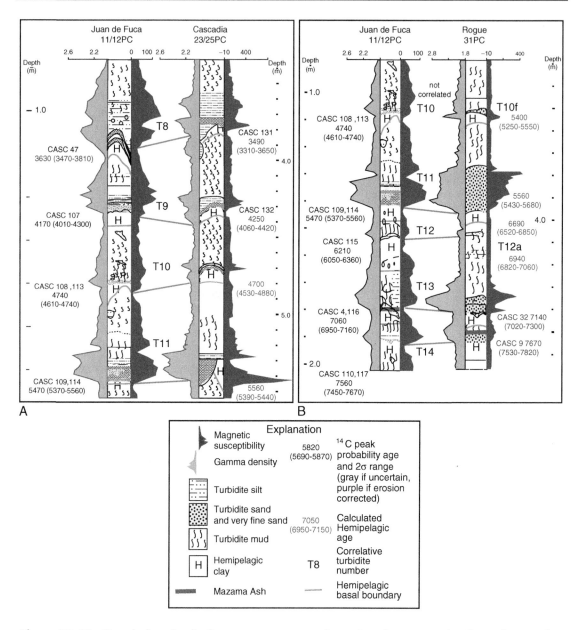

Figure 2B.19: Correlation details from two representative pairs of cores on the Cascadia margin. (A) Events 8–11 in cores from Juan de Fuca Channel (left) and Cascadia Channel (right). Left traces are raw gamma density, right traces are magnetic susceptibility. Lithologic logs are also shown. Note correspondence of size, spacing, number of peaks, and trends of physical property traces between these cores. (B) Similarly displays events T10–T14 in Juan de Fuca Channel (left) and T10d–T14 in Rogue Channel (right). (A) Cores are part of the same channel system, distance along channel = 475 km. (B) Cores are in channels that do not meet, separation distance = 500 km. Note that correlation of longer sections and ^{14}C data show that T10f and T10 do not correlate in (B). Similarly, Mazama ash appears in T14, not T13 in Rogue apron, see text for discussion. Modified after Goldfinger *et al.* (2008). (See Color Insert.)

Another key piece of evidence to address multiple triggering mechanisms is the correlative turbidite sequence from Hydrate Ridge Basin at ~44.5N on the Oregon lower slope (Figure 2B.18). This slope basin is completely isolated from land sources of sedimentation, being surrounded by ridges 500–1800 m above the basin floor that prevent downslope transport into the basin from any source other than the flanks of the ridge itself. The physiography and great depth of the basin eliminate input from storms, tsunami, hyperpycnal flow and other external sources, as evidenced by the absence of Mazama ash. The turbidite record from this key site is correlated to other margin sites on the basis of stratigraphic "fingerprints" and ^{14}C ages. The observed strong correlations to this site comprise an independent test of turbidite triggering at a site where all triggers save earthquakes and self failure are eliminated by the local physiography (Goldfinger *et al.*, 2008, 2009).

The synchroneity of a 10,000 year turbidite event record for 500 km along the northern half of the Cascadia Subduction Zone is best explained by paleoseismic triggering by great earthquakes. The southern Cascadia margin includes correlated additional events, many of which are also correlated to Hydrate Ridge Basin, though there are no channel confluences that can be used to test for synchroneity. The average Holocene great earthquake recurrence was found to be ~500 years, for the northern margin, similar to the onshore rate. Goldfinger *et al.* (2009) report that the recurrence times and averages are also supported by the thickness of hemipelagic sediment deposited between turbidite beds. Using stratigraphic correlation and ^{14}C ages, they report that the southern Cascadia margin can be divided into at least three seismic segments that include all of the northern ruptures, as well as ~22 thinner turbidites of restricted latitude range that are correlated between multiple sites. The southern Cascadia record correlates quite well with the onshore paleoseismic record based on ^{14}C data for the past ~4000 years where the records overlap (Goldfinger *et al.*, 2008, 2009). At least two northern California sites, Trinidad and Eel Canyons, probably also record numerous small sedimentologically or storm-triggered turbidites, particularly during the early Holocene when a close connection existed between these canyons and associated river systems under lowered sea-level conditions.

The combined stratigraphic correlations, hemipelagic analysis, and ^{14}C framework suggest that the Cascadia margin effectively has four rupture modes: 19 full or nearly full-length ruptures; two or three ruptures comprising the southern 50–70% of the margin, 9 or 10 events including the southern 50% of the margin and 9 events restricted to southern Oregon and northern California (Figure 2B.20). The shorter rupture extents and thinner turbidites of the southern margin correspond reasonably well with spatial extents interpreted from the onshore paleoseismic record (e.g. Kelsey *et al.*, 2005, Nelson *et al.*, 2008), supporting margin segmentation of southern Cascadia. The total of 41 events defines a Holocene recurrence interval for the southern Cascadia margin of ~240 years.

Goldfinger *et al.* (2009) report that turbidite physical properties along the Cascadia margin reveal a consistent record of turbidite mass per event along the northern margin for many events. Larger turbidites also have a moderately good correlation with the time interval following each event and are uncorrelated with the preceding time. They infer that larger turbidites likely represent larger earthquakes, and therefore the correlation with following time intervals suggests that Cascadia full margin ruptures may follow a time-predictable earthquake model. The long paleoseismic record also apparently indicates a repeating pattern of clustered earthquakes that includes three Holocene cycles of five earthquakes followed by an unusually long interval of 700-1000 years.

Goldfinger *et al.* (2009) suggest that the pattern of long time intervals and longer rupture for the northern and central margin may be a function of high sediment supply on the incoming plate smoothing asperities and potential rupture barriers. The smaller southern Cascadia segments correspond to reduced sediment supply and potentially greater interaction between lower plate and upper plate heterogeneities.

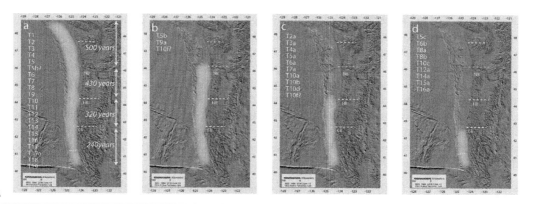

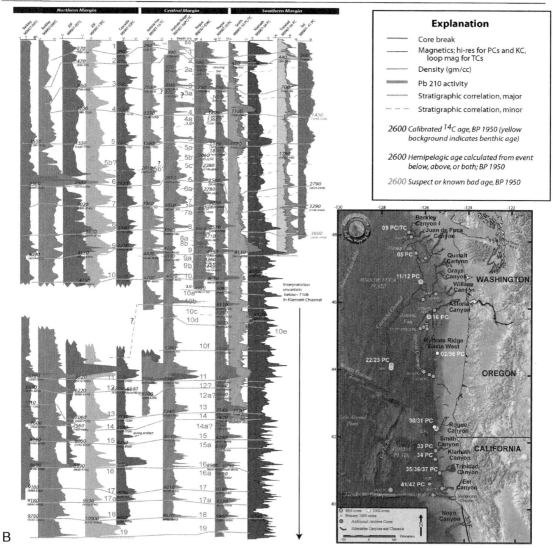

A

B

Goldfinger *et al.* (2009) make comparisons of the frequency of other potential triggering events, including bolide impacts, distal tsunami, storms, and crustal earthquakes and infer that these mechanisms are unlikely to be responsible for the observed record. During great earthquakes, on the other hand, the entire canyon system is affected, a canyon length that can exceed 100 km in Cascadia. The rupture zone also underlies the full length of all of the Cascadia canyons at a shallow depth, typical of other subduction margins, and creating a nearly ideal setting for triggering slope failures. During a great earthquake, the hypocentral distance to the locked fault is never more than between 2 and 10 km from the canyon walls, which likely fail in nearly continuous wall failure during the severe ground shaking of a large earthquake. Peak ground accelerations at such short distances to a great subduction earthquake can be estimated using the attenuation relationships of Atkinson and Boore (1997) and Youngs *et al.* (1997) to between 2 g (Youngs *et al.*, 1997) soil sites and 3.5 g (Atkinson and Boore, 1997) for rock sites. This represents a tremendous suspension and liquefaction force far greater than anything possible from surface ocean waves.

Finally, the recurrence intervals of Cascadia Basin offshore turbidites (Trinidad, Eel, and Mendocino channels excepted) closely match that of the onshore paleoseismic record (Goldfinger *et al.*, 2003a,b, 2007, 2008, 2009). The lack of turbidites overlying the most recent turbidite, dated to within a decade of the 1700 AD Cascadia earthquake indicates that no other triggering mechanism has produced an observable turbidite in the last 300 years, except in Trinidad and Eel Channels with narrow shelves and a local river source. The lack of turbidite triggering in Cascadia Basin by historic El Niño storm and flood events (1964, 1998–1999), and the 1964 Alaskan earthquake tsunami suggest that storm events and tsunami, whether or not sediment is transported to canyon heads, do not generally result in correlative abyssal plain turbidites. The mean peak AMS age of 230 (140–340) cal BP from four channel systems for the youngest turbidite event in Cascadia Channel T1 differs by only 15–20 years from (1) the coastal

---

Figure 2B.20: (A) Holocene rupture lengths of Cascadia great earthquakes from marine and onshore paleoseismology. Four panels showing rupture modes inferred from turbidite correlation, supported by onshore radiocarbon data. (a) Full or nearly full rupture, represented at most sites by 20 turbidites, though with greater uncertainty in southern extent (we include Pleistocene T19 in the figure, but not in the statistics). (b) Mid-Southern rupture, represented by two (1?) events. (c) Southern rupture from central Oregon southward represented by 9 (10?) events. (d) Southern Oregon/northern California events, represented by eight events. Southern rupture limits vary with each event, and many events older than ~5000 years are limited by lack of core older data. Dashed white line offshore indicates reduced confidence in correlations south of Trinidad Canyon. Recurrence intervals for each segment shown in left panel. Each segment includes all full margin events, plus those exclusive to that segment. Rupture terminations are approximately located at three forearc structural uplifts, Nehalem Bank (NB), Heceta Bank (HB), and Coquille Bank (CB). Paleoseismic segmentation shown is also compatible with latitudinal boundaries of Episodic Tremor and Slip (ETS) events proposed for the downdip subduction interface (Brudzinski *et al.*, 2007). These boundaries are shown by white-dashed lines. A northern segment proposed from ETS data at ~48N does not appear to have a paleoseismic equivalent. (B) Correlation plot of Holocene marine turbidite records and ^{14}C ages along the Cascadia margin from Barkley Channel to Eel Channel. All cores are vertically scaled to match Rogue core 31PC which is at true scale. Turbidite ages are shown using probability peaks and averaged where multiple ages at one site are available. Turbidites linked by stratigraphic correlations are shown by connecting lines. Full margin events correlated by using stratigraphy and ^{14}C are shown thicker, local southern Cascadia events are thinner and dashed. Modified after Goldfinger *et al.* (2009). (See Color Insert.)

paleoseismic ages that consistently center about 250 cal BP (AD 1700; Nelson *et al.*, 1995) and (2) the Japanese tsunami evidence showing a date of January 26, 1700 for the youngest great earthquake on the Cascadia Subduction Zone (Satake *et al.*, 1996, 2003). This further validates the synchronous turbidite event record and associated high-resolution AMS radiocarbon ages as a method to provide a long-term paleoseismic record. Temporal correspondence between the onshore and offshore paleoseismic records along the Cascadia margin is quite good, despite a variety of methods and lines of evidence onshore. Within the time rages that the two records overlap, there are few significant discrepancies (Goldfinger *et al.*, 2009). The ties between onshore and offshore paleoseismic data remain limited to radiocarbon timing for all sites except Effingham Inlet on Vancouver Island, which contains turbidites with possible stratigraphic correlatives offshore.

Goldfinger *et al.* (2009) conclude that turbidite systems of the Cascadia Basin are an ideal place to develop a turbidite paleoseismologic method and record because: (A) a single subduction zone fault underlies the Cascadia submarine canyon systems, (B) multiple tributary canyons and a variety of turbidite systems and sedimentary sources exist to use in tests of synchronous turbidite triggering; (C) the Cascadia trench is completely sediment filled, allowing channel systems to trend seaward across the abyssal plain rather than merging in the trench, (D) the continental shelf is wide, favoring disconnection of Holocene river systems from their largely Pleistocene canyons, and (E) excellent stratigraphic datums, including the Mazama ash (MA) and a distinguishable Holocene/Pleistocene boundary (H/P), are present for correlation of events and anchoring the temporal framework in turbidite systems within the northern two thirds of the basin.

2B.4.2.2 Marmara Sea

Correlating turbidites with the historical record is a good way to begin testing a turbidite record for seismic origin if a historical record is available. Considerable effort has been directed toward the Marmara Sea following the 1999 Izmit earthquake to map the submarine North Anatolian Fault (NAF). Concern has been heightened because the fault segment immediately to the west of the 1999 rupture may fail next, and the close proximity of this segment to Istanbul represents a significant hazard to the city. McHugh *et al.* (2006) describe work in the submarine pull-apart basins of the NAF within the Marmara Sea in which CHIRP seismic profiles, multibeam bathymetry and cores were used to test the connections between the turbidite basin fill and the NAF. Unlike channel settings used in Cascadia and the NSAF, the NAF work used cores sited in local depocenters along the fault. The sedimentation rates in the Marmara basins are quite high (0.5–1.0 cm/yr) making possible the resolution of events spaced closely in time. The turbidites in the deep Marmara basins, close to the fault segments are differentiated to some degree from thinner bedded turbidites on the shelf and slope, which McHugh *et al.* attribute to climatic events such as floods, though the earthquake and climatic records are most likely mixed. Thick Holocene deposits (5–20 m) were found on the basin flanks and presumably fail into the basin during earthquakes. In this setting, the method of emplacement and turbidite pathways are not completely clear, though the sedimentary packages thicken basinward, and the fining upward sequences require transport into the depocenters from upslope. Much like the Cascadia and NSAF turbidites, the cores revealed turbidites with multiple fining upward sequences capped by fining upward silt and hemipelagic foram-rich clay (the authors chose to use the term "homogenite" for these deposits though they closely fit the description of turbidites with long mud tails, see Shanmugam, 2006). These deposits were dated with a combination of ^{137}Cs and radiocarbon to search for matches for the historical series of nine earthquakes in 181 AD, 740 AD, 1063 AD, 1343 AD, 1509 AD, 1766 AD, 1894 AD, 1912 AD, and 1965 AD (Figure 2B.21). In the basin depocenters, the thinner bedded turbidites of the shelf and upper slope were not present. In cases such as Cascadia where the historical record is limited or nonexistent, the burden of demonstrating earthquake

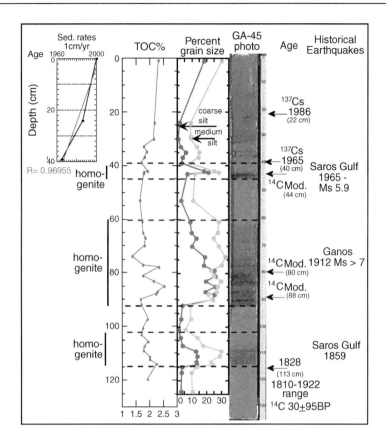

Figure 2B.21: Turbidite paleoseismologic example from the Marmara Sea. Grain size variability ranging from fine-sand to fine-silt and an increase in the total organic carbon (TOC %) of the sediments permit resolution of three homogenite (turbidite) deposits (40–44, 60–92, 100–112 cm). The homogenites are initiated by a sharp basal contact overlaid by multiple sand and silt-size laminae that fine upward to a thick wedge of homogenous fine-grained silt. The homogenites are separated by thin beds of clay (5–10 cm). Short-lived radioisotopes and radiocarbon chronology permit constructing an age model for correlation of the homogenites to the historical record of earthquakes: the large 1912 Ganos event $M_s > 7$ that lead to the deposition of a 30 cm thick homogenite and two smaller events that occurred in the Gulf of Saros. Sedimentation rates of 1 cm/yr were calculated for the upper 40 cm of the core that is apparently undisturbed. From McHugh *et al.* (2006). Reprinted by permission of Elsevier. (See Color Insert.)

origin is a relatively complex series of tests and regional correlations as previously described. In the case where a good historical record is available, the establishment of a good ground truth for seismic turbidites is much simpler. The cores McHugh *et al.* (2006) report in the Ganos, Tekirdağ, and Central Basins appear to record most, but not all, of the local earthquakes on the NAF. The core record also includes some of the events in the Sauros basin, some 60 km distant, bounded by a different fault segment. Cores collected from shallow water near the NAF west of the Hersek peninsula also contained disturbance deposits that were tentatively related to three historical earthquakes in 1509, 1766, and 1860.

Additional work described in Beck *et al.* (2007) describes work in progress on two giant piston cores collected in the Marmara Central Basin and reaches a somewhat more complex conclusion about the sedimentary section and the role of earthquake-generated turbidites from the late deglacial through the Holocene. The long (26 and 37 m) cores collected with this system appear to include the entire marine history of the central basin, as well as a previous lacustrine episode. Beck *et al.* (2007) suggest that while much of the basin fill is likely earthquake generated, they observe several features for which the explanation is perhaps not so straightforward. One of the turbidites is extremely large, and properly called a megaturbidite with an upper homogenite unit ∼8 m thick. This unit roughly corresponds with the transition from lacustrine to marine conditions. They also observe a significant decrease in turbidite frequency at about this same time, ∼16,000 yr BP. Goldfinger *et al.* (2007) also observe occasional turbidites among regional correlatives that are likely earthquake generated that are inexplicably outsized by comparison to others, and to turbidites generated by the maximum regional earthquake, the 1906 San Andreas event. Interestingly, Beck *et al.* (2007) propose that finely laminated units may be the result of bottom seiche currents linked to significant earthquakes. Finally, Beck *et al.* (2007) suggest that observations of planar fluid escape structures may represent liquefaction from ground shaking, as they appear to be spatially correlated to the turbidites in the cores, though they could potentially be related to coring with the large Calypso system.

2B.4.2.3 Northern San Andreas Fault

Using similar methods to their Cascadia work, Goldfinger *et al.* (2007) used 74 piston, gravity and jumbo Kasten cores from channel and canyon systems draining the northern California continental margin to investigate the record Holocene turbidites along the adjacent Northern San Andreas Fault. This fault is offshore or near the coast from San Francisco to the Mendocino Triple Junction and apparently close enough to offshore canyon heads to trigger turbidity currents. The late Holocene turbidite record off northern California was found to pass tests for synchronous triggering and was correlated using multiple proxies between numerous sites from Noyo Channel near the triple junction and the latitude of San Francisco. Preliminary comparisons of the temporal event record based in ^{14}C ages with existing and in progress work at onshore paleoseismic sites show good correlation, further circumstantial evidence that the offshore record is primarily earthquake generated. During the last ∼2800 years, 15 turbidites are recognized, including the one likely generated by the great 1906 earthquake. Their chronology establishes an average repeat time of ∼200 years, similar to the onshore value of ∼230 years. Along-strike correlation suggests that at least eight of the youngest 10 of these events likely ruptured the 320 km distance from the Mendocino Triple Junction to near San Francisco.

The long paleoseismic histories developed for the adjacent Cascadia and NSAF systems allowed Goldfinger *et al.* (2008) to relate the NSAF paleoseismic history to the similar dataset from the Cascadia (Figure 2B.22). They note that the recurrence interval for the NSAF is quite similar to that of the adjacent southern Cascadia margin, where the combined land and marine paleoseismic record includes a similar number of events during the same period. While the average recurrence interval for full margin Cascadia events is ∼500 years, the southern Cascadia margin has a repeat time of ∼220 years during the most recent 3000 year period, similar to that of the NSAF. Comparing these two records in several ways, using offshore data, land data, and the combined land–marine average, they find that 12 of the 15 NSAF events apparently occurred in close temporal proximity to Cascadia earthquakes. There appeared to be a slim temporal lag of ∼0–80 years, averaging 25–45 years, with Cascadia preceding the NSAF, (as compared to ∼80–400 years by which Cascadia events follow the NSAF).

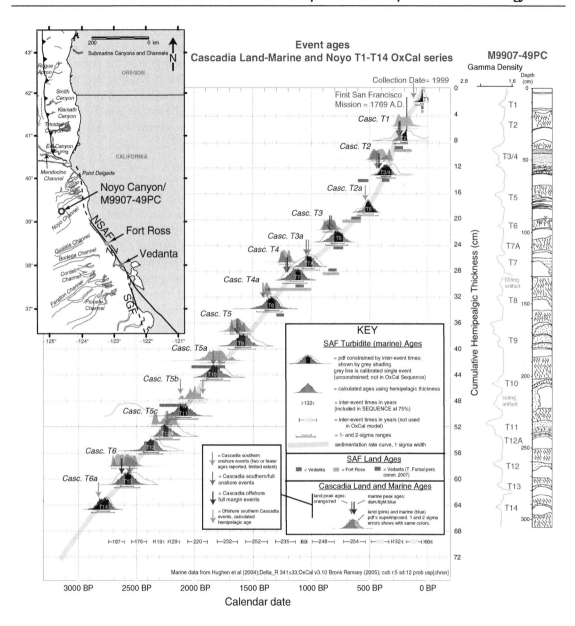

Figure 2B.22: OxCal age model for the youngest 15 events in the NSAF offshore system, and comparison to onshore NSAF ages. Cascadia OxCal PDFs are shown in blue, with lighter blue used where only Hemipelagic ages are available. Land ages from OxCal combines are shown in red. Cascadia mean event ages are also shown with blue arrows for well-dated turbidite events, Purple arrows for hemipelagic age estimates, and light red arrows for onshore paleoseismic events. See text for discussion and tables for data used and criteria, and discussion of temporal relationships. Inter-event times based on hemipelagic sediment thickness (represented by gray segments of NSAF PDFs) were used to constrain original ^{14}C calendar age distributions (gray traces) using the SEQUENCE option in OxCal. Inter-event times were estimated by converting hemipelagic sediment thickness

Based on the observed temporal association, Goldfinger *et al.* (2008) modeled the coseismic and cumulative postseismic deformation from great Cascadia megathrust events and compute-related stress changes along the NSAF to test the possibility that Cascadia earthquakes triggered the penultimate, and perhaps other NSAF events. They concluded that the Coulomb failure stress (CFS) resulting from viscous deformation related to a Cascadia earthquake over ~60 years does not contribute significantly to the total CFS on the NSAF. However, the coseismic deformation increases CFS on the NSAF by up to about nine bars following a typical Cascadia earthquake, most likely enough to trigger that fault to fail in north-to-south propagating ruptures (Figure 2B.22).

2B.4.2.4   Kurile Trench

Marine turbidites as paleoseismic recorders have been investigated along the Japanese islands, primarily in the trench systems and submarine canyons along the eastern coast. Along the eastern Hokkaido forearc along the Kuril Trench, Noda *et al.* (2004, 2008) have investigated the turbidite stratigraphy in Kushiro submarine canyon, offshore Kushiro. The earthquake history in this region during the last few centuries is well known from the historical literature. Two gravity cores (GH03-1033 and GH03-1034) were collected from the bottom of the canyon, and contain a number of turbidites. Sedimentological, geochemical, and micropaleontological data as well as high-resolution seismic data have been used to identify character, provenance, and recurrence intervals of the canyon turbidites. Three tephras from known volcanic events during AD1739, AD1694, AD1667, and AD1663 were used to develop the age model of the cores and recurrence intervals of the turbidites. In the upper canyon, thick mud in the channel suggests that the upper canyon has not been an active pathway during the Holocene. The middle canyon core (GH03-1034) had a source material inferred to be the upper canyon walls on the basis of sand composition and benthic foraminiferal analysis. A recurrence interval of 68 years for the late Holocene is similar to the historical rate of 79.7 years. Individual turbidites were also found to correspond well to the known historical earthquake record (Noda *et al.*, 2004). There are several seismic segments within the Kurile Trench. Noda *et al.* (2008) reported a recurrence interval less than 113 years for another segment along the Kurile Trench.

2B.4.2.5   Nankai Trough

Along the eastern Nankai Trough, Ikehara and Ashi (2005) have observed turbidite sands in cores collected from slope and forearc basins in the Tokai region. Two cores contain 15 and 13 turbidites from two slope basins along the Tokai Thrust, an out-of-sequence thrust in the Nankai accretionary prism. The turbidite ages, determined by ^{14}C dating of planktonic foraminifera suggest that during the last 3000 years, turbidite frequency along the eastern Nankai Trough is 100–150 years, similar to the known intervals of large interplate earthquakes from the historical and archeological records. Ikehara and Ashi

---

between each pair of events to time using the sedimentation rate. Events dated more than once were combined in OxCal prior to calibration if results were in agreement; if not in agreement, the younger radiocarbon age was used in the final model. Five ages are calculated from sedimentation rates where not enough forams were present for ^{14}C dating. The resulting probability distributions (filled black, grey for undated events) are mostly in good agreement with land ages from Fort Ross except for T3–4 and T7a (green lines; Kelson *et al.*, 2006) Vedanta (red lines; Zhang *et al.*, 2006) Bolinas Lagoon and Bodega Bay (Purple lines, Knudsen *et al.*, 2002), and Point Arena (light blue lines, Prentice *et al.*, 2000). Additional Vedanta event is also shown (T. Fumal personal communication 2007). See inset for geographic locations. Figure from Goldfinger *et al.* (2008). Reprinted by permission of the Bulletin of the Seismological Society of America. (See Color Insert.)

(2005) also report occurrence of turbidites in the northern Kumano Trough with a recurrence frequency of ~200–250 years, or about twice as long as the known interval for interplate earthquakes along the eastern Nankai Trough. Turbidites in a core from the Omine Ridge, an outer ridge near the Kumano Trough near an out-of-sequence thrust, suggest a ~1000 year recurrence. The inconsistent results from Nankai point out the importance of spatial coverage to test for earthquake origin among turbidite records, and to test different depositional settings for good deposition and preservation, and the variable results that may be found in different settings within the Nankai and other subduction zones (Ikehara and Ashi, 2005).

2B.4.2.6 Sumatra

The December 2004 Sumatra–India earthquake and tsunami represents an opportunity to catalogue marine effects from a very well-recorded series of events, many of which are unknown or poorly known at present. A suite of piston, gravity, Kasten, and multicores was collected along the length of the Sumatra margin, from the 2004 rupture zone in the north, to the southern tip of Sumatra Island (Patton *et al.*, 2007). Preliminary work suggests that like Cascadia, stratigraphic correlation may be possible in the Sumatra area, supported by numerous tephras with distinct compositional signatures.

Because there were no opportunities for a "confluence test" along the Sumatran margin, their strategy was to densely sample both trench and basin sites to test correlations between these two disparate and isolated site types to test for earthquake origin. Preliminary analysis suggests that the cores contain turbidites most likely generated by the 2004 and 2005 northern Sumatra great earthquakes, and many correllable predecessors. The recent events are represented by a large shallow multipulse event overlain by a smaller single pulse event at the seafloor, with no observed hemipelagic sediment between them. Patton *et al.* (2007) suggest that other turbidites correlate over distinct strike lengths, indicating that seismic segmentation may be resolvable with this dataset. Ongoing ^{14}C and ^{210}Pb dating with stratigraphic correlation will test the origins and connectivity of these and numerous other Holocene turbidites along this poorly known subduction margin.

2B.4.3 Offshore Tsunami Deposits

Evidence of onshore tsunami deposits exists in many forms, many of which clearly have utility as paleotsunami and paleoearthquake records, while others are poorly known and somewhat speculative. Potential modes of tsunami deposition in the marine environment include tsunami-related sedimentation in bays, lagoons, and lakes whose seaward boundaries were overwashed by tsunami waves (lacustrine deposits are discussed more fully in Section 2B.4.4). (We focus here on earthquake-generated deposits at the expense of significant work that has been done in impact generated tsunami, particularly hotly debated work near the KT boundary, e.g., Smit *et al.*, 1996; Keller *et al.*, 2003).

Lesser known than onshore deposits are offshore deposits in open bays, shelves, and forearc basins that result from tsunami passage and backwash. A very few well-documented cases have been reported, and these types of deposits remain to some extent in the realm of speculation. Shiki *et al.* (2008) and Shiki and Tachibana (2008) discuss the conceptual issues and problems surrounding tsunamiites as well as their importance to the geologic record, and relationship to climate and tectonic cycles and events in some detail. Shiki *et al.* (2008) review the features expected in submarine tsunamiites and the sedimentary structures related to various parts of the tsunami wave train. Shiki and Yamazaki *et al.* (1996) discuss a potential tsunamiite in the upper bathyal Miocene section onshore in central Japan that includes cobble imbrication from high-flow velocities and an association with probable shaking evidence, a key discriminator for tsunamiites which are otherwise difficult to distinguish (e.g., Dawson, 1999).

A somewhat similar deposit assigned to shallow water/shoreface depths is reported in the Miocene of Chile (Cantalamessa and Di Celma, 2005). Numerous deposits are reported in the ancient onshore geologic record of other potential tsunamiites, though evidence for tsunami origin is generally not strong. Fujiwara *et al.* (2000) describe tsunami deposits in a drowned valley on the Boso Peninsula of Japan. He proposes a tsunami depositional model based on depositional structures, high-resolution grain size analyses and the taphonomy of molluscan shells and suggests that details of the tsunami waveform may be deduced from the stack of depositional units. Weiss and Bahlberg (2006) performed an analysis of storm and tsunami wave energy and preservation potential along the Australian coast. They used a combination of hydrodynamic modeling and a simplified Hjulstrom–Sundborg diagram and concluded that the most powerful storm and tsunami waves both produce conditions near and at the sea bed that allow the transport of similar sediment grains sizes, up to meters in diameter. The implication is that offshore tsunami deposits in that locality would be reworked by storm waves. For their site-specific study at Brisbane, they concluded that preservation of tsunami deposits is most likely at depths greater than 65 m. Larger tsunami such as those produced by impacts, very large submarine landslides (Goldfinger *et al.*, 2000; McMurtry *et al.*, 2004) may well overcome this problem, as might a selection of localities not subjected to significant storm wave influence.

The few reports of tsunami backwash deposits in nearshore environments are suggestive of potential for preservation, perhaps more subject to special conditions, but possibly offering sites where other modes of preservation are not available. van den Bergh *et al.* (2003) describe a shallow water tsunami deposit from the 1883 Krakatau eruption using textural, compositional and ^{210}Pb geochronological data. The deposit is associated with the 1883 eruption tephra, and thus its origin is relatively clear. The deposit consists of a sandy layer with abundant reworked shell fragments and material apparently locally derived eroded from the seabed. They also note that the deposit included land-derived components when near the coast. Nearshore deposits (<50 m water depth) may also preserve critical information as to wave direction and speed, though these have not been reported to our knowledge. Such evidence might include the preservation of sediment aprons, sand bars, large sediment waves and debris layers deposited during backwash. Also, large objects (boulders, coral blocks, human artifacts) may be dragged or deposited on the seafloor, producing a debris field and other scattered evidence on the seafloor.

An unusual example of a potential hybrid deposit, part tsunami deposit, part turbidite has been described from the Mediterranean seabed. A widespread unit known as a homogenite has been widely described in the Ionian and Sirte abyssal plains and other scattered locations in the central Mediterranean. This deposit, up to several meters thick, is mostly homogenous clay to silt with little or no grading (Cita *et al.*, 1984; Kopf *et al.*, 1998). Cita and Aloisi (2000) describe a pelagic (Type A) and a shallow water (Type B) homogenite. The pelagic deposit, without indications of a shallow water source, has a coarse fraction (sand size) consisting only of planktonic foraminifers (Sironi and Rimoldi, 2005). This deposit has been attributed to a catastrophic eruption of Santorini volcano, which Sironi and Rimoldi (2005) suggest, generated a tsunami during the collapse of the Santorini Caldera. The tsunami in turn is thought to have destabilized mostly hemipelagic marine sediments on shallow ridges such as the Mediterranean and Calabrian Ridges, generating turbidity currents that deposited the homogenite on the abyssal plain. Sironi and Rimoldi (2005) further suggest that this homogenite was then overlain with a megaturbidite (up to 24 m thick) originating on the African continent as a result of the tsunami arrival. Pareschi *et al.* (2006) argue for the same origin for the homogenite, but argue for an origin from Mt. Etna rather than Santorini, and provide tsunami modeling to support the distribution of liquefaction potential to support their model.

While definitive assignment of deep ocean turbidites to a tsunami origin is rare, these papers suggest that such deposits are likely to exist in the geologic record. Along the Cascadia margin, Goldfinger *et al.*

(2009) calculate the potential for tsunami triggering of turbidites on the upper continental slope, and conclude that the potential exists, though correlation with onshore earthquakes suggests that the Cascadia Holocene turbidites are of local earthquake origin.

2B.4.4 Lacustrine Environments

2B.4.4.1 Lacustrine Sediment Pulses Caused by Earthquake-Generated Landslides

Adams (1980) measured sediment loads of rivers in New Zealand immediately following earthquakes and observed an order-of-magnitude increase in sediment load for a period of several months. He correlated increases in load in different areas with the density of earthquake-triggered landslides in those areas and concluded that seismically induced landslides generate large increases in fluvial sediment load, which, in turn, cause increases in sedimentation rates in lakes and oceans. These observations have been corroborated with published observations from earthquakes elsewhere (Adams, 1981; Dadson *et al.*, 2004).

On the premise of these observations, Doig (1986) analyzed organic-free silt layers 0.3–2.0 cm thick in otherwise organic-rich lake sediment in eastern Canada. Using sedimentation rates and radiometric methods, three of these layers were correlated with known earthquakes of AD 1663, 1791, and 1860 + 1870 (two events combined). Two older silt layers were likewise dated and attributed to paleoearthquakes in AD 1060 and 600. Doig (1986) stated that cores from deep lakes likely will yield the best cores for this type of analysis because of lack of bioturbation. He also warned that dating young (a few hundred years) silt layers characterized by lack of organic material can be difficult; he suggested that ^{210}Pb and ^{137}Cs are the ideal radiometric methods for this type of analysis (see details of dating techniques in Chapters 1 and 2A).

2B.4.4.2 Landslide, Turbidite, and Tsunami Deposits in Lakes

A number of lake deposits in various settings have been interpreted as related to seismic shaking, landslides into lakes, submarine landslides, and tsunami overwash into lakes. Lake settings may not offer the constant sediment supply of the offshore environment, but have the advantage of seasonal changes that may be reflected in annual sediment patterns, offering precise chronologies.

Alpine Lakes Beck *et al.* (1996) report evidence of liquefaction and differential compaction with rapid fluid escape (water and/or gas) in the form of ball-and-pillow structures and microfracturing of sediments in Lake Annecy in northwestern Alps, implying a brittle-like behavior of soft, water-saturated, sediment) that they interpret as earthquake induced. Sediment gravity flows in the same lake were also interpreted as likely of seismic origin based on (1) the lack of corresponding sub-aerial landslides and (2) the grain sizes being larger than those found in fluvial input aprons, despite the selection of their drilling sites to avoid fluvial input.

In another alpine lake, Schnellmann *et al.* (2002) interpret five paleoseismic events in the past 15 ka from the evidence of a series of slump deposits in the subsurface of Lake Lucerne. This study identified a stratigraphic "fingerprint" for the sediment deposit associated with the well-described AD 1601 earthquake (Figure 2B.23). The earthquake triggered numerous synchronous slumps and megaturbidites within different sub-basins of the lake, producing a characteristic seismic-stratigraphic linkage between sites imaged with seismic profiling. Four prehistoric events were dated with ^{14}C measurements and tephrochronology on core samples, and used to establish the recurrence period of similar earthquakes, as well as possible tsunami events through the Holocene.

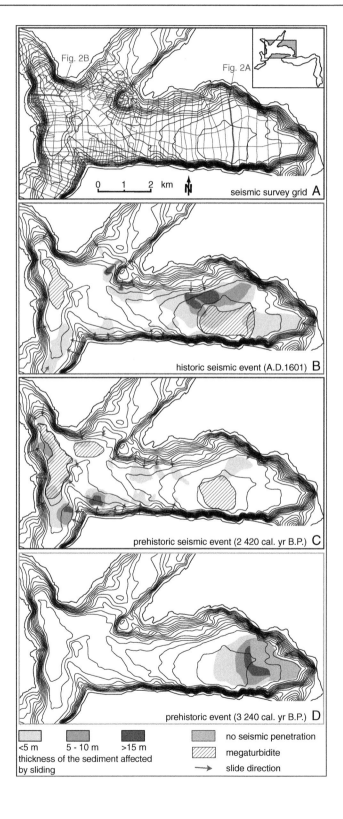

thickness of the sediment affected by sliding: <5 m, 5 - 10 m, >15 m

no seismic penetration

megaturbidite

→ slide direction

A seismic survey grid

B historic seismic event (A.D.1601)

C prehistoric seismic event (2 420 cal. yr B.P.)

D prehistoric event (3 240 cal. yr B.P.)

Other lake records have been described by Carillo *et al.* (2008) in Venezuela, in Lago Icalma, Chile (Bertrand *et al.*, 2008), in Lake Le Bourget in the NW Alps, (Chapron *et al.*, 1999) in Lake Bramant, western French Alps (Guyard *et al.*, 2007), and other localities.

Niemi and Hall (1994) reported evidence for an association between the 1927 Dead Sea earthquake, a submarine slide in the lake bottom, and a ~1 m tsunami apparently generated during this event. Seismic reflection profiling imaged a shallow slide of broad areal extent involving the particularly stable lake sediments. The lack of other potential triggers for this large slide in part was used to infer earthquake origin, along with the eye witness reports of the tsunami originating in the center of the lake, as opposed to a seiche. Eight other similar slides in the same location suggested a repeat time of significant earthquakes on the Dead Sea Fault of several thousand years over the last 20–30 ka. Subsequently, Marco *et al.* (1996) investigated a shallow water record from the Dead Sea Graben, formerly lacustrine, and identified a series of seismic disturbances consisting of pulverized laminae, some in association with fault scarps, indicating earthquakes of $M_w > 5.5$. The recurrence interval from this ~40,000 year lacustrine record is ~1600 years, similar to the Dead Sea record. Migowski *et al.* (2004) report that further investigation of the Dead Sea sedimentary record using varve counting shows that all recent and historical strong local earthquakes could be identified, including the major earthquakes of AD 1927, 1837, 1212, 1033, 749, and 31 BC. A total of 53 seismites were recognized in this study, which also identified long-term patterns of quiescence and greater activity.

Coastal Lakes Tsunami overwash deposits are well known now from several settings including Sweden, Japan, Kamchatka, and Cascadia. They offer long records and continuous sedimentation, much like deeper marine environments, but are more accessible. Deposits in these settings may be mixed with other events, and thus may present some ambiguities, however the same can be said of any off-fault paleoseismology or tsunami deposit.

Bradley Lake, Cascadia Margin Bradley Lake, located close to the coastal dunes of Oregon along the southern Cascadia margin, records local tsunamis and seismic shaking on the Cascadia megathrust (Kelsey *et al.*, 2005). The lake stratigraphy includes 13 landward thinning sand sheets interpreted to be tsunami overwash into the lake on the basis of microfossil analysis. The marine incursions included marked changes in salinity of the freshwater lake. Four additional sediment layers may represent localized turbidity currents from earthquake shaking. The marine incursions had to travel overland to enter the lake, and thus represent a sensitivity test of the magnitude of these tsunamis (Figure 2B.24), which had to be at least 5–8 m above sea level with a duration of at least 10 min. The analysis of the disturbance events in Bradley Lake is analogous to that used for offshore turbidites in that the investigators developed age models for their lake cores, examined sand layers for evidence of basal erosion based on missing section, and uniquely, were able to use the brackish episodes which resulted in varves to establish the interseismic sedimentation rates. The interseismic depositional units included massive muds, but could be used assuming

Figure 2B.23: Slide deposits in Lake Lucerne, Switzerland mapped with high-resolution reflection profiling. Slide deposits related to specific horizons. (A) Grid of 3.5 kHz seismic profiles acquired for this study. (B–D) Distribution and thickness of slide bodies corresponding to three event horizons identified in the reflection profiles. Hachured areas mark extent of megaturbidites directly overlying slide bodies. Bathymetric contour interval is 10 m. From Schnellmann *et al.* (2002) their Figure 3. Reprinted with permission of the Geological Society of America. (See Color Insert.)

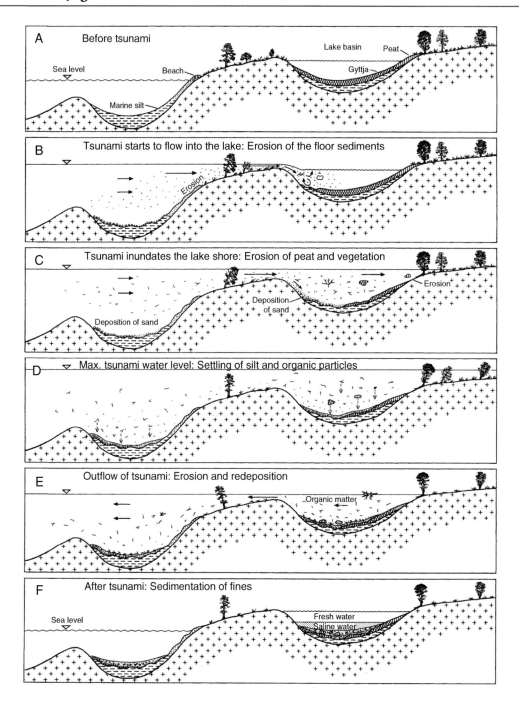

Figure 2B.24: Erosion and depositional model for a tsunami inundating a shallow basin on the sea floor and a lake: (A) before tsunami inundation, normal sedimentation. (B) the tsunami inundates and erodes the shore and flows into the lake where it rips up clasts from the lake floor. Sand is deposited in the marine basin (facies 4 and 5). (C) the tsunami inundates and erodes peat and

the same rates found from the varve-based section, establishing interseismic intervals independent of the radiocarbon data. From these varve-based rates, Kelsey *et al.* (2005) established a sequence of event ages which closely matched the ages based on ^{14}C. Goldfinger *et al.* (2009) conclude that the offshore turbidite record closely matches the Bradley Lake disturbances record for the past ~4600 years, and that both sites have recorded both long and segmented ruptures.

Similarly, stratigraphic, geochemical, and microfossil data were collected from sediments in Laguna Mitla, the Pacific coast of Guerrero, Mexico. The rapid relative sea-level rise, marine inundation, and possible tsunami deposit have been interpreted as evidence of a megathrust earthquake and associated tsunami deposits (Ramirez-Herrera *et al.*, 2007).

2B.4.4.3 The Storegga Tsunami

In an example superficially similar to Bradley Lake, tsunami deposits have been found in coastal lakes adjacent to the Storegga submarine slide off western Norway. Bondevik *et al.* (1997) report distinctive deposits found in small coastal lakes along the Norwegian coast. These lakes, situated from 0 to 11 m above the 7000 year BP shoreline, were sampled with a piston corer and contain a deposit very distinct from the lake sediments in the cores. The base of the deposit is marked by an erosional unconformity which can be correlated around the lake basins. The distinctive normally graded or poorly sorted sand to fine gravel overlies the erosive surface, which shows greater erosion toward the seaward side of the basins. Locally, the sand contains shell fragments and foraminifera. The tsunami deposit is a fining upward sequence with occasional massive sand at the base, which includes the marine fossils. The sand thins and decreases in grain size landward direction. Above the fining upward sand, the sequence includes a coarse organic layer with rip-up clasts. The tsunami unit generally fines and thins upward. One of the most convincing pieces of evidence from this unusual tsunami setting is that the basins show stratigraphic thinning by their elevation from the coast (Figure 2B.24). The basins closest to the paleoshoreline ~7000 BP have several sand pulses separated by organic debris, while successively higher basins (6–11 m above the 7000 year shoreline) have only one sandy unit.

Several basins were investigated that were below the 7000 BP paleoshoreline, but that are now exposed due to postglacial crustal rebound. Bondevik *et al.* (1997) interpret the presence of the tsunami deposit in these basins as well. The character of the deposit in the sub-sea is graded sand beds with occasional organic rich facies between the sand beds. The Norwegian coastal tsunami deposits are linked temporally to the coeval Storegga slide, making this a classic example.

2B.4.4.4 Nankai and Suruga Troughs, Japan

Tsunami deposits have been found along the Japanese coasts in the Nankai and Suruga Trough areas of eastern Honshu. On land, deposits were discovered first at archeological sites, but many of the best

vegetation at the lake shore. Sand brought in by the tsunami is deposited in the lake basin (facies 4 and 5). (D) suspended material such as rip-up clasts, twigs, gyttja, sand and silt settles producing normal graded organic beds (facies 6 followed by facies 7). (E) withdrawal of the wave, erosion and redeposition of the tsunami deposits, organic material carried out of the lake. (F) after the tsunami, deposition of suspended fines in addition to organic matter from reworking of tsunami sediments deposited above the lake. Stages (B)–(E) represent the inundation and withdrawal of one tsunami wave. Basins closer to sea level experienced several waves as is shown by the alternation of sand and organic beds. From Bondevik *et al.* (1997). Reprinted by permission of Wiley Interscience.

deposits are found in coastal lakes (Okamura *et al.*, 2000; Nanayama *et al.*, 2002; Tsuji *et al.*, 2002). Similarly, historical lake tsunami deposits are also found along the Sagami Trough (Fujiwara *et al.*, 2000). Typically these deposits are found in limited sites in the coastal plains and lakes, and do not define landward thinning sheets (Komatsubara and Fujiwara, 2006) as is commonly reported. These deposits resemble the Bradley Lake deposits in that they are typically sandy, fining upward deposits, intercalated with muddy lake sediments, they contain marine fossils including mollusks, nanoplankton, foraminifera, and ostracods (Komatsubara and Fujiwara, 2006 and references therein). These deposits range in thickness from a few centimeter to over 6 m and are sometimes covered with plant fragments. These deposits mostly do not have detailed grain size and stratigraphic analyses, though may correlate to historically recorded earthquakes and in other cases, are dated to within reasonable temporal correlation with historical earthquakes (Komatsubara and Fujiwara, 2006).

2B.4.5   Submarine Landslides Triggered by Earthquakes

Several studies have confirmed the triggering of large submarine landslides and turbidity currents by earthquakes, and numerous others in the geologic record may have been as well. Perissoratis *et al.* (1984) documented a slump covering 15–20 km^2 in the eastern Korintiakos Gulf along the coast of Greece triggered by a series of earthquakes ($M_w = 6.4$–6.7) from 24 February to 4 March 1981. Field *et al.* (1982) documented a sediment flow/lateral spread on a 0.25° slope on the submarine Klamath River delta off the coast of northern California; the feature extends along 20 km of the delta front and is about 1 km long (from scarp to toe). The very low slope and the presence of liquefaction features on the surface both suggest seismic triggering, and repeated bottom surveys before and after the $M_w = 6.5$–7.2 offshore earthquake of 8 November 1980 conclusively linked the landslide to the earthquake. Lee and Edwards (1986) analyzed the stability of four submarine landslides off the coasts of California and Alaska and concluded that three of them required seismic shaking to have triggered failure.

These studies provide the basis for interpreting older submarine landslide deposits in terms of seismic triggering. Examples of other submarine slides and mass wasting deposits include Viscaino *et al.* (2006), who report on turbidites and a submarine slide in the Marquês de Pombal area of the Iberian margin. They report a mixed record of slides and turbidites in which a large landslide deposit observed in acoustic backscatter imagery is not related to the 1755 Lisbon earthquake, but is much older, with ages between ca. 3270 and 1940 yr BP. They found that the deposit more likely related to the 1755 Lisbon earthquake is a thin turbidite.

The majority of historical submarine landslides have been linked to earthquakes, including the well-known Grand Banks earthquake of 1929 which spawned a landslide and associated turbidity current that broke a series of submarine cables downslope, thus recording the direction and speed of travel (Heezen and Ewing, 1952). A near repeat of this event occurred on 26 December 2006, when the magnitude 7.1 Hengchun earthquake was followed by the breakage of eleven submarine cables in the Strait of Luzon, between Taiwan and the Philippines (Hsu *et al.*, 2008).

The causes of submarine landslides may be complex, beginning with a weak depositional sequence that may be climate related, further weakening from a secondary effect such as gas hydrate destabilization, and then final triggering by an earthquake (Masson *et al.*, 2006). Linkages to earthquake triggers for prehistorical events can be problematic as these factors may be difficult to disentangle. The Storegga slide is thought to be just such an event (Bryna *et al.*, 2005). Submarine landslides may occur coincidentally with earthquakes and may also be responsible for enhancing tsunami generation locally, while at the same time making seismologic interpretation difficult, as in the 1998 Papua New Guinea earthquake/tsunami

(Satake and Tanioka, 2003). While equivocal, this and perhaps other similar events may represent slumps where rapid rotation and seafloor offset generate significant seafloor motion and tsunami from a modest earthquake source (Matsumoto and Tappin, 2003).

Numerous other submarine slides have likely earthquake origins, but have not been documented to the level required to establish this linkage. Among the largest known are the super scale slides of the Oregon margin. Using SeaBeam bathymetry and multichannel seismic reflection records on the southern Oregon continental margin, Goldfinger *et al.* (2000) identified three large submarine landslides involving ~8000 km^2, and a volume of ~12,000–16,000 km^3 of the accretionary wedge. The three arcuate slump escarpments are nearly coincident with the continental shelf edge on their landward margins, spanning the full width of the accretionary wedge. Debris from the slides is buried or partially buried beneath the abyssal plain. The ages of the three major slides decrease from south to north, indicated by the progressive northward shallowing of buried debris packages, increasing sharpness of morphologic expression, and southward increase in postslide reformation of the accretionary wedge. The ages of the events, derived from calculated sedimentation rates in overlying Pleistocene sediments, are approximately 110, 450, and 1210 ka. This series of slides traveled 25–70 km onto the abyssal plain in at least three probably catastrophic events, which may have been triggered by subduction earthquakes. The slides would have generated large tsunami in the Pacific basin, possibly much larger than that generated by an earthquake alone. The authors also identified a potential future slide locality with incipient breakaway features off southern Oregon that may be released in a subduction earthquake.

2B.4.6 Coeval Fault Motion and Fluid Venting Evidence

Offshore faults leave evidence of movement and timing in the form of scarps, colluvial wedges, liquefaction, and mass wasting deposits similar to onshore faults, but also in the form of fluid venting. Seafloor evidence of fluid venting is commonly expressed as chemical fluxes (e.g., Gamo *et al.*, 2007) and occasionally as carbonates originating in the commonly methane rich fluids. Carbonates are typically observed as ribbons in fractures, and as "chimneys" composed of annular venting laminae with a vertical orientation (e.g., Goldfinger *et al.*, 1996; Ogawa *et al.*, 1996). ROV and submersible studies, as well as experimental results from flow meters and borehole CORKs (Circulation Obviation Retrofit Kit) in ODP drill holes, suggest such fluid flow is invigorated in the rupture region following a significant earthquake, thus making it potential direct evidence of paleoearthquakes. For example, recent observations at the Nankai Trough Sites 808 and 1173 (Figure 2B.6) illustrate the capture of strain via formation pressure in hydrologically isolated sediment sections. Two deformational events have been captured, the first contemporaneous with a very low-frequency earthquake swarm in the prism, and the second contemporaneous with an earthquake swarm in the subducting plate. The transients reflect both coseismic strain and postseismic relaxation. These are superimposed on a rise in pressure due to local interseismic strain accumulation. Similar observations have been made at other fault sites in Cascadia and Costa Rica using both flow meters and CORKs.

Okamura *et al.* (2005) report the rupture extents of paleoearthquakes in the Sea of Japan can be defined by the extent of scarp activity and fluid expulsion along the fault trace using submersible observations. Quaternary folds and fault zones here comprise several arc-parallel zones, known as the Okushiri Ridge, Sado Ridge, Awashima to Oga ridge, and several others. Historical earthquakes occurred in 1940, 1983, and 1992 along the Okushiri Ridge. Shinkai 6500 submersible dives in the source area of the recent earthquakes widely observed fresh fissures, slope failures, venting and debris on the slope above active segments in the areas of recent earthquakes, and fissures, where similar features were covered with muddy sediments in areas between these ruptures. Evidence of submarine fluid and gas venting directly related to earthquakes has been observed with the Izmit 1999 event (Kuscu *et al.*, 2005) and other localities. Kitamura *et al.* (2002) report a unique sub-aqueous sand blow deposit associated with the 1995 Kobe earthquake.

Acknowledgments

My thanks to Takeshi Nakajima for his review of this chapter, and to Gordon Seitz, John Hughes Clarke, Hannes Grobe, Earl Davis, Celia McHugh, Takeshi Nakajima, Tom Rockwell, Mark Legg, Waldo Wakefield, and Klein-L3 for generous assistance with and permission to use data and figures from their work.

CHAPTER 3

Paleoseismology in Extensional Tectonic Environments

James P. McCalpin

GEO-HAZ Consulting, Inc., Crestone, Colorado 81131, USA

3.1   Introduction

Large extensional earthquakes in the upper crust produce surface deformation recorded by displacements on normal faults, by growth of folds above these faults, and (harder to detect except at coasts) by fault-generated *elastic* and *viscoelastic* crustal stress accumulation and release that produce changes in land elevation over wavelengths of many kilometers. We begin the chapter with a brief regional overview of extensional environments and the structures that characterize them that may be sources for large earthquakes recorded by paleoseismic evidence (Section 3.1). Next we outline the cycle of earthquake deformation on faults and describe some recent historical earthquakes that are important analogs for the study of prehistoric earthquakes.

We then focus on geomorphic (Section 3.2) and stratigraphic (Section 3.3) evidence of coseismic extensional deformation in the shallow upper crust on local and site scales. We focus primarily on dip-slip normal faults that lack major components of lateral slip, although local extension and oblique slip can also occur in strike-slip and compressional environments. The stratigraphic and geomorphic features formed by extensional faulting are commonly easier to see than those formed by compressional or strike-slip faulting, because brittle faulting of surface materials creates sharp-edged scarps that crosscut all preearthquake landforms, and sharp, steep fault planes offset surface deposits.

Next, we address how to date individual paleoearthquakes (Section 3.4), given the unique morphology and depositional environments typical of on-land normal faults. Retrodeformation of normal faults in vertical section (Section 3.5) is the normal technique for interpreting the paleoseismic history. Finally, in Section 3.6 we describe other extensional deformation that can affect the ground surface and deform near-surface deposits, but is not of seismic origin. It is obviously critical for a paleoseismologist to be able to distinguish whether extensional deformation exposed in vertical sections was created by paleoearthquakes, or by some other nontectonic mechanism.

The paleoseismic studies that form the foundation for this chapter in the first edition were performed by the author and colleagues, mainly in the semiarid environment of the western United States. However, modern paleoseismic studies (i.e., studies that utilize both stratigraphic and geomorphic evidence) have now been performed on many normal faults outside of North America, notably in Greece and Italy

---

International Geophysics, Volume 95

ISSN 0074-6142, DOI: 10.1016/S0074-6142(09)95003-3

(papers too numerous to cite individually), China (e.g., Ding, 1982; Zhang *et al.*, 1982; Deng *et al.*, 1984), Israel (Gerson *et al.*, 1993), New Zealand (Beanland *et al.*, 1990), Peru (Cabrera *et al.*, 1987; Schwartz, 1988b), and Russia (McCalpin and Khromovskikh, 1995), to name just a few countries. The climate in New Zealand and Russia is considerably wetter, and in Israel considerably drier, than in the western United States, but nonetheless fault scarps and stratigraphic indicators of paleoearthquakes are very similar in all regions. Although the rates of erosion and weathering of paleoseismic landforms and deposits may vary with climate, their overall genesis and geometry as described in this chapter are felt to be largely independent of climate.

3.1.1 Styles, Scales, and Environments of Extensional Deformation

Crustal extension is typically accommodated by *normal faults*, either singly or in sets of parallel synthetic or antithetic faults. The primary normal fault is a crustal-penetrating fault that may have many kilometers of cumulative throw, and often separates a linear mountain range (upfaulted block, or horst) from an adjacent basin (downfaulted block, or graben) (Figure 3.1). Fault dips in the upper crust are consistently 50°–70°, the result of a horizontal least principal stress combined with Mohr–Coulomb failure of rock with internal friction angles of 20°–40° (e.g., Davis, 1984, pp. 310–311).

At both crustal and smaller scales, the primary normal fault is accompanied by secondary normal faults, either synthetic (dip in the same direction as the primary faults) or antithetic (dip opposite to the primary fault). Almost all secondary faults form on the downthrown side of the primary fault (the hanging wall). Distributed normal faulting creates uplifted blocks (*horsts*), downfaulted blocks *(grabens)*, and rotated blocks or *tilted fault blocks*. Progressive rotation of the crustal blocks bounded by parallel normal faults (the so-called domino style, or bookshelf style of faulting) may tilt the blocks so much that older faults become too gentle to accommodate continued horizontal extension (Proffett, 1977), and new steeper

Figure 3.1: View of the Jackson Hole, Wyoming, graben, the Teton Range horst (left distance), and the Blacktail Buttes intragraben horst (center). The Teton normal fault lies at the base of the range. View is NNW. Town of Jackson, Wyoming at bottom. Photo by J. P. McCalpin.

normal faults develop (Sibson, 1985). [A similar evolution for normal faults that reactivate thrust faults (negative tectonic inversion) is described by West (1992, 1993)]. Tilting progressively increases the dips of footwall strata, leading to *angular unconformities* in basins ranging from crustal scale (tens of kilometers) to microscale (a few meters). Folding does occur near normal faults, but is less common and folds are generally broader and more open than in compressional orogens and fold and thrust belts. The most common type is a drape fold, a gentle monocline that "drapes" over a buried (or blind) normal fault or vertical fault. The second most common fold is a "rollover fold," in which hanging-wall strata are bent downward from their depositional attitude, to be perpendicular to the normal fault plane (Figure 3.2).

Many normal faults are known to decrease in dip with increasing depth (*listric geometry*) based on geophysical and drilling data, but debate is heated on whether listric faults, or the underlying detachment faults that they often sole into, are seismogenic (compare Arabasz *et al.*, 1992, to West, 1993). However, two cases of seismogenic low-angle normal faults are now known. In New Guinea, low-angle normal faults (25°–35°) are clearly seismogenic today over a range of earthquake magnitudes (Abers *et al.*, 1997; Westaway, 2005). In the Basin and Range province, USA, Abbott *et al.* (2001) traced the plane of the 1954 Dixie Valley earthquake fault ($M_s = 6.8$) to a depth of 2.7 km with seismic reflection, and documented that part of the plane dips only 25°–30° (Figure 3.3). Sorel (2000) inferred a still-active detachment fault beneath many higher-angle normal faults in the Gulf of Corinth. Finally, Quaternary movement ($<0.7\ M_a$) was documented on detachment faults in Death Valley (USA) that dip 29°–36° (Hayman *et al.*, 2003), and even younger Quaternary faults scarps may sole into the Canada David detachment (Mexico) which dips about 30° (Axen *et al.*, 1999). Given these observations, it is still somewhat of a mystery why no historic $M_w > 7$ earthquake has displayed a focal mechanism of low-angle normal faulting.

Two types of normal faults also commonly occur in compressional orogens in tightly folded rocks. The first type is bending moment faults, which develop if extension is severe on the axes of folds, typically the crests of anticlines. These faults create structures such as graben on anticline crests. A second type,

Figure 3.2: Normal fault in the Bear Lake graben, Utah, exposed in a roadcut of US Highway 89 just east of Pickleville, Utah. Quaternary (?) cobble gravels (left) are faulted against the red Wasatch Formation (Eocene; right). Fault zone is composed of two major fault planes with beds of rotated hanging-wall strata in between. Note how strata in hanging wall are bent downward (rollover fold) as they approach the fault. Photo by J. P. McCalpin (1991).

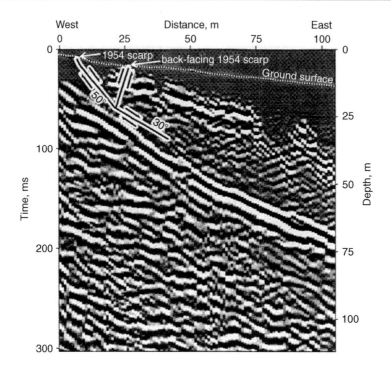

Figure 3.3: High-resolution seismic reflection profile across the 1954 surface rupture of the Dixie Valley earthquake, Nevada. At the ground surface the rupture formed a 20-m-wide graben, created by refraction of the seismogenic fault from a dip of 30° to 50° at a depth of 20 m. From Abbott *et al.* (2001).

less common, is a flexural slip fault with normal sense of slip. Such a flexural slip faults will only be observed on the limbs of overturned folds (on upright folds, flexural-slip faults always have reverse sense of slip).

3.1.1.1   Environments of Extensional Deformation

Extensional tectonic forces affect many varied parts of the earth's surface and result from either: (1) crustal-scale extension related to plate motions, (2) mid-to-upper-crustal extension where extensional secondary faults form related to the major reverse or strike-slip faults, or (3) upper-crustal extension due to gravitational collapse of very elevated terrain (this may include volcanic edifices as well as nonvolcanic mountains). Because 75% of the earth is covered by water, the largest area of extension is associated with oceanic spreading centers, most of which are in water too deep to permit any paleoseismic studies. (However, on-land parts of spreading ridges such as Iceland and the Afar triangle of eastern Africa are covered in Chapter 4.) Likewise, oceanic hot spots and their overlying volcanic islands display normal faulting and are also covered in Chapter 4. In back-arc basins, normal faulting may or may not be associated with volcanism; the latter areas are described in Chapter 4, including the Taupo Volcanic Zone of New Zealand, where considerable paleoseismic work has been performed. Shallow listric normal faulting has also been observed on subduction zone continental margins (McNeill *et al.*, 1997).

This chapter emphasizes on-land tectonic normal faulting not caused by volcanic processes, although there may be volcanic sources in the same area. The largest such regions are broad areas of continental extension (USA Basin and Range province, and northeastern China), followed by narrower intracontinental rifts (East African, Rhine, Rio Grande, Baikal). Another large region of continental extension is high plateaus behind a collisional orogen, such as the Tibetan Plateau north of the Himalayas, or the South American Altiplano (McNulty and Farber, 2002). In these areas crustal extension may involve a component of gravitational collapse toward the topographic margins of the plateau. The smallest areas of tectonic extension are located where secondary, upper-crustal normal faults form related to major strike-slip faults, reverse faults, or folds. This includes strike-slip faults in areas of releasing bends and stopovers (transtension), or where oblique slip is partitioned into parallel but separate strike-slip and normal faults (King *et al.*, 2005). Very localized normal faulting may occur on the crests of anticlines.

3.1.1.2 Segmentation of Normal Faults

Normal faults traces tend to be relatively short (10–50 km), slightly irregular faults that are separated from adjacent normal faults by gaps or stopovers (Figure 3.4). Naming separate normal faults, or subdividing normal faults into "segments" is a rather subjective exercise, since there is no accepted criterion for defining any particular normal fault as a separate named fault, as opposed to a segment of a named fault. Thus, some named normal faults are smaller than segments of other named normal faults, and vice versa. The longest named normal faults, such as the Wasatch fault (Utah) and the Sierra Nevada frontal fault (California–Nevada) are up to 400 km long, but are composed of named separate faults or segments that range from 30 to 65 km long (average about 40 km).

We defer the main discussion of fault segmentation of all fault types to Chapter 9 (See Book's companion web site) (Section 9.4). However, a few observations are relevant here. In large historic earthquakes, normal fault surface ruptures have ranged from as short as 4 km ($M_w < 6$) to as long as 102 km (M_w 7.5; 1887 Sonora, Mexico; Suter, 2006). All of the $M_w > 7$ earthquakes have ruptured multiple faults or fault segments, as noted by dePolo *et al.* (1991). This observation is consistent with the average rupture length of an (historic) M_w 7 normal fault earthquake, which according to Wells and Coppersmith (1994) is 40 km.

The stopping points of surface ruptures are generally (but not always) the gaps and stopovers between the individual fault traces. This is an expectable observation, and argues that rupture-segment boundaries are controlled by whatever subsurface structure is causing faults to lose displacement and die out at the ends of their mapped traces. Knuepfer (1989) states that cross-faults in the footwall are the most common structural indicator of a rupture segment boundary on normal faults, but even those are ruptured through in 60% of the historic cases.

The $M_w > 7$ earthquakes break through at least one segment boundary beyond the one that contains the earthquake epicenter. However, at this point we cannot always predict when such multisegment ruptures will occur, and which adjacent segments they will break. Some historic earthquakes have caused ruptures that traversed the length of one segment and then "broke through" a segment boundary into the next segment, but the displacement in the second segment was a small fraction of that is the first segment, and the rupture did not traverse the entire length of the second segment (e.g., 1983 Borah Peak M_w 6.9 earthquake; Crone *et al.*, 1987). This behavior represents a "leaky" segment boundary.

3.1.2 The Earthquake Deformation Cycle in Extensional Environments

The earthquake deformation cycle for normal faults can be divided into a coseismic phase, a postseismic phase, an interseismic phase, and a preseismic phase. For most on-land normal faults, the only phases that can be measured reliably are the coseismic and postseismic phases. That is because in historic

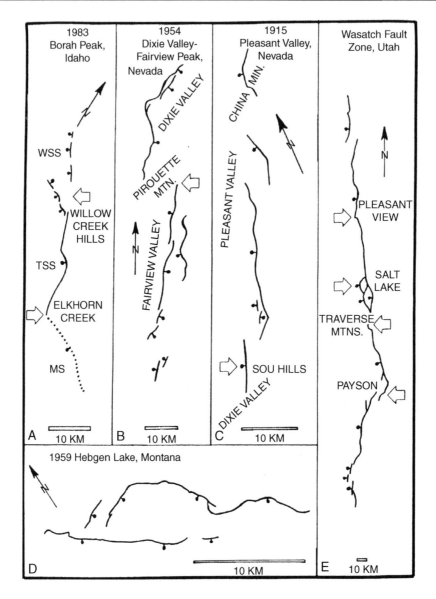

Figure 3.4: Sketch maps of normal fault surface ruptures in the Basin and Range Province, western USA; bar and ball on downthrown sides. Maps (A)–(D) show historic ruptures, map (E) shows the late Quaternary traces of the Wasatch fault zone. Large hollow arrows mark persistent segment boundaries. From Wheeler (1989, p. 437).

earthquakes, the topology of the preceding state is known. For all normal faults that are presently in the interseismic part of the cycle, the preceding topology is not known, so geologic evidence of the deformation cycle is lacking. Contemporary geodesy could in theory illuminate the nature of the interseismic cycle, but deformation rates are typically quite slow on on-land normal faults (compared to plate boundary faults).

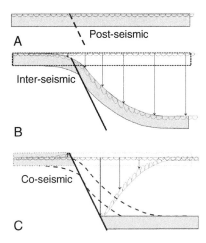

Figure 3.5: Schematic crustal cross sections across a normal fault during three phases of the earthquake deformation cycle. (A) Postseismic phase; the crustal slab (gray) is unbent and undisplaced. The blue wavy line shows a shoreline cut during this phase. (B) Inter-seismic phase; the crustal slab (gray) is bent across the location of fault, which is locked. Dashed box and blue line show geometry from previous phase. The orange line shows the configuration of the shoreline toward the end of the interseismic phase. (C) Coseismic phase; The crustal slab is displaced and elastically snaps back into its original shape. Shorelines cut at the end of the interseismic phase (blue wavy line) are deformed as shown in the orange wavy line.

The best definition of the earthquake cycle on normal faults comes from coastlines, where sea level forms a datum on either the fault footwall, hanging wall, or both. Worldwide, the largest area of normal faulting in contact with coastlines is in the central (Italy and Sicily) and eastern (Greece and Crete) Mediterranean. The effects of normal faulting on coastlines have been intensively studied here (e.g., Stiros and Pirazzoli, 1998; Stiros, 1998a; Stewart and Vita-Finzi, 1999). A schematic model of elastic rebound for a normal fault on a coast (Figure 3.5) shows how shorelines would be deformed in the interseismic and coseismic phases. The basis for this diagram is the geodetic signature of coseismic normal faulting, for example, as measured on the 1983 Borah Peak, Idaho earthquake. In that event, the average vertical displacement across the surface rupture of ca. 1 m was expressed as about 0.1 m uplift of the footwall and 0.9 m subsidence of the hanging wall, relative to sea level. This pattern is remarkably similar to the pattern of inferred long-term deformation across Aegean normal faults proposed by Jackson and McKenzie (1983). As shown in Figure 3.5, in the postseismic phase (immediately after a major earthquake), shorelines formed then [(A) and (B), blue line] will be slowly deformed as the crust bends across the fault during the interseismic cycle. This bending will lower the shoreline on the footwall a small amount as it nears the fault, tilting the shoreline toward the fault (B, orange line). The shoreline on the hanging wall will be lowered and tilted away from the fault, most steeply near the fault, and less steeply farther away (orange line).

In contrast, shorelines that form toward the end of the interseismic period (c, blue line) are carved onto the bent crustal slab. When the normal fault finally slips in a large earthquake, the crustal slab is displaced, and the bent footwall and hanging wall "snap back" elastically into their postseismic shapes. This elastic rebound raises the shoreline on the footwall (C, orange line) about 10% of the net fault displacement, so that near the fault, the shoreline now tilts gently away from the fault. On the hanging wall the opposite happens; the shoreline is dropped ca. 90% of the net fault displacement and now tilts strongly toward the fault.

This simplified model represents only the elastic effects of crustal rebound, and not any delayed viscoelastic effects of sub-crustal material redistributing itself beneath the bent crustal slab, particularly in the postseismic phase. These effects could be significant in areas of normal faulting, where typically the crust is extended, thin, and weak. Nor does this simple model account for fault creep, which is evidently a rare occurrence on normal faults, with the exception of some brief "afterslip" immediately following a large surface rupture. Nevertheless, the simple model forms a framework for detailed paleoseismic studies of normal faults in coastal areas, such as the study of Ferranti *et al.* (2008) in southern Calabria, Italy. For further details on this topic, refer to Scholz (2002).

3.1.3   Historic Analog Earthquakes

Large earthquakes accompanied by historic normal surface faulting form the modern analogs for paleoearthquake studies of normal faults (Table 3.1). This table is identical to the version in our first edition, because in the period 1995–2008, no normal surface-rupturing earthquakes occurred large enough to invite detailed field studies. The earliest studied normal surface faulting is apparently the 5 Feb.

Table 3.1: Well-studied historic normal fault surface ruptures

Date and magnitude	Area/fault	Maximum displacement[a] (m)	Length of rupture[a] (km)	References
a. Ruptures studied immediately after the earthquake				
1954, M_s 6.8	Dixie Valley, Nevada	3.8	45	Slemmons (1957)
1954, M_s 7.2	Fairview Peak, Nevada	4.8	67	Slemmons (1957)
1959, M_s 7.6	Hebgen Lake, Montana	6.1	27	Myers and Hamilton (1964)
1983, M_s 7.3	Borah Peak, Idaho	2.7	34	Crone *et al.* (1987)
1987, M_s 6.6	Edgecumbe, New Zealand	2.9	18	Beanland *et al.* (1989, 1990)
b. Ruptures studied decades after the earthquake				
1857, M_s 7.4	Pitaycachi, Mexico	4.5	75	Bull and Pearthree (1988)
			102	Suter (2006)
1915, M_s 7.6	Pleasant Valley, Nevada	5.8	59	Wallace (1984)
1980, M_s 6.9	Irpina, Italy	1.2	38	Pantosti and Valensise (1990)

[a]From Wells and Coppersmith (1994).

Figure 3.6: Sketch of two normal fault scarps created during the 5 Feb. 1783 earthquake on the Cittanova fault, southern Italy (Sarconi, 1784). The sketch shows the displacement of the Cittanova plain and of the Mercante Road by two vertical scarps (between white arrows), the lower being measured by two members of the Sarconi expedition. The upper (main) scarp is locally affected by landsliding, as shown in the middle of the print. Galli and Bosi (2002) excavated trenches across the fault scarp shown on the left side of the picture. This print is probably the first in the history of earthquake geology to show coseismic surface faulting (vertical scale exaggerated). From Galli and Bosi (2002).

1783 rupture of the Cittanova fault (Figure 3.6), southern Italy, described by Sarconi (1784). Bonilla (1988) and dePolo (1994) suggest that the threshold of normal fault surface rupture is magnitude (M_L or M_s) 5.5 and M_w 6.3–6.5, respectively, based on slightly different data sets and definitions. In the United States our understanding of normal surface faulting is heavily influenced by several $M_w > 7$ earthquakes that occurred in the semiarid Basin and Range Province from 1915 to the present (Figure 3.7). These earthquakes produced normal fault scarps up to 102 km long and 6 m high in alluvium at the base of mountain fronts and are responsible for the topical emphasis in this chapter on fault scarp morphology, degradation, and trenching techniques. Similar scarps occur along normal faults in the Baikal Rift, Tibet, China, South America, and elsewhere, so the phenomena and interpretations described in this chapter have application beyond the United States.

3.2 Geomorphic Evidence of Paleoearthquakes

The primary geomorphic indicator of paleoearthquakes on normal faults is a fault scarp. Normal fault scarps, according to strict definition, vary from mountain fronts thousands of meters high cut on bedrock, to decimeter-scale scarplets that displace Quaternary alluvium and colluvium (Stewart and

Figure 3.7: The surface rupture of the 1915 Pleasant Valley, Nevada, earthquake (M_s 7.6).
(A) A small portion of the 60-km-long surface ruptures at the base of the Tobin Range; view is to the
west. The scarp displaces range-front colluvium on steep slopes, and had widened due to erosion/
deposition in the 70 years since the rupture. Average scarp height here today is 4–5 m. Where the
scarp crosses the canyon mouth at right center, it displaces a low-gradient alluvial fan (see part B).
(B) The 1915 scarp where it crosses the head of an alluvial fan (foreground), and then climbs up
onto the base of the faceted spur (background). Scarp here is 4 m high. There is no pre-1915 scarp
in the alluvium or colluvium here, but there is a weathered bedrock fault plane exposure (from the
penultimate event?) at the colluvium/bedrock contact, directly above the man at center.
Photos taken by J. P. McCalpin in 1985.

Hancock, 1990). In this chapter, however, the term *fault scarp* usually refers to a small escarpment
in unconsolidated deposits created by direct surface faulting, unless otherwise noted. Historic normal
fault ruptures have created parallel fault scarps unconnected to the main scarp (*secondary scarps* as
defined by Bonilla, 1982) ranging from 6 to 95% (mean 49%) as long as the main rupture. Normal
faulting appears to result in more extensive secondary rupture compared to strike-slip and reverse
ruptures (Bonilla, 1970). *Overlaps, step-overs*, and *gaps* are common in normal fault surface
ruptures. In rare cases a broad *swarm* of normal fault scarps may have nearly equal heights (Crone,
1983, p. 24; Crone and Harding, 1984a,b), in which case it is difficult to label one the "main" and
others the "secondary" faults. Scarps in such swarms may have developed in complex space and time
patterns.

3.2.1 Tectonic Geomorphology of Normal Fault Blocks

Studies of large-scale *range-front morphology* are insufficiently precise to identify individual paleoearthquakes, so they fall under the more general heading of neotectonics rather than paleoseismology. However, quantitative geomorphic studies can often suggest a range of vertical slip rates on the normal fault that might be useful in reconnaissance paleoseismic studies. Bull (1984, 1987) used several quantitative measures of range-front tectonic geomorphology (*sinuosity of the range front, valley depth:width ratio*) to define five "classes of relative tectonic activity." Estimated uplift rates for these morphologic classes in arid and semiarid climates are given in Table 3.2. dePolo (1998) developed this approach further and derived an empirical correlation between the height of first-order facts on the range front, and long-term vertical slip rate. Figure 3.8 shows the faceted spurs on several range fronts in the Basin and Range Province, USA. dePolo and Anderson (2000) said that a recognizable basal facet set is typical of faults with uplift rates of 0.10 mm/yr and higher.

However, range-front morphology can be controlled by factors other than uplift rate, particularly by climate, lithology, and structure. In humid climates, where erosion rates are greater than in the semiarid western United States, a more rapid uplift rate may be necessary to maintain youthful-looking range fronts. Bull (1987) notes that rapidly rising (3–8 m/ka) range fronts in humid New Zealand are only in his morphologic activity class 2, whereas in a drier climate like southern California they would probably be in class 1. The morphology of faceted spurs is also strongly controlled by the lithology and structure of the range front. According to Zuchiewicz and McCalpin (2000), the sharpest, best

Table 3.2: Classification of relative tectonic activity of normal fault-block mountain fronts[a]

Classes of relative activity	Piedmont landforms	Mountain-block landforms	Range-front sinuosity[b]	Valley depth/valley width ratio[c]	Inferred uplift rate[d] (m/ka)
1—Maximal	Unentrenched alluvial fan	V-shaped valley in bedrock, U-shaped valley in alluvium or soft bedrock			1.0–5.0
2—Rapid	Entrenched alluvial fan	V-shaped valley	1.1–1.3	0.06–0.53 (mean 0.15)	0.5
3—Slow	Entrenched alluvial fan	U-shaped valley	1.6–2.3	0.2–3.5 (mean 1.5)	0.05
4—Minimal	Entrenched alluvial fan	Embayed mountain front	≥2.5	0.4–3.8 (mean 1.7–2.5)	0.005
5—Inactive	Dissected pediment	Pediment embayment	2.6–4.0	0.9–39.4 (mean 7.4)	≤0.005

[a]Adapted from Bull and McFadden (1977); Bull (1984, 1987)
[b]The sinuous length of the mountain-piedmont junction, divided by the straight-line length.
[c]The ratio of valley depth to valley width at a point 0.5 km upvalley from the mountain-piedmont junction.
[d]Uplift rates in semiarid climates only; in subhumid or humid areas rates may be several times larger for each class.

developed sets of facets exist on range fronts underlain by horizontal to gently dipping sedimentary rocks. The poorest facets exist on igneous/metamorphic rocks. Range fronts with faceted spurs are not solely created by normal faulting; reverse-fault range fronts and even rapidly growing anticlines can also develop faceted spurs (Bull, 2007, p. 80–82).

Late Neogene extension in some regions (Italy, Greece, and US Overthrust Belt) has created Quaternary fault scarps over relatively young normal faults in rolling bedrock hills; such scarps rarely intersect thick surficial deposits (Jackson *et al.*, 1982; Pantosti and Valensise, 1990; Pavlides, 1993; Kokkalas *et al.*, 2007). These relatively immature extensional faults in bedrock terrain are more difficult to study, because (1) Quaternary geomorphic surfaces that might be dated are not present, (2) sedimentation in fault zones is minimal, due to the steep slope of faulted terrain (Figure 3.9), and (3) scarp degradation models cannot be

Figure 3.8: (Continued)

Figure 3.8: Photograph s of range-front faceted spurs in various parts of the Basin and Range province. (A) Faceted front of the Bear River range, Utah, rises 1500 m above the floor of Cache Valley, along the East Cache fault, a Quaternary normal fault of the Basin and Range province, USA. Note the multiple sets of facets, including a basal facet set, which dePolo and Anderson (2000) say starts to form at uplift rates of about 0.10 mm/yr. Paleoseismic studies indicate a latest Quaternary slip rate of 0.07 mm/yr here. (B) Simple facet structure on the west flank of Samaria Mountain, southern Idaho. (C) Range front of the Bear River Plateau created by movement on the East Bear Lake normal fault, southern Idaho-northern Utah. Bear Lake (in foreground) fills the graben formed by the East and West Bear Lake faults. The planar slopes at the range are not facets *per se*, because they are not cut across structure. They are dip slopes on the resistant Nugget Sandstone (Jurassic); so they are "pseudofacets". (D) Reactivated range front in the central Basin and Range province, Nevada. The range from lacks steep first- and second-order facets, and ridges slope gently down to the range front. This morphology indicates an inactive or low-slip-rate range front. However, a low, linear scarp at the range front has displaced all the older Quaternary surfaces (dark tones) and indicates a recent period of renewed displacement. All photos by J. P. McCalpin, 1985–1991.

used if the scarp is entirely underlain by bedrock. The techniques described in this chapter were developed for, and will work best when applied to, fault scarps in unconsolidated (Quaternary) deposits. A few techniques (Section 3.2.1) are applicable to surface fault scarps with bedrock exposed on both sides, and some others (Section 3.3.2) will also work with Quaternary faults that juxtapose bedrock against unconsolidated deposits (such as colluvium).

Quaternary normal faults may also displace planar plateau-like landforms developed on Quaternary eruptive volcanic rocks, such as are common in rift zones. With continued movement over a few million years, these scarps evolve into linear escarpments, in a precursor stage to an (eventual) mountain front escarpment. Two examples studied for paleoseismology are the Hurricane fault, Utah (Amoroso *et al.*, 2004), which has formed an escarpment 150–200 m high across a 0.85 ± 0.06 Ma basalt flow (long-term slip rate of 0.15–0.25 mm/yr), and the Pajarito fault (McCalpin, 2005a), which has formed a 50–130-m-high scarp across the Pajarito Plateau (Figure 3.10), formed by the eruption of the welded

Figure 3.9: Photograph of the trace of the Rock Creek fault west of Kemmerer, Wyoming (Class A fault, USGS ID No. 729). This scarp does not lie at the foot of the range front, but stays high on an erosional bedrock hillslope. Thus, the fault does not displace the Quaternary surfaces in the fore-ground, but only colluvium on steep slopes, except where the scarp crosses canyons mouths such as the one at center. This fault scarp occupies an anomalous topographic position (relative to most Basin and Range normal faults) over its entire 41 km length because it represents an immature fault formed by Neogene extension and tectonic inversion of a preexisting Mesozoic thrust fault. Photo by J. P. McCalpin (1991). (See Color Insert.)

Bandelier Tuff at ca. 1.2 Ma (long-term slip rate 0.1 mm/yr). Other examples are the East and West Klamath Lake faults (70–250 m high escarpments cut in middle Pliocene basalt) and some faults in the East African rift zone. Processes acting on such escarpments are described in Section 3.2.4.

Performing paleoseismic studies on such high fault escarpments is subject to several limitations. First, due to the number of faulting events required to form such escarpments (25–100), the deformation zone is wide and complex. Second, the face of the escarpment is mainly an erosional slope on bedrock. Third, identifiable smaller, single- or double-event fault scarps displacing younger Quaternary deposits are typically absent, but are needed to estimate of displacement per event. Without knowing displacement per event, the only source parameter that can be inferred is long-term slip rate (height of the escarpment divided by the age of the plateau surface). Shorter term slip rates, variation in slip rate, and recurrence intervals cannot be computed, and paleoearthquake magnitude cannot be estimated from displacement per event (only from rupture length). Accordingly, most studies of these large fault escarpments rely on trenches cut across smaller fault scarps that displace younger landforms deposited on or at the toe of the escarpment.

3.2.2   Features of Bedrock Fault Planes and Other Rock Surfaces

Rejuvenation of a range-front fault scarp by repeated normal faulting can expose new *footwall rock surfaces* to weathering. The morphology of these fault surfaces has been described in numerous papers from a structural geology viewpoint (e.g., Stewart and Hancock, 1988; Jackson and McKenzie, 1999), but few of those structural observations can be related to individual paleoearthquakes. Of interest to paleoseismologists is the boundary between the older, weathered bedrock surface exposed before the latest faulting event, and the newer, unweathered fault-plane surface exposed at the time of the latest event, which can be sharp (less than or equal to a few centimeters). Wallace (1984a, pp. A22–A25)

Figure 3.10: The escarpment of the Pajarito fault at Los Alamos, New Mexico, forms the western margin of the Rio Grande rift zone; view is to the south. The escarpment in the foreground is about 120 m high (between white arrows); note buildings of the Los Alamos National Laboratory at the foot of the escarpment (left center). The scarp (between black arrows in distance) displaces the Pajarito Plateau, which is capped by the 1.2 Ma Bandelier Tuff. Therefore, the long-term vertical slip rate over the past 1.2 Ma is 0.1 mm/yr. This scarp varies in structure along strike, from an articulated monocline, to a faulted monocline, to a simple fault scarp; see McCalpin (2005a) for details.

described a 7-m-high bedrock fault plane, the "fresh" lower 3 m of which was exposed by the 1915 Pleasant Valley earthquake, and the upper, weathered 4 m of which was presumably exposed by a similar, previous earthquake (Figure 3.11). The weathered limestone surface contained solution pits and channels up to 1 cm deep and 2 cm wide. In the absence of calibrated measurements on limestone pitting rates, Wallace was unable to determine the age(s) of the earlier 4-m displacement event(s). Similar surfaces exposed by historic and prehistoric earthquakes in the USA are described by Mueller and Rockwell (1995, p. 16) and Pinter (1995, their Figure 6).

Several workers have tried to quantify the surface roughness of fault planes in carbonate rocks exhumed by late Quaternary surface ruptures, and to relate the roughness to exposure age. The earliest was Stewart (1996), who used simple carpenter's tools to prove that roughness increased with height (and thus, age) on the exposed fault plane. However, with the exception of the surfaces exposed during the 1981 Gulf of Corinth earthquakes, he could not independently date surfaces exposed by prehistoric ruptures. Giaccio *et al.* (2003) also made roughness measurements, but found that the greatest roughness coincided with that part of the fault surface that had, at one time, been in contact with the upper 50 cm of soil on the downthrown block (i.e., the active soil profile). Due to water infiltration and organic accumulation in the upper 50 cm of the soil profile, soil-derived weak acids etched the fault surface in a relatively narrow band. Thus, each age segment of the fault plane contained a high-roughness band marking the position of the former soil contact. This model was confirmed by Carcaillet *et al.* (2008), who found that these 30–50 cm-thick, soil-related bands are also rich in major and trace elements, compared to the rest of the fault surface. Major and trace elements also decrease slowly up the fault surface, indicating progressive leaching with greater exposure age. Giaccio *et al.* (2003) concluded that the best method for distinguishing different-age earthquake bands was to enhance digital photographs to bring out color and texture differences.

Figure 3.11: Bedrock fault plane surface of the Tobin fault (Nevada) exposed by the 1915 Pleasant Valley earthquake (lower part, below hat). Rock type is Paleozoic limestone. The upper, darker surface was exposed by the penultimate event. Photo by J. P. McCalpin.

[However, note that quite a few bedrock fault planes have been exposed by quarrying at range fronts, particularly in Mediterranean countries. Quarrying, either ancient or modern, creates an older, weathered (prequarry) fault surface lying above a lower, younger (postquarry) fault surface (e.g., Jackson and McKenzie, 1999), not to be confused with a surface exposed by recent faulting]. Surface-exposure dating of bedrock fault planes using cosmogenic isotopes was pioneered by Zreda and Noller in 1998, and has now been applied to several faults (see Section 3.4.1).

The weathered surfaces of large surface boulders on fault scarps can also display similar discontinuities in weathering. Large clasts exposed on the crest of the (multiple event) Lone Pine, California, fault scarp have *desert varnish rings* that show the ground level on the pre-1872 fault scarp, before it was lowered by scarp crest erosion following the 1872 earthquake. One very large boulder carries two varnish rings, which Lubetkin and Clark (1988) interpret as evidence for three faulting events, the latest of which was the 1872 event. Cation-ratio dating (e.g., Reneau and Raymond, 1991) could conceivably be used to date the varnish rings directly.

3.2.3   Formation of Fault Scarps in Unconsolidated Deposits

When normal faults break to the ground surface during a coseismic surface rupture, the upper few meters to tens of meters of the fault is often rupturing unconsolidated (Quaternary) deposits rather than consolidated bedrock (Figures 3.12 and 3.13). According to eyewitnesses, the scarps form suddenly.

Figure 3.12: **Range-front Holocene normal fault scarps. (A) Oblique aerial photograph of the fault scarp of the Star Valley fault (arrows) at Afton, Wyoming. Vertical surface offset is 11 m across the low-gradient, latest glacial alluvial surface at center and left, and 7 m across the small, postglacial alluvial fan at far right. This scarp was produced by three paleoearthquake displacements of 3–4 m each at about 5540, 8090, and 14,000–16,000 years BP. From Warren and McCalpin (1992); Piety *et al.* (1992); photograph courtesy of L. A. Piety. (B) Multiple-event fault scarp displacing the latest-glacial lateral moraine and outwash plain at McGee Creek, eastern Sierra Nevada, California. Photo by J. P. McCalpin.**

Eyewitnesses 300 m from the 1983 Borah Peak, Idaho surface rupture claim that the 1–1.5-m-high fault scarp formed in about 1 s (Wallace, 1984b). Even closer eyewitnesses only 20 m away from the scarp said once the strong earthquake shaking began, a "sinkhole" 2–3 m deep formed directly downslope of where

Figure 3.13: Extremes of normal fault scarp sizes. (A) Scarp of the 1934 M_L 6.6 Hansel Valley earthquake, Utah. Height here is about 30 cm, maximum height was 50 cm. From University of Utah Seismograph Stations. (B) Scarp of the 1954 M_w 7.2 Fairview Peak, Nevada, earthquake. Scarp height here was greater than 3.3 m in the mid-1985, when this photo was taken. Note the small amount of colluvium shed from this semi-consolidated Quaternary gravel deposit in the 30 years since scarp formation. Photo by J. P. McCalpin (1985).

the fault scarp formed a few seconds later, but that once the scarp formed, the sinkhole disappeared (Pelton *et al.*, 1984).

If the fault plane propagates to the ground surface without steepening, a simple fault scarp is formed. However, due to the lower shear strength and density of unconsolidated materials, the extensional failure plane often refracts to a steeper dip angle as it passes from bedrock to Quaternary deposits. Refraction is caused by a lack of confining pressure near the surface and the existence of a true tensional stress field

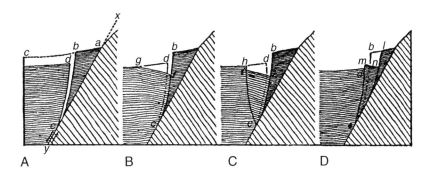

Figure 3.14: Gilbert's theory of formation of complex fault scarps. Close horizontal lines indicate unconsolidated sediments of the hanging wall; wide diagonal lines indicate consolidated bedrock of the footwall. (A) Open tension fissure caused by fault refraction at point *e*. Prefaulting ground surface is defined by *a b c*. The net slip vector is represented by distance *b d*. (B) Back-tilting caused by slumping into the void space. The block of hanging-wall alluvium *g d e* slumps to assume the shape *g f e*. The height of the fault scarp *b f* exaggerates the net slip on the fault *b d*. (C) Graben formation. The alluvial prism *h d e* settles and spreads as to occupy the space *i k e*. The difference between the heights of synthetic (*b k*) and antithetic (*h i*) fault scarps is approximately equal to the net throw *b d*. (D) Step faulting. The alluvial prism *b l e* on the footwall slides down into the tension fissure to assume the position *m n e*. From Gilbert (1890).

(Mercier *et al.*, 1983), resulting in vertical to even slightly overhanging fault scarps at the surface. Refraction creates an *open void space* or fissure (Figure 3.14A), and three types of complex fault scarps form by coseismic failure or slumping of unconsolidated material into the void. These scarp types are a *back-tilted downthrown block* (Figure 3.14B), a *graben* (Figure 3.14C), or multiple *step faults* (Figure 3.14D).

Cross sections through historic normal surface rupture document extensive shattering of surface materials. For example, the main fault rupture of the 1983 M_w 6.9 Borah Peak, Idaho, earthquake at Doublesprings Pass Road included 15 faults and 10 tension cracks in a complex horst-and-graben zone up to 85 m wide. All of the complex fault scarp types described in above (fissures, grabens, back-tilting, step faults) are observed on the Borah Peak rupture (Crone *et al.*, 1987). Both the Borah Peak and the 1959 M_w 7.3 Hebgen Lake, Montana, ruptures displayed small thrust faults (dip 25°–30°) on the downthrown block. The steepening of fault dip (*refraction*) near the surface is generally credited with causing complex scarps, as well as other phenomena such as the increase of fault scarp height at the center of alluvial fans (Slemmons, 1957, p. 373) and the downvalley arcuate "bulging" of fault scarps across stream mouths (Wallace, 1984, p. A14).

It appears that the 1983 Borah Peak ruptures preferentially propagated upward through unconsolidated alluvium along preexisting faults, because trenching showed that most fault strands with displacement in 1983 also displayed displacement in an earlier Holocene rupture (Schwartz and Crone, 1985). This suggests that preexisting faults possessed lower shear strength than did the adjacent unfaulted alluvium. Several laboratory sand-box faulting experiments (Horsefield, 1977; Lade *et al.*, 1984) have shown that local *dilatancy* and *strain softening* occur along laboratory-scale normal fault planes, and that "such fault planes represent planes of low resistive shear strength on which motion will preferentially occur during subsequent deformation stages" (Vendeville and Cobbold, 1988). However, new faults also form in most surface-rupturing earthquakes (see Section 3.2.3.1).

3.2.3.1   Terminology and Measurements of Normal Fault Scarps

Degraded fault scarps in unconsolidated material are composed of several geomorphic components (Figure 3.15; Table 3.3). Slemmons (1957), referring to normal fault scarps, defined a *simple fault scarp* as one on which a single fault rupture had displaced the ground surface without measurable rotation of the ground surface. A *complex fault scarp,* in contrast, was defined by either multiple fault traces or the ground surface being rotated out of its prefaulting gradient. Myers and Hamilton (1964, their Figure 39) propose that a similar fault steepening occurred in the 1959 M_w 7.3 Hebgen Lake, Montana, earthquake in an area of level topography not associated with a range front. On steep colluvium-covered slopes, Witkind (1964, his Figures 21 and 24) observed that the surface fault scarp may be upslope or downslope from the bedrock fault due to slope failure in the overlying prism of colluvium.

The geometric relations between fault displacement, original scarp morphology, and present (degraded) scarp morphology can be complex, but they must be understood to derive consistent fault displacement data from fault scarps. As a general rule, it is preferable to convert *scarp height* values (L, H_2 in Table 3.3) to *surface offset* (SO) values, because the latter does not increase as the scarp broadens with age. A more rigorous approach is to convert SO values to *vertical fault displacement* (H_1) values (if possible), because fault displacement (unlike surface offset) is independent of the slope angle of the faulted surface (*far-field slope*). Vertical displacement values calculated from scarp dimensions can be compared along the length of a fault, regardless of changes in far-field slope or scarp age, and can also be compared with long-term vertical displacement rates. However, any conversion to fault displacement values requires an estimate of fault dip (see next section), and fault planes are usually not visible.

The detailed geometric analysis of normal fault scarps that follows assumes that fault motion at the surface was purely dip-slip, with no oblique component. Detection of an oblique component of slip on fault scarps requires special conditions, either recognizable piercing points on either side of the fault scarp or planar ridge slopes that have been displaced (Chapter 6). In historic ruptures, piercing points have been roads, fences, canals, and other man-made objects, or linear natural topographic features such as channels, debris-flow levees, or erosional ridges that intersect the scarp. For example, the 1983 M_w 6.9 Borah Peak, Idaho, rupture displaced an irrigation ditch 2.9 m vertically and 0.43 m left laterally

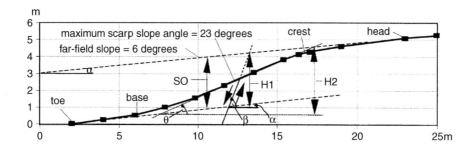

Figure 3.15: Geometry and terminology of a simple fault scarp. Profile constructed by the method of Bucknam and Anderson (1979). Black boxes show ends of measured profile segments; dotted lines show the projections of the maximum scarp angle, upthrown and downthrown surfaces, etc. Terms are defined further in Table 3.3.

Table 3.3: Definitions of fault scarp parameters

Parts of the profile		
Head	H	Edge of the uneroded original upthrown block surface
Crest	C	Point of maximum slope curvature (convex-up) between the scarp head and the steepest part of the scarp face
Face	F	Portion of the scarp profile between the crest and the base
Base	B	Point of maximum slope curvature (concave-up) between the steepest part of the scarp face and the scarp toe
Toe	T	Edge of the uncolluviated original downthrown geomorphic surface
Angular measurements		
Far-field slope	α	Gradient of the faulted geomorphic surface
Maximum scarp slope angle	Θ	Gradient of the steepest part of the scarp face
Fault dip	β	Dip angle of the fault plane underlying the scarp face
Height measurements		
Leveled height	L	Vertical separation between the scarp toe and head, usually obtained by simple leveling. This measurement is typically used in rapid reconnaissance investigations (e.g., Baljinnyam et al., 1993)
Scarp height	H_2	As defined by Bucknam and Anderson (1979), the vertical separation between intersections of the plane formed by the steepest part of the scarp face and the planes formed by the displaced original geomorphic surface
Surface offset	SO	Vertical separation between the projections of the original upthrown and downthrown geomorphic surfaces
Vertical fault displacement (throw)	H_1	Vertical distance between intersections of the fault plane, and planes formed by the displaced original geomorphic surfaces
Net fault slip	ns	Distance, measured on the fault plane, between two points that were originally in contact at the fault plane before faulting. For pure dip-slip motion, equals the distance, measured along the fault plane, between the intersections of the fault plane with the planes formed by the displaced original geomorphic surfaces
Throw on main fault	T_m	Vertical component of fault displacement on the main fault
Throw caused by backtilting	T_t	Vertical component of fault displacement on a main or antithetic fault induced by back-tilting of the downthrown block
Throw on antithetic fault	T_a	Vertical component of fault displacement on an antithetic fault
Net throw	T_{net}	Vertical component of fault displacement across the entire deformation zone, calculated as the difference between synthetic throws (H_1, T_m) and antithetic throws (T_t, T_a), or calculated from the surface offset between projected upthrown and downthrown surfaces

(Crone *et al.*, 1987, their Figure 6). Measurements of centimeter-level precision are typically not possible for prehistoric earthquakes, because (1) no man-made features are displaced by the rupture and (2) linear topographic or depositional features usually cannot be identified to this precision on either side of the fault scarp. On alluvial surfaces, linear channels or ridges may be present on the upthrown block, but often cannot be found on the downthrown block due to burial by colluvium or alluvium. If linear landforms do intersect the fault scarp, they are often so broadened by weathering that measurement errors are on the same scale as suspected lateral offsets, and relations are thus ambiguous. Special geomorphic situations where oblique slip can be measured more accurately are described in Chapter 6.

3.2.3.2   Simple Scarps

A simple fault scarp is initially formed with a near-vertical *free face* (Figure 3.16) in most brittle surface materials; in gravels, original dips are typically $78° \pm 10°$ (McCalpin, 1987a). If the fault plane is vertical, surface offset (SO) equals vertical fault displacement (H_1). For fault dips (β) less than $90°$, surface offset is less than vertical fault displacement, depending on the far-field slope angle (α). Refer to Figure 3.15 and to Table 3.3 for variable definitions.

$$SO = H1(1 - \cot \beta \tan \alpha). \tag{3.1}$$

For a given displacement and fault dip, surface offset is greatest for horizontal far-field slopes ($\alpha = 0$), and decreases to zero as far-field slope approaches fault dip ($\alpha = \beta$). This latter situation is unlikely to occur, except where fault dip (β) is very low and topography (α) is very steep. Fault dips in unconsolidated deposits typically range from $45°$ to $90°$, whereas surface slopes in alluvium and

Figure 3.16: The 2-m-high near-vertical fault scarp of the 1983 M_s 7.3 Borah Peak, Idaho, earthquake. The fault here displaces late Pleistocene (age 15–20 ka?) alluvial fan gravels. Photo taken 36 h after the earthquake by J. P. McCalpin, 100 m south of Doublesprings Pass Road. Note the small amount of colluvium at the base of the scarp. Person at right is 1.8 m tall.

colluvium range from 0° to 35°. The dependence of surface offset on far-field slope should be borne in mind when comparing surface offsets of scarps on steep range front colluvium ($\alpha = 25°$–$35°$) to those of scarps on alluvial terraces or fans ($\alpha = 1$–$10°$). Where fault dip is steep and faulted surfaces are gentle, there is very little difference between fault throw (H_1) and surface offset (SO). With gentler fault dips and steeper faulted surfaces, the difference between H_1 and SO increases.

In the USA we tend to use the definition of "scarp height" proposed by Bucknam and Anderson (1979). This definition (Figure 3.15) is not merely the vertical distance between the scarp toe and head, but is based on the vertical distance between intersections of projections on a measured scarp profile. The scarp height was measured in this way because Bucknam and Anderson were concerned with scarp degradation processes (rather than fault displacement), and needed a measure which reflected particle trajectories on the scarp surface. As a result, the relation between their scarp height (H_2) and vertical fault displacement (H_1) is more complex, because scarp height increases with time as a scarp degrades and broadens (i.e., the maximum scarp angle (θ) decreases) on a surface with some finite far-field slope (α).

$$H_2 = H_1 \frac{\sin\theta \sin(\beta - \alpha)}{\sin\beta \sin(\theta - \alpha)}. \tag{3.2}$$

For example, one can compare the vertical fault displacement (H_1) required to create a 5.6-m-high degraded scarp (H_2) on a gentle versus on a steep surface, by solving Equation (3.2) for H_1. A slightly degraded ($\Theta = 30°$) 5.6-m-high scarp on a gentle surface ($\alpha = 3°$) can be produced by a vertical displacement (H_1) of 5.2 m. In contrast, a more degraded scarp ($\Theta = 22°$) of the same height ($H_2 = 5.6$ m) can be produced on a steeper surface ($\alpha = 15°$) by a vertical displacement (H_1) of only 2 m. By the time the scarp has weathered to a 22° maximum angle (θ only 7° steeper than the faulted surface), there is a considerable elevation difference between the scarp base and crest. In the limit that maximum scarp slope angle (Θ) becomes parallel to far-field slope (α), Equation (3.2) predicts that scarp height (H_2) becomes infinitely large, because $\sin(\Theta - \alpha)$ approaches zero. It may be difficult to recognize a fault scarp once its maximum scarp slope angle (Θ) is within 2°–3° of the far-field slope angle (α).

3.2.3.3 Scarps with Back-Tilting

Rotation of the downthrown block toward the fault scarp (Figure 3.14C) causes scarp height to exceed vertical surface offset (vertical separation); the difference increases as the back-tilted area becomes wider and more tilted. Equation (3.3) relates the additional component of vertical fault displacement (throw) induced by back-tilting (T_t) to the width of the tilted zone (W_f), the angle of tilt (ϕ), and other previously defined measures of scarp morphology (Figure 3.17 and Table 3.3):

$$T_t = W_f \frac{[\tan\alpha + \tan(\theta - \alpha)]\cos\alpha \sin\beta}{\sin(\beta - \alpha)}. \tag{3.3}$$

Using Equation (3.3), one can subtract the portion of throw at the fault scarp due to tilting (T_t) from the apparent throw at the main fault scarp (T_m in Figure 3.17) to calculate the true throw across the fault zone. An example would be a steep scarp ($\Theta = 25°$) with surface offset SO = 25 m, displacing a gentle slope ($\alpha = 5°$), underlain by a fault dipping at 78° (β), fronted by a zone 100 m wide (W) that is back-tilted toward the fault by 3° (ϕ). Using Equation (3.1), we calculate T_m to be 25.5 m; T_t calculated from

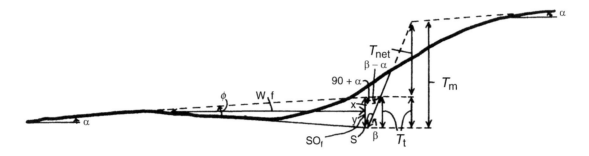

Figure 3.17: Components of a degraded normal fault scarp in which back-tilting of the hanging wall is significant. h1a, apparent scarp height; h1f, component of apparent scarp height due to flexure (back-tilting); h1t, true scarp height after subtracting component hf from h1a; Wf, width of flexed zone; ϕ, angle of back-tilt; sof, vertical surface offset component due to back-tilt; α, ambient slope angle of faulted surface; β, fault dip. From McCalpin, 1983. Reprinted with permission of the Colorado School of Mines Press.

Equation (3.3) equals 5.3 m. Therefore, $T_{net} = T_m - T_t = 25.5 \text{ m} - 5.3 \text{ m} = 19.7 \text{ m}$. Note that 19.7 m only amounts to 79% of the apparent throw (T_m) at the main scarp (25 m). If the back-tilting was not recognized in the field and the height of the main scarp was used to calculate displacement, that value would have overestimated true net throw by 26%. An alternative procedure for calculating T_{net} would be to measure the net SO between the graphical projections of the upthrown and downthrown surfaces, and then calculate H_1 from this net SO value using Equation (3.1).

Back-tilting is usually restricted to within 100–200 m from the fault scarp (Vincent, 1985, p. 76), but depends on the depth at which fault refraction takes place. The important point is that back-tilting of only a few degrees may not be apparent to the investigator in the field, especially if the tilted area is broad. These broad zones of low tilt can cause the scarp height to exaggerate the true net displacement (Figure 3.17), which must be recognized and corrected for. Even more subtle tilting has been documented by geodetic surveys at considerably longer distances from the fault, due to elastic rebound. For example, broad *coseismic tilt* of 100–500 μrad affected the downthrown block of the Borah Peak, USA, fault scarp over a distance of ca. 20 km from the scarp (Stein and Barrientos, 1985). Elastic dislocation modeling (Savage and Hastie, 1966, 1969; Y. Okada, 1985) implies that downthrown block tilt over 6–12 km from the rupture is typical of historic M_w 6.5–7.5 normal surface faulting events. However, for the purposes of measuring displacement on surface ruptures, it is unnecessary to correct for these very small, very broad tilts. One exception is where tilting is restricted to a graben in the rupture zone itself. In that case, angular unconformities in the graben sediments can be used to reconstruct the sequence of tilting and, thus, faulting (Section 3.3.3.4).

3.2.3.4   Scarps with Graben

Graben structures are defined by one or more *synthetic fault scarps* (the largest of which is termed the main fault scarp) and one or more *antithetic fault scarps* (Figure 3.18). The three geometric components of fault displacement in a scarp-graben system are (1) stratigraphic *throw* at the main fault zone (T_m, approximately equal to initial height of the scarp free face), (2) tilt (ϕ) of the graben strata acting over a certain horizontal distance (W) from the main fault, and (3) stratigraphic throw on any antithetic faults

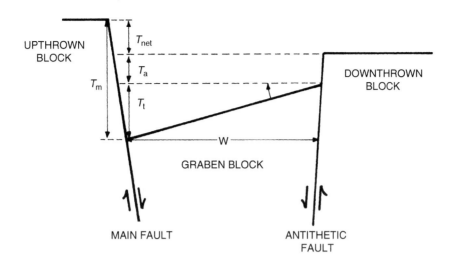

Figure 3.18: Schematic diagram showing the components of displacement in a scarp-graben system, where the graben has been tilted. Terms are defined in the text and in Table 3.3. From McCalpin *et al.* **(1994); reprinted with permission of the American Geophysical Union.**

(T_a; Figure 3.18). If all three components are known for an individual faulting event, the net coseismic throw (T_{net}) for that event is measured as the difference between main fault throw and contributions of throw from tilt ($T_t \cong W \tan \phi$) and antithetic faulting, or

$$T_{net} = T_m - [(W \tan \phi) + T_a] \tag{3.4}$$

An alternative procedure for calculating net vertical displacement is to measure graphically the surface offset between the displaced far-field slopes, and use Equation (3.1) to convert surface offset (SO) to net fault displacement (H_1). If deposition in the graben has been significant, the components of displacement calculated at the main fault (T_m) and antithetic fault (T_a) will yield minimum values (McCalpin, 1983, pp. 42–43), but the net throw across the zone (T_{net}) will remain the same.

Caskey *et al.* (1996) proposed the use of Hansen's (1965) "graben rule" to estimate the depth of, and angle of, fault refraction responsible for creating the graben on the 1954 Dixie Valley rupture (shown in Figure 3.3). According to the rule, the cross-sectional area of the graben trough (without any postfaulting sediment) approximates the cross-sectional area of the void space formed as a consequence of fault refraction, as the downthrown block moves down-dip on a fault plane that shallows beneath the graben. Although there is no unique geometric solution, Figure 3.19 illustrates that, for a given graben geometry (area), the depth at which the dip angle changes is inversely related to the magnitude of the change in dip. If one assumes that the width of the graben reflects the depth at which the dip angle changes, then a fault dip that shallows by 20°–25° (Figures 3.19C and 3.19D) is geometrically most reasonable, and such changes ($\Delta\theta = 25°, 20°$) correspond to subsurface fault dips of 25°–30° E. A fault-dip change ($\Delta\theta$) of only 15° (Figure 3.19E) corresponds to a subsurface dip of 35° E and is less satisfying geometrically because the graben does not appear to have formed across a broad enough zone.

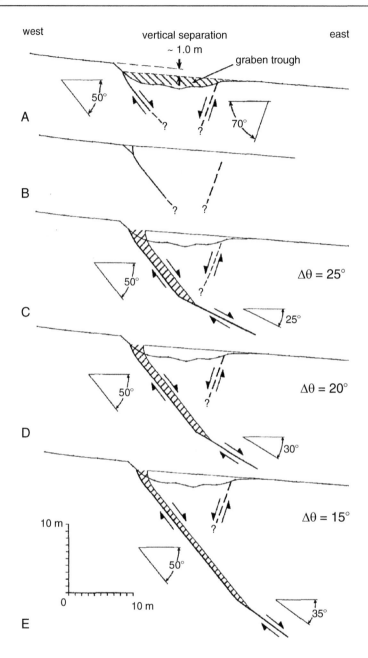

Figure 3.19: Reconstruction of the depth of, and angle of, fault refraction needed to create the observed 20-m-wide surface graben on the 1954 Dixie Valley, Nevada, surface rupture. (A) postfaulting geometry; area of graben trough shown by diagonal lines. (B) prefaulting geometry. (C) equivalent area of graben created as a void space by a 25°-decrease in fault dip; requires the refraction to occur at 9.5 m depth. (D) equivalent area of graben created as a void space by a 20°-decrease in fault dip; requires the refraction to occur at 13.5 m depth. (E) equivalent area of graben created as a void space by a 15°-decrease in fault dip; requires the refraction to occur at 19 m depth. If refraction occurred at 19 m depth, the graben should have been much wider. From Caskey *et al.* (1996); reprinted by permission of Seismological Society of America.

3.2.3.5 Step Faults

Step faults were defined by Slemmons (1957) as parallel fault scarps of similar size and sense of slip created coseismically by collapse of the footwall into a refraction-formed void (Figure 3.14D). If no rotation occurs in the block(s) between step faults, then the net vertical fault displacement across the fault zone is calculated by summing the vertical displacements (H_1) of individual step faults. If block rotation has occurred, the preceding procedure will overestimate net displacement if *backward rotation* has occurred, or underestimate it if *forward rotation* has occurred. To correct for the component of displacement induced by block rotation, one must approximate it as $W \tan \phi$ (as with back-tilted scarps) and either add or subtract that value from the sum of step-fault displacements.

Few ruptures, historic or prehistoric, have been inventoried along their entire length for rupture style and complexity. McCalpin (1983) measured 82 fault scarp profiles at roughly 1.5-km intervals along the 120-km-long Sangre de Cristo fault zone, Rio Grande rift zone. Almost 60% of profiled scarps were simple scarps (as previously defined), whereas graben, step faults, and back-tilted surfaces composed 17, 14, and 10% of profiles, respectively. Crone *et al.* (1987) note that "unusual morphologies" are found along 5–10% of the length of the 1983 M_w 7.3 Borah Peak rupture. Apparently most normal fault surface ruptures produce simple fault scarps; complex fault scarps result from relatively unusual subsurface conditions that lead to fault refraction.

3.2.3.6 Monoclinal Scarps

In contrast to the discrete fracturing that occurs when cohesionless gravels are ruptured, cohesive materials (e.g., finer grained or moister sediments, or some types of bedrock) may form fault-propagation folds such as *monoclines* when ruptured. The crests of the monoclines are often broken by a *crestal tension fissure* that parallels the strike of the monocline (Figure 3.20; also Crone *et al.*, 1987; McCalpin, 2005). The fault scarp face is typically composed of a planar, forward-tilted slab of the prefaulting ground surface. Because a monoclinal scarp does not expose fresh material on a steep free face, it does not progress through the sequential weathering stages described in Section 3.2.4, and thus cannot be dated

Figure 3.20: (Continued)

Figure 3.20: Coseismic monoclinal fault scarps developed in different materials, but with a common overall geometry. A prominent crestal tension fissure is developed on each scarp, and a planar, forward-tilted slab makes up most of the scarp face. (A) Scarp in moist, sandy Holocene alluvium, 1983 Borah Peak, Idaho, fault trace, north of Rock Creek. Fissure in center is 0.9 m wide; note person at upper left for scale. Photograph taken 36 h after the earthquake. (B) Prominent crestal tension fissure (*center*) and forward-tilted slab (*right*) along the 1882 Sonora, Mexico, fault scarp. The faulted Pleistocene alluvial gravels here are cemented to a concrete-like hardness with pedogenic calcium carbonate (caliche) at least 3 m thick, giving the rupture the appearance of a bedrock fault scarp (compare with C). Although the photograph was taken 106 years after the scarp formed, very little material has accumulated in the tension fissure. Rod is 4 m long. (C) Monoclinal fault scarp in Quaternary basalt, Hawaii Volcanoes National Park, Hawaii. Crestal tension fissure is almost completely filled with basalt blocks; note person at upper center for scale. The date of scarp formation is unknown. All photos by J. P. McCalpin.

with diffusion-type techniques (Section 3.4), nor will it create colluvial wedges (Section 3.3.3). Theoretically a monoclinal scarp could be dated from materials that fell into the crestal tension fissure, but no detailed sedimentologic model for fissure coseismic and postseismic sedimentation has yet been published. This is a promising topic for future research.

3.2.4 Degradation of Fault Scarps in Unconsolidated Deposits

Fault scarps formed during paleoearthquakes are soon attacked by weathering and erosional processes. Wallace (1977) presented a conceptual model, based on field observations in the Basin and Range Province, USA, for the degradation of fault scarps in gravelly alluvium and colluvium (Figure 3.21).

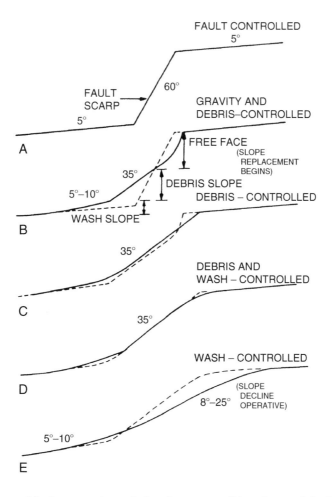

Figure 3.21: Sequence of fault-scarp degradation in unconsolidated materials. To show incremental change, the dashed lines represent the solid lines of the previous profile. (A) Initial scarp; dip is drawn at 60°, although in unconsolidated deposits it is typically steeper. (B–E) Sequential stages of scarp degradation due to gravity processes (B), debris processes (C and D), and wash processes (D and E); see description in text. From Wallace (1977, p. 1269). Reprinted with permission of the Geological Society of America.

These scarps are transport-limited slopes, on which the speed of scarp evolution is limited only by the strength of the transportation process, and not by the supply of transportable material. For a scarp underlain by unconsolidated sand or gravel, the supply of transportable material is effectively unlimited.

The initial fault scarp forms at angles between 60° and 90° (Figure 3.21A), then begins to ravel by mass movement and erosion to build up a basal colluvial debris and wash slope (Figures 3.21B and 3.22). As the *free face* retreats, colluvium accumulates, the *debris slope* grows (Figure 3.11C), and the free face is eventually buried by colluvium (Figure 3.21D). After this stage, wash processes dominate on the scarp, leading to symmetrical slope degradation on the upper half of the scarp and colluvial aggradation on the lower half (Figure 3.21E).

This idealized model assumes that (1) the fault scarp in unconsolidated material stands at greater than the angle of repose after faulting, (2) the scarp is buried mainly by colluvium shed from the free face, rather than by widespread deposition (fluvial, lacustrine, eolian) on the downthrown block, and (3) that erosion and transport of scarp face material are limited only by the strength of geomorphic processes, and not by the availability of loose material (i.e., the scarp face is a *transport-limited slope*; Nash, 1980). Figure 3.23A shows a fault scarp which collapsed into cohesive blocks during coseismic shaking; this scarp bypassed stages (A) through (D) in Figure 3.21, and went directly to the wash stage. At the opposite extreme, the transition from stage (A) to (B) (Figure 3.21) can be retarded by *cementation* of surficial materials exposed in the fault scarp, such as the faulted carbonate-cemented paleosol shown in Figure 3.23B. Fault scarps in Death Valley, California, are cemented with halite and have retained their free faces for thousands of years (L. W. Anderson, personal communication, 1995).

Wallace (1977, his Figure 7) suggested age ranges over which *gravity, debris*, and *wash processes* dominated scarp degradation. Based on observations of scarps produced in Nevada in 1915 and 1954, he estimated that 1000–2000 years would be required to eliminate the free face completely. Subsequent observations have shown that the rate of free face retreat is a complex function of climate and the mechanical properties of the faulted material. For example, along the subhumid, forested 1959 Hebgen Lake, Montana, rupture, the free face has disappeared in most places in only 30 years (Wallace, 1980a). Elsewhere free faces still exist, but are preserved mainly by tree-root cohesion. Free faces created during

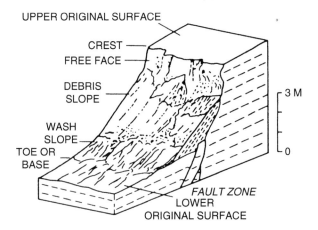

Figure 3.22: Block diagram of a fault scarp in unconsolidated deposits. Terminology is similar to that for degraded fault scarps. From Wallace (1977, p. 1269). Reprinted with permission of the Geological Society of America.

Figure 3.23: Extremes of scarp preservation. (A) Portion of fault scarp in sandy alluvium that collapsed during seismic shaking, 1983 Borah Peak rupture, ca. 150 m north of Doublesprings Pass Road. Uncollapsed near-vertical scarp appears in the distance. Person at left is 1.8 m tall. (B) The 106-year-old fault scarp developed in a cemented carbonate soil horizon (caliche), Pitaycachi fault, Sonora, Mexico, earthquake of 1887 (Bull and Pearthree, 1988). Note the small amount of colluvium that has accumulated at the base of the 2-m-high free face.

the 1981 Gulf of Corinth ruptures disappeared in most places within 23 years, despite the semiarid climate of Greece (Kokkalas and Koukouvelas, 2005; Kokkalas *et al.*, 2007). Along the 1983 Borah Peak, Idaho, fault trace, scarps formed in poorly sorted, matrix-rich colluvium and alluvium maintain nearly vertical free faces (Figure 3.24A). Adjacent scarps that formed in better sorted, matrix-poor stream alluvium had

Figure 3.24: Typical forms of scarp degradation. (A) The free face and debris-facies colluvial wedge of the 1983 Borah Peak rupture, as observed in 1986. Note that the free face is retreating differentially, with more retreat at the top (soil profile). Embayments are also starting to develop in the free face due to minor concentrations in surface runoff at the top if the scarp. Stadia rod is 1.65 m long. (B) Trace of the one of the strands of the Primorsky fault on the western side of Lake Baikal, Russia, north of Olkhon Island. The fault crosses coarse talus deposits and the scarp has broadened far up the talus slope, making an irregularly vegetated band across the center of the photo. In this section of the fault a true free face probably could not exist. Photos by J. P. McCalpin.

lost most of their free faces and had degraded to colluvial slopes of 30°–37° within one day after the earthquake. Crone *et al.* (1987) attribute better free face preservation to increased cohesion from silt and clay, and from the apparent cohesion of oriented angular gravel clasts in colluvium. Fault scarps crossing talus fields soon ravel to the angle of repose of the talus (Figure 3.24B). It thus appears that Wallace's initial 1000–2000-year estimate for free face elimination may be a maximum estimate, applicable only to the slow degradation of cohesive and/or cemented material in the arid climate in which he was working.

The free face/scarp height ratio on historic fault scarps has also been correlated with the mechanical properties of the material exposed in the scarp. Watters and Prokop (1990) show strong positive correlations between free face/scarp height ratios and cohesive strength, peak friction angle, and bulk density of materials composing the 1954 Dixie Valley–Fairview Peak, Nevada, fault scarp. This correlation has been used to date fault scarps (Section 3.4).

After the free face is buried, the rate of scarp decline is affected by both lithologic and microclimatic factors. McCalpin (1983, his Figures 55 and 63) showed that scarps composed of finer gravel had declined to lower slope angles than had scarps of the same age composed of coarser gravel, although the relation between slope angle and grain size had significant variation. Pierce and Colman (1986) demonstrated that late-glacial scarps in the western United States that faced south degraded three to five times as fast as identical scarps that faced north. M. N. Machette proposed that a fault scarp more than 100,000 years old would be obliterated by erosion in arid and semiarid landscapes. This statement became informally known as the "Machette criterion" (cited in Hanks *et al.*, 1984, p. 5787), and has a corollary, that fault scarps with slip rates of <0.01 mm/yr would basically be undetectable in the landscape (a 1 m scarp created every 100 ky is equivalent to a slip rate of 0.01 mm/yr).

A major change to scarp evolution occurs when cumulative displacement becomes so large, that consolidated bedrock is exposed on the scarp face. At this point and beyond, an increasing proportion of the upper scarp profile (i.e., upslope of the fault plane) becomes supply limited rather than transport-limited (e.g., Figure 3.10). As displacement and scarp height increases, three things happen. First, the upper scarp face progressively becomes a stripped erosional slope on bedrock, with only a thin or discontinuous veneer of colluvium. This slope will be steeper than any part of the scarp downslope of the fault, and will develop into a high, steep planar slope underlain by bedrock. Second, the scarp profile downslope of the fault gradually transforms from a site of deposition to a site of transport, as the slope above the fault becomes higher and steeper. Particles that formerly would have stopped downslope of the fault and been deposited on the colluvial wedge, now just keep moving to the scarp toe or beyond. As the scarp grows, these two changes lead to more of the scarp profile lying upslope of the fault. Eventually, the scarp transforms into a bedrock-cored fault scarp with the fault plane nearly at the toe, and a very small colluvial wedge relative to the size of the scarp.

Third, the increasing number of faulting events creates a wider zone of broken rock, and the large scarp height creates enough gravitational potential to start slope instability. This instability may result in landslides, but more commonly, it will result in coseismic fault refraction toward the scarp face (the least confining stress). Fault-bounded slivers of bedrock may begin to tilt (topple) toward the scarp face beneath the scarp profile (e.g., McCalpin, 2005). Tilting will create tension fissures between the tilt blocks, which will further trap colluvium traveling down the scarp face, and "starve" the colluvial wedge at the scarp toe. After hundreds of paleoearthquakes have occurred, these bedrock-cored scarps will develop an erosional geometry typified by range-front facets.

3.2.5   Spatial and Temporal Variations in Surface Displacement

The earthquake surface ruptures that produce fault scarps typically vary in sense and amount of displacement along strike, and also vary in displacement at a given point on the fault between successive paleoearthquakes. To appreciate this range of variability in space and time, we must examine well-studied historic surface ruptures (Table 3.1) as modern analogs to paleoearthquakes.

3.2.5.1   Variability of Displacement Along Strike in a Single Rupture

In a general sense, vertical displacement on historic normal-fault surface ruptures has been greatest at the center and least near the ends of rupture (Figure 3.25). The most symmetrical slip pattern is at Pleasant

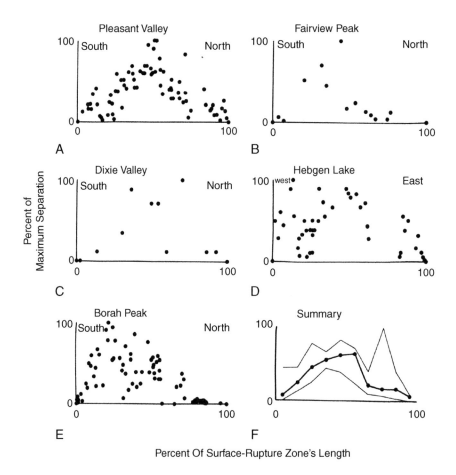

Figure 3.25: Variation in vertical separation along strike for historic normal-fault surface ruptures. All separation data are from fault scarp heights and are expressed as percentages of the maximum separation. Dots connected by a heavy line show median separation, over 10% length increments, from the Pleasant Valley (1915), Fairview Peak (1954), Dixie Valley (1954), Hebgen Lake (1959), and Borah Peak (1983) earthquakes. Light lines show envelope of median separations; anomalously high value at 80% of length is only from Dixie Valley. From Wheeler (1989, p. 435).

Valley, Nevada (despite the fact that it involved four segments), and the least symmetrical is at Hebgen Lake, Montana (two overlapping ruptures summed together). The shorter wavelength lateral variations in vertical surface displacements can be seen in the 1954 Dixie Valley rupture (Figure 3.26).These shorter wavelength variations, measured from fault scarp profiles, may represent the complex response of surficial deposits to rupture rather than spatial variations in displacement along the bedrock fault plane.

Regardless of the origin of such surface slip variations, however, they provide a statistical basis for relating the height of prehistoric normal fault scarps to the magnitude of the causative paleoearthquake (see detailed discussion in Chapter 9, See Book's companion web site). For example, normal fault surface ruptures contain the largest percentage of small displacements along strike (relative to the maximum or mean displacement of the rupture) of all fault types. McCalpin and Slemmons (1998) inventoried 10 well-studied historic normal surface ruptures (Figure 3.27) and were able to combine all their displacement measurements (a total of 559 measurements) by normalizing them to the maximum displacement in each rupture. The normalized-grouped data set shows that the most of the length of normal surface ruptures is comprised of displacements very small in relation to the maximum or average displacement, with 27% of the rupture length displaying displacements between 0 and 10% of the maximum displacement (D_{max}). Over 60% of the rupture length is typified by displacements of <30% of D_{max}. Compared to other fault types, normal fault surface ruptures have longer portions of low displacement that flank the central area of high displacements. The implication of this geometry for paleoseismic studies (explored in more detail in Chapter 9, See Book's companion web site) is that randomly located trench sites on normal fault scarps are likely to encounter displacements considerably smaller than the maximum that occurred in each paleoearthquake.

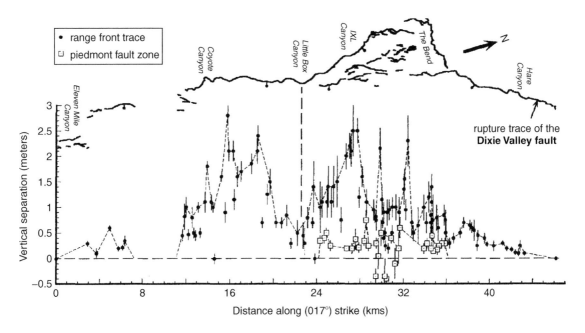

Figure 3.26: Variation in vertical separation along strike during the 1954 Dixie Valley surface rupture, Nevada. Scarp profile measurements were made by Caskey *et al.* (1996) some 35 years after the rupture. Solid dots are from scarps at the range front, open squares are from scarps on the piedmont. Error bars are 2 sigma. From Caskey *et al.* (1996); reprinted with permission of the Seismological Society of America.

10 Normal Faulting Events

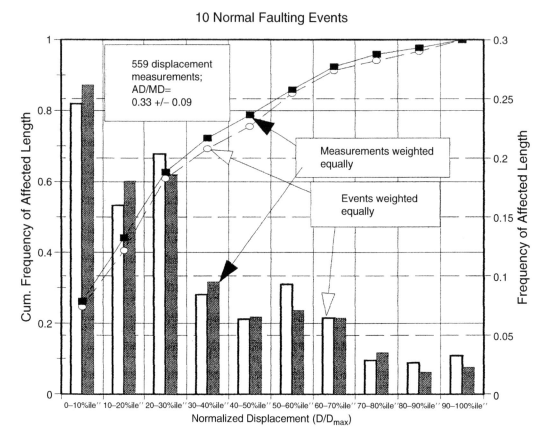

Figure 3.27: Relative frequency (histogram, right vertical axis) and cumulative frequency (lines, left vertical axis) of normalized vertical displacements from 10 historic normal surface ruptures. The field-measured displacements in each rupture were normalized to the maximum (D_{max}) for that rupture, then all 559 measurements were combined into a single dataset. Prior to normalization, displacements were weighted by the length of the rupture they affected, to remove bias from the clustered spacing of measurements along strike. The method is described in Chapter 9 (See Book's companion web site). From McCalpin and Slemmons (1998).

3.2.5.2   Variability of Displacement at a Point

Few historic data exist on the variation of fault displacement and style among repeated coseismic displacements at the same point on a fault, because few fault segments have ruptured more than once in historic time. From paleoseismic evidence, early workers on the Wasatch and San Andreas fault zones, USA (Swan *et al.*, 1980; Schwartz and Coppersmith, 1984) inferred that, at many trench sites, successive Holocene displacements on the main fault plane had been approximately the same size. That inference was supported mainly by measurements of faulted terrace sequences and colluvial wedge thicknesses exposed in trenches. Schwartz and Crone (1985) noted that, in a trench across the 1983 Borah Peak, Idaho, rupture, the 1983 displacements closely mimicked displacement from the prior faulting event in both style and amount. The coincidence of successive displacements is particularly striking because unconsolidated deposits at this location are at least 27–35 m thick (Crone *et al.*, 1987). The observed consistency of displacements through multiple seismic cycles gave rise to the characteristic earthquake model (Chapter 9, See Book's companion web site).

However, later work at some of the same sites revealed variations of about a factor of two in inferred net displacement per event (McCalpin *et al.*, 1994). On the Borah Peak fault scarp, the ratio of 1983 to prior event displacements on 10 fault planes exposed in a trench averages 1.35 but with a large standard deviation (1.23). In other words, on particular fault traces the 1983 displacement ranged from 20 to 350% of the prior event displacement (Schwartz and Crone, 1985, p. 158). Thus, while the net displacement across a complex fault zone may be approximately repeatable from earthquake to earthquake, displacements on individual faults in the zone may vary widely with time. Subsequent to Schwartz and Coppersmith's 1984 paper, which was actually based on very few trench sites, a wealth of data have been collected on displacement repeatability worldwide. These site studies provide evidence for both characteristic and noncharacteristic behavior on normal faults (see Chapter 9, See Book's companion web site).

Some trenches across normal faults (e.g., Machette *et al.*, 1992a, his Figures 10 and 16; McCalpin *et al.*, 1992, their Figure 15) reveal fault strands created in earlier events that were *not* reactivated by later events, as well as "new" fault strands that were created only during the latest event (see Section 3.3). The creation of new fault strands shows that the location of future ruptures cannot be predicted in detail from the pattern of previous ruptures. The correlation between fault patterns in trenches and those created in laboratory sand boxes is a promising topic for future study, and may offer clues to the mechanics of faulting in natural near-surface deposits.

3.2.6 Geomorphic Features Formed by Single and Recurrent Faulting

3.2.6.1 Interaction of Fault Scarps with Geomorphic Surfaces

Most normal fault scarps in rift zones and regions of diffuse extension displace Quaternary geomorphic surfaces such as alluvial fans, river terraces (Figure 3.12A), moraines (Figure 3.12B), or shoreline platforms. When fault scarps cross geomorphic surfaces of different elevation and/or age, changes in scarp height yield important clues as to the history of faulting (see general cases analyzed by Suggate, 1960, and Lensen, 1964a). A common geomorphic setting in extensional regions is fault scarps at a range front displacing the upper parts of alluvial fans. Range-front fault scarps typically intersect fanheads where fanhead incision has created younger-inset-within-older surfaces (Bull, 1991), whereas branch and secondary faults (as defined in Section 3.2) can occur near to or downslope from the fan *intersection point* (Hooke, 1967), where younger deposits lie stratigraphically and topographically on older deposits.
A single-event fault scarp that postdates all geomorphic surfaces, whether upslope or downslope from the fan intersection point, will maintain a roughly uniform height across geomorphic surfaces regardless of their age. Figure 3.28 shows a single-event scarp that is younger than two erosional surfaces (stage a3); note that scarp height remains constant as the scarp ascends from the younger (Pf) to the older (Bf) surface. Near the intersection point, fan deposits of different ages occur at roughly the same elevation, across which the single-event scarp maintains a constant height (stage b3). Although Figure 3.28 shows fill terraces, the surface geometry would be the same for strath terraces.

If periods of terrace formation and faulting have alternated, multiple-event fault scarps will be higher where developed across the older geomorphic surfaces. The increased scarp height on older surfaces merely represents the greater number of cumulative displacements compared to that of younger surfaces. In Figure 3.29, an older fan (Bf) is deposited (stage a1) and later faulted (stage a2). Erosion then results in

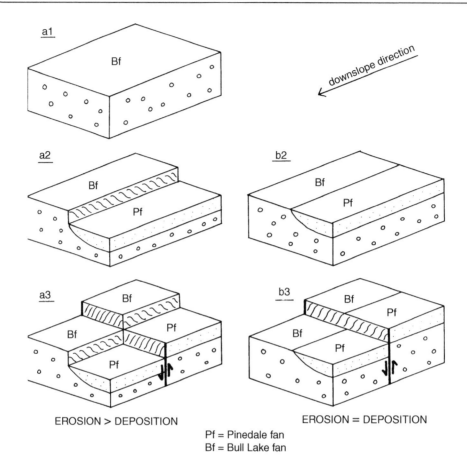

EROSION > DEPOSITION   EROSION = DEPOSITION

Pf = Pinedale fan
Bf = Bull Lake fan

Figure 3.28: Geometry of a hypothetical single-event fault scarp offsetting an alluvial fan terrace sequence. (a1) Deposition of unit Bf. (a2) Channel incision above the intersection point and partial backfilling to create a Pf fill terrace. (a3) Faulting of a height equal to that of the terrace riser. (b2) Channel incision at the intersection point and complete backfilling. (b3) Faulting of equal height to that in stage a3. Note the difference in final geometries (a3 versus b3) depending on whether the faulting occurs above or below the alluvial fan intersection point. From McCalpin (1983, p. 47); reprinted with permission of the Colorado School of Mines Press.

formation of a younger channel, which erodes through the fault scarp (stage a3). Later faulting not only creates a scarp across the younger channel (Pf), but increases the height of the preexisting fault scarp on the older fan (Bf; stage 4). If recurrent faulting occurs near the alluvial fan intersection point, we see a similar abrupt change in fault scarp height (stage b4), but without the incised younger channel. Downfan from the intersection point, only fault scarps younger than the latest period of deposition are preserved at the surface. Earlier displacements offset the older (buried) deposits in the subsurface; such differential displacements can only be analyzed by indirect subsurface methods (Chapter 2).

In the previous example, the stream had incised into both the upthrown *and* downthrown blocks after faulting. Stream incision into the upthrown block after faulting is expected, because the base level of the stream is suddenly (coseismically) lowered by an amount equal to scarp height. This lowering causes a

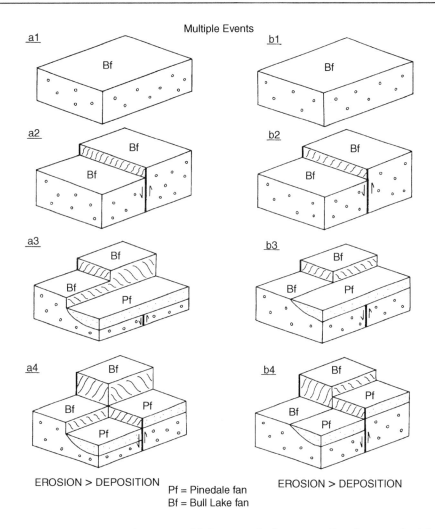

Figure 3.29: Geometry of a hypothetical multiple-event fault scarp offsetting an alluvial fan terrace sequence. (a1) Deposition of unit Bf. (a2) Faulting. (a3) Channel incision into the upthrown and downthrown blocks above the alluvial fan intersection point, and partial backfilling to create a Pf fill terrace. (a4) Renewed faulting of equal height to initial faulting. (b1–b4) Same sequence of deposition, erosion, and faulting at the alluvial fan intersection point, where erosion equals deposition. Note the difference in final geometries (a4 versus b4). The abrupt increase in scarp height between fan surfaces of different age is an indicator of recurrent faulting. From McCalpin (1983, p. 48); reprinted with permission of the Colorado School of Mines Press.

steep *nickpoint* to develop in the stream bed at the fault trace (Figure 3.30A). The nickpoint then migrates upstream, causing progressive incision into the previous floodplain (e.g., at the 1983 Borah Peak, Idaho, rupture; Vincent, 1985, p. 85). An episodic, fault-induced incision into the upthrown block can create *tectonic terraces* that in profile diverge downstream from the modern channel, and abruptly end at the fault scarp. The vertical separation between the projections of these terraces, measured at the inferred fault

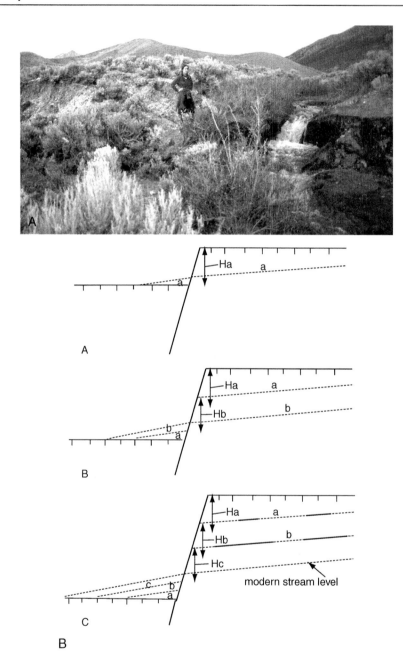

Figure 3.30: Formation of tectonic terraces on the upthrown block of a normal fault. (A) Willow Creek flowing over the freshly created scarp of the 1983 Borah Peak rupture; photo taken 36 h after the earthquake. The stream is flowing on its former floodplain, uplifted about 1 m by scarp formation. The nickpoint had receded less than 1 m from the fault plane at this time. (B) Schematic diagram showing tectonic terraces and their projections used to estimate vertical displacement in individual paleoearthquakes. The prefaulting surface is indicated by vertical ticks. (A) The first

plane, provides a first approximation of the vertical fault displacement in individual paleoearthquakes (Figure 3.30B; also Soule, 1978).

The approximation will be exact if two conditions are met (1) The stream incises the scarp after faulting by the full amount of vertical displacement (i.e., full base level recovery), rather than some percentage of vertical displacement and (2) the terraces are *strath terraces* only, without a significant fill component. Condition 1 will often not be met if an ephemeral stream partly incises the scarp and builds an alluvial fan on the downthrown block. Perennial streams are more likely to incise an amount equal to vertical uplift, because they transport sediment away from the scarp base. Condition 2 will not be met if climatically induced stream degradation and/or aggradation have occurred between faulting events. Nevertheless, tectonic terraces provide geomorphic indicators of individual paleoearthquake displacements that can be measured in reconnaissance, without resort to trenching. If terraces exist in a valley but are not preserved at the fault trace, vertical offsets can sometimes be recognized by projecting terrace profiles to the fault plane. Terrace profiles should be usually be based on elevations of the top of the strath surface, rather than of the terrace surface itself (for reasons explained by Johnson (1944)).

Stream incision into the downthrown block is relatively rare. Where most ephemeral or intermittent streams cross fault scarps, the streams have eroded gullies into the upthrown surface, while simultaneously depositing alluvial fans on the downthrown surface at the mouth of each gully (Chapter 2, Figure 2.2). The fault scarp thus defines a transition zone between local erosion into the upthrown block, and local deposition on the downthrown block. If a stream incises into both the upthrown *and* downthrown blocks after faulting (such as in Figures 3.28 and 3.29, stages a3), it is probable that incision is not due solely to tectonic base level fall, but also results from a nontectonic cause such as *climatic change* in the contributing drainage basin (Bull, 1991). Alternatively, the incision on both sides of the fault may predate the faulting.

Climatically induced fill or strath terraces formed by perennial streams may sometimes be mistaken for tectonic terraces if formed prior to faulting. After faulting of such a terrace suite, the parts of the terraces downstream from the fault scarp may be buried by widespread fluvial (or marsh) aggradation. The resulting geometry is terraces on the upthrown fault block that apparently have no counterparts on the downthrown block. These upthrown block terraces might thus be mistaken for postfaulting tectonic terraces, when in fact they are prefaulting nontectonic terraces (Jones, 1995).

By successively reversing the latest displacement, it is possible to reconstruct step by step the relative sequence of erosion and faulting for any number of displaced geomorphic surfaces (e.g., Lensen, 1968; see also Section 2.3.2.7). In the relatively simple, small-scale field examples described in Figures 3.28

earthquake faults the prefaulting surface by amount Ha, after which the stream incises 75% through the scarp to the level of tectonic terrace a, where it stabilizes; below the scarp alluvial fan bed a is deposited. (B) A second faulting event displaces the scarp by the same amount (Hb), after which the stream again cuts down 75% through the scarp and creates tectonic terrace b; below the scarp alluvial fan bed b is deposited. (C) A third displacement (Hc) occurs, and the stream cuts 75% of the way through the scarp and down to its modern level and deposits fan bed c. Preserved parts of tectonic terraces a and b are shown as solid lines, eroded parts by dotted lines. The number of paleoearthquakes is thus one more than the number of tectonic terraces. Note that paleoearthquake displacement can only be measured from the heights of tectonic terraces if the degree to which the stream incised the scarp after each earthquake is known. Only in the case of complete (100%) incision does the difference in terrace elevations equal fault displacement per event.

and 3.29, two assumptions were made that are often not valid for more complex terrace sequences. First, we assumed that the geomorphic surfaces above and below the fault scarp at a given location are the same surfaces, merely displaced by faulting. If this is true, then the topographic relief between the two surfaces (i.e., the scarp height or surface offset) provides a direct measure of fault displacement. Unfortunately, postfaulting deposition on the downthrown block becomes more common with increasing scarp age, and thus heights of old scarps tend to underestimate cumulative vertical displacement (described later). A second assumption in Figures 3.28 and 3.29 is that of no lateral offset of terrace risers, indicating that fault movement was purely dip-slip. If a small component of lateral slip *had* occurred, terrace risers would be offset laterally, but riser offset provides only a minimum estimate of true horizontal displacement (Chapter 6). In the extensional terrains of the western United States, where oblique movement on historic ruptures has been on the scale of decimeters, faulted Quaternary terrace risers rarely exhibit evidence of lateral offset (within the precision of field measurements on degraded risers).

A good field example of fault scarp/landform interaction is the five geomorphic surfaces at an incised fanhead offset by a compound range-front fault in the Rio Grande rift, southern Colorado (McCalpin, 1987b). Vertical surface offsets range from 1.4 m on the youngest surface to 23.4 m on the highest surface (Table 3.4). If the youngest, 1.4-m-high scarp was created by a single, characteristic event, then simply dividing 1.4 m into the surface offsets of higher scarps would yield a first approximation of the number of events represented by these scarps. However, the fault scarps have variable maximum angles (Θ) and displace slopes of variable gradient (α), so reduction of surface offset (SO) to vertical fault displacement (H_1) is desirable before making any interscarp comparisons (Section 3.2.3). Next, if the smallest scarp was created by *two* faulting events, then the larger scarps may have been produced by twice as many, smaller displacements. Clearly, the first task is to confirm by trenching whether the smaller scarps are products of one or two displacements. Trenching the second smallest (3.8-m-high) scarp at this site confirmed that the 3.8 m of displacement was produced in two faulting events, an earlier one with vertical displacement of 2.2 m, and a later one of 1.6-m displacement.

The number of faulting events of known displacement (1.6–2.2 m) necessary to create the higher, untrenched scarps can now be estimated. Dividing the measured single-event displacements into the calculated net displacements based on scarp profiles suggests that the early Pinedale, Bull Lake, and pre-Bull Lake scarps were created by 4–6, 6–9, and 11–15 events, respectively (Table 3.4). These estimates can be erroneous if (1) earlier displacements at this site were substantially larger or smaller than the 1.6- to 2.2-m displacements observed in the latest two events or (2) geomorphic surfaces at each profile site have been significantly raised or lowered by deposition or erosion postfaulting. Complication 1 cannot be assessed without trenching older scarps; complication 2 is likely for the two oldest profiles, as described below.

It is common in arid and semiarid areas for widespread postfaulting deposition to vertically accrete on the downthrown block, of the same sedimentology as prefaulting deposition on the upthrown block. If this occurs the upthrown and downthrown geomorphic surfaces may look like the same surface, but they are not; the lower surface is younger. In that case, the present scarp height underestimates the vertical displacement that created the scarp, in one or more faulting events. Figure 3.31 shows an example where an initial 2 m displacement occurred (a), scarp colluvium was deposited (b), and then the lower 75% of the scarp was buried by vertically accreting alluvium (c). Later another 2 m displacement occurred (d), resulting in a 2.5-m-high scarp with a colluvial wedge (e). However, the 2.5 m scarp height represents neither the displacement of the first displacement event nor of the combined first and second events. It is an artifact of partial scarp burial which really cannot be related to per-event displacements at all. This fact would become obvious if the scarp was trenched (which would expose the relationships shown in stage e), but it may not be obvious from geomorphic evidence alone.

Table 3.4: Fault scarp measurements and inferred paleoseismic history, Rio Grande rift zone

Type of data	Geomorphic surface	Surface offset (SO) (m)	Vertical fault displacement (H_1) (m)	Cumulative number of faulting events		Age of faulted surface (ka)	Slip rate (m/ka)[a]	Average recurrence interval (ka)[b]	
				1.6-m events	2.2-m events			1.6-m events	2.2-m events
Known from trenching	Holocene	1.4	1.6[c]	1[c]		≤10	0.16	≤10	
	Late Pinedale	3.8	3.8[c]		2[c]	15	0.25		7.5
Inferred from scarp profiles	Early Pinedale	8.9	9.1[d]	5.7[e]	4.1[f]	25	0.36	4.4	6.1
	Bull Lake	13.5	14.0[d]	8.8[e]	6.4[f]	140	0.10	15.9	21.9
	Pre-Bull	23.4	24.2[d]	15.1[e]		≥250	0.10	16.6	22.7
	Lake				11[f]				

[a]Net vertical displacement divided by the age of the surface.
[b]Age of the surface divided by the cumulative number of faulting events.
[c]Measured in a trench.
[d]Calculated from Equation (3.1), assuming β = 75° (measured in trench), α = 5° for Pinedale and younger surfaces, α = 7° for older surfaces.
[e]Net vertical displacement divided by 1.6 m.
[f]Net vertical displacement divided by 2.2 m.

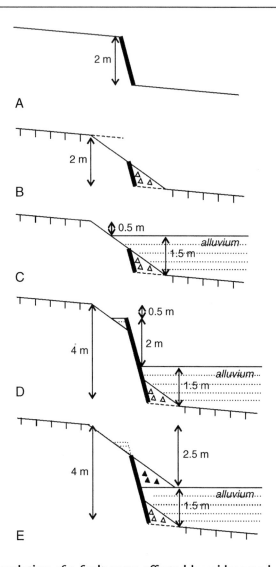

Figure 3.31: Schematic evolution of a fault scarp affected by widespread alluviation on the down-thrown block. (A) initial 2 m displacement of a 5° slope. (B) erosion of scarp and deposition of colluvial wedge. (C) aggradation of 1.5 m of postfaulting alluvium buries the lower 3/4 of the fault scarp. (D) a second 2 m displacement event. (E) erosion of scarp and deposition of second colluvial wedge. The total scarp height (2.5 m) is not equal to either the single-event or double-event displacement. In this case, the best clues that vertical accretion has occurred are: (1) the down-thrown block surface is horizontal, whereas the upthrown surface dips at 5° (but that could also result from back-tilting a single surface), and (2) the soil profile on the downthrown block does not match the soil profile on the upthrown block (vertical lines).

3.2.6.2 Profile of a Compound Scarp

A *compound fault scarp* was defined by Slemmons (1957) as a scarp produced by more than one rupture event [also called a *composite fault scarp* (Stewart and Hancock, 1990) or a *multiple-event fault scarp*]. Compound fault scarps often contain multiple breaks in slope, each of which originated in a separate rupture event. The sequential evolution of these nickpoints is described by McCalpin (1983, his Figure 46) and Figure 3.32 shows a recent example. After the initial scarp formed and degraded to a smooth profile, the scarp in Figure 3.32 was rejuvenated by the 1983 Borah Peak rupture. If *inflection points* are sharp and the scarp profile is essentially composed of linear segments, it may be possible to distinguish single-event displacements on a compound fault scarp even after considerable weathering (Figure 3.33). For example, Haller (1988) used histograms of slope angles versus horizontal distance to emphasize subtle scarp inflections. Distinct multiple breaks in slope on the upper part of a degraded fault scarp may indicate inflection points from earlier faulting events, as suggested by Wallace (1980a, his Figure 12; Wallace, 1984, his Figure 24) for the Madison Range, Montana, and Pleasant Valley, Nevada, fault scarps, respectively. However, it is also possible that these subtle inflections in slope were created by (1) small-scale erosional or depositional events on the scarp face or (2) multiple step faults created during a single event.

Subsurface exposures would show if the inflections were underlain by small-displacement faults or fissures; if so, they are probably primary tectonic features rather than secondary erosional nickpoints.

Figure 3.32: Photograph of a compound fault scarp offsetting alluvial terraces on the 1983 Borah Peak rupture. Rock Creek is barely visible as a line of trees at far right. Person (1.9 m tall) is standing to the left of the crest of the prehistoric fault scarp (C), which had been incised 3.5 m by the stream in the left foreground prior to 1983 (AL). The 2-m-high 1983 fault scarp (unvegetated band at right center) has rejuvenated this older fault scarp and created at new inflection point (C′) that will retreat upslope, eventually regrading the face of the prehistoric fault scarp (area between C and C′). Post-1983 stream incision into the bed of the stream in left foreground (AL) will create a tectonic terrace that will be 2 m high if the stream incises all the way to the bottom of the 1983 fault scarp.

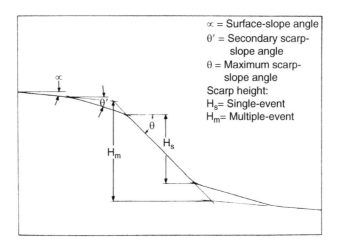

Figure 3.33: Diagrammatic profile of a faceted compound fault scarp, showing how a single-event scarp height (H_s) is graphically determined from the total (multiple-event) scarp height (H_m) and other measures of scarp geometry (see Table 3.3). From Machette (1982); reprinted with permission of the New Mexico Geological Society.

If inflections do *not* overlie faults, however, it might be difficult to distinguish between various erosional origins for the inflections. If the subsurface exposure reveals only a single colluvial wedge, then scarp profile inflections cannot have resulted from multiple faulting events.

Unfortunately, many large fault scarps that are known to be the product of multiple displacements (based on independent geomorphic or stratigraphic relations) do not exhibit multiple breaks in slope. Instead, most of the scarp face is a wide, planar slope at or just below the angle of repose (McCalpin, 1983). On high-slip-rate faults, rejuvenation of the scarp by repeated faulting occurs so often that slope regrading destroys earlier nickpoints (e.g., Wasatch fault zone, Utah; Machette *et al.*, 1992a,b). High, compound scarps that possess large planar faces without nickpoints may indicate short recurrence intervals (a few ka rather than 10s or 100s of ka).

3.3   Stratigraphic Evidence of Paleoearthquakes

Normal surface faulting results in the instantaneous creation of faults, fissures, and tilted beds, and in the delayed response of fault-induced sedimentation. The sequence of paleoearthquakes cannot usually be reconstructed from tectonic or depositional features alone, instead, a combined analysis is required. The key to successful interpretation is to distinguish between tectonic versus depositional features, and to distinguish depositional units that predate faulting from those that postdate faulting.

The concepts presented in this section are derived from studies of many trench exposures of normal faults in the western United States, particularly on the eastern margin of the Basin and Range Province. Detailed trench logs illustrating many of the features described herein can be found in the "Paleoseismology of Utah" series of publications (http://mapstore.utah.gov/ugs/paleosei.htm), as well as in Swan *et al.* (1980), Schwartz and Coppersmith (1984), Forman *et al.* (1989, 1991), Machette *et al.* (1992a), McCalpin *et al.* (1994), and Olig *et al.* (1994).

3.3.1 Characteristics of Near-Surface Normal Faults in Section

Our knowledge of near-surface normal faults comes primarily from paleoseismic trenches, which have been dug in increasing numbers and with increasingly larger dimensions over the past 30 years.

3.3.1.1 Geometry of Faults

When a coseismic normal fault ruptures to the ground surface and creates a fault scarp, it is typically accompanied by secondary faults on both the upthrown and downthrown blocks (Figure 3.34). McCalpin (1987a) compiled statistics on fault geometries as exposed in trenches, in support of crafting land-use regulations in areas of Utah where fault zones were undergoing residential development. As shown in Table 3.5, in unconsolidated material the main (coseismic) fault plane lies roughly beneath the midpoint of the surface fault scarp and dips $77 \pm 10°$ (Figure 3.35). On average, there are two additional secondary faults in the upthrown block within 3 m of the main fault, and four secondary faults on the downthrown block within 13 m of the main fault. These asymmetries arise from fault refraction, as described earlier.

Due to fault refraction in the near surface, normal faults may exceed 90° in dip and assume the geometry of reverse faults (Figure 3.36). However, this does not indicate a compressional stress field, but instead outward bending of the top of the fault toward the scarp face, where confining pressure is least (i.e., a gravitational phenomenon). Because some faults (even the main fault) may bend upward from a normal to a reverse geometry, we recommend avoiding the terms footwall and hanging wall when describing normal faults in vertical section, and to use upthrown and downthrown block instead. Otherwise the reader may become confused as to which side of the fault is the footwall and which is the hanging wall.

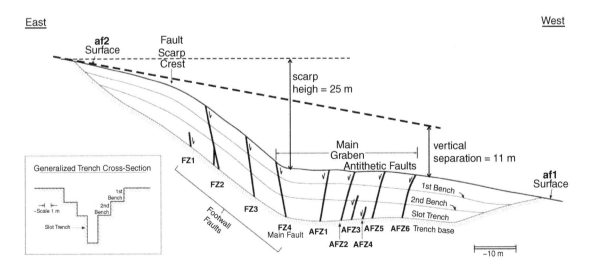

Figure 3.34: Schematic log of structures in the Wasatch fault megatrench on the Provo segment (see photo in Figure 2A.19). The main fault (FZ4) lies at the toe of the scarp, an unusually low topographic position for a normal fault. There are three secondary faults in the upthrown block (FZ1-3) and six secondary faults in the downthrown block, all antithetic (AF1-6). This asymmetry of secondary faulting is typical. Adapted from Olig *et al.* (2004).

Table 3.5: Summary statistics of fault-geometry parameters

Wasatch fault DATA	Main fault[a]		Upthrown block[b]		Downthrown block[c]			
	Position of fault[d]	Apparent dip	No. of faults	Width of deform. zone	No. of faults	Width of deform. zone	% of antithetic[e]	Tilt[f]
No. of Trenches	15	29	29	29	29	29	15	29
Mean	48%	78°	2.8	1.4 m	4.6	14.9 m	14%	9.5°/5.5 m
Std. Dev.	11%	10°	2.0	1.6 m	7.0	26.3 m	13%	16°/13°
Modal Class	35–40%	70–75°	2	0–1 m	1	0–1 m	NC	0–5°
Median	49%	76.5°	2	0.6 m	2	3.0 m	NC	2°
ALL fault DATA								
No. of Trenches	23	40	40	40	40	40	19	40
Mean	45%	77°	3.3	1.8 m	4.1	12.7 m	18%	9.4°/4.8 m
Std. Dev.	14%	10°	3.0	2.4 m	6.7	23.2 m	19%	15.5°/11.6 m
Modal Class	40–45%	70–75°	2	0–1 m	1	0–1 m	NC	0–5°
Median	45%	79°	2	0.7 m	2	3.0 m	NC	2.5°

From McCalpin (1987a).
[a]Fault with the largest displacement.
[b]Includes the main fault and all secondary faults in the footwall.
[c]All structures on the hanging wall.
[d]The position of the main fault plane in relation to the surface fault scarp profile. Measured as the horizontal distance from the base of the scarp to the surface projection of the fault plane, as a percentage of the total horizontal width of the scarp.
[e]The displacement on the largest antithetic fault, as a percentage of displacement on the main fault.
[f]Any tectonic rotation of hanging-wall strata toward the fault (amount of tilt/width of tilted area).

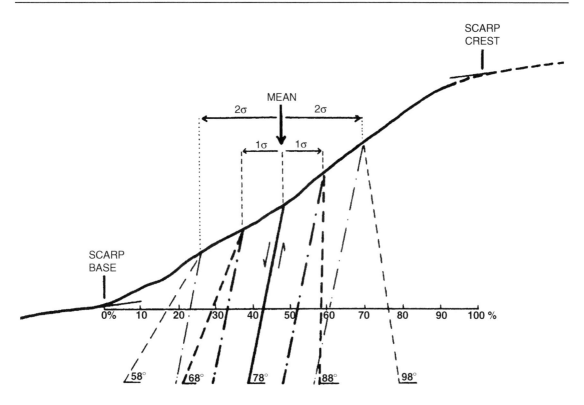

Figure 3.35: Schematic cross section through a normal fault scarp showing the mean position and dip of the main fault plane (solid line at center), with 1 sigma limits on position and dip (thick-dashed lines) and 2 sigma limits (thin dashed lines). Secondary faults are not shown, but their statistics are also given in Table 3.5. From McCalpin (1987).

Where folding is the dominant style of surface deformation, the "blind" normal fault typically underlies the approximate center of the monocline. An example is shown in Figure 3.37. This monocline underwent cultural modification which resulted in ground smoothing and lowering the upthrown block below its assumed original shape (thick-dashed line, a duplicate of the lower contact of folded unit c). The normal fault zone at center is composed of two main traces that bound a forward-rotated structural block. Although these faults displace unit e, they do not vertically displace the top of unit d, having passed from a fault to a fold by that horizon. Due to the ductility of the faulted beds, no tension fissure developed at the crest of the monocline.

3.3.1.2 Architecture of Faults in Section

Normal fault planes in section can vary widely in shape and thickness, depending on the rheology of the faulted material. Cohesive materials tend to develop narrow, sharp fault planes, such as those shown in Figures 2A.33 and 2A.42. Where the upthrown block has a different rheology than the downthrown block, the fault zone may develop two major bounding normal faults with a zone between composed of pieces of relatively intact, but often rotated, strata from either the upthrown block (e.g., Figure 3.36) or the downthrown block (e.g., Figures 3.2 and 2A.44). In the most extreme case there may be multiple fault-bounded "slices" of stratigraphy in the fault zone, with those nearest the upthrown block containing the oldest strata. Very gravelly material seems to yield the most variable fault planes. Where fault planes remain discrete, the main distinguishing feature of the fault plane may be its rotated clast fabric (e.g., Figure 2A.35). However, a more common situation in gravels is the development of a sheared zone, the thickness of which roughly scales with the displacement on the fault (e.g., Figure 2A.39). If fault

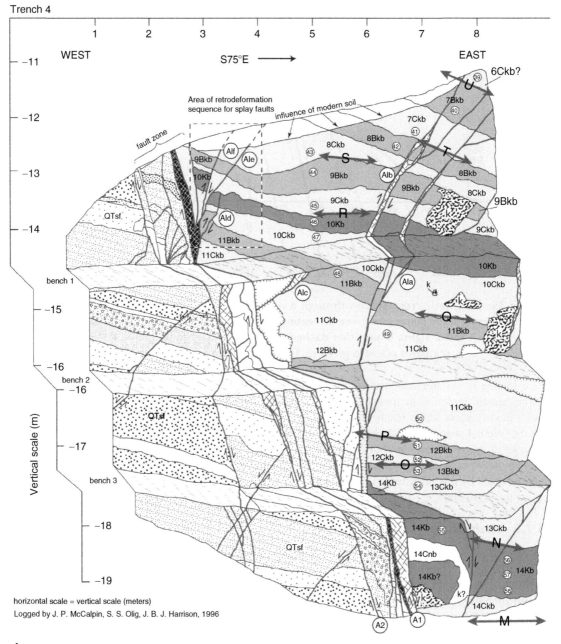

A

Figure 3.36: (Continued)

Figure 3.36: Exposures of the County Dump fault, New Mexico, showing secondary "pseudore-verse" faults that splay off the main normal fault. (A) Log of trench 4. Numbers at left indicate depth in meters beneath the ground surface. The main fault zone (faults A1, A2) displaces middle Pleis-tocene paleosols 7–14 (indicated by shaded horizon horizons at rights) against Pliocene alluvium of the Santa Fe Group (stippled patterns at left). Splay faults A1a through A1f have an apparent reverse geometry. From McCalpin *et al.* (2006). (B) Photograph of trench 4.

refraction has created "space problems," the high degree of extensional stress at the surface will lead to the development fissures and fault-bounded blocks separated by fissures (Figure 3.38).

3.3.2   Distinguishing Tectonic from Depositional Features

Most geologists have no difficulty recognizing faults in bedrock, but faults in unconsolidated deposits are often more subtle and can have many modes of expression (Section 2.3.2.5). Normal fault scarps form in an extensional stress field, so their characteristics in unconsolidated materials differ from those of reverse and strike-slip ruptures. According to Bonilla and Lienkaemper (1991, Table 17) normal faults are more likely to have *fissures filled with rubble*, and less likely to show *gouge, slickensides, breccia, crushing*, or *polishing* compared to reverse or strike-slip faults. If anomalously fine-grained material is exposed along the fault plane in shallow exposures, it is more likely to be entrained clay dragged up along the fault from a subsurface layer (e.g., Eichhubel *et al.*, 2005), than a true cataclastic fault gouge. It is difficult to mistake *erosional* or *depositional* contacts in unconsolidated deposits for normal faults because faults almost always have steeper dips (55°–90°) than erosional contacts. In addition, many clasts along normal fault planes will be rotated such that long axes approximately parallel the fault. Paleoseismologists often refer to this orientation as *shear fabric*, although it has never been reproduced by laboratory shearing experiments. Finally, faults are straighter than most erosional contacts, and often steepen upward.

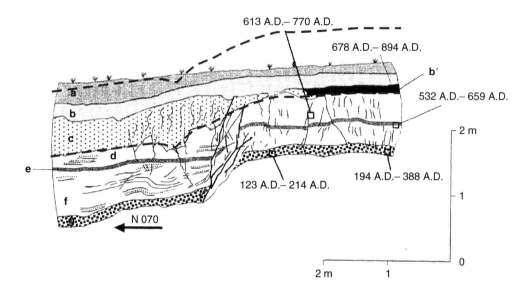

Figure 3.37: Log of Trench 1 across the Bree normal fault, lower Rhine graben, Belgium. This trench wall displays a slightly faulted monocline in well stratified, sandy modern alluvium. The earthquake horizon is inferred to be the unit b (or b′)/unit c contact. The shape of the monoclinal fold can be seen best on the unit c/unit d contact, highlighted by a thick-dashed line. This line has been duplicated at the modern ground surface of the hanging wall to show the approximate shape of the monoclinal surface scarp, prior to erosion of it and the footwall. The maximum slope of the recon-structed scarp was 25°, below the angle of repose, which explains why there is no colluvial wedge on the hanging wall. Log from Meghraoui *et al.* (2000), with dashed lines added by the author.

Despite these criteria, it is often difficult in practice to distinguish sheared *in situ* material from fissure-filling materials or fallen blocks of free face material caught up in a complex normal fault zone. Major construction projects have been delayed and millions of dollars spent to define properly such microstratigraphic relations (e.g., Asquith, 1985). For example, a distinction that must be made at nearly every fault exposure is whether scarp-derived colluvium is in depositional contact or fault contact with upthrown block strata. A depositional contact implies only a single faulting event which is older than the colluvium, whereas a fault contact requires two displacement events—the first to generate the colluvium, and the second to fault it. Due to the tension fissures that are common along normal faults, fault zone stratigraphy is often a confusing assemblage of sheared *in situ* deposits, material that has fallen into fissures in intact blocks, partly disaggregated blocks, and material washed into depressions by running water. Each exposure is different in detail, but Table 3.6 lists some general field criteria for distinguishing between several types of contacts near normal faults.

3.3.2.1   Fissures and Tension Cracks

Fissures and tension cracks (i.e., mode 1 purely extensional cracks) are not unique to the extensional tectonic environments and can also appear in strike-slip (transtension) and reverse (hanging-wall collapse) environments. In recent years, some geologists have encountered fissures of unknown origin in a seismic area and have interpreted them as the result of paleoearthquakes. In a general sense, these fissures are

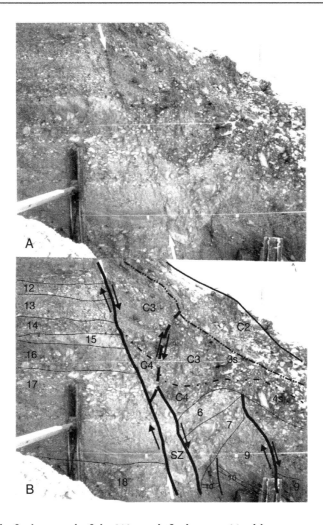

Figure 3.38: The main fault strand of the Wasatch fault zone, Nephi segment, as exposed in the Red Canyon trench (Jackson, 1991). (A) unannotated photo. (B) annotated photo. Units labeled "C" are scarp-derived colluvium; "s" denotes a weak buried soil horizon. Typical features of extensional faulting in gravels are the shear zone (sz), the rotated fault block containing units 6–10, and the refraction of the middle fault to an overhanging dip. Unit C4 may be partly composed of fissure fill.

subvertical, downward-tapering zones bounded by sharp fractures, that are filled with younger sediments, usually from overlying stratigraphic units. These fissures are interpreted as tension cracks that opened coseismically, often immediately before the deposition of the overlying sediment fill. Such an interpretation assumes that the features are coseismic and that the paleoearthquake that created them occurred immediately before the deposition of the overlying sediment. But before we uncritically accept this interpretation, we should apply several observations and criteria to interpreting the so-called "fissures."

Table 3.6: Criteria for distinguishing a normal fault contact from a depositional contact in unconsolidated sediments

Characteristic	Fault contact	Crack fill	Depositional contact
1. Consistence of material	a. Material in fault zone is sometimes softer than that to either side (strain-softening), unless plastic beds have been smeared along the fault. Rotation of clasts and subsequent dilation due to shearing create voids in the fault zone. Rarely, infiltrating water cements the fault zone, making it harder than adjacent materials; groundwater staining is usually obvious	b. Crack fill material is softer than adjacent wall material if it fell into the crack during the latest faulting event, and has not subsequently been sheared. Material hardens with successive shearing events, eventually approaching the hardness of hanging-wall material. Animals preferentially burrow into crack fill, creating many krotovinas	c. Material at the contact is similar in hardness to material on either side
2. Clast orientation	a. Most clast long axes are oriented parallel to fault dip (50-90°), forming a shear fabric	b. Crack fill contains many clasts with vertical or near-vertical long axes, which fell "headlong" into the crack. Most clasts have relatively steep orientations, becoming less steep toward the top of the crack fill	c. Only some clasts are parallel to the contact. Wide variations exist in clast long-axis orientation, with a modal plunge value near the angle of repose for colluvium (ca. 30-40°)
3. Truncation of layers in the abutting colluvium	a. Layers in colluvium maintain constant thickness up to the fault contact; layers are either cleanly truncated by the fault or if plastic, may be drawn out along the fault plane to form a mixed zone	b. Layers are abruptly truncated at the crack margins; there is no mixed material smeared out along crack sides	c. Layers in colluvium steepen within 5-10 cm of the depositional contact and begin to thin. Dip of layers approaches, but does not attain, the dip of the contact. No smearing or mixing of plastic beds occurs at the contact

We should ask these general questions about the fissures:

1. HOW WAS THE VOID SPACE FORMED? (By a tensional opening? By horizontal expansion [as in an ice wedge] or expansion followed by contraction [as in shrink-swell cracking]? By erosion along a zone of weakness? By downdropping along vertical faults, i.e., a graben?).

2. WHAT IS THE VOID SPACE FILLED WITH? (By recognizable parts of an overlying stratigraphic unit? By recognizable parts of an underlying unit? [implies upward injection]. By material derived from the walls of the fissure only? By material totally unrelated to any other in the outcrop?).

3. WHEN WAS THE VOID SPACE FILLED? (Before or after the deposition of the overlying younger sediments? An open tension fissure will be filled with younger sediments. In contrast, a narrow graben will be filled with sediments that are older than the "fissure").

4. HOW WAS THE VOID SPACE FILLED? (Subaqueous or subaerial? Did filling material wash in, fall in, or was downfaulted in?).

Suggested Criteria:

a. What is the lithology and provenance of the fracture-filling material; did it really come from the overlying sedimentary unit, or from somewhere else (such as the bedrock walls of the fissure)?

b. What is the internal fabric and stratification of the filling material? Does it look like it was deposited subaqueously in an open void space (subhorizontally stratified, sorted)? Or, does it have a vertical fabric, as if it fell into an open fissure subaerially? Or, is the fabric completely random, as if it slumped into the fissure as a saturated mass? If the material contains clasts, are the clasts randomly oriented, horizontally stratified, dip inward from the edges, or vertical throughout? If the material contains a finer matrix, what is its fabric?

c. Are the margins of the filling material sheared? Do they have shear fabric that is unique to the edges of the deposit, which does not continue into the center of the deposit?

Structural Setting of Fissures: Fissures are found in several structural settings in extensional environments.

1. Fissures that form along the main fault plane, due to fault refraction, but which do not collapse immediately and remain open after earthquake shaking stops. Such fissures, up to 30 cm wide at the surface, were observed along the main scarp of the Borah Peak earthquake. Subsequent trenching investigations on normal faults elsewhere have documented sets of "nested" fissure fills from different paleoearthquakes up against the main fault, particularly where the fault juxtaposes hard pre-Quaternary bedrock against softer Quaternary colluvium or graben fill (Figure 3.39; see also Gutierrez *et al.*, 2008a). Cross-cutting relationships, and the physical connection of each fissure with a different graben-fill unit, demonstrate that the nested fissures resulted from different paleoearthquakes (see "Fissure-Graben Model" in Section. 3.3.3.2).

2. Fissures that form due to folding, for example, on the crest of a monocline (see Figure 3.20).

3. Fissures on the downthrown block above a refracted fault. These fissures form in response to distributed extension above the refracted fault plane.

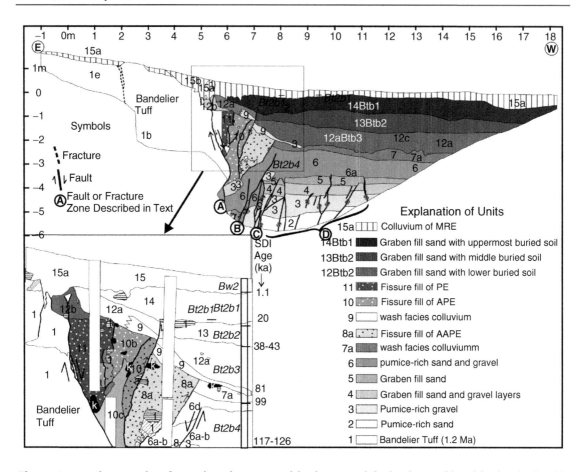

Figure 3.39: Fissures that formed against an antithetic normal fault plane of hard bedrock (in this case, partly welded Bandelier Tuff). On the log of trench 97-4 on the Pajarito fault, Los Alamos, New Mexico, units 11, 10, and 8 represent successively formed and filled, "nested" fissures formed during the penultimate earthquake (PE), the ante-penultimate earthquake (APE), and the ante-ante-penultimate earthquake (AAPE). From McCalpin (2005a).

4. Fissures not near the fault plane. These could represent a response to amplified ground motion, lateral spreading, liquefaction, subsidence, soil shrink-swell, or many other surficial processes. If there is not clear structural connection between these fissures and a coseismic fault plane, nonseismic origins should be considered.

3.3.3   Sedimentation and Soil Formation in the Fault Zone

Coseismic surface rupture creates a severe disturbance to the ambient erosional and depositional processes operating near the fault trace. In normal faulting these vertical elevation changes result in process responses that create a recognizable and predictable stratigraphy. This realization, made in the 1970s (e.g., Swan *et al.*, 1980), meant that a separate line of field evidence was available to supplement observations on tectonic landforms, on which fault interpretations had traditionally been based. Several sedimentologic models have been developed to interpret fault-zone stratigraphy in terms of earthquake occurrence. The most common model worldwide is the colluvial wedge model. However, two other models are described for special conditions related to eolian sedimentation, and upslope-facing scarps.

Because the pioneering work was performed in the semiarid climates of the western United States, there is an unavoidable emphasis in the following discussion on certain types of sedimentation (loess deposition, debris flows, ephemeral sag ponds) typical of arid and semiarid, sparsely vegetated terrain (see Nelson, 1992b; Gerson *et al.*, 1993). Although other climatic regimes experience processes that would require modification of the model described herein (e.g., tree throw, cryoturbation), the observations made in the western United States should apply to a wide variety of climates.

3.3.3.1 The Colluvial Wedge Model

The *colluvial wedge model* is a conceptual model that utilizes the stratigraphy of scarp-derived colluvial deposits to interpret faulting history, much as Wallace's (1977) scarp degradation model did with scarp morphology. Many workers observed that, after formation of historic fault scarps, the loose material exposed on the scarp face (*free face* of Wallace, 1977) falls to the base of the scarp and creates a wedge-shaped deposit of colluvium that overlies the prefaulting surface. Subsequent work on prehistoric scarps (particularly on the Wasatch fault in the late 1970s) showed that the *colluvial wedge* shed from a scarp eventually buried the lower part of the free face, after which the scarp became relatively stable. During stability a soil would form on the colluvium. Subsequent faulting would then lead to deposition of a second colluvial wedge analogous to the first. Multiple faulting should therefore be represented by a succession of vertically stacked colluvial wedges on the downthrown fault block separated by soils, each wedge representing deposition following a surface-rupturing event (Figure 3.40). Japanese workers

Figure 3.40: Photograph of two superposed colluvial wedges (C1, C2) in a trench across the Bear River fault zone, Wyoming (see West, 1993). The fault is indicated by F. Light-colored, gravelly colluvial wedges taper to the right and are under- and overlain by dark organic soil horizons. Upthrown block to left of fault is composed of Eocene claystone. Radiocarbon ages from soils underlying the two wedges indicate faulting at ca. 2.4 and 4.6 ka. Reprinted with permission of the Geological Society of America.

(e.g., Okada *et al.*, 1989) term the pattern of colluvial wedge, over- and underlying soils, and the fault plane as the *D-structure* for its resemblance to that letter of the Roman alphabet. However, they emphasize that the wedge-shaped deposit of colluvium is a less important indicator than the absence of the lower soil on the upthrown fault block, because lenses of colluvium can be deposited at the foot of a steep scarp by nontectonic processes. Pantosti *et al.* (1993) later termed the ground surface at the time of a paleoearthquake as an *event horizon*, although the term *earthquake horizon* is now widely used. An earthquake horizon is stratigraphically defined by either scarp-derived colluvium that buries the prefaulting surface, and/or by unconformities that develop as a result of warping and subsequent deposition. The unconformities typically develop on the downthrown block in fluvial or lacustrine sediments (see Section 3.3.2.2). Therefore, the number of event horizons should equal the number of paleoearthquakes. This simple sedimentologic model has been applied to dozens of faults in the western USA to identify from one to four paleoseismic events (e.g., Forman *et al.*, 1989, 1991; Machette *et al.*, 1992a).

After complete disappearance of the free face, slopewash, rainsplash, and creep processes dominate on the scarp slope. Colluvium deposited by these processes (*wash element* in Figure 3.41) is distinctly finer grained, better sorted, better stratified, and typically richer in organics than debris facies colluvium (Nelson, 1992b). There is often an abrupt contact between debris- and wash-facies colluvium, which is inferred to represent the abrupt disappearance of the free face. If wash facies deposition rates slow sufficiently, the scarp slope will develop a soil profile.

Not every wedge fits the simple model previously described for two reasons. First, the wedge deposited from a simple fault scarp on a subhorizontal geomorphic surface is thickest at the fault and tapers downslope, forming an obvious wedge shape. Colluvium from successive events, however, is deposited on the sloping surfaces of earlier colluvial wedges, which causes later wedges to extend farther downslope from the fault, and be thinner and less wedge shaped than earlier wedges (Ostenaa, 1984; Figure 3.42). In these elongated colluvial deposits from later faulting events the distinction between debris- and wash-facies colluvium is commonly blurred. This loss of distinction between colluvial facies also occurs below fault scarps on steep far-field slopes ($\geq 25°$).

Interpreting the subsurface relations in a trench where colluvium abuts bedrock on a steep slope (e.g., Sullivan and Nelson, 1983, Figures 4 and 5) can be ambiguous. Stewart and Hancock (1988, their Figure 3) claim that the colluvium-bedrock contact will display "tectonic breccias and subordinate slip planes in the Quaternary sediments" if the contact is a fault contact, rather than a depositional contact. Sullivan and Nelson (1983, their Figure 5) argued that because "no evidence exists of shearing along the contacts of the colluvial deposits with the bedrock," and "colluvial units clearly truncate the gouge and breccia of the fault zone," a bedrock fault in Utah had experienced no late Quaternary movement. Criteria for distinguishing fault contacts versus colluvial contacts in trench exposures are described in Table 3.6. Where normal fault scarps are created across very steep ($>45°$) bedrock slopes, the colluvial products (rockfall) may be transported completely off the slope.

Second, sediment traps formed by complex fault scarps may alter the patterns of deposition at the scarp base. Scarps fronted by large tension fissures commonly have abnormally small colluvial wedges for their scarp heights, because much of early colluvium fell directly into the tension fissure. Likewise, much basal colluvium may be required to fill a graben before a transportational slope away from the scarp is established. To further complicate matters, the style of surface rupture at a given site may change with successive events, such that a simple scarp may form in the first event, a graben may form in the second event, and a somewhat wider graben may form in the third event. Deformation occurring during later events can obscure the original geometry of earlier colluvial wedges. The key to correct interpretation is

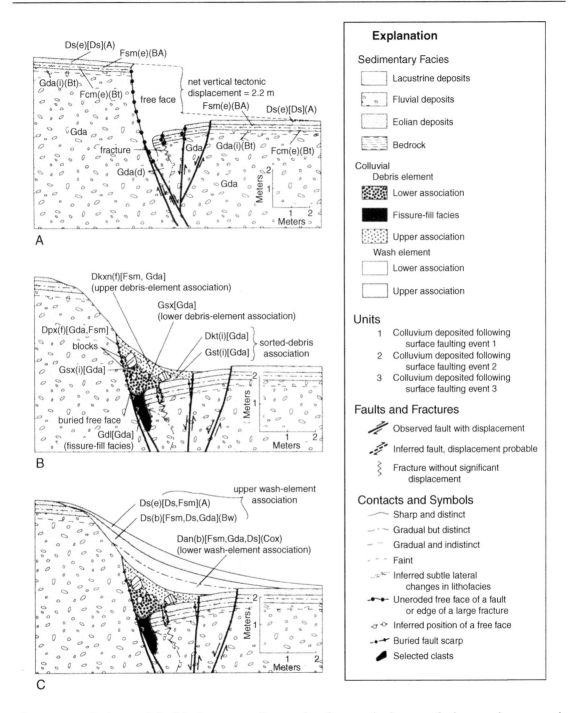

Figure 3.41: Facies model of fault scarp sedimentation from a single-event fault scarp in unconsolidated material. (A) Fault scarp immediately after formation. Diagram shows the basal tension fracture and small-displacement antithetic faults that typically form in unconsolidated sediments.

to reconstruct a sequence of cross sections (retrodeformation, see Section 3.5) which successively reverse the fault displacements and restore deposits to their prefaulting configurations.

The colluvial wedge model states that each surface rupture is followed by the formation of a discrete colluvial deposit. Therefore, simply counting the number of colluvial wedges should yield the number of paleoseismic events at a site. To do this, one must be able to distinguish between fault-scarp-derived colluvial wedges and other types of deposits that are typically present in fault-zone exposures.

Identifying a Colluvial Wedge   A colluvial wedge in a fault-zone exposure can be identified from its contact with the fault, the shape of the deposit, and its sedimentology. The part of the wedge closest to the fault can be either in depositional or fault contact with footwall deposits (for criteria, see Table 3.6). For example, colluvium shed from a single-event fault scarp is in depositional contact with the degraded fault free face (Section 3.2.4). At the base of the colluvial wedge the buried free face overlies a fault contact between prefaulting deposits. The downward transition from a depositional contact (typical dip 40°–60°, no shear fabric) and a fault contact (typical dip 78 ± 10°, with shear fabric) is a key criterion for locating the base of a colluvial wedge. The lower contact of the wedge is an unconformity over the buried prefaulting ground surface; if back-tilting has occurred it is an angular unconformity. The upper contact of the wedge is either the modern ground surface or an unconformity with the next overlying wedge. Colluvium from a multiple-event fault scarp consists of the unfaulted colluvium from the latest event, which overlies the faulted colluviums from earlier events (Figure 3.42).

Sedimentology can also be used to distinguish colluvial wedges from other deposits in a fault zone. Nelson (1992b) proposed a *facies model* for deposition at the base of a single-event normal fault scarp. Two discrete facies of scarp-derived colluvium are defined: *debris facies* and *wash facies* (Figure 3.41). The debris facies is formed by initial spalling of intact blocks and large rocks from the free face, followed by slumping, sliding, and rolling of loose debris to the foot of the scarp (Figure 3.43). The basal colluvium in a wedge usually includes the largest clasts and often contains intact blocks of soil horizons or discrete beds from the upthrown block. The toe of many debris wedges is composed of large clasts with little or no matrix, and represents the largest rocks that have rolled to the tip of the wedge. Nelson (1992b) terms this the *sorted debris facies*. The bulk of the debris facies is usually an unsorted, unstratified deposit which, if rich in clasts, shows a downslope fabric. However, many wedges derived from gravel show internal bedding, with clast-supported layers alternating with matrix-supported layers. These layers have been observed within the post-1983 colluvium at the Borah Peak, Idaho, rupture (Figure 3.43) and are thus inferred to result from seasonal geomorphic processes acting on the free face and do not have tectonic significance (McCalpin and Forman, 1988). Preliminary studies of colluvial clast fabric (McCalpin *et al.*, 1993) show that several orientation subpopulations are present, representing clasts that either slid, rolled, or twirled onto the wedge surface.

No colluvium has yet been deposited. (B) Scarp after deposition of the fissure-fill facies (solid black) and the upper and lower associations of debris-element colluvium (dotted patterns). The lower association contains more intact blocks of free face materials (including soil blocks and vegetation from the crest of the free face) than does the upper association. At the end of stage (B), the near-vertical free face has been destroyed by slope decline. (C) Scarp after deposition of wash-element colluvium. For detailed explanation of facies abbreviations, see Nelson (1992b). From Nelson (1992b); reprinted with permission of the Society for Sedimentary Geology.

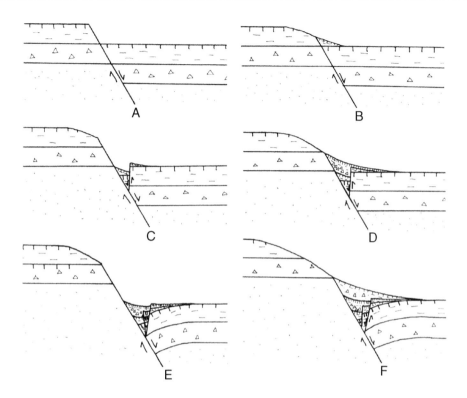

Figure 3.42: Schematic diagrams showing how the provenance of scarp-derived colluvium may vary through successive fault displacements. Vertical ticks indicate soils. (A) First faulting event creates a free face entirely in the unit marked by short dashes. (B) Deposition of the first colluvial wedge, composed exclusively of material derived from the short-dashed unit. (C) Second faulting event creates a basal tension fissure into which the earlier colluvial wedge is dropped. The lower part of the free face at the main scarp is composed mainly of the unit marked by triangles. (D) The second colluvial wedge fills the tension fissure and then prodgrades out onto the downthrown block. The lower part of the second colluvial wedge is derived mainly from gravity and debris deposition from the lower part of the free face and is thus composed of material from the triangle unit. As the scarp continues to backwaste and decline, more colluvial material is derived from the upper part of the scarp (upper short-dashed unit). (E) Third faulting event creates a free face in the dotted unit and a new antithetic fault farther to the right of the earlier antithetic fault, which was not rejuvenated in this event. Most of the free face exposes units marked by dots and triangles. (F) The third colluvial wedge buries the earlier two wedges. The sequence of lithologies in the third wedge roughly parallels the stratigraphic sequence exposed in the scarp face; that is, the basal portion is derived from the dotted unit, and the upper (wash facies) portion is mainly derived from the short-dashed unit.

It is often difficult to distinguish the base of scarp-derived colluvium(s) from the top of the prefaulting deposits of the downthrown block, especially if the latter are diamictons. The most direct approach is to examine carefully the uppermost deposit on the upthrown fault block, and try to identify that same deposit underneath scarp-derived colluvium on the downthrown block. A second check is that the volume of scarp-derived colluvium should approximately equal the volume of material eroded from the scarp crest.

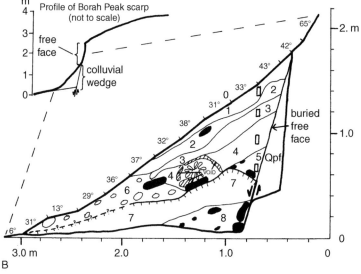

Figure 3.43: The debris-facies colluvial wedge deposited after the 30 Oct 1983 Borah peak, Idaho, earthquake. (A) Photo of the wedge taken in 1987. (B) sketch of colluvial wedge in 1987 (different location than photo); Qpf, Pinedale fan alluvium, age ca. 20 ka; 7, 8, prehistoric proximal colluvium capped by a weak soil (vertical ticks); 6, sorted debris wedge, post-1983; 5, fissure fill, post-1983; 4, earliest debris-facies wedge (1983–1984), including sagebrush that fell from top of free face; 1–3, later debris-facies strata, probably representing spring/summer of 1985 (unit 3), 1986 (unit 2), and 1987 (unit 1). Vertical boxes are thermoluminescence samples; sample 3, probably deposited 2 years prior to sampling, yielded a TL age of 40 ± 4 ka. This result indicates that not only was the sample not re-zeroed during deposition in 1985, but it must have contained 20 ky worth of "inherited" TL signal when it was deposited in the alluvial fan at ca. 20 ka. From McCalpin and Forman (1988).

The area of the eroded material can be estimated by projecting the fault plane upward at a reasonable initial angle (not less than the dip of the fault plane) until it intersects with the projection of the upthrown faulted geomorphic surface. This area should be equal to the area of scarp-derived colluvium on the same trench log. If the eroded area appears significantly larger than the area of colluvium, then the base of the colluvium may still be beneath the trench floor or, alternatively, colluvium could have been removed by lateral stream or shoreline erosion at the base of the scarp. However, the area of colluvium is commonly larger than the inferred area eroded, due to eolian additions to the colluvial wedge (McCalpin, 1983).

3.3.3.2 Other Fault-Zone Facies

Scarp-derived colluvium usually constitutes the largest volume of postfaulting deposits in a normal fault zone, as long as the fault scarp faces downslope, and backtilting and graben formation are insignificant. However, when scarps face upslope, or graben exist, other facies can be significant. The relative volume of these other facies increases with the degree of topographic closure and sediment trapping, for example, on the size and depth of back-tilted areas and graben, or the height of an upslope-facing scarp and the size of the drainage basin it blocks. Topographic traps created by normal fault scarps are generally filled by a combination of *laterally accreting colluvium* shed from the scarp (described previously) and *vertically accreting graben deposits* of various origins. The type of vertical accretion deposits and their deposition rates are mainly controlled by nontectonic factors, such as the geomorphic regime operating on the faulted surface. In arid and semiarid areas, direct airfall loess or eolian sand may fill the depression. Semiarid graben is often filled with sediments derived locally from small gullies that incise the upthrown block. Small alluvial fans form at the mouths of each gully, whereas distal debris-flow sediments spread into the remainder of the graben during severe precipitation events. The intermittent ponds that form during wet seasons may receive vertically accreting sediment from loess or suspended sediment from rare debris flows and stream flows (McCalpin *et al.*, 1994). In more humid climates, perennial ponds in graben (sag ponds) fill with marsh sediments and lacustrine deposits.

Vertically accreting graben deposits interfinger with contemporaneous, laterally accreting colluvium at the toe of the fault scarp (Figure 3.44). The position of the facies interface is determined by the relative rates and loci of deposition of the two facies. Immediately after faulting, colluvium is deposited mainly near the free face. As the scarp broadens with weathering, wash-dominated colluviation (Nelson, 1992a) will push the facies interface away from the scarp. In dry environments wash-facies deposition may prograde completely across the graben between rare periods of fluvial or lacustrine conditions.

Unconformities occur in both the colluvial and graben fill sequences, but their tectonic significance is different. In the colluvial wedge sequence, a soil buried by coarse colluvium usually indicates renewed deposition from a free face after faulting (Swan *et al.*, 1980). In the graben fill sequence, however, renewed deposition atop a soil only signals a change in eolian, fluvial, or lacustrine depositional processes that may or may not be related to faulting. Renewed faulting may create or increase topographic closure in grabens and back-tilted areas. This faulting then sets the stage for renewed graben deposition, but does not necessarily provide for the necessary sediment transport. However, if streams cross the fault nearby and empty into the graben, local stream incision into the upthrown block after faulting should cause a rapid increase in the rate of fluvial/lacustrine deposition in the graben. *Angular unconformities* are almost certainly caused by tectonic deformation and are thus event horizons.

Buried soils in the colluvial wedge sequence can often be traced into the graben where they are buried by lacustrine or marsh sediments that indicate temporary ponding, as described previously. Upsection the graben sediments typically change to a more eolian- or fluvial-dominated facies, indicating that the initial closed depression has filled. Thus, unconformities that are continuous throughout both laterally and

Figure 3.44: Interfingering of scarp-derived colluvium (deposited from left) with graben-fill sediments (far right) beneath the toe of a normal fault scarp. String lines, and tape squares on lines, are 1 m apart. The lower concentration of coarse gravel (between arrows) is the downslope tip of the sorted debris facies of Nelson (1992a). The sorted debris facies is underlain by a matrix-supported soil horizon (S) that predates faulting, and is overlain by finer gravel (UD) of the upper association of the debris element. The upper 0.5 m is generally wash-element colluvium (W), with a concentration of stones at center (a second sorted debris facies) probably induced by human disturbance of the scarp face. On the right margin of the photo fine-grained graben sediments (G) are in contact with the scarp-derived gravels. Photo taken at the middle Sheep Creek trench, Grey's River fault, Wyoming (Jones and McCalpin, 1992; Jones, 1995).

vertically accreting deposits probably represent individual paleoearthquake ruptures (i.e., are earthquake horizons). Additional unconformities in the graben sequence may merely reflect deflection of streams or debris flow into the graben from geomorphic/climatic events that have no tectonic significance. Correlation of individual graben strata to the colluvial wedge sequence becomes important when datable materials are found in the former but not in the latter (McCalpin *et al.*, 1994).

The Eolian Deposition Model   In some arid and semiarid regions eolian sedimentation is so active it dominates over scarp-derived sedimentation even right at the fault scarp, and obscures colluvial wedges (e.g., Personius and Mahan, 2000, 2003; McCalpin *et al.*, 2006). According to Personius and Mahan (2003), on upwind-facing scarps, the eolian/deposits form broad, thin sand lenses of massive to weakly bedded, fine to coarse sand that overlap and bury the scarp between faulting events. The sand lenses occupy a geomorphic position similar to sand ramps that climb up slopes facing the prevailing wind direction. If there are several stacked eolian lenses of different ages, the younger ones occupy successively lower positions on the scarp profile, suggesting that as the scarp grows, it becomes harder for sand lenses to climb up it and bury it. Where these lenses are well developed, there are typically little to no deposits in fault exposures that can be recognized as scarp-derived colluvium.

Personius and Mahan (2003) concluded that these eolian lenses "were deposited rapidly after individual surface-faulting events and that their thicknesses may approximate the amount of throw of individual surface ruptures." Evidence to support this assertion includes: (1) the tops of all the eolian sand lenses or sheets are marked by buried soils and upward fault terminations, and (2) there are no buried soils or significant stone lines *within* the eolian deposits.

On scarps that face downwind, the pattern is different (McCalpin *et al.*, 2006). Sand sheets are deposited on the scarp between faulting events, being slightly thicker on the downthrown side of the fault, and a soil profile develops. When faulting occurs it downdrops the sheet on the hanging wall and preserves it, such that repeated faulting creates a stack of sand sheets/paleosols on the downthrown block (Figures 2A.31 and 3.36). These paleosols are cleanly truncated by the fault plane, such that deposit texture does not coarsen near the fault, nor (usually) do the sheets contain any recognizable material derived from erosion of the fault footwall. These observations indicate that the sand sheet originally extended onto the upthrown block for some distance (i.e., mantled the scarp), was truncated by faulting, and then the part on the upthrown block was eroded away (deflated?).

Over a long period of faulting the eolian/paleosol stratigraphic sequence preserved on the downthrown fault block may accumulate to tens of meters thick. An example is shown in Figure 2A.31, where the sequence is about 17 m thick (County Dump fault, New Mexico). Note that this wedge-shaped deposit contains alluvium as well as eolian deposits and paleosols. Because none of the wedge is composed of scarp-derived colluvium, it is incorrect to call it a colluvial wedge; a more correct term would be fault-angle depression fill. On the nearby Zia fault, part of the fault-angle depression fill is composed of fine-grained paludal (marsh) deposits, indicating period ponding of water against the toe of the scarp.

The Fissure-Graben Model In some normal-fault surface ruptures fissure fills replace colluvial wedges as the primary paleoearthquake indicator. For this to occur, the fissure must be located at the base of the coseismic free face and be large enough to accommodate all the material that erodes from the free face between earthquakes, such that a colluvial wedge never develops. A common situation where such large-scarp-base fissures form is where the upthrown block is composed of pre-Quaternary bedrock, the "free face" is mainly composed of rock rather than unconsolidated material, and the fault refracts enough to form a fissure which does not collapse during the earthquake shaking. As shown in Figure 3.39, repeated faulting under these conditions leads to nested fissure fills.

In the special case where aggradation of the downthrown block also occurs between earthquakes, then successive paleoearthquakes will create a repetitive sequence of paired deposits, consisting of coseismic fissure fills and interseismic graben fill deposits (Figure 3.45). McCalpin (2005) termed this sedimentologic model the fissure-graben model, and suggested it could be used to identify and date paleoearthquakes where scarp-derived colluvial wedges had not formed.

3.3.3.3 Angular Unconformities in Fault Zones

Angular unconformities in normal fault zones are typically associated with either faulting accompanied by back-tilting or the formation of asymmetric graben, or by monoclinal folding or normal drag along the fault plane. In back-tilted areas, the dip of strata may steepen with increasing age and depth, indicating progressive tilting (usually down toward the main fault) by repeated faulting. The differences between the dip of successive strata provide a basis for reconstructing a chronology of faulting that is independent from colluvial wedge evidence, but which can be physically traced (via facies interfingering) to scarp-derived colluvium. When *restoring* graben strata to their pretilt orientations (i.e., retrodeformation

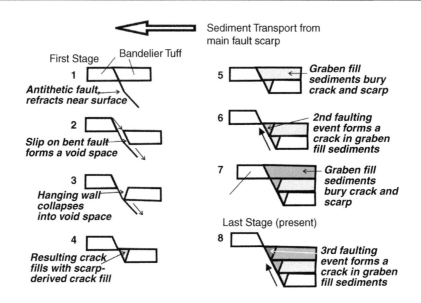

Figure 3.45: Conceptual model of fissure formation and graben filling. The first earthquake (2) creates a fissure (3) which then fills (4). The small volume of material eroded from the bedrock free face is entirely trapped in the fissure, so a colluvial wedge does not develop. The downdropped block is then buried by graben aggradation (5). A second faulting event creates an identical-size fissure in the graben sediments, which then fills (6). The graben again aggrades (7). This process repeats, creating a series of nested coseismic fissures along the main fault plane, and superposed interseismic graben fill deposits in the graben. Paleoearthquake timing is derived from dating samples from graben-fill units; the coarse-grained fissure fills are typically not datable by radiocarbon or luminescence. From McCalpin (2005a).

analysis), one must assume that the laminae within fine-grained deposits (silts, clays) in the graben were initially horizontal. A second assumption is that the initial angle of repose of proximal scarp-derived colluvium has not changed significantly through time; the present surface slope on proximal colluvium (typically 35°–40°) and clast long-axis plunges of 30°–35° thus indicate initial bedding angles in older colluviums.

An example of this type of analysis comes from the Wasatch fault zone, Utah (McCalpin *et al.*, 1994). Measurements used in this analysis include (1) dips of graben lithofacies contacts and (2) dips of the upper and lower contacts of colluvial wedges and dips of clast long-axis fabrics (Figure 3.46). Five of the contacts between units in the graben are defined by the upper contact of laminated or massive silts and clays (dashed pattern) and are presumed to have been originally horizontal. Three other contacts that define the tops of fluvial sands and gravels (circle pattern) are presumed to have been nearly horizontal when deposited along the graben axis. The paleoground surfaces of two distal scarp-derived colluvial units (contacts between soil S1/6A; soil S3/8A) presumably had a slight initial valleyward gradient (2°–4°?) when they became stabilized. By reversing the effects of the latest event and restoring some graben units to horizontal, the sequence of tilting events that accompanied faulting (Figure 3.46) is reconstructed.

Downthrown block strata are occasionally tilted forward (away from the main fault) by faulting, due to fault-propagation folding or frictional drag along the fault plane. If deposits are moist and plastic, strata

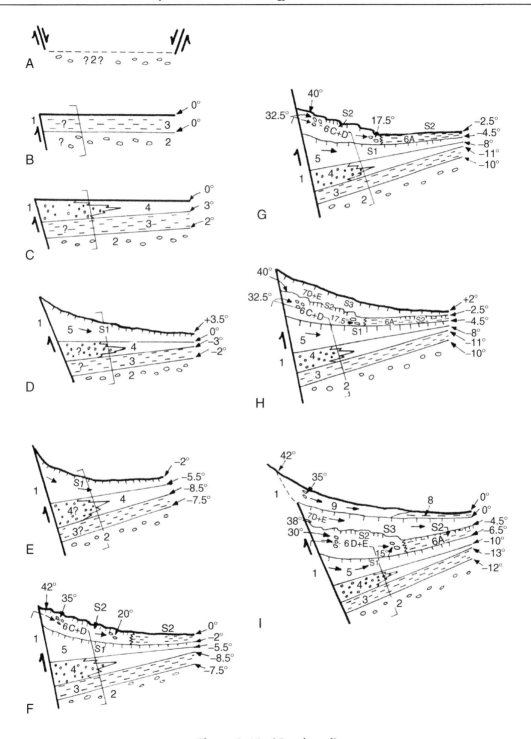

Figure 3.46: (Continued)

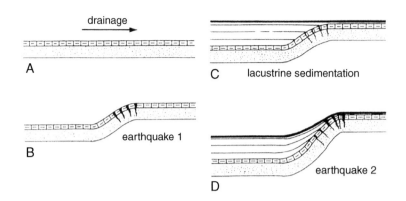

Figure 3.47: Schematic diagrams showing deposition against an upslope-facing scarp in a humid climate. (A) Prefaulting. (B) First earthquake creates an upstream-facing monoclinal scarp split by crestal tension fissures. (C) The stream is dammed to produce a lake, and horizontal lake sediments are deposited over the scarp. The base of the lake sediments is an angular unconformity across the face of the scarp, and a disconformity farther to the left; both are event horizons. (D) The second earthquake rejuvenates the monocline, dragging the lacustrine strata upward and creating new tension fissures at a higher structural level than the first set. Subsequent lacustrine sedimentation upstream of and over the scarp will create a new event horizon like the first. From Pantosti *et al.* (1993); reprinted with permission of the American Geophysical Union.

may be folded into a monocline in which the beds approach parallelism with the main fault plane. This geometry is usually associated with cohesive sediments, for example, moist silts and clays of lacustrine origin. Pantosti *et al.* (1993) analyzed a *monoclinal fault trace* in interbedded lake clays and sands that had experienced several Holocene paleoearthquakes. In that study area the monoclinal scarp faced upstream across a perennial stream. Each faulting event created temporary ponding and lacustrine deposition. Sediments in the scarp-base lacustrine basin were folded during each paleoearthquake and then *onlapped* by horizontal strata (Figure 3.47). The schematic reconstruction of faulting relies heavily on cross-cutting relations between packages of sediment, especially where unconformities are traced into scarp-derived (?) colluvium near the fault plane.

The angular unconformities created by forward-tilting may be useful in paleoseismic interpretation. If downthrown block strata act in a quasiplastic manner (e.g., clayey or silty sands), the monocline may be broken in *domino style* by small thrust faults that dip toward the main fault. In Figure 3.48, a pronounced angular unconformity exists between the youngest scarp-derived colluvium and folded and thrusted older

Figure 3.46: Schematic diagrams showing sequence of faulting events (a = oldest) at Kaysville deduced from graben angular unconformities. Stage (I) reflects the present geometry. Numbers to the right of each diagram show the dip of strata as reconstructed (negative numbers indicate eastward tilt). Unit lithologies: 1, lacustrine sand; 2, alluvial fan gravel; 3, sag pond silt and clay; 4, fluvial gravel deposited in graben; 5,6,7,9, scarp-derived colluvium; 8, sag-pond silt. Colluvial units 6 and 8/9 show interfingering between colluvial and sag-pond facies. From McCalpin *et al.* (1994); reprinted with permission of the American Geophysical Union.

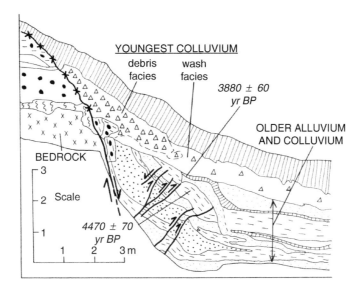

Figure 3.48: Log of trench showing an angular unconformity produced by forward tilt. Vertical line pattern indicates organic soils. Heavy lines show faults, heavy line with Xs shows the buried scarp free face. At lower center, older scarp-derived colluvium (heavy dots) and interfingering alluvium (fine dots and horizontal dashes) have been tilted up to 55° by drag on the main fault plane accompanying the younger faulting event. Forward-tilting caused by fault drag (?) induced several imbricate reverse faults in the older alluvium and colluvium, but these faults are secondary features to the main normal fault plane (*at left*) and do not indicate a regional compressive stress. Following faulting, the youngest scarp-derived colluvium was deposited atop the tilted beds, creating an angular unconformity. Cook Canyon trench, Rock Creek fault, Wyoming (McCalpin and Warren, 1992).

alluvium and colluvium. (These small thrust faults are secondary to the main normal fault and do not indicate a compressional stress regime.) The sequence of events at the trench could be interpreted as follows (1) initial faulting and scarp formation, (2) deposition of scarp-derived colluvium (heavy dotted lens at lower left) and alluvium (lighter dots and dashes) on the downthrown block ca. 4.5 ka, (3) renewed faulting, which folded and faulted units older units near the fault, and (4) deposition of the youngest scarp-derived colluvium ca. 3.8 ka.

3.3.3.4 *Difficult Paleoseismic Evidence: Small-Displacement Faulting at Long Recurrence Intervals*

The stratigraphic expression of small displacements on normal faults with long recurrence intervals can be complex and difficult to identify. The following discussion is based on some recent paleoseismic investigations of this type at Yucca Mountain in southwestern Nevada (e.g., Keefer *et al.*, 2004). Surface displacements commonly are in the range of 5–70 cm on faults with recurrence intervals of 10^4–10^5 years. Several types of deposits may develop following surface ruptures depending on the local structural and depositional setting. Small surface displacement (with or without significant fissuring) produces small and thin colluvial wedges with elongate, finely tapered, and flattened triangular shapes (Figure 3.49) that are difficult to identify uniquely. Wedge recognition is confounded further by the occurrence within thick

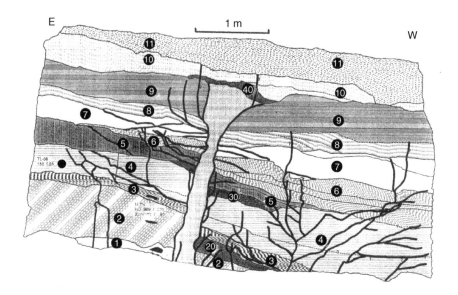

Figure 3.49: Log of Trench 14D on the northern Bow Ridge fault, Yucca Mountain, Nevada. Thick lines show faults and fractures; the upward-flaring patterned area at center is the main fault zone composed of fissure fill. Offset stratigraphic units are differentiated by patterns, with three suspected colluvial wedges marked by the darkest shading. Note how units directly overlying the suspected colluviums thicken across the main fault, whereas higher units maintain more uniform thickness. From Menges *et al.* (1994).

aggradational sequences of poorly bedded, fine-grained eolian or reworked eolian deposits that are difficult to subdivide into lithologic units. Most of the individual deposits in these sequences have broad sigmoidal (flattened S) to downward-thickening shapes and are associated with the initial draping and burial of paleoscarps and their scarp-derived colluvium. These units have no tectonic significance and must be carefully distinguished from true scarp-derived colluvial wedges. In addition, gravelly interbeds in aggradational sequences are deposited by channels localized near the fault zone; these interbeds mimic colluvial wedges, but should not be used directly for paleoseismic interpretation. Identification of event horizons and displacement amounts in such settings ideally should incorporate other criteria such as differential displacements of strata or upward termination of fractures.

3.3.3.5   *Difficult Paleoseismic Evidence: Distributed Faulting on a Large Escarpment*

Large normal-fault escarpments, such as discussed in Section 3.2.1 (and Figure 3.10) present a challenging target for paleoseismic investigations for several reasons. First, the most recent paleoearthquakes may not have formed an obvious small scarp that can be easily trenched (although sometimes this does occur; see Amoroso *et al.*'s 2004 trenching of the 130-m-high Hurricane fault escarpment). Second, an escarpment >50–100 m high is probably underlain my multiple structures of different types (Figure 3.50). Because every paleoearthquake may not have caused displacement on every structure, each structure may record only a partial paleoseismic record. In such case, all structures beneath the scarp must be trenched to retrieve the total paleoseismic history. Third, the escarpments are so high (tens to 100+ m) and steep that trenching the entire scarp may not be feasible, given the available excavating equipment and environmental constraints. Fourth, each of these structures may preserve its

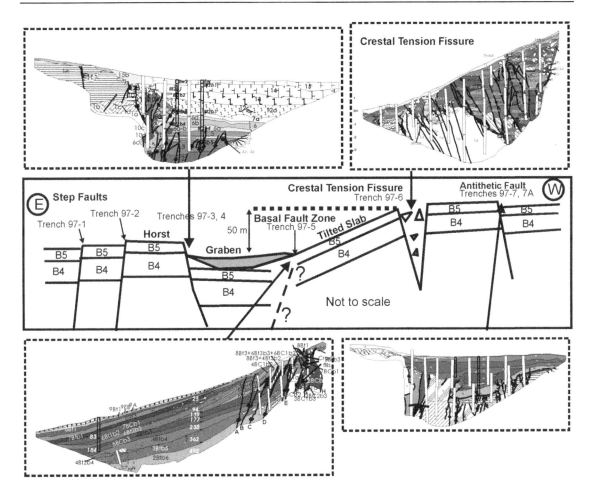

Figure 3.50: Example of a challenging paleoseismic target; the 50-m-high escarpment of the Pajarito fault, New Mexico. Quaternary normal faulting is expressed as a zone of diverse structures up to 2000 m wide (center box) dominated by an articulated monocline and secondary crestal tension fissure, basal fault zone, and graben, plus antithetic and step faults farther away. Boxes with dashed outlines contain trench logs that display the different types of stratigraphic models of each structure. Clockwise from upper left: fissure-graben model; domino-style extension model; fissure-graben model; distributed growth faulting and sedimentary onlap (similar to creeping faults). From McCalpin (2005).

paleoseismic record according to a different stratigraphic model (e.g., colluvial wedge, fissure-graben, eolian). Thus, even simple paleoseismic tasks such as identifying and dating the most recent earthquake (MRE), which normally require only a single shallow, inexpensive trench in other settings, here require a major logistical effort.

An example of trenching a large bedrock-cored fault escarpment is given by McCalpin (2005a). Earlier trenching had been restricted to the easily accessed toe of the 50–130-m-high escarpment. However, the faults exposed there displayed relatively small displacements. During regulatory review of the early trenching, reviewers concluded that (1) it was not clear that the exposed faults constituted the main fault zone, nor (2) that they experienced displacement in the most recent paleoearthquake (MRE). Accordingly, a 2-year project was embarked to trench the entire escarpment to confirm the MRE and to estimate recurrence interval; this effort eventually resulted in 14 trenches. Figure 3.50 shows only four of the trenches, but adequately portrays the complexity in assessing fault escarpments of this size. As always, paleoseismologists must adapt their techniques to fit the local structural and geomorphic setting.

3.3.4   Measuring Displacement on Normal Fault Exposures

The cumulative displacement across a normal fault can be measured relatively simply, from either fault scarp height (Section 3.2.3) or stratigraphically (from the separation of correlative strata across the fault). In contrast, measuring the displacements attributable to each paleoearthquake in a multiply faulted exposure is less simple. In most normal fault exposures, the indicator deposits for individual paleoearthquakes (interseismic deposits such as colluvial wedges, fissures, graben deposits) were formed on only one side of the fault. Thus, they are not displaced by the fault and cannot be used directly to measure displacement during individual paleoearthquakes. To measure those per-event displacements, we rely on indirect indicators associated with the appropriate sedimentologic model (colluvial wedge, eolian, fissure-graben).

3.3.4.1   Displacement Estimates from Colluvial Wedges

Colluvial wedge deposits commonly form only on the downthrown side of the fault, so it is rarely possible to directly measure their displacement across the fault plane. Instead, we use them as indirect indicators of per-event displacement. Scarp degradation models (Section 3.4.3) predict that the maximum thickness of scarp-derived colluvium will be limited to half the height of the free face from which it was shed. Thus, a first approximation of initial scarp height is twice the maximum colluvial thickness exposed in a trench or cut. Ostenaa (1984) suggested that this relation only holds true for colluvial wedges that are deposited on nearly horizontal surfaces. For wedges deposited on steeper surfaces, maximum thickness may only range from a small fraction to 100% of the height of the causative free face (Figure 3.42). Colluvium may also be trapped in rupture complexities such as basal tension fissures, which decreases the volume of the wedge and thus its thickness. In view of the many possible complexities, the rule of thumb "initial scarp height = 2× maximum colluvial thickness" should be used with caution. Typical problems with measuring per-event displacements on normal faults are described by Ran et al. (2003).

In rare cases it may be possible to use *provenance* of colluvial wedges to infer from what part of the upthrown block they were derived and thus estimate displacement. This concept is loosely based on trench observations originally made by Swan et al. (1980) on the Wasatch fault zone, Utah. Figure 3.42 shows three distinct stratigraphic units that are faulted in three vertical increments, each of which is roughly equal to unit thickness. Because of this coincidence, the free face in stage (a) is dominantly composed of the uppermost stratigraphic unit, so colluvial wedge 1 is derived entirely from the upper unit. The second faulting event (c) exposes the middle stratigraphic unit in the free face, so that colluvial wedge 2 is composed of both the middle unit and the upper unit. Likewise, the third faulting event (c) creates a free face composed primarily of the lower two stratigraphic units, which is again reflected in the composition of colluvial wedge 3. The lower part of each colluvial wedge (debris facies) is derived from material exposed in a steep free face, whereas the upper part of the wedge (wash facies) may be derived primarily from stratigraphic units on the crest of the scarp that were eroded during later, wash-dominated stages of scarp degradation (Nelson, 1992a).

3.3.4.2 Displacement Estimates in the Fissure-Graben and Eolian Models

In both of these models, the net vertical displacement in a paleoearthquake is assumed roughly equal the thickness of the interseismic deposit (graben fill, or eolian sand lens). This assumption rests on two beliefs: (1) that normal faulting, by lowering the downthrown block, creates a new volume of "accommodation space" which also constitutes a sediment trap of sorts, and (2) the space is filled with postfaulting deposits until the scarp profile roughly regains its prefaulting shape. As in the colluvial wedge model, these assumptions are only rough guidelines, because it is possible that interseismic deposition might only partly fill the accommodation space before the next earthquake occurs, or conversely, that it might over-fill the space. Thus, the assumptions satisfy Occam's razor as being the simplest explanation that does not require additional assumptions about changes in scarp shape through time.

The only other independent check we can apply to these individual displacements, is that their sum must add to the cumulative vertical displacement of the landform or deposit (e.g., calculated from scarp height). Thus, if we recognize three displacement events on a 6-m-high scarp, the first two of which were approximately the same size and the third twice as large, then estimated displacements of 1.5, 1.5, and 3 m would satisfy those constraints. However, the fact that two of the displacements are cited to the nearest tenth of a meter is a mathematical artifact, and certainly does not imply decimeter precision of the estimate.

3.3.4.3 Displacements Reconstructed from Angular Unconformities

In theory, if successive paleoearthquakes tilt strata on the downthrown block and create angular unconformities, it should be possible to estimate the tilt-related component of displacement in each paleoearthquake by use of Equations (3.3) or (3.4). The fault scarp and tilted graben strata at Kaysville, Utah (Figure 3.46) were subjected to such a simple geometric analysis by McCalpin *et al.* (1994). They estimate T_m (throw on the main fault plane, as approximated by free face height) for paleoearthquakes by multiplying maximum colluvial thickness by an appropriate factor (Table 3.7, columns 2 and 3). The eastward tilt of the graben during the last four paleoearthquakes was estimated from angular

Table 3.7: Fault displacement at the Kaysville, Utah, Trench Site[a]

			Vertical displacements (m)			
Parameter measured	Faulting event	Maximum colluvial thickness	Main fault[b] (T_m)	Tilt[c] (T_t)	Antithetic faults	T_{net}[d] (m)
Displacement per faulting event	5	1.3	3.5	0 or 1.2	1.7–1.9	
	4	2.4	4.8	1.4	1.1 or 0	2.3–3.4
	3	2.3	4.6	3.2	0	1.4

[a]From McCalpin *et al.* (1994).
[b]Estimated as twice the maximum colluvial thickness for events 3 and 4; as three times the maximum colluvial thickness for event 5.
[c]$T_t = W \sin \phi$.
[d]$T_{net} = T_m - [T_t + T_a]$.

unconformities in the graben fill (Figure 3.46). The tilts acted over an estimated horizontal distance of 33 m (W in Figure 3.18), allowing calculation of T_t via Equation (3.4). Throw on the main antithetic fault (T_a; beyond the western end of the trench) was measured by Swan and others (1980) at 2.2 m, based on stratigraphic separations of graben-fill units. Using Equation (3.4) and values for T_m, they estimated the net throw (T_{net}) for each of the last three surface-rupturing events (Table 3.7). The main sources of uncertainty in this particular analysis are (1) whether all of T_a (2.2 m) occurred in the latest event or were partitioned between the two latest events, and (2) what is the appropriate factor (≥ 2) by which to multiply colluvial wedge thickness to estimate T_m.

3.3.5   Distinguishing Creep Displacement from Episodic Displacement

There have understandably been very few studies of creeping normal faults, because such faults do not generate large earthquakes and thus pose no hazard due to ground shaking. However, the few studies that have been performed on normal faults known to creep show that the stratigraphic expression of creeping faults in vertical section is very different from that of episodic, coseismic normal faults. Specifically, creeping faults lack evidence of rapid brittle failure in shallow deposits, even when deposits of that grain size (say, sand or gravel) typically respond to rapid deformation by brittle (rather than ductile) deformation. Thus, creeping faults lack brittle-deformation structures such as fissures, and lack deposits associated with the development of a fault free face (e.g., scarp-derived colluvial wedges, and exotic blocks fallen from a free face; Figure 3.51).

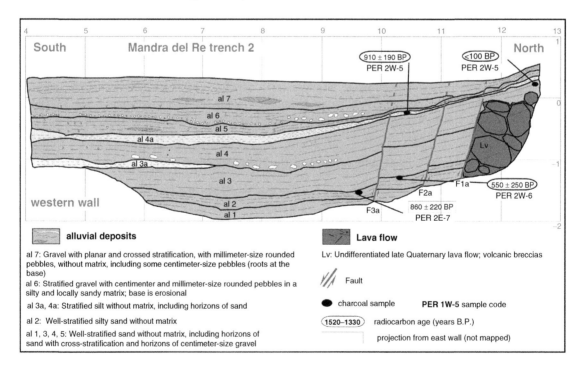

Figure 3.51: Log of a trench across the creeping Pericana fault on Mt. Etna, Sicily. The Pericana fault is dominantly a sinistral strike-slip fault but has a vertical component in many places. From Ferreli *et al.* (2002); reprinted with permission of the Geological Society of America.

Instead, on creeping normal faults strata thicken slightly across faults, and angular unconformities tend to be broad and rather gentle. Strata on the downthrown block tend to onlap and pinch out toward the faults. There is a "continuous and uniform increase in deformation and dislocation along the fault trace as progressively older sedimentary units are encountered down from the surface" (Ferreli *et al.*, 2002). These observations indicate that creeping faults cannot create "free faces," but instead slowly create vertical relief that essentially keeps pace with the ambient erosional and depositional processes. More research is needed to define the unique geomorphic and stratigraphic features caused by creeping faults, both normal and other types. There are practical reasons for this, for example, the need to distinguish seismogenic versus nonseismogenic faults in areas where the two may coexist, such as areas in the western USA affected by both Neogene extensional tectonics and by ongoing (and presumably aseismic) salt tectonics.

3.4 Dating Paleoearthquakes

Paleoearthquakes can be dated directly or indirectly. The most *direct techniques* are dating a fault scarp via scarp degradation modeling, cosmogenic surface-exposure dating, or by quantitative analysis of scarp soils. Although these techniques directly date the formation of coseismic scarps, age uncertainties can be relatively large. *Indirect dating* methods involve bracketing the age of the paleoearthquake by numerical dating of landforms or deposits that predate and postdate faulting. The accuracy and precision of numerical dating may be high, but the ages themselves may not provide close constraints on the age of faulting. Reviews of Quaternary numerical dating methods were cited in Chapter 1 and are not repeated here. This section emphasizes where to collect samples to most tightly constrain the age of faulting, given the occurrence of datable materials. Because radiocarbon dating is by far the most common technique for dating normal fault paleoearthquakes, our examples are generally couched in terms of distribution of organic matter. The reader should consult primary references to see how sampling might differ for other techniques, such as thermoluminescence (Forman *et al.*, 1989, 1991; McCalpin and Forman, 1991; McCalpin *et al.*, 1994) or uranium-series disequilibrium (Peterson *et al.*, 1995).

3.4.1 Direct Dating of the Exposed Fault Plane

In our 1996 edition (p. 97) we speculated that cosmogenic (surface-exposure) dating could be applied to bedrock fault planes exposed by paleoearthquakes. Only 2 years later Zreda and Noller (1998) reported the first use of this technique, using ^{36}Cl to date six paleoearthquakes in the past 37 ka on the Hebgen fault, Montana. The principles of surface-exposure dating are outlined in Figure 3.52, by depicting the depth of shielding from cosmic rays for a hypothetical sample location (white square on the fault plane). Prior to faulting (a) the sample was shielded by 200 cm of hanging-wall overburden, which was slowly thinned by erosion from 12 to 6 ka at a rate of 3 cm/kyr (b). At 6 ka a coseismic displacement of 1.0 m decreases the overburden thickness from 182 to 82 cm (c). A second coseismic displacement at 3 ka (d) exposes the sample. The changing thickness of shielding through time is shown in (f) and forms the basis for interpreting surface-exposure history of any one sample. When combined with multiple samples up-and-down the fault plane, this technique can detect abrupt changes in exposure ages caused by coseismic displacements.

Zreda and Noller's initial application of this method ignited some controversy, since their interpreted chronology included four earthquakes in the past 7 ka, whereas "traditional" geomorphic evidence (fault scarps and tectonic terraces) from the same fault indicated only two Holocene earthquakes. This discrepancy led to the formation of the Hebgen Lake Paleoseismology Working Group (mainly USGS

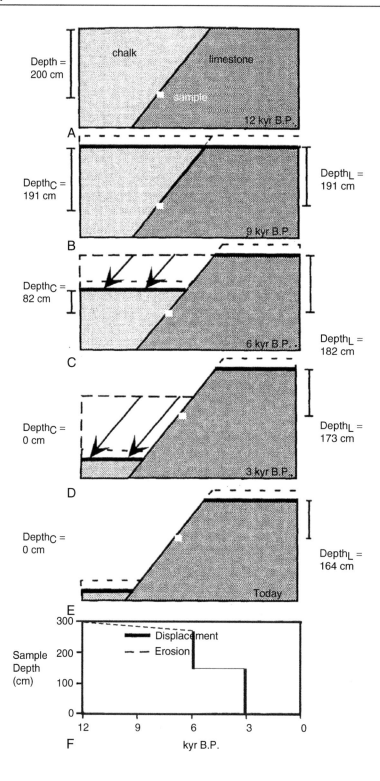

Depth =
200 cm

chalk

limestone

sample

12 kyr B.P.

A

Depth$_C$ =
191 cm

Depth$_L$ =
191 cm

9 kyr B.P.

B

Depth$_C$ =
82 cm

Depth$_L$ =
182 cm

6 kyr B.P.

C

Depth$_C$ =
0 cm

Depth$_L$ =
173 cm

3 kyr B.P.

D

Depth$_C$ =
0 cm

Depth$_L$ =
164 cm

Today

E

Sample
Depth
(cm)

300

200

100

0

Displacement
Erosion

12   9   6   3   0

F   kyr B.P.

personnel) and excavation of two trenches, which ended up confirming by stratigraphic evidence two Holocene and one latest Pleistocene paleoearthquake (Pierce *et al.*, 2000; Hecker *et al.*, 2002).

In the past decade additional fault planes, typically limestone, have been subjected to surface-exposure dating. These include faults in Israel (Mitchell *et al.*, 2001), Greece (Benedetti *et al.*, 2002), and Italy (Palumbo *et al.*, 2004). The Italian Magnola fault was also subjected to detailed chemical analysis by Carcaillet *et al.* (2008) and constitutes the best-studied fault plane so far.

3.4.2 Direct Dating via Scarp Degradation Modeling

Two kinds of techniques based on degradational processes have been used to date fault scarps. The first is the *diffusion dating technique*, which assumes that the symmetrical erosion and deposition occurring in Wallace's wash-dominated stage (stage e in Figure 3.21) can be mathematically simulated by a diffusion-type equation wherein

$$\frac{\mathrm{d}Y}{\mathrm{d}t} = K\frac{\mathrm{d}^2 Y}{\mathrm{d}X^2}, \tag{3.5}$$

where Y is the elevation of points of the surface of the fault scarp, X is the horizontal position of points on the fault scarp surface, t is the time, and K is the diffusion constant.

The diffusion approach makes two assumptions about the erosional modification of fault scarps that (1) the rate of sediment transport is limited only by the strength of the transporting process, and not by the availability of transportable material (i.e., the slope is *transport limited;* Selby, 1993), and (2) the rate of sediment transport is only a function of scarp slope, and not of position on the scarp. The latter assumption requires that any scarps modeled must have degraded mainly by *creep* and *rainsplash*, rather than by wash processes (*sheetwash, rillwash*) which increase in strength downslope. Scarps that do not meet these criteria (e.g., bedrock scarps that are *supply limited*, scarps eroded by running water) should not be dated via the diffusion equation. (Note that the initial gravity- and debris-controlled stages [Figure 3.21A–C] cannot be modeled by diffusion, so the time required for them (30–200 years?; Section 3.2.4) should be added to the diffusion-based estimate of scarp age). Computer programs (Nash, 1980, 2005) and nomographs (Andrews and Bucknam, 1987) based on the diffusion equation yield values of tK for measured fault scarp profiles. Despite early successes in applying this technique to scarps of known age, later work has revealed that the *diffusion constant* (K) is very sensitive to climate, aspect, scarp height (Pierce and Colman, 1986) and grain size (McCalpin, 1983). With the uncertainty in K values induced by the factors listed above, values of tK thus calculated include some uncertainty. As Arrowsmith *et al.* (1998) show, morphologic dating can also be applied to scarps formed by strike-slip faulting.

Diffusion analysis of fluvial scarps of known age near fault scarps can yield a local estimate of K, which can then be divided into the product tK to yield fault scarp age (Begin, 1993; *Enzel et al.*, 1994). Optimally, the fluvial scarps from which K is derived should possess the same orientation, height, parent

Figure 3.52: Schematic diagrams showing how the overburden thickness changes through time over a sample of a fault plane surface (white square), due to slow interseismic erosion and rapid coseismic displacement. See text for explanation. From Mitchell *et al.* (2002). Reprinted with permission of the American Geophysical Union.

material, and vegetation as the fault scarps to be dated. Fluvial scarps (terrace risers) in New Zealand studied by McCalpin (1989a) were asymmetrical, with the lower half of the scarp yielding a smaller product tK than the upper half. These terrace risers were evidently cut incrementally over hundreds or thousands of years, with the toe of the riser being periodically rejuvenated. McCalpin (1989c) thus suggested that while K may be constant on the riser, tK of the upper half of the riser approximately dates abandonment of the upper terrace, and tK of the lower half of the riser similarly dates the abandonment of the lower terrace (as suggested by Knuepfer, 1988; see Section 6.2.1.1). If the ages of the upper and lower terraces are known, K can be estimated and then applied to dating of nearby fault scarps.

Machette (1989) suggested that there are two relatively straightforward methods for calculating fault scarp age. The first is a comparison of scarp heights and maximum scarp slope angles to *empirical regressions* of those two variables from dated scarps, using the graph of Bucknam and Anderson (1979). The second is a calculation of scarp age based on the linear-plus-cubic diffusion model of Andrews and Bucknam (1987), wherein:

$$t = t'(\mathrm{SO})^2/K_0 \tag{3.6}$$

where t is the age of scarp (years), t is the dimensionless scarp age, SO is the vertical surface offset across scarp, and K_0 is the diffusion constant (mass diffusivity) at $0°$ fan slope.

Values for t' are derived from Table 3.8, which relates maximum scarp slope angle (Θ) to the ambient (far-field) angle of the faulted slope (α). The calibrated interval marked on Table 3.8 indicates the range of data derived from independently dated scarps. Values for K_0 have been back-calculated from these dated scarps in the Basin and Range Province, USA, and range from 0.46 to 0.52 $\mathrm{m}^2/1000$ years.

This linear-plus-cubic diffusion model is extremely sensitive to small changes in α and Θ, and should not be used if $\alpha \geq 10°$, or if $\Theta - \alpha < 10°$ (Hanks and Andrews, 1989). For example, scarps degrade much faster on steeper slopes and the runoff process (not modeled in the diffusion approach) becomes significant. The model should only be applied to single-event fault scarps, because only for those scarps can the prefaulting geometry be estimated with any certainty. (However, Hanks and Schwartz (1987) did use the diffusion technique to date the penultimate faulting event at the 1983 Borah Peak rupture by restoring the scarp to its inferred geometry before the 1983 faulting event.) If a multiple-event fault scarp is profiled, diffusion dating will yield a scarp age older than that of the most recent faulting event (Machette and McGimsey, 1983; Mayer, 1984; Colman, 1986).

Table 3.9 shows an example calculation using Equation (3.6) and Table 3.8. Almost all of the uncertainty in the age estimate in Table 3.9 (1309 years) is caused by the inability to measure slope angles in natural terrain closer than $\pm 1°$. By comparison, the uncertainties in scarp surface offset and diffusion constant, while seemingly quite large, contribute only a small amount of uncertainty in age estimate. To minimize the uncertainty in the final age estimate, efforts should be made to minimize the uncertainties in field slope measurements by making multiple traverses along a given fault scarp, and averaging the results. Due to the variability in natural terrain, however, and errors arising from the scarp profiling techniques (Chapter 2), some uncertainty will always be attached to field slope measurements. A second technique for directly dating fault scarps uses the decrease in material *cohesion* on the scarp face through time (due to weathering and soil formation) to estimate scarp age (Ingraham *et al.*, 1980; Watters and Prokop, 1990). The cohesion of material exposed in the post-1872 free face of the Lone Pine, California, USA, fault scarp, is greater than the cohesion of material on the bevels of the post-5000-year and post-10,000-year scarps. By fitting polynomial equations through cohesion values from different-aged scarps,

Table 3.8: Dimensionless age values (t′) for scarp subject to linear-plus-cubic diffusion model and initial scarp slope angle of 31°

Scarp Angle (θ)	Ambient Angle (α)									
	0°	1°	2°	3°	4°	5°	6°	7°	8°	9°
31°	0	0	0	0	0	0	0	0	0	0
30°	0.017	0.019	0.020	0.021	0.023	0.025	0.027	0.029	0.032	0.034
29°	0.026	0.028	0.030	0.032	0.035	0.037	0.041	0.044	0.049	0.053
28°	0.035	0.038	0.041	0.044	0.048	0.052	0.056	0.062	0.068	0.075
27°	0.046	0.049	0.053	0.057	0.062	0.068	0.074	0.082	0.090	0.100
26°	0.058	0.062	0.067	0.073	0.080	0.087	0.096	0.160	0.118	0.132
25°	0.072	0.078	0.085	0.092	0.101	0.111	0.122	0.136	0.152	0.171
24°	0.089	0.096	0.105	0.115	0.126	0.139	0.155	0.173	0.194	0.220
23°	0.109	0.118	0.130	0.142	0.157	0.175	0.195	0.219	0.248	0.283
22°	0.132	0.145	0.159	0.176	0.196	0.218	0.245	0.277	0.317	0.365
21°	0.161	0.177	0.196	0.218	0.243	0.273	0.309	0.352	0.406	0.474
20°	0.196	0.217	0.241	0.269	0.302	0.342	0.390	0.450	0.525	0.621
19°	0.239	0.265	0.296	0.333	0.377	0.430	0.497	0.580	0.685	0.824
18°	0.291	0.325	0.365	0.414	0.473	0.546	0.637	0.754	0.907	1.11
17°	0.355	0.400	0.453	0.518	0.598	0.699	0.827	0.994	1.22	1.53
16°	0.435	0.494	0.565	0.653	0.764	0.904	1.09	1.33	1.68	2.17
15°	0.537	0.615	0.711	0.832	0.986	1.19	1.45	1.83	2.37	3.21
14°	0.666	0.771	0.902	1.07	1.29	1.58	1.99	2.59	3.50	4.99
13°	0.832	0.976	1.16	1.40	1.72	2.16	2.81	3.80	5.43	8.43
12°	1.05	1.25	1.51	1.85	2.34	3.04	4.12	5.89	9.15	16.2

(Continued)

Table 3.8: Dimensionless age values (t′) for scarp subject to linear-plus-cubic diffusion model and initial scarp slope angle of 31° (Cont'd)

Scarp Angle (θ)	Ambient Angle (α)									
	0°	1°	2°	3°	4°	5°	6°	7°	8°	9°
11°	1.34	1.62	1.99	2.52	3.28	4.45	6.36	9.90	17.5	39.1
10°	1.73	2.14	2.70	3.52	4.79	6.85	10.7	18.9	42.3	
9°	2.27	2.88	3.77	5.12	7.34	11.5	20.3	45.5		
8°	3.06	4.00	5.45	7.83	12.2	21.7	48.7			
7°	4.23	5.77	8.29	13.0	23.1	51.8				
6°	6.06	8.73	13.7	24.4	54.8					
5°	9.12	14.3	25.6	57.6						
4°	14.9	26.6	60.0							
3°	27.4	61.9								
2°	63.4									

Table 3.9: Example calculation of fault scarp age, based on Equation (3.7)

Given: scarp surface offset $= 3.3 \pm 0.2$ m; far-field slope $= 3 \pm 1°$; maximum scarp slope angle $= 19 \pm 1°$
From Table 3.7, $t' = 0.357 \pm 0.116$ (see boxed area in Table 3.8)
From Equation (3.6), $t = 0.357 \pm 0.116$ (3.3 $\pm$ 0.2 m)^2/(0.49 $\pm$ 0.03 m^2/1000 year)
Applying the multiplication and division rules for values with unequal standard deviations (Geyh and Schleicher, 1990)
$t^* = (t_1 \times t_2)/t \pm t^* \sqrt{(\sigma_{t1}^2/t_1^2) + (\sigma_{t2}^2/t_2^2) + (\sigma_{t3}^2/t_3^2)}$
$t = 3.888$ ka/0.49 $\pm$ 3.888 ka$\sqrt{0.106 + 0.0037 + 0.0037}$
$= 7.935 \pm 1.309$ ka   or   7935 $\pm$ 1309 years

Watters and Prokop (1990) interpolated the age of an undated scarp from its "parent" and "field" cohesion values. This technique is experimental and has not been applied over a wide range of faulted parent materials or climate zones.

3.4.3 Age Estimates from Soil Development on Fault Scarps

Prior to the development of cosmogenic and luminescence dating methods, paleoseismologists were often forced to use soil profile development to approximate the ages of paleoearthquakes. Although we now have these new methods in our "toolbox" of dating techniques, it is still important to understand what soil profiles tell us about the timing of paleoearthquakes and fault scarp evolution. On fault scarps (as on all slopes) soil profiles thin, thicken, change character, and merge, and these soil-topographic variations (*catenas*) hold clues to the temporal evolution of the scarp (Berry, 1990; McCalpin and Berry, 1996). Observations in numerous trenches show that four soil variations are common (Figure 3.53): (1) The relict summit soil is *truncated by erosion* at the scarp crest; (2) the relict toeslope soil *weakens laterally* as it is traced beneath the colluvial wedge sequence, where it forms the prefaulting soil; (3) soil(s) developed on postfaulting colluvium *merge downslope* with the prefaulting soil, to become the toeslope relict soil; and (4) individual horizons *thin* as they pass over the scarp crest *and then thicken* on the colluvial slope, due to increased infiltration, moisture-holding capacity, and simultaneous deposition and soil formation (*cumulic development*).

The relative soil profile development between the summit and toeslope (relict) soils, the buried prefaulting soil, and the colluvial soils provide a first approximation for faulting history. For example, if faulting occurred soon after stabilization of the displaced landform, then almost all soil formation postdates formation of the fault scarp, so (1) the relict toeslope soil will weaken when traced under the colluvial wedge, until there is no soil at all under the earliest part of the colluvial wedge; (2) soil(s) in colluvium will be very well developed, and will merge gradually with soils above and below the scarp; and (3) the summit soil will show little or no erosional truncation at the scarp crest, since the soil mainly formed *after* the scarp attained its present geometry. Alternatively, if faulting occurred long after stabilization of the landform, then almost all soil formation predates scarp formation, so (1) the buried soil under the colluvial wedge will be nearly as well developed as the full, relict soils above and below the scarp, and "lateral weakening" will be minimal; (2) soil(s) in the colluvium will be weak to very weak; and (3) the summit

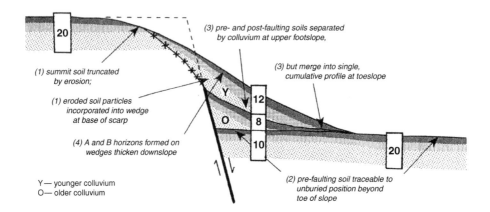

Figure 3.53: Diagram of soil catena relations on a two-event normal fault scarp, showing the four phenomena listed in the text (numbers before phrases). From darkest to lightest, shading represents A, B, and C horizons. Numbers in rectangles are hypothetical profile development index (PDI) values for soils, based on a trench across a faulted 15-ka deposit on the Wasatch fault zone, Utah. The sum of PDI values for the footslope soils (30) is 1.5 times that for the relict summit and toeslope soils (20) due to increased moisture availability and influx of fines on the footslope. Note that the relict soil beyond the toeslope weakens laterally, and is only half as developed beneath the colluvial wedges. If soil formation rates are linear, this implies that initial faulting occurred at about 7.5 ka. Soils on the colluvial wedges suggest that 8/20 of the subsequent geologic time (40% of 7.5 ky = 3 ky) elapsed between the first and second faulting events, and 12/20 of the time (60% of 7.5 ky = 4.5 ky) has elapsed since the second faulting event. From McCalpin and Berry (1996).

soil will be strongly truncated at the scarp crest. The two scenarios above represent end members of faulting chronology, which can be further complicated by recurrent faulting at irregular intervals.

A simple *continuity approach* can yield ratios of soil development times among the various soils, and thus allow estimates of the age(s) of faulting events. This continuity approach requires quantitative measurements for the development of relict and footslope soils, either semiquantitative [such as the *profile development index* (PDI) of Harden, 1982] or quantitative (weight of pedogenic clay or carbonate in the profile, e.g., Machette, 1978). Three assumptions are required for this type of analysis: (1) Soil development rates have been constant over time, (2) soil development rates at all times have been greater on the footslope than above or below the scarp, and (3) the cumulative time for soil formation is the same for the superposed soils under the colluvial footslope as for the stable summit or toeslope soils. If these assumptions are valid, then the development index of the relict soils above and below the scarp should equal the combined development indices of all soils under the colluvial footslope, once those latter indices have been *normalized* to account for their faster development rate due to topographic position. In relation to Figure 3.27, the continuity equation would read:

$$\mathrm{PDI}_r = \mathrm{PDI}_{pf} + (\mathrm{PDI}_Y + \mathrm{PDI}_O)/\mathrm{F}, \qquad (3.7)$$

where PDI_r is the profile development index of *relict* soils above and/or below the scarp, $= 20$ in example, PDI_{pf} is the profile development index of the buried *prefaulting* soil under the footslope, $= 10$ in example,

PDI_Y, PDI_O is the profile development indices of younger (Y) and older (O) footslope soils developed on postfaulting colluvium, = 8, 12 in example, and F is the footslope correction factor.

Using the hypothetical values in Figure 3.53, we may solve Equation (3.7) for F, resulting in $F = 2$. In other words, soils on footslopes in northern Utah developed twice as fast as did soils on stable geomorphic surfaces such as the summit soil. (By comparison, Machette, 1978, derived an F value of 1.25 for soil carbonate in a drier part of New Mexico). One can normalize footslope PDIs to "summit" development rates by dividing by 2, yielding the following PDI values for Equation (3.7): 20 = 10 + 4 + 6. Both sides of the equation represent 15 ka, so the time represented by soils Y, O, and pf must be approximately 7.5, 3, and 4.5 ka, respectively.

If the results of the continuity equation are not compatible with independent age evidence for the faulting events, the cause is most likely variable rates of soil development through time (McCalpin and Berry, 1996).

3.4.4 Bracketing the Age of Faulting by Dating Geomorphic Surfaces

In reconnaissance investigations, the age of scarp-producing paleoearthquakes is typically bracketed between the ages of the youngest faulted and oldest unfaulted geomorphic surfaces. Figure 3.54 shows a

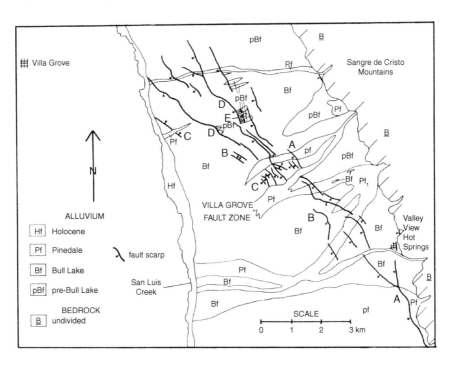

Figure 3.54: Map of fault scarps of various ages, Villa Grove fault zone, Rio Grande rift zone, USA. Based on cross-cutting relationships, the scarps formed over different time intervals; some scarps show a long history of recurrent movement, others appear to be single-event scarps. Ages of scarps: type A, post-Pinedale; B, post-Bull Lake; C, pre-Pinedale, post-Bull Lake; D, post-Pinedale, long history of movement; E, pre-Bull Lake. From McCalpin (1983); reprinted with permission of the Colorado School of Mines Press.

zone of normal fault scarps which have complex relations with fluvioglacial outwash fans dated at 0–10 ka (Hf), 15–30 ka (Pf), ca. 150 ka (Bf), and ≥250 ka (pBf). Scarp A displaces Bf but is truncated by Hf and Pf channels, so it must have formed between 150 and 15–30 ka. Scarp B displaces only Bf, so it can be dated only as younger than 150 ka (i.e., it could be much younger, even younger than scarp A). Most of the scarps between C and D displace Bf but not Pf surfaces, but some scarps (such as D) displace both Bf and Pf surfaces, indicating movement younger than 15–30 ka. Scarp E bounds isolated remnants of pBf and displaces Bf and Pf, suggesting a long history of movement continuing to less than 15–30 ka. Scarp F offsets pBf but not Bf, indicating latest movement before 150 ka.

The temporal resolution of dating paleoevents with bracketing geomorphic surfaces is controlled by the age span between the surfaces. In the example described above, knowing that a faulting event occurred between 150 and 15–30 ka may not be very helpful. Where geomorphic surfaces are created rapidly, resolution is better. In New Zealand, for example, it is common to find 5–10 Holocene stream terraces incised at various levels below the latest Pleistocene aggradation surfaces (e.g., Lensen, 1968; Knuepfer, 1988, 1992). These terraces, formed only a few thousand years apart (see Figure 2A.6), can record fault displacements with better resolution than deposits formed by major climatic changes some 10s or 100s of ka apart.

In the past 15 years it has become more common to date displaced landforms by cosmogenic isotopes to calculate long-term average slip rates. These ages are often called "surface-exposure ages" because they date the time between the abandonment of the geomorphic surface by active deposition (i.e., stabilization of the surface) and the present, a time span during which the surface was "exposed" continuously to cosmic radiation. Studies on normal faults in the USA include Zehfuss *et al.*, (2001), who dated faulted landforms as old as 300 ka by measuring ^{10}Be and ^{26}Al on individual boulders on cinder cones, lava flows, and debris-flow levees of alluvial fans. In the same area of California, Le *et al.* (2007) used ^{10}Be to date five displaced alluvial fan levels (124, 61, 26, 4.4, 4.1 ka) and one displaced rockslide (19 ka) across the southern Sierra Nevada frontal fault, where fault scarps are as much as 41 m high.

These types of "slip rate studies" have advantages and disadvantages. The main advantage is that they permit calculation of mean slip rates over long periods of time (up to 300 ka), comparable to the time span theoretically possible with luminescence dating, but in areas dominated by rapidly deposited, very coarse-grained deposits where luminescence dating would not likely work. Mean interval slip rates can be calculated for the time spans between the dated surfaces, to get an idea of slip rate variability through time. Both mean slip rate and variability are critical inputs to seismic hazard analyses, and also permit comparison with short-term geodetic slip rates measured by contemporary GPS surveys. A disadvantage of a slip rate study is that the displacement-per-event, magnitude, and age of individual paleoearthquakes typically cannot be deduced.

3.4.5   Bracketing the Age of Faulting by Dating Displaced Deposits

Similar relative-dating criteria apply for subsurface exposures as for geomorphic surfaces; faulting is younger than the stratigraphically highest faulted bed, and older than the stratigraphically lowest unfaulted bed. Figure 3.55 shows a late Pleistocene deltaic deposit displaced by a series of normal faults that define a 30-m-wide graben. The faults displace units 1, 2, and 3 and are truncated by units 4, 5, and 6. The faulting thus occurred between deposition of units 3 and 4. Further proof of fault timing is the different geometry of the faulted versus unfaulted units. Units 1, 2, and 3 are laterally extensive and maintain a constant thickness across the fault zone, suggesting they were deposited before any tectonic relief existed. In contrast, units 4, 5, and 6 are restricted to channels that parallel graben faults, and were probably

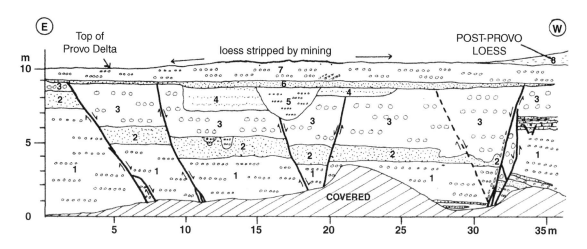

Figure 3.55: Wall of a gravel pit showing truncated normal faults. Because the ages of units 1 and 8 are approximately known from regional 14C chronology, the age of faulting can be roughly bracketed. From McCalpin and Forman (1991); reprinted with permission of the Seismological Society of America.

deposited by streams diverted into the graben axis after faulting. In this example from the Bonneville Basin, the entire delta was deposited (according to the regional ^{14}C chronology; Oviatt *et al.*, 1992) between about 14.5 and 13 ka, with the topset beds (unit 7) being formed about 13.0–13.5 ka. Faulting must have occurred slightly before abandonment of the delta, while active channels still traversed the delta top, but after the bulk of the delta had already formed. Dated strata thus suggest the faulting occurred ca. 13.5 ka.

3.4.6   Bracketing the Age of Faulting by Dating Colluvial Wedges

Many normal faulting chronologies have been reconstructed from numerical ages on colluvial wedge sediments and on the soil horizons that underlie and overlie the wedge. The choice of dating method and sampling strategy are dictated by the characteristics of sediments and soils in the fault zone. For example, past studies in the eastern Basin and Range Province, USA, have used radiocarbon dating ("conventional" gas-proportional counting and accelerator mass spectrometry, or AMS) and/or luminescence dating [thermoluminescence (TL) and optically stimulated luminescence (OSL)] (Forman *et al.*, 1989, 1991; McCalpin and Forman, 1991; Jackson, 1991; Lund *et al.*, 1991; Personius, 1991; Machette *et al.*, 1992a,b; McCalpin, 1994; McCalpin *et al.*, 1994; West, 1994). In hyper-arid environments where organic carbon is rare, soil precipitates (calcium carbonate, silica) may be dated via uranium-series disequilibrium, TL, and cosmogenic isotopes such as ^{36}Cl. Another option is AMS ^{14}C dating of microscopic organic material in layers of rock varnish (Paces *et al.*, 1994; Peterson *et al.*, 1995).

Fault-zone exposures in semiarid regions typically contain *low-organic-content* soils, and bulk samples that span a considerable age range must be collected. In the studies cited earlier, soil A horizons containing ≤5% organic matter required large (1- to 5-kg) samples to yield sufficient carbon for conventional radiocarbon dating. Radiocarbon ages from such bulk samples are termed *apparent mean residence time* (AMRT) ages (see Appendix 2, in Book's companion web site). Machette *et al.* (1992a, his Appendix) discuss application of AMRT ages to faulting chronologies.

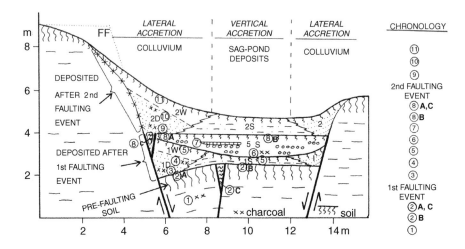

Figure 3.56: Schematic cross section through a normal fault scarp and graben formed by two surface-faulting events; this is a composite section showing features observed in many trenches across the Wasatch fault zone, Utah. Colluvium shed from the main scarp (at left) is divided into debris facies (D) and wash facies (W); both facies grade into coeval sag-pond deposits (1S, 2S). Between faulting events a fluvial deposit (1.5S), containing lenses of gravel (circles), completely filled the graben. Circled numbers indicate potential sites for obtaining radiocarbon samples, and the chronology at right shows which samples most closely constrain the times of the faulting events. Debris facies colluvium is deposited immediately after faulting, whereas sag-pond deposition may be climatically controlled and postdate faulting by decades. Radiocarbon dates from sites 2A and 8A have traditionally been interpreted as the closest maximum limiting ages on faulting, as shown on the chronology at right. However, if radiocarbon ages from soils 2 and 8 are MRT-corrected to reflect the age of soil burial, they become minimum age constraints on faulting, with the closest constraint closest to the fault. Samples from wash facies colluvium (5, 11) and interbedded fluvial deposits (6, 7) do not provide close age constraints on faulting. From McCalpin and Nishenko (1996); reprinted with permission of the American Geophysical Union.

Figure 3.56 shows typical age sampling locations in a normal fault exposure and how sample ages constrain the age of the event horizon, that is, the unconformity between scarp-derived colluvium (units 1D, 1W, 2D, 2W) or graben sediments (units 1S, 1.5S, 2S) and the prefaulting soil (sinuous vertical lines). The ages of the two faulting events are most closely bracketed by samples 2 and 3, and 8 and 9, respectively in Figure 3.30. The age of the event horizon decreases away from the fault due to time-transgressive burial of the soil by laterally accreting scarp-derived colluvium. Thus, samples 4 and 5 provide less close limiting ages on faulting than does sample 3. Scarp-derived colluvium is deposited rapidly (ca. 30–200 years for debris facies, units 1D, 2D), whereas graben sediments may accumulate slowly.

To achieve the closest age constraints, the age of the event horizon (upper soil horizon contact, or UHC) may be extrapolated from the age of several subsamples of the buried soil horizon (Figure 3.57). The A horizons are buried by proximal, coarse-grained, debris-facies colluvium in the western United States (under semiarid climates and scrub vegetation) and display mean age trends with depth of 4.6 ± 1.3 yr/mm (McCalpin and Nishenko, 1996), and this age trend can be used to extrapolate the age of the UHC

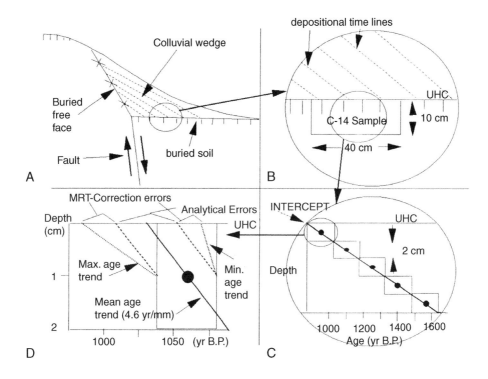

Figure 3.57: Idealized diagrams of a single-event normal fault scarp showing various sources of uncertainty in relating numerical ages to the time of paleoearthquakes. (A) Simplified cross section of the colluvial wedge. Dashed lines show depositional time lines in colluvium. (B) Close-up view of a typical radiocarbon sample, emphasizing the time-transgressive nature of the event horizon [i.e., burial of the upper horizon contact (UHC) of the soil]. (C) Close-up of the 10-cm-thick radiocarbon sample, showing the trend of increasing age with depth and the principle of extrapolating the age of the UHC. (D) Close-up of how age uncertainties are calculated for the event horizon (UHC) immediately above the dated sample. Solid circle shows mean radiocarbon age; surrounding box shows dendro-corrected radiocarbon age, with 2σ limits shown by the horizontal dimension of the box. The vertical dimension of the box indicates sample thickness. The solid line shows a least-squares regression line through the dendro-corrected mean age. The top of the graph represents the UHC of the soil. The total 2σ error range on age of the UHC is composed of an analytical component and a MRT-correction (extrapolation) component. The analytical component is defined by extrapolations of the mean age trend (4.6 yr/mm) from the ±2σ limits on the age of the uppermost 10-cm-thick sample (dashed lines). The MRT-correction components are defined by extrapolations of the maximum age trend (6.7 yr/mm) from the -2σ age limit and the minimum age trend (3.7 yr/mm) from the +2σ age limit (dotted lines).

from 5- to 10-cm-thick soil samples, as shown diagrammatically in Figure 3.57. Age trend with depth is established by dating 1- to 2-cm-thick subsamples (Figure 3.57C). Each of the subsamples, in turn, yields an AMRT age affected by carbon compounds of various ages and molecular weights. The extrapolated age of the UHC contains uncertainties arising from analytical errors on the soil AMRT age and MRT-correction errors arising from the extrapolation procedure itself (spread between the intercepts from the minimum and maximum age trends with depth, Figure 3.57D).

Preferred samples for ^{14}C dating have short age spans (charcoal, wood, peat, shells). If low-organic soils are sampled, one should collect material over the smallest vertical stratigraphic thickness practical. This may involve exposing a large (>1 m^2) surface of the sampled stratum and scraping off a few millimeters of stratigraphic thickness, usually from the top of the soil horizon if this UHC defines the event horizon. Optimal sample weights and pretreatments are described in Appendix 2, in Book's companion web site.

3.4.6.1 Example of Detailed Dating

An exposure of three colluvial wedges and associated soils on the Wasatch fault zone, USA, was closely sampled to examine radiocarbon age variation with depth and organic component, as well as correspondence with TL ages. The radiocarbon ages from an initial set of 10- to 15-cm-thick bulk samples (Figure 3.58, boxes with crosses), can be compared to a later set of 2- to 3-cm-thick decalcified samples

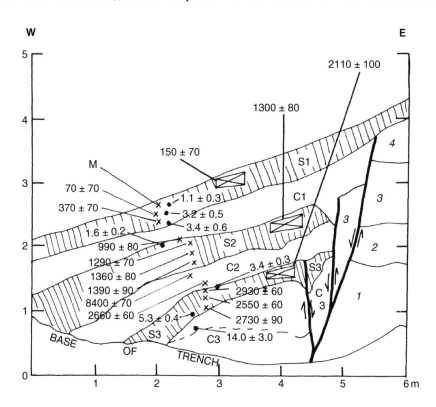

Figure 3.58: Log of trench at Garner Canyon, Wasatch fault zone, USA, showing radiocarbon and TL ages. Large boxes with crosses show correct dimensions of initial bulk radiocarbon samples dated by Machette *et al.* (1992a); ages in large numbers are in ^{14}C year BP Small ×s show ages in ^{14}C year BP of decalcified soils; M = modern (<200 years BP). Solid circles show TL sample locations and total bleach age estimates (ka). Dimensions of ×s and solid circles correctly represent sample dimensions. Soil A horizons S1, S2, and S3 are developed on stony colluvial deposits C1, C2, and C3, respectively. Units 1–4 on the upthrown block are gravelly alluvium. The apparent reversal of radiocarbon ages within soil S3 is anomalous and does not occur in the other carbon fractions. Modified from Stafford and Forman (1993).

($\times$s) dated by AMS. Additional radiocarbon ages on the humic acid and humin fractions of all 12 samples are reported by Stafford and Forman (1993). They found, paradoxically, that the decalcified soil fraction usually dated oldest and the humin fraction youngest for each sample, whereas the reverse is usually true.

For each organic fraction in the A horizons, the trend of increasing age with depth allows extrapolation of the age of the UHC (as in Figure 3.57C). Sufficient carbon for AMS dating was even found in the relatively inorganic colluvial deposits themselves (e.g., C2, Figure 3.58). Age reversals and large disparities occur between the ages of different organic fractions, and indicate that the milligram-size particles of organic matter dated by AMS have complex origins, including reworking of older carbon from sources upslope (e.g., the anomalous 8400 years BP age in deposit C2). However, the basal radiocarbon age from deposit C2 (2660 ± 60 years BP) is similar to the ages from thin samples within underlying soil S3 (2550–2930 years BP), which indicates that some small carbon particles in colluvium accurately reflect the age of colluvium. If the age reversal at the top of soil S3 was not present, it might be possible to bracket tightly the age of faulting between the extrapolated age of the event horizon, and the 2660 ± 60 years BP age of the oldest overlying colluvium. Thus, this method of AMS dating of thin samples appears to hold promise for refined dating of paleoearthquakes, compared to the previous method of bulk sampling. Ages might be refined even further with the use of Bayesian statistics, as described in Section 6.4.

The correspondence of TL ages with radiocarbon ages is best near the top of each soil and worsens downward. This phenomenon has been observed elsewhere (Forman *et al.*, 1989, 1991) and suggests that the dated silt particles contain a higher amount of inherited TL with depth in soil horizons. Light exposure of silt particles, which erases inherited TL, should increase upward in a soil due to slower deposition rates, bioturbation and pedoturbation. For example, TL total bleach ages are only 600 and 470 years older, respectively, than radiocarbon ages at the tops of soils S2 and S3, but become much older with depth (Figure 3.58). Similar very old TL ages were observed at the Borah Peak, Idaho fault scarp, where proximal colluvium deposited at the base of the free face in 1983 and sampled in 1985 (true age 2 years) yielded a total bleach TL age of 40 ± 4 ka (McCalpin and Forman, 1988). By comparison, the alluvial fan gravels composing the scarp free face had an estimated age of ca. 15 ka. The 40-ka apparent TL age of colluvium indicates that the dated silt grains not only escaped exposure to light during colluvial redeposition, but they must have already possessed a large (ca. 25 ka) inherited TL component when deposited at 15 ka in the alluvial fan deposits.

3.4.7 Age Estimates from Cosmogenic Nuclides in Depth Profiles on Fault Scarps

This unique method combines the diffusion dating concept (Section 3.4.2) with cosmogenic dating (Section 3.4.1), to track the cosmogenic exposure of particles as they are eroded, transported, and redeposited on a fault scarp. Phillips *et al.* (2003) point out that, using the diffusion equation alone, "rates of scarp degradation depend strongly on the geomorphic diffusivity, a parameter that is difficult to constrain independently. This difficulty may lead to large uncertainties in the estimated ages of rupture events." Therefore, in their study of the Socorro Canyon normal fault in New Mexico, they combined the accumulation of the cosmogenic nuclide ^{36}Cl to a diffusion model of scarp degradation, to constrain the value of the geomorphic diffusivity and thus determine rupture age(s). Independent stratigraphic evidence from the hanging wall (two colluvial wedges) indicated the scarp had been formed by two late Quaternary rupture events.

First they measured ^{36}Cl accumulated *in situ* in three vertical depth profiles each 4 m deep, located 1.5 m above the fault plane (footwall), 1.5 m below the fault plane (hanging wall), and a control profile 27 m upslope of the fault plane, well above the fault scarp. The control profile penetrated a stable surface not affected by scarp degradation, and showed a simple exponential ^{36}Cl profile from which a depositional age of 122 ± 18 ka was calculated. The footwall profile 1.5 m above the fault trace showed a ^{36}Cl deficit relative to the control profile, indicating net erosion on the upper scarp face, as the diffusion model predicts. The hanging-wall profile showed a ^{36}Cl excess relative to the control profile, indicating net deposition.

They then modeled the accumulation of ^{36}Cl in the vicinity of the scarp, by simulating the erosional redistribution of mass using the diffusion equation. By matching the measured hanging-wall and footwall ^{36}Cl profiles to profiles calculated by the model, they predicted ages of 92^{+16}_{-13} ka and 28^{+18}_{-23} ka for the two rupture events. This technique thus returns results similar to that of the soil catena approach to dating ruptures (Section 3.4.3), but appears to be considerably more precise.

3.5   Interpreting the Paleoseismic History by Retrodeformation

The first version of the trench log forms the basis for the initial interpretation of the style, amount, and age of paleoseismic deformation. Based on the initial log, workers identify the critical structural and stratigraphic contacts (faults and earthquake horizons) that indicate paleoearthquakes, and reexamine them on the trench walls. At this time some attempt is made, in either a rigorous sense or less formally, to reconstruct graphically the sequence of deformation by *restoring stratigraphic units* to their predeformation positions. This procedure is also known as *retrodeformation analysis,* and involves restoring stratigraphic units to their (inferred) original geometries by graphically reversing the sense of displacement on faults. Retrodeformation analysis is based on several basic stratigraphic assumptions, such as (1) the original physical continuity of faulted beds and (2) the original horizontality of the upper contacts of fine-grained beds (Section 3.5.2).

Trench log retrodeformation sequences first appeared in the published literature in the mid-late 1980s (e.g., Asquith, 1985; Meghraoui *et al.*, 1988; Upp, 1989). The more rigorous of these early examples were constructed by cutting the paper trench log along faults and shifting the cut pieces to restore stratigraphy to its prefaulting geometry. Successive deformation stages by then rendered by hand-tracing the restored logs. However, some early sequences were evidently modeled by working forward in time from an assumed initial geometry (*forward modeling)*, rather than backward from the present geometry (*backward modeling*), as indicated by geometric inconsistencies between successive stages.

Retrodeformation sequences appear more frequently in the literature of the 1990s, particularly since 1996. The Italians are the most prolific authors of retrodeformed trench sequences (mainly for normal faults), and some of their artistic drawings even include man-made structures above the ground surface (e.g., Galadini and Galli, 1999). Since the mid-1990s, published sequences have been created by using computer graphics software to manipulate a digitized version of the trench log. Use of computer software enables retrodeformation sequences to be more faithful to the original trench log (i.e., less schematic) and more geometrically rigorous.

The goal of a retrodeformation sequence is to graphically reverse the displacement exposed on the trench wall, in a way that restores stratigraphic relationships that existed prior to faulting. The easiest way to do this is to focus on the paleoearthquake-related unconformities (earthquake horizons) on the trench log. The objective should be to restore these unconformities (paleoground surfaces) to smooth lines. In order to do this, one may simplify the trench log to various degrees as described below.

3.5.1 Types of Retrodeformations

Retrodeformation sequences begin with the present trench log geometry and work backward through time. It eases the drawing process if the starting trench log is simplified somewhat, as described in Table 3.10. Such simplifications are permitted as long as they do not eliminate or reshape any key contacts (such as unconformities) that would change the sequence of events.

3.5.2 Assumptions Used when Restoring Strata to their Prefaulting Geometry

Any retrodeformation sequence should generally honor the laws of superposition, original horizontality, and cross-cutting relationships. Beyond that, there are some useful rules of thumb to follow:

1. Subaqueous strata deposited from suspension (fluvial, lacustrine) are assumed to have been deposited with the upper contact horizontal; the lower contact may have originally had any geometry (undulating) or orientation.

2. Subaqueous strata deposited as bedload are assumed to have been deposited with a subhorizontal upper contact.

3. Subaerial strata (colluvium, slopewash) are assumed to have been deposited with nonhorizontal top and bottom contacts. Debris-facies colluvium (cf. Nelson, 1992b) is assumed to have been originally deposited at the angle of repose for the material (e.g., Figure 3.46, Kaysville trench).

4. Blocks of intact material found in filled tension fissures must have been derived from erosion of a source stratum that was exposed in a nearby fault free face. This stratum may not longer be present on the upthrown block, if it was removed by later erosion.

5. As the lower part of that free face became buried, subsequent material deposited in nearby fissures could only come from strata exposed in progressively higher parts of the free face.

6. If adjacent fault blocks contain similar stratigraphic sequences of subaqueous deposits, but the upper part of the sequence is missing on some relatively uplifted fault blocks, then assume that the preerosion sequence on that block resembled the most complete section on any of the nearby blocks. This rule applies best when, after faulting, subaqueous erosion truncates the differentially uplifted blocks to the same elevation, after which subaqueous deposition of uniform-thickness beds commences on a subhorizontal surface.

7. Rule 6 may not apply if the differentially uplifted fault blocks are buried rapidly after faulting, without the development of an erosional unconformity. In that case, the earliest postfaulting strata will be thinnest on structural highs and thickest in structural lows.

8. An additional complication to Rule 6 is the change of facies near the margins of a fault-controlled depression. Subaqueous strata should be expected to thin and become coarser near the depression margins. Conversely, subaerial strata such as colluvial wedges thicken near the margins of a fault-controlled depression.

Most examples of retrodeformation analysis are two-dimensional; for example, on a vertical trench wall only the vertical components of displacement are restored. Figure 3.59 shows an example of retrodeformation of three normal-faulting events in which beds were faulted, colluvium deposited, and then partly removed by erosion. Attempts to make restorations often point out inconsistencies or

Table 3.10: Types of retrodeformation sequences, as defined by how much the trench log is generalized, and whether there is a drawing block representing every geologic event

Generalization of cross section	
Schematic	Geometry of the stratigraphic units and structures on the trench log is modified significantly in the first stage of the retrodeformation sequence. Modifications are made to both the stratigraphic units and the structures, resulting in a "cartoon-like" present geometry. Stratigraphic units are generalized by combining several thin units into a single thick unit, particularly if the thin units are genetically related. Irregularities and undulations in unit contacts are smoothed out. Soil horizons or other weathering phenomena are deleted. Faults exposed in the trench wall are straightened out, and minor faults and fractures are deleted altogether
Simplified	The only change to the *present trench log geometry* is straightening out the faults on which displacement is to be reversed. Thus, *it* contains all stratigraphic units and minor faults and fractures that were mapped on the trench log. Straightening the faults to straight lines simplifies reversal of movement across them. If the fault lines contain bends, then reversal of fault movement tends to create gaps or overlaps along the fault
Realistic	The faults retain their originally mapped shape during the retrodeformation analysis. In other words, the present *trench log geometry* is identical to the field trench log. If gaps or overlaps are formed when reversing movement along the faults, these gaps/overlaps are either left as-is, or the shapes of the polygons abutting the fault are altered so that gaps or overlaps are eliminated. However, this latter practice violates the conservation of bed area
Temporal Completeness	
Complete	Every deformational, erosional, depositional, and weathering episode that has affected the trench wall is represented by a separate cross section in the graphical retrodeformation sequence. Such a complete sequence is created by first removing the surface soil profile, then removing the uppermost undeformed parent material(s) on which the soil was formed, then reversing the effects of the latest deformation episode. Then, remove all strata displaced by the MRE but not by the PE, remove any weathering that predates these strata, and reverse the deformation that occurred in the PE. This may require restoring parts of the stratigraphic sequence that were eroded between the PE and MRE. Continue in a similar manner back in time
Incomplete	A retrodeformation sequence can be incomplete in two ways. First, some reconstructed events in the chronology may be combined into a single step, such as the deposition of a parent material stratum and the subsequent development of its weathering profile (soil profile). However, combining deposition and soil formation can lead to difficulties in interpretation. Another common combination is to combine the deposition of several conformable strata into a single step. We term a sequence with such combinations an *incomplete retrodeformation sequence*. Second, the sequence may not progress far enough back in time to remove all the deformation events; this we term a *partial retrodeformation sequence*

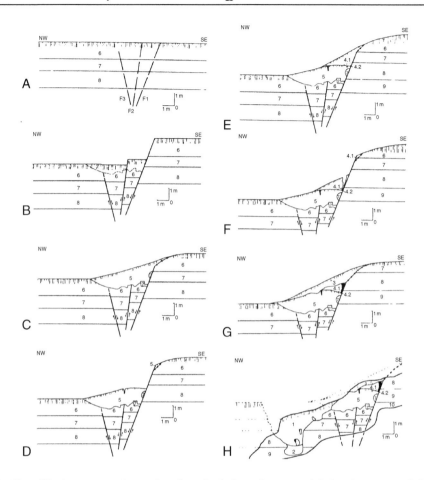

Figure 3.59: Simplified cross sections showing the inferred sequential development of the north wall of the Pole Patch trench, Brigham City segment of the Wasatch fault zone, Utah, USA. (A) Prefaulting stratigraphy. (B) Scarp immediately after the first faulting event; material above dashed line is shattered and eroded. (C) Deposition of first colluvial wedge (unit 5). (D) Second faulting event, displacement on right fault only. (E) Deposition of second colluvial wedge (unit 4). (F) Third faulting event. (G) Deposition of third colluvial wedge (unit 3). (H) Present configuration of trench (within heavy line). An alluvial channel (units 1 and 2) is cut into the toe of the scarp. This type of retrodeformation analysis does not rely on continuity of strata across the fault, except for the first faulting event. Instead it relies on colluvial wedges as event indicators. From Personius (1991); reprinted with permission of the Utah Geological Survey.

impossibilities in the initial trench log, such as correlation of units across faults that is opposite to that required by the inferred sense of slip. As much time may be spent in reconciling the final trench log with a viable retrodeformation sequence, as was spent in making the entire initial trench log. Although each trench log is different, each must pass the test of being restorable without resorting to unreasonable or physically impossible sequences of events.

3.5.3   Accounting for Soil Development in Retrodeformation

Most retrodeformation analyses (such as Figure 3.59) only trace the evolution of stratigraphic units, their erosion (physical subtraction of strata from the cross section) or deposition (addition of strata). However, soils develop through time *within* previously deposited strata, and on long-recurrence faults, there is more time represented by soil profiles in the trench log than in physical deposits. So, in retrodeformation we must be able to "subtract" the effect of soil formation back through the time sequence, separately from removing the host strata of the soil.

Solution: For each soil the soil development index (SDI) is calculated. Then in retrodeformation the SDI value that accumulated during a given time stage may be subtracted without removing its host unit (parent material).

Two underlying premises exist: (1) the SDI value of a soil profile can be increased through time without thickening the stratigraphic unit that hosts the soil, and (2) the SDI value in a soil profile can only be decreased by physical removal (erosion) of the soil and its host strata together.

In order to use SDI as an integral part of the retrodeformation process, *first* one must calculate SDI values in all fault blocks that may have experienced a different uplift-erosion-soil formation history (Figure 3.60). This will often require multiple vertical soil transects in a trench, one for each major structural block in the retrodeformation. *Second*, soils on the fault scarp probably with strong catena effects, and these effects increase the SDI on footslope soils and decrease them on crest soils, relative to the stable surfaces above and below the fault scarp. Since we want to use SDI on footslopes and crests as a measure of soil development time only, we must correct the SDIs on footslopes and crests to what they would have been in a stable setting. For example, McCalpin and Berry (1996) concluded that SDI accumulated 35–37% faster on footslopes than on the lower gradient faulted surface itself. Thus, one would decrease footslope soil SDI values before beginning the retrodeformation "subtraction" steps.

Principles illustrated in Figure 3.60A:

1. All deposits lack soil development at the time of deposition, so their initial SDI = 0

2. Soils will subsequently develop in any deposit exposed at the ground surface, forming a surface soil

3. Due to uplift across the fault, periodic erosion occurs on the upthrown block; this erosion may partially or completely remove both soils and parent materials (deposits) there

4. Because the deposits on the upthrown block are periodically thinned by erosion, soil development may penetrate completely through these thinned deposits and superimpose itself onto an underlying, older soil; this is termed "welding" an upper soil onto a lower one; the original component soils may no longer be recognizable in the "welded" soil profile

5. On the downthrown block, in contrast, periodic deposition occurs. Therefore, soils on the downthrown block are separated by parent materials (deposits) unaffected by pedogenesis (i.e., have an SDI = 0)

6. Due to periodic erosion on the upthrown block, the total SDI there will always be less than the total SDI on the downthrown block

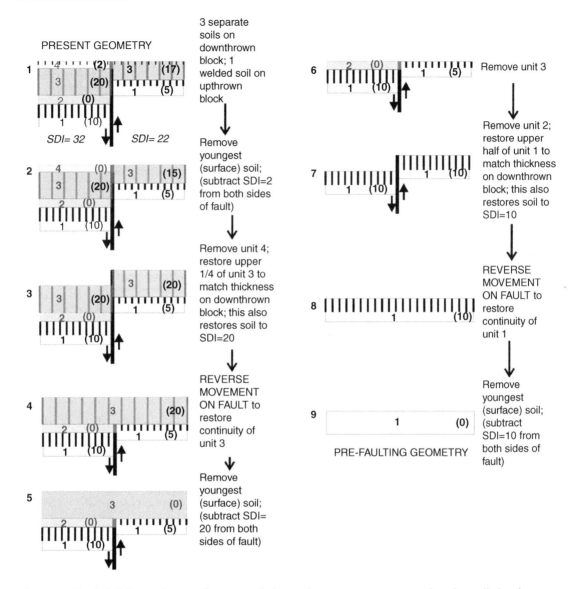

Figure 3.60: (A) Schematic complete retrodeformation sequence, accounting for soil development time. (1) present trench geometry; (2) after removal of the modern soil profile with SDI = 2; (3) after removal of unit 4 (interseismic deposit of the downthrown block) and restoration of the eroded part of unit 3 on the upthrown block, along with its contained paleosol; (4) reversal of the younger faulting event, to restore the continuity of unit 3; (5) after removal of units 3s paleosol, representing SDI = 20; (6) after removal of unit 3; (7) after removal of unit 2 (interseismic deposit of the downthrown block) and restoration of the eroded part of unit 1 and its contained paleosol; (8) reversal of the older faulting event to restore the continuity of unit 1 and its paleosol; (9) removal of the unit 1 paleosol (SDI = 10).

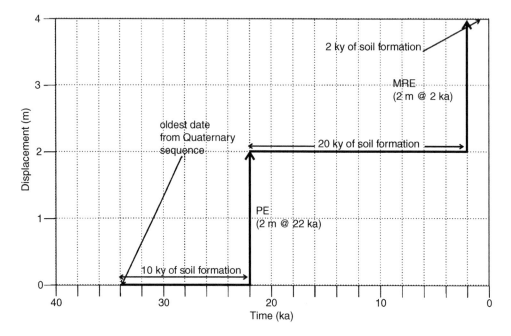

Figure 3.60 (Continued): (B) Slip history diagram derived from the above retrodeformation sequence, assuming that: (1) displacement was 2 m in each faulting event, (2) it took 1 ky to develop 1 SDI unit, and (3) deposition time for units 2 and 4 was negligible compared to soil formation time.

3.6   Distinguishing Tectonic from Nontectonic Normal Faults

In Chapter 2A (Section 2A.4) we argued for the importance of distinguishing tectonic, seismogenic faults from nontectonic and nonseismogenic faults. There are several types of nontectonic and nonseismogenic normal faults that can deform the ground surface and shallow geologic deposits, as described below. Criteria for distinguishing these faults from seismogenic faults are described by Hanson *et al.* (1999), which we summarize below.

3.6.1   Tectonic, but Nonseismogenic Normal Faults

Into this class of faults fall many types of "passive" or secondary faults related to other larger faults and folds. In the extensional environment, the most common of the faults are shallow antithetic and synthetic faults caused by refraction of a larger (seismogenic) fault. These faults may move simultaneously with the seismogenic fault, but they merge with it at such a shallow depth that they produce negligible seismic moment even then. They do not generate $M_w > 5.5$ earthquakes by themselves. The related group of "growth faults" faults may penetrate deeper, but because their displacement arises from differential compaction, the displacement across the fault is greatest at the surface and least at depth. Given the small amount of displacement at depth, and the ductility of the compacting deposits, these faults do not generate $M_w > 5.5$ earthquakes either. The key diagnostic criterion for creeping faults is the lack of brittle-deformation features in the subsurface, and the absence of evidence for past free faces that shed sediment from the scarp.

In compressional environments passive extensional faults are found on fold axes (bending moment faults). This group of faults includes normal faults that form by stretching on the crests of anticlines and monoclines. This includes hanging-wall collapse faults behind active thrust fault tips, such as developed during the El Asnam, Algeria surface rupture of 1980. These normal faults can penetrate only down to the neutral line within the fold, so are generally not deep enough to nucleate $M_w > 5.5$ earthquakes. The key diagnostic criterion for these faults is that they occur in pairs defining grabens, which are restricted to the crests of anticlines or monoclines.

3.6.2 Nontectonic, but Seismogenic Normal Faults

This rather odd class of faults includes mega-landslide blocks on the flanks of volcanic edifices that are underlain by normal faults. Due to the great relief of the edifice, these normal fault planes may penetrate deep enough in the crust to generate $M_w > 5.5$ earthquakes, even though they are basically just huge gravity slides. Examples include the Kalapana Fault on the Island of Hawaii, which generated an M_w 7.2 earthquake on 29-NOV-1975, and similar faults on the eastern flank of Mount Etna, Sicily. The key diagnostic criterion for these mega-landslide faults is the fault trace curvature in plan view (similar to that of landslide headscarps) and their location on the flanks of very large volcanoes.

3.6.3 Nontectonic and Nonseismogenic Normal Faults

This class of faults includes a diverse array of extensional surface processes, some of which are restricted to shallow unconsolidated deposits, and none of which penetrate deeply enough into the crust to generate $M_w > 5.5$ earthquakes.

3.6.3.1 Landslide Faults

At the head and middle of a landslide, the main failure plane has the geometry of a normal fault (see Cotton, 1999). Due to the extension at the head of a landslide, many of the secondary faults related to fault refraction also exist (graben, back-tilting, and step faults). Figure 3.61 shows a trench log across a landslide headscarp in Tertiary volcanic rocks in Utah (McCalpin, 2005b). This landslide scarp contains many of the features found in near-surface tectonic normal faults, including a main normal fault, antithetic fault forming a graben, and extensive network of fissures on the downthrown block. The key diagnostic criteria for landslide headscarps are their plan curvature, their high ration of height versus length, and their location at the head of anomalously hummocky topography with a distinctive bulging toe.

3.6.3.2 Sackung

The sackung (sagging) process describes deep-seated gravitational spreading of mountain ridges and slopes (Chapter 8, Section 8.4.3). This spreading is accomplished primarily by normal faults that outcrop high on mountain slopes (Figure 3.62), and which pass downward into the mountain mass, but rarely are expressed as toe bulges or thrust faults at the base. Due to this asymmetry, Morton and Sadler (1989) termed them "half a landslide." Kinematic models indicate that, because sackung faults are gravitationally driven, they do not extend downward lower than the foot of the mountain ridge. The diagnostic criteria for sackung scarps are their location high on mountain ridges, their orientation perpendicular to the local fall line, and their occurrence as swarms of parallel scarps, usually facing upslope. More detail is given in Hart (2003).

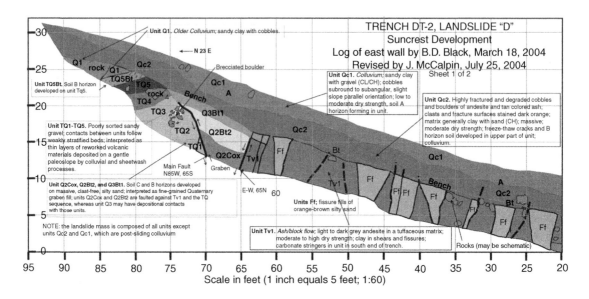

Figure 3.61: Log of a trench through a landslide headscarp, showing the similarities of landslide normal faults to coseismic normal faults. Note the graben at Sta. 65–70, the unusual back-tilted footwall, and extensive fracturing and fissuring of the hanging wall. Based on this exposure alone, it would be difficult to distinguish this fault zone from a seismogenic fault. From McCalpin (2005b). (See Color Insert.)

Figure 3.62: Antislope sackung scarp on Mount Chabenec, Low Tatra National Park, Slovak Republic. This scarp is one in a series of parallel scarps on the southern slope of Mount Chabenec, a series made famous by the pioneering papers of Mahr and Nemcok (1977). Photo by J. P. McCalpin.

Figure 3.63: Four types of earth fissures. (A) circular depressions or "potholes"; (B) surface cracks; (C) fissure gullies, formed by water erosion enlarging surface cracks; (D) fissure with vertical displacement, which forms small scarps (compare to coseismic scarp of Figure 3.13A). From Shipman and Diaz (2008).

3.6.3.3 Subsidence/collapse Faults

Differential subsidence of surficial materials, from a variety of causes, can create normal faults in nonductile deposits. Common causes of differential subsidence are compaction due to fluid withdrawal (Figure 3.63), and collapse or sagging of surficial deposits into subsurface void spaces. Although many subsidence fractures are purely extensional cracks ("earth fissures"), others have a vertical component of displacement (Figure 3.63D). The diagnostic criteria for subsidence-related faults are their association with areas of fluid withdrawal, their often concentric pattern around such areas, or their existence in areas underlain by karstic sinkholes and caverns. Trenching of sinkholes by Gutierrez *et al.* (2009) has shown that sinkhole collapse faults may form as vertical to steep normal faults, but post- and synfaulting sagging of material toward the sinkhole center may rotate them to a steep reverse fault geometry.

CHAPTER 4

Paleoseismology of Volcanic Environments

Suzette J. Payne,* William R. Hackett,[†] and Richard P. Smith[‡]

*Idaho National Laboratory, Idaho Falls, Idaho 83415-2203, USA, Suzette.Payne@inl.gov
[†]Consulting Geologist, 2007 Cherokee Circle, Ogden, Utah 84403, USA, wrhackett@comcast.net
[‡]Consulting Geologist, Nathrop, Colorado 81236, USA, rps3@realwest.com

4.1 Introduction

Faults, tensile cracks, and related extensional features can develop solely as a consequence of magma intrusion. Magma-induced surface faults with co-intrusive displacements of several meters can form aseismically or be accompanied only by shallow, low-magnitude earthquake swarms. *Magma-induced extensional structures* may thus be misinterpreted as coseismic tectonic features (such as normal faults described in Chapter 3), and their seismic potential thereby exaggerated. Special problems exist for paleoseismology in volcano-extensional environments. If traditional methods of assessing slip rates and maximum magnitude are applied without considering the volcanic record or the mechanics of magma intrusion, then the estimated frequencies and magnitudes of paleoseismic events may be erroneous.

Seismicity, surface faulting, magma intrusion, and volcanism are expressed within most tectonic settings, and extension of the brittle crust is globally accommodated by a combination of normal faulting and magmatic processes. The relative significance of these processes varies widely from region to region, is dictated by the balance of magma flux relative to regional-extension rates, and has major implications for the pattern and intensity of world seismicity (Parsons and Thompson, 1991). On a worldwide scale, the pattern of earthquakes coincides very closely with that of active volcanoes, and together these reveal the pattern of lithospheric plate boundaries. One of the most fundamental of all earth processes, the accretion of new lithosphere along the midocean ridge system, occurs through a complex interplay of tectonic and magmatic extension (Bergman and Solomon, 1990; Kong *et al.*, 1992). In addition to its acknowledged role in the formation of ocean crust, there is growing recognition that synextensional magmatism is an ubiquitous characteristic of highly extended continental terrains (Eaton, 1982; Coney, 1987; Lipman and Glazner, 1991; Parsons and Thompson, 1991, 1993; Harry *et al.*, 1993). For example, mass-balance calculations indicate that approximately 5 km of the present crustal thickness in the western Cordillera of the United States was added to the crust of the eastern Great Basin during Cenozoic extension and magmatism (Gans, 1987; Glazner and Ussler, 1989). Seismic and global positioning system (GPS) observations suggest that the process of deep crustal magma intrusion is ongoing in the Great Basin (Smith *et al.*, 2004).

It has long been known that the intrusion of magma produces earthquakes. *Magma-induced seismicity* commonly exhibits spatial migration, focal mechanisms, and waveform characteristics that differ from

International Geophysics, Volume 95
ISSN 0074-6142, DOI: 10.1016/S0074-6142(09)95004-5

those of nonvolcanic sources. The ascent of magma beneath central volcanoes and the lateral injection of dikes beneath volcanic rift zones are commonly accompanied by fracturing of adjacent rocks, producing seismicity (e.g., Brandsdottir and Einarsson, 1979; Klein *et al.*, 1987). *Volcanic earthquakes* (B-type) are typically of low frequency (1–5 Hz), commonly occur without a main shock, and sometimes have non-double-couple mechanisms (Minakami, 1974; Foulger and Long, 1984; Shimozuru and Kagiyama, 1989; Foulger and Julian, 1993; Dreger *et al.*, 2000). These characteristics differ from those of *tectonic earthquakes*, which are commonly of high frequency, include a discrete mainshock-aftershock sequence, and have double-couple mechanisms. A third category referred to as *volcanotectonic earthquakes* incorporates high frequency (5–15 Hz) "tectonic" waveforms and is commonly associated with magma intrusion (Minakami, 1974; Power *et al.*, 1992; Zobin, 2003). Volcano-tectonic earthquakes are so commonly observed that magmatic intrusion must be viewed as capable of altering stresses within large volumes of the shallow and middle crust. Observed seismic and geodetic data associated with magma intrusion show that dike intrusion, propagation, and dilation alter stresses that result in faulting and earthquakes (e.g., Savage and Cockerham 1984; Rubin and Gillard 1998; Owen *et al.*, 2000; Toda *et al.*, 2002; Roman and Cashman 2006; Pedersen *et al.*, 2007).

The purpose of a paleoseismic analysis is to reconstruct the frequency-magnitude history of past earthquakes. Toward this goal, we develop criteria for the recognition of magma-induced, extensional structures, to alleviate their misinterpretation in the paleoseismic record as products of single, large-magnitude earthquakes. In addressing the physical and seismic aspects of magma intrusion, we emphasize *dike intrusion*, which is a widespread process within the upper crust, regardless of magma type or tectonic setting (Emerman and Marrett, 1990; McKenzie *et al.*, 1992). We use this background to describe the growth of discrete volcano-extensional faults and fissures, and the structural development of volcanic rift zones. Our analysis is based on field observations of magma-induced deformation, the contemporary seismicity of active volcanic zones, geodetic remote sensing of dike-intrusion events, and the results of numerical and physical modeling. We assert that volcano-seismic recurrence should be evaluated through careful mapping and geochronology of the coseismic and postseismic volcanic materials and structures. We suggest modifications or alternatives to traditional paleoseismic methods of field excavation and geochronology because the material properties of volcanic rocks and the geometry of volcano-extensional structures commonly differ from those normally selected in paleoseismic investigations. We derive conservative, upper bounds for the maximum magnitudes of volcano-seismic events, using the dimensions and offsets of magma-induced faults. The observational seismicity of active volcanic regions shows that purely magma-induced earthquakes are characteristically of low to moderate magnitude.

Readers may ask why volcano-extensional structures should be included in this book, if they form aseismically or produce only small-to-moderate earthquakes that pose little threat. We offer several reasons. First, many paleoseismological investigations are done within a regulatory context that dictates *all* seismic sources must be considered within tectonic provinces of interest. Second, to assess the significance of volcanically induced earthquakes in a seismic hazard analysis, the role of magma intrusion to accommodate extension in place of regional faulting or in triggering slip along preexisting tectonic faults must be understood. Third, magma-induced structures can be mistaken for the products of single, large earthquakes. Although we are not the first to consider volcanic zones or structures as seismic sources, we offer this chapter as an initial step toward full incorporation of magma-induced seismicity into the larger framework of paleoseismology.

In the decade since this book was first published, new investigations of magmatic processes at active volcanoes and volcanic rift zones evaluate geologic, geodetic, seismic, and geophysical data. New approaches use inversions of observed surface deformation and seismological data to understand and constrain the geometries of magma intrusions and related fault slip. In this revision we (1) incorporate the results of recent field investigations of magma-induced extensional structures, (2) discuss the results of numerical modeling of observational data toward advancing our understanding of magma intrusion and associated faulting, (3) expand the compilation of dike related dimensions, (4) update the tables of maximum magnitudes of dike-induced and magma-related earthquakes, and (5) offer recommendations for future research.

4.2 Volcano-Extensional Structures

Volcanoes grow as much by intrusion as by eruption (Tilling and Dvorak, 1993). Magma intrusion is accompanied by seismicity and deformation, and much of what is known about the internal workings of volcanoes derives from observational seismology and geodesy. Extensional structures are universally associated with volcanism, and range from small *tensile fissures* to major *normal faults* and *caldera-ring fractures*. Because magma intrusion and volcanism are common in extending regions, purely magma-induced structures are likely to be mingled with regional-tectonic structures. Worldwide examples shown in Figure 4.1 and discussed in this section are not comprehensive, but are given as indicators that magmatic processes are important in the tectonics and seismicity of extending regions. The interplay between tectonic and volcano-tectonic sources can be complicated and can vary with time, especially in areas of oblique slip along hotspot tracks, or in transitional regions between strain accommodation by magmatism and normal faulting. Regional-tectonic and magma-induced structures are not always easily distinguished; in Section 4.3 we offer criteria for recognition. Bearing this caveat in mind, we schematically show in Figure 4.2 the great variety of volcano-extensional structures that develop as a consequence of magma withdrawal, intrusion, and eruption.

4.2.1 Worldwide Examples of Volcano-Extensional Structures

A selection of diverse tectonic settings is shown in Figure 4.1, for which magma intrusion is known or assumed to contribute to crustal extension. The eastern Snake River Plain of Idaho (Figure 4.1A) occupies the bimodal volcanic (silicic-mafic) track of the Yellowstone hotspot (Pierce and Morgan, 1992). The region surrounding the eastern Snake River Plain extends by slip on seismogenic normal faults in the northern Basin and Range province with earthquakes reaching magnitudes of $M_w > 7$. In contrast, the eastern Snake River Plain volcanic province has no earthquakes above (local magnitude) M_L 2.0 (Jackson *et al.*, 1993). The low relief of the eastern Snake River Plain reflects the absence of major normal faulting as a mechanism of continental extension. Instead extension is accommodated by emplacement of dikes along volcanic rift zones (Rodgers *et al.*, 1990; Parsons and Thompson, 1991; Smith *et al.*, 1996; Parsons *et al.*, 1998). The major products of eastern Snake River Plain magmatism are isolated rhyolite domes, subsurface basalt-dike swarms, eruptive fissures, basaltic-shield volcanoes, tephra cones, lava fields, *tensile fissures, normal faults*, and *monoclines*, the latter three of which mimic tectonic structures (Hackett and Smith, 1992; Kuntz *et al.*, 1992, 2002). Although the spacing and orientation of volcanic rift zones on the eastern Snake River Plain is similar to that of the normal faults in the northern Basin and Range, reflecting upper-crustal structure and northeast–southwest extension, the rift zones are not colinear with major normal faults in the surrounding Basin and Range Province. The eastern Snake River Plain is unique because it is a region with extensional magmatism, relative aseismicity in comparison to the surrounding Basin and Range Province, the presence of small normal faults within the volcanic rift zones, and the absence of large tectonic faults across the volcanic province.

In the Mono–Inyo region of eastern California (Figure 4.1B), the movement of range-front faults of the Sierra Nevada has apparently been affected by silicic-dike intrusion beneath the Long Valley caldera and

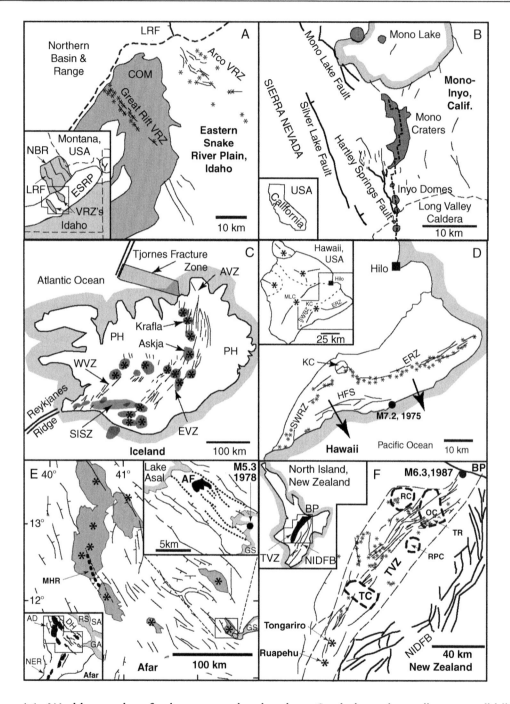

Figure 4.1: World examples of volcano-extensional regions. Symbols apply to all maps: solid lines, fissures and faults related to dike intrusion; heavy solid lines, tectonic faults; asterisk, volcanic vent or volcano; areas with stippled margin, lakes or water bodies. (A) Eastern Snake River Plain (ESRP), Idaho, USA (adapted from Kuntz *et al.*, 1988, 1994). Dark shading, Holocene basalt flows;

the Mono Craters–Inyo Domes. During the past 40,000 years, dike intrusion has accommodated crustal extension in the Mono Craters–Inyo Domes area. Local stress changes due to dike intrusion have caused cessation of movement along the adjacent Silver Lake fault and triggered earthquakes on the colinear Hartley Springs fault (Bursik and Sieh, 1989; Bursik *et al.*, 2003). Magmatic products consist of subsurface rhyolitic dikes, tensile fissures, normal faults, monoclines, rhyolite domes and tuff rings, and isolated basaltic vents. Since 1980, Long Valley caldera has had earthquakes and resurgent uplift of the caldera floor (Hill *et al.*, 2003). The patterns of uplift and seismicity, particularly the occurrence of possible non-double-couple events (Julian and Sipkin, 1985; Dreger *et al.*, 2000), suggest ascent of magma. However, explanations for caldera unrest at Long Valley are equivocal because the caldera is intersected by normal faults of the Sierra Nevada range front, and most earthquakes have double-couple mechanisms indicative of tectonic faulting. We include the Long Valley example because it is an area where magmatic processes are thought to influence local tectonics and seismicity (Bursik *et al.*, 2003).

Iceland is an emergent portion of the mid-Atlantic Ridge where voluminous basaltic-magma production related to a mantle plume has produced several historically active volcanic rift zones (Gudmundsson, 1986; Rubin, 1990; Ryan, 1990). Two transcurrent fault zones, the Tjornes Fracture Zone and South Iceland Seismic Zone, connect the active volcanic rift zones. The transcurrent fault zones have periods of increased and decreased seismicity that are correlated with episodes of dike intrusion (Gudmundsson, 2000; Figure 4.1C). Magmatic features include large central volcanoes with summit calderas, tephra cones, isolated silicic-to-intermediate centers, and extensional structures related to dike emplacement along volcanic rift zones. In the axial region, crustal extension is accommodated largely by dike intrusion

Y, Yellowstone caldera; COM, Craters of the Moon; NBR, northern Basin and Range; LRF, Lost River fault; VRZ, volcanic rift zone. (B) Faults and volcanic features of the Long Valley-Mono Craters area, eastern California (adapted from Bursik and Sieh, 1989; Bursik *et al.*, 2003). Dark shading, rhyolite domes; heavy lines, major segments of the Sierra Nevada range front fault system; heavy dashed line, inferred feeder dike for volcanic rocks. (C) Iceland and adjacent seafloor tectonic elements (adapted from Palmason, 1981; Rubin, 1990). WVZ, west volcanic zone; EVZ, east volcanic zone; PH, pre-Holocene volcanic rocks; Central volcanoes (Krafla, Askja, and others unlabeled) indicated by dark stipple. Tjornes Fracture Zone and South Iceland Seismic Zone (SISZ) are two regions of transcurrent faulting. (D) Island of Hawaii, USA, and Kilauea volcano (adapted from Holcomb, 1987; Peterson and Moore, 1987). MLC, Mauna Loa caldera; KC, Kilauea caldera; ERZ, east rift zone; SWRZ, southwest rift zone; HFS, Hilina fracture system; black dot, 1975 M_w 7.2 Kalapana earthquake. (E) Afar Triple Junction (adapted from Tesfaye *et al.*, 2003; Keir *et al.*, 2006; Wright *et al.*, 2006; Yirgu *et al.*, 2006). Dark shading, volcanic fields younger than 1 m.y.; MHR, Manda-Hararo rift in the Dabbahu magmatic segment, heavy dotted line, 60-km-long dike in 2005; light dashed lines, faults activated by 2005 diking event; GS, Ghoubbet Strait. Index map (lower inset) shows juncture with East African rift system. Black shading, volcanic fields younger than 1 m.y.; AD, Afar Depression; DH, Danakil horst; NER, Northern Ethiopian rift; GA, Gulf of Aden; RS, Red Sea; SA, Saudi Arabia. The Asal rift area, Republic of Djibouti, East Africa (upper inset) shows location of 1978 diking, fissuring, and earthquake event. Heavy dotted lines show fault traces activated during the 1978 episode of dike intrusion, seismicity, and volcanism; AF, 1978 Ardukoba lava flow (black shading); black dot, epicenter of 1978 M_w 5.3 earthquake. (F) North Island, New Zealand, and the Taupo volcanic zone (adapted from Nairn and Beanland, 1989; Cole, 1990; Spinks *et al.*, 2005). TVZ, Taupo volcanic zone; heavy dashed lines, <300 ka calderas: RC, Rotorua; OC, Okataina; RPC, Reporoa; TC, Taupo; NIDFB, North Island Dextral Fault Belt; TR, Tarawera rift; BP, Bay of Plenty.

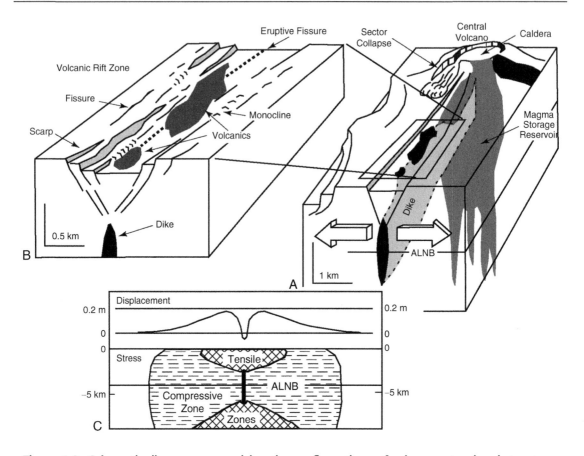

Figure 4.2: Schematic diagram summarizing the configurations of volcano-extensional structures and their relationship to magmatic processes. ALNB, apparent level of neutral buoyancy as described in the text. (A) Brittle deformation associated with upper crustal magmatism. (B) Dike-induced structures along a volcanic rift zone. (C) Results of numerical elastic-deformation model of dike intrusion (modified from Rubin, 1992) are comparable to observed brittle-deformation features of (B). Upper part of diagram shows a vertical-displacement profile above a dike of 1 m thickness, extending from 1 to 6 km deep. Lower part of the diagram shows the compressive and extensional zones that develop around a dike as a result of magma pressure. Panels (B) and (C) adapted from Smith et al. (1996); reprinted with permission of the American Geophysical Union.

into volcanic rift zones that emanate from central volcanoes. Historical seismicity and geodetic monitoring along the volcanic rift zones have contributed greatly to understanding the dynamics of dike intrusion, the formation of co-intrusive extensional structures, and the evolution of volcanic rift zones (Brandsdottir and Einarsson, 1979; Bjornsson et al., 1979; Tryggvason, 1994). Driven by seafloor spreading, older crust is displaced outboard of the active volcanic zones. In these outlying areas, magma supply is small and the volcanic pile is tectonically tilted on regional normal faults, exposing Tertiary dikes and their related extensional structures (Palmason, 1981; Gudmundsson, 1995).

The Hawaiian Islands are along the track of a north Pacific, oceanic-intraplate mantle plume. Although the tectonic settings differ, Hawaiian volcanism has much in common with its Icelandic counterpart, including the growth of large basaltic-shield volcanoes with summit calderas and development of volcanic rift zones. These features are well displayed on the island of Hawaii (Figure 4.1D), which consists of five coalesced Quaternary shield volcanoes, two of which (Mauna Loa and Kilauea) have frequently erupted in historic times. Many cycles of volcanism, intrusion, ground deformation, and seismicity have led to an unrivaled knowledge of the magma dynamics (Decker *et al.*, 1987; Tilling and Dvorak, 1993). Magma ascends along a central conduit beneath Kilauea caldera, where it is stored in a shallow (2- to 5-km deep) chamber, from which dikes intrude laterally into several volcanic rift zones. As a result, the southern part of the island between Kilauea's east and southwest volcanic rift zones is being pushed or is sliding seaward. Normal faults of the Hilina fracture system represent the headwall of a large slump block that is undergoing seaward gravitational failure. GPS velocities and focal mechanisms of deep earthquakes (5–12 km) that are spatially unrelated to the magma-storage system indicate that the entire southern sector of the island is moving seaward, along a decollement of pelagic sediment on which the island grew (Decker, 1987; Wyss, 1988; Gillard *et al.*, 1992; Owen *et al.*, 2000). Strain release in the form of tectonic earthquakes ($M_w > 6$) has occurred between the active Mauna Loa and Kilauea volcano centers and along the southern seaward sector (Ando, 1979; Endo, 1985; Jackson *et al.*, 1992; Arnadottir *et al.*, 1991; Beisser *et al.*, 1994).

Plate divergence encroaches into continental East Africa near the Afar triple junction, where extension is accommodated largely by magma intrusion, which has been observed during two contemporary dike-intrusion events (Figure 4.1E). Geologic mapping and radiometric-age determinations in the Asal-Ghoubbet rift area, Republic of Djibouti, suggest that magmatic activity waxes and wanes while extension occurs continuously across the rift (Stein *et al.*, 1991). During times of low magma input, significant structural and topographic relief accumulates via displacements on tectonic-normal faults. During times of high magmatism, lasting on the order of 10^5 years, volcanic materials fill the structural basins and dike intrusion accommodates much of the extension. The 1978 dike-intrusion event in Asal-Ghoubbet rift area (upper inset in Figure 4.1E) resulted in about 2 m of extension, and was accompanied by both eruption of the Ardukoba lava flow and a (body-wave magnitude) m_b 5.3 earthquake (Ruegg *et al.*, 1984; Stein *et al.*, 1991). An even more spectacular dike-intrusion event took place in 2005 in Afar south of Dabbahu volcano when a 60-km-long dike intruded the upper crust of the Manda-Hararo rift (Figure 4.1E), extending from 2 to 9 km depth and opening as much as 8 m (Sigmundsson, 2006; Wright *et al.*, 2006). New and existing fissures and faults were observed to have horizontal openings as much as 3 m and vertical offsets of up to 5 m along segments up to 3 km long, and the event was accompanied by eight earthquakes with magnitudes between 5.0 and 5.6 (Yirgu *et al.*, 2006; Rowland *et al.*, 2007). Rowland *et al.* (2007) show, with field observations of faults and fissures produced during the event, that all the deformation is consistent with strain distribution induced by dike intrusion at depths as shallow as 2.5 km along the entire rift segment.

The Taupo volcanic zone, North Island, New Zealand is an ensialic, backarc basin that developed landward of the Taupo-Hikurangi subduction system, a zone of regional compression (Figure 4.1F). Oblique slip along the plate boundary is accommodated by extensional upper-plate tectonics, caldera-related silicic volcanism, and inferred magma intrusion in the Taupo volcanic zone (Cole, 1990; Spinks *et al.*, 2005). Regional GPS observations support oblique right-lateral slip within the North Island Dextral Fault Belt and extension in the adjacent Taupo volcanic zone (Wallace *et al.*, 2004). Opening rates in the Taupo volcanic zone inferred from GPS are greater than those observed in the paleoseismic fault record (e.g., Villamor and Berryman, 2001, 2006) implying that other mechanisms such as dike intrusion contribute to extension (Wallace *et al.*, 2004). In the northern Taupo volcanic zone, aligned silicic domes of the Okataina and other volcanic centers indicate the presence of subsurface rhyolite dikes. In 1886, a basaltic dike produced several meters of dilation and was accompanied by fissure fed, explosive volcanism along the Tarawera rift (Nairn and Cole, 1981). These observations suggest that extension is partly accommodated by magma intrusion, although the record of observational seismicity is ambiguous in this regard (Sherburn, 1992a; Bryan *et al.*, 1999; Hurst *et al.*, 2002). Based on field observations during the 1922 earthquake swarm (estimated maximum-magnitude range of 6.0–7.5), Grindley and Hull (1986) tentatively suggest that ground fissuring and earthquake swarms may occur unpredictably throughout the Taupo volcanic zone as a result of basaltic dike intrusion. The absolute uplift of the western Whakatane graben during the 1987 M_L 6.3 Edgecumbe earthquake (Smith and Oppenheimer, 1989) is attributed by Nairn and Beanland (1989) to protracted, late-Quaternary magma intrusion and heating beneath the area.

4.2.2   Central Volcanoes and Calderas

Contemporary seismicity and magma dynamics have been thoroughly investigated at the intraplate volcanoes of Hawaii (Decker *et al.*, 1987) and Iceland (Rubin, 1990; Ryan, 1990), as well as at composite volcanoes situated along convergent margins (Latter, 1989; McGuire *et al.*, 1991). Beneath *central volcanoes*, magma ascends from sources in the lower crust and upper mantle, and is commonly heralded by earthquakes as deep as 35 km. The magma-conduit systems and shallow reservoirs beneath many volcanoes are seismically well defined. Other central volcanoes, such as Nevado del Ruiz, Colombia, do not have single, well-integrated conduit systems or vent complexes near the surface and are only spatially associated with local seismicity (Zollweg, 1990). Magmatic storage in reservoirs beneath central volcanoes and in dikes beneath volcanic rift zones generally occurs worldwide at depths of 2–4 km. Early concepts of lateral dike propagation in the shallow crust emphasized the role of local density equivalence of magma and host rocks as the primary influence on the depth at which magma laterally propagates. This is the level of neutral buoyancy (LNB) of Ryan (1987). Subsequent modeling and analysis of lateral dike propagation have demonstrated the more fundamental role of differential stress normal to the dike plane (Rubin, 1995; Watanabe *et al.*, 1999). In addition, the source of excess magma pressure at the "LNB" is attributed to magma flux from below (Lister and Kerr, 1991) rather than slowed ascent due to the density equivalence of magma and host rock at shallow depth. In Figure 4.2 we therefore identify the level of magma storage and horizontal dike intrusion as ALNB, the "apparent level of neutral buoyancy," following the concepts and terminology of Takada (1989), Rubin (1995), and Watanabe *et al.* (1999).

Calderas are broad collapse depressions, marked by ring faults or broad flexural zones, and have diameters of 1.6–80 km (Lipman, 2000). Ring faults develop as a result of the state of stress in the host rock, magma chamber geometry and pressure, and mechanical properties of the rock. Field observations of ring faults show that some are nearly vertical, mostly inward dipping, dip-slip faults while others are partly shear fractures or faults and partly extension fractures occupied by dikes (Gudmundsson, 2007). Caldera structures occur in virtually all tectonic and volcanic settings. Withdrawal of magma from shallow reservoirs is their fundamental mechanism of formation, regardless of magma type or eruption style. In basaltic systems, withdrawal of magma from subcaldera chambers commonly occurs by the lateral injection of dikes into volcanic rift zones, at shallow depths (<4 km) and may or may not be accompanied by volcanism. In silicic systems, withdrawal occurs by explosive evacuation of the chamber, with eruption and wide dispersal of pumiceous pyroclastic-flow and tephra-fall deposits.

Several calderas have been sites of earthquakes and other signs of historical unrest, including uplift or deflation of their floors due to subterranean flux of magma or hydrothermal fluid (Dzurisin and Newhall, 1984; De Natale and Pingue, 1993). Contemporary seismicity and geodesy are closely monitored at the Quaternary silicic calderas of Yellowstone and Long Valley, USA and Taupo, New Zealand in light of their past eruptive histories (Julian and Sipkin, 1985; Sherburn, 1992b; Langbein *et al.*, 1993; Bryan *et al.*, 1999; Husen and Smith, 2004; Chang *et al.*, 2007) (Figure 4.1A, B, and F). All three calderas have well-defined ring-fracture systems and are intersected by regional-tectonic faults. Their structural complexity therefore confounds seismic interpretations as to tectonic versus magmatic mechanisms. Geodetic and seismic patterns nonetheless indicate a strong interaction between magmatic and regional-tectonic processes (Bursik, 1992; Smith and Braile, 1994; Wicks *et al.*, 1998, 2006; Waite and Smith, 2002). Hampel and Hetzel (2008) use numerical modeling to show that uplift and subsidence of the Yellowstone caldera can lead to slip reversals along the Teton fault to the south. Their numerical results are consistent with changes in geodetic observations along this normal fault, indicating that caldera inflation resulted in normal slip whereas caldera deflation resulted in reverse slip.

Large-scale caldera collapse and its affiliated seismicity have been observed only twice during the past century. Slip along inward dipping ring faults and earthquakes were associated with the 1968 collapse of the summit caldera of Fernandina, a large basaltic-shield volcano of the Galapagos Islands (Simkin and Howard, 1970; Filson *et al.*, 1973). Abe (1992) describes the seismicity associated with one of the century's largest explosive eruptions of silicic pumice, involving the formation of a caldera near Mount Katmai, Alaska, in 1912.

4.2.3 Volcanic Rift Zones

When magma pressure in the upper part of a conduit or reservoir exceeds the strength of the surrounding rocks, bladelike *dikes* propagate outward from the reservoir along self-generated fractures at depths of 2–4 km. These dikes have heights (vertical dimensions) of several kilometers, and lengths that may extend tens of kilometers from the central magma conduit. The dikes orient perpendicular to the direction of least compressive stress, which is influenced by the combined effects of mass loading by the volcanic edifice, and the regional stress field (McGuire and Pullen, 1989; Ryan, 1990).

Earthquakes usually accompany lateral or vertical dike propagation, but their origins are not well understood. Rubin and Gillard (1998) suggest that most dike-induced earthquakes result from slip along suitably oriented preexisting fractures or faults that are within a region of large ambient differential stress. From several volcanic earthquake swarms, Pedersen *et al.* (2007) concluded that the background stress state was more important than the tectonic setting or stressing rate in influencing the levels of seismic energy released during magma movements at three Iceland volcanoes. For several volcanic eruptions, Roman and Cashman (2006) reported focal mechanisms consistent with the orientations of regional stresses for earthquakes produced in association with the lateral propagation of dikes. They also noted focal mechanisms with compressional axes rotated 90° to the directions of regional stresses for earthquakes associated with dilation of dikes. This result is further supported by the observations of Roman *et al.* (2008) who correlated changes in focal mechanisms with different phases of eruptions at Soufriere Hills volcano in Montserrat. Inflation associated with a NE–SW oriented dike centered beneath the vent resulted in a local ∼90° reorientation of the stress field.

Volcanic rift zones are belts of aligned volcanic vents and associated magma-induced extensional structures such as subparallel normal faults, anastomosing tensile cracks, monoclines, and eruptive fissures (Figures 4.2B, 4.3 and 4.4). Volcanic rift zones are always underlain by dike complexes and the largest volcanic rift systems, at the midocean ridges, are thousands of kilometers long by a few tens of kilometers wide. The smallest systems are a few kilometers by several hundred meters in the case of isolated, monogenetic volcanoes fed by single dikes. In Table 4.1, we compile information on dimensions of dikes and *dike-induced extensional structures*, with the intent of portraying their overall dimensions in

Figure 4.3: (A) Fissures induced by basaltic dike intrusion along the southwest rift zone of Kilauea volcano, Hawaii. (B) Small graben with 3-m scarp, formed by dike intrusion associated with the 1983–1990 eruptive episodes of Puu Oo (active vent in background), east rift zone of Kilauea volcano, Hawaii. From Smith *et al.* (1996); reprinted with permission of the American Geophysical Union. (See Color Insert.)

comparison to tectonic features. The dimensions are derived from: outcrop measurements; inversion of GPS, satellite radar, tilt, and leveling data to derive subsurface-dike parameters; observations of co-intrusive seismicity; numerical modeling; and research drilling. The overall lengths of subaerial volcanic rift zones vary from 10 to 100 km, and all are less than 20 km wide. Graben dimensions within volcanic rift zones vary; some are <3 km by <1 km whereas others are 10–20 km long by 3–5 km wide.

Dike lengths are a few kilometers to several tens of kilometers in Table 4.1; some Icelandic dikes and the 2005 Dabbahu Afar dike have maximum lengths of 60 km. Dike basal depths of lateral intrusions tend to

Figure 4.4: Aerial photograph of basaltic dike-induced fissure swarms, Holocene Kings Bowl lava field, eastern Snake River Plain, Idaho. From Greeley and King (1977); photograph courtesy of Ronald Greeley.

be shallow (<5 km), consistent with the 2- to 4-km depth along which dikes propagate (Gudmundsson, 1984a; Ryan, 1987), whereas vertically propagating dikes can have basal depths that extend to deep crustal or upper mantle source regions. Dike dips range from vertical to as shallow as 5–44° (Iceland; Mount Etna, Italy; Long Valley, California; Table 4.1), but most dike dips are greater than 70°. Field observations of dike thicknesses in Table 4.1 range from less than 1–26 m although most mafic dikes are 1–4 m thick. Silicic dikes can have greater thicknesses (Mastin and Pollard, 1988); Table 4.1 lists 8 m for Mono Craters. Geodetically observed dike intrusion episodes permit estimates of dike openings that range from <1–4 m for most, and up to 8 m for the 2005 Dabbahu Afar dike. Associated with dike intrusion are *dike-induced normal faults*, which have dimensions that are controlled by the depth of dike intrusion. Normal faults associated with shallow lateral propagation of dikes have short lengths (<12 km), basal depths (<5 km), steep dips (mostly 70–80°), and co-intrusive offset usually <2 m (Table 4.1).

Geodetic measurements, co-intrusive seismicity, and field observations of surface deformation during intrusive episodes conclusively support the idea that magma intrusion causes surface faulting along volcanic rift zones. Migrating outward at rates of 0.1–6 km/h (Klein *et al.*, 1987; Einarsson, 1991), the propagating dikes incrementally form normal faults and fissures, resulting in swarms of small, shallow earthquakes (Table 4.2). Elastic and inelastic tumescence of the ground surface occurs over a broad area up to 10 km wide, centered on the dike top (Figure 4.2C). Extensional structures such as normal faults, fissures, and monoclinal flexures occur within a narrow, several-kilometer-wide belt centered above the propagating dike, as in the so-called Icelandic *fissure swarms.* Repetition of the process forms subsurface dike swarms and complex extensional structures. Most dike-induced *fault scarps* are <1 m high, but the emplacement of thick dikes or the reactivation of preexisting structures by multiple dike intrusions may produce fault scarps >10 m in height (Table 4.1; Thingvellir, Krafla, Kilauea east rift, and Afar-Dabbahu).

Table 4.1: Dimensions of dikes, faults, and graben produced by dike intrusion in volcanic rift zones

Location[a]	VRZ[b] length/Width (km)	Dike[c]				Normal Fault[d]			Graben[e] length/Width (km)	Type[f]	References
		Length/Basal depth (km)	Height (km)	Dip (°)	Thickness/Opening (m)	Length/Basal depth (km)	Dip (°)	Vertical indiv./Total offset (m)			
Iceland											
Northwest Tertiary dikes		9.6 m/n		70–90	0–23, 4.3 A/n		69 A	n/0.5–25, 5.3 A		O	1
East Tertiary dikes		22 m/n		45–90	0–26, 4.1 A/n		42–90, 69 A	n/0.5–8, 2.7 A		O	2
Thingvellir fissures	12/7.5					11 M/n		n/40 M		O	3
Vogar fissures						0.36–5.7, 1.9 A/n		n/10 M, 2.3 A		O	4
Eyrarfjall-Karastadir dikes		15 A/n		5–90, 69 A	0.02–25, 1.4 A/n		53–89, 75 A	n/0.5–150, 10 A		O	5
Laki, 1783	100/15	27/n						n/0.5–8	n/0.15–0.30	O	6
Krafla, 1975–1984	80/4–10					0.35–3.5/n		n/42 M		O	7
Krafla, 1975								0.05–0.7/n	20/5	O	8
Krafla, 1975		9/4–6	5.5	90	2m/n		70–80	1.0 m/n		S, GM	9,10
Krafla, 1977		n/8.5	7.2		2.6/n	n/2 M	70	1.6 M/n		GM	11
Krafla, 1978		n/6	4.5		1.5/n	6M/n	55 M	1.7 M/n		GM	11

Kilauea rift zones, Hawaii, USA

East, 1968	60/4–6		75	0.5–1/n		n/20m		O, GM	12,13
East, 1980	3/3	3.5	90					S	14
East, 1981		3	82	1 m/n				GM	10
East, 1982	2.5/3	2.5	90					S	14
East, 1983	14.3 M/3	3m	85M	3.3 M/2.1				GM	15
East, 1997	5.1/2.4		76	n/2				GM	16
East, 1997	3/2.5	2	80	n/0.7				GM	17
East, 1984–1986	1.6/2.9	2.5	87	n/1.1 A				GM	18
Southwest, 1971	30/1–3	3.8	89	n/1.8 M				O, GM	10,13,19
Southwest, 1981	2/1	2	72					S	14
Southwest, 1982	4M/9.6	7.6		2.3/n				GM	15
Hawaii, USA									
Koolau Volcano			65–85	0.05–6.7, 0.7 A/n				O	20
Snake River Plain, Idaho, USA									
Arco rift	20/6				3–5/n		3.3/0.7	O	21
Great rift	85/2–8	15–40/n		0.5–1/n				O	22
Great rift	4–14/23–31	22–30		2–12/n				GM	23

(Continued)

Table 4.1: Dimensions of dikes, faults, and graben produced by dike intrusion in volcanic rift zones (Cont'd)

Location[a]	VRZ[b] length/Width (km)	Dike[c]				Normal Fault[d]			Graben[e] length/Width (km)	Type[f]	References
		Length/Basal depth (km)	Height (km)	Dip (°)	Thickness/Opening (m)	Length/Basal depth (km)	Dip (°)	Vertical indiv./Total offset (m)			
Nevada, USA											
Lake Tahoe, 2003		8/33	4	50	1/n					S, GM	24
California, USA											
Medicine Lake	10/2	4M/n								O	25
Long Valley, 1983	16/2–4	n/12	8	30	n/0.4					GM	26
Long Valley, 1989		10/12	10		n/0.1					S, GM	27
Mono Craters, AD1350	25/10	11/n			8/n	8/n			0.7/0.6	O,D	28
Afar, Africa											
Ethiopian rift	60/20									O	29
Afar rift (sa)	15/ 11 RV							n/150 RW		O	30
Afar, Asal 1978 (sa)						4–12/n		0.1–0.7/n	8–10/ 3–4 RIR	O	31
Afar, Asal 1978 (sa)		4/4.5	3.5		n/2.1					GM	30,32
Afar, Asal 1978 (sb)		9/4.5			n/4.1					GM	30,32

Location									Type	No.
Afar, Dabbahu, 2005		60/n			n/3	3M/n		<5/20	O	33
Afar, Dabbahu, 2005		n/9	7		n/8M, 3.5A	2M/1–2.5	65	7M, 2A/n	GM	34
New Zealand										
Tarawera, AD1886	17/5	n/1–2		80					O	35
Tarawera, AD1315		8/n						1–2/n	O	36
Russia										
Tolbachik, 1975–1976		5–8/n							O	37
Japan										
Sakurajima, 1914		7/n			n/20				GM	38
Izu-Oshima, 1986		12 M/12 M	12 M	85–90	n/2.7 M				GM	39
Ito-oki, 1989 (sb)		6/n	6	85	n/0.11				GM	40
Izu Peninsula, 1997 (sb)		5.6/5.2	4.9	70	n/0.38				GM	41
Izu Islands, 2000 (sb)		15/13	5		n/20	8.9–20/n	41–85	0.39–1.4/n	GM, S	42

(Continued)

Table 4.1: Dimensions of dikes, faults, and graben produced by dike intrusion in volcanic rift zones (Cont'd)

Location[a]	VRZ[b] length/ Width (km)	Dike[c]				Normal Fault[d]			Graben[e] length/ Width (km)	Type[f]	References
		Length/ Basal depth (km)	Height (km)	Dip (°)	Thickness/ Opening (m)	Length/ Basal depth (km)	Dip (°)	Vertical indiv./ Total offset (m)			
Nicaragua											
Cerro Negro, 1999					N/0.5–0.9	0.15M/n		0.5M/n		O	43
Italy											
P. de la Fournaise, 2000		1–3/1M	1	67 M	<1/0.59 M					GM, O	44
P. de la Fournaise, 2003				60 M	n/0.3 A					GM	45
Mount Etna, Italy											
S Flank, 1989	0.4 M/1m		0.8		N/1					GM	46

	2.3 M/n	2.4 M	85–90	n/3.5				GM	47
S Flank, 2001									
S Flank, 2002	1/n	3.1	85	n/1.0				GM	48
NE Flank, 2002	6/n	2.1–2.5	44–89	n/3.3				GM	48

Key: Range of values listed unless otherwise indicated by: A, average value; M, maximum value; m, minimum value; n, not available. Volcanic features: sa, subaerial; sb, submarine; RV, rift valley; RW, rift wall; RIR, rift-in-rift.

[a]Location of the volcanic rift zone and the year or age of the dike-injection event, if available. Iceland, Kilauea, Koolau, Snake River Plain, Lake Tahoe, Africa, and Piton de la Fournaise are associated with past eruptions of predominantly mafic magma; Long Valley is and Tarawera may be associated with predominantly silicic magma; all others are associated with magma of mafic to intermediate compositions.

[b]Dimensions of Volcanic rift zone (VRZ) lengths and widths, which contain surficial features resulting from dike intrusion.

[c]Dimensions of lengths, basal depths, heights, dips, thickness, and opening (during dike intrusion) of dikes within volcanic rift zones.

[d]Dimensions of surficial lengths, basal depths, dips, individual vertical offsets (observed during dike intrusion), and total vertical offsets of normal faults within volcanic rift zones produced by dike intrusion.

[e]Dimensions of lengths and widths of graben produced by dike intrusion.

[f]Type of data reported: O, Outcrops; S, Seismicity; GM, Modelling of geodetic observations; D, Drilling.

References: (1) Gudmundsson, 1984b; (2) Gudmundsson, 1983; (3) Gudmundsson, 1987a; (4) Gudmundsson, 1987b; (5) Forslund and Gudmundsson, 1991; (6) Thordarson and Self, 1993; (7) Opheim and Gudmundsson, 1989; (8) Sigurdsson, 1980 and Bjornsson et al., 1977; (9) Brandsdottir and Einarsson, 1979; (10) Pollard et al., 1983; (11) Rubin, 1992; (12) Jackson et al., 1975; (13) Holcomb, 1987; (14) Karpin and Thurber, 1987; (15) Wallace and Delaney, 1995; (16) Owen et al., 2000; (17) Cervelli et al., 2002; (18) Hoffman et al., 1990; (19) Duffield et al., 1982; (20) Walker, 1987; (21) Hackett and Smith, 1992; Smith et al., 1996; and Kuntz et al., 1994; (22) Kuntz et al., 1988, 1992, 2002; (23) Holmes et al., 2008; (24) Smith et al., 2004; (25) Fink and Pollard, 1983; (26) Savage and Cockerham, 1984 and Miller, 1985; (27) Hill and Prejean, 2005 and Langbein et al., 1995; (28) Heiken et al., 1988; Mastin and Pollard, 1988; Sieh and Bursik, 1986; Bursik and Sieh, 1989; and Bursik et al., 2003; (29) Ebinger and Casey, 2001; (30) Stein et al., 1991; (31) Abdallah et al., 1979 and Le Dain et al., 1979; (32) Tarantola et al., 1980; (33) Yirgu, 2006; Rowland et al., 2007; (34) Wright et al., 2006 and Sigmundsson, 2006; (35) Nairn and Cole, 1981; (36) Nairn et al., 2005; (37) Fedotov et al., 1983; (38) Hashimoto and Tada, 1992; (39) Hashimoto and Tada, 1990; (40) Okada and Yamamoto, 1991; (41) Aoki et al., 1999; (42) Nishimura et al., 2001a and Toda et al., 2002; (43) La Femina et al., 2004; (44) Fukushima et al., 2005; (45) Froger et al., 2004; (46) Bonacorrso and Davis, 1993; (47) Bonacorso et al., 2002; (48) Aloisi et al., 2006.

Table 4.2: Maximum magnitudes and focal depths of earthquakes associated with dike intrusion

Location[a]	Rifting event[b] (Year)	Maximum magnitude[c]	Focal depth(s)[d] (km)	References
Iceland				
Krafla fissure swarm	1975–1976	4.5	0–6	1
Krafla fissure swarm	1977	3.8	0–6	2
Krafla fissure swarm	1978	4.1	1–4	3
Hawaii, USA				
Kilauea rift zones				
East	1965	4.4 (M_L)	0–8	4
East	1968	3.3	<5	5
East	1969	2.9	<5	6
East	1976–1977	3.8	<10	7
East	1980, Aug.	3.0 (M_c)	0.5–3	8
East	1980, Nov.	3.1 (M_c)	0.7–4	8
East	1982	3.0 (M_c)	0.5–3	8
East	1999	3.7	nd	9
Southwest	1975	3.0	nd	7
Southwest	1981	3.4 (M_c)	1–2	8
Africa				
Asal, Afar	1978	5.3 (m_b)	0–6	10
Dabbahu, Afar	2005	5.6	1–10	11
Nyiragongo	2002	4.8		12
Nevada, USA				
Lake Tahoe	2003	2.2 (M_L)	29–33	13
Japan				
Izu Peninsula	1989	5.5 (M_{JMA})	<8	14
Izu Peninsula	1997	5.3 (M_w)	5–10	15
Izu Islands	2000	6.4	2.3	16

(Continued)

Table 4.2: Maximum magnitudes and focal depths of earthquakes associated with dike intrusion (Cont'd)

Location[a]	Rifting event[b] (Year)	Maximum magnitude[c]	Focal depth(s)[d] (km)	References
New Zealand				
Taupo Volcanic Zone	1964–1965	4.6	4–8	17
Ruapehu Volcano	1995	4.8 (M_L)	5–20	18
Wyoming, USA				
Yellowstone Caldera	1985	4.9 (M_c)	2–10	19
Mean $\pm$ 1 σ, $n = 23$[e]		4.1 $\pm$ 1.1		This Chapter

[a]Worldwide dike-injection events associated with mafic magma except for: Mono Craters, which is associated with silicic magma; Others, Taupo Volcanic Zone, Ruapehu, Miyake-jima, and Yellowstone, may be associated with silicic or intermediate magma compositions.

[b]An episode of dike intrusion and associated seismicity having a known beginning and end.

[c]Maximum magnitude reported for the dike-injection event. Magnitudes: M_L, Local or Richter; M_c, Coda; M_w, moment; M_{JMA}, Japan Meterological Agency; m_b, Body wave. No definition of magnitude scale was reported for values without magnitude designation.

[d]Depth range of volcanic seismicity and maximum magnitude earthquake associated with the dike-injection event; nd, No data obtained.

[e]Mean and one standard deviation computed based on magnitudes as presented.

References: (1) Einarsson and Bjornsson, 1979; (2) Brandsdottir and Einarsson, 1979; (3) Einarsson and Brandsdottir, 1980; (4) Bosher and Duennebier, 1985; (5) Jackson *et al.*, 1975; (6) Swanson *et al.*, 1976b; (7) Dzurisin *et al.*, 1980; (8) Tanigawa *et al.*, 1981, 1983, Nakata *et al.*, 1982, and Karpin and Thurber, 1987; (9) Cervelli *et al.*, 2002; Dzurisin *et al.*, 1980; (10) Abdallah *et al.*, 1979; Lepine and Hirn, 1992; (11) Yirgu *et al.*, 2006; Ebinger *et al.*, 2008; (12) Kavotha *et al.*, 2002; (13) Smith *et al.*, 2004; (14) Okada and Yamamoto, 1991; (15) Aoki *et al.*, 1999; (16) Nishimura *et al.*, 2001a; (17) Grindley and Hull, 1986; (18) Hurst and McGinty, 1999; (19) Waite and Smith, 2002; This chapter.

4.2.4   Magma-Induced Slope Instability

Steep-sided volcanoes are inherently unstable, and collapse of their flanks is marked by horseshoe-shaped craters, similar in form to nonvolcanic landslide scarps. *Failure and mass movement* are commonly limited to unstable volcanic slopes as a result of loading by eruptive materials or from oversteepening due to shallow magma emplacement (Moore *et al.*, 1989; Borgia *et al.*, 1992, 2000). In some cases, failure has led to massive *landslides* that displace large volumes of seawater resulting in *tsunamis* (e.g., Ma *et al.*, 1999; Bonaccorso *et al.*, 2003). Failure and mass movement can involve entire volcanic piles (as shown on Figure 4.1D), generating large-magnitude earthquakes and associated tsunamis. Along the East Rift Zone of Kilauea volcano, the wedging action of dikes appears to be widening the volcanic rift zone down to a depth of about 8 km. The widening is accommodated by intermittent, seaward movement of the southeastern slopes of the island of Hawaii, along a subhorizontal decollement of seafloor sediment on which the volcano grew (Decker, 1987; Owen *et al.*, 2000). Swanson *et al.* (1976a) used geodetic data to document the accumulation of compressive strain seaward of the volcanic rift zone (schematically depicted by the arrows in Figure 4.1D). In 1975, strain was released in the (surface-wave magnitude)

M_s 7.2 Kalapana earthquake on the south flank of Kilauea volcano (Figure 4.1D). Similar processes may be at work along the lower east flank of the Mount Etna volcano in Sicily. The volcano sets at the northern end of the Malta Escarpment, which separates continental crust to the west from oceanic crust to the east. Recent trenching investigations and geodetic remote sensing suggest that magma intrusion abets extension along the escarpment, resulting in slope instability and eastward displacement of the east flank of the volcano (Azzaro, 1999; Azzaro et al., 2000; Ferreli et al., 2002; Lundgren et al., 2003).

Deformation due to intruding magma (tumescence) and a probable magma-induced M_L 5.1 earthquake contributed to the May 18, 1980, catastrophic slope failure and lateral blast of Mount St. Helens, Washington, USA (Lipman and Mullineaux, 1982). Moriya (1980) identifies several Japanese volcanoes with large avalanche scars, and notes that the long axes of the scars are oriented normal to aligned fissure vents and cones. Siebert (1984) suggests an apparent tendency for composite volcanoes with subparallel dike swarms to undergo large-scale sector collapse and generation of debris avalanches.

4.3   Criteria for Field Recognition of Volcano-Extensional Features

Many of the volcano-extensional structures that we have described, those with clear relationships to central volcanoes and circular structures related to caldera collapse, are recognized using the conventional field techniques of volcanic and structural geology. We do not discuss them further. For landslides, we refer the reader to Chapter 8 of this book. We find it more beneficial to discuss the linear-extensional structures and zones that are related to dike intrusion. These features occur within large extensional regions such as the Basin and Range Province of the western United States and the Afar region of eastern Africa, where they may be juxtaposed with regional-tectonic structures such as range-bounding normal faults. Within this context, dike-induced extensional structures could be interpreted as being of tectonic origin. As we show in Section 4.4.4, a mechanistic distinction between magmatic and tectonic structures can be problematic, because major tectonic structures may be capable of supporting large-magnitude earthquakes, but purely magma-induced structures generally cannot. We develop diagnostic criteria to recognize dike-induced extensional surface features using several different methods that include modeling of dike emplacement and the tectonic geomorphology observed in volcanic rift zones. We also discuss the use of geophysical data and geodetic signals to identify and assess the characteristics of magma intrusions beneath extensional structures.

4.3.1   Results of Empirical and Numerical Modeling

Numerical modeling and scaled-physical experiments of dike intrusion provide a foundation for understanding the origins of dike-induced deformational structures. Numerical modeling assuming elastic media gives information on the relationships between dike geometry, stress and strain distributions, and surface deformation (Pollard et al., 1983; Marquart and Jacoby, 1985; Rubin and Pollard, 1988; Mastin and Pollard, 1988; Rubin, 1992, 1993, 1995; Roth, 1993). In numerical-elastic experiments, dike intrusion produces broad uplift, with narrow subsidence centered above the propagating dike (Figure 4.2C). The locations of displacement (or strain) maxima are a function of the ratio of dike height (vertical dimension of the dike) to dike depth (distance of dike top below surface). Displacement or strain maxima are symmetrical for a vertical dike, but asymmetrical for nonvertical dikes (Pollard et al., 1983). These findings reflect the calculated stress field around the dike (Pollard et al., 1983; Figure 4.2C) and are generally consistent with field and geodetic observations along active volcanic rift zones undergoing dike intrusion, such as those of Iceland and Afar (Rubin, 1992).

Because real-earth deformation is not purely elastic, normal faults and fissures often develop where the tensile zone above the dike top interacts with the earth's surface (Rubin, 1992). Although the regions alongside the propagating dike are under compression (Figure 4.2C), compressional structures are only occasionally observed in the field (Pollard *et al.*, 1983) because the magnitude of the compressive stress is small in relation to the compressive strength of the rocks. In contrast, the tensile stress often exceeds the tensile strength of fractured upper crustal rocks.

Surface faulting and the formation of other inelastic structures have been investigated using *physical-analog models* of dike intrusion (Mastin and Pollard, 1988). Vertical dikes simulated with cardboard sleeves are inserted into boxes filled with flour-sugar mixtures representing the brittle crust. Strain measurements are made during successive stages of dike dilation, including fissure-and-fault development (Figure 4.5). Fissures appear in two symmetrical zones when dilation of the sleeve is about 10% of the sleeve depth and continue to form progressively inward with additional dilation. Most fissures are oriented parallel to the sleeve plane. A shallow *topographic trough* forms above the sleeve top and is flanked by two broad topographic highs, outboard of the fissure zones. When sleeve dilation approaches about 20% of the sleeve depth, dip-slip movement produces *monoclines* and a central *graben*. Subsurface extensional fractures develop, accommodate dip-slip movement, and propagate by connecting with extensional features in adjacent layers and with surface fissures, forming nearly vertical normal faults at the surface but decreasing to 70° dips at depth. Fault depths are shallow; they extend only slightly deeper than the dike top. The total horizontal component of displacement (or opening) is about 60–75% of the dike thickness. Thus, inelastic structures of the physical models are generally consistent with (1) the elastic-strain profiles of numerical experiments; (2) geodetic inversions showing that faults extend only above and ahead of propagating dikes (Du and Aydin, 1992); and (3) field, geodetic, and seismic observations on active volcanic rift zones (Rubin, 1992).

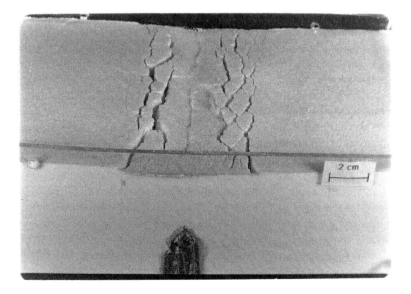

Figure 4.5: Fissures and graben formed by intruding a "dike" of cardboard within a linoleum sleeve into a layered, sugar-flour mixture. Photograph taken at maximum dilation when the width of the linoleum sleeve approached 0.2 times the depth to the top of the sleeve. Mastin and Pollard (1988, p. 13,229); copyright by the American Geophysical Union; courtesy of Larry Mastin.

With reference to Table 4.1, implications of the physical models for geologic field investigations include the following:

1. The magnitude of cumulative fault throw at the surface above the intruding dike and the total horizontal displacement are proportional to the dike thickness, with thicker dikes producing more pronounced graben, normal faults, and fissures. Basaltic dikes are generally 1–4 m thick (Table 4.1). Hence, the dike-induced vertical and horizontal displacements measured in the field for single basaltic dikes should be on the order of several meters. Thicker dikes (rhyolitic dikes can be tens of meters thick) should produce proportionately greater displacement, and this is consistent with the more pronounced structures observed above silicic dikes (with throws of 10 m or more) relative to basaltic dikes (Mastin and Pollard, 1988).

2. Graben width is related to dike depth, such that deeper dikes produce wider zones of extension than shallower dikes. Few dike-induced graben in volcanic rift zones are greater than 2 km wide, suggesting that dikes generally do not induce surface faulting until they are within a few kilometers of the surface. The small offsets and rupture areas of dike-induced faults suggest that the magnitudes of associated seismicity will be small, as discussed in Section 4.4.4.1.

4.3.2   Volcano-Tectonic Geomorphology

Although discrete, dike-induced structures are morphologically similar to nonmagmatic-extensional structures, the bilateral symmetry, the lack of net vertical displacement across graben, the small-scale offsets of individual structures, and the association with eruptive fissures along volcanic rift zones are *diagnostic indicators* of magmatic origin. Perhaps the best criterion for identifying volcano-extensional features is their inferred or demonstrated relationship to cogenetic volcanic materials (Figures 4.2A and B, and 4.4). Within volcanic rift zones, extensional structures commonly occur as a diffuse belt several kilometers wide, and commonly with a central graben that is symmetrically disposed around a central eruptive fissure. Tensional fissures are most abundant (Figure 4.3), indicating that most of the deformation is purely dilatational. Some deformation indicates localized stress variations leading to compressional features including pressure ridges or buckles, strike-slip faults (Rowland *et al.*, 2007), and reverse faults (Gudmundsson *et al.*, 2008). Typically, magma-induced normal faults involve no major displacement or rotation of crustal blocks. Most offsets are less than 1 m, but may reach several tens of meters where thick (silicic) or repeated dike-intrusion events are involved. The 2005 Dabbahu Afar dike event caused slip along preexisting faults as shown in Figure 4.6A (Rowland *et al.*, 2007). Vertical displacements typically vary abruptly along strike, and individual faults are short (hundreds of meters to about 10 km; Table 4.1), commonly grading into monoclines or purely tensional fissures. Monoclinal folds develop above the upper tips of normal faults, which have ascended to depths of 25–50% of their fault lengths but have not yet breached the surface (Grant and Kattenhorn, 2004). Where a vertical normal fault breaches the surface, it breaks though the upper hinge of the monocline. This process results in both throw and dilation (or heave), which generally forms chasms along the base of the fault scarp (Rowland *et al.*, 2007). The limb of the monocline commonly occurs in the hanging wall of the fault (Figure 4.6B). Tensional fissures as well as normal faults can have horizontal openings <1–70 m resulting in chasms tens of meters deep (Gudmundsson, 1987a; Thordarson and Self, 1993; Gudmundsson, 2000).

On the regional scale, extensional magmatism produces diffuse belts of volcanism, fissuring, and subdued normal-fault scarps. Even after millions of years of basaltic volcanism and dike intrusion the terrain is topographically subdued (e.g., Iceland and the eastern Snake River Plain; Figure 4.1A and C, respectively). This is in contrast to extensional provinces that lack substantial magma flux into the upper

Figure 4.6: Two contrasting styles of dike-induced faulting. (A) Scarp of a reactivated normal fault in the Dabbahu rift, Afar, Ethiopia. Features include vertical offset shown by the color contrasts of the basalt rock, thin veneer of recently deposited sediment, little to no horizontal opening along the fault, and abrupt change in displacement along strike. Reprinted with permission of Julie Rowland. (B) Southwest view of the Almannagja normal-fault scarp in Thingvellir National Park, Iceland, where it broke through the surface in basalt along the upper hinge of the monocline resulting in both vertical throw and horizontal opening. The limb of monocline (on left) occurs in the hanging wall of the fault. Sediments and basalt blocks fill the chasm along the fault. Hengill volcano is visible in the background. Reprinted with permission of Simon Kattenhorn. (See Color Insert.)

crust (Parsons and Thompson, 1991), where recurrent faulting is a primary mountain-building process that produces several kilometers of vertical offset and substantial topographic relief.

4.3.3 Geophysical Methods

Because the surficial features of volcanic rift zones result from dike intrusion, geophysical evidence of subsurface dikes will lend confidence to interpretations of magmatic versus tectonic origin. The geophysical signatures of dikes result from physical contrasts between dikes and surrounding rocks, or from changes in properties related to structural offset and are observable with potential-field and electrical methods.

The densities and magnetic susceptibilities of shallow igneous intrusions commonly differ from those of the surrounding country rocks, and the resulting potential-field geophysical anomalies can be used to make inferences on their depths and shapes (e.g., Telford et al., 1990; Blakely, 1996). Electric-field and magnetic anomalies are also used in mapping the configuration of subsurface fluids, which is influenced by volcanic materials and structures (Keller and Rapolla, 1974). Geomagnetic and gravimetric methods have proved particularly useful for understanding the subsurface structure of volcanoes and their intrusive roots (e.g., Yokoyama, 1974; Kauahikaua et al., 2000; Masturyono et al., 2001; Yoshio, 2007).

The densities and magnetic susceptibilities of solidified *mafic* intrusions are generally greater than those of country rocks, including compositionally similar lava flows (Schoenharting and Palmason, 1982; Flanigan and Long, 1987). The growth of secondary magnetite in deuteric envelopes around the intrusions commonly enhances the magnetic contrast with country rocks (Bleil et al., 1982). The volcanic rift zones of Hawaii possess characteristic magnetic patterns, primarily long wavelength, linear, magnetic-low zones, probably depicting rocks that have been chemically and mineralogically altered by hydrothermal fluids at depths greater than 1 km (Hildenbrand et al., 1993). Shorter wavelength, positive anomalies probably reflect slowly cooled, unaltered intrusions. As a result of these factors, linear, symmetrical, and magnetic anomalies, together with geologic information, have been used worldwide for the mapping of mafic dikes and dike swarms within several igneous provinces including: the Columbia Plateau (Swanson et al., 1979); East African Rift System (Halls et al., 1987); Auckland volcanic field (Rout et al., 1993); Canadian Shield (Schwartz et al., 1987); and Basin and Range Province, USA (Zoback and Thompson, 1978; Blakely and Jachens, 1991; O'Leary et al., 2002).

Electrical and other potential-field anomalies, in combination with volcanic- and structural-geologic information, can be used to identify the shallow intrusive masses that commonly underlie volcanic zones. The high-level impoundment of groundwater by dikes (Stearns, 1985) or high gradients of local water table levels due to abrupt changes in hydraulic conductivity (Mundorff et al., 1964; Kuntz et al., 2002) are widespread in areas of intrusion and eruption. Broad (>3-km) self-potential anomalies occur along the volcanic rift zones and beneath the summit caldera of Kilauea volcano (Jackson and Kauahikaua, 1987). Hermance et al. (1984) show that magnetotelluric observations can indicate important physical features associated with caldera structures, including the location and offset of major boundary faults because of the high resistivity contrast between caldera fill and crystalline basement, and the structural control of saline hydrothermal-fluid flow.

4.3.4 Geodetic Remote-Sensing Techniques

Geodetic remote-sensing techniques combined with mathematical methods are powerful tools to quantitatively assess the dimensions, rates, and processes of magma intrusions. Results show that dikes generally intrude as tabular bodies of definable length and height (or depth extent). They can propagate vertically beneath or laterally along volcanic rift zones, change shape through inflation or dilation, and be associated with seismicity and slip on preexisting nearby faults. Dikes are generally positioned directly beneath the diagnostic indicators of surface strain that were discussed previously.

Airborne laser swath mapping (ALSM) or light detection and ranging (LiDAR) surveys when coupled with a GPS base station can produce high-resolution digital elevation models (DEMs) to an accuracy of 5–10 cm vertically and 20–30 cm horizontally, even beneath dense vegetation (Carter *et al.*, 2007). LiDAR can be used to map the surface deformation produced by magma intrusion and, with multiple surveys, track changes in surface features during volcanic activity. Airborne and ground-based LiDAR can be used to comprehensively map and measure the surface deformation, such as the full areal extent of horizontal and vertical offsets, which can then be mathematically modeled to interpret the relationships between the subsurface dike dimensions and surface strain.

Interferometric synthetic aperture radar (InSAR) is another remote-sensing technique that has allowed quantification of the dimensions of intruding dikes and the assessment of associated fault slip. Satellite images acquired before and after a volcanic or intrusive event can be used to map the event's surface deformation. Magma-induced surface displacements modify the distance between the ground and satellite, creating phase changes that can be represented by fringe patterns on an interferogram of the two images (Massonnet and Sigmundsson, 2000). Various mathematical methods (e.g., inverse theory, dislocation sources in elastic half-spaces, and boundary-element methods) have been used to interpret the subsurface deformation from InSAR images in volcanic rift zones and at active volcanoes. For example, Wright *et al.* (2006) and Amelung *et al.* (2007) performed inversion of the InSAR data for the 2005 Afar, Africa and 2002 Mauna Loa, Hawaii events to estimate the dikes' depth extent, dip angle, and amount of opening or dilation. Wright *et al.* (2006) also estimated the amount of slip that occurred on associated normal faults (see Table 4.1). In another example, Froger *et al.* (2004) and Fukushima *et al.* (2005) performed three-dimensional mixed boundary-element modeling of InSAR data for two dike-intrusion events at Piton de la Fournaise volcano, Reunion Islands in 2000 and 2003. In addition to other parameters, they determined that the dikes dipped at angles of 50°–70° (Table 4.1). Lundgren *et al.* (2003) inverted InSAR data to determine that 1993–1995 deformation of the east flank of Mount Etna volcano could be explained by the inflation of an spheroidal magma source centered at 5 km below sea level, and slip along a basal decollement.

In addition to LiDAR and InSAR data, other observational data such as seismicity, GPS velocities, and tilt measurements can be incorporated into the mathematical analyses through joint inversion or as starting parameters for numerical models. Morita *et al.* (2006) used seismicity to constrain the initial geometry of the dike and then performed a time-dependent inversion of GPS data to interpret the evolution of the dike-intrusion process, volumes, and orientations of two dikes during the 1998 earthquake swarm of the Izu Peninsula, Japan. For an earlier dike-intrusion event in Japan, Okada and Yamamoto (1991) used dislocation-source modeling to interpret tilt, leveling and GPS velocity data. They determined that the 1989 earthquake sequence and eruption involved both dike intrusion and slip along a nearby reverse fault.

4.4 Paleoseismological Implications and Methods

The main purpose of this section is to suggest viable approaches to the paleoseismic investigation of magma-induced surface deformation. We therefore do not attempt a comprehensive overview of paleoseismic investigations of tectonic faults in volcanic terrains, but refer readers to other sections in this book. We only present a few examples to illustrate paleoseismic methods and interpretations that are currently being employed in volcanic environments.

Volcanic environments commonly host features, events, and processes of solely tectonic origin. For example, the trenching of tectonic faults in the Acambay graben, a large intra-arc basin within the Trans-Mexican Volcanic Belt (Suter *et al.*, 2001), was investigated by Langridge *et al.* (2000) in an effort to understand the 1912 M_s 6.9 Acambay earthquake and at least four late Pleistocene and Holocene earthquakes with similar surface ruptures to the 1912 event. They determined that large-magnitude earthquakes ($M_w \sim 7$) occur along faults within the Trans-Mexican Volcanic Belt as result of the regional stress field with only a minor influence from volcanism. Using more traditional paleoseismic techniques, Tibaldi and Leon (2000) describe the relationships of Pleistocene–Holocene transpressional tectonic faults to volcanoes in the southern Andes of Colombia. They determined the slip rates of active Quaternary faults and the relationships between the active faults and volcanism. Such phenomena can be investigated and interpreted using the more typical paleoseismic field techniques within tectonic environments described elsewhere in this book.

4.4.1 Excavation

In this section, we discuss selected approaches that may be applied toward the excavation of magma-induced structures associated with dike intrusion. In contrast to the poorly consolidated sedimentary materials that are shed from rising normal faults and that are typically sought for paleoseismic excavation (Chapter 3), dike-related volcanogenic structures and materials are commonly developed in volcanic bedrock, which is composed of strongly lithified materials that do not produce well-developed *colluvial wedges* (Figure 4.6). In volcanogenic structures associated with collapse of volcano flanks where magma intrusion occurs beneath nearby rift zones, colluvial wedges are not everywhere present in the hanging wall, the sedimentary response is completely different from that of coseismic surface faulting, and the scarps in some places evolve through aseismic creep (Ferreli *et al.*, 2002). In addition, many volcanogenic structures are monogenetic (*nonrecurrent*) and have small displacements; these factors further diminish the potential for colluvial wedge development. Therefore, magma-induced structures should be carefully chosen for excavation and age dating. Although our prior illustrations have shown that successful paleoseismic trench excavations can be done in volcanic environments where scarps are developed in unconsolidated volcanic ash and volcaniclastic sediment, we also assert and discuss below that it may be more productive to demonstrate cogenetic or relative relationships with volcanic deposits, rather than to attempt excavation and dating of the deformational features themselves.

Some of the possibilities and problems for dating volcano-extensional structures are shown in Figure 4.7. Tensile fissures can be viable sites for excavation and dating because they are a common type of magma-induced structure and because they are sediment traps. However, their irregular vertical geometry and the potential complexity of their sediment fill preclude traditional back-hoe-trenching operations (Figure 4.7A and B). Hand excavation is required in many cases, proceeding more along the lines of archaeological excavation than paleoseismic trenching. Magma-induced faulting and fissuring are commonly expressed in resistant lithologies such as lava flows. As a result, scarps degrade slowly and contribute blocky debris to noncolluvial-sediment wedges (Figure 4.7D).

In an environment inaccessible to the application of normal paleoseismic field methods, high-resolution seismic-reflection surveys in Yellowstone Lake, USA have provided an alternative to excavation for estimating the ages and displacements of surface ruptures along a submerged extensional fault system within the Yellowstone caldera. Distinctive reflector horizons and coring investigations allowed the estimation

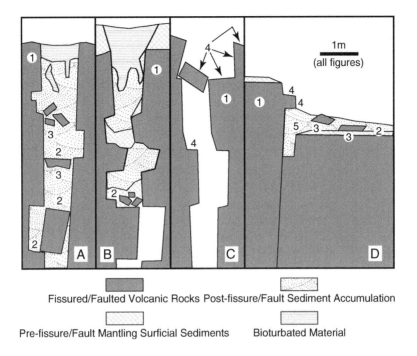

Fissured/Faulted Volcanic Rocks   Post-fissure/Fault Sediment Accumulation

Pre-fissure/Fault Mantling Surficial Sediments   Bioturbated Material

Figure 4.7: Schematic, near-surface cross sections of volcano-extensional structures. Single-deformation events are depicted; recurrent slip or fissuring would produce more complicated schemes. Panels (A)–(D) indicate geometric and lithologic configurations to which geochronometry could be applied: (A) Magma-induced fissure in resistant volcanic rock; fissure fill consists of blocks of volcanic-wall rock and post fissure sediment. (B) Fissure as in (A), but developed within volcanic rock having a mantle of surficial sediment; infilling consists of three components: volcanic blocks, clasts of mantling sediment, and a matrix of post fissure sediment derived from deflation of the surrounding surficial-sediment cover, as well as from distant sources. (C) Magma-induced fissure with no sediment fill. (D) Scarp developed in lithified volcanic rock. Volcanic blocks are of colluvial origin, but the sedimentary wedge is transported material, not derived from scarp degradation. Numbers 1–5 refer to potential geochronometric methods: (1) Conventional dating of deformed volcanic rocks, to constrain maximum age of deformation (e.g., argon-isotopic dating of lava, or radiocarbon dating of charred vegetation beneath the lava). (2) Dating deep portions of post deformational sediment fill by radiocarbon or thermoluminescence, to constrain the minimum age of deformation. (3) Dating the time that blocks have fallen from the fissure or scarp, using thermoluminescence or radiocarbon on sedimentary or organic material beneath the blocks. This gives a conservative minimum age because blocks can fall long after deformation. (4) Use of in situ cosmogenic radionuclides to date pre syn, and postdeformational geomorphic surfaces. (5) Conventional trenching investigation of surficial sediment accumulated at the base of a scarp. From Smith *et al.* (1996); reprinted with permission of the American Geophysical Union.

of displacement per event, slip rate, and recurrence intervals for three surface-rupturing events. In addition, speculations about fault length and down-dip extent provided some earthquake-magnitude scenarios for a seismic hazard assessment (Johnson *et al.*, 2003). Using similar procedures, total vertical-slip rates and long-term extension rates were obtained for submerged rift faults in an Icelandic lake (Bull *et al.*, 2005).

4.4.2   Geochronology

Here we cite some successful applications of geochronology to illustrate the emerging methods that can be used to date volcanic materials, structures, and surfaces (e.g., Grosse and Phillips, 2001). Traditional rock age dating techniques, such as Potassium/Argon (K/Ar) and Argon isotope dating (Faure, 1986) as well as paleomagnetic measurements (Butler, 1991), can be used to determine ages of older volcanic deposits. These techniques, combined with radiocarbon, thermoluminescence (or TL), and other methods for younger volcanic deposits are necessary for assessing recurrence intervals of volcanism. In turn, volcanic recurrence is a basis for estimating the minimum frequency of dike intrusion (e.g., Hackett *et al.*, 2002), and thus periods of volcanic seismicity.

As a demonstration of the precision and accuracy that are achievable with current methods, Argon isotopes have been successfully used to determine the age of eruptive products from the AD 79 Vesuvius eruption as being 1925 ± 69 years, the calendar age of the eruption in 2004 (Renne *et al.*, 1997; Renne and Min, 1998; Lanphere *et al.*, 2007). Bursik and Sieh (1989) use radiocarbon dating to constrain the timing of dike intrusion in the Mono–Inyo area of California (Figure 4.1B), and Kuntz *et al.* (1986) develop meticulous pretreatment procedures for dating Holocene basaltic-lava fields on the eastern Snake River Plain, Idaho (Figure 4.1A). Forman *et al.* (1993, 1996) show that heating of fine-grained sediment beneath basaltic lava flows effectively resets the thermoluminescence ages of the sediment, and "baked" sediment is therefore useful for dating young lava flows. Sediment in fissures or on scarps can potentially be dated by radiocarbon or thermoluminescence, given that the materials are carefully excavated and their genetic relationships are well understood (Figure 4.7). Problems include (1) uncertainty about whether the sample represents early accumulation; (2) inadvertent sampling of collapsed clasts of mantling-surficial sediment that may be substantially older than the deformation; (3) the possibility that deformation occurred in a series of small steps, each of which disrupted the accumulating sediment; and (4) bioturbation by burrowing rodents and insects. On the other hand, rodents leave middens that can be dated by radiocarbon (Betancourt *et al.*, 1991).

Cation-ratio dating of rock varnish is an empirical surface-exposure dating method that can be applied to volcanic rocks (Reneau and Raymond, 1991), and in situ produced, cosmogenic radionuclides are increasingly being used for dating geomorphic surfaces. Volcanic olivine (Cerling, 1990; Kurz *et al.*, 1990) and plagioclase (Poreda and Cerling, 1992) are suitable minerals for cosmogenic helium and neon dating of lava surfaces. Cosmogenic ^{36}Cl is used by Zreda *et al.* (1993) to date a young basaltic-eruption complex, and Jannik *et al.* (1991) use it to establish a chronology of lacustrine sedimentation. We mention these examples because they can potentially be applied to date lava-flow surfaces that have been cut by fissures, lava-flow surfaces on blocks that have been rotated to expose new surfaces during or after deformation, and certain infilling or mantling sediment (Grosse and Phillips, 2001) (Figure 4.7C and D).

4.4.3   Recurrence Intervals

Magma-induced faults may lack sufficient displacement or suitable materials for the development of colluvial wedges (e.g., Figure 4.6A), and many are monogenetic features, unlikely to have undergone recurrent movement. An alternative to the excavation and direct dating of magmatic faults and fissures is to focus on mapping and dating of the associated volcanic materials. The utility of mapping and dating of volcanic materials is demonstrated by Cannon and Burgmann (2001) for the Hilina fault system in Hawaii. The vertical displacements of young lava flows during historic events provided measures of horizontal and vertical displacements per event. Assuming similar displacements per event for paleoruptures in dated prehistoric lavas, the number of events and the recurrence intervals for prehistoric earthquakes were estimated.

Because earthquakes in volcanic rift zones occur as a consequence of dike intrusion during magmatic cycles, earthquake recurrence on volcanic rift-zone faults can be estimated by establishing the recurrence interval of volcanic cycles. Establishing volcanic-recurrence intervals requires thorough knowledge of volcanic processes and the regional patterns of volcanism and takes into account the nature of vent clusters (single dikes can produce many aligned vents). A means of estimating the general proportion of eruptive to noneruptive cycles should also be sought, because not all dikes erupt.

Even when precise and sufficient age determinations are available from volcanic rocks to establish confidently volcanic-recurrence intervals, the information is not analogous to that established by paleoseismic studies of individual normal faults. Rather than being a "one scarp per earthquake" situation, each cycle of recurrent volcanism may involve several dike-intrusion events, and each dike in turn may or may not generate earthquake swarms. Like recurrent faulting, volcanism is episodic. Thus, a viable approach is to estimate volcanic recurrence using relative and absolute chronology and use that interval in estimation of magma-induced seismic recurrence. Conservatism is introduced by decreasing the ratio of volcanic vents per dike intrusion (i.e., assuming one vent per dike-intrusion episode), and by adopting a maximum magnitude that is consistent with the largest measured fault dimensions (see Section 4.4.4.4). Added conservatism is achieved by assuming that each dike-intrusion episode produces a maximum-magnitude earthquake, even though the observed seismicity during numerous dike-intrusion episodes indicates substantial variation in maximum magnitude (e.g., Table 4.2, Kilauea and Krafla volcanic rift zones).

4.4.4 Maximum Magnitude

In this section, we offer methods to assess the maximum magnitudes of earthquakes associated with magma intrusion. We discuss worldwide observational seismology in terms of three geologic settings, and then we compare these data to calculated moment magnitudes based on fault dimensions. Table 4.2 is a compilation of *maximum magnitudes* of earthquakes associated with dike intrusion, mainly along volcanic rift zones. Table 4.3 gives maximum magnitudes of earthquakes at some central volcanoes and calderas, which owing to their greater depth range and structural complexity than purely dike-induced settings, generate larger magnitude earthquakes (Figure 4.8). Table 4.4 lists some maximum magnitudes of "tectonic" earthquakes that may have been triggered by adjacent magma intrusion. In contrast to the seismicity associated with shallow magma intrusion, the large magnitudes of these earthquakes are due to the larger rupture areas of their source faults. We offer reasons for the observation that maximum magnitudes of earthquakes associated with dike intrusion are less than those observed at central volcanoes and calderas and those observed to be triggered by other magmatic processes (Figure 4.8). In Table 4.5, we calculate moment magnitudes from known or reasonably assumed fault dimensions for dike-induced normal faults. In the accompanying discussion, we identify these calculated values as an upper limit of maximum magnitude and compare them to the observational data of Table 4.2.

4.4.4.1 Earthquakes Associated with Dike Intrusion

Maximum magnitudes of observed earthquakes during dike-intrusion events are given in Table 4.2, ranging from M_w 2.2 to 6.4 and largely from the Icelandic and Hawaiian basaltic rift zones. Dike-induced earthquakes in Iceland and Hawaii rarely exceed (magnitude) M_w 5, and maxima at other places are more commonly near M_w 4. Areas such as Africa, Japan, and New Zealand have greater maxima, ranging from

Table 4.3: Maximum magnitudes and focal depths of earthquakes associated with calderas and central volcanoes

Location[a]	Rifting event[b] (Year)	Maximum magnitude[c]	Focal depth(s)[d] (km)	References
Iceland				
Krafla Caldera	1975–1976	5.0	0–4	1,2,3
Heimaey Volcano	1973	4.0	15–25	1
Bardarbunga Volcano	1974	5.0 (m_b)	nd	2
Grimsvotn Caldera	1983	4.0	nd	4
Hekla Volcano	1991	3.0	nd	5
Hawaii, USA				
Kilauea Caldera	1969	3.6	15–35	6
Kilauea Caldera	1976–1977	3.4	<3	7
Africa				
Nyiragongo	2002	4.8 (m_b)	nd	8
La Reunion Island				
Piton de la Fournaise	2002	3.0	<1	9
Nicaragua				
Cerro Negro Volcano	1999	5.2 (M_w)	nd	10
Costa Rica				
Arenal	1968	4.5	nd	11
Galapagos Islands				
Fernandina Caldera	1968	5.1 (M_s)	nd	12
Sierra Negra	2005	5.5 (M_w)	nd	13
Papua, New Guinea				
Rabaul Caldera	1983–1985	5.1 (M_L)	0–4	14
Kamchatka, Russia				
Shiveluch Volcano	1964	5.3	nd	15

(Continued)

Table 4.3: Maximum magnitudes and focal depths of earthquakes associated with calderas and central volcanoes (Cont'd)

Location[a]	Rifting event[b] (Year)	Maximum magnitude[c]	Focal depth(s)[d] (km)	References
Plosky Tolbachik Volcano	1975–1976	5.0	10–20	15
Kliuchevskoi Volcano	1983	3.5	2–5	16
Gorely Volcano	1985	6.0	nd	17
Washington, USA				
Mount St. Helens Volcano	1980	5.1 (M_L)	0–7	18
Alaska, USA				
Katmai Caldera	1912	7.0 (M_s)	nd	19
Redoubt Volcano	1989–1990	2.2 (M_c)	0–10	20
Spurr Volcano	1992	2.2	0–40	21
Mexico				
Paricutin	1943	4.5	<10	22
El Chichon Volcano	1974	3.3 (M_c)	<20	23
Colima Volcano	1991	3.8 (M_c)	<8	24
Colima Volcano	1994	3.0 (M_c)	3–6	25
Chile, South America				
Lonquimay Volcano	1988	4.7 (M_c)	<10	26
Columbia, South America				
Nevado del Ruiz Volcano	1985–1986	3.5 (M_c)	<7	27
Luzon, Phillippines				
Mount Pinatubo Volcano	1991	5.7 (M_s)	nd	28
Indonesia				
Galunggung	1982	3.8	nd	29,30
Merapi	1984	2.5	nd	29,30
Anak Ranakah	1987	4.0	nd	29

(Continued)

Table 4.3: Maximum magnitudes and focal depths of earthquakes associated with calderas and central volcanoes (Cont'd)

Location[a]	Rifting event[b] (Year)	Maximum magnitude[c]	Focal depth(s)[d] (km)	References
Banda Api	1988	4.0	nd	29
Kelut	1990	2.0 (M_c)	nd	31
Italy				
Mount Etna	1974	4.5	nd	32
Mount Etna	1983	3.2	<3	33
Mount Etna	1989	3.3 (M_L)	<4	34
Mount Etna	1991	3.3 (M_L)	<6	35
Mount Etna	2001	3.9 (M_c)	<4	33
Mount Etna	2002	4.2	nd	36
Phlegraean Fields Caldera	1982–1984	4.0 (M_L)	2.5	37
Japan				
Bandai	1888	5.0 (M_L)	nd	38,39
Asama	1910	5.4 (M_s)	nd	40
Sakura-jima	1914	7.0 (M_s)	nd	40
Mount Usu	1910	5.1 (M_s)	nd	40
Mount Usu	1977	4.3 (M_s)	nd	40
Izu Islands	1983	6.2 (M_{JMA})	1–13	30
Oshima	1986	5.1	<10	39
Mount Unzen	1990	4.0	10	41
Miyake-Jima	1962	5.9	1–4	42
Miyake-jima Volcano	1985	6.2 (M_{JMA})	15	43
New Zealand				
Taupo Volcanic Zone	1983	4.3	6–10	44

(Continued)

Table 4.3: Maximum magnitudes and focal depths of earthquakes associated with calderas and central volcanoes (Cont'd)

Location[a]	Rifting event[b] (Year)	Maximum magnitude[c]	Focal depth(s)[d] (km)	References
California, USA				
Long Valley Caldera	1978	5.3 (M_s)	7	45
Long Valley Caldera	1980	6.1 (M_s)	8–10	45
Long Valley Caldera	1982–1983	5.2 (M_L)	<8	46
Long Valley Caldera	1997–1998	4.9	7–8	47
Wyoming, USA				
Yellowstone Caldera	1975	6.1 (M_L)	<6	48
Mean ± 1σ, n= 57[e]		4.5 ± 1.2		This Chapter

[a]Worldwide calderas and central volcanoes selected: Krafla, Grimsvotn, Hekla, Kilauea, Nyiragongo, Cerro Negra, Paricutin, and Piton de la Fournaise are associated with past eruptions of predominantly mafic magma; Taupo volcanic zone, Long Valley, and Yellowstone are associated with predominantly silicic magma; all others are associated with magma of mafic to intermediate compositions.

[b]An episode of volcanism and seismicity having a known beginning and end associated with magma movement unless indicated.

[c]Maximum magnitude reported for volcanic event. Magnitudes: m_b, body wave; M_c, Coda; M_L, local or Richter; M_w, moment; M_s, surface wave; M_{JMA}, Japan Meteorological Agency. No definition of magnitude scale was reported for values without magnitude designation. Magnitudes for Kamchatka volcanoes are derived from: $M_s = (K_s − 4.6)/1.5$; where K_s is the mean energy class determined as the arithmetical mean from short-period S-waves of several stations (Gorel'chik, 1989).

[d]Depth range of volcanic seismicity and maximum-magnitude earthquake associated with the volcanic event; nd, No data obtained.

[e]Mean and one standard deviation computed based on magnitudes as presented.

References: (1) Einarsson and Bjornsson, 1979, (2) Einarsson, 1991, and Bjornsson *et al.*, 1977; (4) Einarsson and Brandsdottir, 1984; (5) Gudmundsson *et al.*, 1992; (6) Swanson *et al.*, 1976b; (7) Dzurisin *et al.*, 1980; (8) Kavotha *et al.*, 2002; (9) Longpre *et al.*, 2007; (10) La Femina *et al.*, 2004; (11) Matumoto, 1976; (12) Filson *et al.*, 1973; (13) Chadwick *et al.*, 2006; (14) Mori *et al.*, 1989; (15) Fedotov *et al.*, 1983; (16) Fedotov *et al.*, 1983; (16) Gorel'chik, 1989; (17) Bulletin of Volcanic Eruptions, 1987; (18) Endo *et al.*, 1981; (19) Abe, 1992; (20) Power *et al.*, 1994; (21) Power *et al.*, 2002; (22) Yokoyama and de la Cruz-Renya, 1990; (23) Medina *et al.*, 1992; (24) Nunez-Coma *et al.*, 1994; (25) Jimenez *et al.*, 1995; (26) Barrientos and Acevedo-Aranguiz, 1992; (27) Nieto *et al.*, 1990; (28) Mori *et al.*, 1989; (29) W. Tjetjep, personal communication, 1993; (30) McClelland *et al.*, 1989; (31) Lesage and Surono, 1995; (32) Guerra *et al.*, 1976; (33) Patane *et al.*, 1984; (34) Barberi *et al.*, 1990, and Bonaccorso and Davis, 1993; (35) Ferrucci and Patane, 1993; Patane *et al.*, 1984; (36) Aloisi *et al.*, 2003; (37) Branno *et al.*, 1984; (38) Yamamoto *et al.*, 1999; (39) Okada, 1983; (40) Abe, 1979; McClelland *et al.*, 1989; Okada, 1983; (41) Umakoshi *et al.*, 2001; (42) Nakamura, 1984; (43) Aramaki *et al.*, 1986; (44) Grindley and Hull, 1986; (45) Julian and Sipkin, 1985; (46) Savage and Cockerham, 1984; (47) Dreger *et al.*, 2000, and Hill *et al.*, 2003; (48) Eaton *et al.*, 1975, and Pitt *et al.*, 1979; This chapter.

M_w 4.6 to 6.4. These larger earthquakes seem to occur in regions where dike intrusion is relatively infrequent, the seismogenic crust is thicker, or dikes have greater thicknesses (also see discussion of Table 4.4).

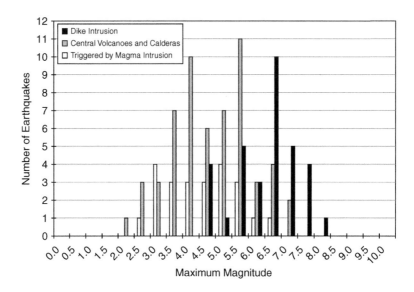

Figure 4.8: Histogram showing the distributions of maximum magnitudes of earthquakes associated with dike intrusion, occurring at central volcanoes and calderas, and triggered by magma intrusion (data are from Tables 4.2–4.4, respectively). The distributions show maximum magnitudes of earthquakes associated with dike intrusion are less than those at central volcanoes and calderas and triggered by magma intrusion. (See Color Insert.)

The mean of 23 earthquake maxima cited in Table 4.2 is M_w 4.1 ± 1.1. Dike-induced earthquakes have low magnitudes for several reasons. Local fracturing of small volumes of adjacent country rock by the intruding magma may not produce large earthquakes because fault area and fault slip are small (Table 4.1). With the exception of deep conduits arising from magma-source regions, dikes tend to propagate at shallow levels (<4 km), and the associated extensional structures are correspondingly shallow (Du and Aydin, 1992). The size of a seismic rupture, and hence earthquake moment, is controlled by the geometrical complexities and variability of accumulated stress levels along the fault structure (Wesnousky, 2006). The structures of faults with small displacements are more complex than presumed preexisting smoother faults that have accommodated many kilometers of cumulative displacement. Small faults that have not accommodated significant displacements over their short life span produce only small earthquakes (Dolan, 2006). Dike intrusion causes incremental surface deformation, which produces new short ("rough") fractures or faults (e.g., Figure 4.6).

Typically, the slopes of the earthquake frequency-magnitude recurrence curves in regions of active dike intrusion are greater than or equal to 1, indicating low effective stress (Bjornsson *et al.*, 1977). In the upper few kilometers of the crust, host rocks contain open fractures occupied by fluids. Such rocks are therefore weak, incapable of supporting large differential stress, and their deformation occurs largely by the opening and closing of fractures. Invoking these properties in their analysis of seismicity near a volcano, Legrand *et al.* (2004) attribute a high b-value of nearly 1.5 as being due to a lack of large-magnitude events. Through a hydraulic-fracture mechanism rather than classical elastic rupture, they explain the absence of high-magnitude events by overpressurized water or gas ejection, with resulting increased normal stress across fault surfaces.

Table 4.4: Maximum magnitudes and focal depths of tectonic earthquakes possibly triggered by magma intrusion

Location[a]	Earthquake date/Volcanic event[b] (Year)	Maximum magnitude[c]	Focal depth(s)[d] (km)	References
Iceland				
Tjornes/Krafla	1975–1976	6.3	nd	1
Laki	1784	7.1	nd	2
Hawaii, USA				
Kilauea east rift	1969	4.7	10	3
Kau, Kilauea	1868	8.0	9	4
Kalapana, Kilauea	1975	7.2 (M_s)	5–6	5
South Flank, Kilauea	1989	6.1 (M_s)	9	6
South Flank, Kilauea	1997	5.5	nd	7
Kona/Mauno Loa	1951	6.9 (M_s)	10–16	8
Kaoiki/Mauna Loa	1962	6.1 (M_L)	nd	9
Kaoiki/Mauna Loa	1974	5.5 (M_L)	5	9
Mauna Loa South Flank	1982	5.6	10–14	10
Kaoiki/Mauna Loa	1983	6.6 (M_L)	11	9,11
Nevada, USA				
Lake Tahoe	2003	4.2 (M_w)	<18	12
Africa				
Nyiragongo	1977	5.2 (m_b)	nd	13
Costa Rica				
Arenal	1987	4.2	nd	14
Chile, South America				
Lonquimay Volcano	1989	5.30 (M_w)	<10	15
Indonesia				
Merapi	2006	6.3 (M_w)	<30	16
Kamchatka, Russia				
	1978	5.2 (m_b)	nd	17

(Continued)

Table 4.4: Maximum magnitudes and focal depths of tectonic earthquakes possibly triggered by magma intrusion (Cont'd)

Location[a]	Earthquake date/Volcanic event[b] (Year)	Maximum magnitude[c]	Focal depth(s)[d] (km)	References
Klyuchevskaya Vol. Group				
Akademia Nauk	1996	6.6 (M_s)	10	18
Italy				
Mount Etna	1991	4.5 (M_L)	<6	19
Japan				
Asama	1916	6.3 (M_s)	nd	20
Izu Peninsula	1930	7.0 (M_{JMA})	nd	21,22
Izu-Oshima	1978	6.8 (M_s)	4	22,23
Miyake-jima	1982	6.4 (M_{JMA})	20	24
On-Take, Honshu	1984	6.8 (M_{JMA})	2	10
Unzen, Kyushu	1984	5.7 (M_{JMA})	7	10
Mount Iwate	1998	6.1 (M_{JMA})	4	25
New Zealand				
Taupo volcanic zone	1895	7.5	nd	26,27
Taupo volcanic zone	1922	7.5	nd	26,27
Taupo volcanic zone	1987	6.3 (M_L)	8	28
California, USA				
Mono Craters	1325 ± 20	6.3 (M_w)	nd	29
Long Valley Caldera	1996	4.3	2–10	30
Long Valley Caldera	1998–1999	5.6	2–10	30
Mean ± 1σ, $n = 33$[e]		6.0 ± 1.0		This Chapter

[a]Worldwide volcanic rift zones, calderas and central volcanoes selected: where provided, name fault system, volcanic rift zone, and caldera or volcano. Krafla, Kilauea, Mauna Loa, Lake Tahoe, Nyiragongo are associated with past eruptions of predominantly mafic magma; Long Valley is and Taupo Volcanic Zone may be associated with predominantly silicic magma; all others are associated with magma of mafic to intermediate compositions.

[b]Date of tectonic earthquake or episode of volcanism (magma movement) which may have triggered the earthquake.

[c]Maximum magnitude reported for triggered earthquakes: Magnitudes; m_b, body wave; M_L, local or Richter; M_s, surface wave; M_{JMA}, Japan Meteorological Agency. No definition of magnitude scale was reported for values without magnitude designation.

[d]Depth range or depth of maximum magnitude earthquake; nd, No data obtained.

[e]Mean and one standard deviation computed based on magnitudes as presented.

Earthquakes associated with extensional-tectonic ruptures commonly nucleate in the midcrust near the brittle-ductile transition and rupture instantaneously (Sibson, 1982; Smith and Bruhn, 1984; Scholz, 1988). At these depths (>10 km), ruptures must have relatively large areas and moments to break the surface. In contrast, faults with several meter displacements and several kilometer lengths form incrementally during shallow dike propagation at rates of several kilometers per hour, and in some instances aseismically rather than in a rapid episode of strain release. Thus, Brandsdottir and Einarsson (1979) observed fault displacements of 1 m and extensive fissuring along the Krafla rift zone of Iceland, but the associated earthquakes did not exceed M_w 4.0. In Iceland and for other recent dike-intrusion events, Pedersen *et al.* (2007) reported the geodetic moment release is usually greater than the seismic moment release by an order of magnitude or more, indicating strain by magma intrusion or other aseismic processes rather than by earthquake ruptures (Einarsson and Brandsdottir, 1980; Stein *et al.*, 1991; Toda *et al.*, 2002; Smith *et al.*, 2004; Wright *et al.*, 2006).

4.4.4.2 Earthquakes at Calderas and Central Volcanoes

Calderas and central volcanoes are the surface expression of long-lived, polygenetic magmatic systems. They differ from more diffuse, less persistent volcanic features such as cinder cone fields, by having deep-seated central conduit systems and by mass loading of the central volcanic edifice. Mass loading and thermal effects are both capable of altering the regional stress field in the vicinity of the volcanoes (e.g., Tibaldi and Leon, 2000).

The compilation of maximum-magnitude earthquakes at calderas and central volcanoes (Table 4.3) suggests that magma propagation and associated deformation may generate somewhat larger earthquakes than simple dike intrusion (Table 4.2). Maximum magnitudes range from M_w 2 (Kelut) to M_w 7 (Katmai; Sakura-jima) and the mean of 57 values is M_w 4.5 ± 1.2 (Table 4.3). Several moderate to large earthquakes are associated with brittle failure of a thin crust above inflating magma chambers (Bardarbunga; Grimsvotn) or deflation of calderas primarily due to withdrawal of magma (Krafla; Fernandina; Katmai). Some of the smaller magnitude earthquakes are associated with summit or flank eruptions of central volcanoes (Redoubt; Spurr; Colima; Mount Etna; Table 4.3). Ito's (1993) analysis of large earthquakes near active volcanoes in Japan shows that all large earthquakes ($M_w > 6$) occurred at distances greater than 10 km from active volcanic centers, where seismicity occurs at greater depths of 5–20 km. Compared to the shallow seismicity (<5 km), subsurface alteration, and surface heat-flow values near centers of active volcanoes, these observations suggest the seismogenic layer is thin resulting in smaller rupture dimensions that produce lower magnitude earthquakes. Seismicity at Mount Etna provides another example of magmatic influences on earthquake magnitudes (Azzaro *et al.*, 2000). Beneath the Mount Etna volcano where it straddles the Apennine normal fault zone and the Malta Escarpment, slip rates are significantly higher and earthquakes are more frequent, of small magnitude, and occur at shallower focal depths than elsewhere along the Apennine fault zone. The observed seismicity may reflect a thinner seismogenic crust in the vicinity of the volcano or instability of the east flank of the volcano due to magma intrusion in nearby volcanic rift zones.

References: (1) Bjornsson, *et al.* 1977; (2) Gudmundsson, 2000; (3) Swanson *et al.*, 1976b; (4) Wyss, 1988; (5) Ando, 1979; (6) Arnadottir *et al.*, 1991; (7) Owen *et al.*, 2000; (8) Beisser *et al.*, 1994; (9) Jackson *et al.*, 1992; (10) Bulletin of Volcanic Eruptions, 1987; (11) Buchanan-Banks, 1987; (12) Smith *et al.*, 2004; (13) Kavotha *et al.*, 2002; (14) Barquero *et al.*, 1992; (15) Barrientos and Acevedo-Aranguiz, 1992; (16) Walter *et al.*, 2007; (17) Zobin, 1990; (18) Zobin and Levina, 1998; (19) Ferrucci and Patané, 1993; (20) Abe, 1979; (21) Nasu, 1935, Kuno, 1954, and Shimazaki and Somerville, 1979; (22) Shimazaki and Somerville, 1979, and Thatcher and Savage, 1982; (24) McClelland *et al.*, 1989; Bulletin of Volcanic Eruptions, 1987; (25) Nishimura *et al.*, 2001b; (26) Grindley and Hull, 1986, and Eiby, 1968; (28) Smith and Oppenheimer, 1989; (29) Sieh and Bursik, 1986; Bursik *et al.*, 2003; Prejean *et al.*, 2002; (30) Hill *et al.*, 2003; This chapter.

Table 4.5: Calculated moment magnitudes from fault dimensions for normal faults in volcanic rift zones

Location	Volcanic rift zone or Episode	Surface-rupture length[a] (Km)	Rupture width[b] (km)	Rupture area[c] (Km²)	Calculated Moment Magnitudes[d]			Observed maximum magnitude[e]	References
					Surface length	Rupture width	Rupture area		
Observed geometries[b]									
Iceland	Krafla, 1978	6.0	1.5	9.0	6.0	4.4	5.0	4.1	1
Africa	Asal, 1978	12.0	1.0	12.0	6.3	4.0	5.1	5.3	2
Africa	Dabbahu, 2005	3.0	2.0	6.0	5.6	4.7	4.8	5.6	3
Japan	Izu Islands, 2000	20.0	8.0	160.0	6.6	6.1	6.2	6.4	4
Estimated geometries[b]									
Iceland	Krafla, 1975	3.5	3.5	12.3	5.7	5.3	5.1	5.0	5
Iceland	Krafla, 1978	6.0	4.0	24.0	6.0	5.4	5.4	4.1	1
Iceland	Thingvellir	11.0	4.0	44.0	6.3	5.4	5.7	na	6
Africa	Asal, 1978	12.0	4.0	48.0	6.3	5.4	5.7	5.3	2
Africa	Dabbahu, 2005	3.0	3.0	9.0	5.6	5.1	5.0	5.6	3
Japan	Izu Islands, 2000	20.0	4.0	80.0	6.6	5.4	5.9	6.4	4
Idaho, USA	Snake River Plain, Arco	5.0	4.0	20.0	5.9	5.4	5.3	na	7
California, USA	Mono Craters, 1350 AD	8.0	11.0	88.0	6.1	6.4	6.0	na	8

		6.1 ± 0.3	5.2 ± 0.7	5.4 ± 0.5
Mean ± 1σ, n = 12[f]				
Mean ± 1σ, n = 36[f]			5.6 ± 0.6	

[a]Maximum normal-fault length observed at the surface in the volcanic rift zone or from a volcanic episode produced by dike intrusion.

[b]For "Observed geometries," dimensions obtained from Table 4.1 where surface length equal maximum value of normal-fault length; rupture width is the depth to the dike top (=dike basal depth-height). For "Estimated geometries," surface length equal maximum value of normal-fault length (Table 4.1); rupture width is based on maximum depth for dike propagation, 2–4 km (Gudmundsson, 1984a; Ryan, 1987). In cases where lengths are less than 4 km, the down-dip width is assumed to be equivalent to its length. The intent is to estimate the maximum area possible. For Mono Craters the down-dip width is from Bursik et al. (2003).

[c]Rupture area = surface-rupture length × rupture width.

[d]Calculated using the following relationships (Wells and Coppersmith, 1994): $M_w = 5.03 + 1.19 \times \log_{10}(SRL)$; $M_w = 4.01 + 2.29 \times \log_{10}(RW)$; $M_w = 4.01 + 1.0 \times \log_{10}(RA)$; where M_w = moment magnitude, SRL = surface rupture length, RW = rupture width, and RA = rupture area.

[e]Observed maximum magnitudes obtained from Table 4.2; na, not available (no instrumental recordings).

[f]Mean and one standard deviation computed using both observed and estimated geometries for surface length, rupture width, and rupture area, and all calculated moment magnitudes combined.

References: (1) Einarsson and Brandsdottir, 1980, and Rubin, 1992; (2) Abdallah et al., 1979; Le Dain et al., 1979; Tarantola et al., 1980, and Lepine and Hirn, 1992; Sigmundsson, 2006; (3) Wright et al., 2006, and Yirgu et al., 2006; (4) Nishimura et al., 2001a, and Toda et al., 2002; (5) Brandsdottir and Einarsson, 1979, 1980, and Pollard et al., 1983; (6) Gudmundsson, 1987a; (7) Hackett and Smith, 1992; Kuntz et al., 1994, and Smith et al., 1996; (8) Bursik et al., 2003.

The 1968 Fernandina and 1912 Katmai earthquake sequences are of particular interest because they were coincident with catastrophic caldera collapse (Table 4.3). In both cases, the calculated kinetic energy releases, based on the volume change due to the gravitational failure of the caldera blocks, generally agreed with seismic-moment calculations (cumulative energy release based on instrumental recordings) (Filson *et al.*, 1973; Abe, 1992). The "tectonic" characteristics and overall conformation of the energy calculations indicate that the release of seismic energy is consistent with down faulting of the observed caldera volumes.

4.4.4.3   *Tectonic Earthquakes Induced by Magmatic Processes*

Maximum magnitudes of tectonic earthquakes that were triggered directly or indirectly by magmatic processes are listed in Table 4.4. The values range from M_w 4.2 (Lake Tahoe; Arenal) to $M_w \sim 8$ (Kau, Kilauea), and the mean of 33 values is 6.0 ± 1.0. These earthquake sequences have characteristics that are typical of tectonic earthquake sequences in nonvolcanic areas, including mainshock-aftershock groups, and *b*-values similar to tectonic swarms. In some cases, the tectonic earthquakes occur along faults that existed prior to the volcanic eruptions (Merapi; Lonquimay; Miyake-jima; Tjornes/Krafla). The coincidence of an M_w 6.3 earthquake in the Tjornes fracture zone (Figure 4.1C) with dike intrusion along the Krafla rift zone suggests a direct, mechanistic relationship.

Several tectonic earthquakes occurred immediately following eruptions or intrusions, and the related seismicity continued for up to several years as a result of accumulated stresses due to repeated magma intrusion (Kilauea; Mauna Loa; Klyuchevskaya; Laki; Mount Etna; Lake Tahoe). At Lake Tahoe, Nevada, USA, dike-related, small-magnitude earthquakes (M_w 2.2; Table 4.2) with a cumulative seismic moment equivalent to a M_w 6.1 event occurred at depths between 25 and 30 km. This sequence triggered upper crustal seismicity at depths <20 km including the 2004 M_w 4.2 earthquake (Table 4.4), which lasted from 3 to 26 months following the dike-related seismicity (Smith *et al.*, 2004; von Seggern *et al.*, 2008). Examples from Hawaii include the 1983 M_L 6.6 Kaoiki, Hawaii, earthquake (Buchanan-Banks, 1987), which involved strike-slip displacement between the active volcanoes Mauna Loa and Kilauea, and was probably generated by differential compressive stress associated with inflation of the magma reservoirs beneath both volcanoes. The 1975 M_s 7.2 Kalapana earthquake occurred on the Hilina fault system (Figure 4.1D) and resulted from protracted dike intrusion along Kilauea's east rift zone (Swanson *et al.*, 1976a). In Japan, inflation of a magma chamber triggered several moderate- to large-magnitude earthquakes (1930 M_{JMA} 7.0; 1978 M_s 6.5; 1998 M_{JMA} 6.1; Table 4.4) on nearby preexisting faults that were optimally aligned for failure (Thatcher and Savage, 1982; Nishimura *et al.*, 2001b).

Several examples of potential magmatic triggering of moderate- to large-magnitude earthquakes have been recognized in paleoseismic trench investigations. Using stratigraphic relationships of volcanic deposits and fault offsets in the paleoseismic record, Bursik *et al.* (2003) estimated the maximum earthquake magnitude of M_w 6.3 for rupture along the Hartley Springs fault triggered by dike intrusion at Mono Craters, California (Table 4.4). From Coulomb stress-change modeling, Parsons *et al.* (2006) concluded that intrusion of a dike at the Lathrop Wells volcano, Nevada, USA decreased the least principal stress for nearby normal faults and this led to one or more *paleoearthquakes* immediately following the Lathrop Wells eruption (Keefer and Menges, 2004).

We preferentially selected tectonic earthquakes with close spatial and temporal associations to magma movement and volcanic eruptions, but we also recognize that tectonic earthquakes may precede and possibly trigger eruptions. Zobin and Levina (1998) attribute the double eruption of the Akademia Nauk and Karymsky volcanoes in Kamchatka, Russia to the 1996 M_s 6.6 earthquake, which occurred 14 h earlier at distances of 9 and 17 km south of the volcanoes (Table 4.4). A sequence of earthquakes is thought to have led to the eruption of Mount Pinatubo. An earthquake of M_w 4.8 occurred within 8 km of Mount Pinatubo, just 4 h after the 1990 M_w 7.7 Luzon earthquake on the Phillipine fault 100 km away. Volcanic activity began at Mount Pinatubo 3 weeks after these earthquakes and culminated 11 months later in the climactic eruption of 1991, accompanied by the 1991 M_s 5.7 earthquake (Table 4.3) (Nostro *et al.*, 1998). At Vesuvius, modeling of Coulomb stress change suggests that earthquakes can promote eruptions by opening suitably aligned near-surface conduits and compressing the magma body at depth. Modeling also shows that withdrawal of magma or eruption from a dike that is oriented parallel to the Appenine normal faults brings these faults closer to failure (Nostro *et al.*, 1998). Analysis of temporal and spatial relationships between earthquakes ($M_w \geq$ 4.8) and volcanic eruptions (volcanic explosivity index, VEI $\geq$ 0) worldwide indicates that earthquake triggering is not the only process involved in promoting volcanic eruptions; other processes include volcano dynamics and regional stress changes related to deep magma storage (Lemarchand and Grasso, 2007).

Instead of inducing earthquakes on adjacent structures, stress imposed by intruding magma may operate in the converse sense: Magma intrusion in extensional terrains is believed to inhibit seismicity by supplanting tectonic normal faulting (Parsons and Thompson, 1991). Thus, Bursik and Sieh (1989) attribute the absence of Holocene oblique-slip faulting along the Silver Lake fault of the Sierra Nevada range front to extensional magmatism of the adjacent Mono Craters (Figure 4.1B). The relative aseismicity and low relief of the eastern Snake River Plain volcanic province, in contrast to the surrounding seismically active, high-relief Basin and Range (Figure 4.1A), are attributed to the substitution of basaltic dike intrusion for normal faulting and low strain rates in the eastern Snake River Plain (Rodgers *et al.*, 1990; Parsons and Thompson, 1991; Hackett and Smith, 1992; Jackson *et al.*, 1993; Smith *et al.*, 1996; Parsons *et al.*, 1998; Payne *et al.*, 2008). The Husavik-Flatery Fault within the Tjornes Fracture Zone (Figure 4.1C) appears to have been locked by increased compressive stresses due to dike intrusion in the Krafla volcanic rift zone 1975–1983. Seismicity along the fault ceased in 1976, but the 1994 M_w 5.5 earthquake at its westernmost end suggests increased differential stress, and hence, unlocking of the fault (Gudmundsson, 2000).

4.4.4.4 Comparison of Moment-Magnitude Calculations to Observational Seismicity

Table 4.5 gives the results of moment-magnitude calculations for normal faults within volcanic rift zones based on surface-rupture lengths, rupture (down-dip) widths, and rupture areas, using the empirical relationships between these fault-rupture dimensions and moment magnitudes (Wells and Coppersmith, 1994). The unweighted mean of 36 moment-magnitude calculations is 5.6 ± 0.6, and as expected this result is greater than the observed earthquake maxima during dike-intrusion episodes (See Table 4.2). In the moment-magnitude calculations of Table 4.5, field observations of surface-rupture lengths are used to assess moment magnitude from the empirical relationships of Wells and Coppersmith (1994). Subsurface-rupture lengths are not used, since it is difficult to determine subsurface-rupture dimension of dike-induced normal faults, based on the typically diffuse seismic swarms associated with incremental fault ruptures (as opposed to tectonic mainshock-aftershock seismicity patterns). Rupture widths are determined from the estimates of down-dip fault widths based on the shallow crustal depths (2–4 km) along which dikes propagate. We assume that normal faults terminate near the tops of the dikes, and possibly extend as deep as 4 km, unless otherwise indicated. The surface fault lengths and down-dip widths are then used to estimate the rupture areas.

Moment-magnitude calculations based on fault-rupture dimensions assume single-rupture events; therefore the results give an upper bound for the moment magnitudes of purely magma-induced earthquakes. This is because the dike-induced faults were likely to have ruptured incrementally, produced many earthquakes, and were formed in the shallow crust, where differential stress and rigidity are lower than at depth. Typical values for midcrustal rigidity are about 3×10^{11} dyne/cm^2 (Hanks and Kanamori, 1979), but shear modulus decreases rapidly within a few kilometers of the surface, and crustal rigidity in the shallow crust where dikes are intruded is therefore likely to be lower. Thus, crustal rigidity values ranging from 0.5 to 1.8×10^{11} dyne/cm^2 are required in order to account correctly for observed deformation volumes (Filson *et al.*, 1973; Mori *et al.*, 1989; Stein *et al.*, 1991).

Wells and Coppersmith (1994) also develop relationships between moment magnitude and maximum displacement, average displacement, and subsurface-rupture length. The displacement–magnitude relationships are not used because deformation associated with earthquakes having moment magnitudes of less than approximately 5.7 may be a secondary effect from ground shaking and not due to primary rupture of the fault (Wells and Coppersmith, 1994). Furthermore, displacements along dike-induced normal faults may not form during single-event ruptures, but are commonly the result of many incremental displacements during dike intrusion, or of multiple dike-intrusion events.

A sound approach to estimating the maximum magnitude from the empirical relationships of moment magnitude and the fault-rupture dimensions shown in Table 4.5 is to apply weighting factors to the rupture dimensions. Rupture dimensions that are best constrained should be given higher weights. A weighted average can then be calculated to derive a maximum moment magnitude.

In using the maximum magnitudes of dike-induced earthquakes as part of a seismic hazard analysis, investigators should consider the context of dike intrusion within the regional-tectonic setting. In the absence of nearby preexisting faults, maximum magnitudes of dike-induced earthquakes may dominate the hazard within an earthquake source zone, depending on the volcanic-recurrence intervals. In contrast, nearby preexisting faults capable of generating large-magnitude earthquakes will almost certainly dominate the hazard, but may be triggered by dike intrusion depending on the magnitudes of regional differential stresses.

It may be appropriate in some volcanic areas to derive site-specific relationships between maximum magnitude and fault-rupture dimensions. The empirical data of Azzaro (1999) for a central volcano show a general linear correlation of surface-rupture lengths with small magnitudes for recent and prehistoric normal-fault ruptures on the flanks of Mount Etna. The earthquake magnitudes range from <5.0 to around 2.5 for surface-rupture lengths from <10 km to 100 m, respectively (Azzaro, 1999; Azzaro *et al.*, 2000; Ferreli *et al.*, 2002). The Mount Etna data show a linear trend with a slope less than that calculated by Wells and Coppersmith (1994) for tectonic earthquakes. Typically, surface ruptures are not produced for tectonic earthquakes of $M_w < 6.5$, owing to their midcrustal nucleation depths. Development of a magnitude-rupture length relationship for a volcano at very low earthquake magnitudes emphasizes the need to account for the interplay between magmatic and tectonic (e.g., slope instability) driving mechanisms and the thinner seismogenic crust within the active volcanic region.

4.5 Conclusions

From the standpoint of natural hazards, magma-induced earthquakes are important because they can cause significant damage and injury in the epicentral region, and they frequently occur prior to eruptive activity. Many earthquakes in volcano-extensional regions are directly related to magma intrusions, particularly to dikes, which propagate along self-induced fractures and cause deformation on extensional faults and fissures. Other earthquakes result from redistribution of the load of the volcano or from caldera collapse. Still others, tectonic earthquakes in the vicinity of volcanoes, are indirectly induced or inhibited by magma intrusion.

Volcanic rift zones are diffuse belts of extensional surface deformation, dominated by tensional fissures and small-offset normal faults, and underlain by dike swarms. Eruptive fissures commonly bisect a central graben, or lie between two sets of anastomosing fissures. Normal faults, monoclines, and fissures developed along volcanic rift zones are individually similar to those formed by nonmagmatic tectonic mechanisms. The aggregate geometry of the surface-deformation features, their spatial and temporal association with cogenetic volcanic materials, and geophysical evidence for shallow magma intrusions are among the criteria for distinguishing magma-induced extensional structures from nonmagmatic ones.

In volcanic extensional environments, recurrence estimates of magma-induced faulting and seismicity should be based largely on integrated approaches of multiple geochronologic methods and mapping of the coseismic volcanic materials, particularly in places where trenching of surface faults or fissures is not possible. Postseismic colluvial wedges do not always form due to resistance of lithified volcanic materials or the growth of scarps by aseismic creep, and thin colluvium may be quickly eroded away due to the small throws of magma-induced faults. Fissures and scarps can sometimes be dated by the innovative use of conventional geochronologic methods, such as thermoluminescence dating of baked sediments beneath cogenetic volcanic materials, or the use of in situ cosmogenic radionuclides to date geomorphic surfaces exposed during deformation.

Coseismic field investigations of the Icelandic, Hawaiian, and Afar volcanic rift zones clearly demonstrate that meter-scale displacements of dike-induced normal faults occur without large-magnitude seismicity. During the last century, worldwide observations of dike-induced seismicity indicate that such earthquakes rarely exceed M_w 5.5 and most have $M_w < 4.5$. Earthquakes at calderas and central volcanoes have somewhat higher maxima (M_w 6.5–7.0), owing to greater depths of origin or to the energy released by fracturing of large volumes of roof rocks atop shallow magma chambers. The maximum magnitudes of earthquakes associated with two caldera-collapse events (Galapagos and Katmai) generally conform to the calculated energy released by down-drop of the observed caldera volumes.

Indirectly related to volcanism but nonetheless important are the large-magnitude (M_w 7+) tectonic earthquakes that can occur on adjacent structures as a result of protracted magma intrusion or direct triggering. Such earthquakes are most likely to occur in nonextensional settings where compressional stress can accumulate near intrusions. Knowledge of magmatic processes and volcanic recurrence should influence the assessment of seismic recurrence along adjacent tectonic structures, but traditional paleoseismic techniques and empirical moment magnitude-fault rupture dimension relationships can be applied to these structures.

Assumptions that might lead to the overestimation of earthquake frequency and magnitude in volcanic rift zones include (1) the association of a dike-intrusion event with each volcanic vent; (2) the association of a maximum-magnitude earthquake with each dike-intrusion event; (3) treatment of the rupture mechanics of normal faulting during dike intrusion as if it were analogous to fault rupture during large-magnitude, tectonic earthquakes; and (4) the use of midcrustal rigidity values in moment-magnitude calculations for shallow, dike-induced events. These overestimates can be rectified by incorporating dike-intrusion and volcanic rift zone concepts into the assessment: (1) the use of geologic information to group volcanic vents into eruptive episodes, based on physical and chronologic relationships; (2) the use of statistical compilations of worldwide, co-intrusive seismic sequences to quantify the probability of a maximum-magnitude event; (3) application of the mechanics of dike intrusion, which shows that most dikes propagate laterally in the shallow crust at rates of several kilometers per hour; dike-induced faults therefore develop incrementally and concurrently with dike propagation, rather than catastrophically; and (4) the use of more reasonable values for crustal rigidity in shallow-crustal environments, based on observed lithologies and seismic velocities in the region of interest.

The concurrent processes of crustal extension, particularly the nature of the interaction among magma intrusion, volcanism, and tectonism, remain an emerging area of inquiry that will occupy geologists and seismologists into the foreseeable future. The influence of local magma intrusion on the seismicity of adjacent tectonic structures is in some places unequivocal, in other places debatable, and in still others yields diametrical consequences: magma intrusion seems capable of inducing, inhibiting, and changing the style of tectonic extension. We recognize three areas for future research: (1) the further development of conceptual models for dike intrusion and related seismicity, based on surface strain revealed by field observations and geodetic remote-sensing data; (2) improvement of our understanding of the observed 2–4 km depth of magma stagnation at the ALNB; and (3) advancement of the methods (such as age dating, paleoseismic field techniques, recurrence estimates of volcanic earthquakes, and assessment of maximum magnitudes) by which the role of magmatic processes can be fully incorporated into seismic hazard analyses.

We offer this second-edition contribution in the same spirit as we did the first, as a step toward the incorporation of volcanic seismicity into the field of paleoseismology. We will have achieved our purpose if this discussion helps paleoseismologists to recognize volcano-extensional structures, to interpret them within the context of magmatic processes, and to advance the field by developing and applying some of the methods we have suggested.

4.6   Information on the Companion Web site

The accompanying web site contains the following information for this chapter: (1) color photographs of those shown in Figures 4.3 and 4.6; (2) brief summaries of the data reported in Table 4.1 for each volcanic event; (3) criteria used to select earthquakes for each table and brief summaries of the magmatic processes associated with the earthquakes for Tables 4.2–4.4; and (4) color histogram of the maximum magnitude distributions shown in Figure 4.8.

Acknowledgments

Our current thinking has evolved as a result of many discussions over many years of collaboration with private and government researchers at the Idaho National Laboratory, where we have attempted to assess seismic hazards within the aseismic, volcano-extensional environment of the eastern Snake River Plain. We are grateful to several individuals for helping us to formulate and focus our ideas, assisting us with compilation of information, and performing reviews of the manuscript: Elizabeth Baker, Kevin Coppersmith, Steve Forman, Ronald Greeley, Mark Hemphill-Haley, Nick Josten, Simon Kattenhorn, Derek Keir, Larry Mastin, Rob McCaffrey, Tom Parsons, Dave Pollard, Dave Rodgers, Julie Rowland, Walt Silva, George Thompson, Tamera Waldron, Peter Wallmann, Don Wells, Ivan Wong, and Jim Zollweg. We are particularly grateful to the editor Jim McCalpin for his valuable assistance with revising this chapter for its second edition. The U.S. Department of Energy (DOE), Office of Nuclear Energy, Science, and Technology funded this work under DOE Idaho Field Office contract DE-AC07-05ID14517.

CHAPTER 5

Paleoseismology of Compressional Tectonic Environments

James P. McCalpin* and Gary A. Carver[†]

*GEO-HAZ Consulting, Inc., Crestone, Colorado 81131, USA
†Carver Geologic, Kodiak, Alaska, USA

5.1 Introduction

Large compressional earthquakes in the upper crust, and even larger plate-boundary earthquakes in subduction zones, produce surface deformation recorded by displacements on reverse or thrust faults, by growth of surface folds, and by changes in the elevation of the land surface. Study of such stratigraphic and geomorphic features yields information about the size and recurrence of large earthquakes that is not available from historic sources in many regions. Coupled with regional-scale knowledge about structure, geophysics, and landscape development, such data allow characterization of seismic fault source zones in regions dominated by compressional tectonics (Chapter 9, See Book's companion web site).

Stratigraphic and geomorphic features formed by active compressive faulting and folding are commonly more diverse, and often more subtle, than those formed by extensional or strike-slip faulting. As a result, reverse faults have often been overlooked near urban areas, even where seismic hazard studies have been carried out. It would not be an exaggeration to say that the most dangerous faults on the planet are subtle reverse faults and blind-thrust faults. This premise was first advanced by Stein and Yeats (1989), but has been underscored by events that occurred since the publication of our 1st Edition in 1996.

During that time period, all the largest and deadliest earthquakes worldwide have been compressional events on reverse faults either previously unknown or not thought to be active (central Taiwan, 1999 (Uzarski and Arnold, 2001); Bhuj, India, 2001 (Jain *et al.*, 2002); Bam, Iran, 2003 (Naeim *et al.*, 2005); Sumatra, Indonesia, 2004 (Bilek *et al.*, 2007); Kashmir, Pakistan, 2005 (EERI, 2006); Sichuan, China, 2008 (EERI, 2008)). Even moderate reverse earthquakes have had major impact, for example, the M_w 6.6 Chuetsu offshore earthquake of 16 July 2007 which occurred near the Kashiwazaki-Kariwa nuclear power plant in Japan, the world's largest NPP (8.2 Gw). The suspected seismogenic "F-B fault" had been identified in the preconstruction hazard assessments (1970s-vintage), but as "inactive" and only 7 km long (IAEA, 2007). This conclusion followed the beliefs of that time, that active folding was accomplished by aseismic, plastic deformation, and that the underlying blind faults were no longer seismogenic. A later

International Geophysics, Volume 95
ISSN 0074-6142, DOI: 10.1016/S0074-6142(09)95005-7

2003 study used modern tools for detecting active folds and blind thrusts, and a modern understanding of blind faulting and fault-propagation folding, and redefined the fault as "active" and 23 km long, a major change. This paradigm shift is probably the most important change in paleoseismology in the past two decades, but unfortunately occurred long after construction of the NPP. The unanticipated M_w 6.6 earthquake of 2007 caused ground shaking that exceeded the plant's design parameters. Although the plant underwent a safe shutdown and insignificant radioactive materials were released, as of this writing 1.5 years later the entire NPP is still shut down, pending a reassessment of seismic hazard.

Also subsequent to 1996, numerous active reverse faults in urban areas have been recognized or characterized for first time, using modern methods of investigation as described in this chapter and in Chapter 2A e.g., the Hollywood fault (Dolan et al., 1997) and the Santa Monica fault system (Dolan et al., 2000) in California; the Seattle fault and backthrusts in densely forested Washington State (Nelson et al., 2003; Sherrod et al., 2004). Previously unrecognized, relatively short (<50 km) active reverse faults have been found in areas of low seismicity, such as the Provence region, France (Chardon et al., 2005); the Pyrenees, France (Alasset and Meghraoui, 2005); the Tagus Valley, Portugal (Vilanova and Fonseca, 2004). In general these "new-found" reverse faults are relatively short, have low Quaternary slip rates, with surface traces obscured by dense vegetation or by cultural smoothing. These hidden faults pose the current greatest challenge for paleoseismology.

5.1.1   Organization of This Chapter

Due to the wide variety of compressional tectonic environments, this chapter is subdivided somewhat differently than the chapters on extensional (Chapter 3) and strike-slip (Chapter 6) environments. Paleoseismic features formed by reverse or thrust faults on land are described in a manner similar to that used in Chapters 3 and 6; that is, geomorphic evidence first (Section 5.2) and stratigraphic evidence second (Section 5.3). Evidence for prehistoric secondary fault displacement (Section 5.6) and coseismic folding (Section 5.8) is described separately. Dating methods are not discussed in a separate section, but are integrated into Sections 5.2–5.8.

The coseismic deformation associated with subduction zone earthquakes is sufficiently unique that it demands a separate treatment in Sections 5.9–5.13. There are several reasons for this organizational distinction. First, the seismogenic megathrust crops out on the seafloor under great water depths, so on-fault paleoseismic evidence (the focus of most on-land fault studies) cannot be directly observed. Second, the main paleoseismic indicator on land masses near subduction zones is regional vertical deformation, usually as evidenced by relative sea-level changes. Third, the presence of paleo-sea-level datums often allows reconstruction of the interseismic as well as the coseismic parts of prehistoric earthquake deformation cycles, something that is generally not possible on continental faults. The best preserved paleoseismic indicators, and the means of interpreting them thus fall more under the classical fields of Quaternary stratigraphy and geomorphology, rather than utilizing the more specialized fault profiling and trenching techniques developed for study of most on-land faults.

5.1.2   Styles, Scales, and Environments of Deformation

Convergent plate margins—the great belts of active tectonism, seismicity, and volcanism that release most global seismic energy (Minster and Jordan, 1987)—are the largest and most widespread compressional tectonic environments. The majority of the earth's major earthquakes have occurred on the largest of these belts, the circum-Pacific "Ring of Fire" (Zhang and Schwartz, 1992) (Figure 5.1). Convergent plate

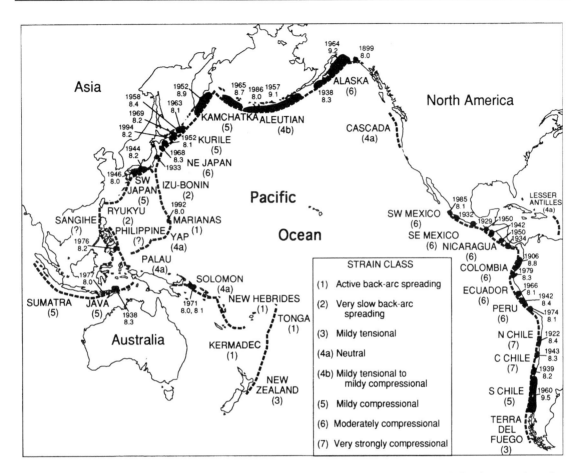

Figure 5.1: Principal Pacific rim subduction zones and some modern great subduction earthquakes ($M_w > 8$). Classification of the subduction zones is based on Jarrard's (1986) comparison of large-scale structure and kinematic characteristics and reflects a range from strongly extensional (Class 1, back-arc spreading) to strongly compressive (Class 7, active folding and thrusting). Most of the modern great subduction earthquakes have occurred along the eastern and northern Pacific where (in general) younger oceanic plates are subducting beneath continental margins. Earthquakes of $M_w < 8$ are common along subduction zones in the southwestern Pacific, where generally older oceanic plates are subducting beneath other oceanic crust, but few great earthquakes have been recorded.

margins can be broadly classed into continent-continent collision, ocean-continent subduction, and ocean-ocean subduction. The first class is described directly below and is studied by the techniques of on-land paleoseismology (Chapter 2A). The second and third classes mainly lie offshore and are studied by offshore paleoseismic techniques (Chapter 2B) and by specialized coastal techniques described in the second half of this chapter.

5.1.2.1 Terrestrial Environments of Compressional Deformation

Yeats *et al.* (1997) proposed a useful grouping of terrestrial compressional deformation zones into four classes:

1. Active fold and thrust belts and reverse-fault ranges that constitute the boundaries between plates of continental crust, in which the boundaries are nearly perpendicular to the direction of relative plate motion. Examples: the great zone of continental collision between India and Asia (Molnar and Qidong, 1984) and the subduction and collision zones in northern Africa, southern Europe, and the Middle East (Meghraoui *et al.*, 1986; Vita-Finzi, 1986).

2. Fold and thrust belts and reverse-fault ranges that occur continentward of the volcanic arc in association with a subduction zone. Examples: northern Honshu, Japan and the eastern foothills of the Andes.

3. Reverse-fault ranges associated with major strike-slip faults, such as those of the California Transverse Ranges, northwest China, and New Zealand. These also include smaller reverse faults such as the Susitna Glacier thrust, which activated at the end of the 2002 Denali, Alaska surface rupture.

4. Isolated active reverse faults in stable continental regions (SCRs), including stable cratonic cores (SCCs) and Phanerozoic accreted fold belts. The former include Precambrian shield areas such as Canada, Scandinavia, and central Australia.

5.1.2.2   *General Style of Deformation in Compressional Zones*

The most common seismogenic structure in compressional tectonic environments is the *thrust fault* (a low-angle reverse fault). Movement on thrusts results in thickening of the vertical section of crust cut by the fault and shortening of the horizontal distance across the fault. In many geologic environments, thrust faults bifurcate as they project toward the surface, producing complex fault systems at the surface. Individual fault splays at all scales separate *thrust-bounded slices* of rock, and these slices are often stacked by repeated fault offset. At depth the splays merge and the dips of the faults decrease until they merge into near horizontal *detachments* below which shortening does not occur through brittle deformation.

At regional scales such systems of thrusts often form long arcuate belts containing generally parallel faults bounding elongate slices of rigid crust. In cross section *imbricate thrust systems* are commonly wedge or lens shaped. Large accretionary fold and thrust systems along convergent plate margins are commonly modeled as the faults in internally deforming wedges, much like a pile of snow being pushed by a snowplow (Davis *et al.*, 1983). Thrust faults are commonly interspersed with folds, particularly folds produced by the movement of thrust sheets over bends in underlying thrusts (*fault-bend folds or ramp folds*) and folds formed at the propagating ends of thrusts (*fault-propagation folds*) (Suppe, 1983) (Figure 5.2). Some fold and thrust belt faults are upward branching splays off of an underlying sole thrust. However, many thrust faults probably root above the sole thrust in upper thrust sheet flexures and in the cores of folds, and thus may not be seismogenic.

There is wide variation in the dips of thrust faults at both shallow and deep levels in the crust. Subduction megathrusts may dip as gently as $10°$ (Section 5.6), but most continental thrusts dip at $30°$ or less. Active thrusts in crustal fold and thrust belts that have dips $>45°$ are termed *reverse faults* or *high-angle reverse faults*. Although many reverse faults are rotated low-angle thrusts with reduced rates of slip, earthquakes on high-angle reverse faults are common in many crustal fold and thrust belts. The dip of many thrusts in complex imbricate systems increases as the accrued slip produces *stacking* or *duplexing* (Suppe, 1985) of thrust-bounded slices. Initially low-angle faults thus become high-angle reverse faults through time. Often the thrusts become inactive as new splays with lower dips form and accommodate slip transferred from the oversteepened older thrusts. Such evolution of imbricate thrust systems results in migration

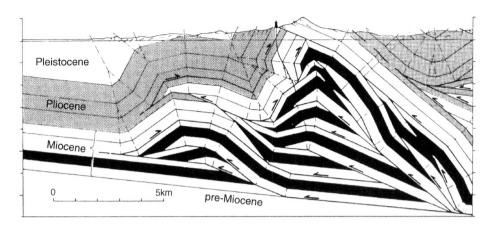

Figure 5.2: Structural interpretation of the Nanliano anticline, southern Taiwan (from Suppe, 1985). Shallow thrusts are often complex and include upward branching faults, dip changes, and dip reversal forming thrust wedges. Long-term displacement stacks the fault-bounded slices, thickening the section and producing permanent uplift. Fault-propagation folds produced at the ends of blind thrusts, and fault-bend folds that form where thrust sheets ride over bends (dip changes) in the faults, have distinct geometry characterized by planar limb segments bounded by sharp hinge lines. At smaller scale, surface thrusts commonly include multiple upward vergent branches, shallow fault-bend and propagation folds, and thrust wedges that are observed in outcrop or in trenches across scarps.

of fault slip from higher to lower splays, and in the periodic generation of new surface traces in the footwall of thrust systems.

Two types of *secondary faults* are associated with thrust faults. The crests of active surface anticlines are commonly cut by normal faults and grabens that trend along the fold crest parallel to the fold axis. These extensional *bending-moment faults* (Yeats, 1987) are secondary to the thrusts that underlay such folds and are believed to form suddenly during thrust-generated flexure of the anticline. Some active anticlines are cut by many closely spaced bending-moment normal faults; in such settings microfaulting is locally pervasive. For example, in the El Asnam, Algeria, thrust earthquake of 1980, a normal fault scarp on the anticlinal crest was more prominent than the thrust trace itself and was initially thought to reflect a normal faulting event (Yielding *et al.*, 1981; Philip and Meghraoui, 1983); similar observations of bending-moment normal faulting accompanying the 1978 Tabas-e-Goldshan earthquake in Iran were made by Berberian (1979). A prehistoric example is given by Beanland and Barrow-Hurlbert (1988) on the reverse Dunstand Fault, New Zealand, where antithetic bending-moment fault scarps comprise 30% of the trace length, often where the thrust trace itself lacks conspicuous scarps. Bending-moment faults are usually short and project to shallow depths, presumably to the level of a neutral stress surface in the flexing block above the thrust. Because they do not deeply penetrate the crust, bending-moment faults are typically not seismogenic.

Sudden bending of thick-bedded sequences of sediments is often accompanied by slip along bedding planes in the fold limbs. This slip generates *flexural-slip faults* that are rooted in the axis of folds and extend through the fold limbs. Where flexure-slip faults intersect the surface, they produce scarps and fault-line features that mimic those produced by tectonic surface faulting. Because incremental growth of many folds takes place in response to coseismic slip on underlying thrust faults, evidence of sudden displacement on flexure-slip faults demonstrates the seismogenic origin and growth of a fold; however, the bedding plane faults themselves are not considered seismogenic.

Crustal thrust and reverse faults seldom occur individually, but instead are generally part of imbricate or overlapping systems made up of multiple faults and folds. Most historic earthquakes in such thrust zones have involved rupture on one of these faults while others in the system have remained quiet. Thus, individual fault structures in larger thrust systems may behave independently. Like all large faults, the regional-scale compressive fault zones are segmented. Possible rupture segments may be reflected by structural discontinuities of the fault, especially where strike-slip tear faults intersect and offset the thrust, or at the ends of overlapping imbricate faults. Other structural indications of possible segment boundaries are sharp changes in orientation of the fault, intersections with branch faults, abrupt changes in dip, and changes in net slip (Knuepfer, 1989). Geomorphic expressions of segmentation may be seen as differences in height of uplifted topography on the hanging wall, or abrupt changes in strike and limits of surface folding. However, few detailed studies have been conducted on the segmentation of shallow intraplate thrust faults.

5.1.2.3 Reverse Faults Expressed by Surface Folds

The advances in understanding fault-propagation folding and its relationship to blind coseismic faulting have been the most significant advance in paleoseismology in the past two decades. The kinematics of fault-propagation and fault-bend folding have been known for many decades, but we have only recently come to recognize the geomorphic expression of active folding (Figure 5.3). The most common expression is warping of Quaternary geomorphic surfaces into anticlines (Figure 5.3B) or monoclines (Figure 5.3C). The recognition of active folding in the landscape is more a topic of tectonic geomorphology than of paleoseismology, and readers are directed to texts such as Burbank and Anderson for the relevant techniques. In this chapter we will concentrate more on the interpretation of geomorphic and stratigraphic features to derive magnitude, displacement, slip rate, and recurrence for the underlying seismogenic fault.

5.1.2.4 Segmentation of Reverse Faults

Defining possible seismic rupture segments for reverse faults requires slightly different approaches, depending on whether the faults are thick-skinned, high-angle reverse faults bounding mountain blocks, or thin-skinned, low-angle thrusts and blind faults. For the former, geometric indicators are used as described in Chapter 9, (See Book's companion web site). For example, range-front morphology can be used to define tectonic "activity classes" based on a number of tectonic geomorphic parameters such as range-front sinuosity, valley depth:width ratios, triangular facets, and so on. The classic work of that type is Bull (2007) on the San Gabriel Mountain front (Figure 5.4). More recent attempts at reverse-fault segmentation have been more robust, relying on a wider array of field data such as slip rates calculated from reverse-fault scarp heights across Quaternary deposits (e.g., Arrowsmith and Strecker, 1999).

One drawback to relying heavily on reverse-fault scarps for segmentation is the tendency of reverse-fault surface ruptures to be discontinuous, with many breaks between short scarps, gaps caused by transitions between faulting and folding, and so on. Applying the same geometric criteria to reverse fault scarps as used for strike-slip and normal faults may lead to relatively short inferred segment lengths, with the possibility of underestimating future surface rupture lengths and earthquake magnitudes. This point was

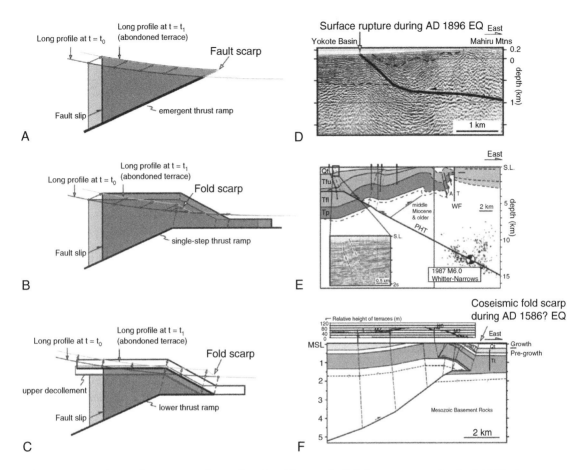

Figure 5.3: Schematic cross sections showing kinematic link between coseismic fold/fault scarps and geometries of underling thrusts in case of (A) emergent thrust ramp, (B) single-step fault-bend fold, and (C) wedge thrust fold. (D) Seismic reflection profile across a coseismic fault scarp during AD 1896 Riku-u earthquake (M_w 7) above an emergent thrust ramp. (E) Cross section across the Puente Hills thrust shows coseismic fold scarp above a ramp flat thrust trajectory. (F) Cross section across Kuwana anticline in central Japan shows AD 1586 coseismic fold scarp above a doubly vergent thrust wedge. Open-headed arrows indicate folding vectors.

emphasized via many case histories by Rubin (1996), who documented that almost all historic $M_w > 7$ reverse-fault surface ruptures ruptured multiple geometric segments (dePolo *et al.*, 1991 made the identical observation for normal faults). Thus, the danger in a place like the San Gabriel range front (Figure 5.4) is that large reverse ruptures will not be confined to individual, 30–35-km-long segments defined geometrically, but will involve multiple segments in ruptures ≥65 km long. Supporting evidence for these multisegment rupture scenarios are the often large displacements per event deduced from geomorphic and trenching studies, displacements that are anomalously large for the length of the segment,

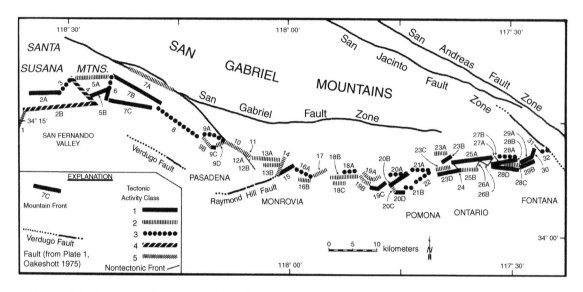

Figure 5.4: Example of segmentation of a range-front reverse-fault zone, based on geomorphic indices; the San Gabriel range front, Los Angeles basin, southern California. The "Tectonic Activity Classes" were defined by Bull (2007) based on range-front sinuosity, valley depth:width ratio, and faceted spur development. Based on these classes, one can deduce very active segments at the eastern (36 km) and western (30 km) ends of the zone, typified by Class 1 range fronts. The central part is dominated by a 35-km-long stretch of Class 2 range fronts, and a smaller western stretch of Class 3 range front. The Class 2 segment may be split by its junction with the Raymond Hill fault. From Bull (2007).

according to empirical equations (e.g., Wells and Coppersmith, 1994). This discrepancy is known as the "short, fat fault problem."

A totally different approach must be made for segmenting blind-thrust faults, as described in Section 5.8.3.

5.1.3   The Earthquake Deformation Cycle of Reverse Faults

The earthquake deformation cycle on terrestrial reverse faults is now known with some precision, based on geodetic surveys after earthquakes in California and more recently, InSAR surveys worldwide (for the coseismic deformation), and detailed surface and subsurface mapping to reveal the geologic structure created by many seismic cycles. The interseismic part of the cycle is less well known, because there has been insufficient time for monitoring it, compared to its length (many hundreds to tens of thousands of years). However, the general character of the interseismic cycle can be deduced if one accepts the conceptual model of Stein *et al.* (1988) that the present geological structure is the sum of deformation during the coseismic and interseismic phases cycle of the earthquake deformation cycle (Figure 5.5).

According to Stein *et al.* (1988) coseismic deformation is well constrained by geodetic observations and elastic dislocation models, to include a sharp spike of hanging-wall uplift and a smaller amplitude of footwall subsidence. Because the cumulative geologic structure also reflects similar perturbations,

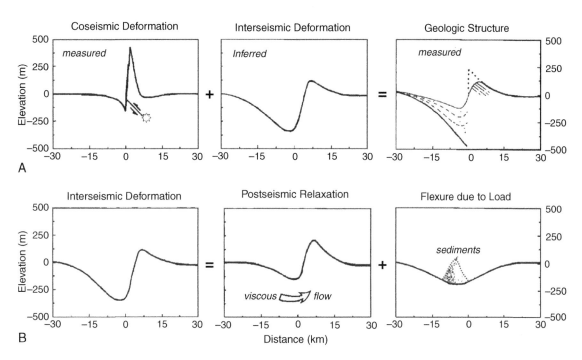

Figure 5.5: The earthquake deformation cycle for terrestrial reverse faults, based on a theoretical 45° dipping reverse fault with 1 km of cumulative slip. (A) Observed geological structure (far right) results from a combination of coseismic and interseismic deformation, as modified by erosion of the hanging wall (dotted line at far right) and deposition on the footwall (dashed lines at far right). (B) Interseismic deformation (far left) results from postseismic relaxation due to viscous subcrustal flow, and isostatic flexure of the elastic crust. The diagram only shows downward flexure due to sediment loading of the footwall, and not the smaller upward rebound due to erosion of the hanging wall. From Stein *et al.* (1988).

it is assumed that the interseismic deformation is also comprised of hanging-wall uplift and footwall subsidence, but with longer wavelengths for both, and a much larger footwall subsidence that hanging-wall uplift (Figure 5.5A, center; and B, far left). The interseismic deformation is thought to be the sum of two mechanisms, postseismic relaxation and flexure (Figure 5.5B). Postseismic relaxation reflects the bending of the brittle crust due to slow flow of viscous subcrustal material from the footwall to the hanging wall. Flexure results from the downward bending of the crustal slab due to a point load of accumulating sediment in the basin. Obviously, the relative amount of these two components will be controlled by crustal thickness and rigidity, the viscosity of subcrustal material, and the sedimentation rate in the basin.

5.1.4 Historic Analog Earthquakes

Well-studied historic thrust-fault surface ruptures form modern analogs for interpreting the paleoseismic record (Table 5.1) and include earthquakes in fold and thrust belts (Algeria, 1980), convergent continental plate boundaries (*collision zones;* Iran, 1978; Armenia, 1988), regions near transpressive bends in plate-

Table 5.1: Well-studied historic reverse and thrust fault surface ruptures

Date and magnitude[a]	Fault/Area	Maximum displacement[a] (m)	Rupture length[a] (km)	References
1952, M_s 7.7	Kern County, California	3.0	57	Buwalda and St. Amand (1955)
1968, M_s 6.9	Meckering, Australia[b]	3.5	37	Gordon (1971) and Gordon and Lewis (1980)
1968, M_s 7.1	Inangahua, New Zealand	1.8	5	Lensen and Otway (1971)
1971, M_s 6.5, M_w 6.7	San Fernando, California[b]	2.5	16	USGS (1971), Bonilla, (1973), and Kahle (1975)
1978, M_s 7.5	Tabas-e-Goldshan, Iran	3.0	85	Berberian (1979) and Haghipour *et al.* (1979)
1979, M_s 6.1	Cadoux, Australia	1.5	15	Lewis *et al.* (1981)
1980, M_s 7.3	El Asnam, Algeria	6.5	31	Yielding *et al.* (1981) and Philip and Meghraoui (1983)
1986, M_s 5.8	Marryat Creek, Australia	0.9	13	Machette *et al.* (1993)
1988, M_s 6.3–6.7	Tennant Creek, Australia	1.1	32	Crone *et al.* (1992)
1988, M_s 6.8	Spitak, Armenia[b]	2.0	25	Philip *et al.* (1992)
1999, M_w 7.6	Chi-Chi, Taiwan	11	100	Ouchi *et al.* (2001)
2005, M_w 7.6	Kashmir, Pakistan	7.0	70	Kaneda *et al.* (2008b)

Bold indicates ruptures that occurred after the 1st Edition was published.
[a]From Wells and Coppersmith (1994).
[b]Described as Case Histories by Yeats *et al.* (1997).

boundary transform faults (California, 1952, 1971, 1983; New Zealand, 1971), and stable continental interiors (Australia, 1968, 1979, 1986, 1988). These historic thrust earthquakes have produced a wide spectrum of shallow structures and scarp morphologies, resulting from both slip on multiple fault planes and folding. For example, small fault-propagation folds and fault-bend folds typically formed in the hanging wall in unconsolidated or semiconsolidated deposits (e.g., Berberian and Qorashi, 1994). These folds occur at all scales and commonly reflect changes in dip of the underlying thrust where it intersects stratigraphic sequences of differing character.

Perhaps due to the complex interaction of faulting and folding, surface displacement on thrust faults tends to be irregular along strike, even more so than for other fault types (Figure 5.6) (see Baljinnyam *et al.* (1993, their Figure 44) for a geometric reconstruction of displacement from fault scarp dimensions). Measurements of fault separation between piercing point landforms neglect the off-fault deformation and provide only a minimum measure of the actual (geodetic) displacement in the shallow subsurface. As in normal surface faulting, the maximum displacement (MD) often occurs in one or more isolated peaks and it far exceeds the median or mean displacement. Unlike the normal and strike-slip ruptures described in Chapters 3 and 6, the thrust ruptures shown in Figure 5.6 do not have long tails of low slip; instead, slip decreases abruptly at the ends of the rupture. Additional detail on this aspect is given in Section 5.2.5. Wells and Coppersmith (1994, Table 3) note that empirical regressions of reverse-fault displacement (average and maximum) on historic earthquake magnitude shows 50% more variance than for other fault types.

One of the best documented examples of combined coseismic folding, thrust faulting, and secondary faulting occurred in northern Algeria during the 10 October 1980 El Asnam earthquake ($M_s = 7.3$). Focal

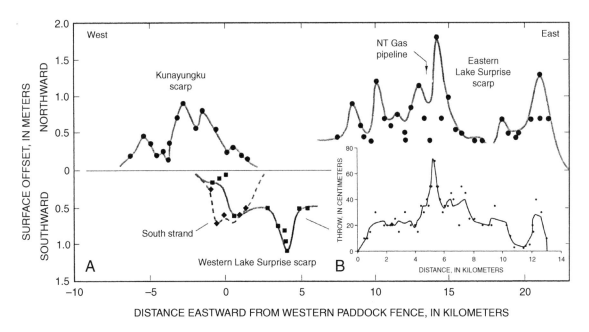

Figure 5.6: Vertical surface offset (or fault throw) along strike in two historic thrust faulting earthquakes. (A) Vertical surface offset associated with the 1988 M_s 6.3–6.7 Tennant Creek, Australia, earthquakes. Solid symbols show field measurements, curve represents smoothed data. From Crone *et al.* (1992). (B) Fault throw associated with the 1986 M_s 5.8 Marryat Creek, Australia, earthquake. Small circles show field data, line represents smoothed data. From Machette *et al.* (1993). Note how maximum offset occurs in a narrow spike and is roughly twice as large as average offset.

mechanism studies indicate that rupture initiated at a depth of about 12 km and extended along a 30-km-long northeast trending northwest dipping thrust fault (Ouyed *et al.*, 1981). The rupture propagated to the surface along the base of low anticlinal hills on the margin of the Chelif Valley (Figure 5.7). Concrete irrigation ditches were offset and shortened, and moletrack scarps formed in cultivated fields and orchards and across the valley floor where the scarp was 3.2 m high (King and Vita-Finzi, 1981). Net slip at depth was estimated to be as much as 6 m based on seismic moment, but thrust offset at the surface was considerably less. The slip deficit appears to have been taken up by the growth of the anticlines in the hills above the rupture surface.

King and Vita-Finzi (1981) observed several types of geologic evidence produced by the coseismic folding. Numerous secondary normal fault scarps up to several meters high formed along the crests of a large asymmetrical anticline in the upper thrust sheet. The normal faults faced both northwest and southeast and bounded long shallow compound grabens aligned in a 1- to 2-km-wide zone, subparallel to the thrust, along the fold hinge line. The normal faults were interpreted to be the result of tension (i.e., bending-moment faults) produced in the upper part of the anticline as it flexed during the earthquake. Long-term growth of the anticline is reflected by increasing bedding dips of successively older Quaternary and late Tertiary sediments. Paleoslip on some of the bending-moment normal faults was indicated by degraded pre-1980 scarps that were reactivated during the earthquake.

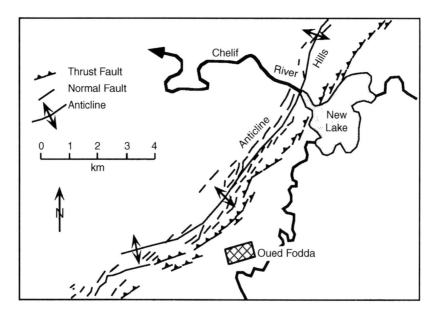

Figure 5.7: Map showing deformation during the 1980 El Asnam, Algeria, earthquake (modified from King and Vita-Finzi, 1981). Surface rupture on the seismogenic thrust produced small discontinuous scarps along the base of the anticlinal hills. More obvious in the field shortly after the earthquake were many normal fault scarps and grabens along the crest of the hills. These normal faults reflect coseismic growth of the anticline above the thrust. Additional evidence of growth of the fold during the earthquake was damming of the Chelif River near the axis of the fold almost 2 km downstream of the surface rupture on the thrust.

A second geologic response to coseismic fold growth was disruption of the Chelif River where it crossed the anticline. Immediately following the earthquake, a lake 5–6 m deep formed upstream of where the river crossed the deformed zone. King and Vita-Finzi (1981) report "the lake was dammed not by the thrust fault scarp, but by a gentle upwarp 1 km further down the valley." The upwarp corresponds with the crest of the anticlines cut by the normal faults. Investigation of stratigraphy near the lake revealed evidence for previous ephemeral lakes filled with paleoflood deposits, presumably formed by paleoseismic displacements similar to the 1980 event. These paleoflood deposits constitute secondary paleoseismic evidence and were later studied in detail (Meghraoui and Doumaz, 1996), who interpreted eight paleoflood units in the past ca. 5.3 ka. Together with fault trench data the paleoflood deposits indicate a slip rate of 0.25–0.37 mm/yr in the past ca. 4 ka and a mean recurrence interval of 720 years, with recurrence as short as 300–500 years in clusters. Without the long record of paleoflood deposits, it would have been very difficult to construct this paleoearthquake chronology from trench data alone.

5.2 Geomorphic Evidence of Reverse Paleoearthquakes

The most direct geomorphic evidence of a terrestrial reverse paleoearthquake is a fault scarp. However, many, large shallow reverse earthquakes are generated by fault rupture that is confined to the subsurface, and no prominent fault scarp may be created (e.g., the 1811–1812 New Madrid, USA earthquakes). Thus, in compressional tectonic environments $M_w < 7$ paleoearthquakes may not be accompanied by any recognizable geomorphic expression of faulting. Yeats (2007) noted that the best-expressed surface ruptures accompanying historical reverse-fault earthquakes are found either in stable continental shields (focal depths 2–7 km), or in fold and thrust belts (focal depth much shallower than the brittle-ductile transition). In contrast, earthquakes nucleating much deeper near the brittle-ductile transition are typically poorly expressed at the surface. Thus he proposed that focal depth controls the development of reverse surface faulting (the same control does not appear to affect normal or strike-slip faults).

If reverse-fault scarps are created, they are typically more sinuous and irregular than for other fault types, being composed of either short, disconnected sections (Gordon and Lewis, 1980; Crone *et al.*, 1992) or producing a continuous but "serrated" rupture trace with zigzags on the scale of meters (photographs in Haghipour *et al.*, 1979). Where thick sequences of late Cenozoic sediments are faulted, the thrusts may reach the surface as imbricate faults, with many discrete slip planes.

Over geologic timescales, repeated reverse faulting typically juxtaposes hard bedrock in the hanging wall against youthful poorly consolidated sediments in the footwall. This contrast in material properties tends to concentrate slip in a narrow zone at the base of a geomorphically distinct, uplifted, and (commonly) folded range front. Here active slope processes, stream erosion, and alluvial fan deposition may act together to obscure the fault trace (Beanland *et al.*, 1986). Newer reverse faults may cut rolling terrain on the hanging wall or gentle depositional slopes on the buried footwall, having propagated through some thickness of unconsolidated sediments.

Paramount in the recognition of past slip events on reverse faults is information on the near-surface structure and kinematics of the fault at the place where the paleoseismic evidence is found. Considerable along-strike variability in structural style and scarp type is common on many reverse surface ruptures, and complex structure is more common than a single simple fault, especially in those depositional environments most likely to record surface faulting events. Because reverse faulting characteristically distributes slip on multiple imbricate faults and as off-fault folding, their paleoseismic study requires measurements across the entire zone of surface deformation.

5.2.1   Initial Morphology of Reverse and Thrust Fault Scarps

The initial form of a reverse-fault scarp can be more varied than for other fault types, due to the varying dip of the bedrock fault (near vertical to near horizontal), the complex nature of mixed faulting and folding, and to the complex response of surficial materials. Gordon (1971) and Philip *et al.* (1992) described eight and seven types of thrust fault scarp morphologies, respectively, based on historic surface ruptures in Australia (1968) and Armenia (1988) (Figure 5.8). The controls on scarp morphology include amount of slip, sense of slip, geometry of the fault(s), properties of surficial materials, and topography (Weber and Cotton. 1980, p. 20; Philip *et al.*, 1992, p. 144).

Steeply dipping (>45°) reverse faults in bedrock produce *simple thrust scarps*, such as those shown in Figures 5.8A and 5.9. Steep faults in brittle unconsolidated materials result in *hanging-wall collapse scarps* (Figure 5.8B), which result when the overhanging scarp collapses, usually during seismic shaking. At lower dip angles, thrust faulting produces *pressure ridges* (Figure 5.8C–F). The type of pressure ridge is dependent on surface material rheology and the magnitude of slip (Figure 5.10). More brittle materials produce more fissuring of the leading edge of the thrust (Figure 5.8C and D), whereas more displacement increases the chance of developing a secondary normal fault in the hanging wall (Figure 5.8D). In plastic surface materials (moist silt and clay, turf), pressure ridges have smoother fronts and may display *backthrusts* (Figure 5.8E) or low-angle pressure ridges (Figure 5.8F), but as thrust displacement decreases below ca. 1 m all pressure ridge types merge into a single type of small *moletrack*. An increasing oblique component of slip results in en-echelon pressure ridges (Figure 5.8G) or oblique tension fissures in pressure ridge fronts (Figure 5.8D).

Hanging-wall ramp folds generated by changes in fault dip at shallow depths (meters to tens of meters) can generate distinct geomorphic *facets* at the top of scarps. These facets resemble the planar bevels in the upper parts of large compound normal fault scarps that sometimes reflect multiple paleoearthquakes. On thrusts, however, the facets may be part of the initial scarp form and independent of the paleoseismic history of the fault and degradational processes acting on the scarp. Some thrust tips form propagation

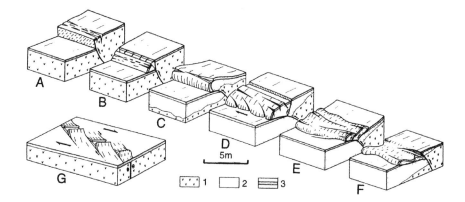

Figure 5.8: Types of reverse-fault scarps produced along the Spitak fault during the 1988 M_s 6.9 Spitak, Armenia, earthquake. (A) Simple thrust (or reverse) scarp. (B) Hanging-wall collapse scarp. (C) Simple pressure ridge. (D) Dextral pressure ridge. (E) Back-thrust pressure ridge. (F) Low-angle pressure ridge. (G) En-echelon pressure ridges. 1, bedrock; 2, soft Quaternary sediments; 3, turf. From Philip *et al.* (1992): reprinted with permission of Blackwell Scientific Publishing Company.

Figure 5.9: Photograph of the fresh overhanging reverse-fault scarp created by the 1971 M_L 6.6 San Fernando, California, earthquake. Faulted material is a weakly consolidated Tertiary sandstone. Large divisions on the horizontal rod are 0.3 m, scarp is ca. 1 m high. Above (A), an overhanging scarp is defined by the fault plane that dips at 65° (compare to Figure 5.11A). Above (B), the overhang has collapsed, resulting in a hanging-wall collapse scarp (Figures 5.5B and 5.11B). Photograph courtesy of Virgil Frizzell and the U.S. Geological Survey. From USGS (1971).

folds at the surface with decreased slip on the fault tip. These folds can also produce multifaceted scarp profiles (Suppe, 1983).

Widely distributed slip generally does not result in the formation of steep scarps, but rather pervasive ground cracking and gentle flexure of the ground surface across a wide fault zone. Microfaulting may occur locally, and mesoscale thrusts may also be present in zones of distributed slip. Where this style of surface thrusting deforms low-gradient terraces the fault trace is expressed as a wide gentle warp of the terrace surface.

5.2.2   Degradation of Thrust Fault Scarps

Surface displacement on a single thrust fault plane initially results in an overhanging scarp, but in unconsolidated deposits such overhangs collapse during or soon after creation. For example, even in weakly consolidated Tertiary sandstone and conglomerate faulted by the 1971 San Fernando, California, earthquake, >50% of the ca. 0.3- to 1-m-high overhanging free face (Figure 5.9) had collapsed after three months (Kahle, 1975, p. 133). The *collapsed tip* of the hanging wall thus creates a free face and a steep debris slope that buries the fault tip. If the free face exceeds the angle of repose in unconsolidated materials, the scarp will progress through successive gravity-, debris-, and wash-dominated degradation stages similar to those described earlier for normal faults (Chapter 3). Many degraded reverse-fault scarps are asymmetric in cross-profile, with the steepest part of the scarp lying downslope of the scarp

Figure 5.10: Types of reverse-fault scarps produced in the 1999 Ch–Chi earthquake, Taiwan. (A) Hanging-wall collapse scarp. (B) hanging-wall collapse scarp (foreground) transitions into a pressure ridge scarp (background). (C) pressure ridge scarp displacing a school running track. (D) complex low-angle pressure ridge scarp in soft and ductile materials. From Chen *et al.* (2001).

midpoint. This asymmetry, caused by repeated overriding of the scarp-derived colluvial wedge (Section 5.3.2), contrasts with the general symmetry of normal fault scarps (Chapter 3). Where repeated movement has generated a high escarpment, reverse faulting may be expressed as a long chain of land slides which obscures the fault trace itself. Such landslides formed along much of the length of the Patton Bay fault scarp on Montague Island during the 1964 Alaskan subduction earthquake (Plafker, 1969a).

5.2.3   Interaction of Thrust Fault Scarps with Geomorphic Surfaces

Youthful surface thrusting is most easily recognized if planar landscape features such as fluvial or marine terraces, alluvial plains, fans, and other constructional geomorphic surfaces are cut by the fault. Thrust faults are often difficult to recognize where the fault crops out in areas of steep or irregular topography. In such settings fault scarps become difficult to distinguish from landslide and sackung scarps (see Chapter 8) and are quickly removed by rapid erosion. Thus, one strategy useful in identifying active thrusts is to examine terraces and other planar geomorphic surfaces for evidence of offset or warping. Because many thrusts reach the surface as broad zones of distributed displacement or folding, the offset may not be concentrated on a distinct scarp and may become apparent only after careful examination of the terrace, or after measuring and plotting long topographic profiles that cross the fault.

5.2.3.1 Fluvial Terraces

The interaction of cyclic fluvial terrace cutting, such as that associated with glacial-interglacial climate changes, and recurrent surface faulting can result in terrace sequences that record the faulting history. If terraces are the result of processes independent of movement on the fault, offset produces vertical displacement of the terrace profile. Paleoseismic events or groups of events are thus recorded by sequentially larger vertical separations on higher regional terraces, in a manner similar to that for normal faults (Chapter 3) (see also Beanland and Barrow-Hurlbert, 1988, Figure 10). Additionally, paleoseismic records are formed when faulting crosses actively aggrading floodplains. Surface offset of vertically accreting overbank sediments produces scarps, mole tracks, secondary faults and fractures, and other fault-line disturbance of the bedded floodplain sediments that become unconformably overlain by subsequent overbank flood sequences. Such paleoseismic events show up as unconformities bounding progressively more deformed flood plain sediments (Section 5.3.5).

Broadly distributed thrusting, surface warping, and folding cannot generally be detected except where planar geomorphic surfaces are deformed. Wide zones of folding, warping, or distributed microfaulting result in obvious disturbance of terraces and in different terrace elevations across the fault. Careful surveys of the terrace and plotting of profiles are often necessary to detect the fault and measure the vertical separation. The elevation differences between the correlative terrace segments on either side of the zone of deformation represent the vertical separation across the fault and provide the best datum for measurement of fault slip, even if a single narrow fault zone cannot be pinpointed in the field.

The displacement created by a single faulting event is easiest to measure from the shape of a single-event fault scarp, such as might be found in the youngest terrace (Figure 5.11). On older terraces the latest single-event displacement may merely add additional incremental relief to a preexisting fault scarp, and possibly spread out over a wider horizontal distance, making it more difficult to separate as an individual displacement event.

5.2.3.2 Marine Terraces

Along uplifted coasts raised marine terraces are excellent geomorphic reference surfaces that record thrust faulting and folding. Paleoearthquakes can be interpreted from differences in the elevation along the shore-parallel profile of faulted late Holocene terraces on rapidly emerging coastlines. Sudden elevation differences in the profile are interpreted to reflect the vertical component of displacement on the fault. Where late Holocene terraces are faulted, reverse or thrust displacement lifts the shoreline and raises the terrace on the upthrown side of the fault. If vertical displacement is great enough a new shoreline is established seaward of the newly raised terrace. Simultaneously the shoreline on the downfaulted side of the fault may be submerged and a new shoreline established landward of its preearthquake position.

Where thrusts intersect emerging coastlines, long-term (late Pleistocene) slip rates can sometimes be determined from raised and faulted glacioeustatic marine terraces (Figure 5.12). If the fault cuts flights of terraces (Figure 5.12A), cumulative displacement can be measured across the fault for each high sea-level stand represented in the terrace sequence (Figure 5.12D). The vertical component of displacement is reflected by the scarp height, that is, the elevation difference across the fault for each of the terraces.

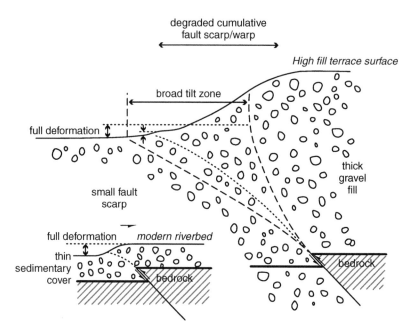

Figure 5.11: Relationship of reverse-fault scarp height to the "full deformation" (vertical displacement), based on observations of the 2005 Kashmir fault scarp. In the river bottom where terraces are young and alluvial cover is thin, the fault scarp height is nearly identical to the fault throw (inset, lower left). On the valley margins where terraces are older and deposits are thicker, the same displacement of bedrock propagates upward and splays into a braoder zone of deformation, creating a small scarplet at the base of the preexisting, multievent fault scarp, and a broad tilt zone. The scarplet height does not reflect the full amount of fault throw. From Kaneda *et al.* (2008b).

A check on the long-term slip rate is possible by comparing the uplift rate determined from the terrace elevation and the age of terraces on either side of the fault (Figure 5.12B and C; see also Lajoie, 1986). The difference in uplift rates (0.77–0.33 mm/yr in Figure 5.12) should match the vertical rate for separation on the fault (in Figure 5.12D, 0.42 mm/yr). Although these slip rates do not define parameters of any individual paleoearthquakes, they do provide a starting point from which to estimate possible combinations of displacement per earthquake, number of earthquakes, and recurrence (see Chapter 9, See Book's companion web site).

5.2.3.3   Coseismic Terraces

Paleosurface rupture can sometimes be represented by anomalous fluvial and marine terraces where a stream traverses the hanging wall of a thrust fault. The formation of nickpoints in rivers or streams on the hanging wall of the Patton Bay fault during the 1964 Alaskan earthquake illustrates the process for forming coseismic terraces (similar to the tectonic terraces described on normal faults, Section 3.2.5.1). If the stream gradient is low, faulting sometimes raises the hanging wall sufficiently to initiate downcutting into the floodplain. The portion of the stream on the footwall may also be lowered enough to cause deposition. Under extreme cases fluvial systems that cross thrusts and are antecedent to the associated fold and thrust mountains have been dammed by fault displacement. Where such processes

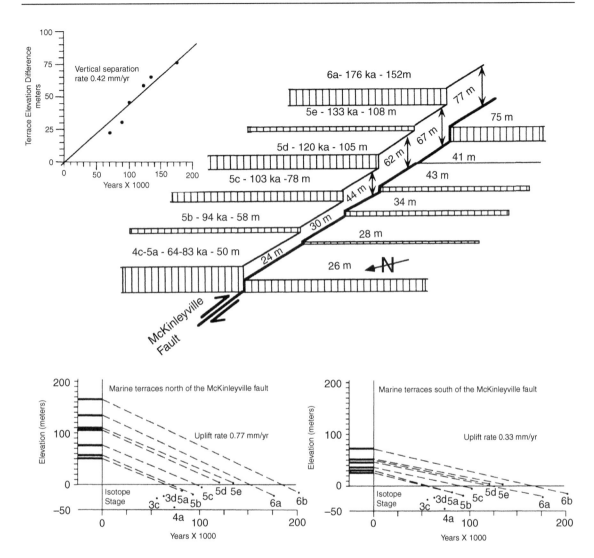

Figure 5.12: Diagram of faulted marine terraces at McKinleyville, northern California. (A) Located in the Mad River fault zone, an upper plate thrust system at the southern end of the Cascadia subduction zone, flights of raised glacioeustatic marine terraces record uplift on either side of the McKinleyville fault. Differences in uplift rates across the fault produce different elevation spacing of terrace treads. The terraces can be distinguished by the soil profiles that reflect a late Pleistocene chronosequence. (B) The vertical separation for each terrace age provides a basis estimate of the fault's late Pleistocene slip rate. (C) and (D) Uplift diagrams showing uplift tie-lines between late Pleistocene highstands (horizontal axis) and present terrace elevation (vertical axis) for the terrace flights on the hanging wall and footwall blocks. The diagrams show relatively constant uplift rates and serve as a check on the age assignments for the terraces (Bull, 1985). From Carver and Aalto (1992); reprinted with permission of the American Association of Petroleum Geologists.

occur they are usually cyclic and produce a record of past fault displacements as raised terraces on the upthrust block and buried sediment layers deposited in response to defeat of the stream (For the response of streams to coseismic folding, see Section 5.5.3).

5.2.4   Slip Rate Studies

Slip rate studies are landform-based paleoseismic studies, the aim of which is to calculate a long-term fault slip rate from landforms displaced by multiple faulting events. They are usually reconnaissance studies on poorly known faults, and precede trenching studies. Slip rate studies can be performed at widely varying spatial and temporal scales. At the gross and small scales, no single-paleoearthquake offsets are usually discernable, so those studies fall more in the realm of tectonic geomorphology (see Burbank and Anderson, 2001). At small and medium scales, long-term slip rates can be calculated spanning tens to hundreds of ka (Table 5.2). The development of cosmogenic dating of arid-zone landforms has enabled many new slip rates to be estimated for faults in places like the western USA and central Asia.

5.2.5   Spatial and Temporal Variations in Surface Displacement

The earthquake surface ruptures that produce fault scarps typically vary in sense and amount of displacement along strike and also vary in displacement at a given point on the fault between successive paleoearthquakes. In order to appreciate this range of variability in space and time, we must examine well-studied historic surface ruptures (Table 5.1) as modern analogs to paleoearthquakes.

Table 5.2: Examples of recent long-term slip rate studies based on landform offsets (reverse-fault scarps or folds over blind thrusts)

References	Location	Landform type	Largest offset or Slip rate	Oldest age	Dating method
Brown et al. (1998)	Tien Shan	Alluvial fans	38 m	40 ka	[10]Be
Thompson et al. (2002)	Central Tien Shan, KYRGYZSTAN	Terraces	3 mm/yr	12 ka, ^{14}C; 170 ka, lum	[14]C, TL, IRSL
Hetzel et al. (2004)	Zhangye thrust, China	Alluvial fans	55–60 m; 0.6–0.9 mm/yr	90 ka	[10]Be, OSL
Benedetti et al. (2000)	Montello anticline, Italy	Terraces	200 m	321 ka	correl. To MIS[1]
Lave and Avouac (2000)	Main Frontal Thrust, Nepal	Terraces	105 m	9.2 ka	[14]C
Ishiyama et al. (2004)	Kuwana anticline, Japan	Terraces	137 m	130 ka	[26]Al, [10]Be

First correlation to Marine Isotope Stages.

5.2.5.1   Variability of Displacement Along Strike in a Single Rupture

In a general sense, vertical displacement on historic reverse-fault surface ruptures has been greatest at the center and least near the ends of rupture (Figure 5.6), but some ruptures are highly asymmetrical, such as the 2005 Kashmir rupture (Figure 5.13), where the MD occurred only 1/6 of the distance from the rupture end. The shorter wavelength lateral variations in vertical surface displacements can be also be seen in this rupture and would probably have been even more pronounced if measurements were taken at closer intervals. As for normal faults, the origin of the short-wavelength variations is unclear. They may represent the complex response of surficial deposits to rupture rather than spatial variations in displacement along the bedrock fault plane.

Regardless of the origin of such surface slip variations, however, they provide a statistical basis for relating the height of prehistoric normal fault scarps to the magnitude of the causative paleoearthquake (see detailed discussion in Chapter 9, See Book's companion web site). McCalpin and Slemmons (1998) inventoried six well-studied historic reverse surface ruptures (Figure 5.14) and were able to combine all their displacement measurements (a total of 193 measurements) by normalizing them to the maximum displacement in each rupture. The number of events (six) and data points (193) is considerably smaller than for normal and strike-slip ruptures, so we are uncertain about this statistical validity of interpreted trends.

Trend 1 is the decrease in scarp heights smaller than 20% of MD. This trend may result because such small displacements are dominantly expressed at the surface by subtle folding rather than faulting, thus no measurable scarp was created. Trend 2 is the increase in frequency of displacements of 90–100% MD, in contrast to the decline in frequency of all other scarps heights slips greater than or equal to 40% MD. This increase appears regardless of the weighting method, so it cannot be ascribed to a single rupture in the six-event set. Examination of the frequencies of the six component earthquakes shows that five of the six events show this tendency. It is possible that workers did not make many measurements on either side of the maximum displacement, in which case that displacement value would have been assigned to a long affected length of the rupture. However, workers typically make dense measurements as displacement increases toward MD, and if anything they omit measurements where displacements are smaller than average. So, we do not know the exact reason for this anomalous frequency pattern. The length-weighted average displacement is 0.38 ± 0.009 MD, or some 15% higher than that of normal faults, which contain a larger proportion of small displacements. The implication of this geometry for paleoseismic studies (explored in more detail in Chapter 9, See Book's companion web site) is that randomly located trench sites on reverse-fault scarps are unlikely to encounter displacements either near the maximum or near minimum that occurred in each paleoearthquake.

5.2.5.2   Variability of Displacement at a Point

Few reliable data exist on the variation of fault displacement and style among repeated reverse surface ruptures at the same point on a fault. The sparse data that do exist come from trench studies where only two or three successive paleoearthquakes can be measured and compared. Clark and McCue (2003) point out that the historic Tennant Creek rupture had a similar displacement to the prior prehistoric event, and McCue *et al.* (2003) make a similar observation for the two latest prehistoric ruptures on the Lake Edgar fault (2.5 m slip in each event). Recent data from the Longitudinal Valley Fault, Taiwan, also show that displacements on that reverse fault have been essentially identical in the past three rupture events on the Rueisuei segment (AD 1951, 1.7 m; AD 1736–1898, 1.6 m; AD 1564–1680, 1.6 m) (Yen *et al.*, 2008).

In contrast, Masana *et al.* (2005) noted noncharacteristic displacements in the two latest ruptures of the Albox fault, Spain, with the latest being measured in decimeters and the prior one at 0.5–1 m. Maruyama

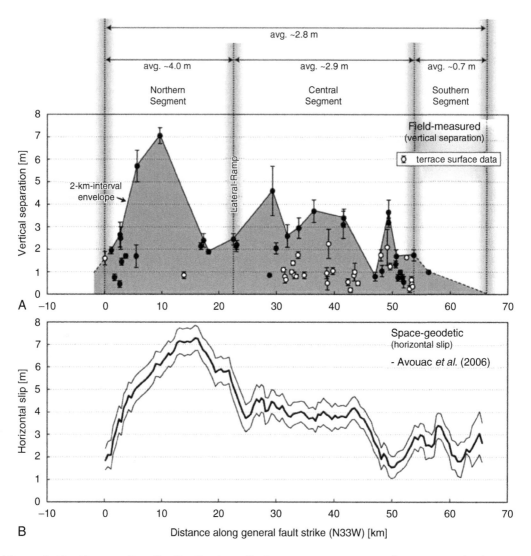

Figure 5.13: Along-strike slip distribution of a large reverse rupture; the 2005 Kashmir event. Compare the length and displacements to those on moderate-magnitude events shown in Figure 5.6. (A) Along-strike distribution of field-measured vertical separation. Open symbols indicate values measured on high fill terraces, whereas estimates at the valley bottoms and elsewhere are denoted by solid symbols. The terrace surface data exhibit systematically smaller vertical separations than those obtained elsewhere, suggesting the absorption effect by thick gravel fill, which is graphically explained in Figure 5.11. Solid line is a slip distribution envelope established by connecting the maximum values on each 2-km-long section along the surface rupture. (B) Along-strike distribution of horizontal slip (thick curve) and associated error (thin curves) deduced from subpixel correlation of ASTER images (Avouac *et al.*, 2006). Redrawn from original data provided by Jean-Philippe Avouac. Overall slip distribution pattern is similar to the field observation although there is a significant mismatch on the southern segment, which is probably the result of diffuse deformation and/or a smaller dip angle of the fault. From Kaneda *et al.* (2008b).

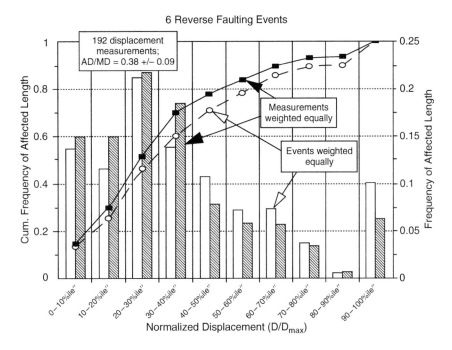

Figure 5.14: Composite normalized frequency data for surface displacements during 6 reverse-faulting events. Average displacement (AD) of the 193 measurements is 0.38 ± 0.09 of the maximum displacement (MD, or D_{max}). Very small displacements (<20% of MD) are less abundant than for normal faulting events (compare to similar figures in Chapters 3 and 6). From McCalpin and Slemmons (1998).

et al. (2007) also document noncharacteristic displacements among the three latest ruptures of an unnamed extension of the active Muikamachi-Bonchi-Seien fault zone, Japan, which ruptured 10 cm in 2004. The two prior paleoearthquakes at the same site had nearly identical displacements of 1.5 m, showing that displacements had varied by a factor of 15 among the latest three events.

Given the small number of studies and their conflicting observations, it is unclear whether most active reverse faults display characteristic earthquake behavior. This situation could be remedied by: (1) performing a comprehensive inventory of data on successive reverse-fault displacements from the published literature and (2) collecting new field data on long paleoearthquake records using "mega-trenches."

5.3 Stratigraphic Evidence of Reverse and Thrust Paleoearthquakes

Reverse surface faulting results in the instantaneous creation of faults, folds and tilted beds, and in the delayed response of fault-induced sedimentation. The sequence of paleoearthquakes cannot usually be reconstructed from tectonic or depositional features alone, instead, a combined analysis is required. The key to successful interpretation is to distinguish between tectonic versus depositional features, and to distinguish depositional units that predate faulting from those that postdate faulting. The concepts presented in this section are derived from studies of many trench exposures of reverse faults in the western United States, as well as worldwide (see citations in text).

5.3.1   General Style of Deformation on Reverse Faults in Section

The simplest stratigraphic expression of a surface reverse fault is a single discrete slip plane between the hanging wall and footwall. If slip is repeatedly concentrated in the same narrow zone, large contrasts in materials are common where older more consolidated rocks are thrust over younger less consolidated surficial sediments. Deformation tends to be concentrated at the lithologic boundary where *gouge* or *cataclastic rock* is produced. *Breccia* and *slickensides* are more common, and *fissures* and *rubble* less common, on reverse faults than on other fault types (Bonilla and Lienkaemper, 1991); evidently near-surface compression across the fault zone promotes grinding and prevents the formation of voids. Where poorly consolidated sediments are in fault contact, usually in the footwall, distributed intergranular shear may accommodate some of the total slip. This distributed slip results in clast rotation and the development of an *imbricate fabric* subparallel to the fault plane.

Thrust fault traces are relatively easy to recognize in exposures unless they flatten to subhorizontal or become parallel to bedding in the footwall, as occurred in well-sorted sand and silt beds along the Dunstan fault in New Zealand (Beanland *et al.*, 1986). In poorly stratified gravel, the fault zone can be up to 1 m wide and defined primarily by *oriented clasts* (Figure 5.15). In eolian sand, the Tennant Creek, Australia, thrust fault was expressed as a 3-m-wide zone of eight parallel thrust faults that had an aggregate net slip of 2.4 m (Crone *et al.*, 1992, pp. A18–A19). The association of multiple, parallel, small-displacement strands with faulted sand was previously noted for normal faults (Chapter 3). If the near-surface deposits

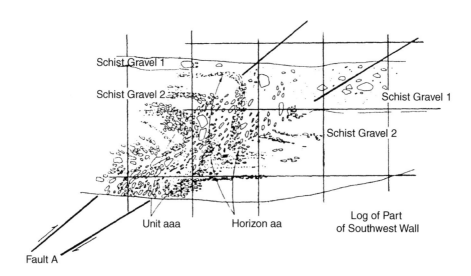

Figure 5.15: Trench log through a reverse-fault zone showing definition of the fault by imbricate clast fabric. Grid spacing is 2 m; clasts are drawn to scale. The zone of reoriented clasts is up to 1 m wide. Note the difference in vertical offsets between the schist gravel 1/schist gravel 2 contact across the fault zone (1.1 m) and between horizons aa and aaa in schist gravel 2 across the fault zone (2.3 m). These differences are indicative of multiple faulting events. Trench log DC 316 on the Pisa Fault, South Island, New Zealand. From Beanland *et al.* (1986); reprinted with permission of the Royal Society of New Zealand.

are clay rich and saturated, ductile deformation can accommodate much or all of the fault slip and fault-propagation folds may form at the thrust tip.

Several trenches have shown that displacement on reverse fault strands may tend to migrate toward the hanging wall over time (Weber and Cotton, 1980; McCalpin, 1989b; Tsukuda *et al.*, 1993, their Figure 9). This geometry results from successive truncation of earlier refracted thrust traces by later fault traces that refract at higher stratigraphic levels (Figure 5.16). The implication is that thrust fault *refraction* occurs at a

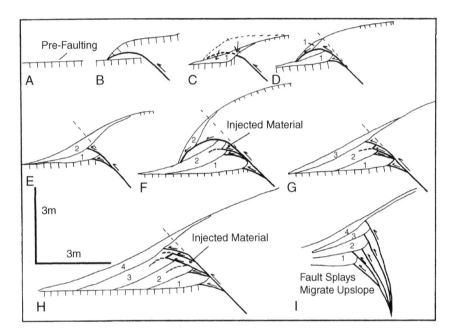

Figure 5.16: An idealized sequence of stages in the evolution of a 45°-dipping reverse fault. Scale at left is approximate. (A) Prefaulting geomorphic surface. (B) First faulting event; fault refracts to a lower angle near the surface and the overhanging fault tip sags and deforms. "Bulldozing" of the tip may occur, but is not shown. (C) Overhanging fault tip collapses onto prefaulting ground surface, creating colluvial deposit 1. (D) Second faulting event. Fault slip propagates upwards on a 45° plane (dashed line), beheads the refracted part of the first-event fault plane, and then refracts to a lower angle at a higher stratigraphic level, forming an overhanging fault tip. Colluvial deposit 1 is truncated and its proximal part is carried up on the hanging wall. (E) The overhanging tip collapses, and colluvial deposit 2 is formed. (F) Third faulting event. The reverse fault again propagates to a higher stratigraphic level on a 45° plane, beheading the refracted fault plane from the second faulting event. In addition, minor movement on the beheaded fault section(s) results in injection of fault breccia into the lower part of colluvial deposit 2. The reactivated part of the refracted fault (dashed line) flattens and loses definition in the colluvium. The proximal part of colluvial deposit 2 is truncated and carried onto the hanging wall. (G) Collapse of fault tip and eventual deposition of colluvial deposit 3. (H) Geometry after a fourth faulting event, tip collapse, and deposition of colluvial deposit 4. More fault breccia has been injected into the base of colluvial deposit 3. All earlier colluvial wedges (1,2,3) are truncated by faulting, only the latest wedge (4) overlies the fault. (I) Alternative geometry where fault refraction occurs at a greater depth and no injection of breccia occurs.

roughly constant depth below the ground surface at a given site, and due to fault-zone deposition, that depth rises stratigraphically with every earthquake. In the idealized geometry of Figure 5.16, each *colluvial wedge* overlies the fault that immediately predates it, and each wedge is truncated at its upslope end by the next younger fault. This simple pattern can be complicated by continued minor displacements on the earlier, *tectonically beheaded faults*, which leads to *injection* of fault zone breccia into the proximal part of the colluvial wedge. Figure 5.17 shows a field example analogous to Figure 5.16, where much of the fault zone material is liquefied sand injected between bodies of fault breccia.

5.3.2 Trenching Techniques

In order to expose the leading edge of reverse and thrust faults in sectional view, trenches are typically excavated perpendicular to the fault scarp (or fault trace) in a manner similar to that used for normal faults (Chapter 3), using similar excavation and shoring techniques. However, the deepest part of a reverse-fault trench must be shifted toward the upthrown fault block, compared to a trench across a normal fault scarp

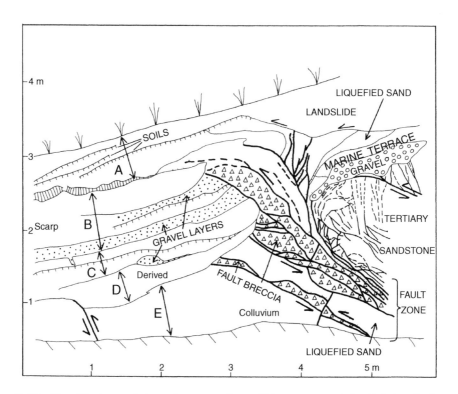

Figure 5.17: Simplified trench log of a natural seacliff exposure of the Frijoles fault, coastal California. Five units of scarp-derived colluvium (A–E) are in contact with a complex reverse-fault zone. The fault zone is composed of fault-bounded slices of coarse-grained fault "breccia" (derived from Tertiary sandstone and marine terrace gravels) and sand injected while liquefied. Note how the lowest faults and breccia zones truncate colluvium (E) but are truncated by colluvium (D). Higher faults truncate colluviums (B) and (C) but are overlain by colluvium (A). Compare to Figure 5.9. Adapted from Weber and Cotton (1980) (Plate XIV).

of similar size. This shift is necessary because reverse faults dip down beneath the upthrown block, and thus the critical paleoseismic features will be found successively farther beneath the upthrown block as they get deeper. This is the opposite geometry to that found on normal faults, where the trench is often not even begun until partway down the scarp face. In many thrust-fault trenches (see trench logs later in this chapter) the deepest part of the trench must be well upslope of the scarp itself, in order to expose postfaulting and interfaulting deposits "run over" by the horizontal advance of the thrust tip. If a geophysical survey is run over the potential trenching site, the subsurface relationships described above will be obvious, and the trench can be configured properly.

The recent trend in trenching reverse faults, as for normal faults, is to dig deeper to expose evidence of more successive paleoearthquakes. Most large trenches have been excavated with the double-benched design. However, in some areas it is logistically impossible to dig such large trenches, and deeper stratigraphic levels must be probed by boreholes, as described in Chapter 2A. Figure 5.18 shows a site where Tucker and Dolan (2001) excavated a 62-m-long, 5-m-deep trench across the geomorphically defined main strand of the Sierra Madre fault, Los Angeles basin. Most of this reverse-fault trace is locally obscured by small landslides, so their trench was excavated along a small canyon in a reentrant in the mountain front, allowing them to cross the entire width of the main active fault zone in an area of active sediment accumulation. After the trench walls were mapped, they backfilled the trench and then excavated a transect of eight large-diameter (70 cm) boreholes (bucket-auger holes) along the length of the trench directly through the back-filled trench in order to define the geometry of the fault and the deformed strata

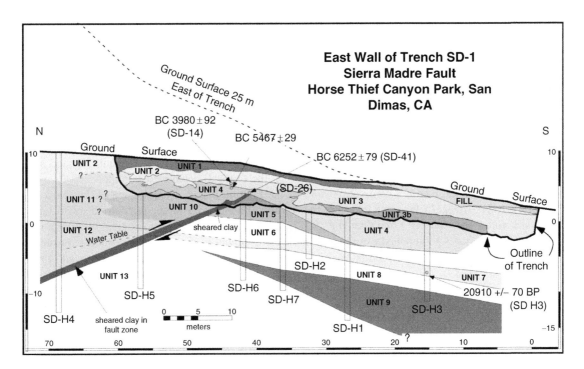

Figure 5.18: Cross section showing east wall of trench SD-1 and large-diameter borehole results from the Horsethief Canyon site. Large-diameter boreholes were drilled directly through back-filled material along original alignment of trench. For clarity, the borehole outlines are not shown where they cross the original trench. No vertical exaggeration. From Tucker and Dolan (2001).

below the trench. The walls of boreholes H1, H3, and H7 were examined directly by lowering a geologist downhole, whereas the remaining boreholes were described by examination of cuttings taken every 25–50 cm of borehole depth.

5.3.3   Structure and Evolution of Reverse-Fault Scarps

Reverse faults (dip >45°) typically form scarps steeper than the angle of repose of faulted materials (e.g., simple thrust scarps and hanging wall collapse scarps, Figure 5.8A and b). Tight fault-propagation folds can also produce steep scarps (Carver, 1987b). Both types of scarps may generate scarp-derived colluvium. This colluvium is deposited by gravity, debris, and wash processes as the scarp degrades by parallel retreat and slope decline, in a manner similar to that described for normal faults (Section 3.2.4). Scarp-derived colluvium exposed in trenches across reverse faults generally fines upward, but colluvial facies have not been characterized in as much detail as for normal faults (e.g., Nelson, 1992b). Nevertheless, some generalizations can be made based on published trench logs.

The basal component of colluvium (lower association of the debris element) is composed of rubble from the collapsed tip of the overhanging fault scarp (Figures 5.9, 5.19A and B). The rubble may include blocks of the prefaulting surface soil horizons as well as pieces of the hanging wall. Blocks of soils and hanging-wall strata typically possess steep or overturned bedding orientations. The fragile nature of many unconsolidated blocks suggest that they were rapidly shed from the overthrust fault tip and were never exposed to significant moisture or weathering. Such rapid burial of soil blocks, engulfment by a matrix, and subsequent preservation can be accomplished by wind and gravity transport of loose material exposed in the scarp face. Prehistoric collapsed-tip rubble was recognized by Bonilla (1973) in trenches across the 1971 San Fernando, California, rupture and was used to date the penultimate faulting event.

As a scarp face continues to degrade, smaller clasts and blocks and loose grains are deposited by gravity and debris processes (upper association of the debris element; Figure 5.19C). After burial of the free face, wash processes should dominate scarp decline and result in deposition of the finer wash element colluvium (Figure 5.19D).

Subsequent fault displacement thrusts the hanging wall over the colluvium on the lower part of the scarp and triggers a new episode of colluvial transport and deposition. Repeated cycles of fault slip and scarp degradation should produce *stacked colluvial wedges* on the footwall, each wedge composed of recognizable facies, that record a complete history of the faulting history over the age of the scarp (Figure 5.19E–H). (*Note*: In Japanese literature the couplet of a colluvial wedge and buried soil is termed the "D structure," see Chapter 3 and Okada *et al.* (1989), for an excellent chronological interpretation of multiple D structures.) Published examples of multiple colluvial wedges in reverse-fault exposures include Weber and Cotton (1980), Sarna-Wojcicki *et al.* (1987), Meghraoui *et al.* (1988), Swan (1988), and Okada *et al.* (1989).

Weber and Cotton (1980) further propose four main criteria to recognize repeated fault movements (see Figure 5.19E–H). First, each earthquake is represented by a colluvial wedge and buried soil. If displacement per event is < 0.3–0.5 m, however, colluvium may not be recognizable and the only evidence of faulting is thickening of soil horizons on the footwall (Weber and Cotton, 1980, p. 33). The second criterion is cross-cutting relations. Colluvial units overlie older faults, but are truncated at their upslope limits by younger faults (Figure 5.16E–I). Within the main fault zone, younger, steeper faults truncate older, flatter faults. Fault geometry often suggests an apparent upward and upslope migration of faulting in sediments of the footwall. Third, wedges of fault breccia are injected into overlying sedimentary layers (colluvium and alluvium) on the footwall (Figure 5.16, stages F, H). Fourth, fault

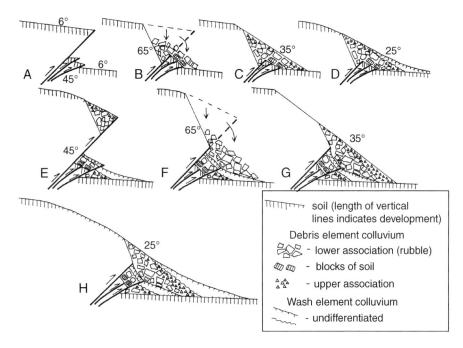

Figure 5.19: Schematic diagrams showing the deposition of colluvium from a two-event reverse-fault scarp. Vertical offset in each event is envisioned to be on the order of 0.5–3 m. (A) Initial faulting (45° dip) of a sloping (6° slope) geomorphic surface with a well-developed soil. Faulting creates an overhanging scarp; note two subsidiary thrusts at scarp base. (B) Over-hanging part of scarp collapses, forming an apron of rubble (compare to Figure 5.6, location B). The lower association of debris element colluvium is composed of blocks of soil and hanging-wall material. (C) Free face retreats and the finer, upper association of debris element colluvium is deposited. The colluvial apron stabilizes at the angle of repose (35°) once the free face is buried. (D) Wash element colluvium is deposited as the scarp declines (shown here at a maximum angle of 25°); a weak soil forms. (E) A second faulting event doubles the height of the scarp. The lower subsidiary fault is slightly reactivated, but the higher one is not. The proximal part of the colluvial wedge is translocated to the hanging wall of the rejuvenated scarp. (F) The overhanging part of scarp collapses. The debris element colluvium deposited after the second event thus includes material recycled from the proximal colluvial wedge from the first event. (G) The free face retreats and upper association debris element colluvium is deposited. (H) Wash element colluvium is deposited. The two faulting events can be deduced from (1) the existence of two colluvial wedges, (2) the fact that the lower wedge is in fault contact, but the upper wedge is in depositional contact, with the hanging wall, and (3) the subsidiary faults terminate upward at different stratigraphic levels.

breccia created by thrust-tip "bulldozing" can be interbedded with fluvial, eolian, or lacustrine deposits. The coarse rubbly deposits formed by breccia injection and by bulldozing of the fault tip may superficially resemble colluvial wedges, but they did not necessarily accumulate on a prefaulting ground surface, and thus their lower contacts are not necessarily earthquake horizons. The possible confusion in differentiating scarp-derived colluvium, injection breccia, and bulldozed debris beneath reverse faults points out the need for a future systematic study of reverse- and thrust-fault zone sedimentology, similar to that performed by Nelson (1992b) on normal faults.

Figures 5.20 and 5.21 show examples of structures and stratigraphic units associated with steep (>45°) reverse faulting. Figure 5.20 shows Tertiary mudstone and sandstone displaced by a zone of subvertical faults, where the uppermost parts of faults near the hanging wall (F1–F4) flatten to subhorizontal and interfinger with scarp-derived colluvium. Cross-cutting relationships follow the conceptual model of Figure 5.16 and indicate at least three displacement events. Figure 5.21 shows a reverse fault dipping about 55° where the leading edge of the fault also flattens to subhorizontal. Before the reverse-fault plane reaches the surface, however, much of the displacement has passed into folding, and as a result no free face was created here and no scarp-derived colluvium was deposited. Instead, the postfaulting deposition is dominated by thin layers of sheetwash which onlap the scarp face. However, even in the absence of coarse debris on the footwall, paleoseismic reconstruction can be based on identifying lenses of finer wash element colluvium and cross-cutting relationships, as was done by Meghraoui et al. (1988).

Lagerbäck (1990, 1992) describes how postglacial (ca. 9 ka) subaqueous reverse faulting in Sweden created a "colluvial wedge" by slumpage and flowage of till, which was later overlain with littoral deposits before Holocene emergence. The wedges are longer, thinner, and more undulatory (Lagerbäck, 1992, his Figure 7) than the subaerial wedges described in this chapter. He also suggests that subaqueous

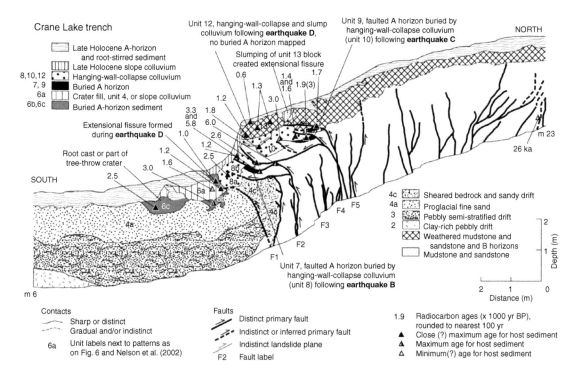

Figure 5.20: Simplified trench log of the west wall of the Crane Lake trench, Washington, USA. Highly weathered mudstone, sandstone, and colluvium were thrust along near-vertical faults up and over latest Pleistocene till (drift), Holocene colluvium, and A horizons during three, or possibly four, late Holocene earthquakes. From Nelson et al. (2003).

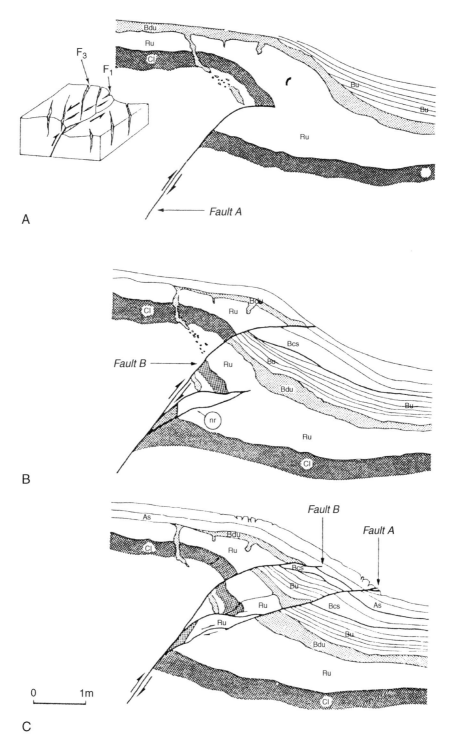

Figure 5.21: (Continued)

faulting liquefied and partially resuspended clasts in the glacial till, thus forming a distinctive "seismically graded till" that can be used as a paleoearthquake indicator. However, neither subaqueous wedges nor graded tills have been described following historic earthquakes, so the mechanics of their origin can only be inferred at this time.

5.3.4   Structure and Evolution of Thrust Fault Scarps

Thrust faults (dip <45°; often 30° or less) typically refract to even lower angles near the surface and form scarps or pressure ridges, the slopes of which are below the angle of repose. In many cases the surface cover of grass, turf, or peat is not broken except for minor tension cracks, even though the scarp may be up to several meters high (Figure 5.10C and D) Weber and Cotton (1980, pp. 28–29) describe a generic scenario that is useful for visualizing the complex process of thrust faulting in near-surface unconsolidated deposits:

> *The hanging wall block is thrust up and over the footwall to form a hump (pressure ridge) or moletrack scarp [Figure 5.22]. The hanging wall block slides out on the former ground surface so that soft sediments and soils are "bulldozed" up in front of the lip of the overthrust block. The leading edge of the overthrust block is a relatively thin wedge of soft sediments that is pushed along the ground surface [e.g., Figure 5.10D]. Eventually, the frictional resistance to movement along the base of the overthrust plate exceeds the strength of the sediments in the overthrust plate, and the movement is transferred from the original shear surface to a second or possibly more shear surfaces that splay off of the main fault break. During one faulting event, it is probable that several fault breaks will splay out from the same main fault. Many of these faults will flatten near the surface and become parallel to bedding on the footwall block. ... The soft sediments of the footwall ... are also deformed by drag underneath the (overriding) plate. Some of the sediments on the footwall block that are initially shoved ahead of the "bulldozing" upper plate do not maintain that position because the fault overrides the brecciated rock as the hanging wall block is pushed farther and farther out over the footwall block. ...*

> *The lip of the overthrust plate itself brecciates and breaks up into masses of mixed rock with a clayey matrix. Consequently, the near-surface portion of the fault zone consists of a wide zone of*

Figure 5.21: Reconstruction of faulting and deposition on the El Asnam thrust fault. (A) Thrust movement on fault A displaces beds Cl and Ru and creates a fault-propagation fold at the surface, defined by soil Bdu (light shading). Postfaulting sediments (Bu) onlap the scarp. (B) Renewed faulting propagates upward past the refraction point of fault A and creates fault B, which also refracts to a lower angle but at a higher level. Fault B, truncates onlapping units Bu and Bcs. A new subsidiary fault (nr) forms near the beheaded part of fault A. (C) During the 1980 earthquake, fault A is reactivated and reaches the ground surface, displacing soil As that covers the entire scarp. Fault B is not reactivated. Due to the small displacements and the tendency to form fault-propagation folds at this site, scarp-derived colluvium is mainly composed of the wash element. From Meghraoui *et al.* (1988); reprinted with permission of the Seismological Society of America.

Figure 5.22: Origin of small moletrack scarps on thrust faults. (A) Moletrack scarp formed by low-angle thrusting during the 2001 Bhuj, India earthquake; white-dashed line follows the scarp crest. Note cow for scale in far distance to left of dashed line. Faulted material here is Quaternary sandy alluvium that contains a hardpan soil profile about 30 cm thick. Moletrack height ranges from 20 to 30 cm; (B) Exposure of the thrust fault beneath the moletrack in a streambank at 23°34.700′N, 70°24.210′E. The 2001 thrust fault (between thick white arrows) dips 13° south and the sense of movement is shown by the thin white arrow. At the bottom of the hardpan layer the fault steepens to subvertical, probably following a subvertical crack caused by ground shaking, and creates a tent structure. From McCalpin and Thakkar (2003).

broken and mixed rock. . . . Another feature present in sub-fault masses of mixed and deformed rock is evidence of local "underthrusting" associated with the "bulldozing." Where jumbled masses of rock material on the footwall block seem to have been pushed ahead of the lip of the thrust, they are sometimes underlain by what appears to be the lip of the over thrust block that has acted in a manner similar to that of a bulldozer blade.

Based in part on these observations, Weber and Cotton (1980, pp. 30–37) proposed a model for recognition of multiple faulting events on thrust faults. The model assumes that the initial scarp form is a pressure ridge made of crushed material, subject to subaerial erosion and weathering. The major processes acting on the pressure ridge are assumed to be sheetwash and rillwash. The following stratigraphic relations should be observed after a single faulting event: (1) a thin deposit of wash element colluvium, (2) this colluvium buries the prefaulting soil, (3) the buried soil is offset by the fault plane, and (4) the soil horizon on the hanging-wall block is thin or absent near the fault and thickens away from the fault.

Figures 5.23 and 5.24 show the spectrum of deformation from low-angle thrust faulting. Figure 5.23 shows a thrust scarp accompanied by the development of a backthrust and a fault-propagation overturned anticline on the hanging wall. The fault zone changes deformation style as it passes upward from brittle cobbly gravel (single planar fault) to semi-ductile sand, (backthrust), and then to highly ductile mud, sand and soil (overturned anticline). Due to the flattening of the thrust fault near the surface, and development of these subsidiary structures, the dip-slip displacement on the thrust fault plane at the surface is too small to create a free face. There is some postfaulting debris overlying the fault tip, but it is

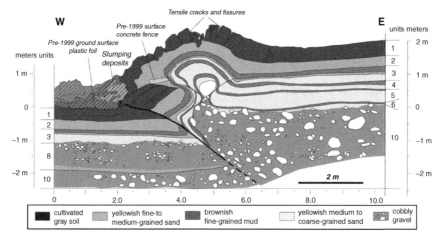

Figure 5.23: Log of the Wufeng trench, 1999 Chi–Chi rupture of the Chelungpu fault, Taiwan. Scarp is 2.5 m high. Trench exposes two distinct sedimentary deposits: a lower cobbly gravel fluvial deposit and an upper overbank sand deposit capped by a cultivated soil. At least three to four fining upward sequences are observed in the overbank sand deposits. Radiocarbon dating of charcoal samples in the overbank deposits (mud layers of Unit 3 and Unit 4) indicated a young depositional age of less than 200 years BP. Flat-lying upper layers (Units 1–3) maintain a uniform thickness across the thrust scarp. In contrast, the underlying layers showed a different depositional facies across the scarp (Unit 8 in the footwall against Units 4–6 in the hanging wall). Deformation structures include a major east-dipping basal thrust and several structures in the hangingwall, including a breakthrough wedge thrust, and a pop-up anticlinal fold bounded by two opposing secondary thrusts. From Lee et al. (2001).

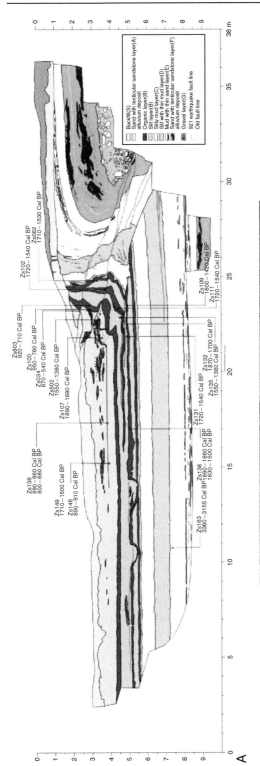

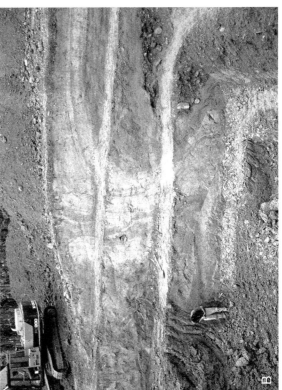

Figure 5.24: (Continued)

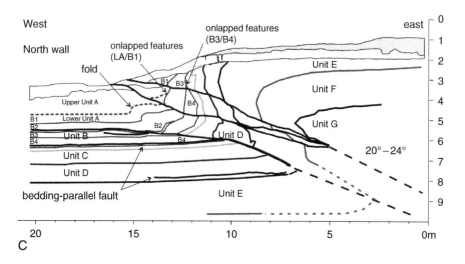

Figure 5.24: North wall of the Chushan trench across the Chelengpu fault, Taiwan. (A) The north wall shows a tight anticline with a steeply inclined forelimb on the hanging wall, and the footwall strata are unfolded. The deformation seems to be a footwall-fixed fault-propagation fold. Note how the lowest (main) thrust plane becomes parallel to footwall strata, and how brittle faulting on this strand constitutes a higher proportion of total deformation that the similar fault in Figure 5.22. (B) Photograph of the fault zone on the north wall. (C) interpretation of structures on the north wall, highlighting onlapped features on the vertical limb of the anticline. These onlaps predate the 1999 rupture and indicate prior topographic/structural relief here. From Chen et al. (2005).

slumping deposits from the oversteepened limb of the anticline, rather than hanging-wall collapse debris. This diagram shows well the interaction of brittle and ductile structures at a thrust fault tip. Figure 5.24 also shows the Chi–Chi rupture, but where the thrusting was more brittle and less plastic.

5.3.5   Stratigraphic Bracketed Offset

Where surface faulting occurs in active depositional settings, individual slip events are often recorded by bracketing strata. Such environments as river floodplains, shallow lakes and ponds, tidal marshland, or aggrading coastlines and dunes are likely environments for preservation of slip events and scarp growth by quick burial. Slip on the main fault (as well as on secondary faults, including normal faults in the hanging wall) produces offsets of preexisting bedding, surface colluvium, and soil. In many places postseismic grading of the scarp removes the constructional microtopography generated by the fault slip before subsequent sediments cover the site. However, occasionally a scarp is buried and details of its morphology are preserved in the stratigraphy.

The unconformity above the faulted sequence records the paleoearthquake and any postseismic modification of the surface trace prior to burial. Unfaulted sediments above the unconformity postdate the fossil earthquake. Where rapid sedimentation has persisted during multiple seismic cycles, a sequence of stacked *unconformity-bounded interseismic sediment packages* may result (Figure 5.25). Faults produced by each earthquake are truncated by the unconformity corresponding to the earthquake that produced them. Faults generated by previous earthquakes are truncated by unconformities and covered by sediments

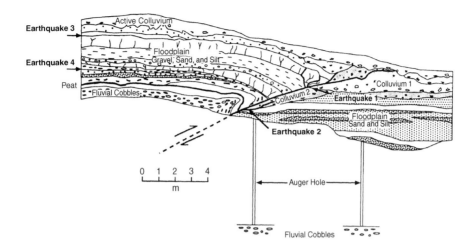

Figure 5.25: Trench log at the Blue Lake site across the McKinleyville fault, showing the effects of repeated thrust faulting in a depositional environment dominated by vertical fluvial accretion. Four paleoearthquakes are inferred. Earthquake 1 (most recent) and earthquake 2 (penultimate) are indicated by overthrust scarp-derived colluvial wedges (colluvium 1 and colluvium 2, respectively). These two events postdate the development of the scarp and the cessation of floodplain sedimentation. Prior paleoearthquakes (earthquake 3 and earthquake 4) occurred during vertical accretion of floodplain overbank sediments. Earthquake 3 is represented by fault splays that displace floodplain sediments below the earthquake horizon and are overlapped by younger sediments. Earthquake 4 (the oldest) is expressed as an angular unconformity that truncates beds that are overturned adjacent to the fault below the unconformity.

lower in the sequence. Faults that have recurrent movement exhibit greater displacement of lower unconformity-bounded sediment sequences. In settings where such sediment sequences and fault relationships are preserved, the deposits laid down across the fault must be thick enough to bury completely the scarp and lap across both the footwall and hanging-wall blocks. Where sedimentation rates cannot keep up with scarp growth, the stacked interseismic sediment packages are preserved only on the downdropped side of the fault, the elevated block is subject to weathering and erosion, and the surface trace is commonly covered by sheets of scarp-derived colluvium.

Surface fault displacement and resulting disturbance of the ground along the fault trace, often coupled with a marked localized increase in surface erosion rate immediately following the earthquake, can produce postseismic pulses of deposition followed by reduced intersesimic sedimentation and soil development. Thus, under ideal conditions unconformity-bounded sediment sequences may reflect these depositional rate changes. For example, coarse high-energy deposits, cut and fill structures, and other indicators of high-energy deposition in the base of a sediment package may grade upward into finer grained parallel stratified material containing a soil at the top.

5.3.6 Fault-Onlap Sedimentary Sequences

The surface expression of many thrust faults in thick unconsolidated sediments is not restricted to a narrow zone of faulting, but rather is commonly distributed across a broad zone on many small displacement faults and accommodated by broad warping and surface folding. Coseismic growth of broad warps in floodplains, marshes, and other *vertically accreting* depositional environments lowers the footwall block relative to the hanging wall and increases the prism available for sediment accumulation. At sites where the vertical displacement elevates the upthrown side of the fault above the limits of deposition (e.g., above the reach of overbank flood flow on a river floodplain or high tide in a tidal marsh), postseismic sedimentation produces layers of sediment that *onlap* the scarp (Carver and Burke, 1989). In many cases these sediments will not overlie any (truncated) surface faults that can be used to bracket faulting events. Restabilization of the landscape following deposition on the downdropped side of the fault is commonly followed by soil formation or peat accumulation (Figure 5.26). Repeated faulting produces *stacked onlap sequences* on the downfaulted side of the fault, each capped by a buried soil or peat layer. The sediments thin as they onlap the warped area along the fault and the soils merge into a single profile on the upthrown side. Abundant plant fossils and charcoal are sometimes present in such deposits and can provide ^{14}C samples useful in limiting the age of faulting events.

5.3.7   Summary of Stratigraphic Evidence for Thrust Paleoearthquakes

The type of stratigraphic evidence created during and after paleoearthquakes on thrust faults is determined by the geometry of near-surface faults and folds and the local depositional environment. Figure 5.27 shows one conceptual view of the continuum of fault-zone stratigraphic geometries. If surface materials are noncohesive and dry, discrete faults form. High-strength materials are typified by single fault strands, narrow scarps, and collapsed fault tips buried by scarp-derived material (lower left). In low-strength materials (e.g., loose sand) wide scarps are formed by distributed conjugate faults (upper left), and no single fault has enough relief to shed a preservable colluvial wedge. In moister or more cohesive surface materials, ductile deformation and folding increase. Sharp fault-propagation folds give rise to colluvial wedges (lower right), whereas open fault-bend folds are often onlapped by vertical accretion deposits (upper right). However, if a broad gentle scarp is formed in a dominantly erosional environment, no paleoseismic indicator strata may be deposited. Although Figure 5.27 is based only on examples from northern California, the general relations shown should apply to any area with similar surface materials and depositional environments.

5.3.8   Distinguishing Creep Displacement from Episodic Displacement

The recognition that reverse faults may creep at the surface was made in the late 1980s for the central Longitudinal Valley fault in Taiwan, which remains the world's most-studied creeping reverse fault. Between 23°N and 23.5°N about 20 mm/yr of differential movement is currently being accommodated by creep across this fault, but in a zone that varies from 100 m to several km wide according to InSAR measurements (Yarai *et al.*, 2006). In the 50-year period between the 1951 M_w 6.2 earthquake and the 2003 M_w 6.8 Chengkung earthquake, this fault experienced creep of 20–30 mm/yr, based on creepmeter measurements (daily), dense campaign GPS surveys near the fault, and land-based surveying since 1989 (Lee *et al.*, 2008). The creep appears to be coming from the upper 0–5 km of the fault plane, which did not rupture in the 2003 earthquake (Wu *et al.*, 2006). However, no trench studies in this part of the fault have addressed the problem of differentiating creep displacement on the fault plane from episodic displacement. In the absence of such studies, we can only assume that the creeping model and criteria proposed for oblique-slip faults by Lienkaemper and Williams (2007; see Chapter 6) should be used to distinguish coseismic displacement from creep displacement in reverse-fault exposures.

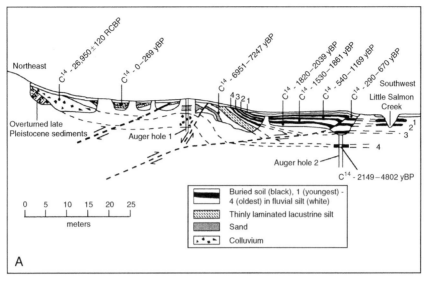

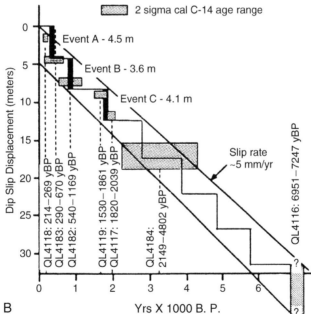

Figure 5.26: (A) Generalized trench log across the west trace of the Little Salmon fault in northern California. The fault reaches the surface in large part as a growth fold above the fault tip. Progressively deformed sediments capped by soils [(1 youngest), to (4 oldest)] onlap the fold and fault tip. These soils are interpreted to record the coseismic growth of the fold and propagation of the fault tip with each slip event. (B) The slip history diagram shows the late Holocene slip rate and paleoseismic history for the last three events represented by the stratigraphy exposed in the trench.

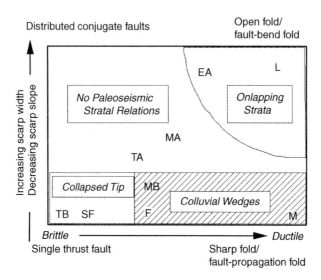

Figure 5.27: Types of stratigraphic evidence of paleoearthquakes associated with various combinations of surface deformation style and scarp morphology along thrust faults. Large bold letters indicate trench logs that show indicated features. Logs in this chapter include M, Mad River fault, School Road trench (Figure 5.28); L, Little Salmon fault, Little Salmon Creek trench (Figure 5.26); MB, McKinleyville fault, Blue Lake trench (Figure 5.25); EA, El Asnam trench, Algeria (Figure 5.21); SF, San Fernando, California, fault scarp (Figure 5.9); F, Frijoles fault exposure, California (Figure 5.17). Other trench logs (MA, McKinleyville fault, Airport trench; TA, Trinidad fault, Anderson Ranch; TB, Trinidad fault, Jager property) are shown in Woodward-Clyde (1980). Modified from Carver (1987b).

The issue of creep on secondary faults (bending-moment and flexural-slip faults) and on blind thrusts is described in other sections of this chapter dealing with those structures.

5.4   Dating Paleoearthquakes

Dating paleoearthquakes on reverse/thrust faults is broadly similar to dating paleoearthquakes on normal faults, as described below.

5.4.1   Direct Dating of the Exposed Fault Plane

To our knowledge no exposed reverse or thrust fault plane has been directly dated in a paleoseismic study, as is now done routinely for normal fault planes (Chapter 3). This is because coseismic reverse-fault planes are rapidly covered by hanging-wall collapse debris and are thus not exposed subaerially for long periods of time.

5.4.2   Direct Dating via Scarp Degradation Modeling

Morphological dating of reverse scarp profiles has been performed using the same diffusion equation technique as for normal faults (e.g., Carretier *et al.*, 2002). The approach assumes that an overhanging

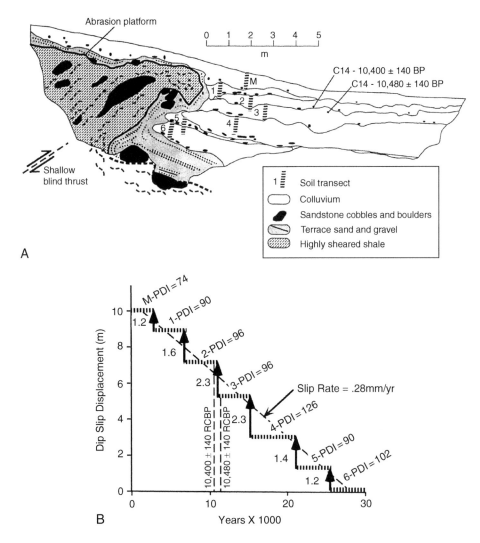

Figure 5.28: (A) Log of a trench across one of the traces of the Mad River fault where it displaces a late Pleistocene terrace near Humboldt Bay, California. At this site, underlain by very ductile highly sheared shaley melange, the thrust is blind and displacement reaches the surface as a sharp overturned fault-propagation fold. The marine terrace platform (heavy line) and overlying terrace sand and gravel are overturned in the forelimb of the fold. Six-stacked colluvial sheets, each with a stone line at the base and containing a buried soil (soil profiles 1–6), extend downslope from the overturned fold limb. A modern active colluvial sheet mantles the scarp at the surface. The colluvium is interpreted to have formed by slope processes on the scarp face between slip events. Coseismic growth of the fold preserved the colluvial layers below the overturned limb and generated new colluvial sheets that buried the downfaulted lower part of the scarp. (B) Diagram of the slip history deduced from the buried soils and two AMS ^{14}C ages on small pieces of detrital charcoal. Soil parameters measured in the field and lab were used to calculate a profile development index (PDI) for each colluvial layer using the methods described by Harden (1982). We assume that the initial (parent material) PDI of each scarp-derived colluvial deposit was equal to the PDI of the underlying soil, because that soil should have been exposed on the scarp crest after faulting and rapidly reworked into the next higher colluvial wedge. The ^{14}C ages are plotted relative to their stratigraphic position and provide rough calibration for the PDI-based slip history.

fault scarp is produced coseismically and then hanging-wall collapse follows rapidly to form an angle-of-repose debris slope. Once this slope is attained then diffusion-type slope decline follows, just as in normal fault degradation modeling. As for normal faults, diffusion-type dating can only yield a scarp age for single-event fault scarps. For multiple-event fault scarps the method yields a vertical slip rate, if one can assume a mass diffusivity value by independent means (Carretier *et al.*, 2002).

5.4.3   Age Estimates from Soil Development on Fault Scarps

Surface weathering profiles (soils) are commonly developed in the upper parts of the colluvial sheets shed from fault scarps. These soils become buried when additional faulting results in the formation of new colluvial sheets. Soil formation on growing thrust scarps is complex and not thoroughly investigated. We know that different soil properties characterize different slope positions on the scarp (Burke and Carver, 1989). The upper parts of scarps are areas of erosion of the uplifted fault tip. Where thrust faults cut upland terrain and are not influenced by off-fault deposition, erosion of the scarp crest is cyclic and driven by coseismic displacement of the fault. Each erosion cycle is initiated by a faulting event and proceeds from rapid gravity-dominated transport of debris off the new fault tip in the immediate postseismic interval, to slow creep of weathered material from the degraded fault tip prior to the next event. Over many seismic cycles, weathering products generated in scarp crest area are removed during downslope transport of the colluvium and the profile development at the crest is minimal.

Below the thrust the colluvial sheets and their weathering profiles are preserved by overthrusting and resulting burial by the new colluvium. The result over multiple seismic cycles is the accumulation of stacked buried colluvium with buried soils reflecting the weathering that accrued on the lower part of the scarp during each interseismic interval (Figure 5.19H). Where these weathered materials accumulate at the toe of the scarp the profiles are thickest and most strongly developed. Below the toe the soil profiles on stacked colluvium may merge into a single soil, as occurs in normal fault scarps (Chapter 3). Within a sequence of stacked soils the profile development generally increases with lower slope position (a similar phenomenon was described for normal faults in somewhat more detail; see Section 3.4.2).

As in normal faults, the age of interseismic deposits on the hanging wall of a reverse fault can be estimated from semiquantitative soil parameters such as profile development index (PDI). Figure 5.28A shows an example of six interseismic deposits in front of a shallow blind-thrust fault (Mad River fault, northern California), each with its own buried soil profile. The development time required for each paleosol was estimated from its PDI and an empirical relationship between PDI and time (Figure 5.28B). At this locality PDI of each successive soil in the stack increased upward, reflecting either (1) steadily increasing recurrence interval with time or (2) progressive recycling of hanging-wall soil profiles eroded and redeposited on the hanging wall. In the slip history diagram the thickness of each interseismic deposit has been used as a surrogate for vertical uplift (displacement) on the blind reverse fault, which is probably on oversimplification.

5.4.4   Bracketing the Age of Faulting by Dating Displaced Deposits

Many reverse/thrust faulting chronologies have been reconstructed from numerical ages on colluvial wedge or onlap sediments, and on the underlying faulted stratigraphy. The choice of dating method and sampling strategy are dictated by the characteristics of sediments and soils in the fault zone, as

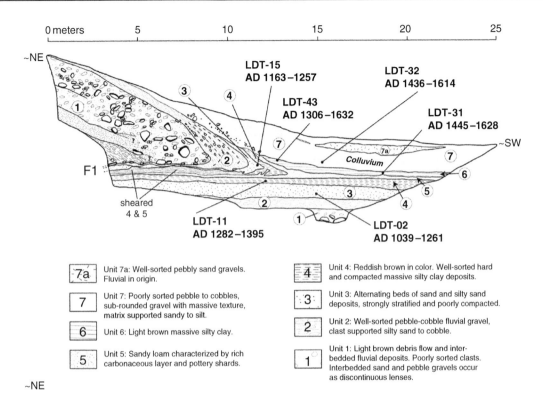

Figure 5.29: Recumbent anticline that has been thrust horizontally over flat-lying strata of the footwall. Units 1–5 are involved in the folding and thrusting, with unit 5 being split by the bulldozing advance of the fold across the ground surface (top of unit 5). The youngest date from the deformed sequence is AD 1282–1395. Unit 6 is the first postthrusting colluvium, dated at AD 1306–1632 and AD 1445–1628. Lal Dhang trench on the Himalayan Frontal Thrust, from Kumar *et al.* (2006).

explained in Chapter 3, Section 3.4.6. Figure 5.29 shows a recumbent fold where the fault plane is difficult to see, because the overturned fold has "bulldozed" its way across the preexisting ground surface, actually splitting unit 5 at the leading edge of the fold. This structure was formed by the processes described above by Weber and Cotton (1980). The trench log shows typical age sampling locations in a thrust fault exposure and how sample ages constrain the age of the earthquake horizon, that is, the unconformity between postfaulting scarp-derived colluvium (units 6, 7) and the prefaulting soil (1–5). The age of the faulting event is most closely bracketed by samples LDT-31 and LDT-43 (postfaulting), and LDT-11 (prefaulting) in Figure 5.29. An even closer maximum age constraint would be the age of unit 5. If taken at face value, these bracketing ages indicate the faulting event occurred after AD 1282–1395 and before AD 1306–1632. These age ranges overlap in the period AD

1306–1395. However, part of the organics dated in unit 6 may have been eroded from a prefaulting soil profile on the hanging wall, and thus may also predate (rather than postdate) the earthquake.

5.5   Interpreting the Paleoseismic History by Retrodeformation

The displacements attributable to individual paleoearthquakes, as well as their timing, are typically estimated from a 2D retrodeformation analysis of the trench log. Reverse/thrust fault retrodeformations fall into two categories: (1) rigid-block retrodeformations, such as used on normal faults, where fault-bounded blocks are restored via translation and rotation and (2) plastic restorations that reverse ductile folding and use line-length balancing or area-balancing techniques.

5.5.1   Rigid-Block Retrodeformations

Figure 5.30 shows a pseudo-rigid block retrodeformation of high-angle reverse faults during three (or four) faulting events. This retrodeformation sequence includes the removal of scarp-derived colluvium after faulting events B (unit 8), C (unit 10), and D (unit 12), as well as the reversal of movement on several strongly curved reverse faults. As for all retrodeformations, the sequence must properly account for all cross-cutting relationships between fault strands and scarp-derived colluviums.

5.5.2   Plastic Retrodeformations

Where low-angle thrust faulting is expressed by parallel folds at the surface, the principles of balanced cross-sections can be used to "unfold" the fold. There are two main techniques used: line-balancing measurement and area-balancing calculation. These two common balancing techniques have been used widely to balance geological cross sections at regional scales (e.g., Ragan, 1985).

Figure 5.31 shows the Wufeng trench log described earlier (Figure 5.23). The difference in elevation of the beds between points across the fault zone is a relatively consistent 2.2 m, which reflects the vertical component of faulting. To estimate the horizontal component, Lee *et al.* (2004) applied two different restoration techniques. Figure 5.31 shows the *line-length balancing method* along the depositional contacts to estimate the amount of the horizontal displacement. Five depositional contacts were divided into line segments, which were separated by faults and tensile fissures. From top to bottom, the summed lengths are 13.2, 13.2, 13.8, and 13.0 m for the bottoms of units 1, 2, 3, and 4, respectively (mean = 13.3 ± 0.3 m). Compared to the postfaulting length of the profile of 10.0 m, this indicates that thrust faulting caused 3.3 ± 0.3 m of horizontal shortening parallel to the exposure.

Area-balancing is a complementary method for estimating the amounts of horizontal displacement across folds. The operative assumption is that the area of an individual bed folded during the earthquake would be the same after the earthquake, assuming no significant change in density of these young sediments. Individual layers are restored to a rectangular shape, with the same thickness as present, assuming no significant pure shear occurred during the earthquake. The length of this restored rectangle is thus equal to the original length of the layer.

For the Wufeng exposure, Lee *et al.* (2004) obtained different horizontal displacements between the line-balancing and area-balancing methods. Area balancing the three uppermost, highly plastic units only yielded 4.8 ± 1.0 m of horizontal displacement. Alternatively, balancing the entire 3-m-thick package of deposits yielded about 2.6 ± 0.3 m of horizontal displacement. The line balancing, on the other hand,

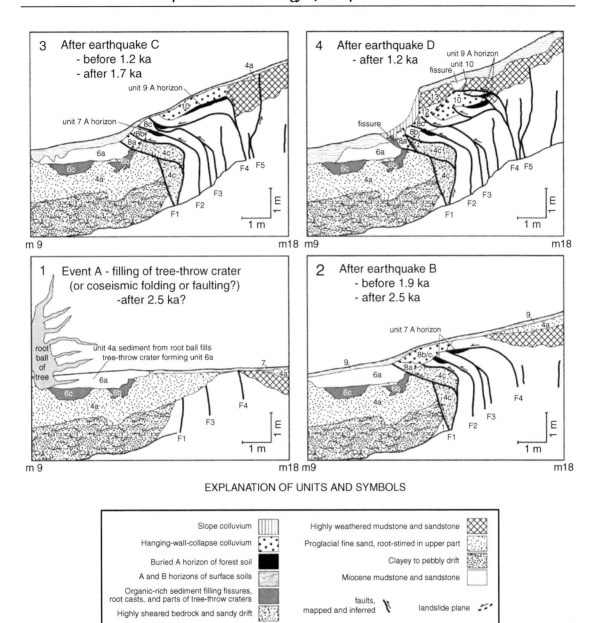

EXPLANATION OF UNITS AND SYMBOLS

Slope colluvium		Highly weathered mudstone and sandstone	
Hanging-wall-collapse colluvium		Proglacial fine sand, root-stirred in upper part	
Buried A horizon of forest soil		Clayey to pebbly drift	
A and B horizons of surface soils		Miocene mudstone and sandstone	
Organic-rich sediment filling fissures, root casts, and parts of tree-throw craters		faults, mapped and inferred landslide plane	
Highly sheared bedrock and sandy drift			

Figure 5.30: Schematic reconstruction of paleoearthquakes at the Crane Lake trench site, Washington, inferred by pseudo-rigid retrodeforming the stratigraphic units. Panel 1 shows how unit 6a may have formed through filling of a tree-throw crater. Alternatively, 6a might have been deposited as a scarp-derived colluvial wedge following folding or faulting of Miocene bedrock, pebbly drift, and proglacial sand (unit 4a) prior to earthquake B. During earthquake B (panel 2), bedrock and proglacial sand were thrust upward and over drift, sand, and root casts and a tree-throw crater, slivering the A horizon of a forest soil (unit 7). The hanging wall then collapsed, burying the soil with the first hanging-wall collapse colluvium (unit 8). During earthquake C (panel 3), most slip probably

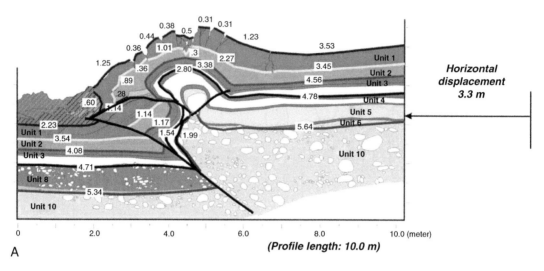

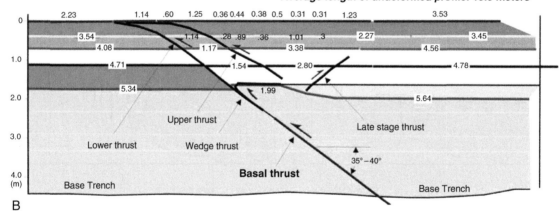

Figure 5.31: Restoration of the Wufeng scarp and pop-up anticline, 1999 Chi–Chi fault scarp, based on measurement of line-lengths of five depositional contacts on the footwall and hanging wall. (A) Numbers indicate line lengths of each segment of the five depositional contacts (m); segments are separated by faults or tensile cracks. The current length of the profile, 10.0 m, is measured from the trench log; (B) Retrodeformed section made by unfolding the line segments shown in (A).

This plastic retrodeformation restores all beds to horizontal (ignores any primary dip) and straightens out the lower thrust slightly and the upper thrust considerably; the late stage thrust and wedge thrust remain unchanged. Fault trajectories are shown by red arrows. From Lee et al. (2004).

occurred on faults F2–F4, truncating the A horizon of a new forest soil developing on the scarp (unit 9), burying the soil with a second hanging-wall collapse colluvium (unit 10). Slip on fault F4 during earthquake D (panel 4) slivered the A horizon (unit 9), faulted the second collapse colluvium (unit 10), and thrust a block of weathered bedrock (unit 13) out over it. The block then slumped, producing the third collapse colluvium (unit 12) and a head-scarp fissure. From Nelson et al. (2003).

yields 3.3 ± 0.3 m of horizontal displacement. Because of large standard deviation, they rejected the 4.8 m estimate and adopted 2.6–3.3 m as bracketing the true horizontal component of displacement.

5.6 Distinguishing Seismogenic from Nonseismogenic Reverse Faults

In Chapter 2A (Section 2A.4) we argued for the importance of distinguishing tectonic, seismogenic faults from nontectonic and nonseismogenic faults. There are several types of nontectonic and nonseismogenic reverse faults that can deform the ground surface and shallow geologic deposits, as described below. Criteria for distinguishing such faults from seismogenic faults are described by Hanson *et al.* (1999), which we summarize below.

5.6.1 Tectonic, but Nonseismogenic Reverse Faults

Into this class of faults fall two types of "passive" or secondary faults related to larger reverse faults and folds. In the compressional environment, the most common secondary faults are bending-moment faults and flexural-slip faults. These faults may move simultaneously with the seismogenic fault, but they root at such a shallow depth that they produce negligible seismic moment even then. They do not generate $M_w > 5.5$ earthquakes by themselves.

Paleoseismic histories derived from secondary faults may be good proxies for events on the underlying thrusts, especially where the underlying seismogenic thrust does not extend to the surface. The paleoseismic indicators for these faults are similar to those for active normal faults and may include the development of colluvial wedges and faceted compound scarps. However, the amount of secondary fault displacement on them may be difficult to relate to slip on the underlying seismogenic reverse fault, as described later.

5.6.1.1 Flexural-Slip Faults

Flexural-slip faults are bedding-plane faults created by differential slip between the strata of synclines and anticlines, due to the requirement that bed length not change during folding. The tighter the fold, the larger the differential displacement between adjacent beds. On the upright limbs of a syncline, flexural-slip faults have a reverse sense of slip, but this changes to an apparent normal sense of slip on overturned limbs (e.g., Figure 5.32). As Yeats pointed out (1986b, p. 67) "there are no known examples of flexural-slip faults formed by aseismic creep. However, there are examples of such faults accompanying earthquakes at Lompoc, California; El Asnam, Algeria [(Philip and Meghraoui, 1983]; and possibly Inangahua, New Zealand [Lensen and Otway, 1971]." Trenching of the Ragged Mountain fault in Alaska in 2006, however, revealed that footwall flexural-slip faults have a creep signature in trenches, whereas the adjacent master thrust scarp and secondary bending-moment normal faults have a signature of episodic displacement (discussed later).

Because flexural-slip faults are upthrown toward the synclinal axis, and modern streams typically flow toward synclinal axes (e.g., Ota and Suzuki, 1979; Rockwell *et al.*, 1984), flexural-slip fault scarps often face upstream and pond local drainage. A classic area for Quaternary flexural-slip faults is the Grey-Inangahua depression in New Zealand, where a swarm of eight parallel flexural-slip fault scarps displaces two fluvial terraces of different ages; scarps are 1.5 m higher on the older terrace (Figure 5.33). The height difference was inferred to reflect the displacement during a single faulting event (Yeats, 1986a).

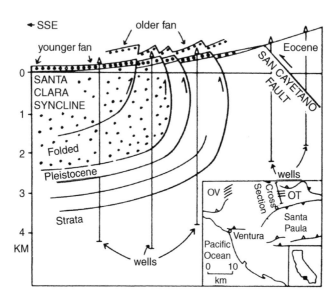

Figure 5.32: Diagrammatic cross section of flexural-slip faults in the Ventura Basin, California. Southward-directed thrusting by the San Cayetano fault (at right) produces tightening of the overturned limb of the Santa Clara syncline (center). Tightening of the fold then produces differential slip between the near-vertical bedding planes on the fold limb, which creates fault scarps at the surface. Note how the older fan has been deformed more than the younger fan. From Yeats (1986b); reprinted from Active Tectonics. Copyright © 1986 by the National Academy of Sciences. Courtesy of the National Academy Press, Washington, DC.

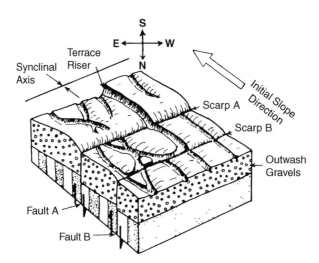

Figure 5.33: Block diagram of flexural-slip fault scarps displacing late Pleistocene outwash terraces at Giles Creek, New Zealand. This diagram shows a surface geometry similar to that of the faulted fans at the top of Figure 5.17. From Yeats (1986b); reprinted from Active Tectonics. Copyright © 1986 by the National Academy of Sciences. Courtesy of the National Academy Press, Washington, DC.

Even larger, multiple-event (?) scarps (up to 12 m high) are found nearby at Blackball, New Zealand. Yeats (1986a,b) noted that the local drainage had ponded against the base of the flexural-slip scarps (Figure 5.33) and argued that such ponding could only be produced by sudden, coseismic rise of the scarp (as opposed to more gradual creep, which would have been unable to defeat the streams). Similar flexural-slip scarps in the Ventura Basin, California (Figure 5.32), displace multiple alluvial fan surfaces dated between 4–5 and 200 ka. Rockwell *et al.* (1984) and Rockwell (1988) used the increasing fault scarp heights and tilting in successively older deposits to reconstruct the chronology of Quaternary folding.

Yeats (1986b) also points out that flexural-slip faults, because they are restricted to folds and their displacement decreases to zero at the fold axis, are probably not seismogenic structures themselves. Rather, flexural-slip faulting results from coseismic (usually synclinal) folding, which itself is caused by coseismic fault displacement on a reverse fault that lies beneath, or adjacent to, the fold. An example of closely spaced reverse-fault scarps, bending-moment faults on the collapsed thrust tip, and flexural-slip faults on a footwall syncline, is the Ragged Mountain fault in Alaska (Figure 5.34). When Figure 5.34B appeared in our 1st Edition, it began a controversy over whether the scarps shown were bending-moment faults on the crest of an anticline, or sackung. Detailed mapping and trenching in 2006 as part of the St. Elias Erosion/Tectonics Project (Figure 5.34A) proved that the scarps were flexural-slip faults in the footwall of the Ragged Mountain thrust. The reverse-fault scarp itself is very hard to identify, with the most prominent range-front landform being an upslope-facing, bending-moment normal fault formed by collapse (breakaway) of the overhanging fault tip. This sequence of normal bending-moment fault, obscure thrust scarp, and flexural-slip scarps (going from hanging wall to footwall) is nearly identical to the sequence formed in the 1980 El Asnam earthquake, albeit in a very different climate.

5.6.1.2 Bending-Moment Faults

Bending-moment faults have also been observed to form in large historic thrust-fault earthquakes, most notably in the 1980 El Asnam, Algeria, earthquake (King and Vita-Finzi, 1981). In some places the normal fault scarps ("extrados fractures" of Philip and Meghraoui, 1983) that formed on the anticlinal crest in 1980 were clearly rejuvenations of earlier, degraded normal fault scarps that had formed in paleoearthquakes. Such well-formed normal fault scarps can be trenched via the techniques of Chapter 3, and the resulting paleoseismic history could then be used as a surrogate (albeit possibly incomplete) history for the underlying, more poorly expressed thrust.

McCalpin and others (in prep.) attempted to reconstruct the Holocene paleoearthquake chronology of the Ragged Mountain thrust, Alaska, by trenching the breakaway bending-moment normal fault in the hanging wall (Figure 5.34A). Their "control" trench across the reverse scarp revealed three faulting events in the past 18 ka. In contrast, the trench across the breakaway scarp/graben revealed only two faulting events in the past 16 ka, and three events in the past 32 ka. Therefore, as expected not every thrusting event was unambiguously expressed as bending-moment displacement in the breakaway zone. However, the stratigraphic/structural indicators of discrete faulting events (colluvial wedges, fissure fills) were actually clearer in the graben trench than in the thrust fault trench, and due to the fact that the graben was a sediment trap, there was more abundant organics for ^{14}C dating and sine sediments for OSL dating. Therefore, the dating control for the latest two ruptures was taken from the normal fault trench rather than from the thrust fault trench.

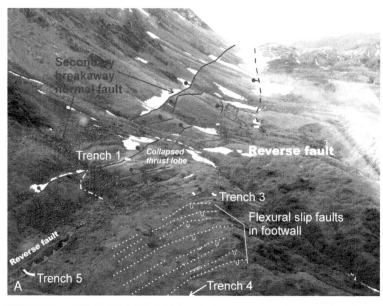

Figure 5.34: Flexural-slip faults in the footwall of the Ragged Mountain fault, Alaska. (A) The reverse-fault scarp (thin line with triangles on hanging wall) is difficult to see, but becomes easier to see when it protrudes out over the footwall, because of the multiple upslope-facing bending-moment scarps that form in the overextended and collapsed thrust lobe. The prominent breakaway normal fault is also a bending-moment fault. On the footwall, overturning of the footwall syncline has formed numerous parallel flexural-slip scarps. Photo by J. P. McCalpin. (B) original photo of the flexural-slip scarps from our 1st Edition, looking in the opposite direction to part (A). Shadows highlight four large, and many small, upslope-facing scarps. Sediments ponded on the upslope side of the scarps displayed evidence of continuous creep displacement, according to criteria described previously in the chapter. Photo by G. A. Carver.

5.6.2 Nontectonic, but Seismogenic Reverse Faults

It is unclear whether any faults of this type actually exist. If they do, they would exist as the toe thrusts of huge gravity slide blocks on the flanks of volcanic edifices. These gravity slides penetrate deep enough in the crust to generate $M_w > 5.5$ earthquakes (e.g., Kalapana slide on the Island of Hawaii, which generated an M_w 7.2 earthquake on 29 November 1975); similar faults exist on the eastern flank of Mount Etna, Sicily. The toe thrusts for these slides lie in deep water and could only be studied via the techniques in Chapter 2B. The key diagnostic criterion for relating these thrusts with gravity slides would be the geomorphology of the on-land part of the gravity slide, with the headscarp trace curvature in plan view (similar to that of landslide headscarps) and its location on the flanks of very large volcanoes.

5.6.3 Nontectonic and Nonseismogenic Reverse Faults

This class of faults includes a diverse array of compressional surface processes, some of which are restricted to shallow unconsolidated deposits, and none of which penetrate deeply enough into the crust to generate $M_w > 5.5$ earthquakes.

5.6.3.1 Landslide Faults

At the toe of a landslide, the main failure plane has the geometry of a thrust fault (see Cotton, 1999). Due to the compression at the toe of a landslide, many of the landforms associated with thrust faulting exist (thrust scarps, pressure ridges). Figure 5.35 shows a trench log across a landslide toe in Tertiary volcanic rocks in Utah (McCalpin, 2005b). The toe thrust shear zone is more than 1 m wide, with the most prominent clayey gouge zone on the footwall side, and several similar internal gouge bands. The shear zone thrusts a block of Tertiary andesite in the toe of a young landslide over an older, compacted Quaternary landslide deposit. The thrusting action induced development of several parallel, sympathetic

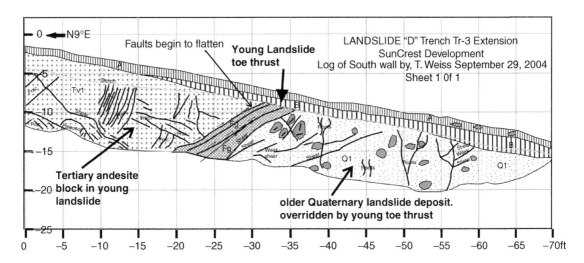

Figure 5.35: Trench log through a late Quaternary landslide toe thrust, illustrating its similarities to tectonic thrust faults. Scale is in feet, no vertical exaggeration. From PSI (2005).

shears in the footwall within 6 m of the toe thrust. The key diagnostic criteria for landslide toe thrusts are their location at the toe of anomalously hummocky topography with a distinctive arcuate scarp at the head; otherwise, they might easily be mistaken for tectonic faults. In fact, because landslides commonly obscure reverse fault traces at the base of steep range fronts, it is sometimes difficult to tell whether reverse faults in range-front trenches (such as at the base of the San Gabriel Mountains, southern California) are tectonic thrusts or landslide toes.

A recent European example illustrates the ambiguity, and the result on seismic hazard assessment. In Portugal, Fonseca *et al.* (2000) interpreted a thrust fault in a trench as evidence for Holocene faulting, which they expected "to have great impact of the seismic hazard assessment of Lisbon." However, they noted that striations on the fault plane indicated north-south compression, instead of the usual northwest-southeast compression observed in western Portugal. In contrast, Cabral and Marques (2000) claim that the thrust fault exposed in the trench is the toe thrust of a landslide, which explains the anomalous compression direction, and conclude the exposed fault has no tectonic or seismogenic significance. Fonseca *et al.* (2001) replied and dispute the landslide origin, but admitted that "conditions near a thrust fault are frequently favorable to landslides." Most of Fonseca *et al.*'s (2001) evidence for a tectonic fault origin is the existence of specific structural features which have been observed in other trenches across known tectonic thrust faults. However, the author has observed most of these structural features also in trenches across landslide toes, something Fonseca *et al.* (2001) do not mention, perhaps because they have not trenched many landslide toes. Thus, at this time the ambiguity remains whether the fault exposed in the Portuguese trench is tectonic and seismogenic, or nontectonic and nonseismogenic.

5.6.3.2 Subsidence/Collapse Faults

Differential subsidence of surficial materials, from a variety of causes, can create reverse faults in nonductile deposits. Common causes of differential subsidence are compaction due to fluid withdrawal and collapse or sagging of surficial deposits into subsurface void spaces. Kinematic models indicate a compressional stress field in the center of a downwarped beam, something observed in areas of severe subsidence associated with fluid withdrawal. Although many faults on the rim of subsidence areas initially form as vertical faults, continued subsidence induces tilting of these faults toward the center, giving them an apparent reverse geometry (Gutierrez *et al.*, in press).

5.7   Hazards Due to Reverse Surface Faulting

In Chapter 9, (See Book's companion web site) we discuss in general how the style of surface rupture has been quantified and then used to define surface faulting hazards to buildings. Recent work on thrust faults by Kelson *et al.* (2001) is worth mention in that respect. They correlated rupture style and building damage on the 1999 Chi–Chi scarp, Taiwan and concluded the following. The narrowest zones of damage were associated with reverse-fault planes that had a constant near-surface dip, regardless of whether surface materials were brittle and formed a hanging-wall collapse scarp (Figure 5.36E) or ductile and formed a narrow pressure ridge scarp (Figure 5.36A). Narrow deformation zones were also produced by reverse faults that created narrow pop-up folds (Figure 5.36D), such as at Wufeng (Figures 5.23 and 5.31). This particular geometry may be the result of near-surface steepening as shown in Figure 5.36D. However, an additional factor may be the presence of thin (<2 m), weak/plastic floodplain sediments over strong/brittle gravels, which caused "decoupling" of the shortening deformation between the two packages.

Where subsurface complexities occurred on the thrust the damage zone was wider, in proportion to the depth of the complexity. This includes shallowing of dip on the main reverse fault (Figure 5.36C), which

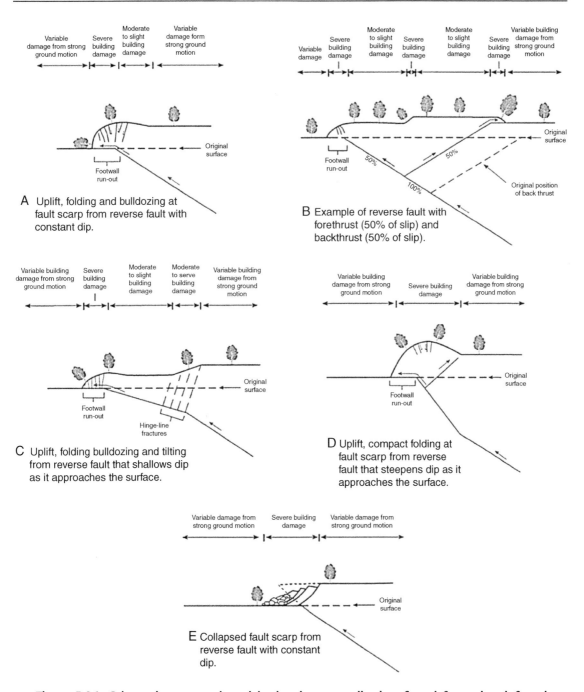

Figure 5.36: Schematic structural models showing generalized surface deformation, inferred subsurface fault geometries, and building damage produced by the 1999 Chi–Chi earthquake. (A) Reverse fault with constant dip. (B) Reverse fault with backthrust. (C) Reverse fault that decreases in dip toward the surface. (D) Reverse fault that increases in dip toward the surface. (E) Hanging-wall collapse scarp. From Kelson *et al*. (2001).

creates a wide zone of footwall runout ("bulldozing"), tilting, and hinge-line fracturing. If a constant-dip reverse fault had reactivated shallow backthrust (Figure 5.36B) the rupture created a broad pop-up structure with severe damage at both margins and in the middle. In both cases, the width of the damage zone could have been predicted if the depth of and dip of the structural complexity had been known. Thus, in theory rupture style could form the basis for land-use regulations, as did models of normal faults described in Chapter 9, (See Book's companion web site).

5.8   Paleoseismic Evidence of Coseismic Folding

The growth of broad coseismic surface folds has been observed during many large compressional earthquakes and is often the only surface expression of thrust displacement on underlying blind faults, the slip on which dies out tens to hundreds of meters below the surface ("blind thrusts"; Stein and King, 1984). Coseismic folding results in subtle horizontal, vertical, and tilting movements of the ground surface that reflect the geometry of the underlying fault and which, given certain conditions, can be preserved as geologic evidence. Detecting these "hidden" seismogenic blind thrusts has become a major applied goal of paleoseismology, particularly in densely populated regions, because they may generate unanticipated future earthquakes.

The theoretical understanding of fault-propagation and fault-bend folds is relatively well established (Suppe, 1983). Dislocation models for rupture on buried thrusts predict general uplift of the area above the rupture surface resulting from thrust transport and internal flexing of the hanging-wall block. The models also predict elastic thinning and subsidence of a broad backstop region behind the downdip limit of fault rupture (Suppe, 1985; Marshall *et al.*, 1991). During large blind-thrust earthquakes these areas undergo vertical displacement of up to several meters for very large earthquakes. The size of these areas of uplift and subsidence varies with the size of the rupture surface, dip of the fault, and depth and amount of slip. Any geometric complexity of the thrust complicates the pattern of surface deformation, as seen on a small scale in Figure 5.36. Abrupt changes in dip of the thrust, or reversals of dip direction (fault wedges), produce upward-propagating axial surfaces in overlying anticlines that grow during fault slip.

In the past two decades paleoseismic study of active surface folds has undergone the greatest advances in techniques and interpretation of any topic in paleoseismology, and deservedly so. Schwartz (1988a, p. 31) remarked three decades ago that "It is unclear, at least at present, how earthquake recurrence intervals can be well constrained in folded deposits or how we can distinguish folds that are surface expressions of seismogenic faults and those that are not. This is certainly an area for future research." The breakthroughs that led to these advances rested partly on technologies barely developed in 1988, such as digital elevation models, free satellite imagery (Google Earth), precise GPS surveying, LIDAR, InSAR radar interferometry, cosmogenic and luminescence dating, and fault dislocation modeling.

5.8.1   *Geomorphic Evidence of Active Surface Folding*

The geomorphic expression of surface folds depends on the subsurface geometry of the blind-thrust fault, the depth to the buried thrust fault tip, and the response of surficial materials to propagation of the buried fault tip. In contrast to primary surface faulting, where evidence of slip on the seismogenic fault is directly observed, coseismic folding often results in deformation that is not concentrated along an identifiable fault zone, but instead includes a large deformed area above the fault rupture (now known from postearthquake InSAR surveys). At sites where the surface processes are sensitive to such elevation changes, paleoseismic evidence of coseismic folding is preserved as local deposition or erosion caused by vertical displacement or tilting, usually associated with the regrading of rivers and streams or shorelines crossing the deformed area. The width of the fold is influenced by the depth at which the blind-thrust transitions into the fault-propagation fold. At very shallow depths folds may be sufficiently narrow that they create *fold scarps* similar to those created by surface faulting.

The geometry of most surface folds is distinctive and includes planar limbs and sharp hinge lines characteristic of fault-bend and propagation anticlines (Suppe, 1983, 1985). The folds are typically asymmetrical with steep forelimbs and gentle back limbs. Multiple hinge lines cut the limbs. Many folds have broad planar crests. Observations from historic earthquakes indicate that growth of these folds occurs suddenly during rupture of underlying thrust as the upper thrust sheet moves across bends in the fault and through the axial planes that propagate toward the surface. In some places these axial planes intersect the surface at distinct hinge lines where the attitude of beds and folded geomorphic surfaces abruptly changes, creating a fold scarp such scarps are often the first indication of active movement on the larger fold.

As anticlines grow in height they also grow laterally (along the axis), and this growth means that the effects of folding on geomorphic processes expands through time. Streams that formerly flowed beyond the anticline tip will eventually be reached by the lengthening fold, and then will have to choose between deflecting around the growing fold tip, or to incising down through the anticline as it grows beneath the stream (antecedence). However, even if the stream can temporarily maintain an antecedent course across the fold, its incision may not always be able to keep pace with the uplift. If it cannot, the stream will be defeated and forced to abandon its valley across the fold, in favor of a new course around the anticline tip. Fluvial surfaces in the abandoned valley will then be deformed by further growth of the fold, making them strain markers. A classic example of this sequence of events is found at Wheeler Ridge, California (Mueller and Suppe, 1997; Mueller and Talling, 1997). Although those studies did not characterize individual folding events, they did determine fold size and slip rate, from which paleoearthquake parameters can be derived (discussed later).

5.8.1.1 Fluvial Datums for Detecting Coseismic Fold Growth

The most common geomorphic datums on which historic or prehistoric folding have been measured are fluvial channels and Quaternary river terraces. The modern analog event that formed the basis for subsequent paleoseismic studies in California was the 2 May 1983 Coalinga, California, earthquake ($M_s = 6.5$; King and Stein, 1983; Stein and King, 1984). This earthquake occurred on a buried reverse fault in central California under the aptly named Anticline Ridge. The fault rupture initiated at a depth of about 10.5 km but did not break to the surface. Resurveys of leveling lines after the earthquake show that the Anticline Ridge grew coseismically by about 0.5 m (Stein, 1983). Stein and King (1984) identified four lines of evidence (mostly geomorphic) for paleoseismic fold growth here: (1) warped stream terraces across the fold; (2) a topographic "hump" where the fold projects across an alluvial plain; (3) increased stream sinuosity upstream of, and decreased sinuosity downstream of, the fold; and (4) clayey sediments upstream of the anticline, suggesting a paleo-lake or -marsh (similar to that formed in the 1980 El Asnam earthquake).

Furthermore, King and Stein (1983) proposed that the mean repeat time between earthquakes could be calculated from the vertical dimensions of the fold crest as

$$T = AU/H, \tag{5.1}$$

where T is the repeat time between uplift events, A is the age of youngest stratum uplifted a distance H at the fold crest, $U =$ is the amount of coseismic uplift of the fold crest during a major ($M_w = 6$–7) earthquake, and H is the structural component of the height of the fold crest.

At Coalinga, they assumed $H = 0$–10 m, $A = 2500$–$10,000$ years, and $U = 0.6$ m, yielding a mean repeat time $(T) = 200$–600 years for morphogenic, blind-thrust earthquakes.

The Coalinga observations and principles were used in early geomorphically based paleoseismic studies elsewhere in California. In his classic study of the Ventura Avenue anticline, Rockwell *et al.* (1988) reconstructed the fold shape from isolated, uplifted fluvial terrace remnants, and deduced vertical folding rates (from 4 to 20 mm/yr) for several discrete time periods in the late Quaternary. In the Los Angeles Basin, Bullard and Lettis (1993) performed what is arguably the first of the modern generation of paleoseismic blind-thrust studies. They used "detailed Quaternary mapping of terraces and surfaces, longitudinal stream and terrace profiles, and construction of a topographic residual map" to measure deformation above the Elysian Park thrust fault. From surface mapping alone they estimated uplift rates of 0.1–0.2 mm/yr and tilting of 0.2–0.5 rad/yr over the past 750 ka, and further inferred that the underlying blind thrust was composed of two distinct 6- to 8-km-long segments with independent deformation histories. Modern geomorphic studies have built on the foundation of these early studies, but have the advantage of technologies and data sources not available to earlier workers.

Folded Terraces of Perennial Streams   A good example of a modern study is that of Benedetti *et al.* (2000) on the Montello anticline, Italy. The anticline forms a whaleback-shaped hill about 15 km long and 5 km wide, with a maximum elevation of 368 m, cored by Mio-Pliocene conglomerates that dip outwards at $10°$–$30°$. The Piave River flows southward out of high topography in the Venetian Alps and had eroded a now-abandoned paleovalley 0.7–1.2 km wide across the anticline. This paleovalley contains seven terraces that lie 11–200 m above the abandoned valley floor, and their ages are estimated via correlation with interglacial high sea-level stands (11–320 ka). The terraces are progressively warped upwards into an anticlinal shape that mimics that of the overall anticline (Figure 5.37), but the shapes are not all self-similar (older terraces warped to a different curve than younger terraces).

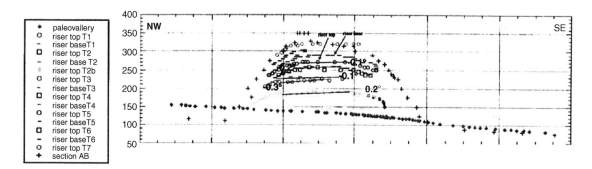

Figure 5.37: Topographic profile of upwarped terrace risers in the abandoned paleovalley crossing the active Montello anticline, northern Italy; vertical exaggeration is 10×. Blue dots show the abandoned valley floor sloping SE. Progressive upwarping of older terraces has resulted from continuing slip on the underlying blind thrust since about 400 ka. From Benedetti *et al.* (2000).

Previous work had shown that the anticline was underlain by a north-dipping ramp and flat thrust system, with the ramp dipping 40°–50°N and its upper tip lying 0.5–2 km below the surface. Benedetti *et al.* (2000) used that data to constrain an elastic dislocation model to account for the observed upwarp of the groups of older and younger terraces. In order to successfully model the observed deformation, they had to assume that the active fault ramp associated with deformation of the older terraces (172–320 ka) dipped 45°N, but then later a shallow 12°-dipping décollement became active and slip shifted to a new 45° ramp 1 km ahead of the earlier one; only this faulting sequence could explain the deformation of all seven terraces (Figure 5.38).

Folded Ephemeral Streams and Piedmonts Many blind faults underlie arid to semiarid areas and range-and-basin piedmonts where there are no perennial streams to form terrace sequences as described above. The signature of blind-thrust folding in such landscapes can be more subtle, especially if fold uplift rates are similar to erosion/deposition rates on the piedmont. Pearce *et al.* (2004) worked on a piedmont affected by both high-slip-rate blind faulting on some structures (which created easily seen anticlines on the piedmont) and low-slip-rate blind faulting on other structures (which merely affected the channel geometry of the ephemeral streams). Upstream of the fold, the ephemeral channel became distinctly braided with an increase in the number of the braid bars and braid index, and an increase in the width-to-depth (w:d) ratio. In the axis of the fold, the ephemeral channels incised, preserving a terrace and effectively increasing the size of braid bars at the expense of the number of bars. Both the braid index and the w:d ratio decreased in the axis of the fold. Downstream of the fold, the w:d ratio and the braid index increased, by an increase in the number of braid bars and a decrease in their size before the channel graded into the distal fan.

While these changes would be detectable by a tectonic geomorphologist, they only tell us about the presence and location of active folds; they cannot be used to derive paleoseismic parameters such as displacement per event, recurrence interval, or slip rate on the underlying blind thrust. However, as discussed later, even knowing the length of an active fold is critical in hazard estimation. In this regard Pearce *et al.* (2004) concluded that the strike lengths of blind thrusts on the San Bernadino piedmont may be as much as 50% longer than shown on published geologic maps, which if true significantly increases the seismic hazard.

5.8.2 Stratigraphic Evidence of Active Surface Folding

Stratigraphic evidence is necessary to confirm that subtle landforms and stream changes seen at the surface are the result of active folding. Such evidence of active surface folding is typically collected in these steps: (1) shallow (0–50 m) geophysics (GPR, seismic reflection, seismic refraction tomography, electrical resistivity tomography) to confirm that strata beneath the suspected fold scarp are in fact folded (i.e., that the scarp is not erosional or some other origin); (2) excavating a fault-perpendicular trench across the fold scarp to the date the latest period(s) of folding; (3) extending the trench interpretation downward with a line of borings up to 15 m deep, which are either logged from core or by lowering a geologist into a large-diameter borehole; and (4) deeper geophysics (500–1000 m) to image the entire fold and its hinge lines, and to try to detect the transition from blind thrust to fold.

These steps can be illustrated by recent work on active blind thrusts in central Japan (Ishiyama *et al.*, 2004, 2007). The Yoro basement-involved fold extends for 20 km in the northern Nobe-Ise Fault Zone and separates the modern Nobi basin in the east from a piggyback portion of the basin and the Suzuka Mountains in the west (Figure 5.39). Shallow high-resolution S-wave seismic reflection data across the

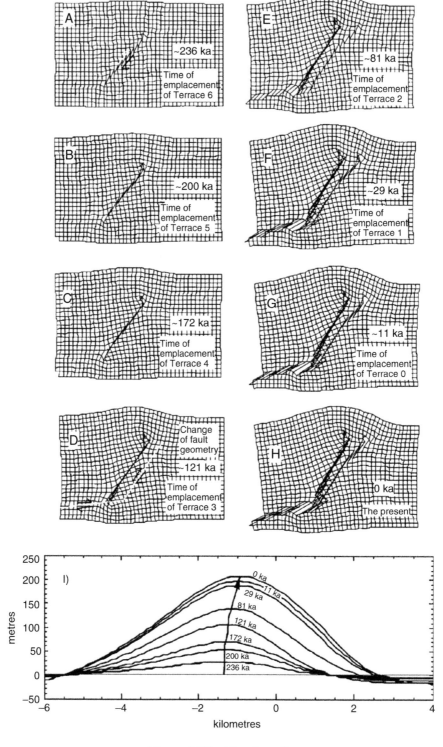

Figure 5.38: (Continued)

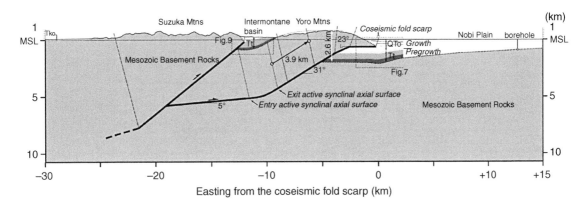

Figure 5.39: Geologic cross section of the Yoro basement-involved fold constrained by the seismic section, geologic mapping, and tectonic geomorphology of its forelimb and backlimb that are linked with bends on the underlying fault trajectory. The close links between the forelimb and backlimb geomorphology and the fault geometry support the wedge thrust model for Yoro basement-involved fold. Abbreviations are Tt, Pliocene-Pleistocene Tokai Group; Tko, Pliocene-Pleistocene Kobiwako Group; and QTo, middle Pleistocene-Holecene Owari Group. From Ishiyama *et al.* (2007).

coseismic fold scarp at Shizu showed very strong and continuous reflectors about 40 m below the ground surface in the basin near the southwest end of the section (Figure 5.40B). The uppermost of these prominent reflectors is interpreted as the top of late Pleistocene alluvial fan deposits are exposed west of Shizu.

To confirm the geophysics they acquired 21 cores up to 5 m long using a "Geoslicer" (see Chapter 2A) and 15 cores up to 7 m long using a percussion core along the seismic line (Figure 5.40A). The boreholes and Geoslicer panels penetrated a late Holocene sequence of delta front and alluvial plain deposits (sand and silt in Figure 5.40B) overlying a prodelta bottomset unit penetrated by deeper boreholes A, B, and C. Within the delta front-alluvial plain package sandy foreset and peaty clay deposits (Figure 5.40A, lower section of unit 2) are laterally continuous and can be traced along the entire 500-m length of the transect, whereas sandy silt (uppermost unit 2) and muddy silt deposits (unit 1) are located only beneath the higher and lower Holocene terraces. The prominent reflectors tied with the basal alluvial fan deposits in the boreholes (15.5 ± 1.2 ka) are clearly folded beneath the east facing fold scarp.

Radiocarbon dating of 18 samples obtained from the cores shows that sediment accumulation has been continuous during the past 10,000 years with no major depositional hiatus. Near-surface stratigraphic relations, radiocarbon dating, and structural restoration of folded strata point to late Holocene and historic fold growth events at this site. Among the events are two recent earthquakes that postdate deposition of the peat layer (AD 120–660) in unit 2. The most recent event occurred in historical time, as suggested by offset of floodplain deposits (unit 1).

Figure 5.38: Dislocation models used to reproduce the observed deformation of terraces 6 (Stage a) through the present valley floor (stage h). The cross sections have no vertical exaggeration in X or Z, but the magnitude of the slip deformation is magnified 5× to make it more visible. (I) Model output curves showing the progressive folding of the river long profile through time, from 226 ka (bottom curve) to present (highest curve). Vertical exaggeration is 10×. From Benedetti *et al.* (2000).

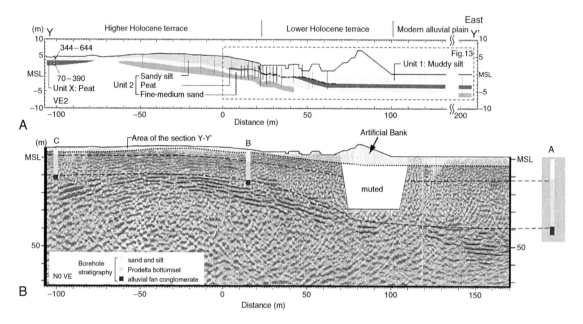

Figure 5.40: Shallow exploration of the Yoro fold scarp. (A) Geologic cross section across the fold scarp of the Yoro fault near Shizu based on Geoslicers and drilled boreholes. (B) A high-resolution S-wave seismic reflection profile (depth migrated) across the coseismic fold scarp. Boreholes tied with prominent reflectors are also shown. Locations of the cross section, seismic section, and boreholes are shown on Figure 10A. From Ishiyama et al. (2007).

Deeper seismic reflection surveys were performed to determine how the small fold scarp was related to the blind thrust and overall fault-propagation fold (Figure 5.41). Growth strata are folded above the propagating tip of the Yoro thrust. These strata collectively define a distinctive shallow synclinal axial surface that bisects the interlimb angles (green line), rather than a reverse fault (red line) that offsets strata at the base of the forelimb.

Based on all the above data, Ishiyama et al. (2007) concluded that the Yoro basement-involved fold has likely formed as a wedge thrust structure above a blind thrust that steepens upward across several gentle bends (Figure 5.42). The dated (15 ka) alluvial fan deposits have been uplifted across the fold scarp about 30 m, yielding a dip-slip rate of 3.5 ± 0.3 mm/yr for intermediate (10^4 years) timescales. At 10^6 year timescales at least 3.9 km of fault slip is consumed by wedge thrust folding in the upper 10 km of the crust, yielding a long-term slip arte of 3.2 ± 0.1 mm/yr. These similar values indicate an essentially constant fold growth rate.

Using techniques described in the next section, Ishiyama et al. (2007) also estimated earthquake magnitude, partly based on the historic $M_w \sim 7.7$ event in AD 1586. Coseismic fold scarps formed during that earthquake can be traced along several en-echelon active folds that extend for at least 60 km, or three times the length of the Yoro fold scarp. They thus concluded that the AD 1586 event was a multisegment earthquake rupture.

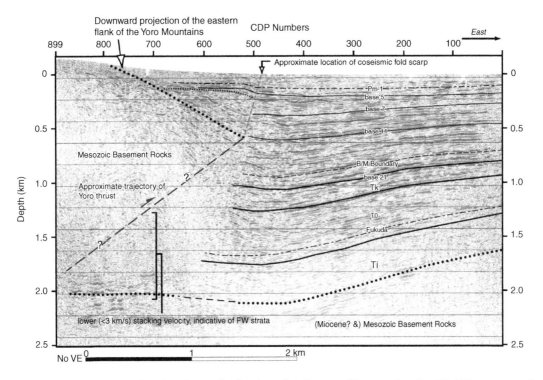

Figure 5.41: Interpreted depth-migrated seismic reflection profile to a depth of 2.5 km across the forelimb of the Yoro basement-involved fold; no vertical exaggeration. Reflectors in the growth strata are labeled; Tk, Komeno Formation; To, Oizumi Formation; and Ti, Ichinohara (all late Pliocene-early Pleistocene). Green-dashed line shows inferred hinge line connecting the blind Yoro thrust with the surface fold scarp. From Ishiyama *et al.* (2007).

5.8.3 Assessing Seismic Hazards from Blind Thrusts

To assess the seismic hazard of blind thrusts, we need to know (1) WHERE is the fault?; (2) HOW BIG will the characteristic earthquake be? (magnitude); and (3) HOW OFTEN will big earthquakes recur? (recurrence interval, slip rate). The steps for deriving these parameters from fold data are slightly different than for faults that break to the surface, and rely more on parameters such as subsurface fault area, stress drop, and seismic moment. Because these parameters are not related to fault scarps, they can also be derived for blind faults offshore (e.g., Rivero *et al.*, 2000).

5.8.3.1 Where?

Previous sections have described how to locate active folds (anticlines), fold scarps, and more subtle geomorphic anomalies formed by active folding over blind thrusts. At a first approximation, well-preserved (or reconstructed) folded Quaternary surfaces can be inverted in an elastic dislocation model to estimate the depth of and dip of the underlying blind thrust. Such estimates of fault location are then confirmed by the stratigraphic techniques described above (trenching, geophysics, drilling).

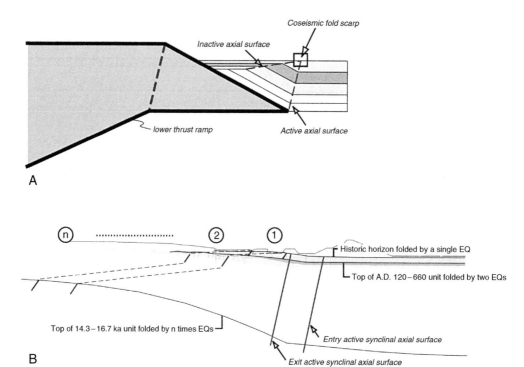

Figure 5.42: Interpretation of combined data for the Yoro thrust. (A) Schematic cross section that explains the surface fold scarp as the trace of an active axial surface above a basement-involved wedge thrust fold. (B) Application of kink band migration across curved hinges to the cross section of the coseismic fold scarp at Shizu site. No vertical exaggeration. Coseismic slip by repeated large earthquakes on the east vergent wedge thrust explains a bed-by-bed growth of the fold scarp, comprised by a more gently dipping and narrower forelimb in the youngest historic horizon produced by the latest earthquake, and steeper and wider forelimbs of older and deeper horizons that have been folded during multiple earthquakes in the past. Circled numbers indicate exit and entry points of inactive axial surfaces associated with the latest two, and "*n* times" earlier earthquakes. From Ishiyama *et al.* (2007).

5.8.3.2   How Big? (M_{max})

In estimating the magnitude of the Characteristic (or Maximum) Earthquake from folds, one cannot use surface rupture length or displacement per event from fault scarps because no fault scarps exist. Instead, one relies on less direct measurements as described below in several alternative methods. The first needed input is a *segmentation model* for the blind thrust, in order to estimate likely subsurface fault rupture length. Yeats *et al.* (1997, p. 342) conclude that many reverse-fault ruptures stopped at lateral ramps in the underlying thrust, for example, the Santa Susana fault with three segment boundaries. At the surface these ramp-controlled segment boundaries may coincide with stepovers of the axes of anticlines, with termination of hanging-wall block mountains, strike-slip cross-faults, and major reentrants in the thrust trace. However, Ishiyama *et al.* (2007) concluded that the AD 1586 $M_w \sim 7.7$ earthquake on the Yoro blind thrust, Japan, ruptured multiple segments as defined by individual surface folds. Such an earthquake thus represents a sort of "worst-case scenario" for seismic hazard managers worldwide, an $M_w > 7.5$ multisegment rupture on a blind thrust.

Once the subsurface rupture dimensions are estimated, there are several alternative methods for estimating earthquake magnitude (STEP 1):

1a.   Use the length of the uplifted/folded zone as a substitute for the surface rupture length in empirical regressions of surface rupture length against magnitude. For example, Grant *et al.* (1999) input the linear extent of marine terrace uplift in coastal California into the regression equation of Wells and Coppersmith (1994) between reverse-fault surface rupture length and magnitude. (This substitution is one of expediency, since the Wells and Coppersmith (1994) surface rupture length data set is composed entirely of ruptures that were NOT blind.)

1b.   Use the area of the seismogenic blind fault plane, based on subsurface information and/or modeling. For example, Shaw and Suppe (1996) used subsurface data to define the area of seismogenic ramps, ramp segments, and decollements (Figure 5.43). They then input that area into the regression equation of Wells and Coppersmith (1994) between reverse-fault rupture area and magnitude. Oskin *et al.* (2000) estimated rupture area in a similar way, but then input it into the regression equation of Dolan *et al.* (1995) between rupture area and magnitude for southern California earthquakes only. These two approaches do not suffer from the expediency mentioned in approach 1a.

1c.   Use an inferred stress drop and the fault area to estimate seismic moment, via the equation $M_o = 16/7\ R^3 \Delta\sigma$, and then input that value into the equation of Hanks and Kanamori (1979) relating seismic moment to moment magnitude ($M_w = 2/3 \log M_o - 10.7$). This approach was used by Myers *et al.* (2003), who assumed a "regional" stress drop of 130 bars, and a circular rupture area with the same diameter as the fault length, to calculate seismic moment from source dimensions.

5.8.3.3   How Often? (Recurrence, Slip Rate)

In theory, estimating the recurrence of the characteristic earthquake can be done in two ways: (1) directly dating displacements or angular unconformities produced by individual paleoearthquakes or (2) dividing the average displacement per event by the long-term slip rate. Due to the difficulty in obtaining good field data on (1), most published studies have used approach (2).

STEP 2: Estimate Slip Rate. Slip rate can be derived in several ways from the amount of deformation of dated geomorphic surfaces overlying the blind thrust (monoclines, anticlines), or the vertical deformation of dated strata in the subsurface. Both approaches yield a long term, average slip rate

2a.   *Simple methods for determining slip rate:* assume that the uplift rate of landforms is identical to the vertical slip rate on the blind fault. If the dip of the blind fault is known, then the vertical slip rate can be related to the net slip rate trigonometrically. This approach was taken by Grant *et al.* (1999). The advantage of this approach is its simplicity.

2b.   *Sophisticated methods for determining slip rate*: use dislocation modeling to invert the observed warping/folding of surface landforms into a "best-fit" fault geometry and net slip amount. This modeling may be quite complex and require a numerical model for fault-bend folding and may only be able to associate the present folded topography to a long-term shortening rate (e.g., Lave and Avouac (2000) for the Himalayan Frontal Thrust; Ishiyama *et al.* (2004) for the Kuwana fault). Then calculate slip rate by dividing the net slip required to produce the observed deformation, by the age of the deformed landform(s).

STEP 3: Estimate the Average Displacement of the Characteristic Earthquake

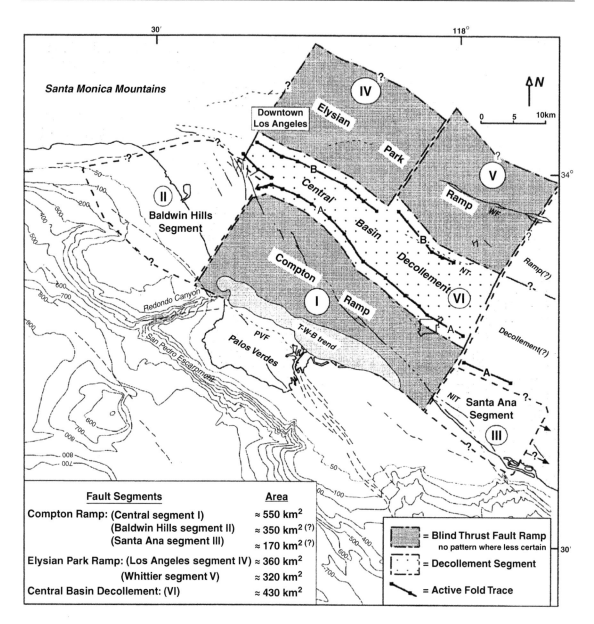

Figure 5.43: Map of the estimated extents of segments of the Compton and Elysian Park blind-thrust ramps and Central Basin décollement. Fold trends: (A) Compton–Los Alamitos trend. (B) Elysian park trend; T-W-B, Torrance-Wilmington-Belmont trend. Offsets in map view of the Compton–Los Alamitos and Elsysian park trends overlie potential segment boundaries of the underlying Compton–Elysian Park blind-thrust system. The Compton ramp consists of a central segment (I) and adjacent Baldwin Hills (II) and Santa Ana (III) segments. The deep Elysian Park ramp consists of the Los Angeles (IV) and Whittier (V) segments and is separated from the Compton ramp by the Central Basin décollement (VI). From Shaw and Suppe (1996).

3a. Given the estimated magnitude, use an empirical regression between magnitude and average displacement per event. For example, Shaw and Suppe (1996) used the regression of Wells and Coppersmith (1994) between magnitude and average displacement for all fault types, arguing that "it was defined by better data and is similar to less statistically significant. . . . relations for thrust and reverse faults."

3b. Given the seismic moment and rupture area, calculate D from equation $M_o = \mu A D$ (Aki, 1966), where D is the average displacement, A is the fault rupture area, and μ is the crustal rigidity.

The seismic moment can be calculated from method 1c, above, or estimated from magnitude by the equation $M_w = 2/3 \log M_o - 10.7$, as done by Oskin *et al.* (2000).

Once the long term, average slip rate is estimated, recurrence interval can be calculated by dividing the average displacement per event (from STEP 3) by the long-term slip rate (STEP 2). A nomograph relating these three seismic characteristics is shown for southern California blind thrusts in Figure 5.44.

5.9 Paleoseismology of Subduction Zones

5.9.1 Introduction

The majority of large compressional earthquakes result from plate convergence at subduction zones. The largest shocks are produced by slip on plate-bounding thrust faults, sometimes called *megathrusts*, and rupture of the entire thickness of the brittle lithosphere. Forearc and backarc thrust belts in the upper plate, and normal and strike-slip faults in the subducting plates also contribute to the seismicity of many subduction zones.

The three fundamentally different configurations of converging plates are (1) convergence between two oceanic plates with subduction of the younger plate beneath the older plate, (2) subduction of an oceanic plate beneath the margin of a continental plate, and (3) collisions between two continental plates. Except for continental collisions, plate convergence results in the formation of *subduction zones* where one plate descends beneath the other and extends deep into the earth's interior. Along most subduction zones the *forearc*, the upper plate between the trench and the volcanic arc, is composed predominantly of accreted marine rocks that are typically strongly tectonized. The trenchward part of most forearcs are active *accretionary complexes*, in which large systems of thrust faults and fault-generated folds accommodate some of the plate convergence.

The principal fault in a subduction zone is the megathrust, the plate-bounding thrust fault that accommodates the relative movement between the plates. At plate scales these huge thrusts are linear or broadly curved to form smooth arcs hundreds or thousands of kilometers long. However, the regional structures and kinematics of megathrusts are complex and vary greatly (Plafker and Savage, 1970). At shallow depths the megathrust separates overlying highly deformed accreted sediments in the tip of the accretionary wedge from underlying subducted marine sediments and oceanic crust. The subducting sediments are poorly consolidated and contain large quantities of water, factors that tend to reduce coupling between the plates and promote *aseismic slip* (Pacheco *et al.*, 1993). As subduction proceeds some of the sediments are scraped off the descending plate and accreted to the tip of the growing accretionary wedge. The sediments that are not accreted move deeper, and dewatering takes place. The water is forced upward along the megathrust and along faults in the overlying accretionary wedge. Dewatering increases the coupling forces and

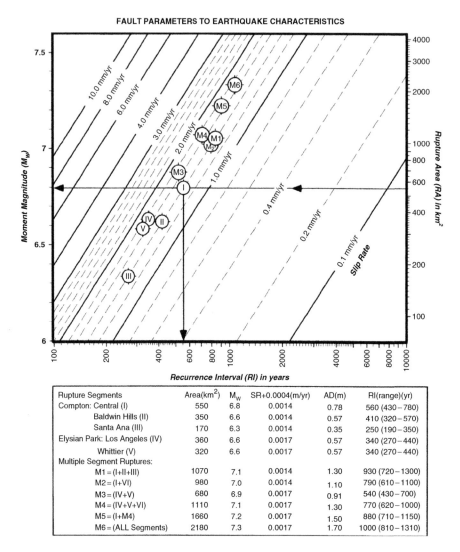

Figure 5.44: (Top) Plot of relationships between fault parameters (area and slip rate) and potential earthquake characteristics (magnitude, repeat time) based on empirical equations relating rupture area to magnitude (Wells and Coppersmith, 1994, all fault types). (Bottom) compilation of potential earthquake characteristics, with repeat time ranges defined solely by ranges in average slip rates. Tabulated estimates of average coseismic displacements (AD) and recurrence intervals (RI) are based on magnitude estimates (M$_w$), rounded to two significant figures. From Shaw and Suppe (1996).

promotes locking that results in *stick-slip motion* between the plates. Roughness of the descending plate, especially large-scale irregularities in the upper surface such as large seamounts or groups of seamounts, submarine plateaus, or fracture zones, creates *asperities* and increases the locking between the plates, promoting seismic subduction. As the oceanic plate descends it transports cold surface sediments down the subduction zone and depresses the temperature along the megathrust.

Slick-slip behavior is possible as long as temperatures remain relatively low, less than 300 to 350 °C (Hyndman and Wang, 1993). Above these temperatures stable sliding prevents the accrual of elastic strain and promotes aseismic subduction. Heat flow measurements above downgoing oceanic slabs and thermodynamic modeling of subduction zones indicate that the 350 °C isotherm generally lies at a depth of 25–40 km.

The seismogenic processes from one subduction zone to another, or from one segment of a long subduction zone to another, vary considerably. Where older oceanic crust is being subducted, plate convergence proceeds by largely aseismic processes, although some of these subduction zones have high levels of seismicity that include abundant small and moderate size earthquakes. In such zones the subducting plate dips steeply and is weakly coupled to the overlying plate. Plafker (1972) classified these as the *Marianas type* of subduction zones. In contrast, subduction zones of the *Chilean type* (Plafker, 1972) are strongly coupled and characterized by seismicity dominated by infrequent very large earthquakes. Chilean-type zones include those that are subducting relatively young oceanic crust at a low dip angle. Subduction zones of the Chilean type that have produced great interplate earthquakes during the last half century include S. Chile ($M_w = 9.6$, 1960), Alaskan ($M_w = 9.2$, 1964), Central Aleutians ($M_w = 9.1$, 1957; $M_w = 8.7$, 1965), and Colombia ($M_w = 8.3$, 1979).

Jarrard (1986) compared 26 parameters for 39 subduction zones and concluded they could be grouped into seven classes (Figure 5.1). Parameters included dimensions, age, and structural characteristics of both the descending slab and the upper plate, as well as the geometry and rate of relative motion between the converging plates. The age of the subducting plate and secondarily the rate of convergence appear to be among the most important parameters in determining the nature of subduction zone seismicity. The largest earthquakes ($M_w > 8.5$) are produced where young plates are subducting at the highest rates (Jarrard's classes 5–7), whereas slower convergence of older oceanic crust (Jarrard's classes 1–4) yields maximum earthquakes in the $M_w = 7–8$ range (Heaton and Kanamori, 1984).

The upper plate accretionary margins of subduction zones are commonly cut by large active faults. Trenchward vergent thrust systems are the most common fault type in upper plate margins, although arc vergent thrusts are also common. These faults reflect permanent strain resulting directly from convergence and demonstrate that plate-bounding thrusts have sufficient strength to transmit motion into the upper plate. Thrust mechanisms for shallow small and moderate magnitude earthquakes in the forearc of some subduction zones indicate that some of these faults are seismogenic, but large historic earthquakes clearly resulting from displacement on upper plate thrusts have been rare, and some subduction zones where large youthful thrusts are present exhibit little or no shallow compressional seismicity. One possible explanation of the apparent lack of large-magnitude earthquakes on forearc thrusts is that these faults represent upward branching imbricate splays from the megathrust and experience displacement only during subduction earthquakes involving slip on the megathrust. Such displacement occurred during the 1964 Alaskan earthquake when large surface displacements developed on the Patton Bay and Hanning Bay faults on Montague Island in Prince William Sound.

High-angle strike-slip faults are also common in some forearc settings, especially where convergence is strongly oblique. In Japan, the Median Tectonic Line, a large strike-slip fault system in the forearc of the Nankai subduction zone, is interpreted to accommodate the oblique component of convergent motion between the Philippine plate and the Asian plate. Although the principal faults composing the Median Tectonic Line have not produced a large earthquake in the past thousand years, a secondary conjugate fault to the Median Tectonic Line produced a M_w 7.0 earthquake near the city of Kobe in January 1995 that resulted in severe damage and more than 5000 fatalities (Comartin *et al.*, 1995).

5.9.2   Segmentation of Subduction Zones

Segmentation of plate-boundary megathrusts in subduction zones is apparent from historic subduction earthquakes. Although the largest of these earthquakes has produced ruptures more than 1100 km long, few have ruptured the entire length of the subduction zone. Thus, great historic subduction earthquakes have usually been restricted to a segment of the convergent margin. In the few localities where multiple seismic cycles have been recorded, the segmentation has differed from one cycle to the next. The long written history of subduction earthquakes on the Nankai trough in southwest Japan includes as many as eight earthquake cycles since A.D. 684 (Figure 5.45) (Ando, 1975; Yonekura, 1975). The great earthquake of 1707 ruptured the entire length of the subduction zone. However, in 1854 the entire zone broke in two separate earthquakes, and in 1944 and 1946 two ruptures covered much of the zone. Based on these and earlier earthquakes, up to four segments can be identified. In most seismic cycles the entire zone ruptured

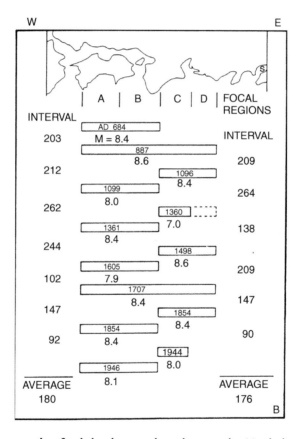

Figure 5.45: Historic records of subduction earthquakes on the Nankai subduction zone along southwest Japan show that five of the seven observed seismic cycles ended with a pair of earthquakes, each extending over about half of the length of the subduction zone. Two cycles ended with single long ruptures that broke most of the zone. Four of the multiple segment cycles produced a pair of earthquakes separated by a few years or less, illustrating temporal clustering of subduction earthquakes also seen on the Aleutian–Alaskan subduction zone during the 1957–1965 period (see Figure 5.1). Modified from Yonekura (1975).

within a time period of less than 3 years. Similar earthquake sequences have occurred along the Mexico and Colombia subduction zones (Thatcher, 1990) and in four circum-Pacific rupture segments described by Schwartz (1999). The great 1960 Chile earthquake broke across two separate rupture segments that had ruptured separately in 1835 and 1837. This behavior of different segmentation in successive seismic cycles may be typical of many subduction zones.

Analysis of *aspect ratios* of rupture length to width for subduction zones earthquakes worldwide shows that most range between 2 and 4, but a few have been as large as 9.7 (1965 Rat Island earthquake, Aleutian subduction zone) (Geomatrix, 1995). Unusually long ruptures associated with giant subduction earthquakes probably reflect domino-like triggering of slip on a series of potential rupture segments anchored on asperities. The length of each segment is probably close to the seismogenic width of the zone. Some convergent margins, notably the south Chile, the Aleutian, and possibly the Cascadia subduction zones, may be characterized by long ruptures that span multiple potential segments. The 1957, 1964, and 1965 subduction earthquakes broke more than 3000 km of the Aleutian subduction zone. In the paleoseismic record the three separate events would not easily be distinguished, and such a series of earthquakes might be misinterpreted as a single giant rupture event.

The ends of rupture segments of some historic subduction earthquakes coincide with prominent structures in lower plates. The 1960 Chile earthquake initiated in the vicinity of the subducted Mocha fracture zone where juxtaposed oceanic lithosphere of different ages is being subducted. The rupture propagated south to the Chile triple junction where the Chile rise intersects the Peru–Chile trench. Fracture zones, groups of seamounts and oceanic plateaus, and subducting plate boundaries between oceanic plate systems are good candidates for termination points of subduction earthquake rupture. Offsets, gaps, or abrupt changes in strike or dip of the Wadati–Benioff zone may also delineate possible segment boundaries (Burbach and Frolich, 1986). Changes in the geometry of the downgoing plate as reflected in the location and offset of aligned arc volcanoes have also been identified as possible segmentation indicators (Guffanti and Weaver, 1988).

Upper plate structures have also been related to seismic segmentation of long subduction zones. The 1957 Aleutian earthquake (M_w 8.6) ruptured 1200 km of the central Aleutian subduction zone, where the architecture of the upper plate includes large strike-slip faults bounding rotating blocks that accommodate the arc parallel component of oblique convergence. The mega-thrust rupture propagated across several of these strike-slip faults (Ekstrom and Engdahl, 1989; Ryan and Scholl, 1989). In 1986 a M_w 8.0 earthquake reruptured a part of the 1957 rupture zone. This rupture may have terminated at one of the strike-slip faults in the upper plate. These earthquakes show that "... regions of high moment release, asperities, vary from earthquake to earthquake ..." (Boyd *et al.*, 1992).

5.9.3   Surface Faulting: Upper Plate Versus Plate-Boundary Structures

During the 1964 Alaskan earthquake slip appears to have been distributed on several imbricate thrust faults in the accretionary wedge as well as the megathrust. Up to 7.8 m of dip-slip displacement at the surface was measured on the Patton Bay fault on the southeast side of Montague Island, and up to 6 m of dip-slip displacement occurred on the Hanning Bay fault on the opposite side of the island (Plafker, 1969a). The marked decrease in uplift trenchward of these faults, which dip landward at 50°–75° at the surface, suggests they probably merge with the megathrust at depth and during the 1964 event they accommodated a significant part of the total interplate slip. Sea-floor scarps and analysis of arrival characteristics of the tsunamis generated by the earthquake suggest the large surface displacements extended southwest of Montague Island to at least the latitude of Kodiak Island. No other surface faulting was found.

Late Cenozoic reverse and thrust faults and folds have been mapped on the sea-floor in the outer part of the accretionary margin of the Aleutian–Alaskan subduction zone, and some of these faults have been identified on land on Hinchinbrook, Hawks, and Montague Islands and on the mainland near Cordova and the Rude River Valley (described later). These faults deform late Pleistocene glacial deposits and landforms, and displace latest Pleistocene marine limit shorelines and Holocene sediments. Most of the known on-land thrusts in the accretionary fold and thrust belt were not active in 1964, but have experienced displacement in the latest Quaternary. Additionally, none of these faults has generated surface displacement of a large earthquake independent of the 1964 earthquake. The likely interpretation is that the fold and thrust belt faults act in concert with the megathrust by partitioning slip from the plate boundary. Probably these faults do not produce large earthquakes and accrue slip independent of the megathrust. The faults in the Rude River–Hinchinbrook Island region that possess latest Pleistocene and Holocene scarps probably reflect previous great subduction earthquakes with slip occurring on different accretionary wedge thrusts than those activated in 1964.

Several of the Cascadia fold and thrust belt faults that come onshore in northern California have experienced late Holocene displacements. The question of whether these faults slipped independent of the megathrust, or accompanied interplate subduction earthquakes, is important to the assessment of the seismic hazard posed by the subduction zone. Seismic refraction and reflection profiling of the seafloor offshore of the Pacific Northwest shows many such faults are present in the outer part of the upper plate, but no seismicity has been associated with these faults. Similar fold and thrust belts are known along other convergent plate boundaries where a young oceanic plate is being subducted; the faults also are characterized by little or no seismicity. The Patton Bay and Hanning Bay faults described earlier are the only well-documented examples of fault rupture of thrusts in the accretionary fold and thrust belts along subduction zones that are similar to Cascadia.

Trenches across the Mad River and McKinleyville faults, two of the principal on-land thrusts in northern California, indicate these faults have experienced several displacement episodes during the Holocene (Figures 5.7 and 5.13). The trench exposures show these faults have long recurrence intervals (several thousand years) relative to the several-hundred-year repeat time of subduction earthquakes suggested for Cascadia from most paleoseismic evidence. If these fault rupture episodes were coeval with slip on the underlying megathrust, only a few of the megathrust earthquakes are represented in the paleoseismic record of these faults. The Little Salmon fault, another of the large fold and thrust belt structures near the south end of the subduction zone, has experienced displacements as large as 7 m repeatedly during the late Holocene. Trenches excavated across the fault where it crosses Little Salmon Creek valley exposed faulted and folded overbank silts and organic soils that record at least three individual displacement events (Figure 5.26). Carbon-14 age estimates for the three paleoearthquakes suggest they occurred about 300, 700, and 1600 years ago (Clarke and Carver, 1992). These age estimates are permissively correlative with three of the four most recent subduction zone paleoearthquakes interpreted from coastal uplift and subsidence evidence and allow the interpretation that the paleoslip recognized on the Little Salmon fault was generated during great megathrust earthquakes similar to the faulting in Prince William Sound during the 1964 Alaskan earthquake. However, the paleoearthquake chronology of any one accretionary wedge thrust is likely to reflect only some of the large or great earthquakes on the megathrust.

5.9.4 Historic Subduction Earthquakes as Modern Analogs for Paleoearthquakes

Large subduction earthquakes differ in scale and process in some important ways from most shallow crustal earthquakes. During subduction earthquakes fault rupture propagates entirely through the brittle lithosphere and produces elastic response of the crust. At the earth's surface above and adjacent to the rupture, large horizontal and vertical displacements accompany the earthquake. The wide rupture resulting from the shallow dip of many megathrusts, the long rupture length (>100 km), and the long recurrence intervals (10^2–10^3 years) for megathrust earthquakes result in much larger rupture areas (10^3–10^5 km^2), displacements (5 to >20 m) and magnitude ($M_w = 7.5$–9.5) than most crustal earthquakes.

The most important feature of large subduction earthquakes is the regional coseismic deformation above the plate-boundary megathrust and the elastic relaxation of the forearc. These regional strains are expressed as vertical changes in land level, with the area above the rupture surface uplifted, and the area between the volcanic arc and the downdip edge of the rupture subsided (Figure 5.46). Since most subduction zones are located along coastlines, these vertical motions are often recorded in the shoreline deposits and landscapes.

Observations of coseismic land-level changes associated with modern subduction zone earthquakes provide the main analog used to identify interplate paleoearthquakes along subduction zones. Affected coasts may rise or fall instantaneously during earthquakes (coseismic uplift or subsidence) and/or change elevation more slowly during *postseismic* and *interseismic* periods (aseismic uplift or subsidence). When great earthquakes ($M_w > 8$) occur at the boundary between the subducting and overriding plates, the region nearest the subduction trench (60–160 km wide) is commonly uplifted; at the same time a parallel zone arcward of the zone of uplift may subside (Figure 5.46) (Plafker, 1972; Ando, 1975; Thatcher, 1984). Local areas in the zone of coseismic uplift can also be thrust upward during slip on imbricate thrust faults or growth of folds within the upper plate (Plafker, 1969a; Plafker and Rubin, 1978; Yonekura and Shimazaki, 1980; Berryman *et al.*, 1989; Page *et al.*, 1989; Ota *et al.*, 1991). Because only large plate-boundary earthquakes produce enduring land-level changes large enough to be recorded in most environments, the precision of these methods limits the threshold of detection of earthquakes along subduction zones to earthquakes of high magnitude. In this sense, the smallest morphogenic earthquakes of some subduction zones may be larger than the largest earthquakes produced by fault zones in some other tectonic settings.

Many of our inferences about past land-level changes during subduction zone earthquakes are based on observations following two of the largest earthquakes of this century, the 1964 M_w 9.2 earthquake in southern Alaska and the 1960 M_w 9.5 earthquake in south-central Chile. Much recent North American research has centered on identifying and dating evidence of prehistoric coseismic land-level change along the coast of western central North America in an attempt to assess the potential for subduction-zone earthquakes on the Cascadia subduction zone (Rogers *et al.*, 1996). Both the 1960 and 1964 earthquakes occurred on tectonically similar subduction zones (Heaton and Kanamori, 1984) situated along midlatitude coasts with landforms, sediments, and climates generally similar to those at Cascadia, so those earthquakes in particular have been studied as modern analogs.

The March 26, 1964, Alaska earthquake ($M_w = 9.2$) resulted from slip on an 850-km-long portion of the Alaskan–Aleutian subduction zone that released stresses accumulated from about 6 cm/yr of convergence between the Pacific and North American plates (Demets *et al.*, 1990). The earthquake was felt over more than a million square kilometers, and caused widespread damage across more than 100,000 km^2. The fault displacement propagated along the megathrust underlying Prince William Sound, southeast

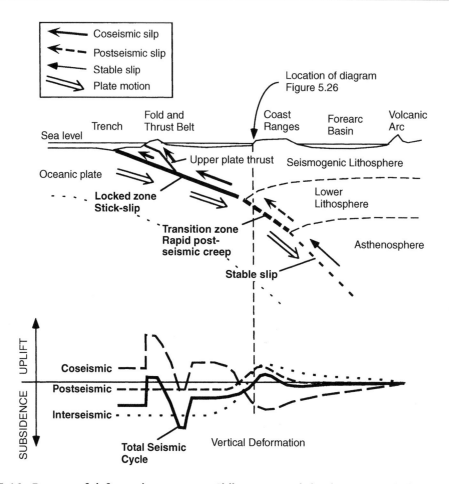

Figure 5.46: Pattern of deformation across a Chilean-type subduction zone (Plafker, 1972) for interseismic, coseismic, and postseismic parts of the seismic cycle. During the long duration interseismic part of the cycle, the locked zones of the megathrust are coupled and the upper plate is carried toward the arc and down with the descending oceanic plate. Compression of the backstop region above the transition zone and the deep stable sliding part of the megathrust generates uplift near the arc. During megathrust earthquakes, coseismic slip on the locked zone produces uplift above the megathrust rupture and elastic relaxation and subsidence between the downdip end of rupture and the arc. Slip on upper plate thrusts can generate localized and permanent uplift and subsidence in the fold and thrust belt. Rapid creep accommodates the slip deficit on the megathrust in the transition zone during the relatively short postseismic interval following the earthquake. This rapid creep produces rapid rebound in the area of coseismic subsidence.

Kenai Peninsula, and the Gulf of Alaska offshore of Kodiak Island in south-central Alaska (Figure 5.47), raising a 150- to 200-km-wide and 800- to 900-km-long part of the floor of the Gulf of Alaska between the trench and the eastern side of Kenai Peninsula and Kodiak Island. An equally large region between the uplifted area and the volcanic arc subsided as much as 3 m. Because the rupture was in part located beneath hundreds of miles of coastline the patterns, styles, and magnitude of coseismic vertical

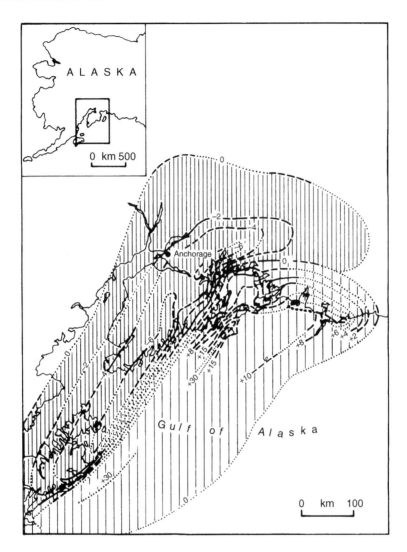

Figure 5.47: Coseismic deformation produced by the 1964 Alaskan earthquake. Regional subsidence (closely spaced lines) and uplift (widely spaced lines) encompassed most of the forearc along the length of the rupture. The axis of maxium regional subsidence and maxium regional uplift were located near the zero isobase (no land-level change). Measurements are in feet. From Vita-Finzi (1986), after data in Plafker (1969a).

deformation were extensively recorded (Plafker, 1969b). Especially useful in mapping and measuring the uplift was the elevation of sessile intertidal plants and animals (barnacle and alga lines) above their life zones. Regional changes in the land level of 1–3 m extended over an area of about 150,000 km², with as much as 9 m of localized uplift on Montague Island along thrust faults activated during the earthquake. Thrust displacement and uplift of the seafloor generated large trans-Pacific tsunamis. Alternating mud–peat and peat–mud couplets, reflecting interseismic sedimentation punctuated by sudden emergence or submergence, were also observed to form (Bartsch-Winkler and Schmoll, 1987).

Subduction earthquakes produce strong shaking over large regions that is commonly of unusually long duration and rich in relatively long periods, factors that favor generation of liquefaction and trigger slope failures. In Alaska ground motion lasted as long as 5 min at some locations and triggered thousands of landslides in the mountains and along the coast. Many large landslides were triggered over more that 150,000 km^2. Some of these involved the collapse of entire mountainsides and minor peaks in the Chugach and Kenai mountains (Plafker, 1969b). Strong shaking caused widespread liquefaction and other forms of seismically induced ground failure in coastal lowlands, river deltas, and along lake margins. Shaking induced submarine landslides that produced >20-m waves in Valdez Arm, Resurrection Bay, and other deep fjords near the epicenter.

The earthquakes of 20–21 May 1960 (M_w 9.5) in south-central Chile produced a pattern of land-level changes generally similar to those in Alaska (Plafker and Savage, 1970; Plafker, 1972). A trenchward belt of coseismic uplift about 100 km wide and nearly 1000 km long raised the sea-floor and several offshore islands as much as 4–6 m, and an adjacent belt of forearc subsidence drowned much of the south-central Chile coast. Strong shaking accompanied the main shock and lasted several minutes, triggering widespread liquefaction, ground failure, and landslides. The main shock was preceded by 12 h by a M_w 8 foreshock and was followed by many large aftershocks that generated additional local strong shaking. Seafloor uplift caused a large trans-Pacific tsunami with 5- to 15-m run-up heights common along the Chilean coast.

Regional coseismic land-level changes reflecting similar patterns of deformation were observed during smaller subduction earthquakes in southwest Japan (1944, M_w 8.0; 1946, M_w 8.1: Ando, 1975), Mexico (1985, M_w 8.1: Bodin and Klinger, 1986), Chile (1985, M_w 7.9: Castilla, 1988), Costa Rica (1991, M_w 7.4: Plafker and Ward, 1992), and northern California (1992, M_w 7.1: Carver et al., 1994b). The principal field evidence of the land-level change associated with these earthquakes was the mortality of intertidal organisms, although small raised terraces were generated in Mexico by the 1985 Michoacan earthquake. These earthquakes were in the magnitude range of 7–8 and resulted in maximum vertical land-level changes in coastal regions of about 1 m, insufficient to produce widespread geomorphic or stratigraphic evidence. In contrast, the 1964 Alaska and 1960 Chile earthquakes produced widespread geomorphic and stratigraphic records where coseismic uplift exceeded 1 m.

The processes of regional coseismic uplift and subsidence, widespread long duration shaking with attendant large landslides, liquefaction, and tsunami, when located along a coast above a subduction zone, produce the unique geologic signature of large megathrust earthquakes. The earthquake causes sudden changes in geologic process, which are typically reflected by an unconformity or sharp contact between pre- and postearthquake sediments, or by landforms formed by contrasting processes and environments. Certain coastline environments, such as salt marshes, river and stream deltas, intertidal wave-cut benches, and reefs are excellent recorders of the sudden vertical changes in elevation along the rupture segment, and act as geological "archives" of subduction earthquakes.

5.9.5   The Earthquake Deformation Cycle in Subduction Zones

Geologic and geodetic observations of vertical movements of the earth's surface along subduction zones show that significant changes in land level take place *between*, as well as during, subduction earthquakes. Vertical movement rates may be especially high during a relatively short (decades long) postseismic period following great subduction earthquakes as displacement propagates downward along the deeper part of the plate interface, and the deeper lithosphere and possibly the upper asthenosphere, compensate for the slip deficit. Tide records for the 25 years following the 1964 Alaska earthquake show uplift of more than 20 mm/yr in the region of principal coseismic subsidence (Savage and Plafker, 1991). Over the same interval, tide stations in the region of coseismic uplift have subsided at rates up to 10 mm/yr. Geodetic observations across the Kenai Peninsula (Cohen *et al.*, 1995) and resurveys of tidal benchmarks across the Kodiak Archipelago (Gilpin *et al.*, 1994b; Savage *et al.*, 1998) also show uplift rates considerably larger than those expected from plate convergence. These high movement rates, if they continue, will compensate for the 1964 coseismic displacements in a century or two, or a much shorter time than the 700-year recurrence interval indicated from paleoseismic studies.

Barrientos *et al.* (1992) likewise show that rapid uplift has occurred over the region of the Chile coast that subsided during the 1960 earthquake, with initial high uplift rates decreasing after about 16 years. They interpret the pattern of vertical deformation as relaxation of the upper lithosphere in response to propagating creep on the downdip extent of the coseismic rupture surface. Aseismic land-level changes over the region of coseismic subsidence in southwest Japan have been deduced from an array of 15 tide gauges for the 50 years (~30% of the historic recurrence interval of 176 years) following the 1944 and 1946 subduction earthquakes (Savage, 1995). Initially high uplift rates that lasted about a decade were followed by relatively linear uplift at a rate about twice that predicted from modeling plate convergence. Most models of the earthquake deformation cycle generalized from data from Japan and Alaska (Chapter 1) show that rates of aseismic movement generally decrease with distance from a fault and with time following an earthquake (Thatcher, 1986a; Savage and Plafker, 1991).

Dislocation models predict interseismic subsidence of the trenchward part of the upper plate above the locked segment of the megathrust (Figure 5.47). The arc-ward part of the upper plate undergoes coseismic subsidence during megathrust earthquakes as the result of elastic thinning behind the megathrust rupture, but then undergoes gradual uplift between earthquakes. This oscillation in land level over repeated seismic cycles helps to explain why most of the regions that undergo coseismic uplift lie below sea level, while regions characterized by coseismic subsidence are mostly above sea level and include coastal mountains and upland areas. Therefore, large-scale landscape characteristics typically reflect the long-term deformation of coastal regions along subduction zones as dominated by interseismic deformation, which is often opposite to the sense of coseismic deformation and the appearance of the postseismic coastline morphology. For example, the prominent shoreline terrace produced by uplift of the Prince William Sound region during the 1964 Alaskan earthquake is the only emergent aspect of that coast's geomorphology; no raised Holocene terraces are found landward of the 1964 terrace. The geomorphic evidence for emergence is present only during the early part of the interseismic cycle (postseismic part). Prior to the 1964 earthquake in Prince William Sound, shorelines were drowned (submergent morphology) and showed little evidence of Holocene uplift.

Stratigraphic and geomorphic features of some coastal environments, such as salt marsh and shallow tidal bays, river and stream deltas, wave-cut platforms, and coral reefs may record both coseismic and interseismic elevation changes and in this way archive evidence of subduction earthquakes. Geodetic measurements of interseismic subsidence and uplift for some subduction zones indicate that rates are great enough to compensate for much or all of the coseismic elevation changes between slip events, so that little long-term elevation change is produced. Permanent coseismic uplift, as indicated by some raised glacioeustatic marine terraces and other emergent features, often represents localized upper plate deformation, usually generated by slip on thrust faults or folds in the accretionary wedge (e.g., Middleton Island; Plafker, 1969b; California coast, see Sections 5.9 and 5.10).

Geodetic and tide gauge measurements, made over periods of many decades before and after some of the largest plate-boundary earthquakes, provide the most detailed data used to develop models of the earthquake deformation cycle (Thatcher, 1986a; Scholz, 2002; Savage and Thatcher, 1992). The cycle begins with *interseismic* strain accumulation, for example, in the upper plate above a locked part of a plate-boundary fault as shown by data from the subduction-zone coast of southwest Japan (Figure 5.48). The more gradual deformation in the interseismic part of the cycle includes short-term *preseismic* and *postseismic* movements that are generally opposite in direction to coseismic deformation at the same sites. Accumulated strain is released through slip on the locked part of the fault during the coseismic part of the cycle (Figure 1.3). Near the fault, deformation tends to be large but decays rapidly with time; deformation farther from the fault is of smaller amplitude and decays more slowly. Throughout the earthquake deformation cycle, steady slip on the fault downdip of the locked part may produce long-term interseismic deformation. Most paleoseismic investigations have focused almost exclusively on the strain release (coseismic) part of the earthquake deformation cycle because deformation during large earthquakes commonly produces landforms or deposits that are more distinctive than those produced by slower deformation during the interseismic parts of earthquake deformation cycles. As discussed in later chapters, inferring deformation rates during the interseismic parts of past earthquake deformation cycles requires various types of geomorphic and stratigraphic markers, the initial shape or elevation of which is known precisely enough to allow recognition of small amounts of deformation. Paleodatums are often produced during brief (geologic) time intervals by climate-driven geomorphic processes (e.g., Bull, 1991).

5.10   Late Quaternary Sea Level

Sea level serves as the physical control on most geological and biological processes in coastal areas, and these processes in turn produce the specific datums used in geomorphic and stratigraphic field studies of subduction zone paleoseismology. Subduction earthquakes are generally recorded by geologic evidence (such as sharp unconformities) that indicate sudden changes in relative sea level. In contrast, interseismic deformation usually appears in the geologic record as gradual changes in shoreline facies and microenvironments. Reconstruction of the relative sea-level changes that result from repeated deformation cycles thus requires detailed information about any late Holocene changes in relative sea level that are independent of local vertical land-level motions (Figure 5.48).

The level of the ocean surface along a shoreline is not static, but is constantly undergoing both short- and long-term changes. Some of these changes represent absolute differences in the level of the water surface relative to the geoid. They include short-term fluctuations such as diurnal tides, periodic storm surges, short-term excursions caused by variations in sea surface temperatures and wind patterns, and other meteorological conditions. These changes in sea level may be local or regional. Longer term rise or fall of eustatic sea level affects all of the world's open oceans and is caused by changes in global ice volumes, glacioisostatic- and hydroisostatic-induced variation in the shape of the geoid and tectonic modification of the shape and size of the ocean basins (Morner, 1976; Clarke *et al.*, 1978). Absolute sea level also varies from place to place due to a variety of local conditions including the latitude, the shape of the coast, and regional or local ocean currents.

Tectonic uplift or subsidence produces apparent changes in sea level along the deformed portion of the coast (see Nelson, 2007). These apparent sea-level changes are the inverse of the vertical movements of the land and can cause responses in shoreline processes that are indistinguishable from those produced by changes in absolute sea level. At any point along a coast the sum of the *absolute sea level* and the *apparent sea level* determines the position of the water surface, or the *relative sea level* (Lajoie, 1986). Paleoseismic methods for identifying subduction earthquakes rely on recognizing apparent sea-level changes by comparing the elevation of past relative sea levels with those expected from the history of absolute sea level.

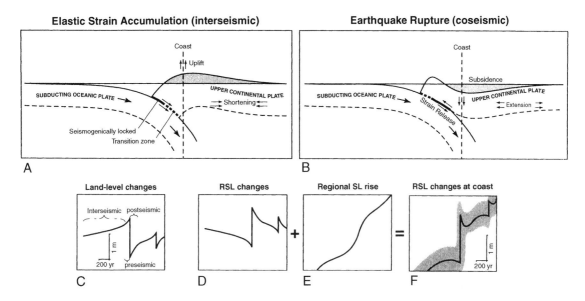

Figure 5.48: Schematic diagrams showing the pattern of (A) interseismic and (B) coseismic deformation associated with a subduction zone megathrust during an earthquake deformation cycle. (C) Land-level changes at the coast during two earthquake deformation cycles of different amplitude (scale approximate). (D) Relative sea-level (RSL) changes produced by the two cycles during a period of no change in regional (eustatic) sea level. (E) A gradual rise in RSL during the cycles that does not include short-term or small-scale changes in local and regional sea level. (F) RSL changes at a coast resulting from the sum of parts (D) and (E). In parts (A) and (B), the solid- and thin-dashed lines mark the megathrust; the thick-dashed line is a transition zone between the locked and plastically deforming parts of the plate boundary. Shading shows the relative amount of uplift and subsidence of the upper plate during the (A) interseismic and (B) coseismic parts of the deformation cycle. The shaded band in part (F) shows ±0.5 m of uncertainty in estimates of former RSLs determined from geomorphic or stratigraphic data. From Nelson *et al.* (1996b); reprinted with permission of the American Geophysical Union.

During the Quaternary period sea levels fluctuated by 100–150 m over intervals of thousands to tens of thousands of years in response to large-scale, long-term climatic changes (Bloom, 1977). Low sea-level stands were produced by withdrawal of water from the global oceans to nourish continental ice sheets and greatly expanded alpine glaciers during each of the Pleistocene glaciations. Glacial minimum sea levels were at least 140 m lower than present. Interglacial sea levels ranged within a few tens of meters of present sea level, with the highest late Pleistocene (stage 5) interglacial sea stands a few meters higher than present sea level. During still stands, coastal processes formed enduring shoreline features, particularly wave-cut terraces, beach berms, and coral reefs that preserved a record of the sea level. Where long-term uplift was sufficient to raise these strand lines above the present sea level, the emergent shoreline markers provide a measure of the net tectonic uplift.

Following the last sea-level minimum about 20,000–15,000 years ago, sea level rose rapidly as the continental ice sheets retreated. Detailed studies of late Holocene sea levels on tectonically stable coasts indicate that *eustatic sea level* has been nearly stable or oscillating within a few meters of present levels during the last 6,000 years (Bloom, 1970; Schofield, 1973). Most studies suggest the recent rise has not been constant, but the shape of the eustatic sea-level curve for the late Holocene is not well established and probably has been slightly different at different latitudes and along different coasts. Some studies of tectonically stable sites suggest a high stand near or slightly above present eustatic levels was reached about 5–6 ka following very rapid early and mid-Holocene rises. A slight decrease in eustatic sea level may have occurred along some coasts in the northern Pacific between 4 and 5 ka (Calhoun and Fletcher, 1994; Grossman *et al.*, 1994; Mason and Jordan, 1994). Minor fluctuations occurred as glacioisostatic adjustments reequilibrated land levels to the interglacial distribution of glacial ice and flooding of continental shelf and coastal margins (Clarke *et al.*, 1978). Estimates of eustatic sea level for the last several thousand years suggest a generally slow rise with the rate accelerating in the past century to 1.5–2 mm/yr (Douglas, 1991).

Landforms and deposits with a known relation to sea level can serve as geologic index points to paleo-sea levels. To be useful in reconstructing the elevation of past sea levels, an index point must have known age, elevation, indicative meaning, indicative range, and tendency (Nelson *et al.*, 1996). "Indicative meaning" is the vertical relation between the index point and the former reference water level (usually mean high tide or mean high water level). "Indicative range" is the vertical range of uncertainty in the index point's relation to the reference water level. Indicative meaning and range differ widely according to the type of index point and the form of the tidal curve (van de Plassche, 1986). "Tendency" reflects whether the index point records an increase or decrease in water level or salinity; movement of marine water toward a site defines a positive tendency.

5.10.1   Sea-Level Index Points along Erosional Shorelines

Along exposed, erosional coasts, emergent shoreline landforms known as *strand lines* provide a record of past sea levels (Lajoie, 1986). For erosional shorelines the principal datum is a *wave-cut shore platform* that commonly forms near the lower intertidal elevation (Bradley and Griggs, 1976; Merritts, 1996). Commonly shore platforms form at or near the level reached by low tides, and platforms grow in width landward as the coastal slope retreats. Shore platforms commonly have microrelief of several meters, particularly where they are cut into heterogeneous rock assemblages, and an initial seaward slope of several degrees (Bradley and Griggs, 1976). Both of these facets introduce uncertainties in relating the elevation of the platform to a past reference water level (Lajoie, 1986).

A variety of high-energy shoreline sediments cover parts of many wave-cut platforms. These deposits include coarse littoral and beach sands and gravels, and boulder and cobble lag deposits in the surf zone and active beach. Soils, terrestrial peats, freshwater marsh and lacustrine sediments, and dunes commonly form on terraces following emergence. The cover sediments overlying wave-cut platforms commonly contain datable material and also stratigraphic records related to local relative sea level (Clarke and Carver, 1992; Merritts, 1996). Terrace cover sediments can vary in thickness and add further uncertainty in assessing the elevation of the terrace surface relative to sea level at the time the terrace was cut.

The shoreward edge of wave-cut platforms, or *shoreline angle* (Lajoie, 1986), provides the best sea-level index point. According to Rose (1981), active shoreline angles around the North Sea have an indicative meaning of +1.5 to 2.0 m above mean sea level, with an indicative range (1σ) of ±0.4 to 0.6 m. By comparison, constructional beach ridges are more affected by large storms and have indicative meanings and ranges of 1.8 ± 0.8 to 3.9 ± 1.4 m (relative to mean sea level). Shoreline angles are often buried by postemergence colluvium shed from the adjacent wave-cut cliff. Detecting the shoreline angle beneath these cover sediments requires drilling or geophysics (Bradley and Griggs, 1976), which induces further components of uncertainty in the elevation of the sea-level index point.

Because elevation changes from deformation during a single subduction earthquake are <2 to 3 m (often <1 m), the rather large indicative ranges of shoreline angles (plus the added elevation uncertainty associated with buried shoreline angles) combine to make shoreline angles only marginally useful in assessment of coseismic vertical movement, except after the largest ($M_w > 8.5$) subduction earthquakes. Often biological features, such as the borings made by sessile intertidal molluscs, have smaller indicative ranges in the decimeter range, and if preserved are better index points than the more ubiquitous but less precise shoreline angles.

On tectonically active coasts, platforms are cut during the interseismic intervals when relative sea level is static or slowly rising. Coseismic uplift raises these platforms, creating emergent terraces if the amount of uplift is large enough to elevate the platform above the high tide level. Along coastlines where coseismic uplift is exceeded by local sea-level rise, shore platforms may be reoccupied repeatedly during successive deformation cycles (e.g., Leonard and Wehmiller, 1992); such reoccupation further complicates using platforms as index points.

5.10.2 Sea-Level Index Points Along Depositional Shorelines

Depositional intertidal environments are usually found in bays and estuaries and along sheltered coastlines, or where active tectonic subsidence has drowned the topography. Where sediments are abundant, such as near large river mouths, sea level is commonly approximated by extensive tidal mud flats, salt marshes, barrier bars, storm berms, beach and dune complexes, and intertidal delta fronts. Such environments provide the most complete record of local sea level and are the most sensitive to small sea-level changes. Because deposition at such sites often continues throughout the seismic cycle, stratigraphic records of both the interseismic and coseismic land-level changes may be preserved. Abundant datable material is commonly preserved in such settings.

Because many depositional processes, as well as the life zones of certain sessile intertidal plants and animals in sheltered shoreline environments, are restricted to narrow vertical ranges, detailed sedimentologic and biostratigraphic analysis can sometimes resolve the magnitude of land-level change with sufficient precision (≤ 0.5 m) to identify subduction paleoearthquakes. Most sensitive are *salt marshes*, which contain distinctive plant communities that colonize tidal flats at or just below the highest tide level (highest high water level or HHWL). These salt-tolerant herbaceous plants occupy overlapping zones commonly less than 1 m in vertical extent (Figure 5.49). Salt marshes are initiated by pioneering plants that colonize the highest part of the *intertidal mud flats*. The mud flats emerge due to sedimentation or lowering of local sea level. Different floras characterize low and high marsh zones. Along the north Pacific coast native low marsh species include *Carix lyngbyei, Distichlis spicata, Triglochin maritimum, Salicorni virginica*, and *Deschampsia caespitosa*. High marsh floral zones also locally contain these low marsh species and also commonly include *Potentilla pacifica, Grindelia stricta, Triglochin concinnum, Jaumea carnosa, Plantage maritime*, and *Orthocarpus castillejoides*. The specific assemblage and vertical range of salt marsh herbs varies regionally and locally, but in general does not exceed about 2 m. Within the salt marsh environment, low, middle, and high marsh subzones can often be recognized by indicative assemblages and index species (Figure 5.49). At sheltered sites the upper limit of these salt-tolerant assemblages is often sharp, with Sphagnum moss, terrestrial herbs and grasses, and vascular trees dominating the flora a few decimeters above the high tide line.

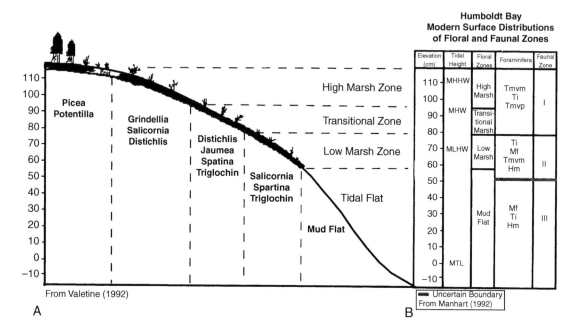

Figure 5.49: Diagram of salt marsh zonation showing the vertical ranges of diagnostic salt marsh plants and foraminifera in Humboldt Bay, northern California. Tmvm, *Trochammina macrescens* v. *macrescens*; Tmvp, *Trochammina macrescens* v. *polystoma*; Ti, *Trochammina inflata*; Mf, *Milliamma fusca*; Hm, *Haplophragmoides manilaensis*. From Li (1992); reprinted with permission of Humboldt State University, California.

At and just below the highest tide level are narrow zones containing distinct assemblages of sessile intertidal foraminiferids and marine and brackish water diatoms (Hemphill-Haley, 1992; Jennings and Nelson, 1992). Microfossil assemblages of these organisms incorporated in the stratigraphy can be used to estimate the elevation of the horizon at the time of deposition relative to the local sea level. Foraminifera indicative of high marsh environments in the north Pacific are typically dominated by *Trochammina macrescens* v. *macrescens, T. macrescens* v. *polystoma*, and *Trochammina inflata*. Low marsh assemblages are locally variable but consistently include *Miliamma fusca*. Diatoms (microscopic unicellular plants) proliferate in marine, brackish, and freshwater aquatic environments, and their siliceous shells, ranging from 10 to 50 μm in size, are commonly abundant in marsh sediments. As with intertidal Foraminiferid, different species of diatoms flourish in microenvironments with restricted salinity and reflect narrow vertical zonation in the upper intertidal level. Quantitative microstratigraphic analysis of sediment samples spaced millimeters apart across mud–peat and peat–mud boundaries is necessary to detect relative sea-level changes (Mathewes and Clague, 1994). Although the indicative range of individual species is often in the decimeter range, overlapping of species assembleges may define an indicative range in the centimeter range (e.g., Hemphill-Haley, 1992; Nelson and Kashima, 1993).

Coseismic land-level changes are reflected by abrupt replacement of indicative assemblages, while gradual changes expected from longer term interseismic motions involve a transition in species composition over centimeters or decimeters of stratigraphic section (showing positive or negative tendency). Biostratigraphic changes in tendency may be identified within lithologically homogenous units, for example, a change from freshwater to brackish diatoms found several centimeters below a sharp peat–mud contact. Such biostratigraphic evidence shows that the rise in relative sea level preceded the sharp lithologic contact, implying a gradual rather than an abrupt sea-level change (Nelson *et al.*, 1996).

5.11   The Coseismic Earthquake Horizon

Most paleoseismic evidence of subduction earthquakes is the direct result of either a sudden land-level change, shaking-induced liquefaction and mass movement, or tsunami run-up along the coastline adjacent to the rupture segment. Each of these coseismic processes results in a wide range of stratigraphic and geomorphic features. Recognition of landforms and stratigraphic horizons representing paleoearthquakes requires detailed microstratigraphic and geomorphic methods to demonstrate the sudden nature of the coseismic processes, and to distinguish evidence of paleoearthquakes from that of nonearthquake events such as storm surges, unusual tides, and other transient marine processes (Shennan, 1989; Nelson, 1992a; Nelson *et al.*, 1996).

Coseismic uplift creates a temporarily emergent shoreline. Emergence is then followed by colonization of the emergent surface by terrestrial plants, accumulation of terrestrial peat and locally derived sediments, and soil development. Geologic evidence of the suddenness of uplift may be preserved by a sharp contact between the active preseismic intertidal surface (either a wave-cut platform along erosional coastlines or beach berms, tide flats, or salt marshes in depositional settings) and the overlying subaerial sediments such as dune sand or freshwater peat. Supporting evidence of sudden uplift includes preservation of assemblages of intertidal mollusks in living position in raised beach and bay sediments (Lajoie, 1986), and borings in raised wave-cut platforms and shoreline boulders containing fossil borrowing clams (*Pholadidae penitella*) (Merritts, 1996). In such cases the fossil shell provides datable material suitable for estimating the age of the uplift event. Along sandy coasts with prevailing onshore wind, sudden uplift can trigger dune building by raising intertidal sandflats above the high water level and promoting dune building that buries the raised shoreline (Carver and Aalto, 1992), preserving datable plants and trees in the dunes. If emergence raises the upper intertidal zone into a subaerial environment that is not conducive to accumulation and preservation of sediment, weathering, soil development, and erosion modify the preuplift surface.

5.11.1   Characteristics of Coseismic Earthquake Horizons

Where coseismic subsidence submerges the coast, the *earthquake horizon* is commonly preserved as a contact between underlying terrestrial or upper intertidal wetland peats and soils, and the overlying lower intertidal mud or sand (Atwater *et al.*, 1995). Unfortunately, alternating mud–peat couplets can form in a variety of ways in addition to cycles of land-level change resulting from subduction earthquakes. Nontectonic variations in local sea level that produce peat–mud sequences include (1) short-term changes in ocean currents and sea surface temperatures, (2) changes in coastline morphology including breaching or growth of barrier spits, (3) migration of tidal channels and delta distributary systems, and (4) gradual local subsidence associated with compaction of thick sediment fills (Shennan, 1989; Nelson, 1992a; Nelson *et al.*, 1996). A critical need in paleoseismology is distinguishing mud–peat and peat–mud contacts that represent coseismic and interseismic land level changes from similar stratigraphic features produced by nontectonic sea-level changes. Nelson *et al.* (1996) propose five general *field criteria* for recognizing abrupt coseismic subsidence of coasts: (1) suddenness of submergence, (2) amount and permanence of submergence, (3) spatial extent of submergence, (4) coincidence of tsunami or liquefaction sands at the earthquake horizon, and (5) synchroneity of submergence.

The *suddenness of submergence* is reflected by the abruptness of the contact between sediments containing evidence of distinctly different elevations relative to sea level. In many bays and estuaries, sudden submergence places vegetated coastal wetlands into newly formed, highly active subaqueous depositional environments where quick burial by intertidal mud and sand is possible. In such settings the above ground stems and leaves of herbs and sedges are entombed in the mud and preserved in upright growth position (Figure 5.50) (Atwater, 1987; Jacoby *et al.*, 1995). The stumps of trees growing along the preearthquake shoreline and in brackish coastal wetlands may also be buried and preserved in intertidal mud following large subsidence events. Fossil plants entombed in growth position, especially trees, provide high-quality material for high-precision radiocarbon dating. Other indicators of sudden submergence are an abrupt change from subaerial to intertidal pollen, diatoms, or foraminifera at the peat–mud contact. Atwater *et al.* (2001) observe that most coseismic tidal deposits preseved in the geologic record probably accumulate in the first decade following the earthquake. Finally, in coseismic peat–mud sequences, peats do not coalesce when traced to the upper edges of tidal wet lands, but remain separated by mud units.

Few nontectonic mechanisms can quickly cause a permanent relative sea-level change of >0.5 m, so the *amount and permanence* of sea-level change is a critical criterion. Field evidence for >0.5-m change in relative sea level includes lithologic (peat–mud) or biostratigraphic successions that bypass one or more intermediate facies or faunal zones, which would have been present if subsidence was gradual (Hemphill-Haley, 1995). If the peat–mud environments brought into contact differ in indicative meaning by >0.5 m, and the change in relative sea level is permanent, a coseismic origin is indicated. Permanence of change is harder to prove, since areas subsided for short terms (hours, days) in regions of high deposition rate can accumulate enough sediment that subsidence may appear "permanent."

Changes in relative sea level due to earthquakes should have a much wider *spatial extent* than vertical changes of similar magnitude that may result from nontectonic mechanisms such as channel or bar migration or river flooding, which should be restricted to individual estuaries. In the Pacific Northwest, earthquake horizons that can be correlated between core holes over an entire estuary (hundreds of meters to a few kilometers laterally) or between estuaries are often considered to be coseismic, whereas thinner peat–mud couplets that pinch out over distances of a few tens to hundreds of meters may reflect nontectonic sea-level changes affecting only part of the marsh (Peterson and Darienzo, 1991).

The *coincidence of tsunami sands or liquefaction features* with peat–mud contacts is a strong independent indicator that the contact is coseismic. However, storm waves can also deposit sand in tidal marshes, so the sedimentology and geometry of sands should be assessed to distinguish storm-related sand from tsunami sand. Coseismic liquefaction features include clastic dikes, sills and sand blow deposits extruded during liquefaction, and ground failure structures including lateral spreads, slumps, and landslides (Chapter 7). Where such features can be stratigraphically traced to buried peat–mud contacts, evidence for coseismic origin is strengthened. Although liquefaction produces enduring deposits in stable geomorphic settings, many liquefaction features (particularly delicate sand blow deposits) are not easily preserved in highly active intertidal or fluvial environments.

Finally, earthquakes are much more likely to create *synchronous submergence* over broad reaches of a coast than are nontectonic mechanisms like bar formation or river flooding, which would occur at different times in every estuary. Eustatic changes in sea level would likely induce gradual changes in relative sea level, of slightly different age and stratigraphic expression, in estuaries with different shapes,

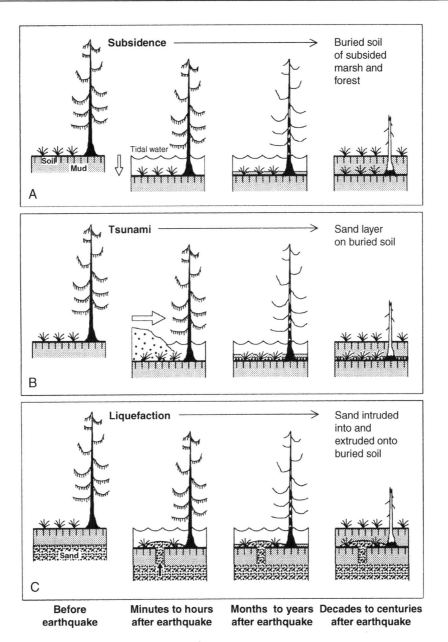

| Before | Minutes to hours | Months to years | Decades to centuries |
| earthquake | after earthquake | after earthquake | after earthquake |

Figure 5.50: Diagram showing three of the principal processes that produce the most diagnostic paleoseismic evidence of subduction earthquakes in coastal environments. (A) Soils or vegetated upper intertidal tidal wetlands that are coseismically subsided into a lower intertidal zone and covered by marine or estuarine sediments. Evidence of coseismic suddenness of the event includes (1) fragile aboveground stems and leaf bases of herbaceous plants entombed in growth position in the marine sediments, (2) tree ring cross-matches that show synchroneity of death for the buried trees, and (3) normal tree ring growth suddenly terminated at tree death. (B) Tsunami sediments deposited directly on a subsided surface. (C) Liquefaction sediments vented directly on a subsided surface. From Atwater *et al.* (1995); reprinted with permission of the Earthquake Engineering Research Institute.

tidal ranges, and sediment inputs. Wide synchroneity of submergence, when combined with suddenness of submergence, is thus good evidence for a coseismic origin. Demonstrating the synchroneity of submergence between noncontiguous sites is limited by the precision of dating methods. Radiocarbon ages of buried peats in Cascadia typically possess 2σ uncertainties of 200–500 calendar years (Nelson, 1992c; Nelson *et al.*, 1996), a range too broad to demonstrate synchroneity. At present, only tree ring dating has sufficient precision to show convincingly synchronous subsidence over broad areas.

The more of the five listed criteria that a contact satisfies, the more likely it is to have a coseismic origin. Nonseismic mechanisms can form peat–mud contacts that might satisfy one, or perhaps two, of the above criteria, but it is unlikely that any nonseismic contact could satisfy more than two of the coseismic criteria.

5.11.2   Earthquake-Killed Trees

Earthquake-killed fossil trees provide unique material for high-precision radiocarbon dating of paleoearthquakes. The dating technique requires samples of wood from earthquake-killed trees that include small numbers of annual rings of known ring position relative to the outermost ring (At water *et al.*, 1991; Nelson *et al.*, 1995). Two or more samples separated by at least a few decades (based on the count of annual rings in the section) are carefully cut from the same root or trunk section. The differential age of the samples obtained from the radiocarbon analysis can be compared with the dendrochronological age difference to verify the precision of the analysis. The technique is especially useful for young samples that yield multiple calibration ages (Figure 5.51). In such cases samples can be collected from trees with many annual rings that produce single calibrated ages, and the sample age can be adjusted by adding the annual ring count from the sample to the outermost ring. For some of the intervals that yield multiple calibrated ages the relative age of samples determined from annual rings (i.e., older versus younger) and the relative ages determined from high-precision radiocarbon analysis can be compared to assess which of the multiple calibration ages is correct. The similarity of difference in the relative ages allows discrimination of which of the several calibrated ages is valid.

The death process and characteristics of earthquake-drowned forests are particularly important to the paleoseismology of subduction zones because tree rings provide the most sensitive time records of subsidence (Atwater *et al.*, 1991). At many places along the Alaskan coast in 1964, the seaward margin of the spruce forest was drowned. Reconnaissance of earthquake-killed trees at Turnagain Arm, Alaska, shows that trees completely submerged below the high water level died during the first growing season following the 1964 earthquake. Trees that were only partially submerged often survived for one or more growing seasons, although their growth was usually retarded. Where only part of the root systems of trees was submerged, only those roots and the branches sustained by those roots were killed immediately, the remainder of the trees continued to grow. Such trees exhibit roots that have different death ages (Jacoby *et al.*, 1995). Fossil forests thus offer one of the best opportunities to test the key criteria of suddenness and synchroneity of submergence.

5.11.3   Tsunami Deposits

Subduction earthquakes generate seismic sea waves (tsunamis) by several processes, including (1) deforming the seafloor and elevating or depressing the overlaying water column, (2) shaking and exciting the water column by long-period seismic waves, and (3) triggering submarine or coastline landslides that displace large volumes of water. Tsunamis generated by submarine land slides can locally reach exceptional run-up heights of many tens to hundreds of meters, as occurred in Lituya Bay, Alaska,

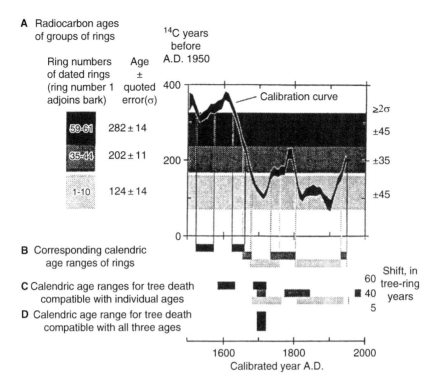

Figure 5.51: High-precision age estimates for paleosubduction earthquakes can be obtained by dating multiple samples of selected ring sequences from an earthquake-killed tree. The radiocarbon ages are matched to the calibration curve and the resulting calenderic ages are corrected for tree death date by ring counts. If the age is correct, the multiple calibrated and ring-count corrected ages for tree death should the same. From Nelson *et al.* (1995); reprinted with permission of Macmillan Journals Ltd.

in 1958 and Valdez, Alaska, in 1964. However, these extreme run-up heights from landslide-generated tsunamis are typically limited to a small part of a coastline, often a single bay, and attenuate rapidly away from the point of origin.

Tsunamis produced from coseismic uplift and subsidence of the seafloor during large subduction earthquakes are commonly regional or transoceanic in extent. Run-up is greatest along the coast adjacent to the rupture and attenuates slowly with distance from the source. The run-up height of waves along open coastlines above the rupture in Alaska (1964) and Chile (1960) generally ranged between 5 and 10 m above normal tide level, with maximum local run-up reaching as much as 30 m in Chile (Plafker and Savage, 1970; Plafker, 1972). The 1960 Chile earthquake caused considerable damage to coastal property in Hawaii and Japan from 4- to 6-m-high waves. The 1964 Alaska tsunami produced run-up heights of 2–4 m at many points along the coast of Washington and Oregon, and waves of almost 5 m at Crescent City in northern California.

Tsunamis produced by regional seafloor deformation during subduction earthquakes typically include trains of waves with periods of several tens of minutes. Successive wave crests arrive along the coast for several hours and create repeated landward surges, followed by seaward return flows of the marine water.

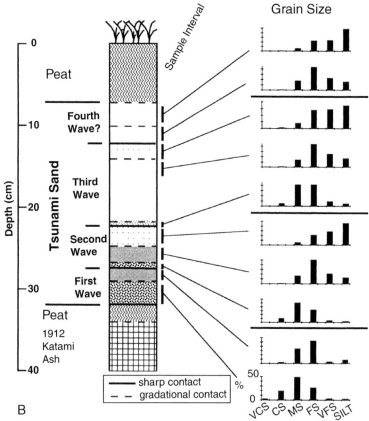

The rise and fall of sea level associated with the arrival of each wave at the coast is typically rapid; water levels can change several meters or more in a few minutes. Some combinations of wave height and coastal morphology produce tsunami bores, breaking walls of water that rush ashore with great force. More commonly run-up takes the form of a turbulent surge of water that flows rapidly landward as the level rises. The incoming surge may last for many minutes. If the wave length is long and the coastal morphology allows water levels on land to equilibrate with the tsunami wave height rapidly, the flow may become static for several minutes during the arrival of the crest, before the flow reverses and flood waters drain. Return flow is also commonly rapid and turbulent, but flow directions are strongly controlled by topography and follow preexisting drainage channels, as opposed to the more landward overland flow associated with the arrival of the wave crest. Debris berms composed of driftwood, broken or uprooted trees, and boats, pieces of buildings, and automobiles are commonly piled along the inland limit of significant flooding. Locally, large dense objects including boulders or blocks of concrete and masonry have been transported by the landward surge of large waves.

From a paleoseismiological standpoint, the tendency for tsunamis to transport and deposit sand and silt in the coastal zone is most important. Thin, often discontinuous, sand sheets have been observed in areas inundated by several modern tsunamis (Chile, 1960; Bourgeois and Reinhart, 1989; Alaska, 1964; Plafker and Kachadoorian, 1966; Clague *et al.*, 1994; Nicaragua, 1992; Bourgeois and Reinhart, 1993; Satake *et al.*, 1993; Japan, 1993; Hokkaido Tsunami Survey Team, 1993; Indonesia, 1992; Yeh *et al.*, 1993). The 1964 Alaska case may be typical. Reconnaissance of tidewater sites on the eastern side of Kodiak Island 30 years after the earthquake shows the waves did not generate preserved geologic deposits along the majority of the coastline (Carver *et al.*, 1994a). However, where environments were favorable for deposition and preservation, the waves are mainly recorded as thin discontinuous sand sheets. These sites are located where abundant sand is present seaward of a coastal lagoon, marsh, or other low-lying deposition site. At most sites the 1964 tsunami is represented by a thin (<10 cm), well-sorted, massive sand lying sharply on the preearthquake surface. At some locations the sand is normally graded and locally contains rip-up peat clasts, scattered pebbles and cobbles, and woody debris. Sand sheets deposited on tidal delta fronts at Kalsin and Middle Bay on Kodiak Island show as many as four upward-fining sequences of well-sorted sand (Figure 5.52). Records of the height and timing of seven waves were obtained from a stream gauge near the mouth of Myrtle Creek at Kalsin Bay (Plafker and Kachadoorian, 1966) and show these deposits were left by waves with 4- to 6-m run-up heights and periods of 35–55 min.

Paleo-tsunamis have been inferred from sand layers in coastal stratigraphy at many localities along the Cascadia subduction zone (Atwater, 1987; Reinhart and Bourgeois, 1987, 1989; Darienzo and Peterson, 1990; Reinhart, 1992; Clague and Bobrowsky, 1994a,b; Kelsey *et al.*, 1994; Hemphill-Haley, 1995) and locations as far apart as Scotland (Long *et al.*, 1989) and Japan (Minoura and Nakaya, 1991; Minoura *et al.*, 1994). The deposits have been found in marsh stratigraphy and in cores from coastal lakes (e.g., Kelsey *et al.*, 2005) and ponds and generally consist of one layer of well-sorted medium and fine sand.

Figure 5.52: (A) Photograph and (B) stratigraphy and grain size distribution of the sand sheet deposited at Middle Bay on Kodiak Island by the tsunami from the 1964 Alaska earthquake. The tsunami at this location included at least six waves with run-up heights of up to about 4–6 m. The deposit is a 25-cm-thick sand sheet composed of up to four upward-fining sequences that are interpreted to have been deposited by successive wave pulses. The mean grain size of each of the four upward-fining sequences also decreases upward, perhaps reflecting winnowing of the available sand by each wave pulse. Tsunami sands at some nearby sites do not show upward fining, but instead are massive.

Some layers are composed of several upward-fining sequences interpreted to represent successive wave pulses, whereas other deposits are massive. Dominey-Howes *et al.* (2006) provide a useful overview of recognition criteria.

Sand beds are also deposited in intertidal and marsh environments by channel migration, river floods, and storms, so some criteria are needed to distinguish tsunami sand from nontsunami sand (Dawson *et al.*, 1991). Field criteria for a tsunami origin would suggest strong landward currents of unusual strength, and an offshore or bayward sand source, and include (1) landward thinning of the sand sheet (Reinhart, 1992), (2) landward fining of the sand, (3) a marine source for the sand, as indicated by marine diatoms (Kelsey *et al.*, 1994), and (4) landward rise of the sand sheet and onlapping of sand onto subaerial deposits or soils (Atwater and Moore, 1992; Clague and Bobrowsky, 1994b).

5.11.4   Coral Atolls and Reefs

The tops of coral atolls and reefs have a fixed relationship with mean sea level and can function as index points in the warmer seas. Although this relationship has been known for decades, work by Kerry Sieh and students in Sumatra between 1996 and 2004 refined the precision with which coral microatolls could be used to infer paleoearthquake characteristics (Zachariasen *et al.*, 1999; Natawidjaja *et al.*, 2003). In the best circumstances, microatolls preserve evidence of relative sea level changes during the slow interseismic part of the cycle as well as the rapid coseismic part, over a large area. This spatial distribution permits construction of deformation profiles both perpendicular to the megathrust and along its strike, a combination not possible with shoreline markers. The pre-2004 work was unexpectedly verified by the occurrence of the 16-December-2004 Aceh-Andaman earthquake (M_w = 9.2, subsurface rupture length 1600 km), which coseismically raised microatolls at its southern end (Meltzner *et al.*, 2006).

5.11.5   Summary of Stratigraphic Evidence for Paleoseismicity

The postseismic and interseismic stratigraphy of coasts above active subduction zones results from a complex interplay between coseismic movement (uplift and subsidence), interseismic movement (uplift and subsidence), the ratio of net coseismic to interseismic movement, and nontectonic factors such as coastal geomorphic processes and deposition rates. Figure 5.53 represents a preliminary attempt to show how these factors combine to produce certain types of stratigraphy, as described in Sections 5.9 and 5.10. The representation neglects factors such as eustatic sea-level changes, so it must be viewed more as an aid to visualization than as a rigorous predictive model. Nevertheless, the diagram does accurately portray the type of stratigraphy encountered at various locations on coasts of the Pacific Northwest.

On exposed, erosion-dominated coasts, net subsidence results in drowning of the coast and submergence of any paleoseismic stratigraphy. Net uplift results in flights of marine terraces veneered with thin littoral sediments. Where coseismic movement is almost exactly matched by equal but opposite interseismic movement, and deposition rates are low, the rocky coasts eventually return to their former positions with respect to sea level, and little geologic evidence may be created from even great paleoearthquakes (e.g., much of the Prince William Sound coast).

On sheltered coasts and embayments with higher deposition rates, various combinations of coseismic and interseismic movement produce peat–mud sequences. Where net subsidence is greater than deposition, the coast will be drowned and paleoseismic stratigraphy submerged. Marine terraces and raised berms form in areas of net uplift, usually areas affected by local upper plate faults and anticlines.

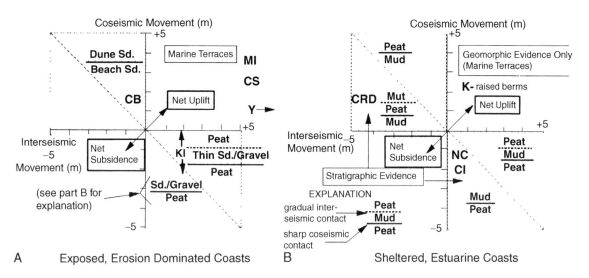

Figure 5.53: Schematic diagrams showing how permutations of coseismic movement (vertical axis), interseismic movement (horizontal axis), and coastal deposition rates (A, B) affect coastal morphology and stratigraphy (diagrammatic sections in various quadrants; see explanation). Shaded areas show the usual fields of coseismic uplift followed by interseismic subsidence (upper left) and coseismic subsidence followed by interseismic uplift (lower right). Dashed diagonal line separates regions of net uplift and net subsidence. Field examples described in the text from Alaska and Cascadia are shown by bold letters. (A) Exposed coasts of low deposition rate. MI, Middleton Island; CS, Cape Suckling; Y, Yakataga–Icy Bay; CB, Clam Beach; KI, Kodiak Island. The first three listed sites have undergone both coseismic and interseismic uplift, hence their marine terrace geometries. (B) Sheltered coasts with high deposition rates. K, Katalla; NC, northern Cascadia; CRD, Copper River delta; CI, Cook Inlet.

Only where coseismic and interseismic movements are opposite in sign and tend to cancel, and deposition rates are moderate to high, will peat–mud sequences form and be accessible for paleoseismic study.

5.12   Paleoseismic Evidence of Coseismic Uplift

At present, the level of paleoseismic investigations for most subduction zones is reconnaissance level at best, and detailed studies have been conducted at only a few localities on those subduction zones that have received the most attention. No subduction zones have been studied in sufficient detail to allow more than a general characterization of the late Holocene paleoeoseismicity, and for those the record extends back only a few cycles. Much of the field evidence interpreted to reflect subduction earthquakes is the result of studies in Alaska and along the Cascadia subduction zone. Although we discuss only North American case histories here, due to our familiarity with the field evidence, similar studies in Japan (e.g., Ota, 1975; Yonekura, 1975; Matsuda *et al.*, 1978; Ota and Yoshikawa, 1978; Maemoku, 1988, 1990) and Chile (Plafker, 1972; Atwater *et al.*, 1992; Nelson and Manley, 1992; Bartsch-Winkler and Schmoll, 1993) describe many of these same phenomena.

5.12.1   Alaska

Paleoseismic evidence of coseismic uplift on the eastern part of the Aleutian–Alaskan subduction zone has been described for the Yakataga coast near Icy Bay (Jacoby and Ulan, 1983), at Cape Suckling (Plafker, 1969b) and Middleton Island (Plafker and Rubin, 1978; Plafker *et al.*, 1992). The principal evidence is flights of raised late Holocene terraces that are interpreted to have been elevated suddenly during rupture of the eastern end the subduction zone. For an overview, see Carver and Plafker (2008).

Yakataga–Icy Bay: The Yakataga coast is situated along the Yakataga segment of the subduction zone which did not break in 1964 (Figure 5.54). Coastal processes along the Yakataga segment are dominantly erosional. Between Cape Yakataga and Icy Bay much of the present shoreline is bordered by wide intertidal wave-cut shore platforms and actively retreating sea cliffs. The lowest emergent terrace is about 8–10 m above the active shore platform, and each of the older late Holocene terraces is elevated 8–10 m above the next (Jacoby and Ulan, 1983; Plafker *et al.*, 1992). Radiocarbon age estimates for the terraces range from about 1000 years for the lowest to about 5000 years for the highest. The terraces are continuous along most of the coast between Icy Bay and Cape Yakataga, a distance of about 60 km.

The flights of raised terraces along the Yakataga coast indicate rapid uplift, probably largely coseismic and associated with rupture of the eastern end of the Aleutian–Alaskan subduction zone. The presence of these bold late Holocene terraces and the lack of historic rupture along the Yakataga of the coast has led some paleoseismologists to identify this segment as the likely site for a future great subduction earthquake (Nishenko and McCann, 1981).

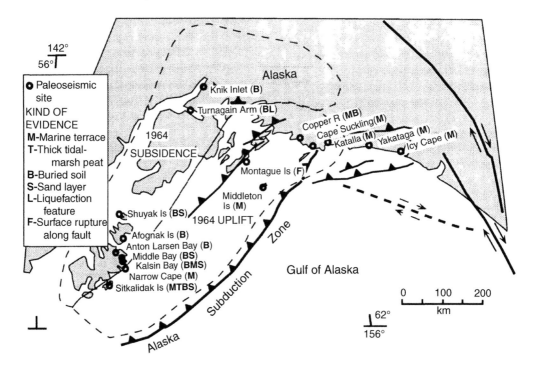

Figure 5.54: Map of the eastern end of the Alaska–Aleutian subduction zone, showing locations of principal sites of paleoseismic evidence mentioned in the text. Dashed lines outline areas of coseismic subsidence (to the northwest) and uplift (to the southeast) in 1964.

Cape Suckling: Cape Suckling, located at the eastern end of the 1964 rupture, was uplifted about 1.5 m in 1964. Evidence of three previous episodes of uplift are present in the form of three raised late Holocene marine terraces at Cape Suckling and on Kayak Island. There terraces are presently at elevations of 6–25 m. Conventional radiocarbon age estimates for the four uplift events along the Yakataga coast indicates that the most recent dates to about 700–900 years ago, with the others occurring within the past 4000 years (Plafker, 1969b).

Middleton Island: At Middleton Island about 3.3 m of uplift in 1964 produced a new terrace (Plafker, 1969b; Plafker and Rubin, 1978; Plafker *et al.*, 1992). The uplift of Middleton Island in 1964 produced terraces that are about half as high as those produced by the last three strain cycles recorded by elevated terraces on the island. Several alternatives have been suggested for this difference between the 1964 event and the paleoseismic record of past events to affect the island. One hypothesis is that much of the slip was partitioned onto the Montague Island thrust system in 1964, but not in previous earthquakes. This explanation is consistent with the lack of obvious evidence for anomalously large amounts of uplift resulting from upper plate thrusting on Montague Island and evidence for repeated late Holocene growth of the Patton Bay, Hanning Bay, or other thrust faults on the islands of northern Prince William Sound. An alternative hypothesis explains the uplift deficit by assuming that future contributions will be made from the Yakataga segment of the subduction zone, and that uplift will increase to the full amount recorded by the pre-1964 terraces when the Yakataga segment ruptures. This hypothesis requires that the rupture surface in 1964 will be reactivated at its eastern end along with the section of the fault that remained locked in 1964.

Katalla: In addition to raised terraces, several of the late Holocene uplift events that affected the eastern end of the Aleutian–Alaskan subduction zone are represented by elevated lagoons, storm berms, and delta tidal flats in the lower Katalla River valley (Figure 5.55). During the late Holocene the Katalla River valley has been the location of rapid coastal progradation. Large storm berms have repeatedly formed across the mouth of the Katalla valley, each seaward of the previous one by hundreds of meters. Small lagoons, tidal flats, and salt marshes developed behind the berms. More than 20 prominent storm berms are preserved in the lower 5–7 km of the valley. The present active beach berm and lagoon has formed since the 1964 earthquake, which caused about 1 m of uplift. The pre-1964 tidal lagoon and salt marsh were elevated above the highest tides and the marsh is being replaced a juvenile spruce forest. Former tidal flats and marsh surfaces behind relic storm berms further inland are separated by elevation differences comparable to the riser heights of the prominent terraces along the Yakataga coast. Stratigraphy of the raised tidal lagoons shows marine mud and salt marsh peat containing fossil salt marsh plants, which are sharply overlain by subaerial sphagnum peat and subaerial plants and trees.

Copper River Delta: The most detailed and best documented paleoseismic record for the eastern end of the Aleutian–Alaskan subduction zone is based on interpretation of sediments accumulated in the extensive delta of the Copper River (Plafker *et al.*, 1992). The delta is located near the eastern end of the coast uplifted in 1964. During the earthquake the delta was raised about 2 m. Prior to the 1964 uplift the delta front included an area of several hundred square kilometers of unvegetated muddy tidal flats and intertidal marsh between distributaries of the Copper River, a large, highly seasonal and very sediment-laden river. The earthquake elevated this region sufficiently to force the coastline seaward by several kilometers and to raise large tracts of the delta front above the reach of the highest tides. These former mud flats and marshes have quickly been colonized by dense vegetation. Preuplift sites of marine mud and sand deposition are now forming postuplift peats and organic soils.

The stratigraphy along the front of the delta consists of 2- to 3-m-thick layers of marine silt interbedded with 10- to 20-cm-thick sequences of peat and organic soil (Figure 5.56). Some of the soils contain roots

Figure 5.55: The lower Katalla River valley, located on a rapidly emerging coast supplied with large volumes of sediment from glacial meltwater rivers, is rapidly prograding and leaving a record of former shorelines as large forested storm berms. Marshes between the berms contain peat with salt marsh plant fossils, capping intertidal mud containing marine shells. The elevation of the berm-marsh sequences increases inland, reflecting repeated episodes of uplift. During the 1964 earthquake the region was elevated about 1 m. Subsequent to uplift, a new storm berm and lagoon have formed seaward of the 1964 beach, and spruce trees have become established on parts of the preearthquake salt marsh.

and stumps of trees that grew while the delta front was emergent. The silt is laminated, with the laminations (each 5–15 mm thick) probably reflecting seasonal deposition of glaciofluvial sediment from the river. These silt layers were deposited on the submerged delta front when the delta was in the "down" position during the latter part of the interseismic interval. The alternation of marine and subaerial sediments shows the surface has oscillated above and below sea level repeatedly during the late Holocene. The peats were formed after uplift events raised the mud flats, as in 1964, and marshes and forests replaced the intertidal environments of the preseismic intervals. A gradual postseismic and eustatic sea-level rise submerged the delta surface, drowned the marshes and forests, and initiated silt deposition. Continued subsidence and deposition formed the thicker silt layers between uplift events. Tide gage records at Cordova, located at the western edge of the delta, show a relative sea-level rise of 9.7 ± 0.5 mm/yr since 1964, a rate that will resubmerge the delta front within about two centuries (Savage and Plafker, 1991).

5.12.2   Cascadia Subduction Zone

Along most of the southern Oregon and northern California coast, raised late Pleistocene glacioeustatic marine terraces record long-term net uplift (Carver, 1987a; Kelsey and Carver, 1988; Merritts and Bull, 1989; Kelsey, 1990; McInelly and Kelsey, 1990). The raised terraces are located along parts of the coast that are closest to the subduction zone deformation front (Figure 5.57), presumably seaward of the zero isobase which probably intersects the shoreline in central Oregon (Peterson *et al.*, 1993). Seismic reflection and refraction surveys of the continental shelf and slope along the Cascadia subduction zone reveal that a 70- to 100-km-wide belt of young thrusts and folds deform the seaward tip of the upper plate (Clarke, 1992; Goldfinger *et al.*, in press). The southern part of the fold and thrust belt intersects the coast and extends onshore in southern Oregon and northern California. Late Quaternary growth of the fold and thrust belt has thus strongly influenced the vertical movements of this part of the coastline and has produced net uplift at some places and subsidence along other parts of the coast. The large differential vertical tectonic motions of the southern Oregon and northern California coasts are in striking contrast to the relatively uniform and small vertical movements during the late Quaternary along the northern Oregon and Washington coasts.

Raised late Holocene marine terraces are present along some parts of the coast in northern California at the south end of the Cascadia subduction zone (Clarke and Carver, 1992; Merritts, 1996). The highest terraces are localized along the crests of thrust anticlines above major faults in the fold and thrust belt. At Cape Mendocino several raised terraces between ~300 and ~6000 years old are present and have been interpreted to have been elevated by interplate earthquakes at the southern tip of the subduction zone (Merritts, 1996). Support for this interpretation comes from the 1992 Cape Mendocino earthquake which was accompanied by as much as 1.4 m of uplift along a 25-km-long section of the coast between Punta

Figure 5.56: (Continued)

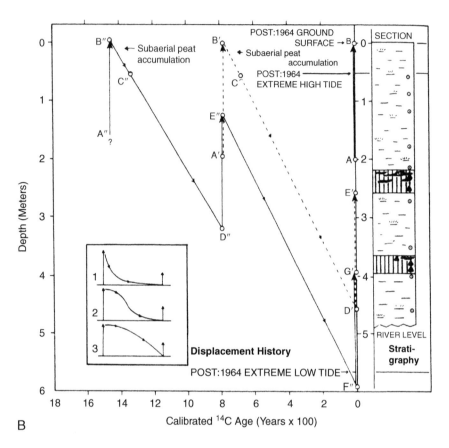

Figure 5.56: (A) Stratigraphy exposed in a channel bank in Alaganak slough on the Copper River delta, Alaska. This area was a tidal mud flat prior to about 2 m of uplift during the 1964 earthquake. The modern vegetation is growing on about 2 m of thinly laminated estuarine mud that buries a 50-cm-thick peat layer (dark layer near the shovel at the bottom of the exposure). The buried peat is composed of remains of the same species that presently grows at the site. Additional peat–mud sequences underlay the exposed section. The peats formed postseismically after emergence such as in 1964. Rapid postseismic subsidence and a rise in sea level have resulted in submergence of the postseismic subaerial peats and burial by intertidal silt, probably as annual couplets averaging about 1–2 cm thick. (B) Stratigraphy and displacement history for the Copper River delta assuming coseismic uplift events comparable to the 1964 uplift and linear interseismic submergence at a rate of about 6 mm/yr. Coseismic uplifts indicated by vertical lines (A–B, A′–B′, A″–B″). Peat accumulation occurs during the interval from coseismic uplift to resubmergence below extreme high tide (post-B, B′–C′, B″–C″), and accumulation of intertidal deposits occurs during the remainder of the interseismic interval (C′–D′, C″–D″). The inset diagram shows three possible nonlinear interseismic subsidence paths. From Plafker *et al.* (1992); reprinted with permission of Springer-Verlag Publishers.

Gorda and Cape Mendocino (Carver *et al.*, 1994b; Prose, 1994a). The location and mechanism for the main shock (M$_w$ = 7.1) and the pattern of aftershocks indicate the causative fault was a low-angle east dipping thrust located at or very close to the plate interface (Oppenheimer *et al.*, 1993).

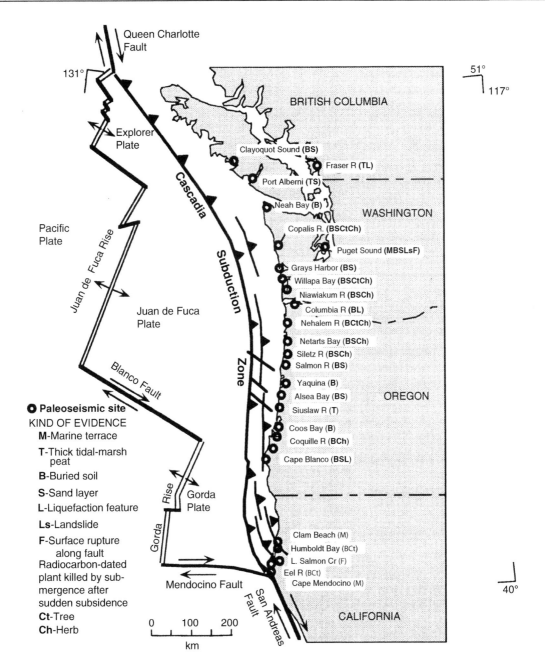

Figure 5.57: Map of the Cascadia subduction zone showing locations of principal sites of paleoseismic evidence for the last event(s) about 300 years ago. Modified from Atwater *et al.* (1995).

Raised late Holocene terraces have also been described in northern California at Clam Beach (Carver and Aalto, 1992; Clarke and Carver, 1992). The broad raised terrace is covered with an upper sequence of dunes that overlie a buried soil with rooted fossil spruce and fir trees (Figure 5.58A). The buried paleosol

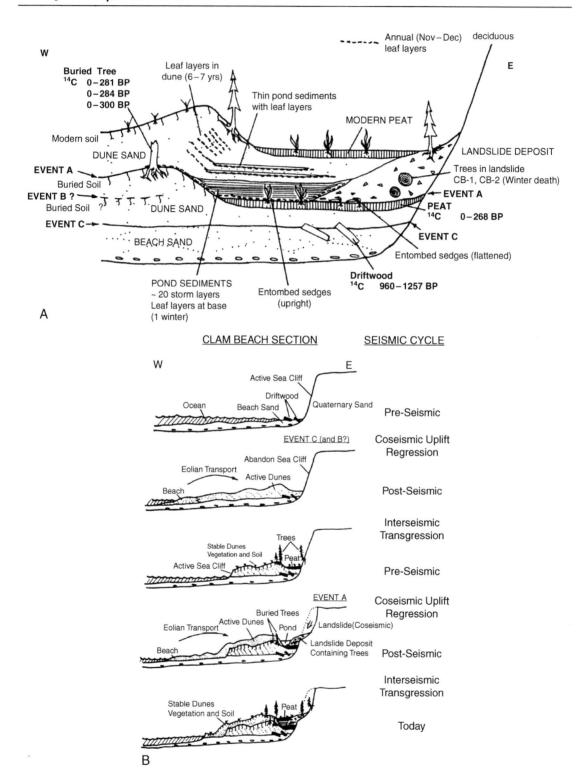

A

CLAM BEACH SECTION   SEISMIC CYCLE

B

Figure 5.58: (Continued)

is developed into a lower sequence of dunes that overlie beach sand containing driftwood. The ^{14}C age for the driftwood is about 1100 years, and the age of the buried trees and soil is about 300 years. The terrace is cut into highly faulted and folded mid-Pleistocene marine sands and muds and the abrasion platform is presently several meters above high tide level. The terrace and cover sediments are interpreted to contain a record of two coseismic uplift events (Figure 5.58B). Each raised the coast enough to regress the shoreline and expose a wide strip of sandy seafloor to strong onshore winds, nourishing dunes along the landward edge of the elevated shoreline. Following each uplift event the supply of sand was diminished by sea-level transgression, and the dunes stabilized. Vegetation was established on the dunes and soil developed during the interseismic intervals.

The occurrence of raised late Holocene terraces along the southern part of the Cascadia subduction zone where the deformation front is closest to the coast indicates that this part of the coast is located above the locked part of the plate interface. In this respect it may be similar to the Prince William Sound region of Alaska. An important aspect of the coastal geomorphology of Prince William Sound is the position in the strain cycle that is considered; the pre-1964 geomorphology was a drowned coastline almost completely lacking Holocene terraces except at the seaward tip of the upper plate very close to the deformation front. The northern California and southern Oregon coast also have raised terraces nearest to the deformation front and a drowned shoreline further inland. This geomorphic pattern may be indicative of a preseismic position in the strain cycle for the Cascadia subduction zone as well.

5.13 Paleoseismic Evidence of Coseismic Subsidence

Stratigraphic evidence of coseismic subsidence has been reported from dozens of sites along the Alaskan coast and bays and estuaries in northern California, Oregon, Washington, and Vancouver Island, British Colombia (Atwater, 1987; Combellick, 1991, 1993; Atwater *et al.*, 1995; Gilpin, 1995). At most sites intertidal mud overlies salt marsh peats. At a few sites in both Alaska and Cascadia intertidal mud containing salt marsh macrofossils and marine and brackish diatoms cover subaerial soils, forest litter mats, and rooted trees, and this unconformity is interpreted as an earthquake horizon.

5.13.1 Alaska

Cook Inlet—Turnagain and Knik Arms: Paleosubduction earthquakes have been interpreted from stratigraphic evidence of relative sea levels during several pre-1964 strain cycles at several sites along the shores of Turnagain and Knik Arms, two large shallow drowned glacial valleys at the upper end of Cook Inlet. The near-shore processes in the parts of the arms are very dynamic and driven by extreme tidal ranges reaching 15 m in some places. Very large amounts of sediment from many glacial melt water rivers

Figure 5.58: (A) Cross section of the shoreline angle and cover sediments on Clam Beach, northern California. The abrasion platform of the raised terrace is presently about 1 m above highest tide level and is covered by a sequence of beach and dune sand containing a buried soil with trees in growth position, and landslide and pond sediments that bury and entomb sedges. (B) The evolution of the terrace stratigraphy is interpreted as resulting from two cycles of coseismic uplift that regressed the shoreline and exposed the seafloor to onshore wind that deposited sand dunes. The landslide was probably seismically triggered and filled the back-dune hollow to create the pond. Leaf layers in the dune above the pond sediments probably reflect annual layers.

drain into the head of Cook Inlet. At low tide large parts of upper Cook Inlet consist of exposed mud flats and extensive salt marshes.

The paleoseismic sites are located along the shoreline in salt marshes and low shoreline areas that were submerged during the 1964 earthquake (Figure 5.59). Prior to the 1964 submergence the vegetated subaerial surfaces were accumulating peat and organic soils. Spruce and willow trees grew around the edges of the marshes. Coseismic submergence initiated rapid sedimentation in places where the former emergent and vegetated surfaces were submerged by postearthquake high tides. At some places the sedimentation was rapid enough to bury the former surface and entomb many of the organic components of the surface (Bartsch-Winkler and Schmoll, 1987).

The composite stratigraphy suggests six to eight subsidence events are recorded in the stratigraphy of the upper Cook Inlet region during the past 5000 years. The age of the penultimate event has been estimated from conventional ^{14}C dating of the outer rings of fossil spruce roots and the upper 2–3 cm of peat from the first buried pre-1964 surface. The ages suggest the event prior to 1964 was about 750–950 years ago

Figure 5.59: An earthquake-killed forest at Girdwood on Turnagain Arm, Alaska, records the 1964 earthquake. This locality experienced about 2 m of subsidence in 1964, flooding the forest marginal to the bay and depositing mud over the former forest floor. The prominent bench extending from beneath the buried 1964 soil to the lower right corner of the photograph is the subsided peat layer and soil resulting from the penultimate event about 800 years ago. Another older buried land surface and associated peat form a subtle bench in the intertidal mud flat along the right side of the photograph near the person in the far upper right corner. Postseismic rebound and sedimentation have raised the subsided land surface into the upper intertidal zone and salt marsh plants have replaced the spruce forest.

(Combellick, 1993). Tide gage and geodetic measurements show the upper Cook Inlet region has been undergoing rapid uplift since the 1964 earthquake (Savage and Plafker, 1991).

Kodiak Archipelago: Stratigraphy reflecting several cycles of submergence and emergence has been described from tidal marshes and shoreline deposits on Kodiak and nearby Afognak and Shuyak Islands near the southwest end of the 1964 rupture (Gilpin, 1995). The sites delineate a transect across the region that underwent coseismic subsidence and the zero isobase in 1964. The axis of maximum subsidence extended along the eastern side of Kodiak Island and amounted to about 2 m. The downwarped region was highly asymmetrical, with the amount of subsidence gradually decreasing to the west across the Kodiak archipelago for nearly 150 km. The downwarp was highly asymmetrical with the axis of maximum subsidence located along the eastern part of the downwarped region. The eastern limb of the downwarp was much steeper, the zero isobase located on the eastern shoreline of Kodiak Island about 20–25 km from the axis of maximum subsidence.

Remeasurement of tidal bench marks and tide gauge records since the earthquake shows that the region that subsided during the earthquake has been uplifting at a rapid rate during the 30 years since the earthquake (Gilpin *et al.*, 1994b). The tide gauge at Kodiak City, which subsided about 2 m during the earthquake, has shown average postseismic uplift of 17.5 ± 0.8 mm/yr (Savage and Plafker, 1991). Other tidal benchmarks in the region of subsidence on Kodiak, Afognak, and Shuyak Islands also have been uplifted since the earthquake in 1964. The recovery during the 30-year postseismic interval averaged about 60% of the coseismic subsidence, and at the southern edge of the 1964 subsidence region the uplift has exceeded the subsidence.

Intertidal deposits in sheltered bays and fjords along the ragged coastline of the Kodiak islands contain sediments that record the 1964 earthquake and several prehistoric subduction earthquakes (Gilpin, 1995). The sites are mostly small salt marshes and delta fronts at the mouths of small rivers and streams. Postsubsidence deposits include thin and discontinuous layers of silt, sand, and gravel that lie on the presubsidence sphagnum peat or soil layer. At many places salt marsh vegetation has become established and is depositing estuarine peat. In contrast to the sediment-rich depositional environments at the Cook Inlet and Copper River, the bays and estuaries on Kodiak Island are typically sediment poor. The resulting stratigraphy is dominated by peat, both estuarine and subaerial. Upper intertidal environments deposit salt marsh peat characterized by the predominance of *Carix lynbgei* with *Triglochin maritima* also common. When the marsh and delta front sites are the upper reach of highest tides, as they were prior to the 1964 subsidence, sphagnum peat is deposited. Volcanic ashes from the Alaska Peninsula volcanoes are interbedded with the peats in the sections.

The stratigraphic record of at least three paleosubduction earthquakes has been interpreted from the marsh stratigraphy in the Kodiak archipelago (Figure 5.60). Carbon-14 age estimates place the penultimate event near 400–500 years BP, and the previous two subduction earthquakes at about 800 and 1300 years BP. At most sites the earthquake horizons representing three earthquakes are closely spaced, only 10–20 cm apart. The close spacing, when considered in light of the rise in sea level over the age of the sediments, indicates that the region experiences more interseismic uplift than coseismic subsidence. Average interseismic uplift rates are as much as 2 m/ka, with much of that occurring in the postseismic interval over a few decades.

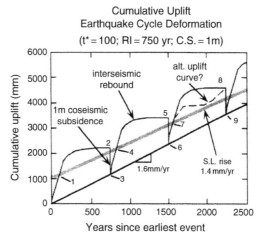

Cumulative Uplift
Earthquake Cycle Deformation
(t* = 100; RI = 750 yr; C.S. = 1m)

Years since earliest event
Numbers correspond to horizons shown on section below

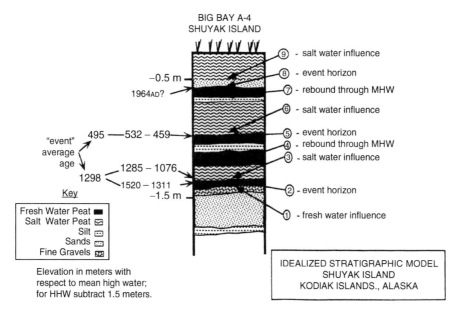

BIG BAY A-4
SHUYAK ISLAND

⑨ - salt water influence
⑧ - event horizon
⑦ - rebound through MHW
⑥ - salt water influence
⑤ - event horizon
④ - rebound through MHW
③ - salt water influence
② - event horizon
① - fresh water influence

−0.5 m
1964AD?

"event" average age
495 —— 532 − 459
1298
1285 − 1076
1520 − 1311
−1.5 m

Key
Fresh Water Peat ■
Salt Water Peat ▨
Silt ▤
Sands ▦
Fine Gravels ▨

Elevation in meters with
respect to mean high water;
for HHW subtract 1.5 meters.

IDEALIZED STRATIGRAPHIC MODEL
SHUYAK ISLAND
KODIAK ISLANDS., ALASKA

Figure 5.60: Paleoseismic interpretation of late Holocene stratigraphy along Skiff Passage on Big Bay, Shuyak Island, Alaska. The sediments consist of terrestrial or fresh water peat alternating with layers of marine silts, sand, and gravel. The upper contacts of the peats are sharp and appear to reflect sudden submergence. The cumulative uplift diagram shows a possible relative sea-level model for the origin of the stratigraphy. From Gilpin *et al.* (1994c).

5.13.2 Cascadia Subduction Zone

The most widespread and compelling evidence of prehistoric interplate earthquakes originating on the Cascadia subduction zone is found in late Holocene stratigraphy in many bays and estuaries from northern California to central Vancouver Island (Atwater, 1987, 1992; Grant and McLaren, 1987; Vick, 1988; Darienzo and Peterson, 1990; Clarke and Carver, 1992; Nelson, 1992a; Clague and Bobrowsky, 1994a). Atwater and Hemphill-Haley, 1997; Kelsey *et al.*, 2002, 2005; Witter *et al.*, 2003. The stratigraphy consists of layers of peat that are interbedded with estuarine sand, silt, and mud (Figure 5.61). The sediments are found along the margins of bays and estuaries where low energy intertidal environments have persisted during the last several thousand years and salt marshes have formed and deposited layers of peat. The peat layers are usually 10–30 cm thick and separated by 0.5–1 m of mud (Figure 5.61A). At many places the upper contact of the peat with the overlying mud records an abrupt change in local depositional environment marked by the submergence of the marsh into the lower intertidal zone and rapid formation of tidal mud flats. In contrast, the change from mud to peat usually appears gradually over many centimeters and is interpreted to reflect the gradual establishment of vegetation on the tidal mud flats as clastic deposition raises the sediment surface into the low-marsh environment.

The peat is composed of the remains of herbaceous plants that indicate the highest tide level to within a few tens of centimeters. Along the Pacific Northwest coast native salt marsh plants that have been used to define this upper intertidal position include *Triglochin meritimum, Salicornia virginica, Jaymea carnosa, D. spicata*, and *G. stricta*. These plants dominate the flora within a meter or less of the elevation of highest tidal inundation where there is little competition with other plants that are not adapted to the saline conditions. Assemblages of these plants are sometimes indicative of vertical subdivisions of the upper inter tidal zone. *S. virginica, T. meritimun, and J. carnosa* are common low and middle marsh species, while *D. spicata* and *G. stricta* are usually restricted to the high marsh (Figure 5.49). The fossilized rhizomes and stem bases of many of the salt marsh plants can be identified in the field. *T. meritimum*, for example, produces distinctive V-shaped rhizomes and large sheathed stem bases that are frequently preserved and easily identified. *G. stricta* has large reddish woody roots that are also easily identified in the sediments. The presence of fossil salt marsh plants allows estimation of the elevation of deposition of the buried peats relative to the elevation of modern marshes. This elevation difference is the total relative sea-level change since formation of the peat and includes the primary and secondary coseismic subsidence, tectonic and nontectonic postseismic vertical motions, and eustatic sea-level changes.

Rooted stems, branches, and leaves of the herbaceous marsh plants have been found entombed in growth position in the basal part of the overlying mud in some bays along the Cascadia subduction zone (Figure 5.61B). The rooted marsh plant stems and leaves extend upward into the overlying mud a few centimeters where many are broken off or have decayed away. However, some are bent over and preserved along bedding surfaces in the mud. The dead leaves and stems of the herbaceous plants are fragile, and if exposed to weathering most decay in a year or less. The preservation of the fragile plant remains in growth position reflects sudden submergence and rapid burial. Where dense marsh vegetation has been buried quickly the peat–mud contact may be gradational over several centimeters. This stratigraphy is inferred to result from coseismic subsidence caused by slip on the Cascadia megathrust landward of the downdip limit of rupture.

Layers of sand interpreted as tsunami deposits directly overlie some of the buried peats at many of the buried marsh sites from northern California to Vancouver Island. The deposits are composed of very well sorted sand, generally fine to medium grain size, and range from a few millimeters to tens of centimeters thick. At Willapa Bay (Figure 5.62) on the southern Washington coast the mineralogy and distribution of

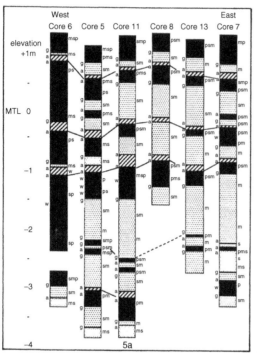

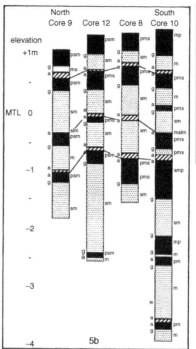

A

B

Figure 5.61: (Continued)

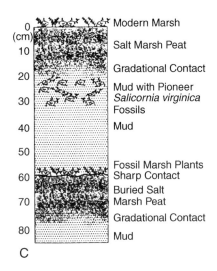

C

Figure 5.61: (A) Core logs and stratigraphic correlations between core sites at Netarts Bay marsh in northern Oregon. Up to seven sudden submergence events are reflected by the peat–mud couplets. Sand interpreted to have been deposited by large local tsunamis lies on the submerged peat surfaces in many cores. The stratigraphy shows local variation from core to core. From Peterson and Darienzo (1988); reprinted with permission of the Oregon Dept. of Geol. and Min. Industries. (B) Photograph of the upper buried peat and overlying mud–peat couplet at Mad River Slough in Humboldt Bay, California. (C) Diagram of the upper peat–mud couplet at Mad River Slough in Humboldt Bay, northern California. The sharp upper contact of the peat includes above-ground stems and leaves of salt marsh herbs entombed in the overlying intertidal mud. The upper part of the mud layer contains fossils of pioneer marsh plants and grades upward into the overlying modern peat. The present salt marsh plant assemblege is the same as that preserved in the buried horizon. From Jacoby *et al.* (1995); reprinted with permission of the Geol. Soc. of Am.

the sand indicates it was transported from the barrier bars and beaches fronting the estuary landward into the head of the bay (Reinhart and Bourgeois, 1987, 1989). The sand sheets in other bays also appear to have been derived from beaches and barrier bars seaward of the marsh sites, suggesting that landward transport occurred as large tsunamis overtopped the barrier bars. Peters *et al.* (2003) have a database of these deposits.

The late Holocene buried marsh sediments around several Pacific North-west bays and estuaries contain layers of salt marsh peat that grade laterally landward into organic rich buried soils. These soils were formed above the reach of the highest tides and supported spruce and on the Washington coast, cedar forests. These forests were submerged into the upper intertidal zone during the paleosubsidence events, and fossil stumps and roots of large trees are entombed in growth position in the overlying estuarine mud at the heads of the estuaries (Figure 5.50). Most of the trees are deeply decayed and missing their outermost rings, but a few have well-preserved annual ring sequences that extend to the bark. The record of growth recorded by these trees indicates that submergence was rapid and synchronous over a large area (Yamaguchi *et al.*, 1989; Atwater and Yamaguchi, 1991). At the south end of the subduction zone dendrochronologic cross-dating ofroots from eight stumps in a buried forest exposed in channel banks of the Mad River slough on the north end of Humboldt Bay shows all died during a 4-year period (Jacoby *et al.*, 1995).

The age estimated for the last subduction earthquake, based on high-precision ^{14}C analysis of earthquake-killed trees at the Copalis River and Willapa Bay in southern Washington (Figure 5.62), is about AD 1700 with a 2σ uncertainty of about ± 20 years (Atwater *et al.*, 1991). Similar analysis of trees interpreted to have been killed by the last subduction earthquake at the Nehaliem River in Oregon (Nelson and Atwater, 1993), and Humboldt Bay in northern California (Carver *et al.*, 1992) also results in a calibrated age of about AD 1700. These estimates are in agreement with the results of mass AMS ^{14}C dating for entombed herbaceous plants from seven bays and estuaries along the Oregon and southern Washington coast (Nelson and Atwater, 1993). The similar ages for the last large earthquake at sites along at least the southern 750 km of the subduction zone leave open the possibility that it was a single giant earthquake in the magnitude range of $M_w = 9$. Alternatively, the paleoseismic evidence may have been produced by a sequence of several great ($M_w > 8$) subduction zone earthquakes that occurred during a few decades or less about AD 1700. Both of these types of rupture sequences have occurred on other subduction zones that are similar to Cascadia (e.g., Figure 5.45).

Conventional ^{14}C ages for earlier paleoearthquakes have been estimated at some sites along the subduction zone (Vick, 1988; Grant *et al.*, 1989; Darienzo and Peterson, 1990; Atwater, 1992; Clarke and Carver, 1992; Nelson, 1992a,c). Samples used for age dating have included bulk peat, rhizomes and stem bases of selected marsh plants including *T. meritimum* and *G. stricta*, and twigs, wood, and charcoal contained in the peat or in the overlying mud. The age estimates suggest between two and five subsidence events have taken place along much of the length of the subduction zone during the last 2 ka. At many sites the conventional ^{14}C ages suggest that, prior to the earthquake ~300 years ago, large earthquakes

Figure 5.62: Prominent ledge of peat buried beneath intertidal mud at Willapa Bay, Washington. This submerged peat layer is locally covered by a thin sand sheet that entombs the above-ground parts of salt marsh herbs rooted in the peat. The sand was probably deposited by a tsunami immediately after the submergence.

also occurred about 700, 1100, and 1600 years ago. The assessment of the number and age of separate paleoearthquakes at different places along the Pacific Northwest coast depends on which evidence is accepted as coseismic and how the paleoseismic earthquake horizons are correlated to define individual rupture segments. Subdivision of the subduction zone into more shorter segments, and acceptance of more of the evidence as coseismic, increases the number of earthquakes but reduces their magnitudes, whereas interpretation of longer segments and fewer events results in larger, less frequent, subduction earthquakes.

5.13.3 *Ambiguities in Characterizing Subduction Paleoearthquakes*

Ambiguities in characterizing subduction earthquakes arise mainly because the dominant evidence is regional geodetic deformation, rather than measurements on the causative fault trace itself (as with on-land faults). Nonseismic mechanisms can cause changes in relative sea level on coasts that mimic those created by all but the largest ($M_w > 8.5$) subduction earthquakes. Thus, there is a high threshold of detection for subduction paleoearthquakes. In addition, there are uncertainties about the lateral extent of paleoearthquakes, due to difficulties in correlating subsidence events between isolated sites where good evidence is preserved, and due to the lack of dating precision. Finally, even those well-documented changes in coastal relative sea level that are almost certainly coseismic could have resulted from either "blind" $M_w > 7$ earthquakes on local upper plate faults, from short $M_w > 8$ megathrust earthquakes, or from long $M_w > 9$ megathrust earthquakes. At a single site, the evidence from all three would appear identical. Due to noncharacteristic earthquakes (Schwartz, 1999) a single site may have been affected by all three types of earthquakes at different times. These ambiguities create major limitations to estimating paleoearthquake magnitude (e.g., Nelson *et al.* [2006, 2008] and Chapter 9, See Book's companion web site).

CHAPTER 6

Paleoseismology of Strike-Slip Tectonic Environments

James P. McCalpin,* Thomas K. Rockwell,[†] and Ray J. Weldon II[‡]

*GEO-HAZ Consulting, Inc., Crestone, Colorado 81131, USA
[†]Department of Geological Sciences, San Diego State University, San Diego, California 92182-1020, USA
[‡]Department of Geological Sciences, University of Oregon, Eugene, Oregon 97403-1272, USA

6.1 Introduction

Strike-slip faults have played a critical role in the development of paleoseismology for several reasons. First, strike-slip faults are often the longest faults on continental landmasses and typically have conspicuous geomorphic expression. Second, many of these faults have long records of seismicity because they pass through populated continental regions and have experienced surface ruptures during large and great historical earthquakes. Third, because coseismic deformation along strike-slip faults is horizontal, subsequent earthquakes (and related interseismic sedimentation and erosion) do not deeply bury, or expose to erosion, traces of earlier events.

Major strike-slip faults that have been assessed for paleoseismicity are typically associated with plate boundaries, such as the San Andreas fault and Queen Charlotte–Fairweather faults (North American/ Pacific plates); the Motagua fault, Guatemala (Caribbean/North American plates); the Alpine fault, New Zealand (Pacific/Indian plates); the North Anatolian fault (Turkish/Eurasian plates); and the Dead Sea transform zone (African/Arabian plates). Other active strike-slip faults are located within the major lithospheric plates and define the boundaries of continental microplates (e.g., central Asia, Mongolia, and China). A final major class of faults, as yet unstudied for paleoseismology, is the numerous submarine transform faults (fracture zones) associated with oceanic spreading centers.

Strike-slip faults by definition displace geologic markers approximately parallel to the earth's surface. Because most young deposits and geomorphic surfaces are planar and form parallel to the earth's surface, this style of displacement poses special problems that are not typically encountered in studying dip-slip faults, which displace deposits and surfaces orthogonal to their original orientation. This chapter focuses on how to overcome this fundamental problem. The geometry of deposits and surfaces being displaced parallel to the earth's surface does provide some advantages, however. Longer records of paleoearthquakes lie closer to the surface of the earth than for dip-slip faults, where evidence of older earthquakes is deeply buried and therefore less accessible to investigation. Also, because strike-slip faults typically produce less relief than dip-slip faults they are less likely to destroy the record by

International Geophysics, Volume 95
ISSN 0074-6142, DOI: 10.1016/S0074-6142(09)95006-9

uplift or erosion, or to disrupt sedimentation, erosion, or other surficial processes that produce stratigraphic and geomorphic markers and preserve the record of events.

We divide this chapter into geomorphic and stratigraphic sections, mainly because the techniques used for each are so different. Historically, geomorphic studies have tended to yield the size of paleoearthquakes, and stratigraphic studies the timing of paleoearthquakes. Determining the size (i.e., net slip) of a strike-slip paleoearthquake through stratigraphic studies was considered difficult, due to slip variability along strike and the need for 3D trenching. However, the increased popularity of 3D trenching and statistical models for slip variability have eased this concern. Likewise, geomorphic studies were limited by the difficulty in dating geomorphic features, but this has changed with the advent of surface exposure dating using cosmogenic isotopes. Currently, geomorphic studies are used mainly to determine long-term fault slip rates and favorable trench sites, and stratigraphic studies for the remainder of paleoseismic characterization.

When looking for sites to carry out paleoseismic investigations, we look for different things depending on whether we wish to date earthquakes or determine their size. The best stratigraphic sites involve rapid deposition, which quickly obscures geomorphic markers, and distributed deformation so that evidence of individual paleoearthquakes is spread out and should be attributable to its component paleoearthquakes (see Section 6.3.1). The best sites for measuring the size of paleoearthquakes and slip rates are regions where (1) slip is localized in a narrow fault zone and (2) many small geomorphic features (gullies, ridges) are constantly being formed, so one may see many examples of offset due to a paleoearthquake, determine its true offset, and distinguish that displacement from those caused by other paleoearthquakes.

Finally, the philosophical theme that we carry throughout this chapter is that one must never rely on a few observations. A single trench, a few geomorphic offsets, or a single numerical age estimate is almost always uninterpretable. Multiple trenches, many geomorphic offsets, and a large number of related dates are necessary to understand paleoearthquakes. We choose our examples in this chapter to illustrate specifically how one combines many observations, rather than interpreting them individually, and to show how one may go astray by not rigorously examining all of the data at hand. Good examples of site studies can be found in the October 2002 Special Issue (vol. 92, no. 7) of the *Bulletin of the Seismological Society of America*, "Paleoseismology of the San Andreas Fault System."

6.1.1   Styles, Scales, and Environments of Deformation

6.1.1.1   Environments of Strike-Slip Deformation

Strike-slip faults exist in many varied parts of the earth's surface and can be classified as follows (after Yeats *et al.*, 1997):

1. Ridge–ridge oceanic transform faults that are perpendicular to, and connect, two spreading ridges. These are the most common type worldwide, but most are submerged.
2. Trench–trench "boundary" transform faults that separate two sections of subduction megathrusts. An example is the Alpine fault, New Zealand.
3. Ridge–trench–boundary transforms, which connect a contractional zone at one end with an extensional zone at the other end. An example is the San Andreas fault.
4. Horizontal extrusion faults, also termed "indent-linked" faults, that form where collisional belts create conjugate strike-slip faults that extrude crustal blocks perpendicular to the collisional direction. An example is the North Anatolian faults in Turkey and faults such as the Altyn Tagh in central Asia.
5. Conjugate pure-shear faults in the hanging wall of subduction zones.

6. Trench-parallel strike-slip faults, which accommodate the oblique component of crustal convergence via slip partitioning. Examples are the Great Sumatran fault, Indonesia, and the Septentrional fault, Dominica.
7. Intraplate strike-slip faults connecting active normal or reverse faults.
8. Intraplate slip-partitioned strike-slip faults.
9. Strike-slip faults associated with lateral margins of crustal-scale gravity failure (mega-landslides) of volcanic edifices (as at Mt. Etna, Sicily).

This chapter emphasizes on-land, tectonic strike-slip faulting not associated with any volcanic processes. The largest such regions coincide with places where plate boundaries come on-land (e.g., San Andreas fault, Alpine fault, Great Sumatran fault) or where parts of plates are extruding perpendicular to crustal collision (North Anatolian fault, faults of central Asia and south China). The smallest areas of strike-slip faulting are located where secondary, upper-crustal strike-slip faults form related to major reverse and normal faults. Very localized strike-slip (or oblique) faulting may occur in stepover zones or bends in normal or reverse faults, and are usually termed tear faults.

6.1.1.2   General Style of Deformation on Strike-Slip Faults: Plan View

The main seismogenic structure in strike-slip or transform tectonic environments is the *strike-slip fault*. The general characteristics of strike-slip (or wrench) faults are reviewed by Sylvester (1988). Where fault strike is parallel to the slip vector, strike-slip faults tend to concentrate deformation along a single linear fault strand (Figure 6.1) that may extend for tens or hundreds of kilometers with only minor changes in strike. If fault strike locally diverges from the slip vector, the dip-slip component of slip on the fault plane increases, resulting in *oblique slip*. Bilham and King (1989) term this degree of strike divergence *segment obliquity*. Although normal (Chapter 3) and reverse (Chapter 5) ruptures commonly have a component of oblique slip, we have deferred the formal discussion of oblique slip to the next section, based on its importance in strike-slip environments.

On right-lateral faults, a bend or stepover to the right induces local extension (*dilation*) and a bend or stepover to the left induces local compression (*contraction*). Crowell (1974) noted that *double bends* (a pair of bends outside of which fault strike is the same) are common along strike-slip faults. On a right-lateral (dextral) fault, *releasing double bends* (such as a bend to the right and then to the left) or stepovers to the right (termed *dilational jogs*) tend to create *transtensional* features such as normal faults, monoclinal folds, rhomboidal grabens, and pull-apart basins (Figure 6.2). Bends and stepovers of the opposite symmetry (*restraining double bends* and *contractional jogs*) induce *transpressive structures* such as pressure ridges, thrust faults, and folds (Figure 6.3). An oblique component of displacement can also be revealed by *en-echelon* fault segments, with (for example) a right-stepping pattern usually indicating left-lateral movement and vice versa. Secondary dip-slip faults and folds may form parallel to straight fault segments or perpendicular near en-echelon steps.

6.1.1.3   General Style of Deformation on Strike-Slip Faults: Sectional View

The main strike-slip fault plane is typically subvertical, with a dip within ca. 10° of vertical where fault strike is parallel to the regional shear direction. With increasing segment obliquity, dip can decrease, in some cases significantly (e.g., Whittier fault, California, decreases from 70° to 20°). Surrounding the main fault plane may be subsidiary fault planes that diverge upward from the main fault (flower structure). In a negative flower structure, subsidiary faults with an oblique normal component steepen toward the

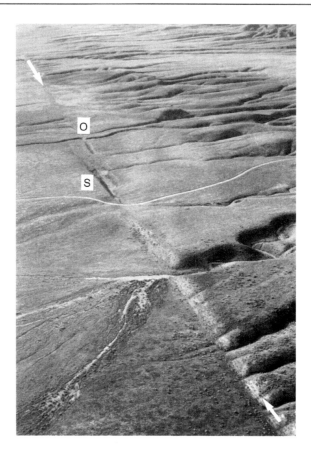

Figure 6.1: Aerial photograph of the San Andreas fault (between arrows) across the Carrizo Plain, California. The linear trace and lack of vertical relief across the fault are typical of many plate-boundary strike-slip faults. Note dextrally offset stream (below "O") and sag pond (to right of "S"). Some strike-slip faults are less linear at this scale, and are composed of alternating tension gashes and pressure ridges oriented obliquely to the main trace (e.g., Baljinnyam *et al.*, 1993). Photograph courtesy of R. E. Wallace and the U.S. Geological Survey; from Wallace (1990).

ground surface (Figure 6.2), whereas in a positive flower structure, subsidiary faults with an oblique reverse component flatten toward the ground surface (Figure 6.3). In both cases, the subsidiary faults may be spaced approximately equally, separating structural blocks of approximately equal dip.

Individual faults in a strike-slip zone are more undulatory in section that normal or reverse faults, and they may change dip often in the space of several decimeters to meters (e.g., Nelson *et al.*, 2000; Dawson *et al.*, 2003, Figure 6). They commonly branch upwards, a phenomenon rare in other stress environments, and may also converge again updip into a single fault strand, something very rare in other environments. Where faults diverge then converge, the intervening space is filled with sheared or rubblized material, or massive sand which may have been injected into the fault plane during liquefaction (e.g., Rubin and Sieh, 1997, Plate 1; Dawson *et al.*, 2003, Figure 14). In transtensional zones faults may transition upwards into fissures that were open at the ground surface, and have subsequently filled with surface-derived deposits

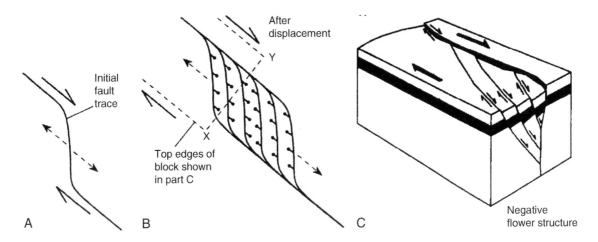

Figure 6.2: Formation of an extensional duplex at an extensional (releasing) bend. Large arrows indicate the dominant shear sense of fault zone; small arrows indicate the sense of strike-slip and normal components of motion on the fault splays. (A) Extensional bend on a dextral strike-slip fault. (B) An extensional duplex developed from the bend in part (A). (C) A block diagram showing a normal, negative flower structure in three dimensions. The block faces are vertical planes along the dashed lines in part (B). From Twiss and Moores (1992); used by permission of W. H. Freeman and Co.

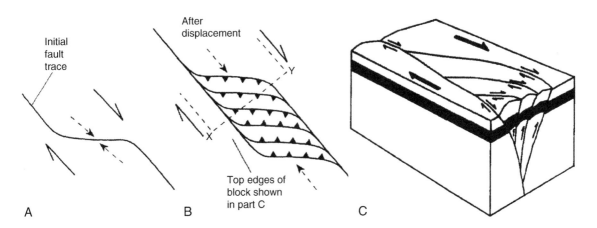

Figure 6.3: Formation of a contractional duplex at a contractional (restraining) bend. Large arrows indicate the dominant shear sense of fault zone; small arrows indicate the sense of strike-slip and reverse components of motion on the fault splays. (A) Contractional bend on a dextral strike-slip fault. (B) An contractional duplex developed from the bend in part (A). (C) A block diagram showing a reverse, or positive, flower structure in three dimensions. The block faces are vertical planes along the dashed lines in part (B). From Twiss and Moores (1992); used by permission of W. H. Freeman and Co.

(e.g., Rubin and Sieh, 1997, Figure 8; Dawson *et al.*, 2003, Figures 9–11). Dieout up and dieout down are also common (e.g., McGill and Rockwell, 1998, Plates 1 and 2).

6.1.1.4 Defining Slip Components

Minor changes in strike on a strike-slip fault often cause major changes in the geomorphic expression of the fault trace, for reasons explained later. Figure 6.4 shows the slip components, and the trigonometric relations among them, for the general case of oblique slip on a nonvertical fault plane. The net slip displacement vector (Δu) in this example represents right-lateral faulting with a reverse component. The three-dimensional slip vector can be partitioned into a *horizontal component* (Δu_h) and a *vertical component* (Δu_v). The horizontal component can be further partitioned into a *strike-slip component* (Δu_{ss}) and an orthogonal *convergent component* (Δu_c). In addition to the three equations in Figure 6.4 at upper right, Baljinnyam *et al.* (1993, p. 29) make the approximation that, for segment obliquity ($\Delta\Theta$) $< 20°$, $\Delta u_{ss} = \Delta u_h$. For larger angles of obliquity, the correct relation should be used:

$$\Delta u_{ss} = \cos, \Delta\Theta(\Delta u_h). \tag{6.1}$$

Baljinnyam *et al.* (1993) quantitatively describe how segment obliquity ($\Delta\Theta$ in Figure 6.4) can result in creation of vertical relief on a strike-slip fault. In their example, the slip vector azimuth is 85°, horizontal slip component (Δu_h) is 6 m, and the fault dips (β) is 55°. For fault segment strikes of 85° (i.e., parallel to the slip vector) $\Delta\Theta = 0$, and thus $\Delta u_c = 0$ and no vertical component results; the result, across low-relief

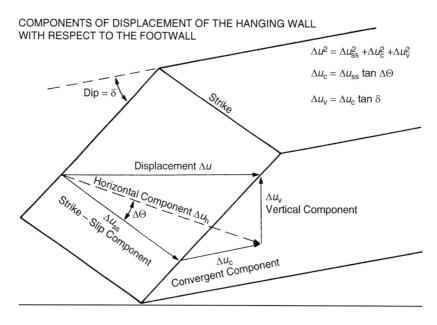

Figure 6.4: Block diagram showing the relationships of horizontal and vertical components of displacement to the slip vector and local strike of the fault. Terms are defined and discussed in the text. From Baljinnyam *et al.* (1993); reprinted with permission of the Geological Society of America.

terrain, would be a fault trace such as that shown in Figure 6.1. For fault segment strikes of 95° and 105°, Δu_c is calculated as 1.04 and 1.75 m, corresponding to vertical displacements (Δu_v) of 1.4 and 2.5 m, respectively, from reverse-oblique faulting. If the latter case was expressed across a low-relief plain with few streams, the lateral component of offset may go unnoticed, and the 2.5-m-high fault scarp could be attributed to pure reverse faulting. For fault segments with the opposite obliquity, for example, an azimuth of 80° compared to a slip vector azimuth of 85°, the equations in Figure 6.4 predict normal-oblique faulting with a vertical component of 0.75 m.

This simple analysis assumes that the earth materials in contact across the fault act as rigid bodies and do not internally deform as the component of convergence increases. For most unconsolidated surficial deposits, this assumption is probably unfounded. For example, Thatcher and Lisowski (1987) report that at many locations less than 70% of the 1906 coseismic slip occurred on the San Andreas fault trace itself, the remainder of the slip appearing as distributed deformation (intergranular shear, subsidiary faulting, and perhaps folding) within 600 m of the fault. McCalpin (1996a,b) compared segment obliquity at 31 sites along a 55-km-long section of the Awatere fault (New Zealand) to the ratio of (prehistoric) horizontal:vertical offset, and found that about one-third of measurements did not obey the relations predicted in Figure 6.4. He concluded that distributed, inelastic deformation was probably responsible.

6.1.2 Segmentation of Strike-Slip Faults

Plate-boundary strike-slip faults are the longest of the on-land faults [e.g., San Andreas, Denali, North Anatolian, Alpine (New Zealand), Great Sumatran], generate the largest earthquakes ($M_w > 8$), and accordingly have the longest rupture segments. However, deducing rupture segments from geometric or geologic evidence is difficult, because large earthquakes on strike-slip faults (as on normal faults) typically rupture through geometric and geologic boundaries. A fuller discussion of fault segment boundaries for strike-slip and other fault types is given in Chapter 9 (See Book's companion web site).

6.1.3 The Earthquake Deformation Cycle of Strike-Slip Faults

The earthquake deformation cycle on strike-slip faults is well documented by observations on the San Andreas fault, USA. Thatcher (1993) divides the cycle into coseismic, postseismic, and interseismic phases (Figure 6.5). Coseismic horizontal deformation is greatest at the fault plane and decreases asymptotically away from it (Figure 6.5B). This pattern is predicted by elastic dislocation theory and accords with observations of displaced/deformed cultural features, such as the straight fence lines displaced and warped by the 1906 San Francisco earthquake. In contrast, the mechanisms responsible for postseismic transient movements and steady interseismic deformation are uncertain. Thatcher (1993) proposes two contrasting models, based mainly on geodetic observations across the San Andreas fault. In the first model ("thick lithosphere model"), the depth of coseismic faulting (D in Figure 6.5A) is much less than the thickness of the strong lithospheric plate. Postseismic movements in this model are caused by episodic slip directly downdip of the coseismic rupture plane, while interseismic deformation is caused by steady aseismic fault slip at even greater depths. In the second model ("thin lithosphere model"), the depth of coseismic faulting (D in Figure 6.5A) is equal to the thickness of the strong lithospheric plate. All interearthquake lithospheric deformation in this model is caused by bulk flow of sublithosphere material. Unfortunately, these two models yield indistinguishable surface deformations, according to Thatcher (1993, p. 14).

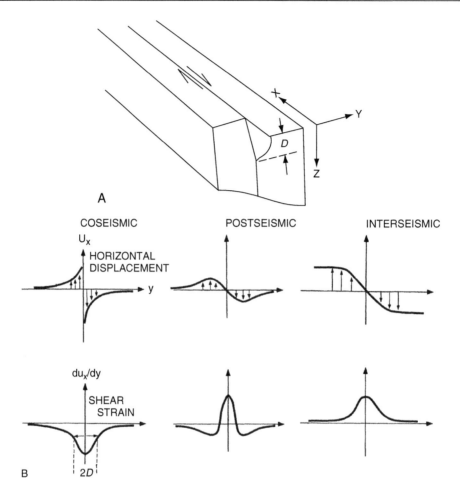

Figure 6.5: Earthquake deformation cycle on strike-slip faults. (A) Fault geometry, showing fault slippage extending from the surface ($Z = 0$) to $Z = D$. In the "thin lithosphere" model, D is equal to the thickness of the elastic lithosphere; in the "thick lithosphere" model, D is less than the thickness of the elastic lithosphere. (B) Schematic plan views of horizontal displacement (top row) and shear strain (bottom row) for the coseismic, postseismic, and interseismic phases of the earthquake deformation cycle. From Thatcher (1993); reprinted with permission of the Instituto Nazionale di Geofisica e Vulcanologia.

6.1.4 Historic Analog Earthquakes

Historical strike-slip surface ruptures are well documented on interplate and intraplate faults (Table 6.1). These coseismic ruptures accompanied great earthquakes as large as M_w 8.6 with up to 18.7 m of displacement over lengths of >400 km, to smaller earthquakes near the threshold of surface rupture (<0.5-m maximum displacement, <10-km rupture length). Several historical ruptures traversed populated areas where lateral offsets of cultural features (roads, fences) permitted precise measurements of horizontal displacement (e.g., 1906 San Francisco; 1972 Managua; 1979 Imperial Valley; 1992 Landers;

Table 6.1: Well-studied historical strike-slip fault surface ruptures

Date and magnitude	Area/Fault	Maximum and (average) displacement (m)	Length of rupture (km)	References
(a) Ruptures studied immediately after the earthquake				
1906, M_w 7.8	San Andreas, California	6.4 (5)	470	Lawson *et al.* (1908)
1957, M_s 7.9	Gobi-Altai, Mongolia	9.4 (4.5)	260	Florensov and Solonenko (1963)
1966, M_s 6.8	Varto, Anatolia, Turkey	0.4 (0.15)	30	Ambraseys and Zatopek (1968) and Wallace (1968b)
1967, M_s 7.4	Mudurnu, Anatolia, Turkey	2.6 (1.63)	83	Ambraseys and Zatopek (1969)
1968, M_s 6.8	Borrego Mountain, California	0.38 (0.18)	31	Clark (1972)[a]
1968, M_s 7.1	Dasht-e-Bayaz, Iran	5.2 (2.3)	80	Tchalenko and Ambraseys (1970)[a]
1972, M_s 6.2	Managua, Nicaragua	0.67	15	Brown *et al.* (1973)
1979, M_s 6.7	Imperial Valley, California	0.8 (0.18)	31	Sharp (1982) and Sharp *et al.* (1982)
1987, M_s 6.6	Superstition Hills, California	0.92 (0.54)	27	Sharp *et al.* (1989)
1992, M_s 7.6	Landers, California[b]	6.1	85	Ebersold (1992) and Sieh *et al.* (1993)
1994, M_w 7.2	Kobe, Japan	2.3	9	Awata *et al.* (1996)
1999, M_w 7.4	**Izmit, Turkey**	**5.2**	**150**	**Rockwell *et al.* (2002)**
1999, M_w	**Hector Mine, California**	**5.5 (2.5)**	**48**	**Treiman *et al.* (2002)**
1999, M_w 7.2	**Duzce, Turkey**	**5.0**	**40**	**Rockwell *et al.* (2002)**
2001, M_w 7.8	**Kunlunshan, Tibet**	**7.6**	**350**	
2002, M_w 7.9	**Denali, Alaska**	**8.8 (4.9)**	**341**	**Crone *et al.* (2004) and Haeussler *et al.* (2004)[c]**
(b) Ruptures studied decades to centuries after the earthquake				
1855, M_s > 8.1	**Wairarapa, New Zealand**	**18.8[d]**	**ca. 145**	**Rodgers and Little (2006)**
1857, M_s 8.3 (M_w 7.9)	San Andreas, California	9.4 (6.4)	360	Sieh (1978b)

(Continued)

Table 6.1: Well-studied historical strike-slip fault surface ruptures (Cont'd)

Date and magnitude	Area/Fault	Maximum and (average) displacement (m)	Length of rupture (km)	References
1905, M_w 8.0	Bulnay, Mongolia	11 (8)	375	Khil'ko *et al.* (1985) and Baljinnyam *et al.* (1993)
1920, M_s 8.5 (M_w 7.9)	Haiyun, China	10.0 (7.25)	240	Zhang *et al.* (1987)
1931, M_w 8	Fu-Yun, China	14.6 (7.38)	180	Baljinnyam *et al.* (1993)
1939, M_w 7.9	Erzincan, Turkey	7.6 (4.7)	327	Barka (1996)
1944	Bolu-Gerede, Turkey	3.6 (1.8)	180	Kondo *et al.* (2005)

Bold indicates studies made after the First Edition was published.
[a]Graphs of displacement along strike are also shown in Thatcher and Bonilla (1989, pp. 389–391).
[b]Surface rupture is shown on videotape (Prose, 1994b).
[c]Contains a digital appendix with photographs of all slip measurement localities.
[d]Largest single-event surface displacement known, for any fault type.

1999 Izmit and Duzce). Figure 6.6 shows two medium-length ruptures (70 and 65 km) where multiple precise measurements of horizontal slip were made. Like normal- and reverse-fault ruptures, slip tends to maintain high levels in the central part of the rupture, with near-maximum slip in very short sections. Slip either decreases rapidly at the end of rupture (left side of Figure 6.6A) or trails off in a long, low-slip tail (Figure 6.6A and B). Pre-earthquake and postearthquake geodetic surveys (e.g., Thatcher and Lisowski, 1987) demonstrate that strike-slip deformation often extends hundreds of meters away from the prominent surface rupture trace, a phenomenon that may explain some of the short-wavelength variability in slip shown in Figure 6.6.

Since the publication of the First Edition, five large strike-slip earthquakes occurred which have been well studied (1999, Hector Mine, California; Izmit and Duzce, Turkey; 2001, Kunlunshan, Tibet; 2002, Denali, Alaska). The latter (Figure 6.7) is a good example of a high-magnitude, 300+ km-long strike-slip rupture. As described later in Section 6.2.3, long strike-slip ruptures such as the 2002 Denali fault are typified by a nearly uniform distribution of offsets of various sizes, with major discontinuities in slip gradient at geometric segment boundaries.

Measurements on historical strike-slip ruptures also document the rate at which slip varies along strike during a single earthquake (Ambraseys and Tchalenko, 1969; Ambraseys and Zatopek, 1969; Sieh, 1978b; Sharp *et al.*, 1982). This rate is termed *slip gradient* and is dimensionless. Slip gradients can be large, for example, the decrease from 4.2 to 1.6 m within 330 m along strike (2.6 m/330 m, or 0.008) in the 1968 Dasht-e-Bayaz earthquake (Ambraseys and Tchalenko, 1969), or the decrease from 6.1 to 3.6 m within 1.4 km along strike in the 1940 El Centro, California, earthquake (Sharp, 1982). The recent detailed measurements on the 1992 Landers, California, rupture have revealed even more extreme slip gradients (McGill and Rubin, 1999). The maximum slip gradient along the Emerson fault was 2×10^{-1}, between two features offset 250 and 535 cm, respectively, located only 12–20 m apart. Maximum slip gradients on the other faults that ruptured in the Landers earthquake were on the order of 10^{-2}, as were the maximum slip gradients for the 1979 Imperial Valley earthquake (Sharp *et al.*, 1982) and the 1987

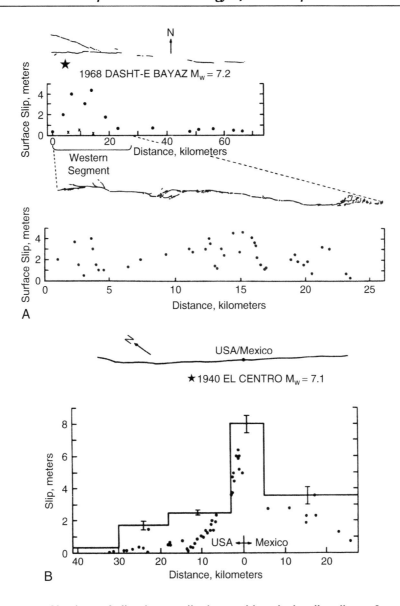

Figure 6.6: Patterns of horizontal slip along strike in two historical strike-slip surface ruptures. Map views show detail of rupture traces; stars indicate epicenters. (A) 1968 Dasht-e-Bayaz (Iran) earthquake. (B) 1940 El Centro, California, earthquake; geodetic estimates of slip are shown by straight lines with one-stranded deviation error bars. From Thatcher and Bonilla (1989).

Superstition Hills earthquake (Sharp *et al.*, 1989), which had smaller maximum and average slip than did Landers (Table 6.1). These slip gradients could be due to changes in the amount of slip across the rupture, or caused by a change from concentrated to distributed deformation. In the latter case, the lateral offset along the main surface rupture trace may appear to decrease, even though the net offset across a broad zone centered on the trace may remain constant.

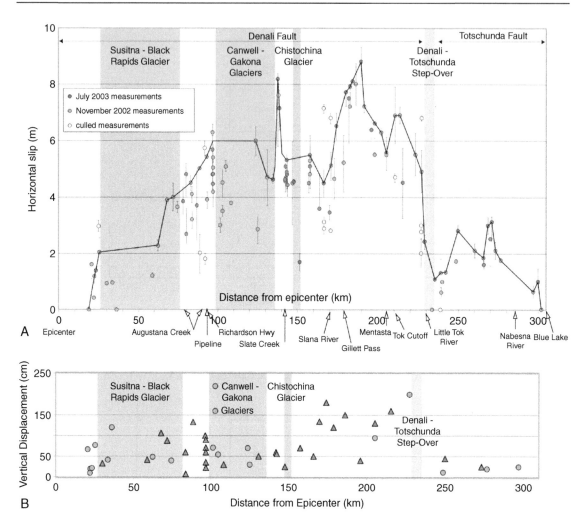

Figure 6.7: Horizontal (top) and vertical (bottom) slip along the 2002 rupture of the Denali fault, Alaska. This diagram is based on 127 measurement locations along the 340 km of rupture. From Haeussler *et al.* (2004).

Great ($M_w > 8$) strike-slip earthquakes in the late 1800s and early 1900s also provided incontrovertible evidence linking fault displacement and earthquakes (e.g., 1855 West Wairarapa, New Zealand; 1872 Owens Valley, California; 1891 Nobi, Japan; 1906 San Francisco, California). The latter gave rise to the elastic rebound theory (Reid, 1910) and to the perfectly periodic model of earthquake behavior (see Chapter 9, See Book's companion web site).

6.2   Geomorphic Evidence of Paleoearthquakes

Active strike-slip faulting produces a characteristic assemblage of landforms (Figure 6.8) including *linear valleys, offset or deflected streams, shutter ridges, sag ponds, pressure ridges, benches, scarps, and small*

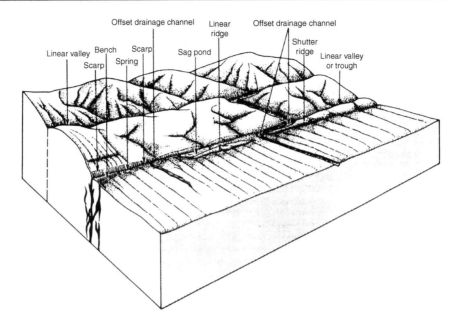

Figure 6.8: Assemblage of landforms associated with strike-slip faulting. From Surficial Geology, Building with the Earth; Costa and Baker (1981). Copyright © 1981 John Wiley and Sons; reprinted by permission of John Wiley and Sons, Inc.

horsts and grabens (Keller, 1986). In many cases, the fault trace is composed of a wide zone of alternating *tension gashes* (extensional) and *moletracks* (compressional) that trend obliquely with respect to overall fault strike. Baljinnyam *et al.* (1993, p. 46) attributed such a wide zone of surface deformation in Mongolia to coseismic rupturing of permafrost in valley bottoms, and noted that the same rupture trace across hills not underlain by permafrost was much narrower.

Strike-slip faults also transport nontectonic landforms laterally, while the erosional and depositional processes forming them continue to operate. This lateral transport causes the most obvious geomorphic effects when faults strike perpendicular to the direction of stream transport. Three landforms are typically used to reconstruct paleoseismic offset histories: fluvial terraces, stream channels, and alluvial fans.

6.2.1 Landforms Used as Piercing Points

In paleoseismology, the main goal of geomorphic analysis is to measure lateral (or oblique) displacements that can be attributed to individual paleoearthquakes. These displacements are then used to estimate earthquake magnitude or seismic moment (Chapter 9, See Book's companion web site). Quaternary displacements across strike-slip faults are typically measured from displaced landforms such as terraces, channels, or fans. Due to the possible oblique nature of the slip vector, field workers need to identify landform elements that intersect the fault plane as *piercing points* on either side of the fault. On gently sloping or flat terrain, the most common piercing point landforms are linear features such as the axes or thalwegs of stream channels, narrow erosional ridges such as interfluves between gullies, bases or crests of terrace risers, or narrow constructional ridges such as debris-flow levees or gravel bars (Figure 6.9). If comparable points on such well-defined landforms can be located in the field and traced to the fault, it is

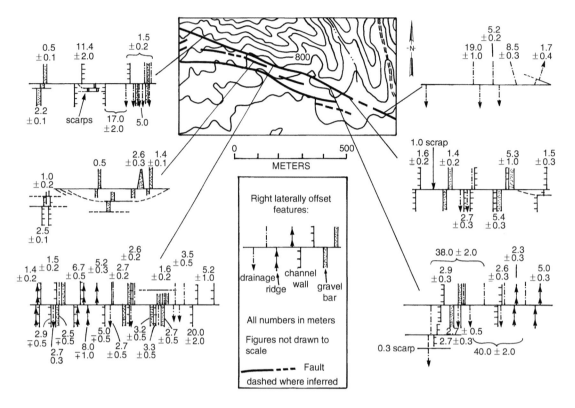

Figure 6.9: Example of an inventory of horizontal offsets for four types of landforms (drainages, channel walls, ridges, gravel bars) typically found in semiarid regions. These measurements can be statistically analyzed to deduce the number of paleoearthquake displacements. From Rockwell and Pinnault (1986); reprinted with permission of the Geological Society of America.

possible to measure the net slip vector directly. In special cases (Section 6.2.1.4) lateral offset can be measured from large landforms even if piercing points cannot be located.

Offset measurements on historic ruptures passing through areas of alluvial fill have detected significant amounts of distributed lateral displacement (bending, drag) away from the fault trace (Rockwell *et al.*, 2002). This type of off-fault deformation occurred on the North Anatolian fault in a zone 5–30-m wide, and averaged 15% of the total lateral displacement, but was up to 40% of total displacement in some places. Thus, measurements of prehistoric lateral offsets from landform misalignments may underestimate the total displacement, if the correlated points across the fault are closer than 30 m.

6.2.1.1 Offset Fluvial Terraces

The initial methods for measuring and interpreting laterally and obliquely displaced fluvial terraces were proposed by Suggate (1960), Lensen (1964a), and Sugimura and Matsuda (1965). Multiple offsets of Holocene river terraces have been extensively documented in New Zealand (Lensen, 1964b, 1968, 1973; Lensen and Vella, 1971; map in Suggate *et al.*, 1978); the most intensively studied locality appears to be the Saxton River terraces across the Awatere fault (Lensen, 1964a; Knuepfer, 1988; McCalpin, 1996b; Mason *et al.*, 2006) and in Japan (summarized by Okada, 1980). In contrast, locations studied since 1996 are mainly in central Asia, Tibet, and southern China.

The classic localities are all located where the strike-slip fault crosses terrace treads and risers nearly perpendicularly. This geometry exists where the strike-slip fault lies in a fault-parallel main valley, and a large tributary enters the valley from one side, where it forms terraces perpendicular to the fault trend (Figure 2A.2). The terraces can be formed by any normal process (base level change, sediment concentration changes, etc.). For example, climatic terraces can be created by base level changes in the main valley (inducing aggradation or degradation in the tributary) or from sediment concentration changes in the tributary watershed upstream of the fault. Tectonic terraces can be created by primary or secondary effects of large earthquakes. If the fault has a dip-slip component, tectonic terraces can be created by incision of the upthrown block (see Chapter 3), although incision cannot exceed the amount of uplift.

Faulted terraces at the classic localities are separated by more vertical distance than the vertical component of faulting there, so they are not primary tectonic terraces. However, the terraces could be secondary tectonic terraces if they were created by changes in sediment concentration caused by an earthquake. Such changes have been observed in New Zealand (Adams, 1980, 1981a,b; Almond *et al.*, 2000) and in Taiwan (fourfold increase measured by Dadson *et al.*, 2004) after the M_w 7.6 Chi–Chi earthquake of 1999. In both cases, an increased sediment concentration was caused by widespread coseismic landsliding in mountain drainage basins, followed by widespread aggradation of river valleys.

If such aggradation affected the floodplain of the tributary watershed more than the larger mainstream watershed, then the aggraded tributary would find itself temporarily above the base level of the mainstream. Therefore, the earthquake-induced aggradation alluvium would eventually be incised as the tributary regraded itself to the mainstream, and this incision would leave the aggraded floodplain preserved as a low terrace. In this model, terraces are formed on the tributary when the slug of earthquake-induced, floodplain aggradation deposits are incised following the earthquake.

The origin of the faulted terraces has not been discussed most in the classic papers, but it bears on this important question; whether terrace formation is temporally related to faulting, or completely time independent of faulting. In the former case, we might expect faulting and terrace formation to alternate, such that there has been one (and only one) earthquake between the formation of each successive terrace. In such a "terrace-dependent" model, every paleoearthquake is represented by differential offsets of terrace treads and risers. Conversely, if terraces are created by climatic causes, there may not be any temporal connection between earthquake timing and terrace formation. In such a "terrace-independent" model, any number of paleoearthquakes (0, 1, >1) can occur between the formation of successive terraces; this results in a very different geometry of faulted terraces than does the terrace-dependent model (discussed later).

Measuring the Offsets of Terrace Treads The terrace tread may contain some linear geomorphic elements that cross the fault, and thus could be used as piercing points. These include channel axes, channel margins, and depositional levees (ridges) (Figure 6.9). The offset of these features can only have accumulated since the abandonment of the terrace tread, assuming that the feature is contemporaneous with the terrace. Although this is normally presumed to be the case, Knuepfer (1988) cites several instances where abandoned channels on a terrace tread are offset much less than the terrace riser below the tread. This telltale geometry indicates that the channel is considerably younger than the terrace tread into which it is cut, and must have been created by tributary flow on the terrace after its abandonment.

Measuring the Offsets of Terrace Risers   Lensen (1968) suggests the following field technique for measuring terrace riser offsets. If offset risers are the same height across the fault, equivalent points of the riser profile (e.g., crests or toes) can be used as piercing points. However, on oblique-slip faults, terrace risers on the upthrown side of the fault are typically higher (and, thus, broader) than risers on the downthrown side. In this case, the riser crest and toe on the higher scarp have probably retreated and advanced (laterally) farther from their original positions than have equivalent points on the lower riser. Lensen (1968) thus suggests using the riser midpoint as the piercing point for measuring lateral offset, based on the assumption that risers erode by slope decline rather than by slope retreat. Knuepfer (1988) argues that even the midpoint may have retreated or advanced, and suggests averaging the lateral offsets between the crests and toes of the offset risers.

The interplay between fluvial processes and strike-slip faulting often produces unique landforms from which one can reconstruct the history of paleoearthquakes, as described in this chapter. Consider the typical geometry (Figure 6.10A) of an active river, riverbank, two terraces (the lower of which, T1, preserves an abandoned channel on its *tread*), and one *terrace riser* crossed by a strike-slip fault. In Figure 6.10B, all landforms are dextrally offset by a constant amount. After faulting (Figure 6.10C) the stream erodes the lateral offset of the riverbank, but the offset is preserved on the abandoned channel and upper riser (R1). Eventually the stream migrates farther to the left and incises (Figure 6.10D), leaving the former floodplain as a new terrace (T2), complete with an abandoned channel. Renewed faulting (Figure 6.10E) dextrally offsets all landforms by the same amount, but on T1 and R1 this new offset adds to the prior offset that occurred in Figure 6.10B. After faulting, the stream once again erodes the lateral offset of the riverbank. This alternation of faulting and terrace formation can continue for many cycles.

This hypothetical scenario emphasizes several critical points in interpreting offset terrace flights. First, terrace treads and their abandoned channels preserve the vertical and lateral components of offset; terrace risers can only record the lateral component of offset. Second, lateral offset of riverbanks is routinely removed by stream erosion after faulting, especially where the riverbank is laterally displaced toward the active channel. In this regard, Bull (1991, p. 238) distinguished the *leading edge* of a terrace flight, where terraces and risers are laterally displaced into the stream's path and the traces of offset destroyed, from the *trailing edge* of a terrace flight, where offset terraces and riverbanks are laterally displaced away from the active channel, and offsets are more likely to be preserved. Third, a terrace riser is abandoned by the river contemporaneously with the abandonment of the terrace *below* the riser (Knuepfer, 1988), not the terrace above the riser as supposed by Lensen (1968). In Knuepfer's model, the age of a riser, and its cumulative offset, correlate with that of the terrace tread below the riser.

From the above points, it follows that heavily scoured terrace risers (usually found on leading edges) record only offsets made after abandonment of the terrace surface below them. This model has been termed the "lower-terrace" model by Cowgill (2007) (Figure 6.11B) and the "strath abandonment" model of Meriaux *et al.* (2005). On unscoured risers (often found on trailing edges) these offsets, which were eroded from leading edge risers, may be partially or totally preserved. In this case, the displacement of the riser has accumulated since the abandonment of the terrace above the riser. This model has been termed the "upper-terrace" model by Cowgill (2007) (Figure 6.11A) and the "fill abandonment" model by Meriaux *et al.* (2005). These two models describe are end-member conditions, where all (or none) of the riser offset is eroded away before the lower surface is abandoned. Intermediate scenarios exist in which only part of the offset is eroded away, approximating Meriaux *et al.*'s (2005) "strath emplacement" model.

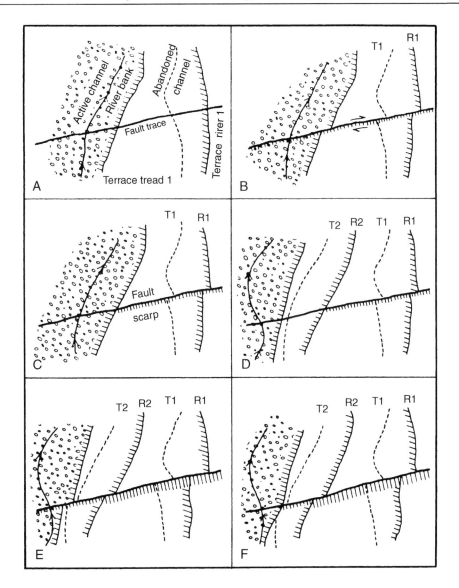

Figure 6.10: Hypothetical plan view of development of a terrace sequence offset by a dextral-oblique fault, upthrown toward the top of each figure. Hachures on terrace risers and fault scarps point to lower surface. (A) Prefaulting geometry. (B) First faulting event dextrally offsets all landforms. (C) Lateral erosion after first faulting event trims the offset riser between T1 and the active channel (patterned). (D) Incision of the active channel and creation of terrace T2. (E) Second faulting event. (F) Lateral erosion after second faulting event. See text for detailed discussion. From Knuepfer (1988).

Thus, the matter of how much riser offset was eroded away before the abandonment of the lower surface becomes a key element in calculating slip rates from faulted terraces sequences. Cowgill (2007) proposes six "indices" for determining this, and whether slip rates should be calculated based on the upper-terrace or lower-terrace models (Table 6.2; Figure 6.12).

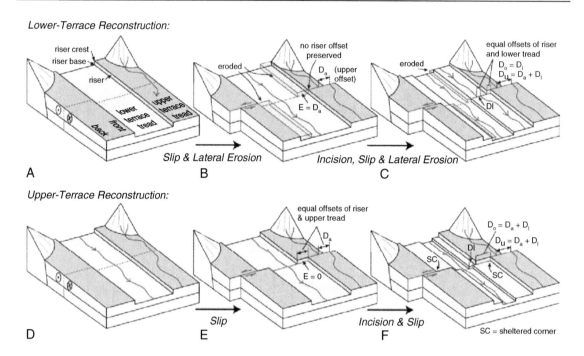

Figure 6.11: Block diagrams showing terrace nomenclature and two end-member models for linking riser offsets with terrace abandonment ages to determine rates of strike-slip faulting. Lower- and upper-terrace reconstructions are shown in panels (A–C) and (D–F), respectively. (A) Initial configuration, showing stream after incision of upper tread but before displacement of lower terrace. Blue line on upper tread is a primary feature that tracks total offset of upper tread (D_u). (B) Lower-terrace model presumes all riser offset is removed by lateral erosion ($E = D_a$) as long as stream occupies lower tread. As a result, stream channel must widen by at least D_a. At this stage $D_u = D_a$ and $D_o = 0$. Gray boxes denote eroded material. (C) After incision of the lower tread, observed riser offset is equivalent to magnitude of slip that postdates lower tread incision ($D_o = D_l$). Total offset of upper tread is larger than D_o and is given by $D_u = D_a + D_l$. (D) Same as (A). (E) Upper-terrace model assumes no lateral erosion of riser ($E = 0$). Thus riser offset and upper tread displacement (D_a) are equal. (F) Riser continues to accumulate slip after incision of lower tread such that $D_o = D_u = D_a + D_l$. From Cowgill (2007).

Morphologic Dating of Risers: An Example from New Zealand   J. P. McCalpin (unpublished data) measured topographic profiles at 10 sites across four risers at the Saxton River terraces, New Zealand, to see if there was morphologic evidence for repeated lateral erosion and "refreshing" of risers. If risers were symmetric (area eroded from crest = area deposited at base), then the entire riser was cut in a single episode, without refreshment. Conversely, if the area deposited was smaller than the area eroded, then some of the deposited material had been removed by lateral erosion of the riser toe. In every case, the measured risers were asymmetric (Figure 6.13). This geometry indicates that the risers were periodically refreshed before the lower terrace was abandoned, and thus the lower-terrace model should be used to calculate slip rates from cumulative offsets of risers.

Offset terrace flights are only interpretable in terms of individual paleoearthquakes if each earthquake is separated from the next by an episode of incision and terrace formation (Figure 6.14). Earlier we

Table 6.2: Tests for determining whether some (or all) of offset was eroded from a terrace riser, prior to abandonment of their lower surfaces (adapted from Cowgill, 2007)

Test no.	Index	Description
1	Riser offset versus inset channel width	Does the offset of the lowest abandoned riser exceed the offset (or width) of the inset channel? If so, there was partial erosion of that riser prior to abandonment of its lower surface.
2	Similarity of riser and tread displacements	If no riser erosion has occurred, then the offset of the riser will match the offset of features on the upper tread (=upper-terrace model). Offsets on the lower-terrace tread will be smaller.
		If complete riser erosion occurred, then the offset riser will match the offset of the lower-terrace tread (=lower-terrace model). Offsets on the upper-terrace tread will be larger.
		Intermediate cases are possible.
3	Morphologic dating of risers	1—to calculate a slip rate from offset of a riser, calculate the age of the riser directly from the diffusion-dating approach (no need to date terrace treads, or determine whether upper-terrace or lower-terrace models are appropriate).
		2—If riser erosion has occurred, the curvature of the lower scarp face will be steeper than that of the upper scarp face. See further discussion in text. A symmetrical riser profile suggests no erosion (=upper-terrace model).
		3—If diffusivity-based ages vary along the length of a riser, it probably has undergone spatially varying amounts of erosion.
4	Riser deflections	Does the angle between the riser and the fault trace change systematically close to the fault, in a way not observed elsewhere? If so, the riser was probably incompletely eroded (=intermediate model).
5	Diachroneity of terrace abandonment	Does the terrace tread below the riser appear to be all the same age, or does it look older nearer the riser? If it looks the same age everywhere, then the stream continued to impinge on the riser until lower-terrace abandonment (=lower-terrace model). If the terrace rises in elevation or looks older near the base of the riser, erosion may have been nil or incomplete (+upper-terrace model).
6	Parallelism of riser crest/base offsets and the slip vector	*If the net slip vector is independently known*: If this vector matches a line linking all riser bases, then the lower-terrace model is used. If this vector matches a line linking all riser crests, then the upper-terrace model is used.
		If slip direction and riser erosion are assumed to have been uniform through time: If this vector matches a line linking all riser bases, then the lower-terrace model is used. If this vector matches a line linking all riser crests, then the upper-terrace model is used.
7	Offset of contemporaneous risers on opposite sides of the stream	The difference between lateral offsets of contemporaneous risers on the leading versus the trailing edges comprises a minimum estimate of the incremental (single-event?) offset that occurred before their abandonment.

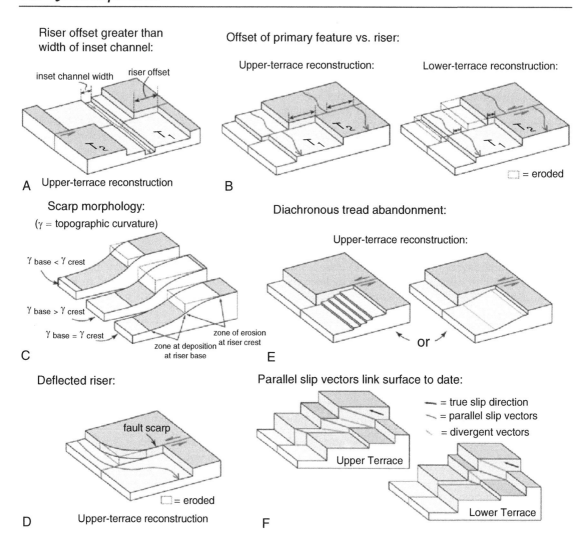

Figure 6.12: Geomorphic indices for evaluating suitability of upper- versus lower-terrace reconstruction. See text for explanation. (A) Magnitude of riser offset relative to width of channel inset into lower tread. (B) magnitude of riser offset relative to displacement of primary features (e.g., relict channels) on tread surface. (C) scarp morphologies. Bottom profile shows a symmetric scarp resulting from diffusion, middle profile shows an asymmetric scarp with a low curvature crest and high curvature base resulting from removal of material from the toe of the scarp during partial riser refreshment, top profile shows an asymmetric scarp with a high curvature crest and a low curvature base resulting from addition of loess and/or the presence of a cohesive unit in the upper terrace. (D) Abrupt truncation versus curved riser at intersection with fault. (E) isochronous versus diachronous tread ages; and (F) parallelism of vectors linking riser crests and bases.

termed this alternation of faulting and terrace formation the "terrace-dependent" model, because it implies some type of causal connection between faulting and terrace formation that exceeds mere coincidence. Where this fortunate alternation of faulting and downcutting has occurred, and no terrace offset was removed by erosion (= upper-terrace model), the offsets of successively older terraces are simply multiples of the single-event offset, if displacement was identical in each earthquake (e.g., Figure 6.14), or somewhat more variable sums if displacement varied among earthquakes (e.g., Figure 6.15). Such uniform offsets are recorded by the two youngest terraces at the Saxton River (New Zealand) of 7.5 m (one event) and 15 m (two events; Knuepfer, 1992; McCalpin, 1996a,b), and on the Wellington Fault (see example later). To create such a consistent geometry, the recurrence interval between earthquakes must be approximately the same, or less than, the interval between terrace formation. Stated another way, the number of terraces must be equal to or greater than the number of earthquakes.

If terrace formation is independent of faulting (terrace-independent model), then terrace formation may occur more or less frequently than faulting. For example, if multiple terraces are formed between earthquakes, then two or more successive terrace treads and risers in the sequence may record identical offsets (e.g., the classic Branch River terraces, Lensen, 1968). In the opposite case, where faulting occurs more often than terrace formation, each terrace tread and riser will be offset by multiple events before it is abandoned. In this case, the difference in offset between successive terrace risers records multiple events. A hypothetical example is shown in Figures 6.16 (where recurrence is variable, but displacement is a constant 5 m per event) and 6.17 (where both recurrence and displacement vary among seismic cycles).

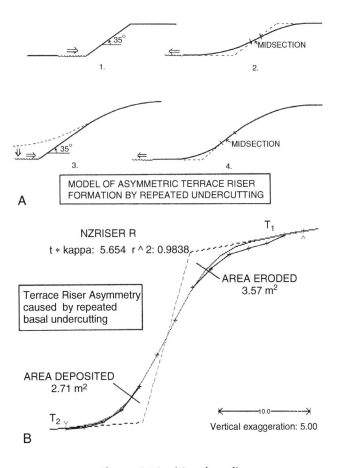

Figure 6.13: (Continued)

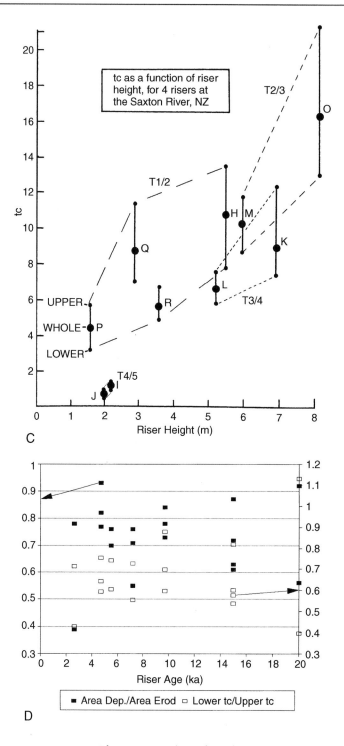

Figure 6.13: (Continued)

Where such a sequence of events has occurred in New Zealand (at the Branch River terraces, among other places), the difference in offset between successive terraces is on the order of 20–30 m, rather than the 6–8 m associated with individual events. In these cases, where earthquakes recur more often than terraces are formed, the number, displacement, and timing of individual paleoearthquakes cannot be uniquely deduced from geomorphic evidence.

Figure 6.18 shows examples of both in-phase and out-of-phase earthquakes and terrace formation, from a single site along the Wellington Fault, New Zealand. Channels on the youngest faulted terrace (T2) are offset 3.7 and 4.7 m, whereas the riser above this terrace shows apparent offset of 7.4 m. Van Dissen *et al.* (1992) conclude that T2 has been offset once and the T2/T3 riser twice, with the 7.4-m offset of the latter representing a minimum value after some (unknown) amount of riser erosion. Thus, subsequent to the cutting of the T2/T3 riser, two faulting events occurred but only one terrace surface preserves evidence of those offsets. The opposite sequence is shown by the consistent offsets (18.0 and 19.0 m) of the risers between T3/T4 and T4/T5. This geometry indicates that terraces were being created more frequently than were faulting events. Such an irregular alternation of terrace formation and faulting events is probably to be expected at most locations, rather than the more idealized sequence of one faulting event following the formation of each terrace. Further details on this site in New Zealand are given in Berryman (1990).

6.2.1.2 Offset Stream Channels

Laterally offset streams were used in early descriptions of active strike-slip faults to document recurrent movement. A classic locality in the United States is the Carrizo Plain segment of the San Andreas fault (Figure 6.1; Wallace, 1968a; Sieh, 1978b; Sieh and Jahns, 1984; Grant and Sieh, 1993, 1994). Wallace (1968a, 1990) recognized that the drainage patterns across the fault zone may have a complex relationship to fault offset. He makes a useful distinction between *stream misalignment* (a purely descriptive term for stream segments that do not align across the fault), *stream diversion* (streams forced to flow parallel to the fault by capture or blockage; also termed *deflection* by others), and *stream offset* (tectonic translation of a stream channel). Stream misalignment may refer to the misalignment of

---

Figure 6.13: Using riser symmetry to determine the appropriate dating model. (A) Conceptual model showing how repeated undercutting creates an asymmetric riser. If the riser is cut in a single erosional episode, it will have a symmetrical profile with the steepest part (midsection) in the center of the profile. If the riser has been refreshed by lateral erosion, it will have an asymmetrical profile with the steep "midsection" closer to the toe. (B) Topographic profile of the riser between terraces T1 (oldest) and T2 at Saxton River, New Zealand (Awatere fault). Vertical exaggeration is 5× to emphasize asymmetry. Dashed line shows inferred initial profile. Solid line with "+" symbols is the field profile, solid line is model profile. Like all profiles at this site, the base of the riser has a stronger curvature than the upper part, indicated that the area of colluvium deposited is smaller than the area of alluvium eroded. (C) Product of "tc" (time multiplied by diffusion coefficient) for the upper half, whole, and lower half of risers at Saxton River, as a function of riser height and riser age. In every case, the lower half of the riser profile yields a younger "tc" (stronger curvature) than does the upper half; the difference increases with riser height, and slightly with riser age (note the small difference for the youngest riser, T4/5). This pattern indicates the lower-terrace model is the appropriate one to use here. (D) Comparison of the area deposited/area eroded (left vertical axis, solid squares) and the lower profile "tc"/upper profile "tc" (right vertical axis, hollow squares), as a function of riser age. The ratio of lower "tc"/upper "tc" ranges from 0.55 to 0.8 for most risers, indicating repeated freshening of risers by lateral undercutting.

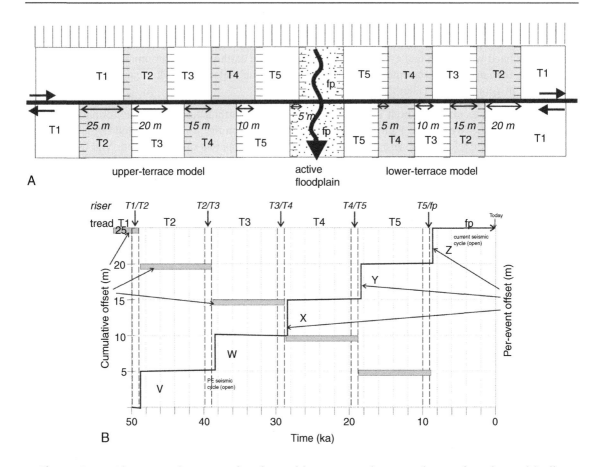

Figure 6.14: Diagrams of terraces that formed between each successive earthquakes, with displacement per event (5 m) and recurrence interval (10 kyr) held constant. (A) Map of terraces. Fault is a thick line running from left to right. Stream flow direction is toward the bottom (curving arrow). Terraces are numbered, with the oldest as T1; fp is the active floodplain. Ticks at top are 5 m apart. Downstream of the fault, risers to the left of the stream (trailing edge) have not been eroded (upper-terrace model). Risers to the right of the stream (leading edge) were fully eroded (lower-terrace model). (B) Slip history diagram of the faulted terraces, comparing times of terrace deposition, riser formation, and faulting. Thick line shows coseismic offsets (vertical sections with letters V–X) and interseismic periods (horizontal sections). Gray boxes show the cumulative offset of each terrace tread and riser, assuming no erosion (upper-terrace model).

sections of a single continuous stream channel across a fault, or the misalignment of a stream on one side of the fault with a disconnected stream segment on the opposite side of the fault. This latter case includes an (often unspoken) assumption that the two segments were once collinear parts of a single stream, which may or may not be true, and has to be supported by some type of evidence.

It is often difficult to determine from surface observations alone whether stream misalignment is caused by diversion, offset, or a combination of both. Human alteration of stream channels can also influence estimates of stream offset. For example, Sieh and Jahns (1984) measured 3.5 ± 0.5 m of lateral offset on

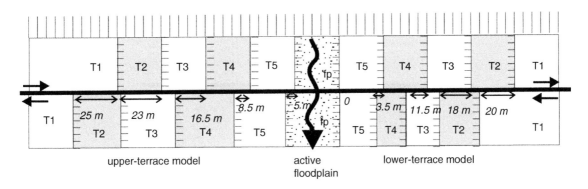

Figure 6.15: Diagrams of terraces that formed between each successive earthquakes, with displacement per event varying but recurrence interval held constant at 10 kyr. Same parameters as in Figure 6.11, but offset is 2 m in the first earthquake, 6.5 m in the second, 8 m in the third, 3.5 m in the fourth, and 5 m in the last. Downstream of the fault, risers to the left of the stream (trailing edge) have not been eroded (upper-terrace model). Risers to the right of the stream (leading edge) were fully eroded (lower-terrace model).

the San Andreas fault near Cholame, California (attributed to the AD 1857 earthquake), whereas Lienkaemper and Sturm (1989) measured 5.7 ± 0.7 m offset at the same site. The difference was caused by 2 m of post-AD 1966 agriculturally induced slopewash deposition in the offset channel, which was not recognized by the earlier workers. The uncertainties introduced by the variable (usually unknown) contribution of diversion and offset to total misalignment can be represented by a quality modifier (good, fair, poor) attached to the lateral offset measurements, as done by Sieh (1978b). Even experts have mistaken stream misalignment for stream offset. For example, 1980s vintage papers on the Karakax fault, Tibet, based on remote sensing imagery described it as a sinistral fault; but later fieldwork revealed it was a dextral fault (Lin *et al.*, 2008).

Offsets of stream channels can be interpreted in terms of individual paleoearthquakes only if the stream re-establishes its course across the fault between each earthquake. Given the tendency of streams (especially intermittent or ephemeral streams) to be diverted laterally along the fault zone, such re-establishment is unlikely to occur after every offset event. Thus, the differences between lateral offsets recorded by adjacent stream channels often reflect multiple faulting events, and amount to tens or hundreds of meters, rather than single-event amounts. For example, the modern channel of Wallace Creek (California) is offset 130 m dextrally by the San Andreas fault, whereas the next (abandoned) channel along strike is offset 380 m dextrally. Both the 130- and 380-m offsets obviously represent many paleoearthquakes, in light of the 9.5-m offset experienced there during the last major earthquake (1857 Ft. Tejon earthquake, $M_w \sim 8$). Sieh and Jahns (1984) conclude that Wallace Creek has only re-established its course straight across the fault twice in the past 13,000 years. The latter re-establishment occurred after the channel section within the fault zone filled up with alluvium, and water was able to spill straight across the fault. The reason for such severe aggradation is unknown, but may have resulted from fault-induced changes is stream gradient, as explained in Section 6.3.1.

Beheaded Streams When intermittent or ephemeral streams are offset, by amounts larger than their channel width, the downstream part of the channel is translated far enough to be totally disconnected from the upstream part. The downstream part thus becomes a "beheaded" stream (Figure 6.19). Where large offsets

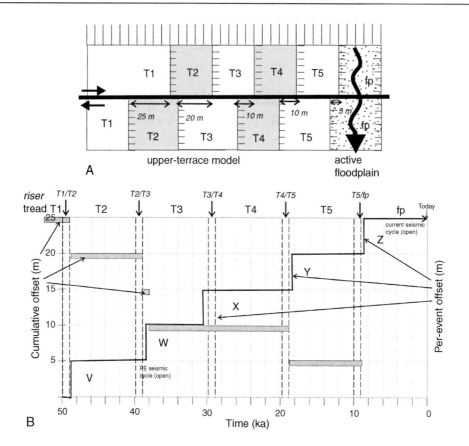

Figure 6.16: Diagrams of terraces that formed at irregular intervals compared to earthquakes, with displacement per event (5 m) held constant. (A) Map of terraces. Same parameters as in Figures 6.11 and 6.12, except that only trailing edge terraces are shown (= upper-terrace model). (B) Slip history diagram of the faulted terraces, comparing times of terrace deposition, riser formation, and faulting. Note that the time period between formation of risers T2/T3 and T3/T4 contains two paleoearthquakes, so the 10-m difference in their offsets represents the sum of two faulting events. Conversely, the time period between formation of risers T3/T4 and T4/T5 contained no earthquakes, so the cumulative offset of those two risers is identical.

have occurred repeatedly, a "master" stream on the upstream part of the fault may have multiple beheaded counterparts on the downstream side, one for each offset event (paleoearthquake). In this case, the per-event offset can be measured by the difference in offset between each pair of beheaded streams, assuming that the master stream re-established a straight course across the fault following each offset event (Figure 6.20).

A corollary hypothesis was proposed by Ferry *et al.* (2007) for the offset of deeply incised gullies in a badland terrain. They assumed that the master gully developed along the trace of the strike-slip fault, and that perpendicular tributary gullies formed opposite each other. They thus reasoned that any misalignment of opposing tributaries represented coseismic offset, although it could also be a primary feature, or a combination of both.

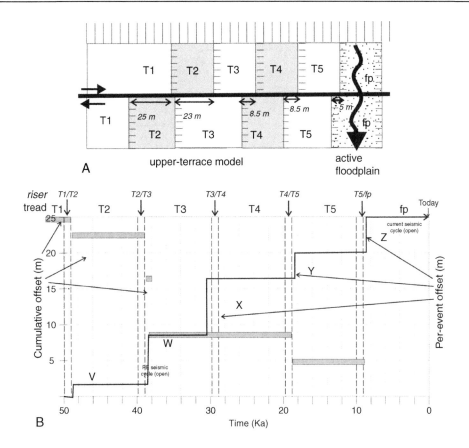

Figure 6.17: Diagrams of terraces that formed at irregular intervals compared to earthquakes, with displacement per event varying as in Figure 6.12, and recurrence varying as in Figure 6.13. (A) Map of terraces. Same parameters as in Figure 6.13, with only trailing edge terraces shown (= upper-terrace model). (B) Slip history diagram of the faulted terraces, comparing times of terrace deposition, riser formation, and faulting. As in Figure 6.13, the time period between formation of risers T2/T3 and T3/T4 contains two paleoearthquakes, and the time period between formation of risers T3/T4 and T4/T5 contained no earthquakes. Due to the variable displacement AND timing of paleoearthquakes, the map pattern of riser displacements (A) is not immediately interpretable in terms of number of events and their displacements, unlike the simple scenario shown in Figure 6.11.

The geomorphology of large perennial streams where they cross strike-slip faults is more complex than for ephemeral streams, because streams are responding to climate changes (resulting in aggradation and degradation) as well as to tectonic transport. The landforms resulting from this interplay of tectonic and climatic forces (channels, fans, and terraces) are typically unique to each site, based on the timing and severity of Quaternary climate changes and the rate of lateral slip. An excellent case history is described by Bull and Knuepfer (1987) and Bull (1991, Chapter 5) from the Charwell River and Hope Fault in New Zealand. However, at this site most landforms were offset by tens of meters (i.e., from multiple faulting events), so only slip rates, and not parameters of individual paleoearthquakes, could be deduced.

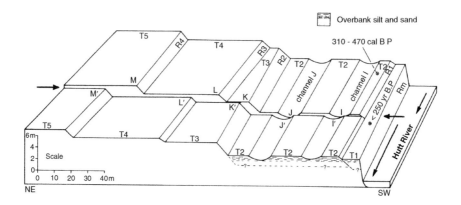

Figure 6.18: Perspective scale drawing of Holocene fluvial channels and terraces displaced by the dextral Wellington fault (between arrows) at Te Marua, North Island, New Zealand. Offsets I–I′, J–J′, K–K′, L–L′, and M–M′ are 3.7, 4.7, 7.4, 18.0, and 19.0 m, respectively. The faulted terraces shown are in a trailing edge geometry, because those on the downstream side of the fault are moving away from the stream. From Van Dissen *et al.* (1992); reprinted with permission of the Royal Society of New Zealand.

6.2.1.3   Offset Alluvial Fans

A third landform that is commonly translocated away from its source is that of alluvial fans. This geometry is similar to that of offset stream channels, except that rather than having an incised channel on both sides of the fault, the geometry here is to have an incised channel on the upstream side of the fault and a matching alluvial fan apex on the downstream side of the fault. For example, Sieh and Jahns (1984) describe a 13-ka alluvial fan at Wallace Creek on the San Andreas fault that has been displaced 475 m right laterally from its source gullies. Matching of the alluvial fan apex (a broad feature) with the suspected source gullies (narrow features) was accomplished by making an isopach map of the alluvial fan, and matching the areas of greatest fan thickness with the mouths of suspected source gullies (Sieh and Jahns, 1984, p. 892). In that regard, the evidence used was a combination of geomorphic and stratigraphic. Similar fan offsets are described by Bull and Knuepfer (1987) and Bull (1991, p. 237) in New Zealand. This technique works best when the fault zone is coincident with the heads of alluvial fans. Due to the poor resolution in defining the apex or axis of most large alluvial fans, this geomorphic technique is not usually useful for measuring offsets of less than 10–20 m, so individual paleoearthquakes often cannot be detected. In contrast, stratigraphic techniques can be more precise (see Section 6.3.5).

6.2.1.4   Offset Landslides

Landslides are a common feature along faults, given that steep slopes often lie on one side of the fault and are periodically subjected to strong ground shaking. Landslides triggered by an earthquake that cross the fault trace will be offset laterally by any subsequent surface ruptures. When the amount of lateral offset is small compared to the width of the landslide, the offset can be measured from misalignments of the margins of the landslide, or from secondary morphological elements on the landslide itself, much like measuring offsets of a moraine. When the amount of lateral offset is large compared to the width of the landslide, or exceeds the width of the landslide, then the source area/landslide head becomes misaligned with the landside deposit/toe. In this case, offset measurements become similar to those on alluvial fans, the source gully of which has been shifted away from the fan deposit/fan apex.

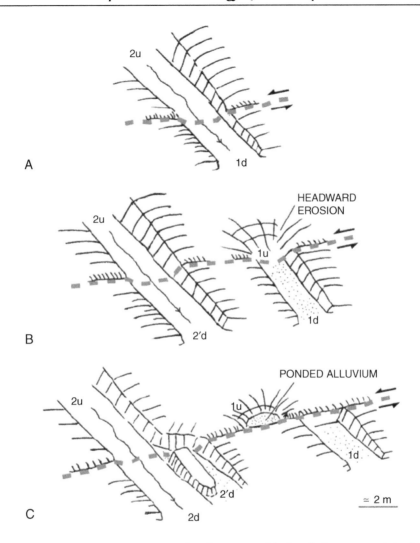

Figure 6.19: Schematic diagrams showing development of beheaded streams. (A) Throughgoing stream before offset; dashed line shows location of sinistral fault. (B) After one large offset event (*offset ≫ channelwidth*), the downstream part of the throughgoing stream (1d) has been transported to the right and "beheaded," and the throughgoing stream later re-establishes a straight course across the fault (2u–2'd). Eventually, the beheaded stream will erode headward (1u). (C) After a second, smaller offset event (offset ∼ channel width), the beheaded channel is further transported to the right, and is beheaded again from its headward-erosion swale. This swale is now dammed by the fault and becomes ponded. The downstream part of the throughgoing channel is shifted to the right (2'd) but not enough to be totally disconnected from the main channel. The throughgoing stream again established a straight course across the fault, by eroding away the offset on the leading edge, leaving a straight channel edge (left side of 2d). From Klinger *et al.* (2000); reprinted with permission of Wiley-Blackwell Publishers.

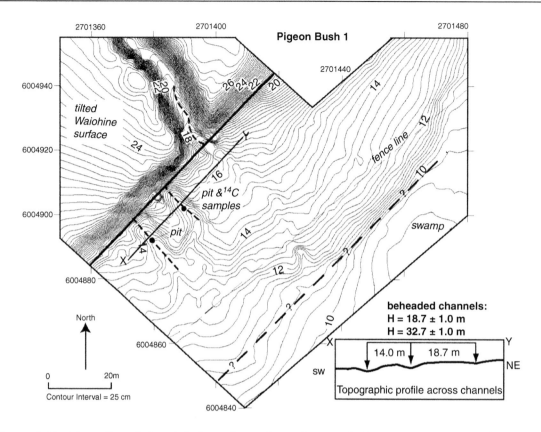

Figure 6.20: Map of two beheaded channels on the Wairarapa fault, New Zealand, showing Pigeon Bush stream (incised stream on "tilted Waiohine surface") and two beheaded channels displaced to the SW (dextral offset). The closer channel was offset 18.7 m by the 1855 earthquake, the largest single-event surface displacement yet measured on a terrestrial fault. Map is based on a survey data set consisting of 1800 points, which are omitted for clarity. Contour interval is 25 cm, elevations are meters above mean sea level, and graticules are NZ Coordinate Grid System (meters). From Rodgers and Little (2006); reprinted with permission of the American Geophysical Union.

Rust (2005) describes a landslide across the San Andreas fault in the Big Bend section on which secondary gullies have been offset 40–46 m laterally. However, the age of the landslide estimated from radiocarbon samples within the deposit (AD 780–1050) implies a slip rate of 34–51 mm/yr, much higher than rates measured north and south of the Big Bend. Thus, it is possible the landslide is older than the dates. This result emphasizes the difficulty of dating landslides, as explained in Chapter 8.

6.2.1.5   Offset Ridges and Valleys

Strike-slip faults that displace high-relief terrain create fault scarps by lateral offset of ridge and valley walls (Figure 6.21). In a general sense, lateral displacement of a ridge will create scarps that face in opposite directions if the net slip vector is inclined closer to horizontal than are the sideslopes of the ridge. This geometry is reflected in Figure 6.21 where the net slip vector is nearly horizontal and ridge sideslopes have a gradient of ca. 30°. On the left side of the ridges shown, the scarp (light tones) faces

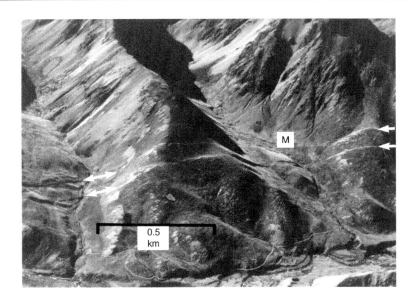

0.5
km

Figure 6.21: Oblique aerial photograph of fault scarps along the right-lateral Awatere fault, New Zealand. Two glaciated valleys are at left and right center. Fault scarps formed by dextral offset of valley sideslopes face toward the mountains (away from viewer) on the right side of each valley wall (light tones), and away from the mountains (toward the viewer) on the left side of each valley wall (shadowed scarps). Note the general lack of vertical relief where the faults cross the flat valley floors, and the dextral offset of the elongated hill of ground moraine below "M."

upvalley, away from the viewer. On the right side of each ridge, the scarp (in shadow) faces downvalley (toward the viewer). On the flat crest of the ridge, there is little to no height to the scarp. In contrast, if the net slip vector has a large vertical component, such that it is more steeply inclined than the ridge sideslopes, scarps on opposite sides of the ridge will face in the same direction but will have different heights.

Where ridge sideslopes are relatively planar and the fault trace is roughly perpendicular to the ridge crest, the net slip vector can be measured graphically even if sharp piercing points cannot be found. Figure 6.22 shows the reconstruction of right-lateral/normal displacement of an erosional ridge along the 1954 Fairview Peak, Nevada, fault scarp. This ridge has planar sideslopes of 23° but its crest is too broad to comprise a precise piercing point. Oblique displacement resulted in a valley-facing fault scarp with vertical surface offset of 1.3 m on the left (south) flank of the ridge but with 4.0 m of valley-facing vertical offset on the right (north) flank. There is only one unique net displacement vector that will result in these two scarp heights on a ridge with 23° sideslopes. That vector can be found by drawing a topographic profile of the ridge parallel to the fault on paper, and on a transparent overlay. One then shifts the profile on the overlay with respect to the profile on the underlying paper (in the same direction as inferred fault slip, without any rotation) until the apparent height of fault scarps on both ridge flanks is equal to that observed in the field. This graphical technique, applied to the profile in Figure 6.22, results in an estimated horizontal slip component of 3.0 m and a vertical slip component of 2.3 m (Figure 6.22C). The estimated vertical component should match that measured from a standard topographic scarp profile perpendicular to the fault scarp along the crest of the ridge (Figure 6.22B). In this example,

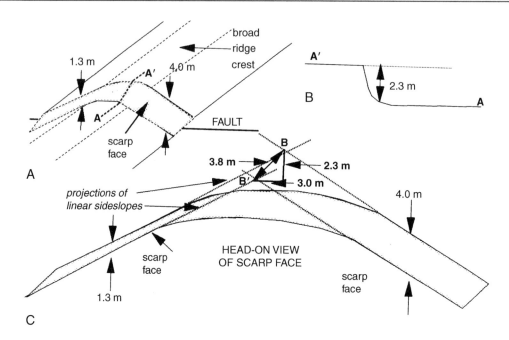

Figure 6.22: Diagrams showing an obliquely offset ridge and a crude method of measuring slip components, based on an example from the 1954 Fairview Peak, Nevada, fault scarp. (A) Perspective drawing of a broad-crested ridge offset in a dextral-normal sense. Scarp height on the left side of ridge = 1.3 m, scarp height on right side = 4.0 m. Ridge sides are linear and slope at 23°. (B) Topographic profile along the ridge crest perpendicular to the fault scarp. Scarp height = vertical surface offset = 2.3 m. (C) Head-on view of scarp face. The lines defining the crest and base of the fault scarp have identical shape, and are shifted laterally and vertically with respect to one another until the observed scarp heights on ridge sideslopes are attained. Graphical projections of linear sideslopes above and below the fault scarp define points B and B', respectively. The vertical offset between B and B' is measured graphically as 2.3 m, the same as the field measurement (see part B). The projection method further indicates a lateral offset of 3.0 m and a net slip (double-headed arrow) of 3.8 m at this site. Uncertainties in this method arise from the variable ridge cross-sectional shape above and below the fault, and errors from line projection.

the oblique-slip vector was inclined more steeply than the 23° ridge sideslopes, so scarps on opposite sides of the ridge face in the same direction.

This technique was originally formulated by Peltzer *et al.* (1988) as a more general case, and used for larger-scale offsets of landforms resulting from multiple offset events (Figure 6.23).

6.2.2   Using Lateral Offsets to Calculate Long-Term Slip Rates

Measuring the lateral offsets caused by paleoearthquakes on strike-slip faults differs in many ways from measuring displacements on dip-slip faults (Chapters 3–5). On the latter structures, displacement could be estimated from fault scarp height or from thickness of scarp-derived colluvium. Strike-slip faulting, in contrast, displaces the landscape horizontally, and vertical relief along the fault trace (if it exists) is

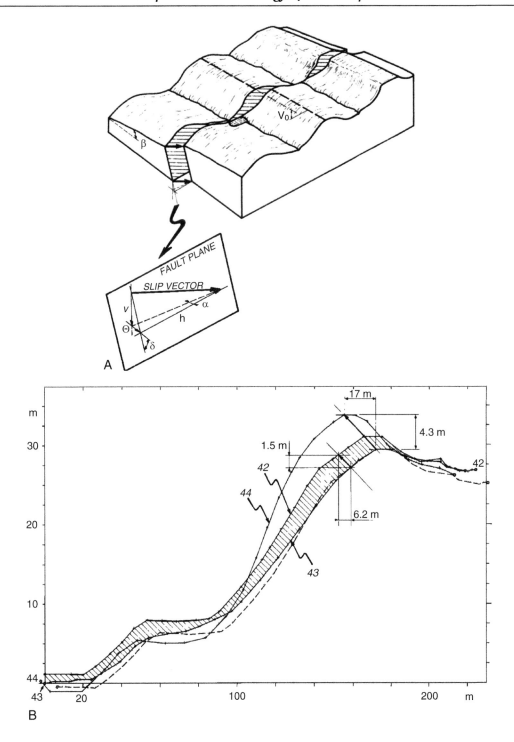

Figure 6.23: Method of Peltzer *et al.* (1988) for measuring the net slip vector from offset erosional landforms. (A) Schematic sketch of offset crests and channels. Dashed line represents a transverse profile along offset ridge crests. The net slip vector *V* has three components: horizontal and parallel

commonly the result of differential erosion or local variations in strike that are not relevant to the regional rupture. Horizontal displacements can only create vertical relief under two conditions (1) if fault strike varies sufficiently from net slip direction to form transtensive or transpressive structures and (2) if the fault trace laterally shifts topography with considerable relief.

Due to erosional smoothing in and near the fault zone, it is usually impossible to trace a linear landform right to the fault trace itself. Instead, the landform will be traced as close as possible to the fault and then projected to the fault plane from either side. The uncertainty in measuring lateral displacements thus has two components, the first arising from the preservation of the landform and its correlation across the fault (qualitative), the second from locating correlative points on the landform and then projecting them to the fault trace (quantitative). [Sieh (1978b) used a rating system (excellent, good, fair, poor) to indicate the uncertainty of the first type.] Excellent and good designations reflect the absence of complicating secondary faults, little or no lateral warping, sharp offset expression, and clearly interpretable geologic and geomorphic features. Uncertainties of the second type are expressed by assigning a plus-or-minus value to every measurement, reflecting the probable bounds of error. To support measurements of lateral offset made with tape measures, Sieh (1978b), McGill and Sieh (1991), and Grant and Sieh (1994) advocate compiling large-scale topographic maps of the displaced landforms with contour intervals as small as 10 cm (see Chapter 2).

6.2.2.1 Slip Rate Studies

Slip rate studies are landform-based paleoseismic studies, the aim of which is to calculate a long-term fault slip rate from landforms offset by multiple events. They are usually reconnaissance studies on poorly known faults, and precede trenching studies. Slip rate studies can be performed at widely varying spatial and temporal scales, and good examples of several scales are described by Hubert-Ferrari *et al.* (2002) for the North Anatolian fault:

Gross scale: displacements of tectonic terrains, tens to hundreds of km; terrains dated in Myr

Small scale: displacement of large river/valleys and drainage basins, km to tens of km; but normally one cannot precisely date erosional topography

Medium scale: latest glacial moraines, flights of glacial and postglacial river terraces, tens of m to hundreds of m; tens to hundreds of ka; dating by cosmogenics, luminescence, and ^{14}C (e.g., Brown *et al.*, 2002; Lasserre *et al.*, 2002; Meriaux *et al.*, 2005)

---

to the fault (*h*); horizontal and normal to the fault (*e*); and vertical (*v*). δ is the dip of the fault plane; *α* is the angle between the local strike of the fault and the horizontal projection of the slip vector; β is the surface slope angle (nearly identical to the ridge crest plunge); V_o is the vertical offset of the ridge crest profile slopes. Note that in this drawing, the net slip vector dips more steeply than any of the surface slopes, so the scarp formed faces to the lower right at all locations. (B) Topographic profiles parallel to the fault trace along the top (42) and base (43) of the scarp (hatched) and on the crest of the ridge north of the fault (44). Maximum error on point position is ±10 cm. Vertical exaggeration is 4×. Dashed line shows the scarp top profile displaced so that culminations along it fit those along the base profile. Overlapping zones (shaded) may result from accumulation of post-MRE colluvial deposits on the downthrown block. Restoration of displacement implies 6.2 m of left-lateral slip and 1.5 m of vertical slip. From Peltzer *et al.* (1988); reprinted with permission of the American Geophysical Union. (See Book's companion web site)

Large scale: small landforms such as debris-flow levees, terrace risers, small gullies/intermittent streams, m to tens of m; Holocene; dating mainly by ^{14}C

At the gross and small scales, no single-paleoearthquake offsets are usually discernable, so those studies fall more in the realm of tectonic geomorphology (see Burbank and Anderson, 2001). At small and medium scales, long-term slip rates can be calculated spanning tens to hundreds of ka (Table 6.3). The development of cosmogenic dating of arid-zone landforms has enabled many new slip rates to be estimated for faults in places like the western USA and central Asia.

At medium and small scales, with good preservation, it is possible to isolate and measure displacements from individual paleoearthquakes (see previous descriptions of offset stream channels, terrace risers, etc.).

Slip Rates from Offset Terraces/Risers Slip rates can be calculated using either the upper-terrace or lower-terrace models, as previously described. The choice of model determines which terrace tread age to divide into the offset of the riser. The terrace treads can be dated by several techniques, with the most common being cosmogenic surface-exposure ages (^{26}Al, ^{10}Be, or ^{36}Cl). However, the difference between slip rates calculated by the upper-terrace and lower-terrace models decreases as the number of earthquake events increases (Figure 6.24). The difference between the offsets of multiply offset risers (uneroded vs totally eroded) in these respective models remains only the single-event offset (5 m in examples of Figures 6.14–6.18), regardless of the magnitude of the riser's cumulative offset. Thus, as the cumulative offset increases, this single-event offset becomes a progressively smaller proportion.

Table 6.3: Examples of recent long-term slip rate studies based on landform offsets, listed in order of increasing oldest (landform) age

References	Location	Landform type	Largest offset or slip rate	Oldest age	Dating method
Van der Woerd *et al.* (1998)	Kunlun fault, China	Terraces	33 m	2.9 ka	^{26}Al, ^{10}Be
Lasserre *et al.* (2002)	Haiyun fault, China	Moraine	200 m	11 ka	^{26}Al, ^{10}Be
Meriaux *et al.* (2005)	Altyn Tagh fault, Tibet	Six terraces	225 m; 18 mm/yr	38 ka	^{26}Al, ^{10}Be
Le *et al.* (2009)	San Jacinto fault, USA	Stream channels	500 ± 70 m	47 ± 8 ka	^{10}Be
Frankel *et al.* (2007a)	Death Valley fault zone, USA	Alluvial fans	4.2–4.7 mm/yr	63–70 ka	^{10}Be, ^{36}Cl
Brown *et al.* (2002)	Karakorum fault, India	Terraces	40 m	90 ka	^{10}Be
Matmon *et al.* (2005)	San Andreas fault, USA	Alluvial fans	42 ± 9 mm/yr	413 ka	^{26}Al, ^{10}Be
Hubert-Ferrari *et al.* (2002)	North Anatolian fault, Turkey	Terraces	85 km; 6.5 mm/yr	13 Ma	Correlation, ^{14}C

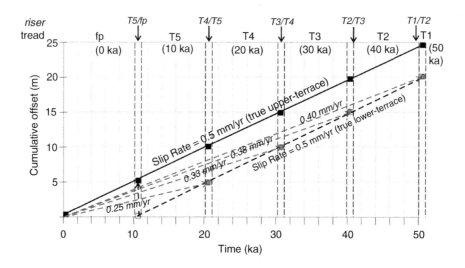

Figure 6.24: Comparison of slip rates calculated from the upper-terrace (trailing edge) and lower-terrace (leading edge) models, based on riser offsets from Figure 6.12. The thick solid line shows the true slip rate from displacement of uneroded risers (trailing edge) using the upper-terrace model, and the thick dashed line shows the identical slip rate calculated from the eroded risers on the leading edge, using the lower-terrace model. Gray dashed lines show spurious slip rates calculated by incorrectly applying the upper-terrace model to the eroded riser offsets on leading-edge terraces. Likewise, spurious slip rates would result from applying the lower-terrace model to the uneroded risers of the trailing edge terraces.

6.2.3   Spatial and Temporal Variations in Surface Displacement

To interpret paleoearthquake displacements on strike-slip faults, we need to understand the pattern and amount of variability in those displacements, both along the strike of the rupture in each event (spatial variation) and from rupture to rupture (temporal variation). The best way to understand the spatial variations is to examine the displacement variability in well-studied, historic strike-slip ruptures (e.g., those listed in Table 6.1). The best way to understand the temporal variations is through trenching studies which measure the displacement at a point in successive paleoearthquakes.

6.2.3.1   Variability of Displacement Along Strike in a Single Rupture

In a general sense, horizontal displacement on historic strike-slip surface ruptures has been greatest near the center and least near the ends of rupture (Figures 6.6B and 6.7), although in some ruptures offset is greatest near the end of the rupture and decreases away from it (Figure 6.6A). Shorter-wavelength variations in offset do exist, but tend to be less marked than in normal or reverse ruptures, particularly in the longer ruptures with higher average displacements (compare Figures 6.6 and 6.7 to slip graphs in Chapters 3 and 5). In addition, strike-slip ruptures tend to have a more uniform distribution of offsets of all sizes, compared to other rupture types. Figure 6.25 shows the frequency distribution of normalized horizontal slips for the 2002 Denali, Alaska surface rupture (the raw slip measurements are shown in Figure 6.7). The most common offsets along strike were between 50% and 60% of D_{max} ($D_{max} = 8.8$ m). The scarcest offsets were those $<10\%$ of D_{max}, but that may reflect a lack of measurements or interest

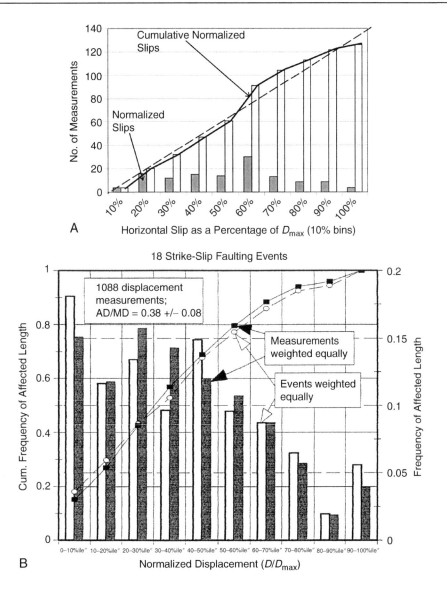

Figure 6.25: Variation of slip along strike in a typical high-magnitude, historic strike-slip rupture; the 2002 rupture of the Denali fault, Alaska (D_{max} = 8.8 m, rupture length = 341 km). (A, B) Individual frequency (black bars) and cumulative frequency (solid line) of normalized horizontal displacements, binned in 10%-ile classes. The 2002 rupture contained an abnormally high percentage of displacements in the range of 50–60% of D_{max}. Dashed line depicts a uniform frequency distribution of normalized displacements. Data from Haeussler *et al.* (2004). (B) Cumulative frequency (line) and discrete frequency (histogram) of normalized horizontal displacements, binned in 10%-ile classes, for a combined data set of 1088 displacement measurements from 18 well-studied strike-slip ruptures. Average displacement (AD) averages about 38% of maximum displacement (MD). Compared to other types of ruptures, strike-slip ruptures have a more uniform frequency

in the low-slip tails. Otherwise, there is a remarkably uniform distribution of normalized slips; this pattern is very different from that observed for normal faults. Along strike the displacements did display some short-wavelength variations, but they tended to have relatively small amplitude, compared to other types of ruptures. If these shorter-wavelength variations represent the complex response of surficial deposits to rupture (as suggested for normal and reverse faults, see Chapters 3 and 5), it suggests that large-displacement ruptures such as this one ($D_{max} = 8.8$ m) are less affected by surface deposit rheology, than are smaller ruptures.

Figure 6.25B shows that some of the displacement characteristics of the 2002 rupture are shared by most large strike-slip ruptures. The normalized horizontal displacement data from 18 well-studied ruptures was combined into a single data set (1088 measurements) by McCalpin and Slemmons (1998). In frequency space, strike-slip ruptures tend to have a rather uniform frequency of various-size displacements less than about 70% of D_{max}; in other words, the cumulative frequency curve is nearly a straight line, only flattening at the very large normalized displacements. This is a very different pattern than shown by normal and reverse ruptures, the former of which has an overabundance of small displacements (<30% of D_{max}) and very rare occurrence of displacements >80% of D_{max}. Thus, strike-slip ruptures, particularly the large ones, tend to have more uniform displacements along strike with less short-wavelength variation. One implication of this geometry for paleoseismic studies (explored in more detail in Chapter 9, See Book's companion web site) is that randomly located trench sites on strike-slip faults are likely to encounter displacements nearer the maximum that occurred in each paleoearthquake, than similar trenches on normal or reverse faults.

6.2.3.2 *Variability of Displacement at a Point*

Based on a limited number of studies, strike-slip ruptures tend to have relatively low variability in displacement at a point on the fault from earthquake to earthquake, compared to other types of faults. This repeatability of displacement per event was noticed early in the development of paleoseismology from data on the San Andreas fault, and formed part of the basis for the "characteristic earthquake" model of Schwartz and Coppersmith (1984).

Since that time, there have been additional studies aimed at quantifying the variability in displacement-per-event through multiple seismic cycles at a single trench site. At this time, the study of Liu-Zeng *et al.* (2006) appears the most definitive. They opened a latticework of trenches across offset channels on the central San Andreas fault and were able to correlate the channels across the fault on the basis of their elevations, shapes, stratigraphy, and ages. The three-dimensional excavations allowed them to locate accurately the offset channel pairs and to determine the amounts of motion for each pair. They found that the dextral slips associated with the latest six rupture events were (from oldest to youngest) 5.4 ± 0.6, 8.0 ± 0.5, 1.4 ± 0.5, 5.2 ± 0.6, 7.6 ± 0.4, and 7.9 ± 0.1 m. This series of displacements has a mean of 5.9 m, sigma of 2.5 m, and coefficient of variation of 0.43. If the smallest (anomalous) displacement is omitted, the displacement series is more regular (mean of 6.8 m, sigma of 1.4 m, COV of 0.20). These numbers bracket the COV of 0.3 cited by D.P. Schwartz (personal communication) as his approximate criterion for characteristic behavior.

---

distribution of different-size offsets, particularly those smaller than 70% of D_{max}, although the exact values depend on whether all measurements are weighted equally, or all ruptures are weighted equally (ruptures contained different numbers of measurements). All data from McCalpin and Slemmons (1998); see complete report on the companion web site, Chapter 9 (See Book's companion web site).

Interestingly, Liu *et al.* (2006) found that smaller offsets in the series followed the shorter seismic cycles and larger offsets followed longer seismic cycles, supporting the slip-predictable model of strain release for that part of the San Andreas fault. However, it should be noted that if any of the displacements attributed to a single event were actually the sum of two events, then the same pattern would arise. Other recent studies showing repeatable strike-slip displacements over the past two seismic cycles include Rubin and Sieh (1997) on the Emerson fault, California (part of the 1992 Landers rupture), and Benson *et al.* (2001) on the Awatere fault, New Zealand.

6.2.4 Reconstructing Individual Earthquake Displacements

Horizontal separations measured between piercing point landforms could be the result of one or many paleoearthquakes, depending on the age of the landforms and the recurrence time of the fault. Numerous workers (e.g., Wallace, 1968a; Sieh, 1977; Rockwell and Pinnault, 1986; Rockwell, 1989; McGill and Sieh, 1991; Trifonov *et al.*, 1992) have suggested that individual paleoearthquakes can be identified from a *frequency histogram* of lateral offsets. Such histograms commonly show clusters or groups of similar displacements (e.g., 3, 6, 9, and 12 m), which are then interpreted to be the cumulative slip associated with a discrete number of paleoearthquakes. McGill and Sieh (1991) contend that such an interpretation rests on two assumptions. First, new geomorphic features must form during every interseismic period and some of these must survive up to the period of observations. If earthquakes and erosional processes are scattered in time, this assumption is reasonable. However, if earthquakes or erosional events are clustered in time, no interearthquake landforms may be formed and the combined displacements of two paleoearthquakes may appear as one large displacement in the geomorphic record (a possible explanation of the anomalous MD in the 1857 Ft. Tejon rupture). Second, one must assume that observed displacements are coseismic (or closely postseismic) rather than the result of continuous creep. Both of these assumptions can be validated on many faults by historical observations and other paleoseismic studies.

McGill and Sieh (1991) and McGill and Rubin (1999) also used historically observed slip gradients as an indirect criterion to distinguish between landforms offset by one versus two paleoearthquakes. For example, if two terrace risers only 10 m apart were offset laterally by 2.7 ± 0.7 and 5.6 ± 0.7 m, that indicates a single-event slip gradient of 3×10^{-1}. Therefore, they would consider any prehistoric landform offsets that yielded equal or lower slip gradients to be arguably the result of a single rupture. Maximum slip gradients measured on the 1992 Landers, California, earthquake were of this magnitude. Given such high observed slip gradients, Weldon (this chapter, First Edition) concluded that it would be very difficult to use the amount of offset of any landform, even when compared to that of adjacent landforms, to infer the number of displacement events it had experienced.

In contrast, Rockwell *et al.* (2002) believe that slip gradients in most large strike-slip ruptures are much smaller than those proposed by (1) McGill and Rubin (1999) for historic earthquakes and (2) several authors for prehistoric earthquakes. In other words, displacement does vary laterally somewhat, but not at extremely high-slip gradients. First, they argue that the high-slip gradients observed in Landers arose from displacement switching from one parallel fault strand to another, not from a monotonic decrease on a single strand. The basis for this conclusion is Rockwell *et al.*'s work on the 1999 Izmit and Duzce ruptures, where they measured total offset across the entire fault zone via long across-fault profiles. These profiles measured the net 1999 offset on all parallel fault strands plus distributed deformation between strands, at a given point on the rupture. When these net offsets were compared along strike, there was some lateral variation, but not as extreme as that cited at Landers (i.e., extremes of 1.8 m/100 m and

<1 m/10–15 m). These slip gradients are similar to those measured by Treiman *et al.* (2002) on the Hector Mine, California, rupture of 1999. Second, they argue that some very high-slip gradients interpreted for prehistoric ruptures could better be explained if the smaller offset were a single-event offset and the larger one a two-event offset.

6.2.4.1   Quantitative Analysis of Multiple Lateral Offsets

In our First Edition, the state of the art was described by McGill and Sieh (1991), who measured 74 offset geomorphic features over a distance of 27 km along the Garlock fault in Pilot Knob Valley, California. This study was distinctive because it not only involved a large number of offsets which are rated for quality, but also included quantitative uncertainties associated with each offset measurement and therefore permitted a reasonably rigorous statistical analysis. The values of the offsets appear to be clustered (Figure 6.26), suggesting multiple ruptures, and all of the proposed clusters span the area of coverage. They interpreted each of the six peaks in Figure 6.26 as resulting from six separate paleoearthquakes, with cumulative offsets as 3.4, 5.3, 8.6, 11.8, 15.9, and 18.0 m. This inference is supported by the approximately integral increase in each peak's offset, that is, 3.4, 5.3 (3.4 + 1.9), 8.6 (5.3 + 3.3), 11.8 (8.6 + 3.2), and 15.9 m (11.8 m + 4.1 m).

This relatively simple interpretation assuming consistent offset through time was complicated somewhat by the observed offset patterns during the 1992 Landers earthquake, in which the ruptures displayed not only short-wavelength variations in offset (common to all ruptures), but also long-wavelength (km-scale) variations (McGill and Rubin, 1999). They noted that each of the four major faults that ruptured during the 1992 Landers earthquake display a multimodal distribution of offset, due to the long-wavelength variations (Figure 6.27). They note:

> *In some cases the different peaks in the histogram correspond to particular segments of the fault, suggesting that if a short enough fault segment is considered, a single rupture may in fact produce a single-peaked histogram. The 5.6-km-long segment that produced the bimodal histogram (shown in Figure 6.26B) can be broken down into 1.5- to 2-km-long segments, some of which have histograms with one dominant peak, although they still have minor peaks at other offset values. Thus, for at least some parts of the rupture the multimodal nature of the histograms is produced by km-scale variations in slip. The fault lengths over which nearly unimodal histograms are observed are so short, however, as to be of little use in deciphering the number of prehistoric earthquakes associated with a set of offset geomorphic features.*

Kilometer-scale variations in slip can also be seen in Figures 6.6, 6.7, and 6.25.

For paleoseismology, the effect of km-scale, long-wavelength slip variations is that on one part of a prehistoric fault scarp, a 4-m height may represent a single event (MRE), whereas on another part where the MRE was only 2 m, a 4-m scarp would represent the MRE and the PE. Such trends can easily be seen by constructing a slip-along-strike diagram on which displacements on deposits of different age are connected by lines (e.g., Figure 9.15).

To remove the ambiguity in number of offset events (on prehistoric ruptures) induced by such km-scale variations in slip, McGill and Rubin tried three techniques. First, they subtracted a three-point running average from all offset measurements, and plotted that histogram. If that histogram is multimodal, it

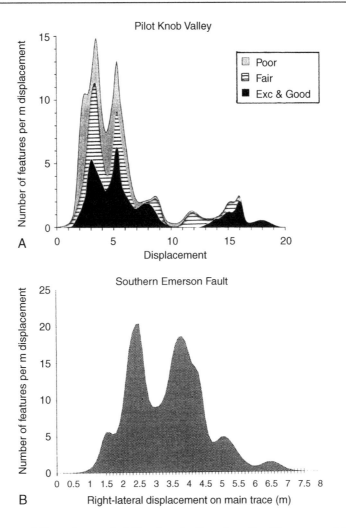

Figure 6.26: (A) Summed Gaussian probability density functions for 62 geomorphic offsets across the Garlock fault zone in Pilot Knob Valley, California. Shading indicates the quality of offset estimates used. The six peaks were interpreted by McGill and Sieh (1991) to represent the cumulative slip associated with six paleoearthquakes. From McGill and Sieh (1991); reprinted with permission of the American Geophysical Union. (B) Summed Gaussian probability distributions of lateral offsets along a 5-km length of the Emerson fault created during the 1992 Landers, California, earthquake. Note the bimodal distribution resulting from a single faulting event. From McGill and Rubin (1999).

argues for multiple earthquakes; however if it is unimodal, it could still represent one, or more than one, earthquake.

Second, they compared the slip gradient between two prehistoric offsets to the largest slip gradient documented in any historic strike-slip rupture (e.g., their gradient of 0.22 observed on the Emerson fault). For example, five pairs of offset landforms in Pilot Knob Valley have slip gradients between 0.26 and 1.05, which greatly exceed any known slip gradient in a single event. The weakness of this method is that

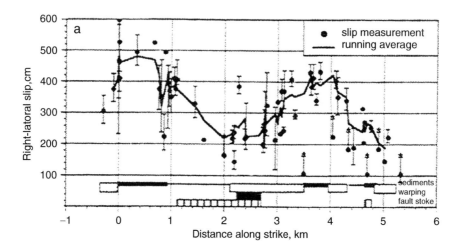

Figure 6.27: Right-lateral offset as a function of distance along strike, Emerson fault, 1999 rupture. Measurements shown here include slip on all fractures within a zone several tens of meters wide and centered on the main fault trace. Measurements that do not span the entire width of the deformation zone are regarded as minimum values and are shown with triangles at the upper ends of the error bars. The solid line is a running average over three points, excluding the minimum values. From McGill and Rubin (1999).

slip gradients may not be scale-invariant over horizontal distances ranging from a few meters (such as measured on the Emerson fault) to tens of meters to kilometers. McGill and Rubin note:

> *The observed slip gradient of 0.09 between offsets that are 18 m apart cannot be used as a precedent for a 9-m change in slip in a single earthquake over a 100-m distance along strike.*

Thus, they admitted to ambiguity in using this criterion to separate single- versus multiple-earthquake offsets.

Third, they investigated the change in slip as a function of distance between the points, plotting the raw input values against each other rather than simply using their quotient (slip gradient). Plotting the change in slip versus distance between points for all the Emerson fault offset measurements produces a cloud of points which are bounded by an "envelope of precedence," i.e., the largest amount of change historically observed at a given distance between points (Figure 6.28). When the five offset pairs from the prehistoric Pilot Know Valley rupture are superimposed on this plot, three of them clearly exceed any of the observed historic values, and are thus interpreted by McGill and Rubin (1999) as representing multiple earthquakes. The two remaining points (triangles in Figure 6.28) plot slightly above the envelope, and their number of earthquakes is still ambiguous.

6.3   Stratigraphic Evidence of Paleoearthquakes

The stratigraphy and deformation of unconsolidated sediments in strike-slip fault zones provides the best evidence for determining the number of and age of paleoearthquakes, and in some cases (Section 6.3.4) the offset in individual events. Much of the progress in reconstructing paleoseismic histories from

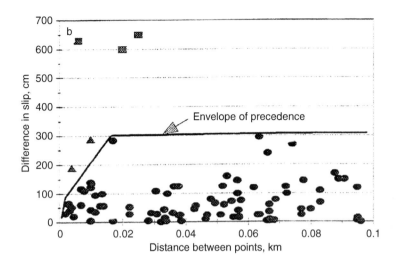

Figure 6.28: Difference in slip as a function of distance between offset measurements along the Emerson fault rupture of 1992, for all measurements within 100 m of each other. All measurement pairs, not just adjacent ones, are plotted. The "envelope of precedence" indicates the most extreme change in slip that has been documented for any given distance between points. Squares and triangles represent measurement pairs along the Garlock fault in Pilot Knob Valley that have slip gradients larger than the largest slip gradients along the 1992 Landers rupture. From McGill and Rubin (1999).

stratigraphic evidence followed Sieh's (1978a) widely read paper on fault-zone stratigraphy and numerical dating on the San Andreas fault. These microstratigraphic techniques, coupled with fault-zone trenching technology, have permitted far more detailed paleoseismic analyses than could have been performed from geomorphic measurements. However, the success of stratigraphic investigations depends heavily on the choice of sites for excavation, as explained later. In addition, stratigraphic features produced outside of the fault zone, mainly from ground failure, are also useful for identifying paleoearthquakes and are discussed in Chapters 7 and 8.

6.3.1   General Style of Deformation on Strike-Slip Faults in Section

The general features of strike-slip fault zones in section were described in Section 6.1.1.3, and include the main fault plane, fault shear zones, subsidiary faults, flower structures, and folds. If subsidiary faults (flower structures) are lacking and only the main fault planes exists, repeated faulting may create a chaotic shear zone along the main fault that is basically uninterpretable from a paleoseismic standpoint; fortunately, this appears to be a rare occurrence. Instead, most paleoseismic interpretations of strike-slip faults are based on subsidiary faults, fissures, folds, unconformities, and sand dikes that emanate from the main fault plane but diverge from it upward. Rockwell and Ben-Zion (2007) show that, where the thickness of sedimentation is small between ruptures, the upward-flaring secondary structures from several paleoearthquakes may overprint each other (Figure 6.29) and give the impression of a broad zone of deformation. In contrast, where interearthquake deposition is thicker it can be seen that the flower

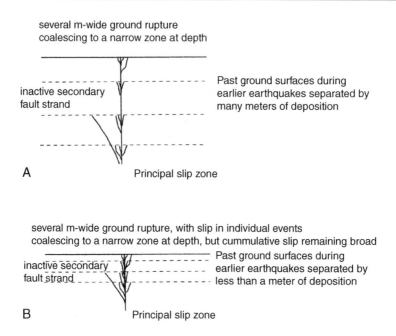

several m-wide ground rupture
coalescing to a narrow zone at depth

inactive secondary
fault strand

Past ground surfaces during
earlier earthquakes separated by
many meters of deposition

A   Principal slip zone

several m-wide ground rupture, with slip in individual events
coalescing to a narrow zone at depth, but cummulative slip remaining broad

inactive secondary
fault strand

Past ground surfaces during
earlier earthquakes separated by
less than a meter of deposition

B   Principal slip zone

Figure 6.29: Evidence for the high localization of slip in strike-slip earthquakes. Schematic representations of two types of superposed structures. (A) Where the thickness of deposition is large between surface ruptures, it may be possible to see the width of the slip zone collapse downward into a narrow zone. (B) Where the thickness of deposition is small between surface ruptures, it is likely that the upward flowering of slip toward the free surface will give the appearance of an overall broader zone. From Rockwell and Ben-Zion (2007).

structure from each paleoearthquake is separate and discrete, and all merge into a single main fault that accommodated all the horizontal displacement over many seismic cycles. This pattern is also found to a lesser extent in transtensional or transpressional bends or stepovers, although there the oblique nature of slip may lead to slip partitioning which widens the zone of deformation (e.g., the Wrightwood site on the San Andreas fault; described later).

6.3.2   Sedimentation and Weathering in Strike-Slip Fault Zones

Previous paleoseismic investigations (e.g., Sieh, 1978a; Fumal *et al.*, 1993) have emphasized the importance of sedimentation and weathering in strike-slip fault zones for reconstructing paleoearthquake histories. The best sites for distinguishing individual paleoearthquakes are where (1) essentially continuous deposition has occurred in the fault zone throughout the time period of interest, concomitant with faulting, and (2) multiple fault strands occur in a wide zone. The second phenomenon cannot always be inferred from surface evidence, but sediment traps are relatively easy to locate from fault-zone geomorphology. In these traps, deposits often continue to accumulate after each displacement event, creating the multiple unconformities and crosscutting relationships from which each displacement event may be distinguished. In the semiarid terrain of the western United States, Holocene sediments typically accumulate in two geomorphic settings along strike-slip faults: sag ponds and intermittent stream channels. Because running water may have removed part of the stratigraphic record in these settings, along with its paleoearthquake evidence, one should always address the completeness of the record during interpretation (e.g., Dawson *et al.*, 2003).

6.3.2.1 The Sag Pond Environment

The best depositional environments for preserving paleoearthquake evidence are relatively low-energy environments where sediments accumulate episodically in thin strata, separated by weathering profiles, organic soils, or peats (e.g., the Pallett Creek site: Sieh, 1978a; the Glen Ivy Marsh site: Rockwell *et al.*, 1986; the Wrightwood site: Fumal *et al.*, 1993, 2002; Weldon *et al.*, 2002; the Hog Lake site: Rockwell *et al.*, 2004). In semiarid and subhumid regions, these conditions are often found in sag ponds fed by unconcentrated slopewash, minor rillwash, or small ephemeral streams.

Sag ponds along strike-slip faults typically occupy structural depressions created by transtension and normal faulting, which are found in minor releasing steps or bends (e.g., the Pallett Creek and Glen Ivy Marsh sites on the San Andreas fault, California). Such features are readily observed on aerial photographs. More rarely, sediments are trapped when shutter ridges partially or completely block ephemeral drainages that flow perpendicular to fault strike, creating marshes or swamps (e.g., Hall, 1984). Alluvial fan deposition in the fault zone can also block fault-parallel drainage and create marshes.

Alternating subaqueous (fluvial/lacustrine) deposition and subaerial exposure typically give rise to a stratigraphic sequence of finely stratified sand, silt, and clay (in beds a few centimeters to tens of centimeters thick), interbedded with soil A horizons or thin peats. The organic content of peats and soils is maximized if the sag pond does not completely dry out during normal dry seasons. In arid regions playas represent a similar depositional environment in closed depressions, although playas are larger than sag ponds and not entirely created by fault movement.

Trenching has been successful in very wet marshes that were drained by dewatering wells (Rockwell *et al.*, 1986) or lakes drained by deliberate breaching (e.g., Hog Lake; Rockwell *et al.*, 2004). Alternatively, recent stream incision from natural or man-made causes can dewater sag ponds (e.g., Sieh, 1978a). Thus, paleoseismologists should search for sag ponds that fill with water and receive fine-grained, distal fluvial deposition from small catchment areas in the infrequent very wet years, but exist as marshes or swamps the rest of the time.

Not every sag pond meets these criteria. Tectonic depressions occupied by larger, through-flowing gullies are often subject to channel scour and lateral erosion (see next section), and may be too well drained and dry between rare depositional events to develop peat or rich organic soils. Sag ponds fed by streams subject to debris flow may fill with thick debris containing minimal organic material. At the opposite extreme, perennial sag ponds are typically sites of continuous fine-grained lacustrine deposition. Trenches in such settings (which are possible only if the sag pond has been drained) have often revealed massive deposits of dark gray clay with few stratigraphic markers (Rockwell *et al.*, 1986). However, sometimes you get lucky. The stratigraphy in the Hog Lake sag pond (Rockwell *et al.*, 2004) is exceptional, dominated by nearly continuous quiet-water deposition of finely laminated clayey silt with some interbeds of well-sorted sand (see Section 6.3.4.8). The depositional environment and the general absence of bioturbation because of the high groundwater conditions permit cm-scale resolution of individual units. Further, the strata contain abundant organic-rich "peat-like" deposits that contain abundant seeds, detrital charcoal, freshwater pond snails, and in situ reed growths, all of which have yielded stratigraphically consistent radiocarbon ages.

6.3.2.2　The Intermittent Stream Environment

Several workers (Wallace, 1990; Sims, 1994) have speculated that the complex stratigraphy of nested cut-and-fill channels in streams that cross strike-slip faults are caused by predictable geomorphic processes, and might be related to individual paleoearthquakes. For example, Wallace (1990, pp. 17–19) describes the hypothetical evolution of a straight channel that formerly crossed the fault at right angles, and was later offset by right-lateral strike slip (Figure 6.30):

The strike slip partly or temporarily dams the stream, causing upstream alluviation at C. A fresh fault scarp is formed in the vicinity of A, and successive offsets expose new scarp areas to the left of A. The dam at B is eroded, and the alluvium deposited earlier at C is dissected. As offset progresses further, the channel segment along the fault line between B and A continually elongates, thus lowering the channel gradient more and more. Because of this decreasing gradient, alluvium is deposited upstream from A to and beyond C, and eventually the stream, having difficulty maintaining a channel along that elongate course, spills across the fault trace and creates a new channel more nearly in alignment with the segment upstream from the fault.

J. D. Sims refined Wallace's idea into a "tectono-sedimentological process–response model" (Figure 6.31) and tested the model at the Phelan Creek site on the San Andreas fault, California. They divide streams crossing the fault into an *upstream reach* (upstream of the fault), an *along-fault reach* (essentially parallels the fault), and a *downstream reach* (downstream of the fault). The nearly right-angle bends between the upstream, along-fault, and downstream reaches are termed the *upstream* and *downstream bends*, respectively. According to their model, each faulting event lengthens the along-fault reach and decreases its gradient. In response, the stream aggrades in the along-fault reach in an attempt to increase its gradient back to its prefaulting gradient. Deposition begins at the upstream bend, due to partial damming. The extent of damming can be assessed by comparing the ratio of the average channel width to the average lateral offset per earthquake. For example, a 9-m coseismic offset would completely dam any stream channel <9-m wide, creating a topographic depression and causing ponding and marsh or

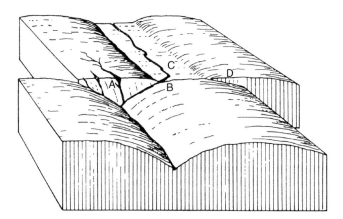

Figure 6.30: Schematic block diagram showing general features and conditions produced where a stream channel is offset by dextral fault slip. See text for discussion of points A–D. From Wallace (1990).

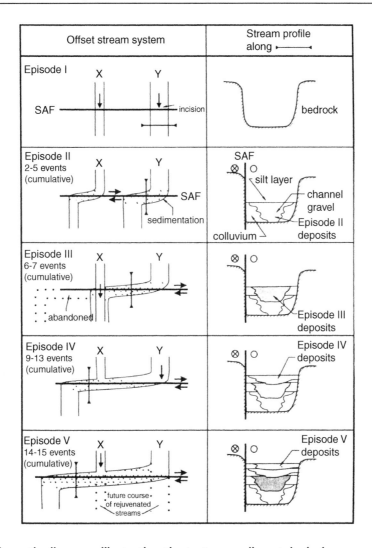

Figure 6.31: Schematic diagrams illustrating the tectono-sedimentological process–response model for strike-slip faults. Left side shows plan views of two intermittent streams successively offset dextrally in multiple faulting events. Right side shows cross sections through the along-fault reach of the stream (except in episode I). Time progresses from top to bottom (i.e., from episode I to episode V). Deposits related to each multiple-event episode are shaded in the right-side diagrams. At Phelan Creek, subunits within deposits of a given episode were assumed to represent individual paleoearthquakes, based on the coincidence of radiocarbon ages with other trench sites. From Sims, 1994).

lacustrine deposition. The same 9-m offset of a >9-m-wide channel would only create channel diversion, inducing fluvial aggradation due to gradient decrease but no ponding.

Each faulting episode forces the along-fault reach to readjust its slope abruptly and a new depositional cycle is initiated (Figure 6.31). The offset history would thus be represented by a sequence of stacked channel fills, in a geometry similar to that of stacked colluvial wedges against normal-fault scarps. Eventually, successive depositional episodes fill the along-fault channel to nearly bankfull, reducing the freeboard and leading to channel overtopping during seasonal flooding. The overtopped channel spills straight across the fault, becomes incised, and a new cycle is initiated.

J. D. Sims and coworkers document that stacked channel fills do exist in along-fault reaches at Phelan Creek. However, it is difficult to prove that every one of these fills is offset induced (rather than climatically induced) at this site because few subsidiary faults exist that would give an independent confirmation (via upward terminations or other structural indicators) that each channel fill formed soon after a paleoearthquake. Instead, Sims notes that the age of each channel fill unit or subunit closely matches the ages of paleoearthquakes dated elsewhere on the same fault segment (e.g., Prentice and Sieh, 1989; Sieh *et al.*, 1989; Fumal *et al.*, 1993; Grant and Sieh, 1994) by the use of more traditional structural indicators such as upward terminations (Section 6.3.3). At this point their model cannot be accepted as proven, but further research is certainly warranted to determine the conditions under which stacked channel fills consistently replicate paleoearthquake sequences.

As a cautionary note, we acknowledge that multiple episodes of sedimentation and erosion from nontectonic causes (storms, fires, stream capture) certainly affect fault zones. For example, 70 cm of fan sediments containing six soil horizons have accumulated in the San Andreas fault zone at the Bidart fan, California, since the last surface-rupturing earthquake (AD 1857). The latest episode of deposition is clearly associated with storms (Grant and Sieh, 1994), and by analogy the six soils and five intervening deposits are all storm related. Thus, workers should be cautious in inferring paleoearthquakes based solely on the complex record of erosion and deposition within strike-slip fault zones (similar caveats were made for normal-fault grabens in Section 3.3.2.2).

6.3.3 Trenching Techniques

There are basically three approaches used to generate the necessary three-dimensional data to characterize strike-slip paleoseismic events. They are (1) to excavate *multiple, closely spaced trenches* orthogonal to the fault, generally one at a time and (usually) backfilling to avoid collapse, (2) to *excavate progressively* a volume of the fault zone by digging successive exposures orthogonal to the fault, and (3) to excavate one or more trenches *perpendicular to the fault* (locator trenches), then excavate two trenches *parallel to and on each side of the fault (fault-parallel trenches, to locate piercing lines)*, and finally to progressively excavate the area between the fault-parallel trenches. The case histories that we discuss later include all of these techniques, but we briefly describe each approach here, discuss some of the advantages and disadvantages of each technique, and suggest what problems are best addressed with each approach. In every case, the strategy must evolve as the excavation develops, to incorporate what is learned as work progresses.

In all three approaches, one or more trenches perpendicular to the fault (*locator trenches*) are excavated to locate the fault precisely and to guide future excavations (Figure 6.32A). We summarize the sequence of excavations at Wrightwood, California (Figure 6.32B) to show how the approach evolved as we learned from each excavation. First, natural streambank exposures of the fault (at Swarthout Creek) were cleaned and peat and wood were dated to confirm the age of local deposits. Trench 1 (a locator trench) was excavated to span the entire fault zone. It was clear that there were at least two fault traces, defined by lush vegetation and subtle scarps (subsequently named the main and secondary fault zones). Numerous small, previously unsuspected, structures were revealed in the secondary fault zone. Because the main fault strand displaced only sediments that were several thousand years old in trench 1, and

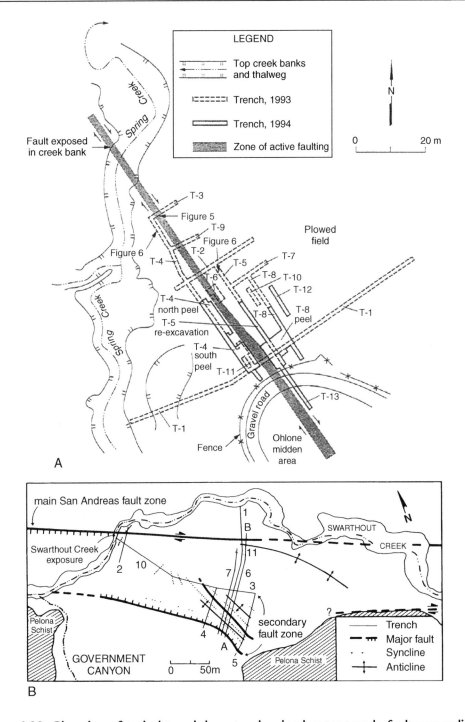

Figure 6.32: Plan view of typical trench layouts, showing locator trench, fault-perpendicular trenches, and fault-parallel trenches. (A) Site plan of trenches on the San Andreas fault, San Francisco Peninsula segment, at Woodside, California (Hall *et al.*, 1999). They centered

displayed very complex repetitive deformation, we excavated trench 2 in the hope of exposing a younger (and less complexly deformed) section. The younger section was found, but the secondary fault zone did not extend to trench 2. Trenches 1 and 2 were later connected by trench 10 (Figure 6.32) and, once connected, the results were subsequently integrated and published (Fumal *et al.*, 1993).

Part of the site, involving the secondary fault zone, was excavated using *multiple parallel trenches*. First, trench 3 was excavated perpendicular to trench 1, away from obvious deformation, to establish the lateral extent of the young stratigraphic section; next, two trenches (4 and 5) were excavated parallel to trench 1, across the fault zone, to establish the continuity of the structure. All of these trenches were mapped (both walls) and individual units could be carried through the entire "fork-shaped" network of trenches. Finally trenches 6 and 7 were excavated, successively to avoid collapse, 1 m from and on each side of trench 1. Thus, we had 10 exposures across the fault zone (2 from each of trenches 4–7, and 1), and six exposures in a 5-m stretch (involving trenches 1, 6, and 7). Each trench was logged and then carefully buried after logging to preserve the grid, marker flags, and nails, for future reference.

The greatest advantage of the multiple parallel trench approach is the preservation of mapped surfaces (though a later, deep excavation destroyed a portion of four exposures). Overall, about 90% of the mapped exposures at Wrightwood can be re-excavated and checked; in fact, we have often re-exposed surfaces to check relationships or simply to tie in the reference grid to new trenches. We feel that it is very important to allow future critical investigators the ability to reinterpret the structures or resample the section. This approach is also quite safe, because one always can work in shored trenches, which is commonly not possible with progressive excavations or fault-parallel excavations. Disadvantages include the inability to have more than one trench open at a time, the time involved in cutting and refilling trenches, the space it takes, and the fact that the exposures must be at least a meter apart, which makes measuring small lateral offsets difficult. However, as discussed elsewhere, it is very unlikely that a single site will yield both recurrence and slip data, so we recommend this approach when seeking recurrence data at a suitable site.

Another approach is to *excavate progressively* a volume of sediment, usually by repeatedly cutting back and mapping a free face that progresses orthogonally to the fault. While we have employed this technique on the Elsinore fault and the main fault strand at Wrightwood at Swarthout Creek (Fumal *et al.*, 1993), the results are not documented in detail in the literature, so we use an example from Pallett Creek (Sieh, 1984); it is, in our opinion, an excellent example of this approach. A 50-m-long section, 5-m deep, and 15-m wide was progressively excavated with walls mapped on average every meter; additional vertical and subhorizontal exposures were also mapped, where critical relationships were found. An upper 2.5-m exposure was first excavated orthogonal to the fault and extended progressively. After suitable room was established, a second, lower, 2.5-m exposure was excavated that

their locator trench (T1) on the 1906 rupture fault trace as mapped after the rupture. Their main focus was on fault-parallel trenches that intersected fault-normal channel deposits. They traced these deposits by systematically removing ∼1-m-wide vertical sections, in essence "peeling" the trench walls until the deposits approached the active fault zone. After each peel, they cleaned and logged the new trench wall exposure and recorded the location and orientation of the margins and thalweg of the individual buried stream channel deposits. Peels extended to within ∼2 m of the axis of the active zone of faulting, where shears and en-echelon cracks from the 1906 rupture rendered the trench walls unstable. From Hall *et al.* (1999); reprinted with permission of the American Geophysical Union. (B) Map of the Wrightwood paleoseismic site on the San Andreas fault zone, California, showing location of trenches (numbered), major fault zones (thick lines), and associated folds. From Fumal *et al.* (1993), modified by the addition of new trench locations.

progressively "followed" the first, as excavation continued, leaving a bench between the two exposures. Exposures were surveyed, gridded, and photographed; critical units and relationships were labeled and mapped, and then a new wall was cut. From these data, isopach and structure contour maps were constructed (Sieh, 1984). A similar progressive excavation study by Wesnousky *et al.* (1991) is described in Section 6.3.4.

The major advantage associated with this approach is the rapidity with which the data are collected, making such three-dimensional reconstructions possible. An additional advantage is seeing the units and structures evolve as the exposure is cut back; one may stop at any place, such as where key information is preserved, and map horizontal surfaces; to locate exact piercing points. The major drawback is the destruction of the volume excavated and thus the impossibility of rechecking or reinterpreting the results. While Sieh (1984) only excavated a small fraction of the young section deformed by the fault, the exposures recorded and interpreted now only exist on photographs and maps. In our experience, photographs are often difficult to interpret, especially by individuals unfamiliar with the site, and with time photographs become progressively more difficult to interpret even by the original investigator, as first-hand knowledge of the site fades. However, in situations where one wishes to recover the deformation associated with individual earthquakes, which was the goal of the Sieh (1984) study, this is probably the best approach. Note, however, that the events characterized by Sieh (1984) had already been discovered in multiple trenches and natural exposures (Sieh, 1978a), and that the deformation determined by the 3D trenching is only about one-third of the total deformation across the fault (Salyards *et al.*, 1992).

The third approach is *trenching parallel to and on each side* of the fault. Typically two or more locator trenches are cut across the fault to establish its exact position, and then trenches subparallel to the fault are cut some distance away from the fault to locate potential markers to trace to the fault. Subsequent trenches are cut close to the fault on each side, leaving a thin *septum* containing the fault. Piercing lines can be located on each side of the fault on the thin septum, and their trends estimated from the two walls of the trench parallel to the fault and, if available, additional trenches farther away from the fault. Because this technique provides relatively few exposures of the fault, it is best suited for measuring offset (Section 6.3.5), rather than determining the stratigraphic horizon or age of the earthquake. However, discrete increases in the downward displacement can be recognized this way, providing information about the stratigraphic level of earthquakes.

A variation of this approach, which is most commonly applied in Japan, is to make a U-shaped, T-shaped (Tsukuda and Yamazaki, 1984), or *rectangular excavation (open pit)* that provides exposures both parallel to and perpendicular to the fault (see Figure 2.11). Typically the Japanese slope the walls, so that they have a 45° exposure. The advantages of this approach include the ability to project onto both vertical and horizontal views (Figure 6.33); the walls are more stable than in a narrow and deep "California-style" trench, especially in soft and wet sediments (Yoshioka *et al.*, 1993), one does not need shoring, and because the walls often meet at the bottom of the excavation one has a continuous 3D exposure, rather than a floor. The major disadvantage is the greatly increased amount of material that must be removed; these exposures are very expensive and time consuming to create, and a large volume of the critical material near the fault is lost. In addition, trench logs as drawn often have different vertical and horizontal scales (e.g., Sato *et al.*, 1992, pp. 562–563) which must be rectified by projecting onto a vertical plane.

In summary, we recommend multiple parallel trenches for the recognition and dating of paleoseismic events. As discussed elsewhere, we do not recommend trying to determine both the timing and offset at a single site, but if that is the goal of the project, the second technique, progressive excavation of the fault zone is probably the best. If one is most interested in the offset of one or a few features, the final method, fault-parallel trenching, is the best approach. There are, of course, many possible modifications and

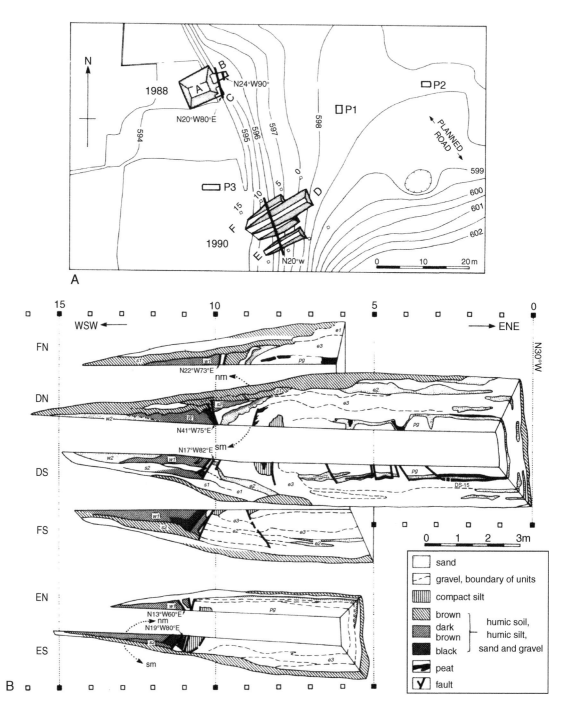

Figure 6.33: Examples of Japanese open-pit trenches and logs made from them. (A) Detailed topography and geometry of trenches across the Gofukuji fault. Fault planes are shown by heavy lines. Shaded areas are the sloping sidewalls of the trench. Elevations are in meters. (B) Simplified plan views of walls of trenches, D, E, and F. Trench D is inset within wider trench F. Each wall is named after the trench name (D, E, F) plus north (N) or south (S). From Okumura *et al.* (1994); reprinted with permission of the Seismological Society of Japan.

hybrids of these three basic strategies that could be adapted to particular projects; one should remain flexible and modify the approach as one learns more about the specific case with each excavation.

6.3.4   Stratigraphic Indicators of Paleoearthquakes

Paleoearthquakes are typically recognized in exposures of strike-slip faults from six general types of evidence (1) upward termination of fault displacement, (2) abrupt changes in vertical separation of strata as faults are traced upsection or downsection, (3) abrupt changes in thickness of strata or of facies across a fault, (4) fissures and sand blows in the stratigraphic sequence, (5) angular unconformities produced by folding and tilting, and (6) colluvial wedges shed from small scarps (Figure 6.34). In section, strike-slip faults tend to define narrower deformation zones than do dip-slip faults (mean widths of 5.5 and 12.1 m, respectively; Bonilla and Lienkaemper, 1991). Fissures and voids are common (58% of trenches inventoried), similar to normal faults but unlike reverse faults, which is indicative of transtension or of low confining pressures at shallow depths. Conversely, features indicating strong compression in the near-surface environment (gouge, breccia, slickensides) are less common on strike-slip faults (observed in 4–15% of trenches inventoried; Bonilla and Lienkaemper, 1991, Table 17).

6.3.4.1   Upward Fault Terminations

Upward fault terminations (Figure 6.34A) are probably the most commonly cited evidence for paleoearthquakes (e.g., Sieh, 1978a). Upward terminations are generally most effective in identifying the latest faulting event, because subsequent ruptures often follow the same plane. However, a single upward termination at a given stratigraphic level is tenuous evidence for a displacement event (Grant and Sieh, 1994), for reasons given next.

First, Bonilla and Lienkaemper (1991) record that, where the ground surface at the time of rupture is known, 73% of principal and secondary strike-slip faults *die out upward* and do not reach the ground surface at the time of the earthquake (see Chapter 2). The depth below the ground surface at which faults *dieout up* varies from a few centimeters to >2 m, with a mode at 15–30 cm (i.e., an asymmetrical distribution; Bonilla and Lienkaemper, 1991, Figure 15, Table 10). Thus, about three quarters of strike-slip fault strands die out before they reach the ground surface at the time of the earthquake, but the depth of dieout up may be predictable only in statistical terms. Second, strike-slip ruptures are often composed of a series of en-echelon steps (Figure 6.35), where slip is being transferred from one fault to another. Faults often dieout up in the step and terminate at different stratigraphic levels from exposure to exposure, even though the faults were all formed at the same time. Third, shaking, *freeze–thaw*, and *wet–dry cycling* commonly make fractures propagate above the event horizon after faulting. It is very common to see small, hairline cracks propagating up into what were probably previously unfaulted units.

In summary, upward terminations of rupture must be interpreted carefully and should only be used when the termination is consistent at many locations or in association with other indicators, like scarp colluvium or fissures. To be confident of a distinct paleoearthquake, upward terminations must exist at the same stratigraphic horizon at several locations in the trench and on both walls. Nelson *et al.* (2000) used a graphical technique to show how many faults terminated at various stratigraphic positions, which supported their interpretation of two faulting events (Figure 6.36).

6.3.4.2   Downward Growth in Displacement

Abrupt *downward increases in displacement* on a fault often indicate multiple faulting events, as described in Section 2.2.2.5 for dip-slip faults. However, these increases can also result during a single

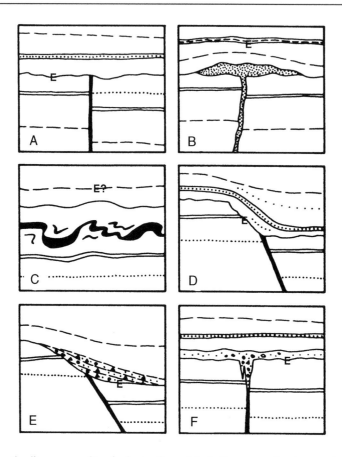

Figure 6.34: Schematic diagrams of typical stratigraphic indicators of paleoearthquakes in strike-slip environments. In all examples strata have been offset by a single event, either along faults (thick black lines) or sand dikes (stippled pattern). The event horizon (*E*) most closely approximates the time of faulting. (A) Fault terminates upward against an unconformity (the event horizon). This abrupt termination contrasts with gradual dieout up (see Chapter 2). (B) Sand dike feeds an injected sill of sand that has folded overlying strata (dashed line). The event horizon is at the contact of the folded and unfolded strata, *not* at the top of the sill. (C) Deformed horizon overlain and underlain by undeformed strata. Identification of the event horizon is complicated by ambiguity in the position of the ground surface at the time of soft-sediment deformation (see Chapters 7 and 8). (D) Fault underlies an eroded scarp and is overlain by unconformable strata; the unconformity is the event horizon. (E) Fault underlies a scarp that has been buried by scarp-derived colluvium. This is the typical geometry encountered on normal or normal-oblique faults (see Chapter 3). (F) Fault grades upward into a fissure that has been filled with material from an overlying unit (dots and circles). The fissure may have opened prior to the deposition of the dotted/circled unit, in which case the event horizon is the basal contact of that unit. Alternatively, the lower part of the dotted/circled unit may have existed at the time of fissuring and pieces may have fallen into the fissure, after which deposition of the unit continued. In that case, the event horizon is within the dotted/circled unit. From Allen (1986a,b); reprinted with permission from Active Tectonics. Copyright © 1986 by the National Academy of Sciences. Courtesy of the National Academy Press, Washington, DC.

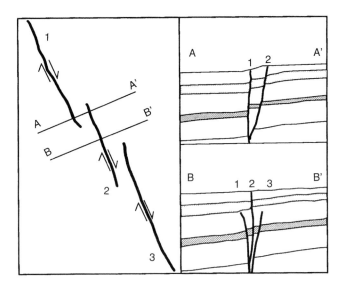

Figure 6.35: Schematic diagrams showing how transfer of strike-slip displacement across en-echelon steps leads to upward terminations of fault strands at different stratigraphic levels. Plan view at left, cross sections at right. In section A–A′, fault strands 1 and 2 are expressed at the surface. In section B–B′, only fault strand 2 reaches the surface. The projections of flanking faults 1 and 3 exist in the subsurface, but die out upward before reaching the ground surface. Faults 1 and 3 formed at the same time as fault 2, but they might be misinterpreted as being older if the principle of upward termination is too strictly applied. For a discussion of the "dieout-up" phenomenon, see Section 2.3.2.5.1. From Rockwell (1987).

faulting event if certain conditions are met. First, the vertical component of slip on a strike-slip fault may decrease naturally upsection, in a manner similar to dieout up. Second, if faulted strata are lenticular (i.e., they thicken and thin along strike), lateral displacement will result in variable vertical separations of contacts at the fault plane. Depending on the shape of lenticular (or previously folded) strata, vertical separations may either increase or decrease upsection, or do both at different stratigraphic levels. *When faulted strata are neither horizontal nor planar, changes in vertical stratigraphic separation do not prove changes in strike-slip displacement.* If sufficient 3D data exist, the shape and offsets of individual stratigraphic units can be traced, and can permit recognition of the stratigraphic horizon of individual earthquakes. Third, discrete downward increases in offset can also be caused by faults merging, rather than from cumulative displacement events. From struggling with many examples like this, we conclude that working on zones where the deformation is localized to a few meters for thousands of years (at least for a fault as active as the San Andreas) is a waste of time. One should look for places where the deformation from individual events is separated as widely as possible.

6.3.4.3 *Downward Increase in Thickness/Facies Contrasts*

Another line of evidence for multiple events is a *downward increase in thickness or facies contrasts* across the fault. Such reconstruction can be done in principle by finding the facies transition or the same thickness of a given unit on each side of the fault, usually accomplished by the progressive excavation technique described in Section 6.3.3 (e.g., Sieh, 1978a, 1984). However, even this technique becomes difficult where strata have been deformed in multiple events or faulted by multiple intersecting strands.

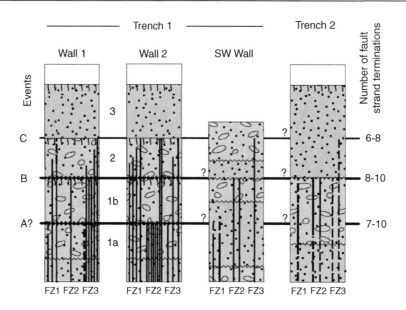

Figure 6.36: Graphical representation of fault upward terminations and their associated strati-graphic levels. Data for faulting events A?, B, and C in trenches 1 and 2 of Nelson *et al.* (2000). Indistinct or inferred parts of strands are dashed. The range in the number of terminations for each faulting event (right edge of figure) is the number of distinct strands and the total number of strands. Question marks indicate uncertainty in the correlation of faulting events. Although as many strands were mapped terminating near the position of event A? as for any other event (7–10), all but one of the distinct strands terminating at this position were in a single fault zone (FZ2, wall 2). For that reason, they did not insist that event A? was a separate faulting event. From Nelson *et al.* (2000); reprinted with permission of the Seismological Society of America.

6.3.4.4   *Fissures and Sand Blows*

We consider *fissures*, either filled with overlying material (Figure 6.34F) or filled with extruded sand (*sand blow*), to be very good indicators of the stratigraphic position of an earthquake. Although many are quick to point out that liquefaction can be caused by other distant faults and therefore be misleading, we are aware of few (if any real) examples of sand blows in paleoseismic excavations being clearly related to earthquakes on some other fault. In every case at the Wrightwood trench site, fissures are associated with fault ruptures. At the Pallett Creek site, Sieh (1978a) identified most of the earthquakes by the stratigraphic position of sand blows and subsequently (Sieh, 1984) showed that those horizons were associated with discrete increases in deformation.

The best cases are when fissures or sand blows are associated with actual fault displacement, such as is shown in the particularly good example of Figure 6.37. In that figure one can see where we cut into the fault revealing a fissure, partially filled with well-sorted sand that was probably extruded as a sand blow, but little if any sand flowed onto the surface. The vertical separation across the fault can be seen by the white sand layer within darker peat.

Sand blow deposits that were vented, and thus lay upon the ground surface at the time of rupture, must not be confused with *sills*, which are injected below the ground surface (Figure 6.34B). In the former

1 METER

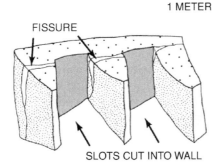

FISSURE

SLOTS CUT INTO WALL

Figure 6.37: Photo of southeast wall of the gully at Wrightwood, California, showing faulting and a large fissure filled with peat and sand that formed along the axis of the fold during event W4. Slots were excavated into the wall to provide 3D exposure of this fissure. Note that white arrow on right points to gray sand that was injected into the fault zone. Yellowish, poorly sorted, organic fissure fill can be seen above the gray sand and at the white arrow to the left. Event W4 faulting displaces peat 135c but not 135d. From Fumal *et al.* (2002).

case the event horizon lies beneath the extruded sand (see Chapter 7), whereas in the latter case (Figure 6.34B) the event horizon (*E*) lies at some distance above the sill, where strata deformed upward by sand intrusion are unconformably overlain by undeformed strata. Detailed descriptions of sills are given in Chapter 7. While caution should be exercised, as in interpreting any data, sand blows and fissures appear to be one of the most reliable indicators of paleoearthquakes in strike-slip fault zones.

6.3.4.5 Angular Unconformities

Angular unconformities produced by folding or tilting constitute relatively unambiguous paleoseismic indicators (Figures 6.34D and E). When poorly consolidated sediments are folded during coseismic strike-slip displacement, an angular unconformity (i.e., an event horizon) is formed when sedimentation resumes. Angular unconformities, typically having wide areal extent, do not suffer from the complications (induced by merging fault traces) that affect paleoseismic indicators restricted to fault planes. In addition, there are few nontectonic processes that can create angular unconformities. The main source of ambiguity associated with angular unconformities is whether folding can be caused by aseismic creep. However, if creep is operative during sedimentation, each successively lower stratigraphic unit will be incrementally more folded. In most exposures (e.g., the Wrightwood trenches), strata are folded in discrete packages, and discrete periods of folding can be inferred from the technique of "unfolding" the folds (see Section 6.5).

6.3.4.6   Colluvial Wedges

The final stratigraphic indicator of paleoearthquakes is the *colluvial wedge* (Figure 6.34E). Production of colluvial wedges in a strike-slip environment requires the creation of a fault scarp containing a free face, typically the result of segment obliquity and the accompanying vertical component of displacement (Section 6.1.1.1). The evolution and sedimentology of colluvial wedges are discussed in detail in Section 3.3.2.1. Due to the small height of most oblique-slip or secondary dip-slip fault scarps in strike-slip fault zones, colluvial wedges are generally thin and colluvial facies are poorly differentiated.

6.3.4.7   Collapse Features

Collapse features were initially described by Clark (1972), who observed them along the surface rupture of the 1968 Borrego Mountain earthquake, and who contrasted them with open fissures formed coseismically. In section, collapse pits have relatively steep margins except near the top where the margins shallow, probably the result of the pit walls eroding back to what was a paleoground surface. The collapse pits/voids are typically filled by two types of material, a basal blocky rubble, and overlying bedded silt and clay. The former is composed of identifiable blocks of strata adjacent to the filled void space. The latter bedded strata vary in thickness from less than 1 cm to as much as 20 cm, forming a 1.5-m-thick sequence. Clark (1972) proposed a conceptual model for pit formation that begins with the formation of fractures by an earthquake. Postfracturing storm events lead to the percolation of surface water down the fractures, which erodes the fracture walls, widens the fracture, and creates a subvertical, lens-like subterranean void. The void eventually collapses upward to the ground surface that exists at the time, filling with the blocky rubble facies, and creating a depression which over fills with sediment of various types (dry gravity fall, dry ravel, sheetwash flooding, etc.). For example, bedded units are probably deposited in storm events, when sediment was settled out of suspension from floodwaters. Collapse features near fault ruptures have been documented by Zellmer *et al.* (1985), Pampeyan *et al.* (1988), McGill and Rockwell (1998), and Dawson *et al.* (2003).

However, others have observed that collapse features can also form nonseismically, by flood water percolating down faults or fractures *not* related to an earthquake (e.g., desiccation cracks, toppling cracks near the edges of arroyos, differential subsidence cracks caused by fluid withdrawal, etc.) (Zellmer *et al.*, 1985; Pampeyan *et al.*, 1988). Because the initial cracking event may not be coseismic, the potential exists to misinterpret a collapse feature as a stratigraphic indicator of a paleoearthquake (similar to a coseismic tension fissure), when in fact it may have developed along a nonseismic crack and not be temporally related to any paleoearthquake. Thus, there are two questions to answer concerning these filled subsurface voids (1) were they originally fissures open to the ground surface, or voids created by subsurface piping? and (2) was the crack exploited by piping formed coseismically, or nontectonically?

Dawson *et al.* (2003) proposed several sedimentologic and structural criteria to answer the above questions. Diagnostic criteria that the voids were progressively enlarged in the subsurface (rather than being open fissures filled from the surface) include (1) missing stratigraphic section in the bottom of the filled void, (2) erosional unconformity at the base of the void, (3) discordant dips of blocks in the blocky rubble facies, and (4) anomalous upward terminations. Criterion 1 indicates a "missing" volume of material (between 30 and 70 cm in their case) that cannot be accounted for by mere fault translation, vertical or horizontal. This suggests that sediment was removed, probably by subterranean water erosion (piping). Criterion 2 hinges on the presence of an erosional unconformity and its interpretation as the bottom of a piping channel, through which water flowed and from which many of the stratigraphic units in the void area were eroded. This erosional contact may also cut across several depositional units and form an angular unconformity. Criterion 3 (discordant stratal dips within rubble blocks) suggests collapse of the roof of the subterranean void, where the higher blocks in the rubble facies came to rest at a different apparent dip than the lower ones. Criterion 4 (anomalous upward terminations) is predicated on all the upward fault terminations *within* the filled void being at a stratigraphic levels not observed elsewhere in the trench. The implication is that these anomalous upward terminations have been eroded downward from their true original position(s).

Dawson *et al.* (2003) observed that all collapse pits in their trenches occurred at a stratigraphic level coincident with paleoearthquake horizons identified from other more reliable (less ambiguous) lines of evidence, such as upward terminations. They thus concluded that (1) the rubble-filled voids in their trenches were mainly created in the subsurface by a piping and collapse process, rather than as open fissures, and (2) the fractures exploited by piping were coseismic, and thus (3) their enlargement and filling followed closely after the surface-rupturing event. At our current level of understanding, collapse structures should probably not be used as the sole evidence to support an inferred paleoearthquake, in the absence of more reliable lines of evidence.

6.3.4.8 Examples from the Hog Lake Trench Site, San Jacinto Fault, Southern California

In this section, we show trench logs containing many types of stratigraphic indicators at a single trench site. The Hog Lake sag pond site is advantageous for paleoseismic study, because (1) sedimentation (mainly subaqueous) was essentially continuous throughout the late Holocene, preserving evidence of each paleoearthquake; (2) fault planes are sharp, clearly visible and localized; and (3) stratigraphic units are thin, well-defined, and organic rich and thus datable by 14C. Due to the subaqueous history of the lake, the most common stratigraphic indicators of paleoearthquakes are (1) upward fault terminations; (2) mismatch of bed thickness and sedimentology across the faults, due to strike-slip motion; (3) fissure fills; (4) growth section (thickened beds on the downthrown side); (5) folding; and (6) angular unconformities. Some stratigraphic indicators typical of subaerial trench sites, such as colluvial wedges, are absent.

Figure 6.38 shows a composite trench log 3.5-m high and 6.5-m wide that spans the main late Holocene deformation zone of the San Jacinto fault. There are 13 earthquake horizons exposed in this wall (marked by small yellow tags). Figure 6.39 shows a close-up of the evidence for the latest three earthquake ruptures (from youngest to oldest, earthquake horizons E1–E3). Figure 6.40 shows horizons E3–E5; Figure 6.41 shows horizons E6–E14; and Figure 6.42 shows horizons E16–E18.

6.3.5 Measuring Lateral Displacements from Stratigraphic Data

There are two general methods for measuring lateral displacements in trenches. The first can be employed in fault-perpendicular trenches, and uses the vertical separation of strata across faults, plus measurements of the rake of the net slip vector on the fault plane (slickensides), to estimate the lateral slip

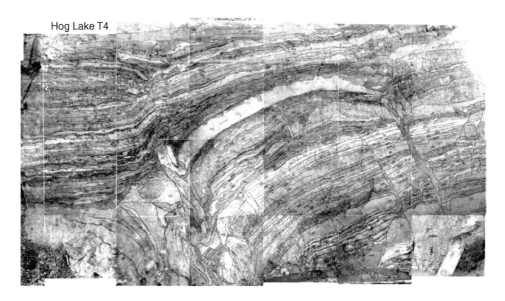

Hog Lake T4

**Figure 6.38: Photomosaic log of trench 4 at the Hog Lake sag pond; individual photos are 1 m².
In the main fault zone at left center, slip during the latest three paleoruptures was restricted to the
mm-thick red line in the upper half of the log, above the thick light-toned bed. Displacement in
earlier events (below the light-toned bed) occupies a slightly wider zone, but still localized in a zone
less than 1.5-m wide. From Rockwell *et al.* (2004).**

component by trigonometry. This method is not always possible, because slickensides are not always
created (e.g., in gravelly deposits) or preserved. An additional complication is that the vertical separation
of a contact across a strike-slip fault is a function of the dip of the contact, as well as the orientation of
the net slip vector on the fault. For example, pure horizontal fault motion displacing a dipping contact
will create a vertical separation in a plane perpendicular to the fault. An identical vertical separation
could also be created by oblique slip on a horizontal contact. Thus, to solve for lateral slip via the
trigonometry, one must know the orientation of the net slip vector, the true dip of the displaced
contact, and the strike of the contact as opposed to the plane of the trench wall.

The more commonly used method is to measure the lateral displacement directly on a stratigraphic
piercing point such as a buried stream channel. Using progressive excavation of fault-parallel trench walls
inward toward the fault zone, buried channels are traced into the fault zone from both sides. In early
studies, the logged fault-parallel walls stopped at some small distance from the fault zone (∼1–2 m), and
the trend of channel margins on each side was linearly projected to the fault plane to estimate offset.
However, later trenching studies showed that within 1–2 m of the fault, a channel might curve to flow
parallel to the fault, and using the linear projection method that component of misalignment would be
lumped together with any fault offset (see two case histories, below).

Difficulties arise when the size of the channel is comparable (or greater) than the offset associated with a
single event, where there are multiple fault traces, and where there is relief near the fault (such as a scarp),
which makes it very difficult to know the original geometry of the channels. In addition, if the fault-
parallel walls are too far apart and do not extend all the way to the fault plane, it very difficult to see if the
stream may have been diverted to flow parallel to the fault.

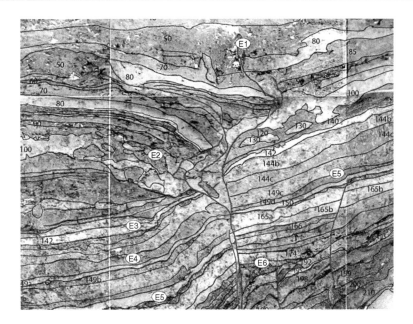

Figure 6.39: Photomosaic of the south wall of trench T4, cut 3, at the Hog Lake trench site, showing evidence for the latest five paleoearthquakes E1–E5; individual photos are 1 m². Note the repeated section of unit 80 (uppermost light-toned bed) by event E1 (the most recent earthquake), which can only be resolved with strike slip. Penultimate earthquake E2 is only represented by liquefaction and disruption of unit 100, overlain by a bedded section. Earthquake E3 is represented by an angular unconformity and folding, as are earthquakes E4 and E5, but E5 is also represented by an upward termination on a secondary fault at far right. From Rockwell *et al.* (2004).

A modern study of this type for the San Andreas fault is that of Liu-Zeng *et al.* (2006), who were able to correlate six buried paleochannels of different age across the fault, by progressively excavating fault-parallel trench walls at 0.5-m intervals, decreasing to 0.2-m spacing in the fault zone. Their six paleochannels were misaligned by 7.25–7.42, 15.1–15.8, 20.7 ± 0.15, 22.0 ± 0.2, 30.0 ± 0.3, and 35.4 ± 0.3 m, respectively. In this case, trenching showed that most channels had never been deflected to flow along the fault, and thus misalignment was almost 100% attributable to lateral offset.

To illustrate this issue more clearly, we discuss three recent three-dimensional trenching projects across strike-slip faults, in which misalignment of a buried stream channel was wholly, partly, or not at all due to lateral offset by single or multiple earthquakes.

6.3.5.1 *Whittier Fault Example (Most of Misalignment Caused by Diversion)*

The Whittier fault zone near Los Angeles, California, crosses a stream with about 50 m of right-lateral misalignment. Trenches were initially excavated across the fault to determine the width and characteristics of the fault as well as the general stratigraphy of the channel environment (Figure 6.43). Five buried alluvial channels were identified in the subsurface, and range in age from historical (complete with rusted iron pipes and wire) to early Holocene. Channels were differentiated based on their stratigraphic position and composition. One channel (Q4) in the middle of the sequence contains

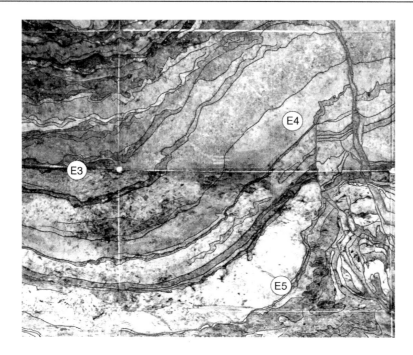

Figure 6.40: Photomosaic of the main fault zone on the south wall of trench T2, cut 2, at Hog Lake, showing evidence for paleoearthquakes E3–E5; individual photos are 1 m². Note the angular unconformities associated with earthquakes E3 (upper left center) and E5 (lower right center). Earthquake E4 is represented by a clear upward termination (upper right center) and mismatch of units across that terminated fault. The stratigraphic thickness between each earthquake horizon is greater on the left side of the main fault zone than on the right side, indicating growth strata associated with each paleoearthquake. From Rockwell *et al.* (2004).

abundant, distinctive tar clasts (a natural tar seep occurs just upstream) that allowed for a unique correlation across the fault. All channels were traced into and across the fault zone to resolve slip. Excavations close to the fault were carried out by hand to avoid obscuring important relationships.

Channel Q4 was exposed in the initial trenches (T4, T17) with a trend at a high angle to the fault. We believed that channel Q4 would directly intercept the fault, and it would be a simple matter to measure its offset across the fault. However, as is clear in Figure 6.43, the channel flowed along a paleoscarp within centimeters of the fault, and where the channel crossed the fault, only about 1.9 m of slip was resolved. If two trenches had been placed parallel to the fault, even as close as 1-m upstream and downstream from the fault zone, the channel would have been projected into the fault to estimate slip of nearly 20 m, an order of magnitude greater than the true displacement of the Q4 channel.

In a similar example, Wesnousky *et al.* (1991) used the progressive excavation technique to log 39 consecutive face positions when tracing a subsurface channel that was misaligned 12 m across the San Jacinto fault, California. Their structure contour map drawn on the base of the buried channel identified three pairs of piercing points which yielded lateral offsets of 4.5, 4.2 ± 0.7, and 5.4 ± 0.6 m, respectively. Thus, only about half (ca. 5 m) of the 12 m of apparent right-lateral misalignment of the 2-ka channel can be attributed to faulting; the remainder was due to diversion of the stream parallel to the fault.

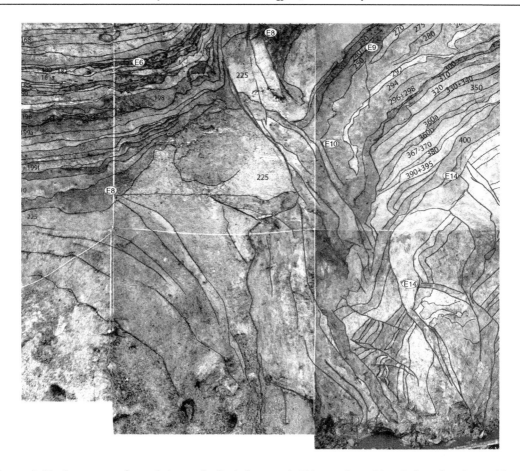

Figure 6.41: Lower portion of the main fault in trench T4, cut 3, at Hog Lake, showing evidence for paleoearthquakes E6–E14; individual photos are 1 m². Note the warping down of unit 225 and the onlap of unit 215 (left center, overlies unit 225) resulting from event E8. Also note the filled fissures from events E8, E9, and E10, and the upward terminations and angular unconformities associated with events E6, E7, E8, E9, and E14. From Rockwell *et al.* (2004).

This study is one of the most detailed published examples of the incremental trenching technique and is highly recommended to interested readers.

6.3.5.2 Rose Canyon Example (Most of Misalignment Caused by Tectonic Offset)

On the Rose Canyon fault (San Diego, California), most of the geomorphology associated with active faulting was destroyed by urban development in the 1960s. Based on predevelopment aerial photographs, the site of greatest potential for preserving the most recent offset now lay beneath a parking lot. Three borings were first conducted to determine the depth of artificial fill beneath the parking lot (about 2 m) and the depth to the water table (about 4.5 m), providing a 2.5-m unsaturated zone in which to trench (Lindvall and Rockwell, 1995).

After a locator trench (T1 on Figure 6.44) determined the precise location and width of the fault zone and the general stratigraphy and age relations, two fault-parallel trenches (T2, T3) were emplaced about 5 m

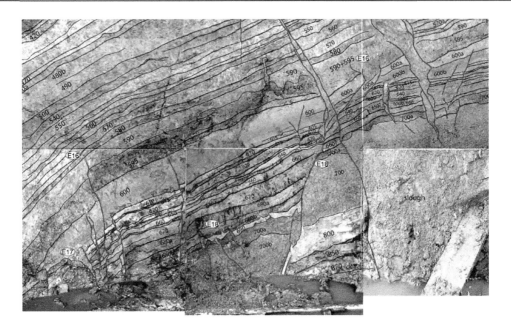

Figure 6.42: Older earthquakes E16–E18 recorded along an older, presently inactive strand in trench T4 at Hog Lake. Earthquake E16 is represented by upward terminations of several fault strands. Earthquake E17 ruptured a significant fault (lower left) less than a meter from the currently active main strand (just out of the field of view to the left), and has an associated filled fissure. Units across this strand are mismatched. Several other fault strands also terminate at this level (unit 600). Earthquake E18 is represented here by the upward termination of a significant strand, as well as major lithologic contrasts across the fault of correlative units, indicating significant strike slip. There also appears to be growth strata adjacent to the primary older strand, although this could also be explained with significant strike-slip and mismatch of strata. From Rockwell et al. (2004).

away from the fault to search for buried channels that could be traced into and across the fault. Only one distinct, gravel-filled channel was located east of the fault in trench 2. The trend of the channel was determined to be roughly perpendicular to the fault and was therefore a potentially useful slip indicator.

The artificial fill was then removed to the top of the original ground surface. Because the channel was small, about 0.5 m across, all subsequent excavations were performed by hand to avoid destroying critical relationships; a photograph of the excavation is included as Figure 6.45. The channel was sequentially exposed in numerous exposures into and through the fault zone (Figure 6.44). All channel exposures as well as important stratigraphic horizons were surveyed as they were exposed to avoid loss of spatial data. The channel was preserved in each fault block in such a way that when the excavations were completed, geologists could walk through the excavations and see the displaced channel segments.

Figure 6.46 shows the inferred reconstruction across the many fault traces. Unfortunately, the stratigraphic unit containing the channel in the northwesternmost corner of the excavation had been graded out when the site was developed in 1960. Consequently, only a minimum slip was determined, and therefore a minimum slip rate was calculated. Nevertheless, it was shown in this case that the channel maintained a relatively straight course across the entire fault zone and true slip could be determined.

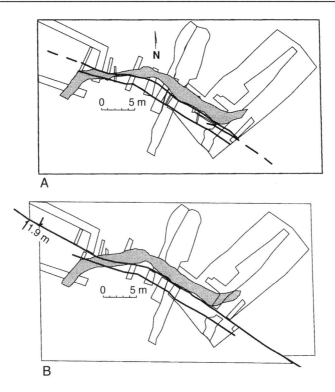

Figure 6.43: Plan view of faults (thick lines), buried channel Q4 (stippled pattern), and trenches (outlined by thin lines) across the Whittier fault, California. (A) Present geometry of the buried channel Q4. The channel is only offset at far left. (B) Restored geometry of buried channel Q4 before the latest fault displacement, based on channel margins as piercing points. Total restored slip is 1.9 m dextral.

The three field examples described emphasize the necessity to resolve completely the channel geometry into and across a fault in order to resolve slip, either for a slip rate or as an estimate of earthquake slip in discrete events. Projection of linear features only a few meters distant into the fault zone can produce substantial errors in slip estimation, if the geometry of the features within the fault zone itself is not known. This is especially true if the paleochannel trends nearly parallel to the fault, for example, in Rockwell *et al.* (2009b). In special cases, even small alluvial fans may have sharp enough boundaries that the isopach method can reveal single-event or two-event displacements, given enough control points. An example is shown by Rockwell *et al.* (2009a) on the North Anatolian fault, where the geometry of the fan deposit and its disconnected feeder channel were measured by progressive excavation, resulting in 1500 data points for thickness contouring. At that location a plan-view retrodeformation showed the fault had offset the feeder channel and fan apex dextrally by 9.1 m in two faulting events.

The preceding examples required substantial amounts of field time to allow for careful tracing of linear features into the faults without their destruction or loss of data. This final point is brought forward because, in most cases, not all structural and stratigraphic relationships are immediately apparent during the early phases of a trench project and fault exposures may be logged without fully understanding them. It is recommended to proceed cautiously when laterally excavating a strike-slip fault because many relationships may be lost during the excavations. Trenching, by its very nature, is destructive. Many sites are unique along a particular section of a fault and there may be only one opportunity to collect the data.

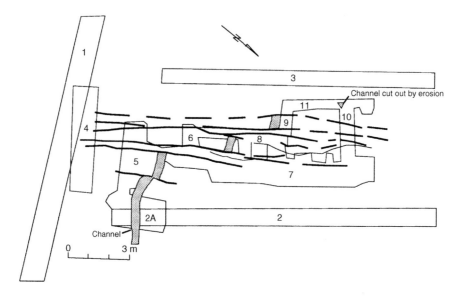

Figure 6.44: Plan view of trenches (thin lines), faults (thick lines), and buried channel (stippled pattern) across the Rose Canyon fault, San Diego, California. Trenches 1–3 were excavated from the surface of a parking lot, whereas trenches 4–11 were hand dug from the floor of a large excavation (not shown) that removed 2 m of artificial fill from beneath the parking lot.

Figure 6.45: Vertical photograph of trenches 5–11 at the Rose Canyon site. Trenches 5 and 11 are labeled for reference. Note aluminum hydraulic shore (light-colored object above and left of "11") for scale; its total length is 2.8 m. North is to lower right corner. Note numerous benches within excavation. For further details, see Lindvall and Rockwell (1995).

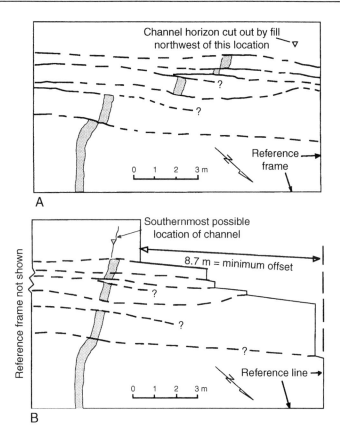

Figure 6.46: **Plan-view retrodeformation of the buried channel across the multiple strands of the Rose Canyon fault. (A) Present configuration of the buried channel exposed by trenching. The open triangle represents the farthest position to the NW that the unit containing the channel is preserved. (B) Restoration of the channel by 8.7 m. This restoration assumes only brittle slip along five fault strands and therefore represents a minimum (dextral) offset. For further details, see Lindvall and Rockwell (1995).**

Careful work will require time, and the work is not complete until every stratigraphic unit and structural relationship is understood.

In theory, offset buried channels could also be detected by detailed fault-parallel geophysical profiles, although this approach seems only to have been tested to date by Ferry *et al.* (2004) on the 1999 Izmit, Turkey rupture trace of the North Anatolian fault. They surveyed one fault-perpendicular GPR line and four fault-parallel GPR lines, two on each side of the fault, and projected channel margins linearly to the fault. As shown in Figure 6.47, the upper channels (purple and green) were consistently deflected 4.5–4.9 m in a right-lateral sense, which is approximately twice the amount of strike slip observed in 1999. From the companion trenching study, these channels were known to be embedded within that part of the stratigraphic section that has been subjected to only two surface ruptures (Rockwell *et al.*, 2001). Similarly, the lower channels (orange and red) were deflected by 6.7–7.4 m when projected across the main fault and were thus interpreted to have been subjected to three surface ruptures. If this "geophysical trenching" technique could be tested and validated at many sites, it might eventually replace the rather tedious and expensive fault-parallel trenching now employed.

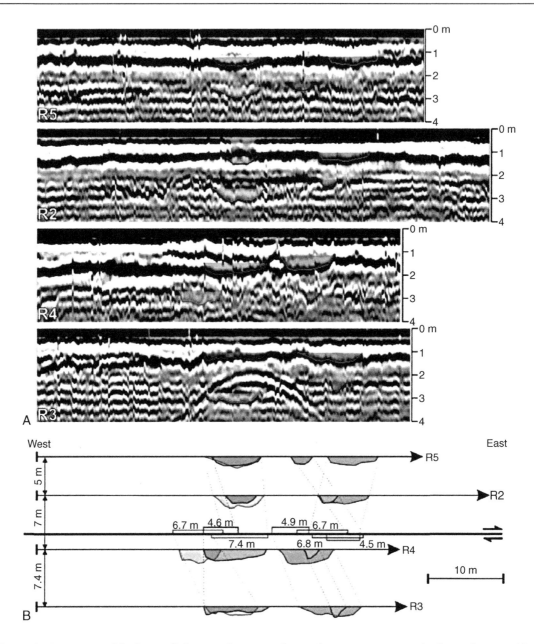

Figure 6.47: A test of fault-parallel ground-penetrating radar surveys as a substitute for trenching, North Anatolian fault. (A) Parallel GPR profiles R5, R2, R4, and R3 collected along main 1999 surface rupture at site 1. Four channels (highlighted in different colors) may be consistently identified on all sections. Depths are given in meters (with respect to top of each profile). (B) Map of occurrence of channels with respect to fault (thick black line with arrows). Channel margins are projected onto fault plane to indicate their horizontal cumulative displacement. From Ferry *et al.* (2004).

6.3.6 Distinguishing Creep Displacement from Episodic Displacement

Strike-slip faults have been documented to creep at the surface, based on the displacement of cultural features and on geodetic measurements (e.g., Thatcher, 1979). Creeping strike-slip faults include the central San Andreas, Calaveras, and Hayward faults (California) and the Ismetpasa section of the North Anatolian fault. However, Cakir *et al.* (2005) note that:

> *In some cases, faults are thought to creep throughout the seismogenic layer at a rate comparable to the geologically determined slip rate, and cannot therefore generate large earthquakes (e.g., central San Andreas fault; Thatcher, 1979; Burford and Harsh, 1980). In other cases, fault creep appears to take place within a shallow depth interval and/or at a rate slower than the overall slip rate, and hence does not prevent the fault from producing moderate-to-large-size earthquakes (e.g., southern and northern Hayward fault; Lienkaemper and Williams, 1999; Schmidt et al., 2005).*

Where creep may have occurred on near-surface fault planes, is it possible to differentiate it from coseismic offsets when mapping trench walls or other vertical exposures? This question has practical significance for determining slip rates and recurrence on faults that experience both coseismic faulting and postseismic/interseismic creep, such as the Hayward fault. Lienkaemper *et al.* (2002) and Lienkaemper and Williams (2007) proposed several criteria for distinguishing coseismic displacement from creep displacement in trenches on the Hayward fault, based on their own observations and previous work by Stenner and Ueta (2000), Kelson and Baldwin (2001), and Ferreli *et al.* (2002).

Features unique to coseismic displacement: (1) fissure fills; (2) colluvial wedge deposits; and (3) packages of blocky scarp colluvium deposited adjacent to the main fault trace (Figure 6.48). Note that these are features common to normal-fault scarps as well.

Ambiguous evidence: (1) liquefaction (can be caused by seismic shaking from the trenched fault, *or* from another fault nearby); (2) upward terminations of faulting and abrupt changes in deformation between units. If creep accumulates over several decades as both slip and folding without deposition, the subsequent burial of the deformed surface could yield both upward terminations and undeformed strata overlying strata that have been deformed, that would be indistinguishable from those formed by coseismic faulting; (3) fault plane overlain by a series of small, upward fanning fractures that extend to the surface with little displacement; interpreted by Fenton *et al.* (2006) to be the result of fault creep (Maacama Fault). However, similar features could arguably be formed by a noncharacteristic earthquake on the trenched fault.

Features unique to creep displacement: (1) no evidence of postevent scarp retreat from location of fault plane; (2) absence of colluvial wedges or any other scarp-derived sediments; and (3) relatively continuous and uniform increase in deformation across the fault of progressively older stratigraphic units.

6.4   Dating Paleoearthquakes

The usual approach to dating individual paleoearthquakes on strike-slip faults is to bracket the earthquake by dating the youngest datable stratum deformed by the quake and the oldest datable stratum that overlies the earthquake horizon. Dating methods commonly used on fault scarps (direct dating of the exposed fault plane, dating of the scarp landform by degradation modeling, soil profile development,

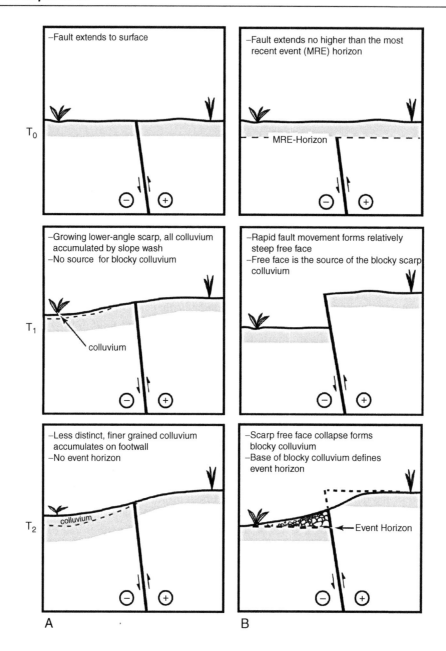

Figure 6.48: Comparison of stratigraphic indicators formed by creep movement on a strike-slip fault (A) versus those formed by coseismic, episodic displacement (B). Time proceeds from T_0 to T_2. The topographic relief shown could be created either by oblique-slip faulting or by pure horizontal faulting across a nonhorizontal ground surface. From Lienkaemper *et al.* (2002).

or cosmogenic isotope profiles; see Chapter 3) are rarely used, due to the scarcity of scarps and the uncertainty about the prelatest-faulting geometry of the scarp.

In bracketing a paleoearthquake by nondisplaced/displaced strata, one may attempt to date the earthquake horizon directly, by shaving off the few millimeters of organics that formed immediately prior to or after the earthquake, and dating the samples very precisely (e.g., Sieh *et al.*, 1989). We have shown (Biasi and Weldon, 1994a,c) that neither of these techniques are optimal because they make little or no use of the information in the age of all the super- and subjacent horizons and are sensitive to potential contamination.

Laboratory ^{14}C ages are normally converted to calendar ages to correct for fluctuations in atmospheric ^{14}C over time (e.g., the freeware program OxCal; Figure 6.49). The probability distributions for calendar ages are often multimodal and in a vertical sequence, will overlap between adjacent samples. However, the age variance of these individual ages can be "trimmed" by considering that (1) the stratigraphically higher unit must be younger than the lower one and (2) sedimentation rate or soil development rate can constrain the time possible between ages based on their stratigraphic separation. Usually the result of applying these two corrections is to shift probability to one peak of a multimodal distribution and reduce the variance.

This method was applied to earthquake ages and interevent times on the southern San Andreas fault by Biasi *et al.* (2002) and Scharer *et al.* (2007), and has become a de facto standard technique for processing large suites of ages stratigraphic successions.

The reason precise dating is so critical for strike-slip faults is that earthquake recurrence may be irregular on them, and an irregular pattern would not necessarily be detected with imprecise dates. For example, Rockwell (2008) dated 150 ^{14}C samples at the Hog Lake trench site on the San Jacinto fault, southern California, and subjected the ages to trimming based on stratigraphic position (using OxCal). The suite of earthquake ages shows that in the past 3.8 ka the fault has switched from a quasiperiodic mode of earthquake production, during which the recurrence interval is similar to the long-term average, to clustered behavior with the interevent periods as short as 20–30 years (e.g., five surface ruptures between about AD 1025 and 1360). Intercluster periods with no ruptures last as long as 550 years, commonly followed by a one to several closely timed ruptures. Similar behavior is documented on the San Andreas Fault at Wrightwood, where a cluster of seven surface ruptures occurred between AD 500 and 1000, the same period during which only one rupture is recognized at Hog Lake. Conversely, during the San Jacinto earthquake cluster of AD 1025–1360, earthquake production at Wrightwood was suppressed. These observations suggest that earthquake recurrence is strongly influenced by fault interaction (see discussion in Chapter 9, See Book's companion web site), a conclusion not possible without refined dating.

6.5 Interpreting the Paleoseismic History by Retrodeformation

Published retrodeformation sequences of strike-slip faulting (also called palinspastic restorations) have been performed in plan view (examples shown previously) and in sectional view (described here). The most common sequences are 2D reconstructions based on the logged vertical walls of fault-perpendicular trenches. Such fault-perpendicular retrodeformations attempt to restore the original continuity of faulted strata by successively reversing the vertical component of slip events on each fault. The spectrum of these vertical retrodeformations has two end members (1) trenches in transpressional or transtensional reaches, where most or all of the displacement is dip-slip, and (2) trenches in nearly pure strike-slip areas, where the observed vertical separations across faults are caused either

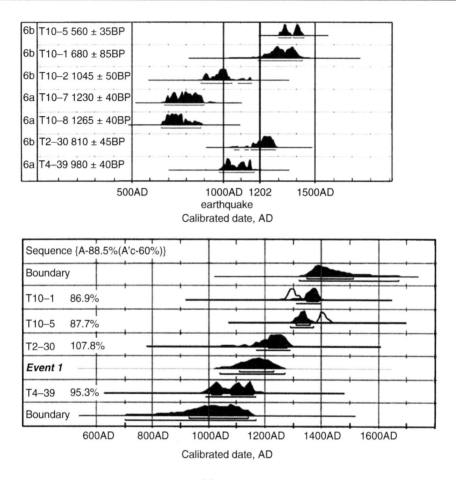

Figure 6.49: Example of trimming calibrated ^{14}C ages based on stratigraphic order. Top: calibrated date distribution for samples from trenches T2, T4, and T10 on the Jordan Gorge fault (Marco *et al.*, 2005). Bottom: trimmed probability density functions for radiocarbon dates that constrain the timing of the penultimate event at the Bet-Zayda paleoseismic site. The dates were trimmed with Bayesian statistics in OxCal and the probability density function for the earthquake age (*Event 1*) is then calculated from the radiocarbon ages. Note that the historical 1202 earthquake falls within the probability distribution, and is in fact the only historical earthquake in the vicinity that can possibly fit the age distribution. From Marco *et al.* (2005).

by a small vertical component of movement (relative to horizontal slip) or by pure strike-slip movement on nonhorizontal contacts.

In the first case, the critical features used in the retrodeformation are identical to those found in dip-slip reverse and normal faults (e.g., colluvial wedges, fissures, fault-propagation folds, etc.). In this case, the principles of retrodeformation are identical to those described in Chapters 3 and 5. One can deduce both the timing of the paleoearthquake(s) and the vertical displacement. However, the vertical displacement that is reversed for each slip event will almost certainly be less than D_{max} of the strike-slip rupture (which will probably be found in an area of pure strike slip), and also will be difficult to relate quantitatively to D_{max} or D_{avg} of the rupture. Thus, a 2D retrodeformation in strongly transtensional or transpressional zones is useful in deducing earthquake recurrence, but not slip rate or displacement for the strike-slip fault. Examples of dip-slip-based retrodeformations on strike-slip faults include Rubin and Sieh (1997), the San Andreas fault at Wrightwood (Weldon *et al.*, 2002), and Gomez *et al.* (2003).

In the opposite end-member case, the vertical separations visible on the trench wall are caused either by a small vertical component of movement relative to the horizontal component or by pure strike-slip movement displacing nonhorizontal contacts. In this case, the primary evidence of paleoearthquakes is upward terminations, mismatch of units, angular unconformities, fissures/collapse structures, liquefaction dikes, and other structures related to strike-slip faulting. These features will again yield a chronology of paleoearthquakes, but perhaps not any useful information about displacement per event or slip rate. Some have suggested that net slip during a paleoearthquake could be calculated given a measured vertical separation on a fault-perpendicular trench wall, combined with a known rake of the net slip vector measured from slickensides on the fault plane. This is theoretically true, but only if (1) strata are horizontal in the vicinity of the trench wall, (2) the rake can be measured accurately, and (3) rake remains constant in both space and time. Small changes in the rake will induce large changes in the calculated net slip vector, particularly at low rake angles. Stratal dip and rake are likely to vary parallel to the fault at most trench sites, so the method only yields a crude estimate of the net slip vector.

To date, no papers have published a true 3D retrodeformation of strike-slip faulting, in which the true net slip vectors are reversed for each slip event in 3D block diagrams.

6.5.1 Retrodeforming the Trench Log

The stepwise construction of 2D vertical cross sections, backwards in time from the present, follows the same general principles for strike-slip faults as for normal and reverse faults. A simple example is shown by Rubin and Sieh (1997, Plates 1 and 2), who simply reversed the 1992 displacement on multiple strands of the Emerson fault that ruptured in the Landers, California earthquake. This style of retrodeformation is similar to that used for normal faults, and treats each fault-bounded block as a rigid entity that is moved up or down along faults without shape change (although limited block rotation is permitted).

A more sophisticated treatment of vertical deformation was performed by Fumal *et al.* (2002) on the San Andreas fault, who show a retrodeformation sequence that removes folding as well as faulting. Their trench 4 at the Wrightwood, California trench cluster contains obvious folding (Figure 6.50). However, it is difficult to tell by mere inspection if some layers have been folded together or how many folding events might be represented.

For this reason, they restored the trench sections in this manner. First, they overlaid the trench log with a rectangular grid of points having a spacing of approximately 15 × 15 cm. The part of the log within each column of points was then successively moved vertically until a particular contact was restored to the same shape as a template ground surface (initially, the modern ground surface). This process removed the vertical component of deformation, but not the horizontal one. In a few instances, they removed the slip on discrete faults by moving an entire block up along the fault plane to match units across the fault, as in a normal retrodeformation of rigid, fault-bounded blocks. They then continued this process, restoring the upper contact of each significant debris flow deposit to the shape of the present ground surface. Below a certain stratum, it became obvious that a ground surface template with a lower slope angle was

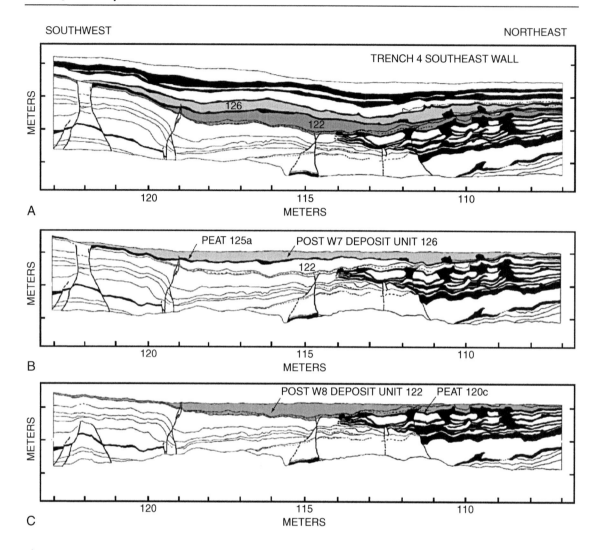

SOUTHWEST   NORTHEAST

TRENCH 4 SOUTHEAST WALL

A

METERS

120   115   110

METERS

PEAT 125a   POST W7 DEPOSIT UNIT 126

B

METERS

120   115   110

METERS

POST W8 DEPOSIT UNIT 122   PEAT 120c

C

METERS

120   115   110

METERS

Figure 6.50: Example of a 2D vertical retrodeformation that unfolds folded beds. (A) Log of portion of southeast wall of trench 4 at Wrightwood, California. Black is peat; all other units are debris flows. (B, C) Log retrodeformed to show restoration of ground surface following deposition of debris flows subsequent to folding during events W7 and W8. In each restoration, the peat layer that was at the ground surface at the time of the particular earthquake is indicated and the debris flow deposited immediately following the earthquake is shaded. Folding from event W8 is subtle in most exposures but in trench 4 it is clear that events W8 and W7 are separate folding events. Note that slip during event W8 was principally on faults near meter 115 and meter 119. In contrast, slip during event W7 occurred on a highly curved fault near meter 111–112 and on faults between meter 121 and 123. The axis of folding during event W8 (thickest part of unit 122), near meter 115, corresponds with a slight high resulting from event W7 folding. From Fumal et al. (2002).

appropriate, which they used for the subsequent retrodeformation stages. They concluded that "the particular choice of a ground surface template is somewhat arbitrary and is not critical," and the patterns of successive folding were not sensitive to the ground surface template used. By the end of the retrodeformation process, they concluded that each of the seven significant debris flow deposits in trench 7 were deposited after a separate earthquake, because they were all folded in different patterns and by different amounts. As expected, the debris flows accumulated to a greater thickness in coseismic synclinal folds and thinner on anticlinal folds.

6.6 Distinguishing Seismogenic from Nonseismogenic Strike-Slip Faults

From a seismic hazards perspective, we need to be able to deduce whether a strike-slip fault, whether exposed on the land surface or in an exposure, is seismogenic or not (defined in hazards literature as being able to release an $M_w > 5.5$ earthquake). This question is related to whether a fault creeps at the surface or not, but is broader since it encompasses the depth of the fault and the origin of the fault. The discussion below follows that of Hanson *et al.* (1999) and similar sections of Chapters 3 and 5.

6.6.1 Tectonic, But Nonseismogenic Strike-Slip Faults

It is possible for a strike-slip fault to be tectonic (i.e., deeply penetrates the crust) but either consistently (or occasionally) nonseismogenic. An example of a persistently nonseismogenic (but tectonic) fault is the creeping section of the central San Andreas fault near Hollister. Faults that slip both nonseismically (creep) and coseismically include the Hayward and Calaveras faults in California. For additional details, see Hanson *et al.* (1999).

Table 6.4: Criteria for recognizing a landslide-created strike-slip fault from a seismogenic, tectonic fault

Type of evidence	Criterion indicating a landslide origin
Seismological and geological	1—fault not associated with historic earthquake activity (not compelling)
	2—fault is associated with a volcanic edifice mega-landslide
Structural/geologic	3—fault is associated with a mappable landslide (Landslide Inventory Map)
	4—fault is creeping (usually compelling, because creeping tectonic faults are rare)
Geomorphic	5—fault is perpendicular to the slope
	6—fault exists only where there is topographic relief, and stops where the relief stops
Microstatigraphic	None are compelling

6.6.2   Nontectonic and Nonseismogenic Strike-Slip Faults

The more common type of strike-slip fault to be nonseismogenic is a nontectonic fault that does not penetrate to crustal depths. Such faults are found on the lateral margins of landslide, where (looking downslope) the right margin is a site of dextral shear and the left margin of sinistral shear. Strike-slip faults on landslide margins create landforms that closely mimic those found along seismogenic strike-slip faults (Fleming and Johnson, 1989; Johnson and Fleming, 1993; Gomberg *et al.*, 1995). In very large landslide such as found on the flanks of volcanic edifices (Hawaii, Eissler and Kanamori, 1987; Mt. Etna, Sicily), it may not be clear whether the marginal strike-slip faults are crustal-penetrating or not, and thus, whether they are seismogenic. Cotton (1999, p. A48) states that "The world-wide geologic record contains excellent examples of faulted rock materials which are the product of large gravity sliding events along detachment surfaces that are regional in extent and have subsequently undergone deformation and deep erosion. Only with extensive field studies can these large fault surfaces on nontectonic origin be accurately recognized." He also proposes criteria for recognizing nonseismogenic strike-slip faults related to landsliding (Table 6.4). Cotton (1999) emphasizes that in vertical exposures near the surface these two fault types may be indistinguishable, so the critical evidence will often be found by surface mapping, specifically, landslide inventory mapping.

CHAPTER 7

Using Liquefaction-Induced and Other Soft-Sediment Features for Paleoseismic Analysis

Stephen F. Obermeier

Emeritus, U.S. Geological Survey, EqLiq Consulting, Rockport, Indiana 47635, USA

7.1 Introduction

This chapter focuses on the methodology for determining whether observed sediment deformation had a seismic shaking or a nonseismic origin. Sediment deformations have myriad manifestations, from manifold causes, so the discussion here focuses on those most relevant to a paleoseismic study. This chapter also focuses on small-scale features, with typical dimensions from millimeters to a few meters. Large-scale features, such as those involving huge submarine slumps, are discussed in Chapter 2B.

Emphasis is placed on features developed from the process of *liquefaction*, which is the transformation of a granular material from a solid state into a liquefied state as a consequence of increased *pore-water pressure* (Youd, 1973). The discussion encompasses various manifestations of liquefaction-induced deformation in deposits laid down in fluvial and relatively shallow subaqueous settings, and the application of criteria for establishing an earthquake origin.

The systematic study of paleoliquefaction is a young discipline. Accordingly, some of the physical parameters that control liquefaction effects in the field are not completely understood. Still, the principles and methodology for conducting paleoliquefaction studies are sufficiently advanced to warrant their routine application in paleoseismic studies. The methods discussed here for conducting paleoliquefaction studies were developed largely out of necessity in the United States, where the historical seismic record is particularly short.

Finding features caused by paleoliquefaction is not always easy, because later deposition may have covered them and thus they will only be observed in natural exposures such as actively eroding streambanks, or in man-made cuts. Geophysical methods, including electrical resistivity and electromagnetic induction (Wolf *et al.*, 1998, 2006) and ground-penetrating radar (Liu and Li, 2001; Al-Shukri *et al.*, 2006), have been refined sufficiently to be used with some success to locate buried liquefaction features. For additional details on shallow geophysical exploration techniques, see Chapter 2A.

International Geophysics, Volume 95

ISSN 0074-6142, DOI: 10.1016/S0074-6142(09)95007-0

Paleoliquefaction studies are useful to engineers and planners because of the *high shaking threshold* required to develop liquefaction features. The threshold is a horizontal acceleration on the order of 0.1 g for strong earthquakes, even in highly susceptible sediment (Ishihara, 1985, p. 352; National Research Council, 1985, p. 34). Worldwide data on historical earthquakes show that features having a liquefaction origin can be developed at earthquake magnitudes as low as about 5, but that a magnitude of about 5.5–6 is the lower limit at which liquefaction effects become relatively common (Ambraseys, 1988). (Earthquake magnitude, M_w, is used rather loosely as either moment magnitude or surface-wave magnitude, whichever is larger. And on a related matter, it should be recognized that the shaking levels for an earthquake of magnitude, M_w, strongly depend on the specific tectonic situation—for example, shaking levels from a subduction earthquake are typically much lower than for a crustal earthquake, owing to large differences in strength of the rock in the rupture zone as well as distances from the source to sites of interest. Shaking durations can also differ greatly, with subduction earthquakes typically having much longer durations than their crustal counterparts.)

Liquefaction has been severe and widespread in many earthquakes worldwide, and its effects occur in many forms. Some noteworthy reports discuss the effects in the region of the 1811–1812 New Madrid, Missouri, earthquakes (Fuller, 1912); in the region of the 1886 Charleston, South Carolina, earthquake (Dutton, 1889); in northern California (Youd and Hoose, 1978); in the vicinity of the 1964 Alaska earthquake (U.S. Geological Survey Professional Papers 542–545); various earthquakes in Japan (O'Rourke and Hamada, 1989) and Italy (Galli and Ferreli, 1995); and the earthquake of 1897 in India (Oldham, 1899). However, the deformational effects of liquefaction have rarely been illustrated and discussed in vertical section, which is the view most useful for paleoliquefaction studies and to which this chapter is devoted. Noteworthy accounts of sectional view observations are reported by Sieh (1978a), Amick *et al.* (1990), Audemard and de Santis (1991), Tuttle and Seeber (1991), Clague *et al.* (1992), Wesnousky and Leffler (1992, 1994), Tuttle (1994), and Sims and Garvin (1995). Numerous photographs showing both seismic and nonseismic soft-sediment features in various field settings, as well as discussion of different types of features caused by liquefaction are given in Obermeier (1998a) and Obermeier *et al.* (2005). Findings of exceptional interest in these reports are discussed herein where appropriate. The principal basis for this discussion of liquefaction effects is the author's observations of liquefaction effects at diverse geologic and geographic settings.

Seismic lique faction effects described in this text are caused mainly by *cyclic shaking* of level or nearly level ground. Primary seismological factors contributing to liquefaction are the *amplitude of the cyclic shear stresses* and the *number of applications* of the shear stresses (Seed, 1979b). These factors, respectively, are related to field conditions of shaking amplitude (i.e., *peak acceleration*) and *duration* of strong shaking. Both peak acceleration and duration generally correlate with the earthquake magnitude within a specific tectonic setting. Engineering methods for evaluating variable and irregular cyclic stress applications typical of real earthquakes yield results generally thought acceptable for engineering analysis (Seed *et al.*, 1983; Green, 2001; Youd *et al.*, 2001; Green *et al.*, 2005; Olson *et al.*, 2005a), providing that shaking amplitude–time records can be reasonably bracketed. These same methods can be used for paleoseismic analysis where there is a widespread distribution of liquefaction features. And for many field situations, there are independent techniques for crosschecking any analysis of strength of prehistoric shaking. This chapter briefly summarizes these techniques.

Pseudonodules and other such small-scale features caused by the *plastic deformation* or *flowage* of very soft muds and freshly deposited cohesionless sediments (often referred to as *syndepositional features* or *soft-sediment deformations*) are also discussed at some length. The purpose of the discussion is to critique the state of the art for interpreting the origin of these features, as well as note additional types of observations and data that can be relevant. Liquefaction is not required to deform muds and extremely loose, freshly deposited cohesionless sediments, although a high pore-water pressure can be involved.

The geologic literature is replete with articles attributing an earthquake origin to deformed muds and convoluted sands (see discussion by Lucchi, 1995). Keep in mind, however, that very weak sediments are also commonly deformed as a result of other geologic processes such as loading during rapid sedimentation, localized artesian conditions, slumping, or large waves. In addition, small-scale soft-sediment deformations often form at such low levels of seismic shaking that the shaking poses no engineering hazard. Thus the usefulness of these features for hazard assessment normally is more limited, and interpretations of origin are more often equivocal.

Still, in recent years many paleoseismic studies have been made throughout southern Europe and the eastern Mediterranean, where the long historical record has proven useful in associating seismicity with soft-sediment deformations in lacustrine deposits. These investigations have contributed greatly to the development and verification of techniques for interpreting the origin of soft-sediment deformations in many lacustrine settings (e.g., Marco and Agnon, 1995; Hibsch *et al.*, 1997; Rodriguez-Pascua *et al.*, 2000; Schnellmann *et al.*, 2002, 2006; Bowman *et al.*, 2004; Monecke *et al.*, 2004, 2006). Also, studies of deposits in ancient marine settings have contributed much (e.g., Pope *et al.*, 1997; McLaughlin and Brett, 2004). I draw heavily upon their findings and interpretations in my critique of assessing the origin of soft-sediment deformations.

The chapter also describes some features formed by chemical weathering and features deformed by a periglacial environment, which can mimic those of earthquake origin. Tests are suggested for interpretation of origin.

This chapter is intended primarily for geologists and to a lesser extent for engineers. Some few terms, though, are used in their geotechnical engineering context because of the lack of geologic equivalents for semiquantitative description of certain sediment properties. Most of these terms are given in Table 7.1, which relates the state of compactness (i.e., relative density) of sand to descriptors such as "very loose" to "very dense." In a similar sense, the term "clean" sand refers to a sand with no silt or clay, or any bonding material. The term *liquefaction susceptibility* refers to the ease with which a saturated sediment liquefies and is described with qualifiers such as "very low" or "very high" (Youd and Perkins, 1978).

7.2 Overview of the Formation of Liquefaction-Induced Features

It is the application of shear stresses that causes a buildup of pore-water pressure, thereby leading to liquefaction of saturated *cohesionless sediment*. For seismically induced liquefaction, these shear stresses are due in most field situations to the upward propagation of *cyclic shear waves* (although in special cases

Table 7.1: Relative density of sand as related to standard penetration test blow counts

No. of blows per ft.	Relative density
0–4	Very loose
4–10	Loose
10–30	Medium or moderate
30–50	Dense
<50	Very dense

From Terzaghi and Peck (1967).

compression waves or surface oscillations have shear stresses sufficient for liquefaction). Cohesionless sediments that are loosely packed tend to become more compact when sheared. Continued cyclic shearing can cause the pore-water pressure to increase suddenly to the static confining pressure, leading to large strains and flowage of the water and sediment. No appreciable change in volume of the deposit is required for this change in state from a solid-like to a viscous, liquid-like material (i.e., liquefaction). The process is driven by the breakdown of the packing arrangement of grains during shearing.

Liquefaction during earthquake shaking commonly originates at a depth ranging from a few meters to about 10 m in alluvial deposits, in which the uppermost meter or so is slightly weathered and thereby more resistant to liquefaction than at depth. Liquefaction typically takes place only where the sediment is completely saturated. Where sediments are in a subaqueous setting, and the sediments have not experienced any effects of weathering, liquefaction can originate within the uppermost centimeters, particularly where the thickness of cohesionless material is on the order of decimeters or more. And even where a much thinner cohesionless sediment is in proximity to the top, yet is capped by a thin (order of a centimeters or so) layer of lower permeability that acts prevent dissipation of pore-water pressure between cycles of shearing, liquefaction can develop during the cycles of shearing.

The zone of liquefaction during shaking depends on the relationship between the cyclic shear stresses generated by the earthquake and the stress required to initiate liquefaction in the sediment (Figure 7.1). Development of liquefaction is increasingly difficult with depth, because the higher initial vertical effective stress (total overburden pressure minus static pore-water pressure) applied by the overburden greatly increases the shearing and deformation resistance of the sediment.

Sediment *vented* to the ground surface provides the most conspicuous evidence of liquefaction at depth. Water from the zone of high pore-water pressure must escape upward to cause venting. A water-sediment mixture typically erupts suddenly and violently to the surface, through pre-existing holes or through fractures opened in the capping material in response to liquefaction. In exceptional situations, the mixture spouts as high as 6–7 m, especially where flow is concentrated into holes and cracks through an overlying *fine-grained cap* (Dutton, 1889; Fuller, 1912; Housner, 1958). Water, sand, and silt can continue to flow to the surface as *sand volcanoes* for hours after earthquake shaking has stopped. Sediment is left behind on the ground surface in the form of cones, often called *sand blows* or *sand boils* (Figure 7.2). The cones of sand can be as much as a meter in height and tens of meters in width. Sediment left behind in the vents through the cap forms *clastic dikes*. Clastic dikes can also develop by sediment being forced into fractures that extend only partway through the cap.

The increased pore-water pressure during ground shaking can be manifested in other ways. The high pore-water pressure decreases the shear strength of granular strata at depth. These strata can then fail in shear even where the ground surface is inclined as gently as 0.1–5% (Youd, 1978; Youd and Bartlett, 1991). Huge masses of overlying soil can shift horizontally in the form of laterally moving, translational landslides (called *lateral spreads*; Figure 7.3). Separation between individual blocks is commonly as much as 2–3 m where shaking has been especially strong. This separation tends to be largest near streambanks or scarps in alluvium, even if only a few meters in height, because these breaks in slope reduce the resistance to lateral movements. Near-vertical fissures typically abound in lateral spreads, whereas (rotational) slumping only occurs in proximity to a streambank or steep slope, as illustrated in Figure 7.3. Figure 7.4 illustrates a variation of lateral spreading (termed *surface oscillation*) that takes place on level ground far from breaks in slope. Here, oscillation of the ground above the liquefied zone forms blocks separated by fissures.

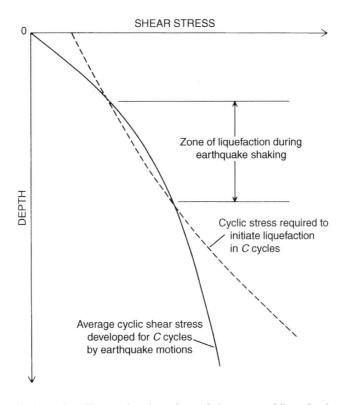

Figure 7.1: Schematic depiction illustrating location of the zone of liquefaction that is frequently induced during earthquake shaking, in a field setting having a shallow-water table and slight degree of weathering of sediment near the surface. C, number of shaking cycles. From Seed and Idriss (1971); reprinted with permission of the American Society of Civil Engineers.

Liquefaction of granular deposits normally leads to surface cracking and formation of localized depressions because of densification of liquefied sediment after high pore-water pressures dissipate, even where there is no evidence of venting of sand and water to the surface. These settlements can be as much as 0.25–0.5 m where thick sands liquefy severely (Tokimatsu and Seed, 1987).

On slopes that exceed about 5%, liquefied sediments can trigger huge landslides (sometimes called *flow failures*) that can flow as much as tens to hundreds of meters (Seed, 1968). Ground disruption can be so severe that it is difficult to establish what the surface geometry was prior to failure (Chapter 8). The strength properties of materials in the failure zone can be changed greatly. Flow failures along streambanks and hillsides can also originate by static (nonearthquake) mechanisms, however. Determination that prehistoric landslide movement on a steep slope was seismically triggered generally requires complex engineering testing and analysis that is difficult to perform in the best of circumstances, regardless of whether liquefaction was involved (Chapter 8). Therefore, sites on level ground are best for distinguishing a seismic from a nonseismic origin.

Clastic dikes that formed on level to nearly level ground are the primary source of data used for paleoseismic interpretations. Very important factors controlling the development and density of dikes

Figure 7.2: Small sand blows near the town of El Centro from the 1979 Imperial Valley, California, earthquake. The sand blows were produced by a mixture of sand and water that spouted to the surface; they provide evidence of extensive liquefaction at depth. White scale in foreground is 20-cm long. Photograph courtesy of R. F. Scott.

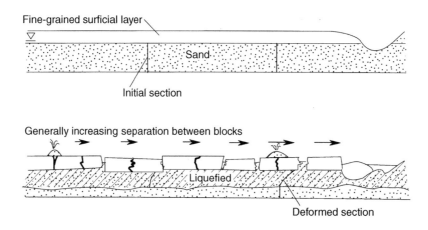

Figure 7.3: Vertical section of lateral spread before and after failure. Liquefaction occurs in the crosshatched zone. The surface layer then moves laterally down the gentle slope, breaking into blocks bounded by fissures. Sand is vented to the surface through some fissures, but other fissures are only partly filled. The blocks can tilt and settle differentially with respect to one another. From Youd (1984a).

include not only the compactness and thickness of sediment that liquefied but also the properties of any cap layer, including thickness, strength, degree of weathering, etc. Such dikes almost certainly form solely in response to *hydraulic fracturing* of the cap in many field situations, but dike development can be enhanced greatly by the cap simply being *pulled apart in tension at sites of lateral spreading as well as by strong oscillatory shaking at the surface*.

Sills in the form of horizontal clastic intrusions beneath fine-grained strata, as well as within the fine-grained cap, are commonly found at sites of clastic dikes. Dikes may not cut to the surface where roots

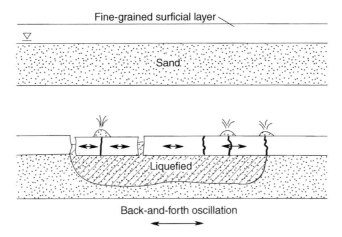

Figure 7.4: Vertical section showing fissures and blocks resulting from oscillation of level ground far from breaks in slope (figure from Youd, 1984a). Ground oscillation can result from fundamentally different mechanisms driven by surface waves (Youd, 1984a) or body waves (Pease and O'Rourke, 1995). In the mechanism driven by surface waves, liquefaction in the zone marked by diagonal lines decouples the surface layer from the underlying liquefied layer. The decoupled layer vibrates at a different frequency than the underlying liquefied layer; as a result, fissures form between oscillating blocks. In the mechanism driven by body waves, ground oscillation becomes severe in response to a resonant frequency effect in which horizontal displacement in the cap is amplified with respect to that of the underlying liquefied sand.

have greatly enhanced the tensile strength of the cap; here, sills can form abundantly beneath the root mat (Clague *et al.*, 1992; Obermeier, 1994b).

7.2.1 Process of Liquefaction and Fluidization

The process of seismically induced liquefaction of saturated granular sediments has been studied extensively and is reasonably well understood (Seed, 1979b; National Research Council, 1985; Castro, 1987; Dobry, 1989). Figure 7.5 illustrates the typical field situation on level ground. A liquefiable sand layer is overlain by a thin, nonliquefiable stratum, and the groundwater table is shallow. Earthquake-induced shear stresses propagate through the sand and cause shear strain of the sediment structure. (Shear strain is the angle in radians, γ, shown in Figure 7.5.) Because grains attempt to move into a denser packing arrangement relatively quickly during back-and-forth shear straining, water in the voids does not have time to escape. The pore-water pressure can thereby increase. In medium and looser sand deposits (Table 7.1), stresses at the grain contacts can approach zero, and concurrently the pore-water pressure carries the weight of the overburden. The first time this occurs is sometimes referred to as *initial liquefaction* (National Research Council, 1985, p. 12). A large loss of strength can take place once this condition is approached. In very loosely packed sand, the shear resistance can decrease rapidly to near-zero values, and blocks on very mild slopes can move large distances. The combination of elevated pore pressure with large loss of strength is referred to as *liquefaction* (Youd, 1973). Only a few cycles of shearing can suffice to achieve this state in loose sand. In more densely packed sand the pore-water

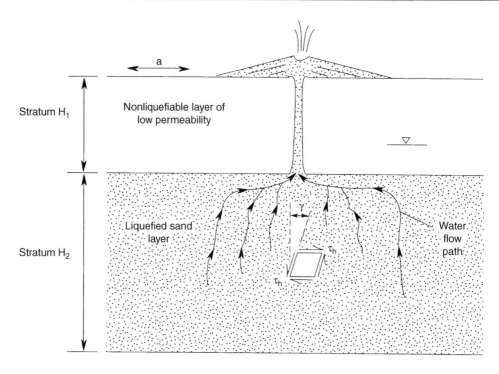

Figure 7.5: Vertical section showing typical sediment relations, seismic loading condition, and water flow paths involved in formation of sand blows; γ, shear strain (angle in radians); *a*, horizontal acceleration; τ$_h$, shear stress induced by horizontal acceleration.

pressure rise is less sudden, and the strength loss is less severe; thus, any increase in strain is less dramatic and is limited. Very densely packed sands can develop elevated pore-water pressures, but not high enough to cause liquefaction. For the purpose of this text, the condition of highly elevated pore-water pressure with a significant loss of strength is referred to "liquefaction" unless otherwise noted.

The cyclic strain in sands that is required to induce liquefaction is usually quite small. Ten cycles of back-and-forth shearing generally suffice if the shear strain exceeds a small number, commonly 0.3%. For earthquakes having long durations and many cycles, the *critical shear strain* can be as low as 0.04% (Dobry, 1989). Following liquefaction, *densification* occurs after water is expelled from the sediment. The vertical settlement caused by sediment densification often is quite small, being less than 2–3% of the height of the stratum that liquefied (Castro, 1987, p. 175).

Only minor disturbance to the original stratification may take place in the portion of the stratum where only initial liquefaction has been approached or occurred. The effects to bedding are often virtually indistinguishable to the unaided eye. Closer inspection may show that platy minerals such as mica and clay are reoriented from their original flat-lying position (and can form structures known as *dishes*, shown later in Figure 7.26), and thin laminations of finest constituents are warped. Small, steeply inclined, flame-like sand-rich structures known as *pillars* (see Figure 7.26) can form where water collects beneath a slightly less permeable lamina and then locally penetrates through the lamina to winnow out silt and other very fine constituents. This winnowing takes place by a process often referred to as *fluidization*.

Fluidization occurs when flowing water exerts sufficient drag or lift to momentarily suspend grains of sediment. When a fluid is forced vertically through a layer of cohesionless sediment at a rate sufficient to cause fluidization, the layer expands rapidly, porosity increases, and the sediment ceases to be grain supported and becomes fluid supported. Fluidized flow typically destroys original bedding and fabric, at least locally. Fluidization can result from a number of mechanisms, including seepage caused by compaction of underlying sediments, seepage from artesian springs, or seepage from deposits liquefied either from static or from earthquake forces (Section 7.5). Silts and muds are much less prone to forming fluidization effects than coarser, sand-rich sediments (Dzulynski and Smith, 1963).

Reoriented minerals and small deformation structures similar to those caused by liquefaction-induced fluidization also occur as syndepositional features in many environments, particularly where great thicknesses of sediment are rapidly laid down in a delta-like setting (e.g., Lowe and LoPiccolo, 1974; Lowe, 1975; van Loon, 1992), which makes interpretation of a seismic or nonseismic origin uncertain. In contrast, larger features caused by fluidization from seismic liquefaction are easier to interpret because, in many field situations, sand-filled dikes whose widths exceed several centimeters and whose heights exceed a meter or so cut weathering horizons or other strata that are obviously much younger than the source zone for the dikes. This relation generally eliminates syndepositional processes from the list of possible causes for formation of the dikes.

Many variables control the formation of large fluidization features caused by seismic liquefaction. The influence of some of the most important variables is fairly well understood. The main elements are illustrated in Figure 7.5, where the base of stratum H_2 lies on an impermeable base. Assume stratum H_2 liquefies. Water tends to flow upward by two mechanisms: relief of the high pore-water pressure and reconsolidation. Reconsolidation causes densification as it progresses from the bottom of stratum H_2 upward (e.g., Scott and Zuckerman, 1973; Scott, 1986). Water expelled from the zone that liquefied during shaking can accumulate beneath a low-permeability capping layer to form a *water-rich zone* (Liu and Qiao, 1984; Elgamal *et al.*, 1989; Fiegel and Kutter, 1994; Kokusho, 2003); this zone in turn probably supplies much of the water and sand that vents to the surface through breaks in the cap. Sediment may also be vented to the surface from greater depth, where liquefaction first developed. Venting can occur during the time of strong shaking or be delayed by as much as a few minutes following very strong shaking (Kawakami and Asada, 1966). The increased pore-water pressure in the underlying sand–water mixture can most easily break through to the surface where the cap is thin (<1–2 m). Characteristics of the source stratum that enhance the liquefaction–fluidization process are (1) a thick, loose sand that, once liquefied, provides a large volume of water available for upward flow and (2) a permeability that is high enough to allow water to flow quickly to the base of the cap but that is not so high as to dissipate excess pore pressures between seismic cycles of shearing (Castro, 1987, pp. 177–179; Dobry, 1989). This simple model fully explains most field observations. However, seemingly contradictory and unexplained manifestations of the liquefaction and fluidization process are also encountered in the field. For example, sand dikes cutting gravel layers much more permeable than the sand dikes have been reported by Tuttle *et al.* (1992).

7.2.2 Factors Affecting Liquefaction Susceptibility and Effects of Fluidization

The most important factors controlling the occurrence of liquefaction and subsequently the development of dikes and sills are (1) grain size of the source bed, (2) relative density (i.e., degree of compactness) of the source bed, (3) depth and thickness of the source bed and overlying strata, (4) age of the sediments, (5) characteristics of any overlying fine-grained cap, (6) topography and nature of seismic shaking, (7) depth to water table, and (8) seismic history. The following synopsis is from an expanded discussion of these factors in Obermeier et al. (2005, pp. 226–228). Also not discussed herein are secondary factors affecting liquefaction susceptibility such as grain shape, sediment fabric, weak grain-to-grain bonding in sediment, and static horizontal stress conditions; those factors are summarized in articles by Mitchell (1976, p. 244) and Seed (1979b).

Liquefaction-induced features often form readily in sand and silty sand. Figure 7.6 illustrates the sizes and gradations generally most susceptible. However, the thickness of potentially liquefiable sediment and the characteristics of adjoining strata can be highly relevant to liquefaction. As noted above, a particularly critical issue can be whether any buildup of pore-water pressure is permitted during cycles of shearing. Field examples illustrated in Figure 7.7 show the role of sediment relationships. Figure 7.7A is a case where no dikes have formed in a layer of clean, medium-sized sand, about a decimeter in thickness, underlain by a much more permeable sandy gravel. Yet dikes formed nearby where the sand thickness was several decimeters. The regional levels of peak acceleration were likely on the order of several tenths g, and the paleoearthquake magnitude was $M_w \sim 8$–9. The probable explanation for the absence of dikes associated with the thinner sands is that any increase in pore-water pressure caused by cyclic shearing dissipated, during shearing, into the underlying highly permeable zone.

A similar example showing the role of adjoining strata is in Figure 7.7B, where dikes and other liquefaction effects have formed by the liquefaction of sand, centimeters in thickness, in thinly bedded sediments. Such relations have been found in lacustrine deposits at widespread locations.

Even though large liquefaction-induced dikes have been documented in gravel (Andrus et al., 1991; Meier, 1993), the threshold magnitude ($M_w \sim 7$, crustal tectonic setting) is much higher than for sand (Valera et al., 1994). Exceptionally loose silt deposits can also form sizable "sand" blows (Youd et al., 1989), but even a small amount of clay greatly diminishes the ability of sediment to liquefy and flow, especially if the clay has significant plasticity (Obermeier, 1996). Liquefaction is unusual in sediments containing more than 15% clay (Seed et al., 1983). Criteria for evaluating liquefaction of clay-bearing sediments are given in Boulanger and Idriss (2007), but often some uncertainty lies in role of clays on liquefaction susceptibility. Simple field tests for relating fines content to susceptability are given in Appendix 3 in Book's companion web site.

The state of *relative density* of cohesionless sediment has a very large influence on liquefaction and flow. Materials of low-to-moderate relative density (Table 7.1 shows the commonly used engineering measures) are most susceptible to liquefaction. Susceptible sediments are usually late to mid-Holocene in age (Youd and Perkins, 1978), although sediments as old as about 200,000 years have liquefied (Obermeier et al., 1990).

Liquefaction under level or nearly level ground conditions generally originates in strata located a few meters to 10 m beneath the surface, but reported depths can be on the order of a few centimeters in a subaqueous setting (Sims, 1973; Rodriguez-Pascua et al., 2000), and the maximum reported depth is greater than 20 m (Seed, 1979b). Susceptible beds are usually ≥ 0.3–1.0 m in thickness, but Tuttle and Seeber (1991) describe an 8–10-cm-thick sand bed that liquefied and formed a dike exceeding 2.5 m in height. And, I have observed that it is not unusual that a sand bed at thin as several centimeters can suffice to form small dikes, millimeters in width and centimeters in height.

Whether sand vents to the surface is normally determined by the ratio of cap thickness (H_2 in Figure 7.5) to thickness of the source stratum (H_1 in Figure 7.5; see Ishihara, 1985). Dikes generally do not extend to the surface when the cap thickness exceeds 10 m and often are severely restricted for thicknesses of more than 5 m.

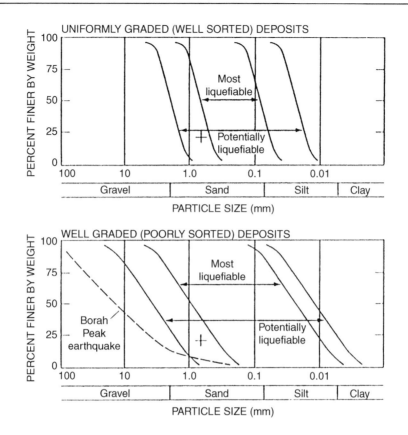

Figure 7.6: Gradation curves showing grain sizes and gradations generally most susceptible to liquefaction; from Tsuchida and Hayashi (1971). Line for the Borah Peak, Idaho, earthquake ($M_w = 7.3$) shows gradation of coarsest natural deposit verified to have produced severe liquefaction effects at the ground surface; from Hardman and Youd (1987); sediment of essentially the same grain size gradation also liquefied in the Wabash Valley paleoearthquake of 6100 years BP (Pond, 1996, Figure 2.9), of $M_w \sim 7.3$.

Effects of seismic shaking typically are topographically amplified near abrupt relief such as streambanks, leading to greatly increased ground breakage with venting (Fuller, 1912; McCulloch and Bonilla, 1970). Still, far from any topographic incisions, dikes can form in response to very strong surface oscillations (T. L. Youd, written communication, 1994).

Liquefaction susceptibility is influenced strongly by depth to the water table. For example, increasing the depth from the surface to 5 m or less commonly reduces susceptibility from high to moderate, and increasing the depth to 10 m reduces susceptibility to nil.

The seismic history of sediment can influence its ability to *reliquefy* during earthquakes many years later. An occurrence of severe liquefaction can cause significant densification of source sediment (Castro, 1987), except possibly within a thin zone just beneath a fine-grained cap. This thin, loose zone is thought to form by water and sediment flowing up from the liquefied zone (Elgamal *et al.*, 1989; Dobry and Liu, 1992). In subsequent earthquakes, liquefaction severity may be diminished because of densification of

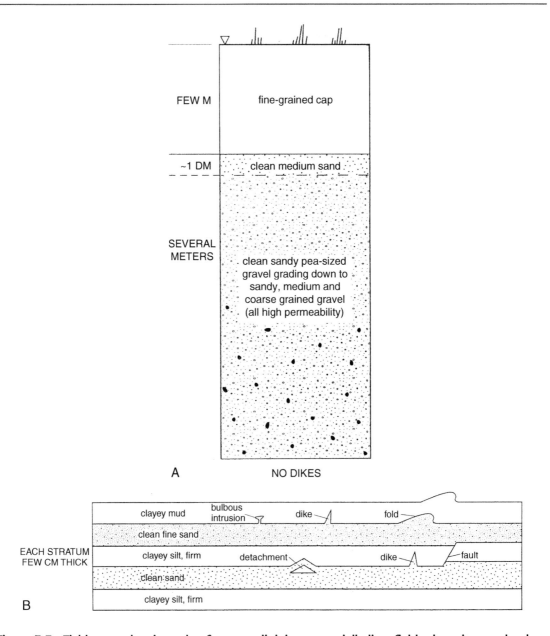

Figure 7.7: Field cases showing role of strata adjoining potentially liquefiable deposits, on development of liquefaction effects. (A) Example of highly permeable stratum beneath thin stratum of clean sand. (B) Example of thin clean sand strata bounded by impermeable strata.

most of the source stratum. Worldwide observations show, though, that liquefaction from widely timed earthquakes commonly recurs at the same site even to the extent of using the same dike for venting (Kuribayashi and Tatsuoka, 1975; Youd and Hoose, 1978; Youd, 1984b; Obermeier et al., 1990; Tuttle et al., 1992).

Field observations of the influence of the preceding factors are discussed throughout this chapter, as case history examples.

7.3   Criteria for an Earthquake-Induced Liquefaction Origin

The following set of criteria can establish whether observed sediment deformation was caused by seismically induced liquefaction:

1.   The features have sedimentary characteristics that are consistent with an earthquake-induced liquefaction origin; namely, there is evidence of an upward-directed hydraulic force that was suddenly applied and was of short duration.
2.   The features preferably have sedimentary characteristics consistent with historically documented observations of the earthquake-induced liquefaction processes, in a similar physical setting. In addition, preferably there is more than one type of feature commonly caused by seismically induced liquefaction. Such features include dikes, sills, vented sediment, lateral spreads, and some types of soft-sediment deformations.
3.   The features occur in groundwater settings where suddenly applied, strong hydraulic forces of short duration could not be reasonably expected except from earthquake-induced liquefaction. In particular, the possibility of an origin from artesian conditions or nonseismic landsliding must be ruled out.
4.   Similar features occur at multiple locations, preferably at least within a few kilometers of one another, in similar geologic and groundwater settings. The regional pattern of size and abundance of features should be consistent with a pattern of shaking associated with an earthquake.
5.   The evidence for age of the features supports the interpretation that they formed in one or more discrete, short episodes that individually affected a large area and that the episodes were separated by relatively long time periods during which no such features formed.

Determining the age of features is generally done by bracketing when liquefaction features formed, chiefly by radiocarbon analysis in combination with other methods such as soil pedogenesis, cultural archeology and regional stratigraphy, or less frequently dendrochronology. Excellent examples using these techniques are found in Munson and Munson (1996) and Tuttle (2001). Most recently, two other methods have proven extremely useful for quantitative determination of earthquake age: optically stimulated luminescence (OSL) and thermal luminescence (TL) (Lian, 2007). These methods determine when sand- or silt-sized silicate grains were last exposed to sunlight. Furthermore, age determination using these methods can extend much farther back in time than ages based on radiocarbon analysis, and the accuracy and sensitivity of these methods is generally much better than radiocarbon analysis. OSL can be used to determine ages ranging from modern to as far back as 150–200,000 years, and techniques are available for routinely determining the validity of the age.

Determining the regional pattern of size and abundance of suspected liquefaction features may be critical to the interpretation of origin. Preferably at least 20–30 km (cumulative) of fresh exposure should be examined. Such a regional approach to a paleoliquefaction study helps eliminate the possibility that nonseismic processes (Section 7.5) are responsible for creating the features, and it helps to develop a sense of the various processes that deform sediments locally within the region.

Whereas the five criteria above can be applied in some studies, there are other settings where the suspected liquefaction features are so sparse or are in unusual field settings that make it questionable to assign a seismic origin. For these situations, engineering testing and analysis can help greatly in assessing origin, through evaluation of liquefaction susceptibility at side-by-side sites, where small features are present at one site but none are nearby. The small size is important to interpretation if it represents a marginal occurrence of liquefaction. For a seismic origin, a site with features should be more susceptible to liquefaction than the adjoining site with no features (Green *et al.*, 2005; Olson *et al.*, 2005a).

The next section illustrates application of the five criteria listed above, and presents examples of why understanding the local geologic setting is critical to interpretations.

7.4   Historic and Prehistoric Liquefaction—Selected Studies

In this section, I describe earthquake-induced liquefaction features in four geologic–geographic settings: coastal South Carolina, the New Madrid seismic zone of Missouri, the Wabash Valley seismic zone of Indiana and Illinois, and coastal Washington State (all in the United States). The latter two areas have not experienced historically documented liquefaction. The discussion emphasizes the role of the local geologic setting in deducing an earthquake origin. Special consideration is given to eliminating artesian springs and nonseismic landsliding as possible sources of observed deformations. Where possible, the magnitudes of the prehistoric earthquakes are estimated by comparison with historical liquefaction-producing earthquakes in the region.

7.4.1   Coastal South Carolina

The strongest historical earthquake in the southeastern United States took place in 1886 near Charleston, South Carolina ($M_w \sim 7.0$–7.2). Clear evidence of seismotectonic conditions in the region is lacking, which prompted searches for prehistoric liquefaction features. Liquefaction evidence has since been found for many strong prehistoric earthquakes (Obermeier *et al.*, 1987; Amick *et al.*, 1990). Results of the searches are shown in Figure 7.8. The figure shows the approximate boundary of the 1886 earthquake meizoseismal zone; sites where swarms of liquefaction features described as explosively erupting *craterlets* were formed in 1886 (Dutton, 1889; see Figure 7.9 herein); and sites where liquefaction features predating 1886 were found. The prehistoric liquefaction-induced features are mainly ancient craterlets that are now filled.

None of these pre-1886 *craters* have an expression on the ground surface that can be discerned by on-site surface examination or on aerial photographs. These features are seen only in sectional view (i.e., walls of excavations). At most of the sites shown on Figure 7.8, at least three or four pre-1886 craters are exposed within a few hundred meters of one another.

The physical setting of the region within 50 km of the South Carolina coast is conducive to widespread liquefaction. That region is known locally as the "low country" because it has low local relief (1–3 m) and low elevation (0–30 m), and because vast expanses are under water much of the year. Most of the Carolina low country is covered by a 5–15-m-thick blanket of unconsolidated Quaternary marine and fluvial deposits, which lie on semilithified Tertiary sediments. The Quaternary deposits primarily occur as a series of well-defined, temporally discrete, interglacial beaches and associated back-barrier and shelf deposits that form belts subparallel to the present shoreline. Increasingly older beach deposits are progressively farther inland and at higher altitudes. Most beach deposits consist of clean, medium- to fine-grained sand.

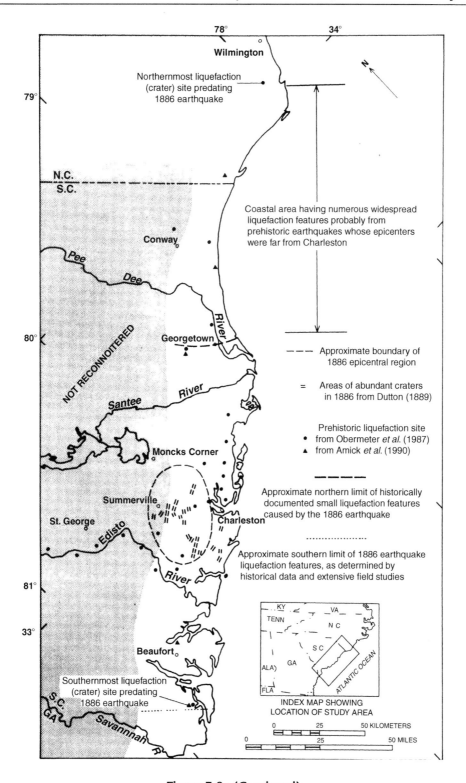

Figure 7.8: (Continued)

Most of the craters discovered were in the ancient beach deposits. The search for liquefaction features was generally restricted to the beach deposits younger than about 250,000 years and older than about 80,000 years (Figure 7.8). Sand in these deposits is loose at many places. Normal depth to the groundwater table in these sediments is about 1–2 m, even in topographically elevated regions. Deposits older than about 250,000 years have such a low susceptibility to liquefaction (due to the effects of chemical weathering) that the likelihood of their liquefying has been extremely low during the late Pleistocene and Holocene. Deposits younger than about 80,000 years generally have such a high groundwater table that exposures are very limited.

Tabular sand dikes were discovered in fluvial terraces and in back-barrier environments where the cap is much richer in clay. Sills were observed only rarely. The following discussion concentrates on the craters, because the section below discussing liquefaction effects in the New Madrid seismic zone adequately deals with characteristics of tabular sand dikes and sills, which are the types of liquefaction-induced features most often encountered where a thick clay-rich cap overlies source sand beds.

7.4.1.1   Characteristics of the Craters

Figure 7.9 shows moderate to large craters produced by the 1886 earthquake. Examination of photographs taken in 1886 shows that surficial sheets of vented sand around crater rims normally had thicknesses of about 15–20 cm. The maximum reported thickness of vented sand was 1 m, and the maximum crater diameter was about 6 m (Dutton, 1889). Almost all craters that predate 1886 have a morphology and size comparable to the 1886 craters of Figure 7.9 except that the craters are now filled with sediment.

Figure 7.10 is a schematic cross section through a typical prehistoric crater site on a beach ridge. According to eyewitness observations in 1886, "craterlets are found in greatest abundance in belts parallel with (beach) ridges and along their anticlines" (Peters and Herrmann, 1986). Thus the locations of many prehistoric crater sites that have been discovered are consistent with historical observations.

The crater sites are located where weathering has imposed a strong soil profile on the ancient beaches (Figure 7.11). Near the surface, a thin A horizon (organic matter and several percent sand) overlies a thin, very light gray E (eluviated) horizon; the E horizon overlies a thick, weakly cemented, black Bh horizon (humate-enriched sand containing a few percent clay) that grades down rather sharply into a variably thick, light-colored B–C horizon (transition zone between B and C horizons). The B–C unit grades down into C horizon sands (parent material). The craters cut the solum and the C horizon. Within the filled crater are well-defined zones of sediment. The fill materials are fine- to medium-grained sand and clasts from the Bh, B–C, and C horizons of the host, and there is sand from depths much below the exposed C horizon. The walls of the crater are commonly smooth and sharply defined when viewed closely, especially in the lower part.

Figure 7.8: Coastal portion of South and North Carolina containing liquefaction sites. Unshaded onshore region is predominantly marine deposits younger than about 240,000 years. Numerous ancient beach ridges lie in this unshaded region. Shading denotes region of older marine deposits that was not reconnoitered, except locally. Younger fluvial sediments occur locally. All liquefaction sites along the Edisto River are in fluvial sediments. Almost every liquefaction site shown represents an area where numerous liquefaction features are exposed in a network of drainage ditches several kilometers in length. Index map shows coastal region searched.

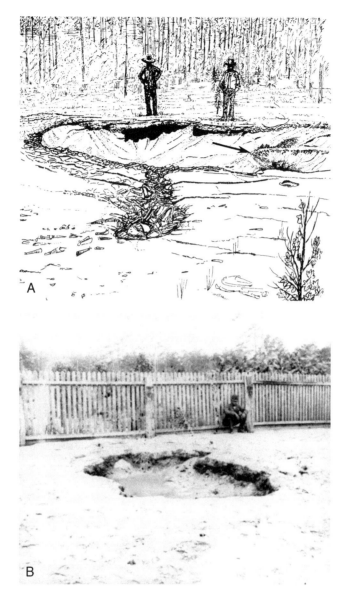

Figure 7.9: Craters produced by the 1886 Charleston, South Carolina, earthquake. (A) Sketch from a photograph of an 1886 crater (sand blow at Ten Mile Hill, near the present Charleston airport). Note that the crater contains sand sloughing toward the lowest parts and that there is a constructional sand volcano located in the right part of the crater (at arrow). The crater is surrounded by a thin blanket of sand partly veneered with cracked mud. (B) Photograph of typical crater produced by the 1886 earthquake. Note the thin blanket of ejected sand around the crater and sand and clasts of dark soil within the crater. Photograph from the archives of the Charleston Museum.

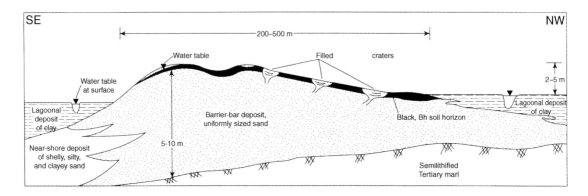

Figure 7.10: Schematic vertical section of representative barrier showing sediment types, ground-water table locations, filled craters, and Bh (humate-rich) soil horizons. Modern shoreline is located southeastward. Lagoonal clay deposit at left is younger and lower in elevation than the barrier-bar deposit.

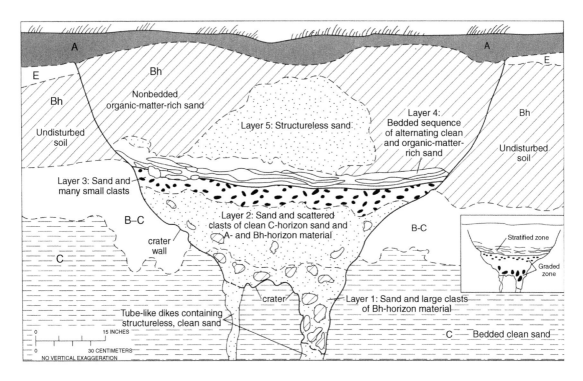

Figure 7.11: Schematic vertical section of filled liquefaction crater that forms a bowl in three dimensions. Letters correspond to soil horizons. The filled crater in this figure long predates the 1886 Charleston earthquake, on the basis of thickness of Bh horizon in the filled crater. Inset shows zonation within the crater.

The Bh horizon of the laterally adjoining undisturbed host generally is abruptly thicker than the Bh horizon on the filling in the crater. With increasing age, the Bh horizon of the filled crater is thicker, more clay rich, and has better developed soil structure. Craters older than about 5000 years have Bh horizons that approach having the thickness and development in host sediment enclosing the craters.

The filled craters are characterized by a *sequence* of five layers. Layer 5 (Figure 7.11) is a structureless, gray, humate-enriched and cemented sand, which overlies a thinly (2–3-mm) laminated sequence of alternating light- and dark-colored sands (layer 4). The lamina typically is discontinuous and irregular in thickness. The dark color is due to humate staining. The basal bed sharply overlies a medium-gray structureless sand (layer 3), which contains many small clasts of Bh material and wood. Layer 3 grades down into a structureless sand zone (layer 2) containing many intermediate-sized clasts (1–5 cm in diameter). Layer 2 grades down into layer 1, which contains densely packed intermediate-sized (1–5-cm) and large-sized (>5-cm) clasts of Bh material in a structureless sand matrix; the large clasts have diameters exceeding 25 cm in many filled craters. Dikes containing clean sand are present beneath the bowl. The dikes are tube-shaped in plan view.

The filled craters are interpreted to have formed in the following phases (1) a large hole is excavated at the surface by the violent upward discharge of the liquefied mixture of sand and water; (2) a sand rim accumulates around the hole by continued expulsion of liquefied sand and water; (3) sand, soil clasts, and water are churned briefly in the lower part of the bowl, followed by settling of the larger clasts and formation of the graded-fill sequence of sediment (layers 1–3); and (4) the crater is intermittently filled by adjacent surface materials to form the thin stratified-fill sequence (layer 4) during the weeks to years after the eruption. Layer 5 is in the strongly bioturbated zone and thus has no stratification. The sand blanket ejected from the crater is indistinguishable in the field from the surface and near-surface (A, E, and Bh) soil horizons, because the blanket has been incorporated within these soil horizons.

The presence of friable, angular clasts of B–C, C, and Bh horizon material in the graded-fill portion (layers 1–3) is consistent with a short-lived, churning type of upwelling from the vent. Water commonly flows for a day or so from vents, on the basis of worldwide observations. The violent, boiling phase is much shorter in duration. Hence, the presence of friable clasts argues against a long-term artesian spring origin; such a spring would abrade, round, and disaggregate the clasts. In addition, springs are very unlikely to form in the topographic–geologic setting at some of the crater sites (Figure 7.10).

An earthquake origin for the craters is also supported by the presence of sand-filled tabular fissures, whose overall shape and dimensions strongly suggest that they are *incipient craters*. Fissures such as those shown in Figure 7.12 are rather common in the epicentral zone of the 1886 earthquake. The figure shows a V-shaped fissure connecting to a tube-like dike with sand transported upward from depth. The tabular fissures in the V-portion widen with depth until they connect to a single, near-vertical, large, sand-filled tube. The V-shaped fissures probably represent the early phase of development of craters; upward forces, however, were too weak to excavate the overlying material.

It is possible that liquefaction produced craters because of a fortuitous combination of sediment properties. The source beds that liquefied were exceptionally susceptible to liquefaction; they are loose (in the engineering sense), fine-grained, uniformly sized, and free of clay (Martin and Clough, 1990). These properties would cause the source beds to liquefy abruptly and, once liquefied, the sand–water mixture would flow readily. I suspect that the liquefied sand strata quickly migrated laterally to a hole such as that left by a decayed root. The sudden application of an upward force around the hole caused the formation of a V-shaped crack. The liquefied sand vented violently because of its exceptional ability to flow. The V-shaped cracks developed because overlying sediment is isotropically cemented with humate,

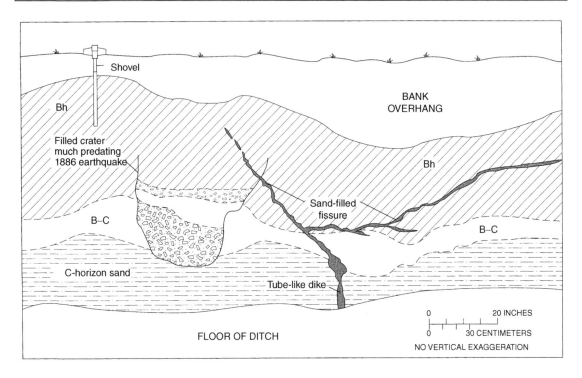

Figure 7.12: Schematic vertical section showing V-shaped, sand-filled fissures interpreted as resulting from liquefaction during the 1886 Charleston earthquake. Sand in the fissures came from a depth of 6 m, on the basis of grain size and mineralogy. Fissure cuts soil horizon developed on filled crater predating 1886.

has no pronounced planes of weakness, and is very brittle; the process is similar to formation of a conchoidal fracture in an isotropic, brittle medium such as glass, when struck by a rock.

Liquefaction-induced craters are common during earthquakes worldwide. Good examples for various earthquakes in Japan are shown in articles by Kawakami and Asada (1966) and Iwasaki (1986). Craters were especially prevalent during the M_w 7.5 Niigata earthquake of 1964; soils around Niigata typically are sand rich all the way to the ground surface (Katayama *et al.*, 1966), similar to the Charleston region where craters formed. For the northeastern United States, sketches of prehistoric seismically induced craters in host sand deposits are to be found in Tuttle (1994). An example of a filled crater at a site where the surficial material is very weakly cemented, in the New Madrid seismic zone, is shown by Obermeier (1998a). Similar features have been found in Brazil (Berezza *et al.*, 2005) that in sectional view have gravels concentrated at the base of a bowl-shaped region, much like that of a crater, probably from fluidization of a gravelly sand in response to seismic liquefaction.

Craters can also form in a clay-rich cap. In Alaska during the great earthquake of 1964, craters having a relatively small diameter (many ~1 m) formed at sites of violent venting, which apparently eroded deep holes (the craters) through a clay-rich cap; some "craters" (actually surface depressions) as much as 10 m in diameter and extremely shallow (25–40 cm) also formed (Reimnitz and Marshall, 1965). These very large "craters" apparently formed as sand vented to the surface and undermined a clay-rich cap, thereby making a swale. In Argentina, Youd and Keefer (1994, pp. 227–229) have shown that pre-existing holes through a clay-rich cap later led to erosion of large craters in response to liquefaction during a M_w 7.4 earthquake. Craters also formed in a clay-rich cap in the Nile River Valley during the M_w 5.9 Dashur, Egypt, earthquake of 1992 (Elgamal *et al.*, 1993). No mechanism for formation of the craters in Egypt can be demonstrated because of the lack of geologic and geotechnical data at the crater sites. However, the regularity of alignment of the craters suggests that man-made holes led to formation of the craters.

From reports of craters formed at worldwide sites, it can be deduced that craters are of two fundamentally different origins (1) those that erupt violently and concurrently excavate a hole (i.e., crater) in the ground and (2) those that erupt much less violently and have a hole eroded from the flowing mixture of sand–water. The violent type likely forms where there is an impermeable, brittle cap that overlies very loose, liquefied sand. The erosional type likely forms where the cap is not so impermeable, or is much more friable, or overlies sand that is less prone to being suddenly liquefied.

7.4.1.2 Prehistoric Seismicity

Many filled craters in South Carolina contained small twigs and bark from trees. This woody matter is concentrated along the contact between layers 3 and 4 (Figure 7.11) and obviously fell into the open pits soon after they formed. Twigs and bark from various craters yield radiocarbon ages of approximately 600, 1250, 3200, 5150, and older than 5150 years, documenting five prehistoric earthquakes in the Charleston area (Obermeier *et al.*, 1987; Amick and Gelinas, 1991; Talwani and Schaeffer, 2001).

An estimate of the magnitude of prehistoric South Carolina earthquakes is provided by comparison of their liquefaction effects with worldwide observations, and also by comparison with observations of liquefaction in South Carolina in 1886. Data from the 1886 earthquake furnish a database for the regional development of craters, as well as their size and abundance. Worldwide data show that features having a liquefaction origin can be developed at magnitudes as low as $M_w \sim 5$ but that a magnitude of about 5.5 is the lower limit at which liquefaction effects are relatively common (Ambraseys, 1988). The source sands that produced craters in coastal South Carolina commonly are highly susceptible to liquefaction and flow; because of this susceptibility, one might suggest that a low-magnitude earthquake could have produced the prehistoric craters. However, numerous prehistoric craters in the Charleston area, many having diameters in excess of 3 m, clearly are too large to have been the result of marginal liquefaction. Such large diameters suggest that the earthquake that produced them was much stronger than M_w 5–5.5. In addition, the prehistoric craters that formed 600 and 1250 years ago extend along the coast at least as far as the craters produced by the M_w 7.0–7.2 earthquake of 1886.

Interpretations of prehistoric earthquake magnitudes must account for liquefaction susceptibility. Principal variables are water table depth and the compactness of the source sands. The water table is presently very shallow, being less than 1–2 m below the ground surface. Almost certainly the water table has been essentially unchanged for the past few thousand years at many of the crater sites (Amick *et al.*, 1990). Just prior to the 1886 earthquake, the Charleston area was experiencing an extraordinarily wet period (Taber, 1914, p. 126), so water table conditions were optimal for production of liquefaction features. *Standard penetration test (SPT)* data also show that the source sands are so loose as to liquefy readily. It is not unusual that the sand deposits at the liquefaction sites have SPT blow counts as low as 10 or less (Martin and Clough, 1990). It is difficult to conceive of any mechanism that would have made the sands much more compact when the prehistoric earthquakes occurred. In summary, the liquefaction susceptibility was high at many places when the 1886 earthquake struck.

We noted previously that craters having ages of 600 and 1250 years extend along the coast at least as far as craters of the 1886 earthquake extend. A comparison of the size (diameter) of the craters shows that those formed 600 and 1250 years ago are larger than the 1886 craters, both in the vicinity of Charleston and far away. Considering all of these factors suggests that these prehistoric earthquakes were at least on the order of the M_w 7.0–7.2 event in 1886.

Paleoliquefaction evidence for the event that took place 3200 years ago has been found only in the vicinity of Charleston. The existence of abundant, exceptionally large craters for this event might suggest that the earthquake was exceptionally large, but the limited size of the affected area suggests otherwise. The absence of craters far from Charleston might be explained alternatively by a lower sea level and thus a lower water table level, and by a generally drier climate during this part of the Holocene (Amick *et al.*, 1990). Absence of the 3200-year-old craters far from Charleston might also be explained by an exceptionally shallow earthquake, in which energy attenuates rapidly within a short distance. For craters having an age of 5000 years or older, there is a greatly diminished chance for preservation of organic material that can be dated with accuracy, so it is difficult to evaluate the magnitude of such old events.

Some of the craters far to the north of the Charleston area, in the vicinity of the South Carolina–North Carolina border (Figure 7.8), have ages different from those of craters to the south. This difference suggests another epicentral region in the vicinity of the state boundary.

7.4.2   New Madrid Seismic Zone

The 1811–1812 sequence of earthquakes in the central United States consisted of four very strong earthquakes ($M_w \sim 7$–8) within a 3-month period. Six aftershocks had magnitudes of 6–7 (Hamilton and Johnston, 1990). Epicenters of the strongest 1811–1812 earthquakes probably were distributed along a fault zone exceeding 100 km in length (McKeown *et al.*, 1990). These epicenters, in combination with continuing seismicity, define the New Madrid seismic zone (Figure 7.13).

The meizoseismal region for the 1811–1812 earthquakes was centered in a huge area of alluvial lowlands. Prominent effects of liquefaction extended over an area of 10,000 km^2 in the lowlands and plainly were visible on the ground in recent years (Figures 7.14 and 7.15). Large areas have more than 25% of the surface covered with vented sand more than a meter thick (Obermeier, 1989).

The alluvial lowlands is an area of very low relief, thick strata of fine and medium sand at shallow depth, a very high water table, and a clay-rich cap. The sand strata generally are moderately compact. The lowland is made up largely of late Wisconsinan braid-bar terraces that formed in floods of glacial meltwater carrying great quantities of sand. Thickness of sand beneath the terraces generally exceeds 30 m, and at most places the sands are capped with clay-rich strata interbedded with thin sand and silt strata having a total cap thickness of a few meters. Much of the alluvial lowland is also occupied by large areas of Mississippi River meander-belt deposits, which were laid down during Holocene time as insets into the braid-bar terraces. Most of the meander-belt deposits consist of point-bar accretion topography of arcuate ridges and swales, abandoned channels, and natural levees. Many abandoned channels are filled with as much as 30 m of soft clay and silt. A cap of montmorillonite-rich clay at least a few meters thick lies on meander-belt sediments at most places. Overall, the alluvial lowland region is quite susceptible to the formation of earthquake-induced liquefaction effects during strong shaking.

Reports made shortly after the 1811–1812 earthquakes noted great multitudes of sand blows, linear fissures as deep as 6 m and hundreds of meters long, craters many meters in diameter, and lateral spreads as long as hundreds of meters (Penick, 1976). Individual and coalesced sand blows and some long linear

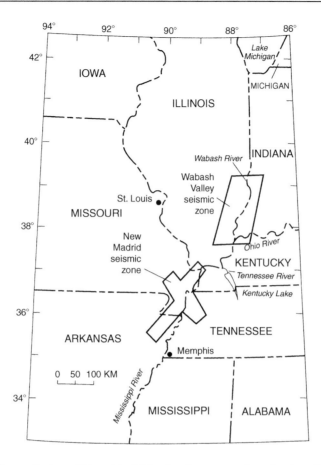

Figure 7.13: Approximate limits of New Madrid seismic zone and Wabash Valley seismic zone. New Madrid seismic zone is the source area of the 1811–1812 earthquakes and continues to have many small earthquakes and a few slightly damaging earthquakes. The Wabash Valley seismic zone is a weakly defined zone of seismicity having infrequent small to slightly damaging earthquakes.

fissures through which sand vented are the only features that are still readily visible on the ground surface. Great numbers of intruded dikes and sills can be seen in walls of deep (>3–4-m) drainage ditches.

Also within the 1811–1812 earthquakes meizoseismal region are many sedimentary features of unknown or nonseismic origin. Mainly, the features formed as nonseismic sand boils, mima mounds, or deformed mud.

7.4.2.1 Characteristics of Venting and Fracturing at the Ground Surface

Even though sand that was vented to the surface by the earthquakes of 1811–1812 is still visible today, most evidence of venting has been obliterated by agricultural practices. However, even small features were abundant at the time of a field study by Fuller (1912). Individual sand blows induced by the 1811–1812 earthquakes typically are dome-like accumulations of clean sand on the ground surface. Fuller (1912, p. 79) noted that "the normal blow is a patch of sand nearly circular in shape, from 8 to 15 feet

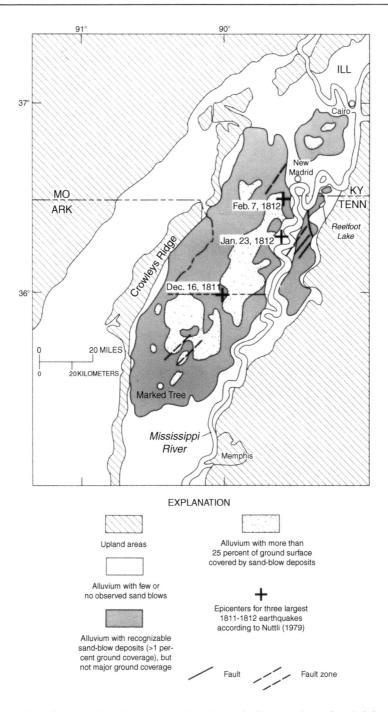

Figure 7.14: Regions having abundant vented sand, excluding modern floodplains, in the New Madrid seismic zone (from Obermeier *et al.*, 1990). Sand was presumably vented in response to the 1811–1812 earthquakes. Severe liquefaction also occurred locally beyond the areas shown on the map, especially along streams west of Crowley's Ridge (Fuller, 1912). Also shown are the approximate epicenters for the three strongest 1811–1812 earthquakes and major faults and fault zones.

Figure 7.15: (Continued)

0 ⊢─────────────────────────────┤ 1 MILE
0 ⊢──────────────┤ 1 KILOMETER

Figure 7.15: Aerial photos that show long fissures (dikes) through which sand vented (light-colored linear features) and also individual sand blows (light-colored spots) formed by liquefaction during the 1811–1812 New Madrid earthquakes. (A) A portion of the Manila, Arkansas, 7.5-min ortho-photo quadrangle. Fissuring and venting took place in braid-bar deposits of latest Pleistocene age and in younger Holocene point-bar sediments. Note how fissures formed parallel to the scrolls of point-bar deposits. Note also the abundance of fissures in the upper part of the right side of the photo. These fissures of lateral spreading origin have formed near a break in slope, where the terrace that is underlain by braid-bar deposits is adjacent to the slightly lower floodplain level of point-bar deposits. Isolated sand blows in upper left formed by hydraulic fracturing. (B) Aerial photo taken about 40-km north of part (A). Extensive fissuring and venting in braid-bar deposits of latest Pleistocene age. Cap thickness (about 6 m) is relatively uniform. Topographic relief is about 1 m throughout the region. Note severe fissuring over a width of at least 3 km. Long fissures probably formed by surface oscillations, because relief seems too low for lateral spreading.

across, and 3 to 6 inches high." Such small sand blows as Fuller described can rarely be found today. The sand blows now visible range from about 0.3 to 0.7 m in height at the center and thin to a feather edge at a distance of 5–20 m from the center.

Sand vented to the surface by the 1811–1812 earthquakes is obvious on aerial photographs, where the sand vented onto the dark clay-rich cap. The vented sand dries more rapidly than clay during seasonal drying, making a tonal contrast on the photograph. Figure 7.15A and B illustrates the contrast and also illustrates how venting has taken place at irregular and somewhat erratically spaced intervals. The photographs also show that extensive venting took place through approximately linear dikes that are more-or-less parallel. These patterns of erratic spacing and parallelism generally reflect small differences in site characteristics. One of the most obvious is cap thickness. Typically, venting in point-bar deposits has taken place along the highest, thinnest part of the meander scroll. Where the cap is thicker, sand blows tend to be less abundant but larger. The wider spacing between dikes apparently causes more concentrated flow to the surface.

Long, linear dikes, commonly with exceptionally large quantities of vented sand, also tend to develop parallel to topographic declivities along streams and scarps. Dikes here have formed in response to lateral spreading movements, which generally take place more readily near the declivities (Figure 7.15A). Wide (>1-m) dikes having lengths of many hundreds of meters are not unusual. While the widest dikes tend to be close to the declivities, they also may develop many hundreds of meters away. Fuller (1912, p. 49), for example, states:

> In the sand-blow districts the spacing of (lateral spreading) fissures varies from several hundred feet down to less than 10 feet. ... In the case of the large (several meters wide) fault-block (lateral spread) fissures the spacing is greater, several hundred feet often intervening between the cracks, while the space between them may be half a mile or more. Isolated cracks of this type are not uncommon.

The direction of shaking during the earthquake probably had a very secondary influence on orientation of the largest dikes at most places in the meizoseismal region of the 1811–1812 earthquakes (Obermeier, 1989). It is the local geologic–topographic setting that is of predominant influence. Cap thickness and proximity to streambanks and abandoned meanders are most important. Such important influence of the local setting has also been shown in a report about the 1964 Alaskan earthquake (M_w 9.2) by McCulloch and Bonilla (1970, Figure 46) and is emphasized by Oldham (1899) for the great earthquake of 1897 in India, so the observed effects in the meizoseismal region of the 1811–1812 earthquakes seem typical. Still, localized extension and compression of the ground surface, which may relate to the direction of strong surface shaking, may be relatively common at places (Oldham, 1899, p. 99; Obermeier *et al.*, 2005).

The aerial photograph in Figure 7.15 indicates enough randomness in dike orientation that most vertical exposures will intersect many dikes. This is especially relevant because searches for paleoliquefaction features are often made in banks of ditches or rivers, which may not be oriented optimally to intersect dikes.

7.4.2.2 Characteristics of Sand Blows and Dikes in Sectional View

Most sand blows of the 1811–1812 earthquake have a well-defined set of internal relations and stratigraphy, shown in a somewhat idealized version in Figure 7.16. The main *feeder dike* occurs beneath the central part of the dome. The basal few centimeters of sediment that vented onto the original ground

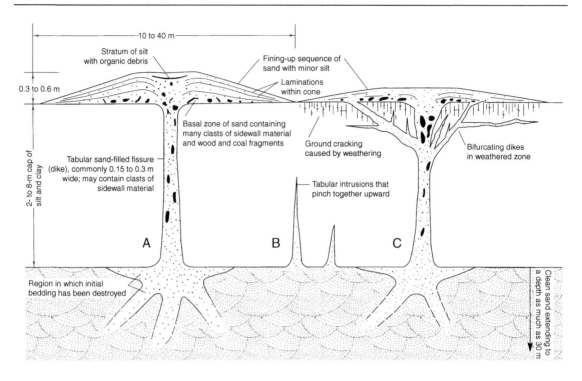

Figure 7.16: Schematic vertical section showing dikes cutting through overbank silt and clay strata and the overlying sand-blow deposits. (A) Stratigraphic details of sediment vented to the surface. (B) Dikes that pinch together as they ascend. (C) Characteristics of dikes in fractured zone of weathering, in highly plastic clays. Situations shown are encountered in many places in meizoseismal zone of the 1811–1812 New Madrid earthquakes.

are generally a fine to medium sand with a slight to moderate amount of silt, containing scattered centimeter-long round to irregular clasts derived from the underlying clay-rich strata cut by the dike. Sediment along the basal few centimeters grades up within a few centimeters to coarser sand with less silt containing numerous irregular 1–5-cm-long clay-rich clasts. The clasts are encased in clean, medium- to coarse-grained sand. The clasts are largest and most plentiful near the feeder dike. The basal part of the sand-blow deposit also contains numerous 1–3-cm-long rounded lignite fragments and other low-density materials vented to the surface. These low-density materials originated from the source sand. Clasts from the cap occur almost exclusively in the lowermost quarter of vented material. Above this in the larger sand blows, away from the vent, is a much thicker zone (tens of centimeters) of very clean, generally medium- to coarse-grained sand that is nearly structureless except for rather muted sediment laminations. Higher yet, the sand grades upward to mainly medium-grained. Here there are weakly to moderately developed planar to wavy laminations of fines and sand generally a few millimeters in thickness, which gently dip down and away from the central part of the sand volcano (Figure 7.16A). Where sediment vented into swales on the ground surface, this sequence may be capped by a silt-rich stratum with organic debris, 0.5 to several centimeters thick; this cap may also contain multiple very thin (1-mm-thick) clay-rich layers (Saucier, 1989). The organic matter in the cap consists of small pieces of charcoal and wood. The organic debris and thin clay-rich layers formed in swales located both above the vent (Figure 7.16A) and in depressions far from the mound of vented sediment.

Closest to where dikes vented onto the ground surface, are well-defined strata that dip steeply into the dike (Figure 7.16A). These strata typically contain the coarsest sediment vented. Next to these strata, in the lower and central part of the sand blow, there may be evidence of shearing and disruption caused by the forceful expulsion and boiling of sediment and water.

The overall *upward-fining sequence* of vented sediment, from the basal clast-bearing sand to the uppermost organic matter-bearing stratum, represents the transition from the turbulent violent eruption very shortly after initial venting to the final ebbing flow to the ground surface. The planar and wavy laminations probably represent weak variations in flow from the dike.

Many of the larger sand blows have more than one vertical sequence of the vented sediment, as illustrated in Figure 7.16, with no intervening pedological development between sequences. This has been interpreted by Saucier (1989) to represent more than one discrete episode of venting, closely spaced in time. Such an interpretation accords with the historic observation that there were at three very strong earthquakes (M_w 7–8) that took place in the region in the years 1811–1812, and that the epicenters of all three were relatively close to one another, within tens to a hundred kilometers.

Dikes that formed in the clay-rich cap of the meizoseismal region of the 1811–1812 earthquakes typically are sand-filled fissures that are steeply dipping (60–90°) and planar in plan view. In vertical section, dikes having widths exceeding several centimeters commonly are spaced from several meters to hundreds of meters apart. Smaller dikes often occur at much closer spacing.

Dike widths range from millimeters to several meters. Many of the widest "dikes" are sand-filled fissures that were almost certainly caused by lateral spreading. Figure 7.17 shows a feature probably having such

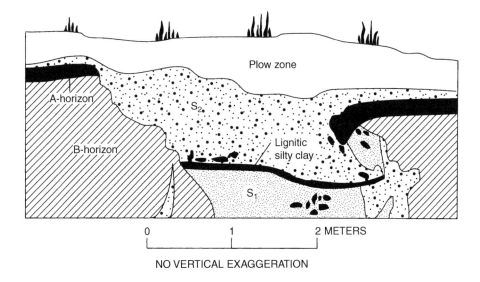

Figure 7.17: Sketch of vertical exposure in a ditch in the meizoseismal zone of the 1811–1812 New Madrid earthquakes showing evidence for about 1.5 m of lateral spreading. The S_1 sand was emplaced during lateral spreading. The lignitic, silty clay layer next was laid down on S_1 sand. Later, S_2 sand was vented to the surface, burying the lignitic silty clay and S_1 sand. Note that the sidewalls are approximately parallel. From Wesnousky and Leffler (1992); reprinted with permission of the Seismological Society of America.

an origin; the lignitic, silty clay stratum in the sand indicates the location of the top of the sand (S_1) that flowed into the opening of the lateral spread. The upper sand (S_2) probably was vented later during a following earthquake. (In contrast, often little or no sand vents to the ground surface from the large space between blocks opened by lateral spreading.) The tendency of sidewalls of many of the larger sand dikes throughout the 1811–1812 earthquakes' meizoseismal zone to be parallel to one another in vertical section also indicates a pulling-apart origin during lateral spreading.

Dikes 15 cm or less in width are very common. Dikes in this width range normally narrow upward as illustrated in Figure 7.16A. The tapering may represent downwarping of the ground surface in response to sand at depth having been vented to the surface. At almost all places, even isolated sand blows have vented through small, vertically planar dikes. The smallest dikes pinch together as illustrated in Figure 7.16B.

Within the uppermost meter or so, dikes intruding the weathered portion of the clay-rich cap may branch irregularly upward into many smaller segments (Figure 7.16C). Possibly, pre-existing planes of weakness related to weathering processes cause a single large dike to branch into many small members. The clay cap typically contains a large percentage of montmorillonite. During dry years, desiccation cracks extend a meter or more in depth. Pedogenesis has also developed a strong soil structure (pedons) in a thick B horizon near the surface.

Where larger dikes branch extensively (Figure 7.16C), there may be only minor evidence of venting onto the ground surface. Apparently, this network is effective in dissipating the energy of the flow. Locally though, venting has excavated the highly fractured zone, leaving behind a widened dike at the top. This excavated zone may contain many clay clasts mixed irregularly in a matrix of sand.

Dikes that cut through the fine-grained cap generally are filled with a loose mixture of fine and medium sand and a minor amount of silt. Clasts of clay, some as long as 20 cm, may also occur but generally are not abundant. Elongate clasts tend to be parallel to sidewalls. The clasts were derived from the sidewalls and transported up the dike. The mixture of sand, silt, and clasts has a sharply defined contact with the sidewalls. Weak laminations within the sand and silt may parallel the sidewalls. Crosscutting, steeply to vertically oriented zones of sand and silt within the dike are also commonplace. These probably represent episodes of venting during separate pulses or venting from different source zones at depth.

The finest constituents (fine sand and silt) have been winnowed from dike fillings at some sites of venting. *Winnowed zones* within the dike are commonly tubular and as much as several centimeters in width. Locally, winnowing extends several meters down into the dike. This winnowing probably took place by water flowing up through the dike during final phases of water expulsion, following initial emplacement of sand in the dike.

Many dikes that pinch together upward have a large proportion of silt and clay mixed with the sand near the top. Often, it is unclear whether the silt and clay have invaded the dike by pedogenesis or whether the silt and clay were constituents of the originally intruded sediment. Dikes that taper upward (without reaching the ground surface) are generally of limited usefulness for paleoliquefaction studies, owing to difficulties in determining when the dikes formed.

Many variations of the relations shown in Figure 7.16 exist in various field settings. One of the most common variations is a large amount of *downwarping* of the cap toward the dike. This downwarping tends to be most pronounced where a large amount of sand has vented to the surface. It is not unusual that the cap be downdropped by more than 0.5 m on one or both sides of the dike and that the cap be otherwise faulted or severely deformed near the dike (e.g., Wesnousky and Leffler, 1992, Figures 9, 12, and 14).

Scattered small tubular dikes can also be found in clay caps of the 1811–1812 earthquakes' meizoseismal region, especially in thin very soft strata immediately beneath the cap. Also, holes that originated from decay of tree roots or from excavation by crayfish are ubiquitous and doubtlessly were used as the conduits for small tubular dikes. These tubular holes through the cap had a very minor role as conduits for venting, though, as compared to steeply dipping planar dikes. However, these holes possibly were preferred paths during the early phase of venting, and thereby controlled where hydraulic fracturing developed tabular dikes. Small holes with walls defined by angular breaks, and which have a tortuous upward path, are also commonly observed conduits for sand venting during the 1811–1812 earthquakes. Similarly, Audemard and de Santis (1991) report another field example, in Venezuela, where venting was localized in tubular (crab) burrows; this venting occurred in response to limited liquefaction during moderate earthquakes.

Beneath the fine-grained cap, the dikes typically extend steeply downward into liquefied sand strata. For an isolated sand blow in a clay-rich cap, these dikes in the sand are typically tubular in plan view. Where shaking and liquefaction have been severe, resulting in many tabular dikes cutting through the cap and significant venting of sand, the dikes beneath the cap can be either tabular or tubular in plan view.

The preceding discussion has focused on characteristics of dikes that cut through a clay-rich cap that varies in consistency from very soft to brittle. Sims and Garvin (1995) present excellent descriptions and detailed drawings of tabular dikes that cut interbedded clay and sand strata during the 1989 Loma Prieta, California, earthquake (M_w 7). The dikes and sand blows that they describe generally are much smaller and represent less forceful venting than those of the 1811–1812 earthquakes, yet many of the sediment relations in the dikes and sand blows in both regions are quite similar.

Locally, the cap in the 1811–1812 earthquakes' meizoseismal region is a very weakly cemented sand containing only a minor amount of silt and clay, with some slight bonding from oxides of iron and manganese. Liquefaction features in these areas appear to have been mainly large open craters, similar formed during the 1886 Charleston, South Carolina, earthquake. In contrast to the filled craters in South Carolina, very large clasts of host sediment generally are not present in the craters in the New Madrid region. Apparently, the host sand typically is too friable to form large clasts. Obermeier (1998a, Slides 57 and 58) illustrates an example of a filled crater with clasts from the weakly cemented cap in the New Madrid region.

7.4.2.3 Characteristics of Sills in Sectional View

Combinations of dikes and sills are also common within a nonliquefiable cap. Where sills are abundant, dikes may also be plentiful. Sills form preferentially at three locales (1) along the base of the cap; (2) along bedding and other horizontal planes of weakness in the cap; and (3) beneath dense, strong root mats. These locales are illustrated in Figure 7.18.

Laterally extensive sills as thick as 0.5 m are commonplace along the base of the cap, and exceptionally are as thick as a meter. An intruding sill can dome an overlying flexible, clay-rich cap having a thickness of as much as a few meters. Sills are also common within the cap where original horizontal sedimentary structures and planes of weakness have not been destroyed by weathering. Sills especially tend to form irregularly within thin beds of silt or sand sandwiched between clay-rich beds. Small branches from the main sill commonly intrude into the overlying more clayey stratum, forming more sills and dikes. A thin (less than 0.5 m) clayey stratum above the sill can be shattered for meters into irregular clasts, 1–10-cm across. Clay-rich clasts can abound within the sill. The clasts can be transported many meters horizontally, but clasts that detach and founder vertically several centimeters are also common

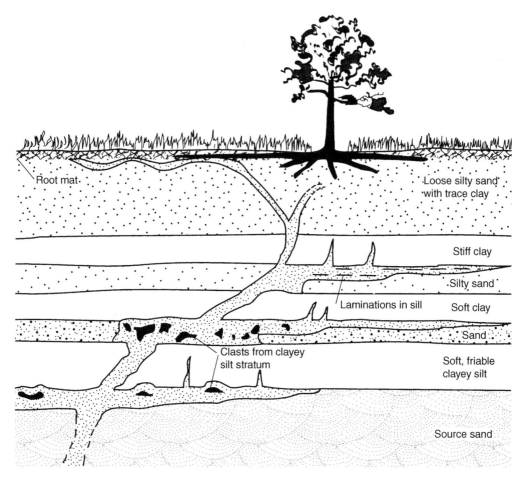

Figure 7.18: Schematic vertical section showing where sills form preferentially. Note that thin sills can extend great distances horizontally, especially where the overlying cap is thin. Such severe sill development as shown in the figure is typically accompanied by large sand blows.

(see Figure 7.18). The generally angular shape of clay-rich clasts in the sills indicates brittle fracturing of the host stratum. Despite such fracturing, clasts and their hosts commonly are so soft as to easily permit several centimeters of penetration by thumb pressure.

Sills at a depth of about a meter or less in a clay-rich cap can be quite wavy in vertical section (Obermeier, 1998a, Slide 23). These sills can thin and thicken dramatically within a horizontal distance of a meter or less and produce blister-like bulges on the surface. In plan view, these bulges can range from circular to very elongate. Sills can be as thick as 0.7 m near the surface. Such large thicknesses are less common at greater depths, except locally along the base of the cap. Sills very close to the ground surface generally seem to have formed beneath root mats.

Figure 7.19A shows a commonly observed field example in which near-vertical dikes are connected to a laterally extensive sill. The sill has formed within a clay cap that has only incipient horizons of weakness.

Figure 7.19: (Continued)

Figure 7.19: Sand dikes and sills exposed in a nonliquefiable cap of silt and clay; the dikes and sills are interpreted as having originated by liquefaction during the 1811–1812 New Madrid earthquakes. Outcrop is the bank of a ditch in the meizoseismal area of the 1811–1812 earthquakes. (A) Overview and line drawing of typical small dikes and small sills. The sill extends far beyond the photograph and is at least 25-m long. Rectangle shows area of part (B). (B) View showing details of layering in sills. (C) Very close view showing details of layering in sill. Sill consists of fine- and medium-grained sand with some silt- and sand-sized lignite. Black bands are laminae of small pieces of lignite.

The sill is at least 25-m long but is only 10-cm thick. The sill essentially follows a single horizon in sectional view. The type of internal layering seen in this sill is commonplace. Individual laminae are composed of small pieces of lignite or fine-grained sand and silt (Figure 7.19B and C). Also common at this site and elsewhere are structureless sills containing many *clay clasts* in a sand matrix, as well as sand with graded bedding in which the larger clay clasts are concentrated along the base.

Laterally extensive sills also tend to form in the upslope direction along the base of a cap, where the base dips appreciably. Such sills occur preferentially along the base of a clay-filled abandoned channel. In this case, a sill typically extends upslope to where the cap is thinner, where the sill feeds into a steeply dipping dike that vents onto the surface.

The exposure illustrated in Figure 7.20 shows another common field example where sand dikes and sills cut the upper liquefied sands and the lower portion of the nonliquefiable cap. In the upper part of the liquefied sand (bed A, Figure 7.20A), small dikes branch out from a large dike (feature 1), cut through the basal bed of the cap (bed B) at horizontal intervals of 0.5–1 m, and extend upward about 0.5–1 m. A few dikes and sills intrude to much higher elevations (features 2–4). At exposures nearby (not shown), dikes extend to the surface. Sand has vented to the surface to produce many large sand blows in the field adjoining the outcrop of Figure 7.20.

The dikes and sills shown in Figure 7.20A as features 1–4 generally contain clean, medium-grained sand. The edges of the intrusions are sharp in clay-rich beds such as beds C, D, and G. Locally, intrusions fracture the very soft clay with sharp angular turns and breaks that follow a haphazard path (features 2–3). Edges of intrusions are generally less distinct in beds of clean, permeable sand. Dikes commonly widen and terminate upward as flame-shaped structures (feature 4) in a permeable sand bed (bed H). Clay clasts may be present in lower portions of the structures.

Figure 7.20D shows at left-center a small dike having an irregular, more-or-less circular shape in plan view; the sediment within the dike is more silt-rich than in other dikes of feature 1. In vertical view, the irregular dike has fed into a feature that is essentially bulbous in shape, with an overall shape in three dimensions much like that of a mushroom. Within the meizoseismal region, such small intrusions are relatively commonplace near the base of the cap, where the basal portion of the cap is a soft sandy silt or very silty sand. The diameter of the dike feeding into the bulbous portion is typically less than a centimeter, and the sediment within it is generally very rich in silt with a slight amount of highly plastic clay (montmorillonite). Nearby there are commonly large sand blows and dikes, made up of clean medium-sized sand. Thus, the presence of the small bulbous intrusions almost certainly reflects a relatively low ability of the silt-rich sediment that liquefied to form the bulbous intrusion (perhaps because of its low liquefaction susceptibility or a low permeability), rather than a low strength of seismic shaking.

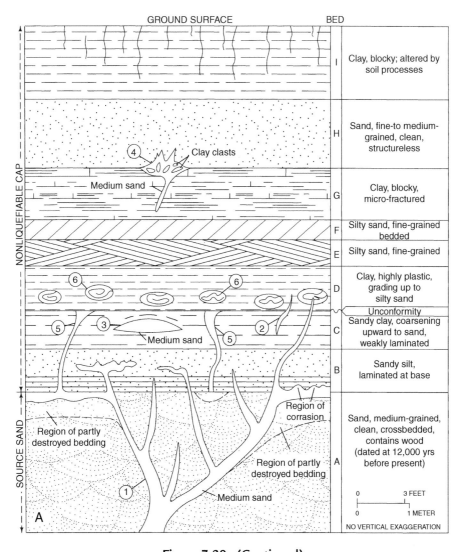

Figure 7.20: (Continued)

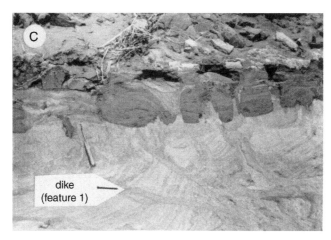

Figure 7.20: (Continued)

Figure 7.20: Schematic vertical section showing Holocene sediments (nonliquefiable cap and underlying Wisconsinan age source sand that liquefied). Exposure is in a ditch in the meizoseismal region of the 1811–1812 New Madrid earthquakes. Earthquake-induced intrusions cut section at many places. (A) Schematic diagram of stratigraphic relations and liquefaction-induced features (numbered 1–5). Feature 1, dikes of medium-grained sand that cut cap and source sand; features 2 and 3, intruded dikes and sills of massive, clean, medium-grained sand; feature 4, dike and flame structures of medium-grained sand containing and large clasts from bed G; feature 5, dikes of medium-grained sand, truncated unconformably; feature 6, pseudonodules collapsed in bed D. (B) Photograph of beds A–D showing sand intrusion (feature 3). Knife is 12-cm long (B–E). (C) Photograph of dikes (feature 1) cutting source (bed A) and bed B. (D) Photograph of plan view of bed B showing intrusion structures caused by feature 1; the plan view is in the area of the knife that is oriented vertically in part (B). (E) Photograph of part of feature 4, showing clay clasts in sand matrix intruded into bed H.

The sill at the base of bed B, at the extreme right-hand side of Figure 7.20, has an irregular contact along the top. Here the sill has *corraded* the base of the friable bed B. Small intact pieces of bed B have sunk into the sill, probably attesting to a very water-rich condition in the sill at the time of its intrusion. Such destruction of friable beds by sills is commonplace throughout the meizoseismal region, especially along the base of the cap.

In regions of only marginal development of small dikes, sills along the base of the cap and within the cap are commonly observed at many places, and where present may help to serve as independent evidence for a seismic liquefaction origin. Even beyond the region of small dikes, small sills are often found.

Features labeled 1–4 in Figure 7.20 are interpreted to be earthquake-induced because (1) they are widely distributed over tens of kilometers; (2) they contain dikes and sills commonly as wide as 15 cm that suggest intrusion by large volumes of water-saturated sediment; (3) they contain clean, medium-sized sand containing large angular clay clasts (which is evidence of forceful intrusion); and (4) artesian conditions and aseismic landslides are unlikely at these sites.

The dike shown as feature 5 has an uncertain origin (Figure 7.20A). Here, three small dikes that were truncated at the contact of beds C and D were exposed in a 25-m section along the ditch but were not found in other nearby exposures of beds C and D. The dikes may represent seismic liquefaction that occurred prior to the 1811–1812 earthquakes. However, they contain a large percentage of silty fine-grained sand, which does not suggest forceful intrusion at this site. Possibly, they resulted from springs that formed near the base of a streambank or as slump-related features that formed soon after initial deposition of the host sediments. Feature 6 (pseudonodules) is discussed in a later section.

Flame-shaped structures such as those of feature 4 have developed at many places within thick sand beds in the meizoseismal region of the 1811–1812 earthquakes. Flame-shaped structures commonly have widths ranging from millimeters to about a third of a meter. Apparently the upper sand beds do not liquefy and, because of their relatively high permeability or higher density, perhaps in combination with an unsaturated condition, permit the pressurized water from beneath to dissipate within the large volume.

A rather uncommon deformation feature (not shown here), but probably related to seismic liquefaction, involves plastic deformation of silt and clay along the base of the cap. In this case, the basal 10–20 cm of the cap contains a convoluted mixture of severely disturbed, plastically deformed silt and clay, intruded by sand. Such *convolutions* probably take place only where extremely soft silt and clay lie directly on liquefied sand. An outstanding example of such convolutions of almost certain seismic shaking origin from the region of the New Madrid earthquakes is shown in Obermeier (1998a, Slide 63). Overall, it is generally very difficult to definitively assign a seismic origin to convolutions in fluvial terraces with shallow-water tables, such as those of the New Madrid region, because of the widespread presence of small, localized artesian pressures in swales that occur almost annually. In contrast to plastically deformed convolutions of nonseismic origin, almost all seismically induced liquefaction features in the region have sharply defined tabular breaks in plan view, which have been filled from beneath with fluidized sediment.

7.4.2.4 Paleoliquefaction Studies

Systematic paleoliquefaction studies have been undertaken by many researchers during the past 15 years throughout the very large meizoseismal region of the 1811–1812 earthquakes, which extends a few hundred kilometers from north to south (Figure 7.14). As mentioned previously, this region is comprised almost completely of large terraces of late Wisconsinan and early Holocene age where the water table appears to have been very shallow since the terraces were formed (Wesnousky and Leffler, 1992, 1994). Thus, if very strong earthquakes occurred since early Holocene time, liquefaction features should be present in the geologic record. Definitive evidence for liquefaction predating 1811–1812 was first found in the northern part of the New Madrid seismic zone (see Figures 7.13 and 7.14) near Reelfoot Lake (Russ, 1979) and at a site nearby (Saucier, 1991b). Russ (1979) found that three earthquakes have induced liquefaction during the past 2000 years and, on that basis, suggested a recurrence interval of about 600 years for liquefaction-producing events. Saucier (1991b) estimated an average recurrence interval of 470 years for liquefaction-producing events in the past 1300 years. Later extensive studies by Tuttle (1999, 2001) and Tuttle *et al.* (2002a, 2005) have shown that at least six liquefaction-producing paleoearthquakes have struck in the region during the past 4500 years; and at least two are inferred to be comparable in strength to the M_w 7–8 earthquakes of 1811–1812 on the basis of sizes of sand blows and the regional extent over which the features formed. Tuttle *et al.* (2005) also confirmed the findings of previous researchers regarding return period for the large paleoearthquakes during the past 1500 or so years. The large sizes of the many of the older paleoliquefaction features suggest earthquake magnitude(s) much larger than the minimum required to trigger liquefaction. Regionally, this threshold magnitude is about M_w 5.6 based on historical observations of liquefaction-producing events in the terraces of the New Madrid seismic zone (Pond, 1996; Olson *et al.*, 2005b).

7.4.3 Wabash Valley Seismic Zone

Many small to slightly damaging earthquakes have occurred throughout the region of the Wabash River valley of Indiana and Illinois in the past 200 years. Seismologists have long suspected that the weakly defined Wabash Valley seismic zone (Figure 7.13) could be capable of producing earthquakes stronger than the largest of record ($M_w \sim 5.8$). Those suspicions have been confirmed by a search for paleoliquefaction features, which has resulted in discovery of at least seven large prehistoric, Holocene earthquakes.

Alluvium commonly as thick as 10–30 m overlies bedrock throughout much of the study area. The main search region, the Wabash River valley, contains expanses of low glaciofluvial terraces of late Pleistocene age. These terraces are mainly braid-bar deposits of gravel and gravelly sand. Inset into the Pleistocene terraces are slightly lower Holocene floodplains of finer point-bar sediment. The sand and gravel deposits of both braid bars and point bars typically are overlain by a 2–5-m-thick alluvial cap of sandy to clayey silt. Bordering the valley are extensive plains of silt and clay that contain patches of clean sand, laid down in slackwater areas during glaciofluvial alluviation. The water table is presently shallow (<3 m) and appears to have been shallow over large areas much of the time following glaciofluvial alluviation, on the basis of depth of carbonate leaching and B horizon soil development in sandy and silty alluvium. This combination of a relatively shallow-water table and widespread sand-rich deposits with an overlying fine-grained cap has provided an excellent opportunity for liquefaction features to form throughout much of the Holocene.

7.4.3.1 Field Techniques

A study of the field techniques used by Munson and Munson (1996) and secondarily by Obermeier (1998b) to distinguish the various paleoearthquakes in the region is highly recommended for those undertaking a paleoseismic study in fluvial deposits, where streams are actively cutting through the deposits. Their methods of data collection enhanced the utilization of radiocarbon, archeological, stratigraphic, pedological, and spatial data for age determination of the various earthquakes, and their methods were also conducive to locating epicentral areas.

In particular, the chief method of collecting data was by examination of actively eroding streambanks from a boat, which permitted viewing the features through much of their height. The streambanks were commonly 3–5-m high and clean of debris, thereby greatly facilitating a search for dateable materials. And viewing the liquefaction features in section permitted detailed measurements of dike location and width, as well as locations where no liquefaction features were present—all of which are important for the engineering back-analysis of the strength of paleoearthquake shaking (discussed later). This sectional view also permitted the investigators to identify dikes that pinched together and never vented to the ground surface.

The field methods described in Munson and Munson (1996) and Obermeier (1998b) were also extensively used for successful paleoliquefaction studies in the New Madrid seismic zone, noted in the preceding section. And these field methods were used to first verify seismic shaking from a great subduction earthquake that struck the Pacific Northwest coast of the U.S. in 1700 AD, discussed below.

7.4.3.2 Characteristics of Liquefaction Features

Hundreds of dikes were found in Holocene point-bar deposits and in late Pleistocene glacial outwash and slackwater deposits. Figure 7.21 shows the locations searched, sites where dikes were discovered, size (width) of widest dike at a site, and spatial limits of dikes from various earthquakes of different ages.

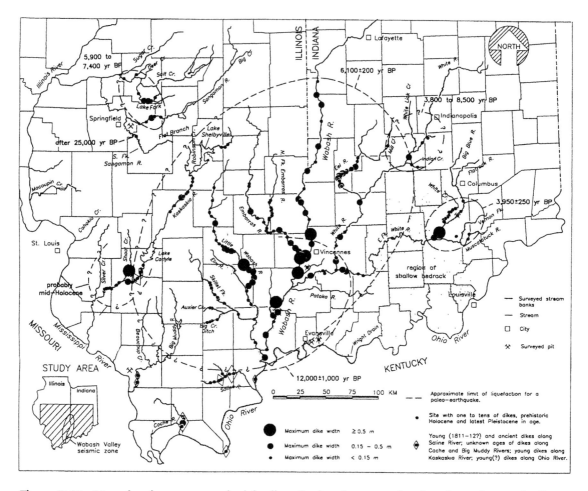

Figure 7.21: Map showing area searched for liquefaction features, showing sites where paleolique-faction features (mainly dikes) were discovered, and showing regional limits of liquefaction for different earthquakes. About 10% of the length of the rivers searched has freshly eroded exposures. Only exceptionally are there no fresh exposures of mid-Holocene or older sediments within a 20-km length of river; no suitable exposures are in the (dotted) region of shallow bedrock on the map. Liquefaction sites on the map generally denote exposures with numerous dikes. Some sites in the southern part of the study area have unweathered dikes near the surface, probably induced by the 1811–1812 New Madrid earthquakes. Dike widths on the figure were measured at least 1 m above the base of the dike. Sites with dikes having a width >0.7 m are shown for the 6100-year-BP earthquake. Note the core region of exceptionally large dikes around Vincennes. Modified from Munson *et al.* (1997) and Obermeier (1998b).

Virtually all sites shown on the figure have more than one dike, and many sites have more than 10. Almost all the liquefaction sites are in actively eroding banks of rivers, which were about the only areas searched. The dikes are steeply dipping, tabular, and typically connect at depth to a sand to gravelly sand source beds, some with significant silt but virtually no clay. Many smaller dikes pinch together as they intrude the overlying silty cap. Dikes filled with sand containing some gravel and silt are very common. While mainly sand was vented, large quantities of vented gravel also occurred commonly (Figure 7.22). Many dikes contain an upward-fining sequence of coarsest material. Also within many dikes are sharply defined, vertically intertwined and intersecting zones containing distinctly different grain sizes, which in places can be traced to different source strata at depth. Where dikes cut through a thick clay cap without pronounced horizontal planes of weakness, it is not unusual that the dike width does not exceed a centimeter throughout a height of as much as 4–5 m. But, widths as much as 15 cm or more throughout the height of the dike are very common and widespread in the region. Locally, though, in the meizoseismal regions for the various paleoearthquakes, dikes can be much wider and have widths of as much as 0.7–2.5 m.

Sediment vented to the surface extends as much as 40 m in diameter. Thicknesses of vented sediment of 0.15–0.2 m are not unusual. Most vented sediment fines upward and laterally, especially if gravel was vented. The vented sediment at sites bordering the Wabash River generally lies on a paleosol and is buried beneath a 1–3-m thickness of overbank deposits, as illustrated in Figure 7.22. On slightly elevated terraces, where flooding has been rare, the vented sediment has been incorporated into the surface soil.

At some sites, there is good evidence for more than one pulse of venting. This evidence is best shown at sites near Vincennes (Figure 7.21), where both surface sand blows and their dikes are largest for the event of 6100 years BP. The vented sediment is manifest as two upward-fining sequences. The lower pulse of vented sand fines upward to a thin silt layer; the upper pulse has abundant sand and gravel that also fines upward. Here, there is no evidence of a significant hiatus between pulses.

The interpretation that sediments were vented from the dikes onto the ground surface rather than intruded as sills is based on several lines of evidence. Sills would cut irregularly across sedimentary horizons at some places (see Figures 7.18 and 7.19), rather than always being confined to a single horizontal layer as in the Wabash Valley. Sills also would tend to follow the contact between sand and clay strata rather than lie on a paleosol, and sills probably would have some dikes branching up.

7.4.3.3 Ages of Dikes and Epicentral Locations

Dating at widespread sites shows that at least seven large paleoearthquakes have struck between 3950 and 12,000 years BP. Many of the liquefaction features in Figure 7.21 resulted from a single earthquake 6100 ± 200 years BP (Munson and Munson, 1996). Very large dikes with extensive sand blows from this earthquake are centered near Vincennes, demonstrating a proximate epicenter.

The next strongest earthquake in the Wabash Valley took place about 12,000 years ago, but the paucity of exposures in Pleistocene deposits precludes determination of a regional pattern of dike sizes. Other smaller paleoearthquakes have also been identified, all located beyond the modern concentration of small earthquakes that define the Wabash Valley seismic zone (shown in Figures 7.13 and 7.21). In the southernmost part of the search area are a few small, essentially unweathered dikes very near the ground surface. These dikes clearly are very young, and because of their proximity to the New Madrid seismic zone are suspected to be from the New Madrid earthquakes of 1811–1812.

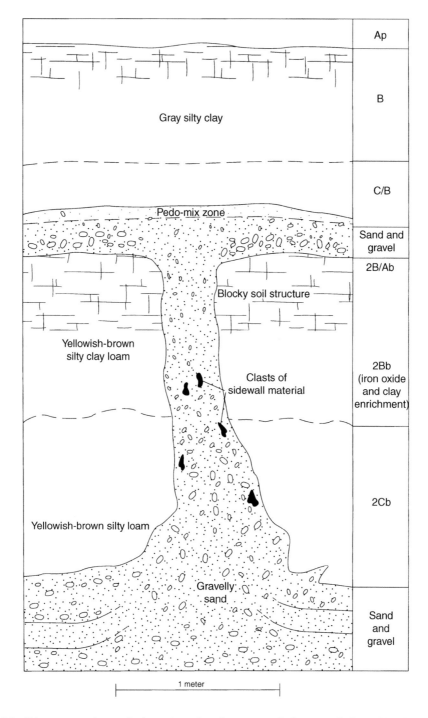

Figure 7.22: Diagrammatic vertical section showing general characteristics of buried sand- and gravel-filled dikes along the Wabash River. Source beds are Holocene point-bar deposits or late Wisconsinan age braid-bar deposits overlain by much finer overbank sediment. Sediment in source beds beneath dikes shows evidence of flow into dikes. Gravel content and size decrease upward in dikes at many places. The column on the right side of the figure contains pedological descriptions of host materials.

Epicentral locations of the paleoearthquakes (defined for this purpose as region of strongest seismic shaking, essentially analogous to macroseismic epicenters) were approximated from the regional pattern of dikes widths. It was found that using either maximum dike width or sum of dike widths, for fixed increments of exposure, worked equally well providing that conditions for amplification of shaking above bedrock were reasonably constant (Munson and Munson, 1996). The conceptual basis for this method is that the largest dikes develop from lateral spreading, and that the width of lateral spreads is essentially independent of thickness of fine-grained cap (Bartlett and Youd, 1992; Obermeier, 1998b). This independence arises because the horizontal forces involved in lateral spreading are much larger than the strength of the cap.

The assumption that the region of largest dike widths was also the region of strongest seismic shaking has been confirmed by engineering back-analysis of the strength of shaking, for the earthquake centered about Vincennes (Green *et al.*, 2005).

7.4.3.4 Evidence for Seismic Origin

All aspects of the Wabash Valley dikes can also be observed in the meizoseismal region of the 1811–1812 New Madrid earthquakes, which has a physical setting generally similar to that of the Wabash Valley. An earthquake-induced liquefaction origin is interpreted for the dikes of the Wabash Valley for the following reasons, considered in combination:

1. The dikes widen downward or have walls that are parallel (agreeing with a lateral spreading origin).
2. Dikes are approximately linear in plan view and exhibit strong parallel alignment in local areas.
3. The dikes vented large quantities of sandy sediment to the surface.
4. Material in the dikes fines upward and was transported upward.
5. Bedding in some source beds is homogeneous, and the contact of source beds with overlying fine-grained sediment is highly disturbed in some places.
6. Flow structures project upward from the source zones into the bottom of the dike.
7. Many dike sites are in flat and topographically elevated landforms, located at least several kilometers from any high, steep slopes that might have existed at the time the dikes formed, and therefore could not have been induced by nonseismic landsliding.
8. Other nonseismic mechanisms such as artesian springs that could produce similar features are not plausible at many dike sites because of the lack of topographic relief and the local geologic setting.
9. The size and abundance of the dikes along the Wabash River, the area where data are most complete, generally decreases with increasing distance from a core region of largest dikes (Figure 7.21).
10. Large regions in the same geological setting, with liquefiable sediment, have been searched far north of the Wabash Valley seismic zone and have no dikes.

Sills both within the fine-grained cap and beneath it are sparse in comparison to sill development in the New Madrid seismic zone, except for the earthquake of 12,000 years BP. For that paleoearthquake, the depth to water table was typically much shallower than for others in the Wabash Valley region, being at most a few meters in depth at many places, similar to the conditions at the time of the 1811–1812 in the New Madrid earthquakes.

7.4.3.5   Paleoseismic Implications

Historical earthquakes in the Wabash Valley, with magnitudes as high as $M_w \sim 5.8$, have not been reported to have caused liquefaction. Undoubtedly, the paleoearthquakes in the Wabash Valley far exceeded the magnitude of any historical events because of the large areal distribution of liquefaction effects and the large size of some of the dikes.

Two independent methods have been used to estimate the magnitude of the paleoearthquake centered near Vincennes, both yielding $M_w \sim 7.2$–7.5. The method used by Green *et al.* (2005) involved back-calculating the strength of shaking at many widespread sites, and determining which suite of back-calculated values agreed best with those predicted for various values of M_w. The method used by Olson *et al.* (2005b) yielded a best estimate of M_w 7.3. Both methods involved assumptions regarding seismological parameters, owing to the lack of seismological data on large earthquakes in the region. The method used by Green *et al.* (2005) required assumptions about ground motion amplification from bedrock to the surface, and the method used by Olson *et al.* (2005b) was based on a method utilizing the farthest distance of liquefaction features from the epicenter of the largest earthquake of the nearby 1811–1812 New Madrid earthquakes, of $M_w \sim 7.5$–8 (the current best estimate for these earthquakes). The fact that different assumptions were used by the two methods, yet estimates of the value of M_w were very close to one another yields significant confidence in the interpretation.

All smaller paleoearthquakes in the Wabash Valley region were almost certainly well in excess of M_w 6, based on the regional extent of their liquefaction effects (Olson *et al.*, 2005b).

7.4.4   Coastal Washington State

Tidal marshes buried in coastal Washington and nearby coastal Oregon record episodes of sudden submergence accompanied by tsunamis during late Holocene time, which have been ascribed to great ($M_w \sim 8$–9) earthquakes on the basis of the large region along the coast that appears to have submerged simultaneously (Atwater, 1987, 1992, 1996; Darienzo and Peterson, 1990; Nelson, 1992a; Rogers *et al.*, 1996). These inferred earthquakes are presumed to have originated by rupture along the thrust fault where the Juan de Fuca oceanic plate is being subducted beneath the North America continental plate, that is, the Cascadia subduction zone (Figure 7.23B). However, no direct evidence had been discovered to corroborate that seismic shaking accompanied the episodes of submergence and no strong Cascadia earthquake has occurred during the time of written history in the Pacific Northwest, some 200 years. Modern seismicity on the subduction zone is limited to scattered, small earthquakes, none with thrust mechanisms in the region of sudden submergence.

Atwater (1992, 1996) inferred at least two occurrences of coseismic subsidence when great earthquakes struck the coast of Washington, including the region around the Columbia River valley, during the past 2000 years. Strong evidence indicates that one event was about 300 years ago, and less widespread evidence suggests that another event occurred between 1400 and 1900 years ago. The portion of the thrust fault that ruptured and provided energy for seismic shaking was most likely a small distance offshore (a few tens of kilometers), on the basis of the location of the subsided zone (Atwater, 1987), heat-flow data (Hyndman and Wang, 1993), and strain data (Savage and Lisowski, 1991).

The inferred earthquakes would be expected to have caused such strong shaking as to have produced abundant liquefaction features near the coast, even in sediments having moderate-to-low susceptibility. To verify occurrence of strong shaking, I initiated a search for liquefaction features in cutbanks of islands in the Columbia River.

7.4.4.1 Columbia River Features

Many large islands were searched between the towns of Astoria and Portland (Figure 7.23). These islands originated as braid bars on a grand scale. The islands are flat, poorly drained, and swampy. Large portions are submerged during the highest tides. Strong currents and wave pounding are severely eroding many islands and as a result have sculpted clean, vertical banks as high as 2 m, which extend from water level to the top of the banks. Significant areas are also being cleaned in plan view by tides.

The banks of the islands between Astoria and Longview expose mainly soft clay-rich silt deposits (Figure 7.24). Age at the base of the exposed clay-rich cap is less than 1000 years and more than 600 years on most islands on the basis of radiocarbon ages of fossil marsh plants (genus *Scirpus*) found in growth position and now just above the level of low tide.

Regional stratigraphic control of sediments exposed on the islands is excellent. About 1.5 m below the top of the banks is a tan horizon with a thickness of a few centimeters. This horizon is exceptionally rich in volcanic ash. About 10–15-cm lower is a blue-gray horizon, generally several centimeters thick, also rich in ash. Very locally, rounded pumice clasts as large as 5 cm in diameter occur in the lower ash horizon. Chemical analysis shows that the ash and pumice have minerals and elements identical with those of an eruption from Mount St. Helens in AD 1480–1482 (C. D. Peterson, written communication, 1992). Therefore, the radiocarbon ages on fossil marsh plants and the ash data show that the sediments are old enough to record liquefaction associated with the 300-year-old downdropping event but probably are not old enough for the event of 1400–1900 years ago.

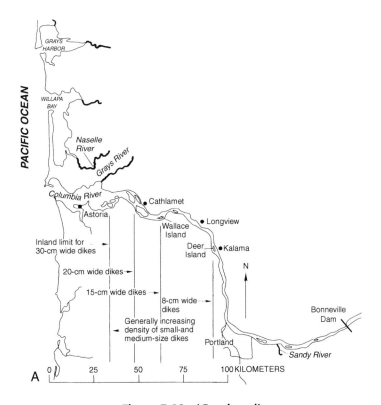

Figure 7.23: (Continued)

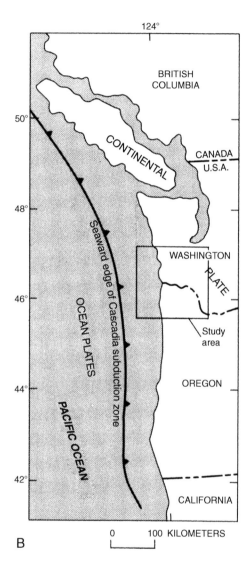

Figure 7.23: (A) Map showing that part of the Columbia River where banks of islands were searched for paleoliquefaction features. These islands have ages between 600 and 1000 years at most places. Sands beneath islands are fine to medium grained and generally are at least moderately susceptible to liquefaction. Maximum dike width is measured at least 1 m above the base of the dike. (B) Index maps shows Cascadia subduction zone.

At many places, sand is exposed immediately beneath the clay-rich cap. In general, alluvium that makes up the islands, generally thick, fine- to medium-grained sand, probably exceeds 100 m in thickness. Conditions on many islands are nearly ideal for the formation of large liquefaction-induced features. Not only is the cap thin, but the groundwater table has almost certainly been within a meter or so of the ground surface since the islands formed. The tidal range at these islands is about 2–2.5 m, and high tides inundate parts of the islands and doubtlessly have done so for at least several hundred years.

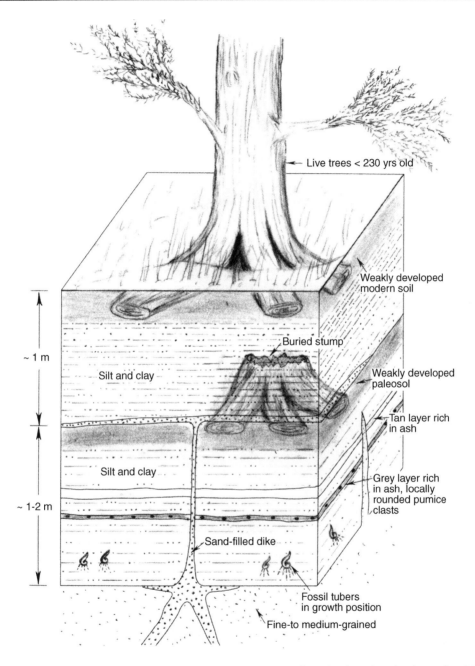

Figure 7.24: Block diagram showing typical field relations at liquefaction sites in the Columbia River islands. A sand-filled dike cuts through a 1–2-m thickness of soft silt and clay with a weakly developed soil at the top. The dike connects to a thin sand sheet on the soil horizon that is buried by a 1-m thickness of silt and clay. Tubers at widespread sites collected in their growth position near the base of the stratum cut by dikes have radiocarbon ages ranging between 600 and 1000 years. Widths of top of dikes are only several millimeters at most places.

Hundreds of dikes have been found along 9 km of vertical banks in scattered islands upstream as far as Deer Island (Figure 7.23A). At some places, the tabular nature of the dikes is exposed in plan view. Maximum dike widths and abundance of dikes tend to decrease in the upstream direction. Figure 7.24 illustrates relations observed at many islands. A thin sand sheet lies on a weakly developed, very soft soil that is about 1 m below the present surface. Locally, the upper few centimeters of the soil are contorted by small (centimeter-sized) folds and other soft-sediment deformations. The sand sheet is 1–4 cm in thickness and is as wide as 10 m. The sheet connects to a nearly vertical, narrow planar dike that widens downward markedly and connects to sand beneath the clayey cap. The width of the uppermost 5–15 cm of almost all dikes is only several millimeters or less. The width near the base of the cap is generally less than a few centimeters. Where pits were dug along the bottom of the cap, flow structures in the sand could be observed going into the base of the dikes. Pits also exposed sills as thick as 0.1 m running along the base of the cap. Rarely were sills observed to have intruded into the cap.

All dikes are interpreted to have been caused by the coastal downdropping earthquake event about 300 years ago for the following reasons (1) the radiocarbon ages of sticks along the surface of venting agree with the 300-year-old downdropping event; (2) ages of trees (determined from tree rings) rooted in sediment above vented sand have maximum age values (about 200 years) that are reasonable for the 300-year-old downdropping event; (3) dikes generally increase in abundance toward the coast; and (4) maximum dike sizes (widths) increase toward the coast. (The fact that the dike widths increase dramatically downstream, toward the coast and into an increasingly estuarine setting, eliminates the possibility that the dikes were caused by wave pounding from river flooding.) The 1-m thickness of silt and clay above the vented sand is interpreted to have been deposited following the regional downdropping from the subduction-zone earthquake 300 years ago. This 1-m thickness agrees well with the estimates of coastal tectonic submergence (Atwater, 1987, 1992, 1996; Darienzo and Peterson, 1990).

The 1–4-cm-thick sheet of vented sand may be exceptionally thin because of the tidewater action or because of subaqueous venting. The surface of venting is submerged at high tide. Tidewater flows relatively fast in this area, so any large cones of sand initially vented to the surface could have been beveled off and the sand scattered over a large area.

Many dikes are so narrow at the top as to be hardly distinguishable. This same relation can also be observed infrequently in dikes in the Wabash Valley and in dikes in the meizoseismal region of the 1811–1812 New Madrid earthquakes. The cause for this pinching is not known but may represent venting of a very water-rich mixture of sand and water through a very soft cap that closes partially after venting. The widest dikes (as much as 30 cm) on the Columbia River islands are interpreted to have formed by lateral spreading, because the sidewalls appear to be parallel throughout their height.

It was observed in the field study that sills were common occurrences in 8-cm-diameter core samples taken beneath the cap of Figure 7.24 in thinly layered strata of sand and silt, even where no dikes or sills were observed in the cap. Others have suggested that the properties of the soft silt cap on the islands may have prevented dikes from forming extensively; they suggest, instead, that only sills formed as a result of severe liquefaction over large regions. This suggestion is based on thin sills and sill-like features along the base of the cap that have been observed in samples collected in 8-cm-diameter tubes. I believe that such an interpretation is unlikely because the mechanics of forming sills by severe liquefaction over a large region, without also forming dikes, does not seem plausible. In order for sills to form over a large area, the cap must be lifted to provide space for the intrusion. The force required to lift the cap must equal the weight of the cap. The simplest of calculations shows that the hydraulic uplift pressure required to counteract the weight of a 1–2-m cap is much higher than the tensile strength of a soft cap. Furthermore, dikes formed abundantly in the New Madrid seismic zone in very soft capping material, in response to the

1811–1812 earthquakes. Therefore, dikes should develop, if liquefaction occurred. There are almost certainly other mechanisms whereby small sills would form without liquefaction; for example, due to seismic shaking at levels less than required for liquefaction, fluidization alone can lead to sill formation (discussed later in the section about estimating strength of paleoearthquakes).

The dikes described above were found before Satake *et al.* (1998) reported evidence for a small tsunami in Japan, on 26 January, 1700 AD, and interpreted it to be from a great subduction Cascadia earthquake. However, it was the liquefaction effects noted above that proved a seismic shaking event had occurred at that time. The occurrence of the tsunami implied nothing about levels of shaking, and indeed whether any shaking occurred.

Over the past decade, many investigators have found evidence of tsunamis in the form of our-of-place sediments along the coast of Washington–Oregon, thereby demonstrating that large tsunamis have struck repeatedly in Holocene time (Peters *et al.*, 2003).

7.4.4.2 Strength of Prehistoric Shaking

Subduction earthquakes can have very large variations in shaking characteristics, offering the possibility of an especially long duration of shaking at very low frequencies. Such uncertainties cause difficulty in interpreting the strength of prehistoric shaking. Still, significant conclusions can be drawn for the earthquake of 300 years ago. Small dikes possibly with venting in the Columbia River islands appear to go inland as far as 90 km. These dikes very likely formed at an acceleration level on the order of 0.1–0.2 g (Obermeier, 1994a; Obermeier and Dickenson, 2000). These accelerations accord with both theoretically and statistically derived accelerations from seismological models for the scenario of an $M_w \sim 8$ or larger subduction earthquake slightly offshore (Geomatrix, 1995).

The lack of abundant wide dikes throughout islands of the lower Columbia River valley also supports the seismological models noted earlier, which predict that exceptionally strong shaking should not extend very far onshore. Even though severe erosion of some islands has probably removed evidence of dikes wider than 30 cm at some places, there are many locales where erosion probably has been slight. (Also note that the widest observed dikes, 30 cm, are quite small in comparison to the width of dikes commonly found in the meizoseismal region of the 1811–1812 New Madrid earthquakes or in the Wabash Valley.)

Another plausible model consistent with observed liquefaction effects in the Columbia River islands is a subduction earthquake having a long duration of shaking at moderate-to-low peak accelerations, in combination with an exceptionally low dominant vibration frequency (perhaps less than a few hertz). The fine sands that underlie the Columbia River islands are typically so extensive and have such low permeability that pore pressures would not dissipate between cycles of shaking, even at an extremely low frequency. In addition, the thick alluvial deposits of the Columbia River valley probably strongly amplify bedrock accelerations, regardless of vibration frequency (Dickenson *et al.*, 1994). Therefore, the sand deposits of the Columbia River should be susceptible to forming liquefaction and fluidization features during a long duration of shaking, even at low-to-moderate accelerations and exceptionally low frequencies.

On a related matter, the abundance of small sills at sites in the islands, which have been observed in relatively undisturbed samples by using a "geoslicer," have been interpreted as having been induced by liquefaction and thereby indicating high levels of strength of shaking (Takada and Atwater, 2004). However, as noted previously, the presence of sills does not necessarily indicate an occurrence of liquefaction.

The level of bedrock shaking from the subduction earthquake of 1700 AD is difficult to interpret from study of liquefaction features in the Columbia River islands, alone, because of large uncertainties in amplification of bedrock motions as they went through the thick column of soft sediment underlying the islands. As a result, the best approach for estimating bedrock shaking is from study of liquefaction effects in much smaller streams in the region, where thin alluvium hosts any liquefaction features. Using this approach, Obermeier and Dickenson (2000) showed that relatively low levels of shaking were the norm, regionally.

7.4.4.3   Ancient Marine-Terrace Features

Many ancient fluidization features have been identified in late Pleistocene marine-terrace deposits in the region. The fluidization features can be seen for a span of 500 km in cliffs along the coast from central Washington to near the California–Oregon boundary (Peterson *et al.*, 1991; Peterson, 1992; Peterson and Madin, 1997). The features are of particular interest because they support the possibility of a long continuing record of subduction-zone earthquake shaking near the coast. The methodology for interpretation of a seismic liquefaction origin is instructive.

Source beds for the fluidization features include beach sands and sandy gravels, and lagoonal sands. Clastic dikes are as much as 5 m in height. Dikes are filled with clean sand or gravelly sand at almost all places. Dikes are as wide as a meter in scattered locales. Some dikes have penetrated upward into dune sands or have cut through lagoonal muds and peat. Sills are particularly abundant. Sills commonly extend beneath lagoonal muds and peats; small, steeply dipping dikes branch off from these sills at many places and cut up into thin (less than 0.5 m) strata of low permeability at the surface. The largest sills are as much as a meter thick. Even thin sills can extend over considerable lateral distances.

The possibility that these fluidization features were caused by wave action that induced liquefaction must be considered because the terrace deposits were laid down under shallow marine or shoreline conditions. Storm waves can impose significant shear stresses on the ocean-bottom sediments, even where the water depth exceeds 60–70 m (E. C. Clukey, written communication, 1992). Wave-induced cyclic shear stresses are thought to cause liquefaction in sands and granular deposits in a manner analogous to seismically induced liquefaction (Nataraja and Gill, 1983; Owen, 1987; Chowdhury *et al.*, 2006). The action of storm waves pounding on beaches also seems plausible as a mechanism for forming fluidization features. For the fluidization features in the marine terraces of coastal Oregon and Washington, though, the mechanism of wave-related liquefaction probably can be eliminated at some sites because dikes extend up into dune sands where wave action seems unlikely. Additionally, some dikes and sills cut lagoonal deposits at places that probably would have been protected from wave action. Significant artesian pressures at these lagoonal sites are also implausible. Thus, a seismic liquefaction origin seems probable for some of the features along the coast.

7.5   Features Generally of Nonseismic or Unknown Origin

Deformation features in unconsolidated sediments can form by many nonseismic processes (e.g., van Loon, 1992; Lucchi, 1995). Figure 7.25 provides a graphical overview of how various seismic and nonseismic processes can create certain features. Note that some features are never produced seismically, whereas others (e.g., warped and folded bedding) can be induced by either seismic or nonseismic mechanisms. In the following section, I briefly survey the voluminous literature of soft-sediment deformation, making special note of common features that might be confused as resulting from seismic

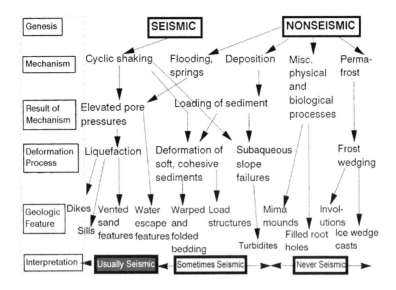

Figure 7.25: Graphical explanation of how various geological features of sediment deformation can be caused by either seismic or nonseismic causes. Much of the ambiguity in genesis arises because a given *deformation process* or *result of mechanism* can be caused by either seismic or nonseismic mechanisms. Features at the left side of the diagram are usually caused by earthquakes, whereas features on the right side are never caused by earthquakes.

shaking. Assignment of an unambiguous origin to isolated occurrences of such features is often impossible without determining their regional pattern.

7.5.1 Terrestrial Disturbance Features

Artesian conditions cause abundant nonseismic sand boils (vented sand volcanoes) to form in lowlands near the levees of the Mississippi River, USA, every few years (Kolb, 1976). The sand boils are typically restricted to a belt within 0.5–1 km of the levees, which aids in interpretation of origin. The cone-shaped, external form of the sand vented to the surface by these sand boils generally is very similar to that of a solitary sand blow caused by earthquake-induced liquefaction. Internally though, there can be significant differences, especially for the larger vented features. Going upward in artesian pressure-induced sand boils are rhythmic, planar strata of silty sand to clean sand that dip away from the central part of the sand boil; the individual strata are almost invariably much more sharply defined, more uniform in thickness (typically about 1 cm), and have a more narrow range of grain sizes than sediment vented in response to seismic liquefaction. Especially relevant for paleoseismic studies is that the sand boils have vented through dikes that are almost always more-or-less circular, and whose diameters are commonly a few decimeters, but can be as much as a meter. Fortunately for paleoseismic studies, the regional pattern of seismically induced liquefaction features generally should become evident as earthquake magnitude increases, which reduces the likelihood of confusing these features with nonseismic sand boils (Li *et al.*, 1996).

Slumps along streambanks are commonplace in both seismic and nonseismic conditions. Translational sliding blocks can develop in special nonseismic field settings and can resemble seismically induced lateral spreads. However, sliding blocks of nonseismic origin rarely extend far back from a free face, as commonly occurs for lateral spreads of seismic origin in which dikes can extend back hundreds of meters from the free face or incision (e.g., see Figure 7.15A). I believe that generally, it is not worthwhile to attempt to assign seismic or nonseismic origin to prehistoric slumps along actively eroding streambanks; there are too many uncertainties about the physical setting when such an ancient slump formed. For example, questions crucial to interpretation include the initial ground slope and the possibility of a streambank having been undercut by an eroding stream. These questions are virtually impossible to resolve. Slumps in loose, fine sands are especially difficult to interpret because they may be prone to static liquefaction even on relatively gentle slopes (Lade, 1992). Methods to interpret the seismic or nonseismic origin of landslides in geologic settings other than eroding streambanks are discussed in Chapter 8.

Thrown trees can excavate pits that resemble liquefaction-induced craters. Sediment filling in these pits typically does not have the orderly progression of clasts found in liquefaction-induced craters. Pits excavated by thrown trees sometimes can be distinguished from liquefaction-induced craters by the absence of feeder dikes or sediment from depth. Thrown trees can also form tabular breaks as roots are pulled through a fine-grained cap. The breaks can be filled with sand, gravel, or other sediment dragged into the break. However, the dragged sediment tends to be arranged haphazardly and have a much larger range of grain sizes than an intrusion formed by seismic liquefaction.

Mima mounds (also called prairie mounds or pimple mounds) have had many origins attributed to them, most commonly to animal burrowing, but also to seismicity (Berg, 1990). Mima mounds in the meizoseismal region of the 1811–1812 earthquakes are domes less than 30 m in diameter and 1-m high. In many upland areas, though, mima mounds are formed on nonliquefiable deposits and therefore are not of earthquake origin (Saucier, 1991a). Mounds on alluvial lowlands can be identified as not resulting from earthquake liquefaction if excavation shows an absence of vents to connect the mounds to underlying source beds (Fuller, 1912, p. 80).

7.5.2   Features Formed in Subaqueous Environments

This section focuses on soft-sediment deformations formed in subaqueous environments whose sizes are relatively small, commonly millimeters to centimeters to as much as a few meters. Until very recently, it has been very difficult to impossible to interpret the origin of such features from this type of setting, owing to a lack of proven techniques. This problem has been largely alleviated on the basis of findings from many field studies undertaken during the past 10–15 years, where historical records of seismicity have been associated with the deformations.

Many types of soft-sediment deformations can form in subaqueous settings, from a multitude of causes. And almost all can be either gravity- (i.e., static-) or seismically induced. Processes involved in nonseismic formation can be driven by rapid deposition (e.g., syndepositional factors), fluidization, liquefaction, storm waves (breaking), nonbreaking waves, strong currents, slumping, artesian pressures, and other mechanisms. Whereas all the criteria invoked in Section 7.3 for determination of a seismic liquefaction origin (i.e., features having sedimentary characteristics consistent with a seismic origin, and conforming to sudden, synchronous, widespread development around a central area) also apply for soft-sediment deformations in a subaqueous setting, the additional potential causative mechanisms must also be considered. Lucchi (1995) has shown numerous examples of features that can be of either seismic or nonseismic in origin, and also presents site details relevant to formation of such features. Wheeler (2002) has given an excellent philosophical discussion of requirements for interpreting a seismic origin.

7.5.2.1 Processes and Associated Deformations

Numerous types of soft-sediment deformations in subaqueous settings are discussed in articles by Lowe and LoPiccolo (1974), Lowe (1975), Reineck and Singh (1980), Allen (1982), Mills (1983), Jones and Preston (1987), and Einsele *et al.* (1991). Montenat *et al.* (2007) have presented numerous examples throughout Europe and northern Africa, in deposits generally of Tertiary and early Pleistocene ages, although some interpretations of origin are made on the basis of morphology alone. A particularly relevant discussion of the mechanics of subaqueous formation of dikes and sills in a marine setting, also applicable to features in a fluvial setting, is given by Jolly and Lonergan (2002).

In the following, I attempt to delineate the main classes of soft-sediment deformations and also relate them to triggering causes noted above. Unless otherwise stated this discussion focuses on features that develop on level ground, which would be so flat-lying as to preclude any influence of slope in their formation. Level ground conditions are less than about two degrees for most gravity-induced slumps (Lowe, 1975), and for sheet-like failures can be as low as about half a degree (Allen, 1982).

Syndepositional deformations very commonly involve intrusion of silty or sandy material down into finer-grained, muddy sediments that are extremely soft, and often are also highly sensitive (i.e., they become essentially viscous upon being sheared beyond a small threshold strain). The intrusion is driven by shear failure of the underlying material, caused by rapid and uneven loading from above, followed by gravity-induced sinking into the viscous mud. Additionally, the rapid loading can elevate the pore-water pressure in the underlying materials, thereby contributing to their shear failure or to fluidization.

Commonplace features of syndepositional origin are bulbous intrusions known as *load structures, load casts,* or *load-casted features* and include several variants. *Pseudonodules* form when overlying sandy or silty sediments become detached and sink to become isolated kidney-shaped bodies encased in the underlying mud. In *load-casted ripples,* sandy or silty intrusions form because of the unequal loading of migrating ripples of sand on a mud substratum. Load-casted ripples show progressively deformed radial internal lamination caused by the rotation of the ripple crosslaminations as the ripples sink (Dzulynski and Walton, 1965, pp. 146–149). *Convolute bedding* is manifested as more-or-less regular folds that develop either throughout or confined to the upper part of a single sedimentary unit (Allen, 1982). Similar to load structures in mud are *ball-and-pillow structures* in which kidney-shaped bodies of sand, locally slightly silty, have foundered into a cleaner sand; another setting associated with these structures occurs where thin limestone strata overlie more muddy strata, with foundering into the muddy strata (see Pettijohn and Potter, 1964, Plate 100A). Ball-and-pillow structures are abundant in many glaciofluvial deltas, where huge volumes of sand and silty sand have been rapidly deposited.

Elevated pore-water pressure in combination with fluidization cause water-escape features that include *dish structures, pillars,* and *convolute laminations* (Figure 7.26) (Pettijohn and Potter, 1964; Lowe and LoPiccolo, 1974). Dishes appear to form as upward-moving water locally flows in sediment beneath very thin laminations (mm scale) of lower permeability, and this flow creates short, dish-shaped features. Pillars are circular columns or sheet-like zones of structureless or swirled sand, sometimes bounded by dark laminations, that cut steeply through sand ranging from structureless to laminated. Figure 7.26 shows field relations commonly observed between dish structures, pillars, and convolute bedding, and these features and relations are often found in proximity to clastic dikes at sites of earthquake-induced liquefaction.

On slopes, features such as *sheet slumps, warped beds* (Allen, 1982), and *recumbent folds* (Allen and Banks, 1972; Owen, 1987) are well represented in the geologic record and in many cases are clearly of

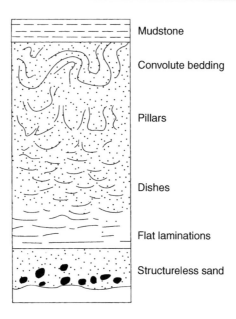

Mudstone

Convolute bedding

Pillars

Dishes

Flat laminations

Structureless sand

Figure 7.26: Schematic vertical section showing commonly observed vertical sequence of the sedimentary structures of convolute bedding, pillars, and dishes. Example shown is in thick sandstone beds. From Lowe and LoPiccolo (1974); reprinted with permission of the Society for Sedimentary Geology. I have observed this sequence of structures in sediments exposed in the banks of a river near the meizoseismal zone of the 1886 Charleston earthquake.

nonseismic origin, but they can also be of seismic origin. For example, Audemard and de Santis (1991) observed that during a moderate earthquake, warping developed within the uppermost 0.2 m of sediment, where mud overlaid a thin sand stratum. Severe warping of the ground surface is commonly observed where very strong ground shaking has caused liquefaction in thick sand deposits, as for the great New Madrid earthquakes of 1811–1812 (e.g., Fuller, 1912).

Liquefaction commonly occurs in subaqueous settings. The mechanics of liquefaction is unaffected by the depth of overlying water, even if hundreds of meters. The liquefaction-induced features are the same as those described for subaerial conditions, and include ground fracturing with associated dikes and sills as well as a multitude of types of plastic deformations, many which can be described as *hydroplastic* or *ductile* owing to smearing of fine-grained portions without completely obliterating the bedding. Good examples of various types of such deformations have been shown in liquefaction experiments in the laboratory by many researchers (e.g., Nichols *et al.*, 1994, Figure 4; Nichols, 1995, Figure 1).

Storm waves can induce liquefaction of the seabed from two mechanisms: back-and-forth shearing (causing conventional liquefaction), and *momentary liquefaction* caused by pulsating pore-water pressure within sediment in response to changes in pressure from overlying waves, either breaking or not (Owen, 1987; Chowdhury *et al.*, 2006). In recent years, many engineering methods have been developed for evaluating wave-induced liquefaction (e.g., see discussion in Chowdhury *et al.*, 2006), but field verification is typically lacking—which can cause uncertainty in interpreting origin in some field settings, such as shallow to moderately deep marine situations where wave heights can be very large. Certainly, liquefaction can be induced from very large ocean waves, to depths of many tens of meters. Fortunately for paleoseismic interpretations, in many lacustrine settings, and especially for smaller lakes, such very large waves cannot develop.

Chowdhury *et al.* (2006) found from experimental studies that even partially saturated sands are prone to being liquefied in their uppermost parts from pulsating storm waves, in a field setting corresponding to a very shallow marine condition such as proximity to a beach. Similarly, Dalrymple (1979) showed that even relatively small waves caused slumping and flowage of liquefied sand to depths of 0.15–0.35 m along the crest and upper stoss side of ebb megaripples. Wave action from storms has been interpreted as causing both load structures and downward-penetrating dikes in hummocky cross-stratified silts (Martel and Gibling, 1993); the dikes pinched together downward and some were quite large, being as wide as 40 cm at the top and penetrating as much as 5 m.

Shear stresses from bottom currents can also induce liquefaction and fluidization effects (Lowe, 1976; Herbich, 1977). The critical bottom current velocity required to trigger liquefaction can be greatly exceeded during large hurricanes (e.g., Pope *et al.*, 1997, p. 495); however, in apparent contradiction Pope *et al.* also point out that studies of modern hurricane effects in marine shelves do not report any evidence of such features. Still, one must consider the possibility of soft-sediment deformations from currents in both marine and lacustrine situations.

The possible influence of wave action in producing soft-sediment deformations is typically evaluated by searching for deposits having characteristics associated with wave action, such as graded bedding, hummocky crossbedding, or symmetrical ripples with sharply peaked anticline structures (e.g., Bowman *et al.*, 2004). To evaluate the possibility of storms or wave action, a search should also be made for effects such as localized destruction of bedding within the uppermost parts of megaripples (Dalrymple, 1979), and for downward-penetrating dikes (e.g., Martel and Gibling, 1993). Additionally, one must consider the possibility of seismically induced waves, in the form of a tsunami in marine situations or a seiche in lakes.

Factors in Development of Load Structures Load structures are commonly observed in the geologic record, and interpreting their origin is frequently an integral part of assessing the role of seismicity. Load structures develop most often where a sand-rich sediment has been laid rapidly upon a mud, causing foundering of the sandy material into the muddy host. Virtually the same structures can develop from seismic shaking (Figure 7.25). Factors relevant to interpretation of origin can include the size of the features, the rate of the deposition of foundered material, and the properties of the foundered material in relation to those of the host. Schematic drawings of several types of load structures in various stages of development are shown in Obermeier (1996, pp. 51–54).

Having a large size for soft-sediment load structures, in conjunction with widespread distribution, has been suggested as good evidence for a seismic origin in a field study by Moretti and Sabato (2007). Moretti and Sabato reasonably argue that such relationships were strongly suggestive of a seismic origin because other mechanisms such as wave loading or very rapid deposition from a nonseismic event had been eliminated.

Having a relatively large size for load structures, in combination with a slow rate of deposition for material that foundered, is sometimes used as the basis for interpreting a seismic origin, especially where small load casts along the base of the foundered material are rare (e.g., McLaughlin and Brett, 2004). Such an argument seems reasonable where numerous sites are available for observation, regionally.

The properties of the host can also be relevant to interpretations. For example, Moretti *et al.* (2002) interpreted a seismic origin where a coarse sand had foundered into a silty sand, largely on the basis that the silty sand would have been too strong to permit foundering without being weakened by elevated pore-water pressures. Sources of excessive pore-water pressure from nonseismic origins were eliminated, thereby supporting a seismic liquefaction origin. Somewhat similarly, a seismic origin was deduced for features found in the New Madrid seismic zone where detachments from the base of a silt stratum had foundered into a much coarser, silty sand stratum (Obermeier, 1998a, Slide 63).

Another factor that can be very important in development of load structures is plasticity characteristics of clay minerals. For example, McLaughlin and Brett (2004) showed that in ancient limestone strata, seismic shaking probably had triggered the ball and pillows in K-bentonite-bearing, fine-grained sediment. Such clays can become greatly softened upon small shearing. Relating the possible role of plasticity, by means of simple engineering tests such as Atterberg limits (discussed in any elementary text on soil mechanics), is not often done in studies of the origin of soft-sediment deformations, but possibly should be done more often.

Soft-Sediment Deformations, and Liquefaction and Fluidization Whether seismic liquefaction should be accompanied by recognizable fluidization effects can be highly dependent upon the particular sedimentary situation. Grain sizes as well as the vertical sequence of sediments can be critical to development of fluidization effects. For example, sediments such as silts and muds are not prone to forming fluidization features such as ball and pillars (Dzulynski and Smith, 1963), probably owing to their low permeability and any small cohesion.

A laboratory study by Moretti *et al.* (1999) demonstrated that liquefaction commonly induces fluidization effects in poorly sorted sediments, where sediments of lower permeability overlie those that liquefied; in contrast, they found no recognizable "water discharge" (i.e., fluidization) features where medium-coarse sands were liquefied without an overlying permeability barrier. This finding of Moretti *et al.* that seismic liquefaction of sand can occur without collateral development of recognizable fluidization features has also been noted by Lowe (1975, p. 165).

On a related matter, I suspect that load structures frequently may develop with only very minor development of fluidization effects, in response to seismic shaking-induced buildup of pore-water pressure, but not liquefaction. As noted in a previous section, buildup of pore-water pressure takes place progressively with continuing application of cycles of shearing in all but the loosest of sands (which can liquefy virtually catastrophically).

I suspect that the presence of any recognizable fluidization features can also be highly dependent on details of bedding, such as very thin laminations of fine sediment. Such laminations should significantly enhance any effects of fluidization.

The possible role of the rate of deposition of sediment on the development and nature of fluidization effects has been pointed by Lowe (1975), who stated, "Rapidly deposited [thick] sands may deform by fluidization [from compaction of underlying sediments, without liquefaction], but where this is the dominant process there should be evidence of water escape structures in the form of dish and pillars." And according to Lowe, fluidization of prolonged duration in such a setting (as in a proglacial delta), and especially in very young sediments, should be expected to develop pipe-like vents. However, my field observations indicate that another variable in formation of fluidization effects is related to "aging," in the engineering sense. (This type of aging imparts a slight strengthening to the sediment, typically within months to a few years, by mechanisms thought to be readjustment of grains and molecular forces (Mesri *et al.*, 1990; Mitchell, 2008). Aging is very commonly observed in the field, in almost all sediments.) Where aged, seismic liquefaction frequently develops numerous tabular as well as tubular dikes that cut through in host sands, much like the tabular dikes shown in Figure 7.20C; these tabular dikes can be relatively large, as shown in the figure. Large-diameter tubular dikes (tens of centimeters in diameter) can also be very commonplace in seismically liquefied, aged host sands.

Brenchley and Newall (1977) have interpreted a seismic liquefaction origin based mainly on the presence of ball-and-pillow features that had sunk into relatively thicker sand strata, 1–2 m in thickness, with no effects of fluidization such as tubular or tabular dikes; no significant translational movements of the larger mass had taken place. Even if the interpretation by Brenchley and Newall of seismic origin is correct, I suspect that the absence of fluidization effects in their sedimentary setting generally admits only the possibility of a low level of seismic shaking, especially in view of a paucity of horizontal deformations (as I discuss below).

7.5.2.2 Paleoseismic Criteria and Selected Field Study Examples

In this section, I recommend papers about field studies that are especially useful to the student on the basis of clear writing and logic, and illustrations and photographs. Many other excellent papers are available, but I have chosen only the few needed to achieve understanding for the conduct and interpretation of a field study.

Proving a regional, coeval formation of plastic and hydroplastic features in fluvial deposits is frequently difficult, because of the lack of lateral continuity of strata through a large region. Proving such coeval development in lakes, especially deep lakes, can be much more tractable. Deep lakes in which sedimentation has been so slow as to produce thin strata of widespread lateral continuity, and which have depths so great as to preclude effects of waves and currents, can be especially useful for paleoseismic studies because many candidates can be eliminated from formation of soft-sediment features (Lucchi, 1995). Additionally, in deep lakes often there may be many highly relevant features that develop, which are either detectable by seismoacoustic profiling (Ouellet, 1997) or so small they can be collected with conventional sampling tubes (e.g., Monecke *et al.*, 2004).

For the deep lacustrine field setting outlined just above, there is one type of deformation, *fault-graded bedding*, which is almost uniquely of seismic origin for level ground conditions. The fault-graded bedding (of Seilacher, 1969) commonly involves minor faulting, with offsets of millimeters to centimeters, in conjunction with other bedding disruptions. Many variations are possible, depending on factors such as strength and duration of shaking, and sediment properties. Idealized depictions are illustrated in Figure 7.27. Figure 7.27D shows the most complete manifestation of fault-graded bedding (modified from

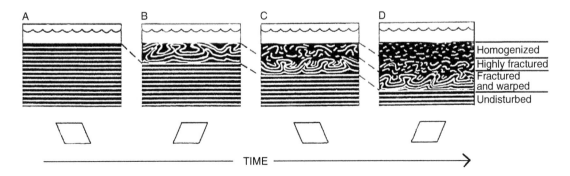

Figure 7.27: Schematic sketch showing the evolution of deformations in a laminated sequence subjected to continuing back-and-forth seismic shearing. Modified from Rodriguez-Pascua *et al.* (2000).

Rodriguez-Pascua *et al.*, 2000 to also show thrust faulting at shallow depth), with a fourfold sequence going from the top down (1) completely homogenized (soupy) zone, (2) severely disrupted (rubble) zone, (3) microfaulted and warped zone, and (4) undisturbed zone. Rodriguez-Pascua *et al.* noted all variations of fault-graded bedding in Figure 7.27 in their instructive study of lakebed deposits. In some field situations only the top two zones develop, creating what Agnon *et al.* (2006) refer to as *intraclast breccias.* Agnon *et al.* clearly showed that the brecciated zones were related to nearby faults.

In some field situations, only minor thrust faulting or recumbent folding develops, as shown in Figure 7.27B. Sims (1973), for example, observed low-amplitude recumbent folds in the uppermost 4–5 cm of flat-lying lake sediments (mud with very fine sandy laminae and partings), and convincingly related the folds to a nearby magnitude 6.5 earthquake. A back-and-forth orientation of thrusting, as shown in Figures 7.27B–D, has also been interpreted as indicating a seismic shaking origin (Lignier *et al.*, 1998).

I have noted above my opinion that having the uppermost sediment be so deep as to be beneath the influence of wave action is relevant to interpreting a seismic origin, for fault-graded bedding features. This requirement is added because I am unaware of any data showing whether or not waves can induce such features, although I suspect that the field conditions for their formation from waves would be unusual and probably exceptionally so. Given this caveat, having a regional occurrence of fault-grading features can be important for interpreting origin of the features, because a widespread occurrence would eliminate the likelihood of any other mechanisms, such as nonseismic slumping.

Another illustrative example of fault-graded bedding is given by Ringrose (1989a), where features with small faulting offsets were found at widespread exposures in sediments of a late Quaternary glacial lake, in Scotland. The sediments consist mostly of laminated silt and sand. Deformation features developed in two episodes about 10,000 years ago. The features are restricted to well-defined horizons of very large areal extent. Figure 7.28 shows how the styles vary from most severe (Figure 7.28A) to marginal (Figure 7.28D). The most severe deformations were centrally located in a geographic sense.

Another instructive paper, by Hibsch *et al.* (1997), discusses plastic–elastic deformations of seismically induced features found in thinly bedded lacustrine deposits in Ecuador.

The features shown in Figure 7.28 can also be useful in the study of *varves* from glacial melting origin. In eastern Canada, Adams (1982) reported the presence of thin contorted zones within silt and clay varves of early Holocene age. Synchronous contortions occur at widely scattered glacial lake sites, with a crude regional pattern with greatest deformations at the center. Thus, a paleoseismic origin was interpreted. Morner (1985) has reported a similar type of finding in glacial lake varves of early Holocene age in Sweden; association of deformed varves with tectonic faulting, fracturing, and slumping led to a seismic interpretation.

Widespread *turbidites* are very common in marine sediments (see articles in the journal edited by Cita and Lucchi, 1984; also Chapter 2B). It is often tacitly assumed that such a distribution strongly suggests a seismic origin. For flat-lying turbidites in a deep marine setting, though, a seismic origin may be assigned from widespread synchronous features such as small folds, and especially by small recumbent or back-and-forth folds or shears such as in Figure 7.28B.

Features illustrated in Figure 7.28 appear to be relatively commonplace on the basis of my survey of the literature. However, not all sediments are amenable to being deformed in such a manner for a variety of reasons, including the nature of seismic shaking and nature of the sediments. But distinctive features may still be found.

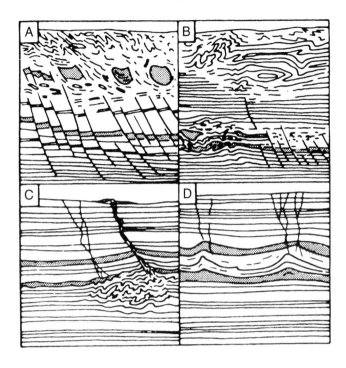

Figure 7.28: Schematic depiction of sediment deformation structures observed by Ringrose (1989a) in late Quaternary lake sediments. Sediment consists of laminated silts and sands (stippled). Parts (A–D) represent change from most severe deformations to slightly distinguishable deformations. Each box is about 2 m across.

For example, sediments having thin strata (even as thin as a few cm) of liquefied sand or silt can have small detachments from overlying cohesive sediment that visibly sink into the liquefied sediment. Also, small sills and hydroplastic and ductile deformations can form in this setting. Good photographic examples from tube samples taken from a lake in Switzerland are shown by Monecke *et al.* (2004, Figures 11 and 12), and from laboratory experiments by Nichols *et al.* (1994) and Nichols (1995). Many other researchers report such deformations (e.g., Rodriguez-Pascua *et al.*, 2000). A cautionary note that I offer, though, is that the constraints under which small-scale deformations develop in thin strata, such as those in reports noted above in this paragraph, are not well defined. Based on unpublished data by me and Prof. Scott Olson, after examining tens of meters of tube samples from many widely scattered sites in deep marine settings, plus my own field examination of hundreds of kilometers of fluvial sediments in river banks, in very differing geologic and tectonic settings, I can only conclude it is not possible to predict occurrence of features in thin sand and silt beds such as those reported above, even in locales of strong shaking (i.e., accelerations of at least several tenths g, in very large to great earthquakes). Thus, an absence of such small-scale features should not be interpreted to indicate an absence of strong paleoseismic shaking, in many field situations. Still, I suspect that the presence of small-scale features in thin strata can be highly suggestive of seismicity, especially if brittle fracturing has occurred in cohesive material bounding the thin sand or silt beds, because fracturing implies a large pore-water pressure was suddenly applied to the cohesive material (Obermeier *et al.*, 2005).

Where thick, flat-lying sand strata have been convoluted regionally, and both underlying and overlying cohesive strata that also have been severely deformed, a seismic interpretation can be interpreted for level ground conditions in a deep-water setting (Olson and Obermeier, 2007). Applying a somewhat similar logic, Roep and Everts (1992) interpreted a seismic origin to deformation structures that extend upward, only, from sand strata with many fluidization effects, whereas no deformations went down into underlying muds; Roep and Everts noted that depositional overloading usually affects underlying muds.

A widespread development of submarine slumps can be evidence for seismic shaking, and is cited as evidence in many studies (e.g., Schnellmann *et al.*, 2002, 2006; Karlin *et al.*, 2004; Monecke *et al.*, 2004, 2006; Strasser *et al.*, 2006). But, a cautionary note is that slumps can also develop from wave action during large storms, as commonly occurs in the Gulf of Mexico during hurricanes (Bea *et al.*, 1983; Lee and Edwards, 1986). Also, widespread (rotational) slumping can be induced from sudden drawdown of the water level (whereas lateral spreading, extending far back from a free face, would not be expected from rapid drawdown).

Pope *et al.* (1997) have proposed tests to evaluate a syndepositional versus seismic origin for very low-angle sheet-like slumps found in conjunction with deformed ball-and-pillow beds; a seismic origin is indicated if there are random axial-planar traces, there is lateral interfingering with undeformed beds, and if there is widespread development of the features in both shallow- and deep-water conditions. Using these tests throughout a huge region, hundreds to thousands of square kilometers in areal extent, in association with studies of faulting, Pope *et al.* showed that ancient marine sediments were probably extensively deformed by seismic shaking. I suspect, though, that ball-and-pillow features and very low-angle sheet-like slumps reported by Pope *et al.*, alone, do not indicate strong seismic shaking, especially in view of the low threshold of shearing require to develop ball-and-pillow features in many settings. For strong shaking I would anticipate other features should be present, such as clastic dikes, fault-graded bedding, or readily discernible fluidization features such as those illustrated in Figure 7.26.

In a study of ball-and-pillow features and associated highly deformed strata, Greb and Dever (2002) found opposing orientations of pillows within the same mass, with axial planes oriented close to the horizontal, suggesting seismic wave oscillations; Greb and Dever also clearly associated huge, largely downward-penetrating masses of sand, indicating large-scale load casting (termed *flow rolls*) with nearby faulting. These relations of opposing axial orientations and large penetrating masses, near a fault, seem to me to be strongly supportive for a seismic source.

Finally, it should be apparent that any evaluation of a seismic source for soft-sediment deformations must consider a multitude of alternate causative triggers. For determination of the triggering source for deformations in lacustrine and marine deposits, I emphasize that one should attempt to determine to the extent possible the regional development of the features. And, one might consider examining both the marine deposits and any nearby fluvial deposits, where entirely different processes could be operative.

7.5.3   Features Formed by Weathering

Distinguishing liquefaction-induced features can be difficult where weathering effects are severe. A wide variety of features produced by chemical weathering in the southeastern United States mimic those caused by earthquake liquefaction (Obermeier *et al.*, 1990). The boundary between the E (eluviated, or bleached) and B soil horizons is commonly abrupt and irregular and is characterized by narrow, near-vertical *pedogenic tongues* of the white E horizon that penetrate downward into the B horizon. Locally, tongues of

E horizon sand extend more than a meter into a thick, red to brown, clayey B horizon. Tongues of this size and shape can give the impression of fractured and brecciated ground and might be mistaken for liquefaction features unless examined carefully. Pedogenic tongues can range in morphology from tubular (Gamble, 1965) to planar (defining B horizon polygons; Nettleton *et al.*, 1968a).

Another category of pedogenic feature that might be confused with earthquake-induced liquefaction is the BE' horizon, which forms from the progressive chemical destruction of a clay-rich Bt (argillic) horizon (Daniels *et al.*, 1966; Nettleton *et al.*, 1968a,b; Steele *et al.*, 1969). Sharp contacts between the leached quartz sand areas and the much more clayey material may suggest to the uninitiated that this feature formed by some type of ground disruption, possibly ancient liquefaction. Examples of both phenomena just described are diagrammed in Obermeier (1994b, Figures 36 and 37).

7.5.4 Features Formed in a Periglacial Environment

A summary of features that can form as a result of freezing are given in Obermeier (1996b), and are briefly reviewed below. Two classes of features produced in a periglacial (freezing–melting) environment, loosely defined as (1) *involutions* and (2) *ice-wedge casts*, can resemble those having an earthquake-induced origin. Involutions are surficial manifestations of frost-related stirring (*cryoturbation*) and are often characterized by distortion and mixing of the roughly uppermost meter of sediments. Ice-wedge casts are downward-pinching, planar, nearly vertical features that originated by thermal contraction of frozen ground.

Vandenberghe (1988, p. 182) lists six types of involutions in terms of symmetry, amplitude to wavelength ratio, and pattern of occurrence. The features range from individual folds of small amplitude and large wavelengths (resembling slightly warped bedding caused by earthquake shaking) to intensely convoluted forms having amplitudes generally between 0.6 and 2 m (again resembling earthquake-induced convoluted bedding); load structures, diapirs, and dikes are also common. The genesis of involutions is probably related to three main process categories: load casting during melting, pressures in water trapped between freezing fronts, and pressures and heaving caused during freezing (Vandenberghe, 1988). Involutions are associated with ground patterned by freezing (*patterned ground*) and ice-wedge casts at many places, however, which aids interpretation.

Ice-wedge casts (Black, 1976), also called *ground-wedge pseudomorphs* (Harry and Gozdzik, 1988), result from ground cracking caused by contraction of frozen ground. Wedges of ice then form in the cracked ground. Later, when permafrost melts, host sediment slowly replaces the melting ice and forms an ice-wedge cast. An earthquake-induced origin can be incorrectly assigned to ice-wedge casts, especially in clean granular host deposits. Strong vertical alignment of infilling sediment commonly takes place, especially in narrow casts. This vertical alignment superficially resembles the effects of upward-flowing water such as winnowing out of finer grain sizes and fluidization. However, a liquefaction origin is almost unequivocally eliminated if there are no feeder dikes going from the host to sediments in the cast. Field studies in the northeastern United States (Stone and Ashley, 1992) have shown that the uppermost meter of a cast is often composed of fine sand and silt, including pebbles polished by wind abrasion, rather than sediment from depth vented to the surface. Stone and Ashley also found that air photographs can reveal a well-developed polygonal network of wedge structures, thereby demonstrating an almost certain permafrost origin. If the dike sediment contains magnetic minerals, Levi *et al.* (2006) suggest that passively filled fractures such as these can be distinguished from liquefaction dikes by the anisotropy of magnetic susceptibility.

7.6   Estimation of Strength of Paleoearthquakes

Interpretations of the strength of paleoearthquakes can be made using several independent techniques. First, a very crude association exists between severity of shaking, as measured by the modified Mercalli intensity (MMI) scale, and the threshold for formation of liquefaction effects and soft-sediment deformations such as pseudonodules and recumbent folds. In a second approach (magnitude-bound method) applicable to field situations where the regional extent of liquefaction from a paleoearthquake can be estimated, a probable minimum magnitude can be determined. A third method involves back-analysis of strength of shaking using an engineering-based procedure such as the cyclic stress method of Seed *et al.* (1983), or the energy-based method of Green-Mitchell (Green, 2001). Using these procedures, limits can be placed on accelerations that formed the liquefaction features in many field situations, which can then be used to estimate earthquake magnitude by comparison with estimates from seismological models (e.g., as shown in Green *et al.*, 2005).

The methods discussed next for estimating prehistoric earthquake magnitude first require locating the epicentral region. I suggest using the regional pattern of maximum dike widths as an easy means to approximate this region. Where field data are adequate, a preferable measure is the sum of dike widths normalized to the amount of outcrop (Munson and Munson, 1996). Maximum dike width is used rather than density of dikes because the density generally is highly sensitive to cap thickness (Ishihara, 1985; Obermeier, 1989). Lateral spreading, alternatively, is insensitive to cap thickness (Bartlett and Youd, 1992).

7.6.1   Association with Modified Mercalli Intensity

MMI value is a qualitative measure of earthquake-induced damage (Wood and Neumann, 1931). The scale ranges from I to XII, with I representing the level at which shaking may be felt slightly and XII representing total destruction. MMI of about VI seems to be the threshold for widespread development of small-scale soft-sediment deformation features such as folds, pseudonodules, contorted laminations, and recumbent folds (Sims, 1975; Rodriguez-Pascua *et al.*, 2000; Monecke *et al.*, 2004).

Although liquefaction effects have occurred at MMI values as low as V and VI (Keefer, 1984), the lowest intensity at which liquefaction-induced features can become common is a value of VII, where highly susceptible deposits are present (National Research Council, 1985, p. 34). Values of VIII–IX are generally required before liquefaction-induced ground failure becomes severe enough to cause damage to buildings. A serious shortcoming of use of MMI values as a measure of the strength of shaking of prehistoric earthquakes is caused by the very crude association with ground failure effects. Furthermore, any association of soft-sediment features with MMI is fraught with uncertainty because of the unknown properties of the sediment when the features formed. Despite this major shortcoming, attempts have been made to associate plastic deformations with severity of shaking and in turn earthquake magnitude (e.g., Hibsch *et al.*, 1997; Rodriguez-Pascua *et al.*, 2000). Still, associating intensity effects with distance from the energy source may offer the ability to place very crude limits on earthquake magnitude (Galli and Ferreli, 1995).

7.6.2   Magnitude Bound

Figure 7.29 shows the distance from the epicenter to the farthest observed liquefaction effect at the ground surface (plan view), such as venting of sand or ground fissuring. The data are from worldwide earthquakes in a wide variation of tectonic and alluvial settings. (A similar curve for Italian earthquakes within the

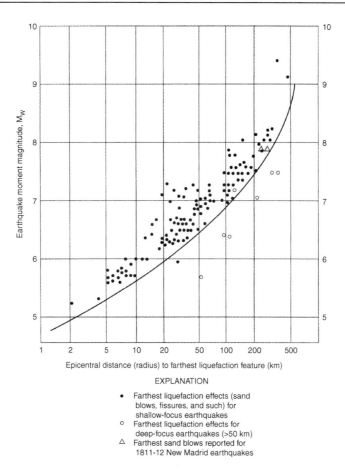

Figure 7.29: Relationship between earthquake moment magnitude (M_w) and distance from earthquake epicenter to the farthest liquefaction effect (venting to the surface or ground fracturing), with bound suggested by Ambraseys (1988). Bound is for estimating minimum value of magnitude. Curve separates data from earthquakes worldwide that had focal depths <50 km from those having depths >50 km.

past millennium, based on 317 liquefaction cases, has been developed by Galli (2000).) The sites of liquefaction range from having thin to thick alluvium. Amplification of bedrock accelerations is probably negligible to small for thin alluvium, whereas for thick alluvium, and especially soft clays or loose sands, the amplification is undoubtedly as high as 2–2.5 at places (Idriss, 1990).

Liquefaction susceptibility is also doubtlessly very high at many of the farthest sites in Figure 7.29 (Ambraseys, 1988). Therefore, using the bound in the figure should provide a minimum estimate of magnitude of paleoearthquakes, especially in regions where conditions are less than optimal for forming liquefaction effects.

The farthest sites of liquefaction effects on the ground surface (in plan view) often cannot be determined in a paleoliquefaction search (normally in sectional view). However, it has been my experience that this farthest distance can be reasonably approximated if a very large amount of exposure is searched in sectional view. A lower limit magnitude thus can be established by using the plot in Figure 7.29. Two illustrative examples are given on the basis of my field experience, one for the Charleston earthquake of 1886 and the other for the New Madrid earthquakes of 1811–1812. Both examples are regions of widespread moderate-to-high liquefaction susceptibility and probably at least moderate amplification of bedrock accelerations. Using the farthest liquefaction effects observed (90–100 km) at the time of the 1886 Charleston earthquake as a point on the outer bound of Figure 7.29 yields an estimated minimum magnitude of about M_w 6.8; current best estimates for magnitude are $M_w \sim 7.0$–7.2. I have found small liquefaction features (unweathered dikes extending almost to the surface) as far as 100 km from the epicenter of the 1886 earthquake that were almost certainly caused by that earthquake. In a similar vein for the 1811–1812 New Madrid earthquakes, I have found liquefaction features (unweathered dikes extending nearly to the surface) in widely spaced regions as far as 250–275 km from the epicenter of what was probably an earthquake of $M_w \sim 7.8$ (current best estimate). Farthest effects reported in 1811–1812 were the same distances in these widely spaced regions. Using Figure 7.29 yields a magnitude of 7.5–7.7. What these findings suggest is that if effects of liquefaction are observed at the surface at the time of the earthquake, in the same region there should be an abundance of dikes that pinch together only slightly below the ground surface—and that some could be found in a search for liquefaction features. (Keep in mind, though, that small, pinching-up dikes that do not approach the surface may extend much farther from the meizoseismal region.) In summary, my experience for the two earthquakes above, based on farthest discovered paleoliquefaction features that approach the ground surface (within 1–2 m), indicates that the estimates of magnitude using Figure 7.29 are not outrageously lower than actual values, providing field conditions are favorable for liquefaction and that sufficient exposures are available at distance from the likely epicentral region.

The outer bound line in Figure 7.29 can sometimes be adjusted, in regions of historic liquefaction, to account for influence of the local setting on farthest development of liquefaction effects. An excellent example for developing an appropriate calibration is given by Olson *et al.* (2005b).

It is not unusual to see curves such as Figure 7.29 be used for estimating the earthquake magnitude required to produce widespread soft-sediment features such as ball and pillows, in lacustrine or marine settings. I do not believe that this use is valid for two reasons: (1) the curves were developed using features observed at the ground surface in plan view that resulted from liquefaction—which is not the causative mechanism for many soft-sediment features and (2) there is no means to realistically assess the sediment properties when the soft-sediment features formed.

7.6.3   Engineering-Based Procedures

A means to circumvent the problem of not being able to locate the distal effects of liquefaction required for the magnitude-bound method is provided by an engineering-based procedure such as the method of Seed *et al.* (1983, 1985), referred to as the "cyclic stress method." This method was developed to provide engineers with an estimate of the shaking threshold required to produce surface manifestations of liquefaction during future earthquakes (Figure 7.30). The method (and its updated versions, noted below) can be adapted to paleoseismologic studies. The method is based on worldwide observations following many earthquakes, chiefly crustal earthquakes in California and Japan. The curves of Figure 7.30 are intended to indicate where liquefaction will suffice to produce scattered occurrences of effects of level ground liquefaction, such as limited venting, small ground openings (several centimeters) at the surface, minor lateral spreading, or noticeable settlement or warping of the ground surface. Sites of wide lateral spreads, especially near free faces, have been excluded because of the possibility of enhanced ground breakage with copious venting.

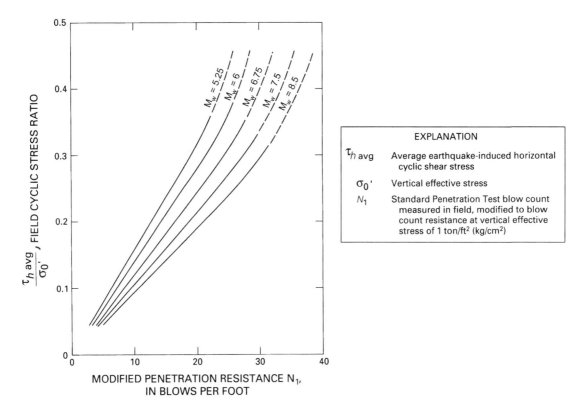

Figure 7.30: Curves for the method of Seed *et al.* (1983) used to evaluate the potential occurrence of liquefaction with accompanying venting of sand on appreciable ground cracks at a site on level ground. Curves are for clean sand deposits (average diameter >0.25 mm) and for different earthquake magnitudes (5.25–8.5). Points above and to the left of curves show conditions having high potential for liquefaction.

To use Figure 7.30, the relative density of the source deposits is evaluated in situ by the SPT blow count method (Table 7.1). An adjusted SPT blow count value (the N_1 value of Figure 7.30), which is the value corrected for site conditions of depth of the water table and static overburden stress conditions of the source stratum, is then related to the shaking required to cause venting. Figure 7.30 shows the boundary curves for different earthquake magnitudes. The curves relate the N_1 value in clean sands to the field cyclic stress ratio, which is defined as the ratio of the average earthquake-induced horizontal cyclic shear stress (τ_{havg}) to the vertical effective stress (σ'_{vo}). The field cyclic stress ratio due to earthquake shaking is computed from the following equation (Seed *et al.*, 1983):

$$\frac{\tau_{\mathrm{havg}}}{\sigma'_{\mathrm{vo}}} = \frac{0.65 A_{\mathrm{max}} \sigma_{\mathrm{vo}} r_{\mathrm{d}}}{\sigma'_{\mathrm{vo}} g}, \tag{7.1}$$

where A_{max} is the peak horizontal acceleration at the ground surface, σ_{vo} is the total vertical (overburden) stress on the sand at a particular depth, σ'_{vo} is the effective vertical (overburden) stress (total stress minus pore-water pressure) on the sand, r_d is a stress reduction factor ranging from a value of 1 at the ground surface to a value near 0.9 at a depth of about 10 m, and g is the acceleration of gravity. The curves of Figure 7.30 for various magnitudes are based on the premise that the expected duration (i.e., number of cycles) of strong shaking is longer for increasingly higher magnitudes.

In recent years, it has been found that there are many factors that influence the location of the curves of Figure 7.30, and some suggested modifications as well as limitations are reported in Youd *et al.* (2001). Even in its modified versions, the method is intended to be slightly conservative for predicting where liquefaction will develop. The method is thought most applicable for crustal earthquakes, of $M_w < \sim 8$, and so using it otherwise has significant uncertainties.

If the intent is to reverse the process and back-calculate the maximum ground acceleration that probably occurred at a paleoliquefaction site, then the engineering conservatism in the cyclic stress method (updated or not) could lead to a value that may be too low, at least for the types of earthquakes from which it was developed. Whether the value is too low depends on the constraints on liquefaction and venting at the site, and these constraints depend greatly on factors such as cap thickness and strength, and depth to the water table. If all factors are favorable, the cyclic stress method probably provides a good estimate of the strength of shaking at sites where only minor venting or ground failure occurred. Favorable conditions would include a cap thickness of less than a few meters (Ishihara, 1985; Obermeier, 1989), a water table within a few meters of the surface, and a cap that is not greatly strengthened by a mat of roots. Sites of marginal liquefaction also offer a good likelihood that the SPT blow count (N values) of the source bed have not changed substantially by the occurrence of liquefaction. Paleoliquefaction sites where severe liquefaction took place should be avoided for back-calculations, because these sites can yield only estimates of minimum shaking levels. Severe liquefaction can be indicated by large sand blows or strong warping of the cap. Sites of wide lateral spreads, especially near free faces, should be avoided because these sites may yield back-calculated values that are too low.

The curves of Figure 7.30 do not fully account for field conditions that are not conducive to venting, such as the presence of a very thick cap or a mat of peat in the cap. Elimination of sites having these unfavorable conditions would likely increase the probability. A common misconception is that the strength of clay in the cap has a major bearing on whether venting occurs. However, the pore-water pressure that develops along the base of the cap during liquefaction normally greatly exceeds the tensile strength of the cap sediments; any extensive liquefaction should be adequate for hydraulic fracturing of the cap and subsequent dike formation. Even where the cap is very soft, field observations show that a brittle mode of fracturing almost always occurs, rather than formation of plastic intrusions, when liquefaction is severe.

To use the cyclic stress method for estimating the strength of paleoseismic shaking, bounds must be placed on depth of the water table at the time of the earthquake. Maximum depth to the water table sometimes can be bracketed by observing the highest regional level at which dikes cut the base of the cap. The water level almost certainly was at least this high; if the water level had been much lower, the excess pore-water pressure probably would have been dissipated into granular, permeable sediment along the base of the cap.

The energy-based method of Green-Mitchell (Green, 2001) is theoretically more valid than the cyclic stress method. The method can be used to back-calculate the same parameters as the cyclic stress method, by using seismological attenuation parameters for bedrock motions in a specific tectonic setting. The disadvantage of the method is that it has not been evaluated using field occurrence of liquefaction to the same degree as the cyclic stress method. Still, by using various attenuations relationships, good approximations of strength of shaking and earthquake magnitude can sometimes be confidently back-calculated as shown by the example for the Vincennes paleoearthquake in the Wabash Valley (Green *et al.*, 2005).

Any back-calculation should account for liquefaction severity (i.e., marginal, moderate, severe, or no liquefaction) as shown by Green *et al.* (2005). Techniques for picking specific sites for back-analysis of strength of shaking, as well as a critique of the cyclic stress and Green-Mitchell methods, are discussed in detail in Olson *et al.* (2005a).

Sites hosting dikes have served as the primary basis for developing the above engineering procedures. And sites hosting dikes have also served as a principal source of data for back-analysis in paleoseismic studies. Recently, though, it has been proposed by some that sites having sills can also be used for back-analysis, with the premise that even small sills are manifestations of liquefaction. Whereas large, thick sills almost certainly result from liquefaction, small sills may often the result of fluidization, without complete development of liquefaction, on the basis that they commonly develop much farther from the epicentral region than dikes, especially for earthquakes of prolonged duration and shallow-water table (Obermeier and Olson, 2007). Dikes, and especially dikes from hydraulic fracturing (which are the principal dike sites that are recommended for back-analysis), develop from *suddenly* applied high pore-water pressure, which is often a much higher threshold than required for fluidization alone.

7.6.4 Overview of Estimates of Magnitude

It is obvious that the curves in Figure 7.30 do not yield a unique solution to both acceleration and magnitude. Only possible combinations can be determined. The estimate of magnitude preferably should be based on estimates of accelerations from seismological models and statistical analysis, in conjunction with an estimate of magnitude based on the observed areal distribution of the paleoliquefaction features.

Factors that control the regional extent (span) of liquefaction features and sizes of liquefaction features include liquefaction susceptibility and amplification of bedrock shaking, as well as seismological factors of focal depth and shaking frequency of the earthquake, and stress change (drop) in the rock of the rupture zone at the time of the earthquake. Possible shaking characteristics are myriad. A shallow hypocentral depth can result in more severe shaking in the meizoseismal area than that caused by a deeper earthquake. Far from the meizoseismal area, however, shaking may be less severe for the more shallow earthquake than for a deeper one. A higher stress drop should cause accelerations to be higher both in the meizoseismal area and far away. Clearly, any interpretation of prehistoric magnitude needs to be calibrated as much as possible to the local tectonic setting in order to select the most reasonable parameters.

An accounting for all the factors above was implicit in the analysis of prehistoric liquefaction features centered about Charleston, South Carolina, discussed previously. There, I concluded that some prehistoric earthquakes were at least equal in strength to the 1886 earthquake ($M_w \sim 7.0–7.2$). Seismological factors were indirectly considered by showing that the size and span of features predating 1886 exceeded those of the 1886 event, both in the meizoseismal region of the 1886 earthquake and far away. In addition, near Charleston, the fact that features that predate 1886 are more abundant and larger than elsewhere helped to define the region of strongest shaking.

The South Carolina example, for which one can compare historical and prehistoric liquefaction effects to estimate past earthquake magnitudes, is exceptional. At most places such extensive historic liquefaction data from a very strong earthquake do not exist for the same seismotectonic setting. Estimates of magnitude elsewhere must be based on a more numerical- or geological-based analysis.

7.6.5   Negative Evidence

The absence of liquefaction features (negative evidence) also plays an important role in assessing prehistoric earthquakes. Where depth of the water table can be bounded through time, and the susceptibility of potential source deposits can be estimated, the lack of liquefaction effects can be used to place limits on the maximum levels of prehistoric ground shaking. There is no well-defined procedure for determining the length of outcrop that must be searched for liquefaction features to support such a negative conclusion. Some uncertainty in interpretation also arises because shaking probably attenuates from the energy source in a somewhat variable (stochastic) manner. These variations often have no readily identifiable basis. Similarly, both the size and abundance of liquefaction-induced features can have large variations within a local area. For example, in the meizoseismal area of the 1811–1812 New Madrid earthquakes, within a given length of outcrop there may be hundreds of dikes and sills of all sizes. Yet, in an outcrop of the same length nearby, where conditions for liquefaction appear to be about the same, an order of magnitude fewer features can be found.

Dikes from lateral spreading, at least near streambanks, often appear to form easier than from other mechanisms. Paleoseismic studies in which I have been involved suggest that at accelerations up to about 0.25 g, lateral spreading near streambanks is the dominant mode of ground failure. Severe bank erosion can later remove most of this evidence of paleoliquefaction. Thus, I suggest that even where sediments in a typical fluvial setting in the eastern and central U.S. (i.e., where many ages of sediments are exposed), which have been at least moderately susceptible to liquefaction through time, 10 or more kilometers of fresh exposure should be searched to find effects from accelerations as low as 0.1–0.2 g. Preferably, the contact of the base of a fine-grained cap with potentially liquefiable sand can be observed at many places, to search for even small dikes and sills. The search is best conducted when the water table is very low.

CHAPTER 8

Using Landslides for Paleoseismic Analysis

Randall W. Jibson

U.S. Geological Survey, Golden, Colorado 80401, USA

8.1 Introduction

Most moderate to large earthquakes trigger landslides (Figure 8.1). In many environments, landslides preserved in the geologic record can be analyzed to determine the likelihood of seismic triggering. If evidence indicates that a seismic origin is likely for a landslide or group of landslides, and if the landslides can be dated, then a paleoearthquake can be inferred, and some of its characteristics can be estimated. Such paleoseismic landslide studies thus can help reconstruct the seismic shaking history of a site or region.

Paleoseismic landslide studies differ fundamentally from paleoseismic fault studies. Whereas fault studies seek to characterize the movement history of a specific fault, landslide studies characterize the shaking history of a site or region irrespective of the earthquake source. In regions that contain multiple seismic sources and in regions where surface faulting is absent, paleoseismic ground-failure studies thus can be valuable tools in hazard and risk studies that are more concerned with shaking hazards than with interpretation of the movement histories of individual faults. In fact, paleoseismic studies in some parts of the world typically rely more on landslides and sackungen than on surface fault ruptures (Solonenko, 1977a,b; Nikonov, 1988a,b).

As discussed in this chapter, the practical lower-bound earthquake that can be interpreted from paleoseismic landslide investigations is about magnitude 5–6. This range is comparable or perhaps slightly lower than that for paleoseismic fault studies. Obviously, however, larger earthquakes tend to leave much more abundant and widespread evidence of landsliding than smaller earthquakes; thus, available evidence and confidence in interpretation increase with earthquake size.

Paleoseismic landslide analysis involves three steps: (1) identify a feature as a landslide, (2) date the landslide, and (3) show that the landslide was triggered by earthquake shaking. This chapter addresses each of these steps and discusses methods for interpreting the results of such studies by reviewing the current state of knowledge of paleoseismic landslide analysis. Only subaerial landslides are discussed in this chapter; submarine landslides are dealt with in Chapter 2B.

International Geophysics, Volume 95

ISSN 0074-6142, DOI: 10.1016/S0074-6142(09)95008-2

Figure 8.1: Madison Canyon landslide, triggered by the 1959 Hebgen Lake, Montana, earthquake (M_w 7.1). Strong shaking caused 28×10^6 m^3 of rock to slide into the canyon, which dammed the river and created a lake more than 60 m deep (Hadley, 1964). Slide scar at left is 400 m high, debris is as thick as 67 m in valley axis, and slide debris traveled 130 m up the right valley wall. Twenty-eight people were killed by the slide. [Photograph courtesy of J.R. Stacy, U.S. Geological Survey Photographic Library (photo no. 209a).]

8.2   Identifying Landslides

Landslides include many types of movement of earth materials. In this chapter, the classification system of Varnes (1978) is used, which categorizes landslides by the type of material involved (soil or rock) and by the type of movement (falls, topples, slides, slumps, flows, or spreads). Other modifiers commonly are used to indicate velocity of movement, degree of internal disruption, state of activity, and moisture content (Varnes, 1978).

Identifying surface features as landslides can be relatively easy for fairly recent, well-developed, simple landslides. Older, more degraded landslides or those having complex or unusual morphologies can be more difficult to identify. Several excellent summaries of approaches to landslide identification and investigation have been published (Schuster and Krizek, 1978; Záruba and Mencl, 1982; Brunsden and Prior, 1984; McCalpin, 1984; Turner and Schuster, 1996), and the details need not be repeated here.

In general, landslides are identified by anomalous topography, including arcuate or linear scarps, backward-rotated masses, benched or hummocky topography, bulging toes, and ponded or deranged drainage. Abnormal vegetation type or age is also common.

Earthquakes can trigger all types of landslides, and all types of landslides triggered by earthquakes also can occur without seismic triggering. Therefore, an earthquake origin cannot be determined solely on the basis of landslide type. However, some types of landslides tend to be much more abundant in earthquakes than other types. For example, Solonenko (1977a,b) described some common characteristics of landslides triggered by earthquakes in the former Soviet Union. In a more comprehensive study, Keefer (1984) ranked the relative abundance of various types of landslides from 40 major earthquakes throughout the world (Table 8.1). Overall, the more disrupted types of landslides are much more abundant than the more coherent types of landslides. Keefer (1984) also observed that most earthquake-induced landslides occur in intact materials rather than in pre-existing landslide deposits; thus, the number of reactivated landslides is small compared to the total number of landslides triggered by earthquakes. Keefer (1984, 2002) described typical characteristics of various types of earthquake-triggered landslides and their source areas. In general, slope materials that are weathered, sheared, intensely fractured or jointed, or saturated are particularly susceptible to landsliding during earthquakes.

Sackungen (ridge-crest troughs) are a somewhat controversial type of ground failure that have been related, in some cases, to seismic shaking (Jibson *et al.*, 2004). Sackungen are identified by one or more of the following: (1) grabens or troughs near and parallel to ridge crests of high mountains, (2) uphill-facing scarps a few meters high that parallel the topography, (3) double-crested ridges, and (4) bulging lower parts of slopes (Varnes *et al.*, 1989).

Table 8.1: Relative abundance of earthquake-induced landslides

Abundance	Landslide type
Very abundant	Rock falls Disrupted soil slides Rock slides
Abundant	Soil lateral spreads Soil slumps Soil block slides Soil avalanches
Moderately common	Soil falls Rapid soil flows Rock slumps
Uncommon	Subaqueous landslides Slow earth flows Rock block slides Rock avalanches

Data from Keefer (1984). Landslide types use nomenclature of Varnes (1978) and are listed in decreasing order of abundance.

8.3   Determining Landslide Ages

Paleoseismic interpretation requires establishing the numerical age of a paleoearthquake. In the case of earthquake-triggered landslides, this means that dating landslide movement is required. Several methods for dating landslide movement can be used; some are similar or identical to those used for dating fault scarps (as discussed in other chapters), while others are unique to landslides. Most of the methods discussed here are simply applications of numerical dating techniques discussed in Chapter 1, which contains general descriptions of sampling and testing procedures.

Different types of landslides could be datable by different methods, depending on a variety of factors such as distance of movement, degree of internal disruption, landslide geometry, type of landslide material, type and density of vegetation, and local climate. Ideally, multiple, independent dating methods should be used to increase the level of certainty of the age of landslide movement (Johnson, 1987).

8.3.1   Historical Methods

Some old landslides could have been noted by local inhabitants or could have damaged or destroyed human works or natural features (e.g., Whitehouse and Griffiths, 1983). In some parts of the world, potentially useful historical records or human works extend back several hundreds or thousands of years. For example, a prehistoric encampment at Mam Tor, in Derbyshire, England, was partly destroyed by a landslide (Johnson, 1987). The encampment was first occupied about 3000 years BP according to archaeological studies (Jones and Thompson, 1965); this date provides an approximate maximum age of the landslide. In the United States, few historical records exceed 300 year in length, but some of these could still be useful. In a paleoseismic investigation of landslides possibly triggered by the 1811–1812 New Madrid earthquakes, Jibson and Keefer (1988) reported that oral accounts of local inhabitants helped establish minimum landslide ages in the 1850s, which helped bracket absolute ages. Also, grave markers on landslide masses, datable roads and trails whose locations clearly show that they either postdated or predated landslide movement, disturbed stone fences or other property markers, and other human works can potentially bracket or definitively date landslide movement (e.g., Jibson and Keefer, 1988).

For fairly recent events, comparing successive generations of topographic maps or aerial photographs can bracket the time period in which mappable landslides first appeared (e.g., Jibson and Staude, 1992; Jibson et al., 1994a,b).

8.3.2   Dendrochronology

Dendrochronology can be applied to date landslide movement in several ways (Hupp et al., 1987). At the simplest level, the oldest undisturbed trees on disrupted or rotated parts of landslides should yield reasonable minimum ages for movement (Jibson and Keefer, 1988; Logan and Schuster, 1991; Williams et al., 1992). On rotational slides that remained fairly coherent, pre-existing trees that survived the sliding will have been tilted because of headward rotation of the ground surface; if both tilted and straight trees are present on such landslides, the age of slide movement is bracketed between the age of the oldest straight trees and the youngest tilted trees (Fuller, 1912). Using this simple application of dendrochronology to date coherent translational slides is more difficult because trees can remain upright and intact even after landslide movement. On all types of landslides, trees growing from the surface of the scarp will yield minimum ages of scarp formation, from which the age of slide movement can be interpreted.

In some cases, trees killed by landslide movement will be preserved and can thus yield the exact date of movement. For example, Jacoby *et al.* (1992) dated trees beneath the surface of Lake Washington near Seattle that were drowned by landsliding into the lake. They were able to date the landslide movement from the preserved tree-ring records and from radiocarbon dating of the outermost wood.

A more sophisticated application of dendrochronology involves quantitative analysis of growth rings. For trees that have survived one or more episodes of landslide movement, such analysis can be used to identify and date reaction wood (eccentric growth rings), growth suppression, and corrosion scars, which might be evidence of landslide movement (Hupp *et al.*, 1987). Reconstruction of movement histories by such dendrochronologic analysis has been documented successfully in several areas (e.g., Terasmae, 1975; Reeder, 1979; Jensen, 1983; Bégin and Filion, 1985; Hupp *et al.*, 1987; Osterkamp and Hupp, 1987; Stefanini, 2004). Some landslides block stream drainages and form dams that impound ponds or lakes. Inundation of areas upstream from landslide dams can drown trees that can be dated dendrochronologically (Logan and Schuster, 1991).

8.3.3 Radiometric and Cosmogenic Dating

Radiometric dating (most commonly using ^{14}C) can be used in a variety of ways to date organic material buried by landslide movement (Stout, 1977). Landslide scarps degrade similarly to fault scarps, so colluvial wedges at the bases of landslide scarps might contain organic material that can be retrieved by trenching or coring and dated radiometrically. Fissures on the body of a landslide, particularly near the head where extension can take place, can also trap and preserve organic matter. If the landslide mass is highly disrupted, as in rock or soil falls or avalanches, then some vegetation from the original ground surface might have become mixed with the slide debris; such organic material excavated from slide debris can be dated radiometrically (Burrows, 1975; Whitehouse and Griffiths, 1983; McCalpin, 1989a,b, 1992; Aylsworth *et al.*, 2000). At the toes of landslides, slide material commonly is deposited onto undisturbed ground; if this original ground surface can be excavated beneath the toe of a slide, buried organic material from this surface can be dated to indicate the age of initial movement.

Sag ponds commonly form on landslides, and organic material deposited in such ponds can be dated radiometrically. Organics at the base of the pond deposits should yield reliable dates of pond formation (Stout, 1969, 1977; McCalpin, 1989a,b).

Vegetation submerged from inundation of areas upstream from landslide dams can also be dated radiometrically. Schuster *et al.* (1992) dated the emplacement of rock-avalanche dams by ^{14}C dating the outer few rings of drowned trees protruding from landslide-dammed lakes and detrital wood and charcoal in lacustrine deposits that formed behind a landslide dam. Similarly, landslides into lakes can submerge and kill vegetation that can be dated radiometrically (Jacoby *et al.*, 1992).

Rock-fall and rock-avalanche deposits can be dated cosmogenically if the surface of the deposit has been relatively stable since the time of emplacement. This method measures the amount of time that specific types of mineral grains have been exposed to cosmic radiation. The assumption is that a significant proportion of the material on the surface of a rock-fall or rock-avalanche deposit was newly fractured and exposed when the landslide occurred. This method has produced consistent ages (considered minimum landslide ages) for landslides in the Argentine Andes (Hermanns *et al.*, 2001).

8.3.4   Lichenometry

Lichenometry—analysis of the age of lichens based on their size—has been used to date rock-fall and rock-avalanche deposits (Nikonov and Shebalina, 1979; Oelfke and Butler, 1985; Nikonov, 1988a,b; Smirnova and Nikonov, 1990; Bull *et al.*, 1994, 1995; Bull, 1996a,b, 2003; Bull and Brandon, 1998). By measuring lichen diameters on rock faces freshly exposed at the time of failure, numerical ages can be estimated by assuming that lichens colonized the rock face in the first year after exposure. Because rock-fall and rock-avalanche deposits typically include abundant rocks having freshly exposed faces, numerous samples generally can be taken to create a database for the statistical analysis required by lichenometry. Lichenometric ages must be calibrated at sites of known historical age or by comparison with other numerical dating techniques. Lichenometric dating is subject to considerable uncertainty, however, because several decades can elapse before lichens colonize a fresh rock exposure, and lichens might never colonize unstable landslide deposits on very steep slopes (Oelfke and Butler, 1985).

8.3.5   Weathering Rinds

For a given climate and rock type, measuring the thickness of weathering rinds can be used to date when rocks were first exposed at the ground surface (Chinn, 1981; Knuepfer, 1988). For rock falls and rock avalanches and for other landslides whose movement exposed rock fragments at the ground surface, measuring the thickness of weathering rinds can be used to date landslide movement (Whitehouse and Griffiths, 1983; McCalpin, 1989a,b, 1992). Determining which rock surfaces were initially exposed at the time of landsliding can be difficult, but if a sufficiently large number of samples can be measured, consistent statistical results of predominant ages that relate to landslide movement can be obtained.

8.3.6   Pollen Analysis

Analysis of pollen in deposits filling depressions on landslides can yield both an estimated age of initial movement and, in some cases, a movement history through time (Franks and Johnson, 1964; Adam, 1975; Tallis and Johnson, 1980; Dietrich and Dorn, 1984; Johnson, 1987; Skempton *et al.*, 1989). Such analyses assume that sediment deposition and incorporation of pollen occur immediately following landslide movement and that local climatic and vegetation variations can be accounted for. Pollen samples from the buried ground surface beneath the toes of landslides also have potential for use in dating landslide movement.

Landslide activity that alters sedimentation patterns and vegetation cover in an area also has been detected and dated using pollen analysis (Dapples *et al.*, 2002; Glade, 2003). This approach does not necessarily date a specific landslide but rather an area where landslide activity has occurred.

8.3.7   Geomorphic Analysis

Landslides are disequilibrium landforms that change through time more rapidly than surrounding terrain. By analyzing the degree of degradation of landslide features such as scarps, ridges, sags, and toes, relative ages can be assigned to various landslides (Schroder, 1970; McCalpin, 1984, 1986; Crozier, 1992). For example, McCalpin and Rice (1987) analyzed 1200 landslides in the Rocky Mountains and assigned each of them to one of four relative age groups based on morphology. Numerical age ranges for these groups were estimated based on correlation with other landslides in the Rocky Mountains that have similar morphologies and surface-clast weathering and for which ^{14}C dates were available. Although the classification scheme of McCalpin and Rice (1987) was developed for the Rocky Mountains, similar schemes could be developed for other areas (Wieczorek, 1984).

Another example of relative dating by geomorphic analysis was developed in New Zealand by Crozier (1992), who identified distinct age groups of landslides based on degree of definition of landslide features, soil development, tephra cover, stream dissection, preservation of vegetation killed by movement, and drainage integration. Ranges of numerical ages for these groups were estimated by dating organic material retrieved from representative landslides from each group.

Jibson and Keefer (1988) concluded that because a large group of landslides in the New Madrid seismic zone all appeared to have the same degree of geomorphic degradation, these landslides were contemporaneous. Other types of evidence (Jibson and Keefer, 1988, 1989, 1993) were then used to link the synchronous ages of these landslides to triggering by the 1811–1812 New Madrid earthquakes.

Models of fault-scarp degradation also have potential application in landslide dating because landslide scarps should behave similarly to fault scarps. Several approaches to morphologic fault-scarp dating have been proposed (e.g., Bucknam and Anderson, 1979; Nash, 1980; Mayer, 1984), all of which require calibration for various parameters such as climate and scarp material. Scarp degradation commonly is modeled as a diffusion process, in which degradation rate varies in time and is a function of slope angle, which represents the degree to which the scarp is out of equilibrium with the surrounding landscape (Colman and Watson, 1983; Andrews and Hanks, 1985; Andrews and Bucknam, 1987).

Christiansen (1983) used sedimentation rates to date landslide ages. An ancient landslide moved over alluvium deposited by the North Saskatchewan River, Canada, and part of the landslide was buried by continued deposition. The rate of alluvial deposition was determined by ^{14}C dating to be fairly uniform at about 2.4 mm/yr. By measuring the depth of the landslide shear zone below the present surface of the alluvium, an age of about 4000 years BP was estimated.

Johnson (1987) discussed some other geomorphic methods to date landslide movement, including correlation of landsliding with specific periods of fluvial downcutting or aggradation and correlation with known limits of ice sheets.

Analysis of soil-profile development is also a potential tool for dating landslides. New soil profiles will begin to develop on disrupted landslide surfaces. If such surfaces can be identified, dating the newly developed soil profile will indicate the age of movement (Small and Clark, 1982; Birkeland, 1999; Knuepfer, 1988; Birkeland *et al.*, 1991).

8.4   Interpreting an Earthquake Origin for Landslides

Interpreting an earthquake origin for a landslide or group of landslides is by far the most difficult step in the process, and methods and levels of confidence in the resulting interpretation vary widely. This section summarizes several basic approaches that have been documented to interpret the seismic origin of landslides.

8.4.1   Regional Analysis of Landslides

Many paleoseismic landslide studies involve analysis of large groups of landslides rather than individual features. The premise of these regional analyses is that a group of landslides of the same age, scattered across a discrete area, were triggered by a single event of regional extent. In an active seismic zone, that event commonly is inferred to be an earthquake. Such an interpretation could be justified in areas where landslide types and distributions from historical earthquakes have been documented and can be used as a standard for comparison. In areas where such historical observations are absent, assuming an earthquake origin for landslides of synchronous age is much more tenuous, primarily because large storms can trigger widespread landslides having identical ages and spatial distributions.

Differentiating between such groups of storm- and earthquake-triggered landslides might be possible using a statistical approach to characterize the distribution of steep slopes in a region. Densmore and Hovius (2000) argued that storm-triggered landslides form most commonly near the bases of slopes and thus tend to form landscapes characterized by steep inner gorges. Earthquake-triggered landslides, on the other hand, tend to form more uniformly across the entire reach of slopes; corresponding landscapes generally lack well-developed inner gorges. These landscape patterns provide supportive evidence of the possible seismic origin of landslides in a region, but they cannot be used to definitively determine the origin of any specific landslide.

Crozier (1992) cited six criteria to support a seismic origin for some landslides in New Zealand; these criteria can be applied generally: (1) ongoing seismicity in the region, which has triggered landslides; (2) coincidence of landslide distribution with an active fault or seismic zone; (3) geotechnical slope-stability analyses showing that earthquake shaking would have been required to induce slope failure; (4) large size of landslides; (5) presence of liquefaction features associated with the landslides; and (6) landslide distribution that cannot be explained solely on the basis of geological or geomorphic conditions. Obviously, the more of these criteria that are satisfied, the stronger the case for seismic origin.

Russian scientists were the first to analyze the distribution and ages of landslides in seismic zones for paleoseismic analysis. Nikonov (1988a,b) estimated that analysis of landslides in a region can detect earthquakes having magnitudes greater than 6.5 and that epicentral zones can be located within about 10 km. Analysis of fault features is considered preferable for epicentral location and magnitude estimates; analysis of landslides is preferable for age determination (Nikonov, 1988a,b). The method developed by the Russians (Nikonov, 1988a,b) involves complementary studies of fault-related features and shaking-induced features in a known seismic zone. The premise of the approach is that large earthquakes in mountainous areas trigger many landslides, and that the number, size, and areal extent of the landslides are proportional to the size of the earthquake (Solonenko, 1977a,b). Many landslides in a seismic zone are dated either by the radiocarbon method or lichenometry; if one or more groups of landslides cluster in both space and time, then an earthquake origin is inferred (Nikonov, 1988a,b). Each age cluster is interpreted to define a different paleoearthquake. Generally, no criteria other than synchronous age are used, so the seismic origin of these landslides is, to a large degree, simply assumed. An earthquake origin is more certain in cases where landslide ages match ages of local fault features and where the types of landslides correspond to those documented in historical earthquakes (Solonenko, 1977a,b). Based on historical observations that large, deep-seated landslides are triggered only within modified Mercalli intensity (MMI) isoseismals VII–IX, only large earthquakes that triggered large, well-preserved landslides have been interpreted from such landslide studies (Nikonov, 1988a,b). This method was applied to rock-avalanche deposits in the epicentral region of the 1907 Karatog and 1949 Khait earthquakes (both $M_w \approx 7.4$) in Tadjikistan (Nikonov and Shebalina, 1979). Lichenometric ages from young-looking deposits near the epicenter of these two earthquakes correlate with the 1907 and 1949 earthquakes, respectively. Lichenometric dates from the older parts of the deposits suggest an earlier earthquake about 200 years before the study.

Aylsworth *et al.* (2000) documented several large landslides scattered across several drainages in eastern Canada that clustered in age. They postulated an earthquake origin based on several factors in addition to contemporaneity: (1) historical earthquakes have triggered similar landslides; (2) the failures occurred in sensitive marine sediment, material known to be susceptible to failure during seismic shaking; (3) the slides occurred during a drier climatic period, when rainfall triggering would be less likely; and (4) landsliding occurred long after channel abandonment, the period when fluvial processes would have been most likely to have caused failure.

Tibaldi *et al.* (1995) analyzed the distribution of landslides triggered by the 1987 Ecuador earthquakes ($M_w = 6.9, 6.1$) and compared this distribution with locations of known faults and recent earthquake epicenters. They found good correlation between the elongation of the landslide distribution and the location and dimensions of the seismogenic faults in the area; thus, they concluded that this method could be used to reconstruct the geometry of seismogenic faults in other areas where synchronous landslide distributions can be mapped.

Bull *et al.* (1994) used lichenometry to date numerous rock-fall deposits and rock-fall scarps near the Hope fault on South Island, New Zealand. Recent ages of deposits were linked to historical earthquakes, and older deposits were interpreted to have been triggered by previous earthquakes.

Adams (1981a,b) used landslide-dammed lakes in New Zealand to identify paleoearthquakes. He examined 17 historical landslide-dammed lakes and found that 15 of them formed during earthquakes; he therefore concluded that a seismic origin reasonably can be inferred if several synchronous prehistoric landslide dams cluster in an area. Perrin and Hancox (1992) later confirmed that most landslide dams in parts of New Zealand were, indeed, seismically triggered. Adams (1981a,b) estimated magnitudes of prehistoric earthquakes by comparing the areal extents of landslide dams of a given age with areal extents of landslide dams in historical earthquakes. He dated a group of prehistoric landslide-dammed lakes on South Island, New Zealand, using three types of samples: (1) woody detritus in the debris of the landslide dams, (2) standing trees drowned by the lakes, and (3) submerged soil horizons cored beneath lake sediment. His results indicated an earthquake of magnitude 7.4 in about AD 1650. Adams (1981a,b) indicated that such analyses could identify earthquakes of $M_w \geq 6.75$ that occurred within the past few hundred or thousand years.

Schuster *et al.* (1992) used a similar approach to date prehistoric rock avalanches that dammed streams in the Olympic Mountains in Washington State. Synchronous dates for several such avalanches indicate a common triggering event at about 1100 years BP, which they argued was a large earthquake. Several lines of evidence for seismic triggering were cited: (1) the rock that failed is not known to have failed historically either during large storms or in moderate earthquakes; (2) more than 40% of a recent inventory of worldwide rock avalanches that formed landslide dams were formed by earthquake shaking (Costa and Schuster, 1991); and (3) in New Zealand, the distribution of landslide-dammed lakes approximates the distribution of shallow earthquakes having magnitudes 6.5 or greater (Perrin and Hancox, 1992).

Jacoby *et al.* (1992) used dendrochronology to date prehistoric landslides that moved into Lake Washington near Seattle. They were able to correlate the tree-ring records from these landslides directly with a tree buried in a tsunami deposit elsewhere in the region. Thus, they inferred an earthquake origin for the Lake Washington landsliding since it was synchronous with a deposit of more certain seismic origin.

Jibson and Keefer (1989) used a regional analysis based on both spatial distribution and synchronous age. They used discriminant analysis and multivariate regression to analyze the geographic distribution of three distinct types of landslides along bluffs that extend more than 300 km through the New Madrid seismic zone. Field evidence indicated that landslides of two of the three types (old coherent slides and earth flows) were synchronous and could have ages consistent with triggering in the 1811–1812 earthquakes there; landslides of the third type (young rotational slumps) appeared much younger and unrelated to seismic activity. The bluffs were divided into segments 762 m (2500 ft.) long, and the percentage of the length of each segment covered by landslides of the three types was measured for use as the dependent variable in the statistical analyses. Independent variables measured for each segment included slope height, slope angle, stratigraphic thicknesses of various units, slope aspect, and proximity to the estimated hypocenters of the 1811–1812 New Madrid earthquakes. Discriminant analysis showed that bluffs having old coherent slides and earth flows are significantly closer to the estimated hypocenters of the 1811–1812 earthquakes than bluffs without these types of slides (Jibson and Keefer, 1989). Bluffs having young slumps showed no such correlation. Multiple regression analysis, which simultaneously combined all factors, showed that the distribution of old coherent slides and earth flows correlates strongly with proximity to the hypocenters of the 1811–1812 earthquakes, as well as with slope height and aspect (Jibson and Keefer, 1989). Again, young slumps showed no such correlation with earthquake-related independent variables. The results of these statistical analyses thus showed that old coherent slides and earth flows in the New Madrid seismic zone are spatially related to the 1811–1812 earthquake hypocenters and thus probably formed in those earthquakes. This type of analysis can be used only in areas where landslide locations can be correlated with well-defined seismic source zones.

8.4.2   Landslide Morphology

Some landslides have morphologies that strongly suggest triggering by earthquake shaking. For example, stability analyses of landslides on low-angle basal shear surfaces show that they generally form much more readily under the influence of earthquake shaking than in other conditions (Hansen, 1965; Jibson and Keefer, 1988, 1993). Landslides that formed as a result of liquefaction of subsurface layers are also much more likely to have formed seismically than aseismically (Seed, 1968). Perrin and Hancox (1992) indicated that slides that form as a result of intense rainfall are more fluid and tend to spread out more across a depositional area, whereas seismically induced landslides could have a blockier appearance and a more limited depositional extent in some cases. None of these criteria is definitive, but the types and characteristics of landslides described previously do suggest seismic triggering and can be used as corroborative evidence of earthquake triggering.

Solonenko (1977a,b) described several types of earthquake-triggered landslides documented in the former Soviet Union, some of which had morphologies that he argued could be unique to seismic origin. His descriptions of such landslides can be condensed into six types: (1) subsidence of areas tens of square kilometers in extent by the opening of fracture systems in very large ($M_w > 8$) earthquakes; (2) collapse of slopes and mountain spurs crossed by active faults; (3) toppling of steep mountain peaks; (4) translational or rotational sliding of topographic benches covering several square kilometers; (5) rock falls and rock avalanches having abnormally long runout distances, including extreme runout events that might have moved on an air cushion; and (6) "ground avalanches and flows," where thick deposits of weak sediment, such as loess, collapse and flow large distances even on nearly level ground.

Landslide size is also cited widely as evidence of seismic triggering (e.g., Whitehouse and Griffiths, 1983; Nikonov, 1988a,b; Crozier, 1992). Size commonly is inferred to be a factor because of observations of large landslides in past earthquakes (Solonenko, 1977a,b; Whitehouse and Griffiths, 1983; Nikonov, 1988a,b). In areas where large landslides have been documented in historical time to occur only during earthquakes, the large size of prehistoric landslides could suggest seismic origin and could even be used to infer the relative size of the triggering earthquake (Nikonov, 1988a,b); very large landslides commonly are triggered by longer duration and longer period shaking, which generally relate to larger magnitude earthquakes.

Multiple lines of evidence strengthen an argument for seismic triggering. For example, Philip and Ritz (1999) argued that a huge landslide in the Gobi-Altay region of Mongolia was seismically induced because of its large size; its proximity to a major, active fault; and its low-angle basal shear surface.

Remember, however, that landslides of all sizes form in the absence of earthquake shaking in a wide variety of environments. And Naumann and Savigny (1992) reached an opposite conclusion from their analysis of the stability of several rock avalanches in British Columbia, Canada. They showed that the larger slides analyzed were more susceptible to failure from increased pore-water pressure than from earthquake shaking and that earthquakes are more likely to trigger smaller rock falls (Naumann and Savigny, 1992).

Although earthquake-induced landslides generally fail during or immediately after seismic shaking, some earth flows and debris slides/debris flows have a delayed response and begin movement days after an earthquake. Jibson *et al.* (1994a,b) list six known instances of such activity: three involved slow-moving earth flows and three involved fast-moving debris slides/debris flows. The three earth flows—the Kirkwood earth flow, following the 1959 M_w 7.1 Hebgen Lake, Montana, earthquake (Hadley, 1964), and the Chordi and Zhashkva landslides, following the 1991 M_w 7.0 Racha, Republic of Georgia, earthquake (Jibson *et al.*, 1994a,b)—were similar in size (250–500 m wide, 1000 m long), began moving 3–5 days after the earthquake, and moved 30–70 m in the 3–4 weeks following the earthquake. Large (0.2–2.5 $\times$ 10^6 m^3) debris slides/debris flows occurred 13 days after the 1906 M_w 8.25 San Francisco, California earthquake; 3 days after the 1949 M_w 7.1 Tacoma, Washington, earthquake; and 2 days after the 1983 M_w 7.0 Borah Peak, Idaho, earthquake (Jibson *et al.*, 1994a,b). Several investigators (e.g., Keefer *et al.*, 1985; Wood, 1985; Rojstaczer and Wolf, 1992) have speculated that delayed landslide movement is caused by increased groundwater flow arising from either locally increased permeability (due to coseismic fracturing) or increased pore pressure. The length of the delay might represent the time needed for groundwater levels in the slide (or pore pressures on the slide plane) to lower the factor of safety (FS) below 1.0. In addition, Chleborad (1994) stated that a small failure high on the slope loaded the head of the Tacoma Narrows, Washington, debris slide, leading to massive failure 3 days later after the 1949 earthquake. Because most delayed landslides are initiated by groundwater changes that could just as well have been induced over a longer period of time, they are, at best, ambiguous indicators of paleoseismicity.

In summary, landslide morphology and size can, in some circumstances, be used as corroborative evidence for seismic triggering, but only when a clear link between a specific morphology or size and earthquake triggering is observed.

8.4.3 Sackungen

Sackungen are geomorphic features in mountainous areas that are characterized by ridge-parallel, uphill-facing scarps; double ridge lines; and troughs or closed depressions along ridge crests. While topography and gravity clearly influence the ridge-parallel geometry of sackungen, several different processes for their origin have been proposed, including gravitational spreading due to long-term creep (Bovis, 1982; Varnes *et al.*, 1989; Chigira, 1992), stress relief due to deglaciation (Beck, 1968; Radbruch-Hall, 1978; Bovis, 1982), faulting (Cotton, 1945; Cotton *et al.*, 1990; Johnson and Fleming, 1993), strong shaking

(Solonenko, 1977b; Clague, 1979, 1980; Ponti and Wells, 1991), or a combination of factors (Johnson and Cotton, 2005). It now appears that sackungen can form under a variety of conditions or combinations of conditions because sackungen have been documented to have formed in different ways in different tectonic and geologic settings (McCalpin, 1999).

Bovis (1982) and Varnes *et al.* (1989) studied sackungen in seismically quiescent parts of western North America and argued that movement stems from long-term, gravity-driven creep, although both studies mention tectonism as a possible contributor in some cases. Modeling of stresses in ridges has confirmed that gravitationally driven creep in weak, foliated rocks can form such features (Savage and Swolfs, 1986; Savage *et al.*, 1986; Savage and Varnes, 1987; Pan *et al.*, 1994). Trenching studies of sackungen in Colorado and British Columbia, both relatively aseismic areas, revealed progressive fold deformation supportive of a creep mode of failure (McCalpin and Irvine, 1995).

Earthquake shaking has been invoked frequently as a cause of sackung formation (e.g., Jahn, 1964; Salvi and Nardi, 1995). Beck's (1968) investigations of sackungen in New Zealand concluded that earthquake shaking was the most likely trigger of movement, primarily because the sackung topography appeared stable over long periods of seismic quiescence and because sackung were abundant in seismically active areas there. Tabor (1971) suggested that earthquake shaking might play a minor role in sackung formation in the Olympic Mountains of Washington. Russian workers have long believed that most, if not all, sackungen are created during earthquakes (Solonenko, 1977a,b; Nikonov, 1988a,b; see photographs in Khromovskikh, 1989, pp. 270–271). Several recent postearthquake investigations have documented fresh sackung features (e.g., Dramis and Sorriso-Valvo, 1983; Cotton *et al.*, 1990; Ponti and Wells, 1991; Blumetti, 1995; McCalpin and Hart, 2003; Jibson *et al.*, 2004), including those detected via InSAR and later confirmed in the field (Moro *et al.*, 2007). These studies suggest that the sackungen formed as a result of strong shaking, but in some cases they formed in the immediate vicinity of, and parallel to, the seismogenic faults, which suggests that fault-related tectonic deformation as well as strong shaking might contribute to their formation (Ponti and Wells, 1991; Johnson and Fleming, 1993; Jibson *et al.*, 2004).

Sackungen have been directly associated with both active and inactive fault traces. For example, McCleary *et al.* (1978) concluded that some troughs and uphill-facing scarps in the Cascade Mountains of Washington were tectonic reactivations of old faults. The origin of extensive and complex ground cracking in the vicinity of the fault trace of the 1989 Loma Prieta, California, earthquake remains controversial: Ponti and Wells (1991) argued that most of the observable ground cracking was caused by downslope movement driven by prolonged seismic shaking; Cotton *et al.* (1990) proposed that the deformation was caused by coseismic bending-moment slip on bedding-plane faults; and Johnson and Fleming (1993) believed the cracking was directly related to surface fault rupture. In the French Alps, Hippolyte *et al.* (2006) argued that some sackung features were seismically induced reactivations of old bedrock faults.

An even more complex situation arises when evidence shows that a single sackung feature has formed by a combination of processes. A landslide and related sackung features in southern California were shown to have had multiple episodes of movement, some related to seismic shaking and some to periods of climatically induced increased groundwater levels (Johnson and Cotton, 2005).

Thus, sackungen can form in a variety of tectonic and geologic environments and can form by several different processes. This obviously makes paleoseismic interpretation of sackungen difficult and commonly tenuous. Criteria for establishing the seismic or nonseismic origin of sackungen have been proposed with the aim of differentiating between features indicating abrupt, episodic movement versus those indicating gradual, continuous movement (McCalpin and Irvine, 1995; Psutka, 1995; McCalpin,

1999, 2003). McCalpin and Hart (2003) proposed seven criteria that included stratigraphic, geomorphic, and structural evidence: (1) evidence of continuous deformation of sediments suggests a nonseismic origin; (2) sackung deformation events that are contemporaneous with other regional paleoseismic features could be coseismic; (3) if sackungen overlie a steeply dipping crustal fault zone that has a net displacement much larger than the scarp height, the fault could be active; (4) gravity-driven sackungen tend to occur in swarms and be shorter, less continuous, and arcuate, whereas tectonic scarps tend to be longer, more continuous, singular, and straighter; (5) height-to-length ratios of gravity-driven sackungen are much greater than those of tectonic faults; (6) an asymmetrical fault zone having a sharp upper boundary and transitional brecciated lower boundary is more likely to be a sackung than a tectonic fault; and (7) subsurface deformation zones of tectonic faults can occur in any spatial relation with the modern topography, whereas subsurface deformation zones of sackungen are closely related to modern topography.

No single criterion is sufficient to unequivocally prove the seismic or aseismic origin of a sackung feature. And a seismic origin could have resulted from strong shaking, primary tectonic faulting, sympathetic faulting on a feature other than the seismogenic fault, or a combination of these factors. Also, a single sackung feature could have had both seismic and aseismic episodes of movement. Therefore, paleoseismic interpretation of sackungen is generally quite challenging and in many cases impossible. The difficulty is exemplified by two recent papers on antislope scarps in the Alps, where one group correctly identified the scarps and grabens as sackung (Hippolyte *et al.*, 2006), whereas another group, using four oversimplified geometric criteria, identified most of their scarps as active tectonic/seismogenic faults (Persaud and Pfiffner, 2004). In some cases, however, evidence for abrupt episodes of movement that can be linked to a contemporaneous seismic event can provide valuable paleoseismic evidence.

In many cases, identifying a sackung as seismic will rely on temporal criteria (criteria 1 and 2 of McCalpin, cited above), rather than static geometric/structural criteria. This means the movement history of the sackungen must be dated somehow, or the origin of the sackungen will remain ambiguous (as in Gutierrez *et al.*, 2005). Most sackung ages to date have come from radiocarbon dating of deformed trough sediments (e.g., McCalpin and Irvine, 1995; McCalpin and Hart, 2003; Gutierrez *et al.*, 2008a,b). However, it is also possible to date the emergent bedrock "fault" plane using cosmogenic isotopes such as ^{10}Be (Hippolyte *et al.*, 2006), in a manner identical to that used on normal faults (Chapter 3).

8.4.4 Sediment from Earthquake-Triggered Landslides

Earthquake-triggered landslides can profoundly affect alluvial systems by denuding slopes, generating large amounts of disrupted sediment that will move into the alluvial system, and physically disrupting drainage systems. The commonest types of landslides triggered by earthquakes are shallow, highly disrupted slides in unconsolidated surficial material (Keefer, 1984), and the deposits of these types of landslides tend move quickly into stream drainages. Thus, earthquakes can deposit large pulses of sediment into alluvial systems, which can (1) lead to the creation of new fans, (2) cause widespread aggradation of channels, (3) provide material for subsequent deposition on fan surfaces by debris flows and hyperconcentrated flows, and (4) affect the overall development of the fan surface on the long term (Keefer, 1999). Thus, landslides triggered by earthquakes can leave evidence in the depositional record of alluvial systems.

These types of effects were documented by Keefer and Moseley (2004) following the 2001 M_w 8.4 earthquake in southern Peru. They proposed the term *shattered landscape* to describe areas experiencing widespread ground-failure effects including abundant landslides, pervasive ground cracking, microfracturing of surficial hillslope materials, collapse of drainage banks over long stretches, widening of hillside rills, and lengthening of first-order tributary channels. They concluded that this shattering of the landscape (1) increases the capacity of channels to carry runoff by enlarging upstream channels and (2) detaches large amounts of loose slope material, which increases the amount of sediment available for transport and deposition.

Satterlee et al. (2001) compared and contrasted flood and debris-flow deposits attributed to a fourteenth century flood to deposits dated to shortly after an $M_w \sim 8.4$ earthquake in AD 1604. As compared to the flood-related deposits, the postearthquake deposits (1) were abnormally thicker, (2) contained the largest clasts, (3) contained the highest percentage of coarse clasts, and (4) contained more angular clasts. They proposed using these criteria to identify deposits that originated after significant earthquakes. Keefer et al. (2003) applied these criteria to propose a paleoseismic interpretation of debris-flow and flood deposits in southern Peru related to El Niños (climatic events causing above-normal rainfall) following great earthquakes. They inferred four earthquakes in the last 38,000 years.

8.4.5   Landslides That Straddle Fault

In some areas, landslides have formed on slopes immediately above fault traces, and the slide mass has extended across the trace (Hunt, 1975; Morton and Sadler, 1989). Subsequent surface movement of such a fault would offset the landslide mass and allow estimation of fault slip rates if the slide could be dated. This approach does not require that the landslide be seismically triggered because the paleoseismic interpretation is based on postlandslide fault offset of the landslide mass. However, landslides triggered in the immediate vicinity of active faults commonly are seismically triggered (Burrows, 1975).

8.4.6   Precariously Balanced Rocks

Brune (1996) proposed using precariously balanced rocks as crude paleoseismoscopes. The premise of this approach is that areas containing precariously balanced rocks indicate the absence of strong earthquake motions since the precarious rocks developed; paleoseismic interpretations can be made by estimating the peak accelerations required to cause toppling and the length of time the rocks have been precarious (Anderson and Brune, 1999). The shaking required to topple precarious rocks has been estimated using analytical and numerical modeling, physical modeling using shaking-table tests, and field experiments on actual precarious rocks (Shi et al., 1996; Anooshehpoor and Brune, 2002; Anooshehpoor et al., 2004). Brune (1999) defined precarious rocks as being capable of being toppled by peak accelerations of 0.1–0.3 g; rocks requiring 0.3–0.5 g were defined as semiprecarious.

Precarious rocks have been dated cosmogenically and by analysis of rock-varnish microlaminations. Results of dating studies have shown that precarious rocks in various locations in southern California, are >10.5 ka; rocks at Yucca Mountain, Nevada, are >10.5 to >27.0 ka (Bell et al., 1998).

Brune (1996, 2002) studied several areas around epicenters of large historical earthquakes in California and Nevada and found few or no precariously balanced rocks; he concluded that such rocks had been toppled by previous earthquakes. He also investigated areas of undocumented seismic potential and suggested that those areas where balanced rocks were present had not experienced ground shaking exceeding about 0.2 g in the past several thousand years, the estimated time needed to form the balanced rocks.

A study of the 1991 Hector Mine, California, earthquake (M_w 7.1) appeared to confirm some aspects of the precarious-rock methodology (Brune, 2002). Some previously documented precarious rocks (Brune, 1996) were toppled by the earthquake. A nearby strong-motion station recorded ground accelerations of about 0.2 g, similar to the estimated toppling accelerations from previous studies. Trenching studies indicated that the last earthquake on that segment of the fault occurred more than 10 ka, which is consistent with the estimated ages of the precarious rocks (Brune, 2002).

Other studies have shown significant inconsistencies between results of precarious-rock studies and other seismic hazard models (e.g., Brune, 1996; Anderson and Brune, 1999; Stirling *et al.*, 2002a,b). Although several possible reasons for these inconsistencies have been suggested, there is no consensus regarding the reasons for these inconsistencies and, therefore, of the validity of the results of the precarious-rock studies. In this context, several caveats regarding precarious-rock studies should be kept in mind:

1. Large uncertainties exist in the required toppling accelerations owing to both the geometric complexity of the rocks and the complexities of 3D ground motion.

2. Not all precarious rocks in a given area will be toppled by the estimated threshold ground shaking. Studies of overturning of tombstones in Japan have shown that a given threshold acceleration will overturn only a fraction of a group of seemingly identical tombstones (Ikegami and Kishinouye, 1950; Omote *et al.*, 1977). Therefore, finding some precarious rocks in an area does not necessarily mean that the area has not experienced the threshold ground shaking.

3. Establishing the age of precarious rocks does not necessarily determine the minimum time since a certain level of shaking has occurred because the toppling acceleration of a given rock will have been continuously changing as the rock has evolved into a precarious state. For example, a precarious rock with an estimated age of 20 ka and a present-day toppling acceleration of 0.2 g might have had a toppling acceleration of 1.0 g at 20 ka, 0.5 g at 10 ka, etc.; therefore, it cannot be concluded that the present-day toppling acceleration has not been exceeded in 20 ka.

8.4.7 Speleoseismology

Speleoseismology is the investigation of earthquake records in caves. Such records can include broken speleothems (stalactites, stalagmites, soda straws, etc.), cave-sediment deformation structures, offset along fractures and bedding planes, simple rock falls, and coseismic fault displacement (Becker *et al.*, 2006). Before an earthquake origin can be inferred, all other possible causes of the disturbance must be ruled out. Such causes include human or animal disturbance, water flow, ice movement, debris flow, and sediment creep. Accounts of possible earthquake effects in caves date back at least to the early twentieth century, and formal studies of possible earthquake effects in caves began appearing in research journals in the 1960s. The large majority of speleoseismological studies have been conducted and published in Europe (see Forti, 2001; Becker *et al.*, 2006).

Forti and Postpischl (1984) were the first to detail a systematic method for analyzing the toppling of stalagmites for paleoseismic interpretation. By measuring and dating tilting and collapse of many stalagmites in a region, they differentiated sudden (seismic) versus gradual movements and local versus regional causes. Tilting and collapse events can be dated by analysis of radiometrically determined speleothem growth rates; uranium-series isotopes can be analyzed to date speleothems precisely within the 0–500 ka range (Pons-Branchu *et al.*, 2004). By modeling stalagmites as simple inverted pendulums, Forti and Postpischl (1984) estimated the minimum ground shaking necessary to cause collapse using pseudostatic engineering analysis. Numerous studies followed their methodology and attributed broken speleothems to earthquakes (see Forti, 2001; Becker *et al.*, 2006).

Lacave *et al.* (2000, 2004) used numerical and physical models to examine the ground shaking that would be required to break and topple various types of speleothems. They measured the natural frequencies and damping characteristics of speleothems and the peak ground accelerations (PGAs) necessary to break them. They concluded that the natural frequencies of most speleothems are between 50 and 700 Hz, well above the range of seismically generated ground motion (0.1–30 Hz). The only exceptions are so-called soda straws, long slender speleothems that can have natural frequencies as low as 20 Hz. Their studies also indicated that most speleothems would require ground accelerations in excess of 1 g to cause breakage; some very long, thin soda straws, however, could be broken at accelerations as low as 0.1–0.2 g. They concluded that only exceptionally long, thin speleothems having weak sections are likely to break during earthquakes. And even in a well-documented strong earthquake, only about 2% of such structures broke (Gilli, 2004). Lacave *et al.* (2004) concluded that neither the presence of broken nor unbroken speleothems could be used to accurately quantify ground accelerations, but that they might be used for case studies of recent earthquakes where the population of broken and unbroken speleothems is known both before and after the earthquake.

Re-evaluation of several earlier studies concluded that many broken speleothems previously attributed to earthquake shaking were probably broken by other processes (e.g., Gilli, 1999, 2004; Kempe and Henschel, 2004).

8.4.8   Summary

A wide variety of methods for interpreting the seismic origin of landslides has been developed and, in some cases, successfully applied to paleoseismic analysis. Virtually all of the methods summarized in this section have one aspect in common, which is stated explicitly in most papers: the seismic origin of the features being interpreted remains tentative and cannot be proven, because in each case a nonseismic process could have produced the observed features. Circumstantial evidence for seismic triggering ranges from very strong to extremely tenuous. Indeed, on the latter end of the spectrum, the reasoning can be rather circular: an earthquake origin for a feature is assumed and then an earthquake origin is interpreted and concluded from analysis of that feature. Any paleoseismic interpretation of a feature is limited primarily by the certainty with which seismic triggering can be established. The following section addresses this dilemma by describing an approach to assess directly the conditions leading to failure of individual landslides.

8.5   Analysis of the Seismic Origin of a Landslide

The most direct way to assess the relative likelihood of seismic versus aseismic triggering of an individual landslide is to apply established methods of static and dynamic slope-stability analysis (Lee and Edwards, 1986; Crozier, 1992; Jibson and Keefer, 1993). The first step in such an analysis involves constructing a detailed slope-stability model of static conditions to determine if failure is likely to have occurred in any reasonable set of groundwater and shear-strength conditions in the absence of earthquake shaking. All potential nonseismic factors must be considered; these might include processes such as fluvial or coastal erosion that oversteepens the slope or undrained failure resulting from rapid drawdown (for slopes subject to submersion). If aseismic failure can reasonably be excluded even in worst-case conditions (minimum shear strength, maximum piezometric head), then an earthquake origin can be inferred.

Dynamic slope-stability analyses can then be used to estimate the minimum shaking conditions that would have been required to cause failure. In the sections that follow, a method for conducting such an analysis is described using an example from the New Madrid seismic zone summarized from Jibson and Keefer (1993) and updated with more recent data and modeling approaches.

8.5.1 Physical Setting of Landslides in the New Madrid Seismic Zone

The New Madrid earthquakes of 1811–1812 are the largest historical earthquakes in the central and eastern United States. Three events of large magnitude occurred on 16 December 1811, 23 January 1812, and 7 February 1812, and thousands of aftershocks of moderate and small magnitude occurred in the months between and after the principal shocks. The sparse population in the area at the time and the paucity of eyewitness accounts makes it difficult to estimate the magnitudes and locations of these events. Magnitude (M_w) estimates of the 16 December event range from 7.3 to 8.1, for the 23 January event from 7.0 to 7.8, and for the 7 February event from 7.5 to 8.0 (Johnston, 1996; Hough *et al.*, 2000; Bakun and Hopper, 2004). Figure 8.2 shows the epicentral zone of the 1811–1812 earthquakes and the estimated sources of the three major events. Current seismic hazard analysis of the region indicates that the 1811–1812 events probably occurred on three interrelated faults (Wang *et al.*, 2003). Although these faults have no consistent surface expression, they can be located approximately using recently monitored earthquake locations, regional geophysical studies, and reports of geologic effects from the 1811–1812 events (Figure 8.2).

The 1811–1812 earthquakes triggered many large landslides (Fuller, 1912) along the bluffs that form the eastern edge of the Mississippi River alluvial plain in Tennessee and Kentucky (Figure 8.2). Many of these landslide features remain visible along the bluffs, and one of these landslides was analyzed to determine if a seismic versus nonseismic origin could be established with a reasonable level of confidence.

The bluffs in the study area are not, for the most part, active river banks and thus are subject to landsliding related to fluvial erosion in only a few locations. The bluffs stand as high as 70 m above the alluvial plain of the Mississippi River and therefore are not subject to landsliding from conditions such as rapid drawdown because the bluff is never inundated by flooding to a significant height. The average height of the bluffs in this area is 35 m, and slope angles range from a few degrees to almost vertical, but typically are 15–25°.

The base of the bluffs throughout most of the area is formed by as much as 45 m of shallow-marine clays and silts of the Eocene Jackson Formation (Conrad, 1856). Lying unconformably on the Jackson Formation is as much as 20 m of Pliocene alluvial gravel and sand of the Lafayette Gravel (McGee, 1891; Potter, 1955). The bluffs are capped by 5–50 m of Pleistocene loess lying unconformably on the Jackson Formation and Lafayette Gravel. The average thickness of the loess in the area is about 15 m.

A translational block slide about 11 km north of Dyersburg, Tennessee, referred to as the Stewart landslide, was chosen for detailed analysis (Figure 8.2). This landslide is representative of coherent block slides in the area, which previous research (Jibson and Keefer, 1988, 1989) indicated were probably triggered by the 1811–1812 earthquakes. Figure 8.3 shows a profile of the Stewart landslide.

8.5.2 Geotechnical Investigation

Four rotary drill holes were placed along the line of profile to determine the bluff stratigraphy and to procure soil samples for geotechnical testing (Figure 8.3). Standard penetration testing (SPT) yielded

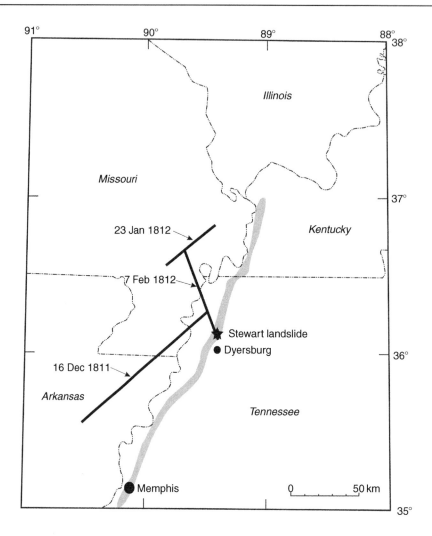

Figure 8.2: Map of the epicentral region of the 1811–12 New Madrid earthquakes. Bold lines show approximate locations of the likely causative faults of the three principal earthquakes in the sequence (Wang *et al.*, 2003). Shaded area shows bluffs forming the eastern edge of the Mississippi River alluvial plain (width exaggerated for visual emphasis). Star shows location of Stewart landslide.

split-spoon samples, which typically were heavily disturbed by the sampling process and were used primarily for determining index properties, such as grain size, plasticity, water content, and color. Several 13-cm-diameter undisturbed piston cores were procured to measure soil unit weight and shear strength, both needed for limit-equilibrium stability analysis. Jibson (1985) described the sampling methods in detail.

Shear strength can be characterized in different ways to model different types of failure conditions. In aseismic conditions, drained or effective shear strengths are used because pore-water pressures are assumed to be in static equilibrium. During earthquakes, soils behave in a so-called undrained manner because excess pore pressures induced by the transient ground deformation cannot dissipate during the

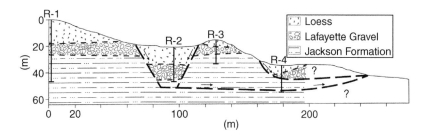

Figure 8.3: Cross section of Stewart landslide showing subsurface stratigraphy (identified from drill holes designated R-1 through R-4) and diagrammatic representation of failure surfaces. Undisturbed stratigraphy is shown at R-1. [Modified from Jibson and Keefer (1993).]

brief duration of the shaking; therefore, undrained or total shear strengths are used to model seismic failure conditions. Drained (effective) shear strengths were measured using two methods: (1) direct shear in which the rate of strain was slow enough to allow full drainage and (2) consolidated–undrained triaxial (CUTX) shear in which pore pressures were measured to allow modeling of drained conditions (Jibson, 1985). Undrained (total) shear strengths were measured primarily by CUTX tests. CUTX test results were supplemented by vane-shear and penetrometer data and correlation with SPT blow counts where undisturbed samples were unavailable.

8.5.3   Static (Aseismic) Slope-Stability Analysis

Figure 8.4 shows an idealized model of the prelandslide bluff in drained conditions, appropriate for modeling static (aseismic) stability. The bluff is 45 m high as measured from the profile (Figure 8.3). Undisturbed bluffs adjacent to the Stewart slide have slopes of about 20° and have simple, uniformly sloping faces. Geotechnical properties of the stratigraphic layers in the model were assigned using the results of the shear-strength tests; layers where no shear-strength tests were performed were assigned strengths based on stratigraphic and index-property correlation with layers where strengths were measured (Table 8.2).

Lack of data made detailed modeling of groundwater conditions along the bluffs difficult. Therefore, several potential groundwater conditions were modeled that effectively bracket the most and least critical conditions that are physically possible (Figure 8.5). Because of the local topography and hydraulic properties of the bluff materials, the most critical condition modeled (Figure 8.5, condition 1) is a more critical situation than can realistically exist in the bluffs and thus provides a worst-case bounding condition. The most likely groundwater condition was also modeled: a water table sloping upward from the base of the bluffs to the top of the Jackson Formation, and a second water table perched on the relatively impermeable Jackson that saturates the Lafayette Gravel.

The STABL computer program (Siegel, 1978) was used to determine the stability of the modeled bluff in aseismic conditions. STABL searches for the most critical failure surface by randomly generating circular, wedge, and irregular slip surfaces and calculating the factor of safety for each generated surface. Factor of safety (FS) is the ratio of the sum of the resisting forces or moments that act to inhibit slope movement to the sum of the driving forces or moments that tend to cause movement. Slopes having FS greater than 1.0 are thus stable; those having FS less than 1.0 should move. The program plots the 10 most

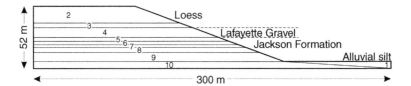

Figure 8.4: Idealized model of pre-landslide bluff at Stewart site in drained conditions, appropriate for modeling aseismic conditions. Table 8.2 shows soil properties for each designated layer in the computer model. [Modified from Jibson and Keefer (1993).]

critical surfaces of each type and their FS. The geometry of the actual failure surface (shown diagrammatically in Figure 8.3) was estimated by locating weak or disturbed layers by drilling and by analysis of the surface geometry of the landslide. The FS for this surface was calculated using the simplified Janbu method (Siegel, 1978) for each groundwater condition.

Determining the stability of the bluffs from the FS requires judgment. Gedney and Weber (1978) recommended that engineered slopes have safety factors between 1.25 and 1.50 for the type of analysis used here. Because of the high density of good-quality geotechnical data, this range is used as the criterion to evaluate slope stability: Between FS 1.00 and 1.25, slopes are considered to be marginally stable; between FS 1.25 and 1.50, slopes are considered to be stable; and above FS 1.50, slopes are considered to be very stable.

The results of the stability analyses are summarized in Table 8.3. The lowest FS in the most critical groundwater condition is 1.32, which indicates that the bluff at the Stewart site is stable in aseismic conditions even in the most critical groundwater condition. In the most likely groundwater condition (sloped and perched), the minimum FS is 1.82, indicating a very stable bluff. The FS of the estimated actual failure surface in the most likely groundwater condition is 1.88.

The analysis shows that an artesian piezometric surface tens of meters above ground level at the top of the bluff would be needed to reduce the FS to 1.0. Such an artesian condition is impossible because (1) the regional geology and topography preclude such a condition because the top of the bluff is 30–70 m above the alluvial plain, and no topographically higher artesian recharge area exists and (2) a piezometric surface high above the bluff-top that dips steeply to the base of the bluff is physically unrealistic.

Figure 8.6 shows the locations of the most critical slip surfaces of various shapes and of the actual slip surface. All the surfaces have grossly similar shapes, but the most critical computer-generated surfaces all lie well above the actual failure surface. This disparity suggests that the sliding did not take place under drained, static conditions.

The rather high FS values, even in unrealistically high groundwater conditions, and the disparity between the most critical computer-generated slip surfaces and the actual surface indicate that it is highly unlikely that the existing landslide at the Stewart site formed in aseismic, drained conditions.

8.5.4   Dynamic (Seismic) Slope-Stability Analysis

Jibson and Keefer (1993) evaluated the seismic stability of the bluff using the dynamic displacement analysis developed by Newmark (1965), which is used widely in engineering practice (Seed, 1979a,b;

Table 8.2: Geotechnical properties of layers in models of drained (aseismic) and undrained (seismic) conditions

Layer[a]	Unit weight (kN/m³)	Friction angle (°)	Cohesion[b] (kPa)
Drained (aseismic) model			
1	19.3	34	12
2	19.6	35	24
3	19.8	30	21
4	22.0	32	15
5	18.1	15	98
6	18.1	20	86
7	17.3	12	34
8	18.1	20	108
9	18.1	15	98
10	18.1	12	120
Undrained (seismic) model			
1	19.3	0	400
2	19.6	0	220
3	19.6	0	761
4	19.6	0	220
5	22.0	0	187
6	22.0	0	212
7	18.1	0	121
8	18.1	0	170
9	17.3	0	115
10	18.1	0	239
11	18.1	0	151
12	18.1	0	180
13	18.1	0	208

Data from Jibson (1985).
[a]Layer numbers refer to layers shown in Figures 8.4 (drained) and 8.7 (undrained).
[b]Undrained shear strength is treated as cohesion in the undrained model.

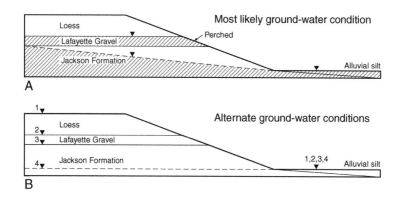

Figure 8.5: Groundwater conditions modeled in the slope-stability analyses. (A) Most likely groundwater condition; saturated zones shown by cross-hatched pattern. (B) Piezometric surfaces of four alternate groundwater conditions are shown by inverted triangles numbered 1–4. [Modified from Jibson and Keefer (1993).]

Jibson, 1993; Jibson *et al.*, 2000; Jibson and Jibson, 2003). Newmark's method models a landslide as a rigid-plastic friction block that slides on an inclined plane. The block begins to slide when a given critical (or yield) base acceleration is exceeded; critical acceleration is defined as the base acceleration required to overcome basal shear resistance and initiate sliding on the inclined plane. The analysis calculates the cumulative permanent displacement of the block as it is subjected to the effects of an earthquake acceleration-time history, and the user judges the significance of the displacement. Laboratory model tests (Goodman and Seed, 1966; Wartman *et al.*, 2003, 2005) and analysis of actual earthquake-induced landslides (Wilson and Keefer, 1983; Pradel *et al.*, 2005) have confirmed that Newmark's method can fairly accurately predict landslide displacements if slope geometry and soil properties are known accurately and if earthquake ground accelerations can be estimated using real or artificial acceleration-time histories. Detailed treatments of how to conduct a Newmark analysis on landslides in natural slopes have been published elsewhere (Jibson, 1993; Jibson and Jibson, 2003).

Newmark (1965) showed that critical acceleration is a simple function of the static factor of safety and the landslide geometry; it can be expressed as

$$a_c = (FS - 1)g \sin \alpha, \tag{8.1}$$

where a_c is the critical acceleration (in terms of g), the acceleration of Earth's gravity; FS is the static factor of safety; and α is the angle from the horizontal (hereafter called the thrust angle) that the center of mass of the potential landslide block first moves.

The algorithm first developed by Wilson and Keefer (1983) and modified by Jibson and Jibson (2003) is used to apply Newmark's method. The algorithm consists of a two-part integration with respect to time (1) the parts of the selected acceleration-time history that lie above the critical acceleration of the landslide block are integrated to yield the velocity of the block with respect to its base and (2) the velocity curve is then integrated to determine the cumulative permanent displacement of the block. Displacements estimated using this method are referred to as Newmark displacements.

Table 8.3: Static factors of safety from stability analyses of the Stewart landslide in drained and undrained conditions

Type of failure surface	Location of piezometric surface				
	Base of bluff[a]	Top of Jackson Formation[b]	Top of Lafayette Gravel[c]	Top of bluff[d]	Most likely[e]
Drained stability analyses					
Circular	**1.90**	**1.66**	**1.61**	1.35	**1.82**
Irregular	1.95	1.69	1.64	**1.32**	1.87
Wedge, layer 5	4.06	3.98	3.76	2.83	4.03
Wedge, layer 6	4.24	4.03	3.80	2.79	4.23
Wedge, layer 7	2.46	2.28	2.14	1.47	2.45
Wedge, layer 8	3.81	3.39	3.23	2.51	3.72
Wedge, layer 9	2.83	2.48	2.38	1.88	2.71
Wedge, layer 10	2.40	2.10	2.03	1.68	2.25
Actual surface	1.96	1.73	1.67	1.40	1.88
Undrained stability analysis					
Circular	1.72	1.72	1.64	**1.99**	1.62
Irregular	**1.64**	**1.64**	**1.55**	2.16	**1.53**
Wedge, layer 5	2.81	2.81	2.50	3.59	2.49
Wedge, layer 6	3.23	3.23	2.96	3.84	2.93
Wedge, layer 7	2.19	2.19	1.99	2.81	1.97
Wedge, layer 8	3.18	3.18	3.05	3.57	3.02
Wedge, layer 9	2.00	2.00	1.89	2.41	1.87
Wedge, layer 10	1.99	1.99	1.88	2.25	1.87
Actual surface	1.74	1.74	1.66	2.12	1.65

Results from Jibson (1985). Most critical surface for each groundwater condition is shown in bold type.
[a]Figure 8.5B, piezometric surface 4.
[b]Figure 8.5B, piezometric surface 3.
[c]Figure 8.5B, piezometric surface 2.
[d]Figure 8.5B, piezometric surface 1.
[e]Figure 8.5A.

Conducting a rigorous Newmark analysis requires knowing the critical acceleration of the landslide and selecting one or more earthquake acceleration-time histories to approximate the earthquake shaking at the site. The critical acceleration of a potential landslide can be determined in two ways: (1) for relatively

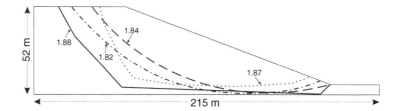

Figure 8.6: The three most critical slip surfaces and their factors of safety (FS = 1.82, 1.84, 1.87) for aseismic, drained conditions at the Stewart site in the most likely groundwater condition. Solid line shows location of actual failure surface (FS = 1.88). [Modified from Jibson and Keefer (1993).]

simple slope models where material properties do not differ significantly between layers, Equation (8.1) can be used to estimate the critical acceleration, and (2) for more complex slope models that include layers having complex geometries or widely differing material properties, the critical acceleration should be determined using iterative pseudostatic analysis, where different seismic coefficients are used until the static FS reaches 1.0. The seismic coefficient yielding a FS of 1.0 is the yield or critical acceleration (Abramson *et al.*, 1996). The slope model used for analysis of the Stewart landslide is simple enough that the critical acceleration can be estimated using Equation (8.1); this requires determining the static FS in undrained conditions and the thrust angle.

8.5.4.1 Undrained Static Factor of Safety

A layered model of the bluff in undrained conditions was constructed to model failure in seismic conditions (Figure 8.7). Undrained shear strength is treated as a single numerical quantity that is represented in the analysis as cohesion, and the friction angle is taken to be zero (Lambe and Whitman, 1969). Undrained shear strengths used in the model (Table 8.2) were measured directly in the laboratory, as described in Section 8.5.2. Because undrained strength depends in large part on consolidation stress, layers of roughly similar thickness were constructed that reflect the increase in shear strength with depth even for relatively homogeneous materials.

STABL was used to generate potential failure surfaces and to determine the most critical failure surface in the same manner as described in Section 8.5.2 for the aseismic stability analysis in drained conditions. Table 8.3 summarizes the results of the undrained slope-stability analyses using this model. The lowest FS is 1.53, which shows that the bluff is statically stable in undrained conditions. Figure 8.8 shows the locations of the most critical slip surfaces for the most likely groundwater condition. All the slip surfaces, including the actual failure surface, plot very close to one another and have similar factors of safety. Both circular surfaces have large radii and approximate planar basal shear surfaces, as is expected from the shape of the actual shear surface (Figure 8.3). The fact that the most critical computer-generated surfaces closely parallel the actual failure surface indicates that the model of the bluffs is reasonable and that slope failure is more likely to have occurred in undrained conditions than in drained conditions.

8.5.4.2 Thrust Angle

The thrust angle (α) is the direction in which the center of gravity of the slide mass moves when displacement first occurs. For a planar slip surface parallel to the slope face (an infinite slope),

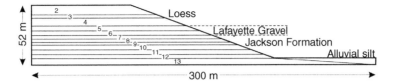

Figure 8.7: Idealized model of pre-landslide bluff at Stewart site in undrained conditions, appropriate for modeling seismic conditions. Table 8.2 shows soil properties for each designated layer in the computer model. [Modified from Jibson and Keefer (1993).]

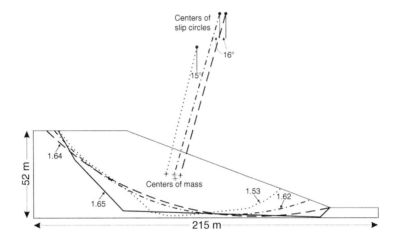

Figure 8.8: The three most critical slip surfaces and their factors of safety (FS = 1.53, 1.62, 1.64) for static, undrained conditions at the Stewart site in the most likely groundwater condition. Solid line shows estimated location of actual failure surface (FS = 1.65). Geometric constructions to determine thrust angles also shown. [Modified from Jibson and Keefer (1993).]

the thrust angle is the slope angle. For rotational movement on a circular surface, Newmark (1965) showed that the thrust angle is the angle between the vertical and a line segment connecting the center of gravity of the landslide mass and the center of the slip circle.

Figure 8.8 shows geometric constructions of the thrust angles for the two circular failure surfaces and the circular approximation of the irregular surface. Thrust angles for these surfaces all are 15–16°. The thrust angle of the actual surface is difficult to estimate because of its irregular shape and consequent complex movement. An average inclination of the actual failure surface was calculated by weighting the inclinations of the line segments forming the actual surface by their relative lengths. This yielded an average inclination of 16°, consistent with the other generated surfaces.

8.5.4.3 Critical Acceleration

Critical accelerations were calculated based on a thrust angle of 16° and on the factors of safety (FS 1.62, 1.64) of the two circular slip surfaces in the most likely groundwater condition. These slip surfaces most closely coincide with the actual surface and have the lowest factors of safety. Equation (8.1) yields critical accelerations of 0.17–0.18 g for these input values. A critical acceleration of 0.175 g is used in the subsequent analyses to calculate the Newmark displacement.

8.5.4.4   Earthquake Acceleration-Time History

The hypothesis being tested is that the Stewart landslide was triggered by the 1811–1812 earthquakes; therefore, earthquake acceleration-time histories must be selected to approximate the shaking conditions from the 1811–1812 earthquakes at the Stewart site. Estimating various ground-motion characteristics of the 1811–1812 earthquakes at the Stewart site and comparing these estimated characteristics with those of existing earthquake records provides a basis for choosing an input ground motion. Uncertainties regarding magnitudes, locations, and seismological characteristics of the 1811–1812 earthquakes, and the paucity of strong-motion records from large-magnitude earthquakes in stable cratonic interiors, make this selection process difficult. Estimating various seismological properties in different ways facilitates estimation of ranges of values that bracket the properties needed to select a suite of records. Key properties to consider in estimating ground motions and selecting records include magnitude, source distance, peak ground acceleration, Arias (1970) intensity, and shaking duration.

As stated in Section 8.5.1, there is considerable disagreement regarding the magnitudes of the 1811–1812 earthquakes because of differences in the characterization of intensity reports, stress drops, fault lengths, local site effects, and other important properties (Nuttli, 1973; Johnston, 1996; Hough *et al.*, 2000; Bakun and Hopper, 2004). Table 8.4 shows the range of magnitudes estimated in the recent literature (Johnston, 1996; Hough *et al.*, 2000; Bakun and Hopper, 2004); estimates span 0.5–0.8 magnitude units for the three principal earthquakes.

Estimating source distance is likewise difficult because of uncertainties regarding the locations of the 1811–1812 events and disagreement about how to estimate those locations. Various estimates of epicentral locations have been made (Nuttli, 1973; Johnston and Schweig, 1996; Bakun and Hopper, 2004), but these estimates differ by tens of kilometers in some cases. Most currently used attenuation equations use closest distance to the fault rupture rather than epicentral distance, therefore estimating the location of the fault-rupture zones is the best approach for determining distance measurements. Figure 8.2 shows diagrammatically the estimated locations of the causative faults of the 1811–1812 earthquakes, based on recently recorded seismicity, geological and geophysical studies, and damage patterns from 1811–1812. It appears most likely that the 16 December 1811 event occurred on the southern fault

Table 8.4: Estimated shaking conditions at the Stewart landslide site from the three principal 1811–1812 New Madrid earthquakes

Earthquake	M_w^a	r^b (km)	PGA^c (g)	$D_{5-95\%}^d$ (s)	D_N^e (cm)
16 Dec 1811	7.3–8.1	15	0.76–0.99	19–31	37–81
23 Jan 1812	7.0–7.8	60	0.09–0.16	21–34	0
7 Feb 1812	7.5–8.0	1	1.45–1.48	13–17	117–122

[a]Moment magnitude range from Johnston (1996), Hough *et al.* (2000), and Bakun and Hopper (2004).
[b]Closest distance to fault-rupture surface.
[c]Peak horizontal ground acceleration estimated using Campbell's (2003) attenuation model (Equation 8.2).
[d]Duration required to build up the central 90% of the Arias intensity, estimated using Equation (8.4).
[e]Median values of Newmark displacement from rigorous analysis of the records in Table 8.5 for the ranges of PGA shown.

Table 8.5: Strong-motion records used to model shaking conditions at the Stewart landslide from the 16 December 1811 and 7 February 1812 earthquakes

Earthquake	Record component	M_w	r (km)	PGA (g)	$D_{5-95\%}$ (s)
16 December 1811 earthquake					
Cape Mendocino 1992	RIO-270	7.1	19	0.39	15
Cape Mendocino 1992	RIO-360	7.1	19	0.55	11
Chi–Chi, Taiwan 1999	CHY080-000	7.6	7	0.90	22
Chi–Chi, Taiwan 1999	CHY080-270	7.6	7	0.97	22
Chi–Chi, Taiwan 1999	CHY101-000	7.6	11	0.44	27
Chi–Chi, Taiwan 1999	TCU045-000	7.6	24	0.51	11
Chi–Chi, Taiwan 1999	TCU045-270	7.6	24	0.47	11
Chi–Chi, Taiwan 1999	TCU072-000	7.6	7	0.40	24
Chi–Chi, Taiwan 1999	TCU072-270	7.6	7	0.49	22
Chi–Chi, Taiwan 1999	TCU074-270	7.6	14	0.60	12
Chi–Chi, Taiwan 1999	TCU078-270	7.6	8	0.44	26
Chi–Chi, Taiwan 1999	TCU079-000	7.6	10	0.39	27
Chi–Chi, Taiwan 1999	TCU079-270	7.6	10	0.74	24
Chi–Chi, Taiwan 1999	TCU084-000	7.6	10	0.42	23
Chi–Chi, Taiwan 1999	TCU084-270	7.6	10	1.16	15
Chi–Chi, Taiwan 1999	WGK-000	7.6	11	0.48	25
Duzce, Turkey 1999	DZC-270	7.1	8	0.54	11
Duzce, Turkey 1999	VO-000	7.1	8	0.97	13
Duzce, Turkey 1999	VO-090	7.1	8	0.51	13
Tabas, Iran 1978	DAY-TR	7.4	17	0.41	12
7 February 1812 earthquake					
Cape Mendocino 1992	CPM-000	7.1	9	1.50	6
Cape Mendocino 1992	CPM-090	7.1	9	1.04	10
Chi–Chi, Taiwan 1999	CHY028-000	7.6	7	0.82	6
Chi–Chi, Taiwan 1999	CHY080-000	7.6	7	0.90	22
Chi–Chi, Taiwan 1999	CHY080-270	7.6	7	0.97	22

(Continued)

Table 8.5: Strong-motion records used to model shaking conditions at the Stewart landslide from the 16 December 1811 and 7 February 1812 earthquakes (Cont'd)

Earthquake	Record component	M_w	r (km)	PGA (g)	$D_{5-95\%}$ (s)
Chi–Chi, Taiwan 1999	TCU079-270	7.6	10	0.74	24
Chi–Chi, Taiwan 1999	TCU084-270	7.6	10	1.16	15
Duzce, Turkey 1999	BOL-000	7.1	18	0.73	9
Duzce, Turkey 1999	BOL-090	7.1	18	0.82	9
Duzce, Turkey 1999	VO-000	7.1	8	0.97	13
Landers 1992	LCN-000	7.3	1	0.79	14
Landers 1992	LCN-275	7.3	1	0.71	13
Tabas, Iran 1978	TAB-074	7.4	3	0.71	16
Tabas, Iran 1978	TAB-344	7.4	3	0.81	16

segment, the 23 January 1812 event on the northern segment, and the 7 February 1812 event on the central transverse segment (Johnston and Schweig, 1996). The first two earthquakes are thought to have been strike-slip events, the last a thrust event. Thus, I measured distances from the Stewart landslide to the closest points on each of these fault segments assuming that the rupture reached the surface (Table 8.4). This yields source distances that in general are much smaller than epicentral-distance estimates used in previous analyses of this landslide (e.g., Jibson and Keefer, 1993; Jibson, 1996). For example, most studies place the epicenter of the 16 December event at the southern end of the southern fault segment (Figure 8.2), whereas the Stewart landslide is near the northern end of that segment.

PGA is estimated using empirical attenuation equations that depend on some combination of magnitude and source distance for a given site condition and focal mechanism. Several attenuation equations for midcontinent earthquakes have been published, and their predicted ground motions differ significantly (Atkinson and Boore, 1995, 2006; Frankel *et al.*, 1996; Toro *et al.*, 1997; Somerville *et al.*, 2001; Silva *et al.*, 2002; Campbell, 2003; Tavakoli and Pezeshk, 2005). For the purposes of this exercise, I chose Campbell's (2003) equation because its predicted ground motions lie near the center of the range of the various models. For the conditions under consideration, the equation takes the following form:

$$
\begin{aligned}
\ln \text{PGA} = {} & 0.0305 + 0.633\,M_w - 0.0427(8.5 - M_w)^2 \\
& -1.591 \ln\sqrt{r^2 + [0.683\exp(0.416\,M_w)]^2} + (0.000483\,M_w - 0.00428)r,
\end{aligned}
\tag{8.2}
$$

where PGA is the mean of the two horizontal components of peak ground acceleration (in terms of g), M_w is the moment magnitude, and r is the closest distance to the fault rupture (in km). PGA values for the conditions under consideration are listed in Table 8.4.

A measure of shaking intensity developed by Arias (1970) is useful in seismic hazard analysis and correlates well with distributions of earthquake-induced landslides (Harp and Wilson, 1995; Jibson *et al.*, 2000). Arias intensity is the integral over time of the square of the acceleration, expressed as

$$I_A = \frac{\pi}{2g} \int_0^{t_f} [a(t)]^2 dt, \tag{8.3}$$

where I_A is the Arias intensity (in units of velocity), $a(t)$ is the ground acceleration as a function of time, and t_f is the total duration of the shaking. Arias intensity thus accounts for both the amplitude and duration of the shaking and is a very useful measure of the total shaking content of an earthquake record. Wilson and Keefer (1985) were the first to develop a predictive equation to estimate Arias intensity as a function of magnitude and distance, but it was based on a rather small data set. Travasarou *et al.* (2003) developed an empirical regression equation based on a much larger data set (1208 records from 75 earthquakes), but the data were exclusively from interplate regions, where strong-motion attenuation is significantly greater than in intraplate regions. This yielded Arias intensity estimates that were unrealistically low for the conditions under consideration here; therefore, Arias intensity is not used for this comparison.

Estimating the duration of strong shaking can be done in several ways because several measures of duration have been proposed. The duration measure proposed by Dobry *et al.* (1978) as the time required to build up the central 90% of the Arias intensity (commonly referred to as Dobry duration or $D_{5-95\%}$) has come into common usage and will be used for this analysis. Dobry *et al.* (1978) developed an empirical regression equation to predict duration as a function of unspecified earthquake magnitude using 26 earthquake records. The availability now of much larger data sets facilitates development of updated regression models. I used the data set of Jibson and Jibson (2003), which contains 2160 strong-motion records from 29 earthquakes ($5.3 \leq M_w \leq 7.6$) and a full range of site conditions, to construct a regression model to predict $D_{5-95\%}$ as a function of magnitude and fault-rupture distance:

$$\log D_{5-95\%} = 0.252 \, M_w + 0.196 \log r - 0.784 \pm 0.187, \tag{8.4}$$

where $D_{5-95\%}$ is in seconds and r is in kilometers. This model's R^2 value is only 49%, but the coefficients are all significant above the 99.9% confidence level, and the model standard deviation (the last term in the equation) is not excessively large. Durations estimated using this equation are shown in Table 8.4.

Selecting a suite of strong-motion records that have characteristics similar to those listed in Table 8.4 allows us to estimate the Newmark displacements of the Stewart landslide that would have occurred as a result of the shaking from the three principal earthquakes in the 1811–1812 sequence. The software developed by Jibson and Jibson (2003) facilitates parameter searches on a collection of 2160 strong-motion records to select appropriate records, and rigorous Newmark analyses can then be performed on the selected records. The software allows scaling of PGA values; this allows selection of a larger suite of records that has a broader range of PGA values. Standard practice dictates that PGA values can be scaled within a factor of 2 and yield acceptable results; thus records having PGA values within a factor of 2 of the estimated PGA values at the site were considered for inclusion in the analysis.

The estimated PGA from the 23 January 1812 earthquake at the Stewart site is 0.09–0.16 g (Table 8.4), which is below the estimated critical acceleration of 0.175 g. This means, within the estimated parameters of this analysis, that the 23 January event would not have generated shaking strong enough to have caused significant displacement of the Stewart landslide. Therefore, no rigorous analysis of the 23 January event was performed.

Table 8.5 lists the strong-motion records selected for analysis of the 16 December 1811 and 7 February 1812 earthquakes. Fewer records were selected for the 7 February earthquake because the extraordinarily high PGA values estimated for this event at the Stewart site (Table 8.4) limit the number of records within the appropriate scaling range.

8.5.4.5   Estimation of the Newmark Landslide Displacement

The software of Jibson and Jibson (2003) was used to conduct rigorous Newmark analyses of the strong-motion records listed in Table 8.5. For the 16 December 1811 and 7 February 1812 earthquakes, the software scaled each of the strong-motion records to the upper- and lower-bound PGA values indicated in Table 8.4, which thus yielded a range of displacements. Table 8.4 shows the median values of Newmark displacement for the bounding PGA values for the two groups of strong-motion records used to model the two largest 1811–1812 earthquakes. Displacements for the 16 December 1811 earthquake are 37–81 cm; those for the 7 February earthquake are 117–122 cm. Displacements in successive earthquakes are additive, which means that the cumulative displacements of just these two earthquakes would total 154–203 cm.

The significance of the Newmark displacements must be judged in terms of the probable effect on the potential landslide mass. For example, Wieczorek et al. (1985) used 5 cm as the critical displacement leading to failure of landslides in San Mateo County, California; Keefer and Wilson (1989) used 10 cm as the critical displacement for coherent slides in southern California. Detailed analysis of landslides triggered by the 1994 Northridge, California, earthquake showed that most landslides occurred by the time 10–15 cm of Newmark displacement had accumulated (Jibson et al., 2000). When displacements in this range occur, previously undisturbed soils can lose some of their strength and be in a residual-strength condition. Static factors of safety using residual shear strengths can then be calculated to determine the stability of the landslide after earthquake shaking (and consequent inertial landslide displacement) ceases.

The soils sampled and tested at the Stewart site all showed significant strength reductions during strain in both drained and undrained conditions (Jibson, 1985); residual strength generally was reached after shear displacements of about 0.5 cm (for silts and sands) to 6 cm (for clayey silts). Therefore, even modest displacements would have at least partially reduced the soil shear strength and thus would have reduced the critical acceleration of the landslide in future earthquakes. The relatively large cumulative displacements (~2 m) modeled for the earthquakes of 16 December and 7 February events almost certainly would have reduced soil shear strengths to near residual levels. For all groundwater conditions, all static factors of safety for the Stewart slide calculated using residual shear strengths in both drained and undrained conditions were less than 0.8 and in most cases were less than 0.4 (Jibson, 1985). Therefore, if the bluff materials reach residual strength, continuing postseismic displacement is highly likely. Displacements of ~2 m thus would reduce the shear strength of the bluff materials to a residual-strength condition and probably would lead to catastrophic failure.

Of course, incremental deformation in the early earthquakes in the sequence would tend to partially reduce the shear strength and, consequently, the critical acceleration, which would make the landslide more susceptible to additional displacement in the later earthquakes. Thus, the estimated 2 m of displacement in the current analysis should be considered a minimum estimate. The repeated shaking of the bluffs by three large earthquakes (and the far more numerous moderate earthquakes) and the reduction of the critical acceleration of the partially failed landslide mass leave little doubt that the very large displacements of the Stewart landslide would have occurred during the entire 1811–1812 earthquake sequence.

8.5.4.6 Summary

In summary, static stability analyses of drained conditions indicate that failure of the Stewart landslide in aseismic conditions is highly unlikely. Dynamic analysis shows that shaking conditions similar to those in 1811–1812 would have induced large displacements that probably would have led to continuing postseismic failure. Further analysis (Jibson and Keefer, 1993) showed that no earthquakes since 1812 could have triggered the observed landslide movement. The results of these analyses are consistent with results from field and regional studies (Jibson and Keefer, 1988, 1989), which indicated that the ages and regional distribution of landslides similar to the Stewart slide are consistent with triggering in 1811–1812. Datable material needed to determine the precise age of landsliding at the Stewart site could not be recovered; thus, the analytical approach outlined here was crucial in linking the landslide to the 1811–1812 earthquakes. Considered together, these studies strongly support such a conclusion.

The reliability of the results of an analysis such as this obviously depends on the amount and quality of input data and the appropriateness and accuracy of the modeling approach used. As Clark and Cole (1992) pointed out, obtaining samples that accurately reflect the shear strength along a failure plane is very difficult, particularly in cases where reactivated landslides having well-formed basal shear surfaces are being analyzed. In such cases, using minimum shear-strength estimates is generally appropriate because the material along the pre-existing shear surface is probably at residual level (Clark and Cole, 1992).

8.5.5 Analysis of Unknown Seismic Conditions

The procedure described in the last section was used to test the hypothesis that an individual landslide was triggered by an historical earthquake whose magnitude and location have already been estimated. The goal of most paleoseismic investigations, by contrast, is to detect and characterize prehistoric or undocumented earthquakes whose effects are recorded in the geologic record. Therefore, a more general procedure for paleoseismic landslide analysis is required.

If static stability analyses clearly indicate that failure in aseismic conditions of a landslide of unknown origin is highly unlikely, then an earthquake origin can be hypothesized on that basis alone. A dynamic analysis can then be used to estimate the minimum shaking intensities necessary to have caused failure. Such an approach requires a general relationship between critical acceleration, shaking intensity (which can be characterized in various ways), and Newmark displacement. Several such relations have been published (Watson-Lamprey and Abrahamson, 2006; Bray and Travasarou, 2007; Jibson, 2007; Saygili and Rathje, 2008).

For example, in the case of the Stewart landslide, if we knew nothing of the shaking history of the site, the minimum earthquake shaking intensity could be estimated using any of a number of equations from the studies just cited. For this example, consider the following equation from Jibson (2007):

$$\log D_{\mathrm{N}} = 0.215 + \log \left[\left(1 - \frac{a_{\mathrm{c}}}{a_{\mathrm{max}}} \right)^{2.341} \left(\frac{a_{\mathrm{c}}}{a_{\mathrm{max}}} \right)^{-1.438} \right] \pm 0.510, \tag{8.5}$$

where D_{N} is the Newmark displacement (in cm), a_{c} is the critical acceleration, a_{max} is the peak ground acceleration, and the last term is the model standard deviation; the ratio of a_{c} to a_{max} is commonly referred to as the critical acceleration ratio. Applying this equation requires judgment regarding the amount of Newmark displacement (the critical displacement) that would reduce shear strength on the failure surface to residual levels and lead to catastrophic failure. As discussed previously, critical displacements of about

10 cm are probably realistic for this type of slide, based on previous studies (Wieczorek *et al.*, 1985; Wilson and Keefer, 1985; Keefer and Wilson, 1989), laboratory shear-strength testing of soil samples from the site (Jibson, 1985), and field studies of landslides in the region (Jibson and Keefer, 1988). Insertion of a displacement value of 10 cm into Equation (8.5) yields a critical acceleration ratio of 0.275. For a critical acceleration (a_c) of 0.175 for the Stewart landslide, this would yield a PGA of 0.64 g as a minimum required to trigger enough displacement to cause general failure.

The PGA value from such an analysis can be used by itself as a basis for hazard assessment, or it can be used to estimate various magnitude/distance combinations of a possible triggering earthquake. For example, Equation (8.2) would yield a minimum magnitude of ~6.2 for a PGA of 0.64 g and an assumed fault-rupture distance within 10 km of the Stewart landslide. Although this magnitude is considerably lower than those estimated for the 1811–1812 earthquakes, it would provide a reasonable lower-bound magnitude in the absence of other information. If more than one landslide of identical age were similarly analyzed in an area, magnitude and location estimates could be optimized by using the larger required source distances between two or more separate sites.

Equations that use other parameters are available and could be applied similarly. For example, a minimum threshold Arias intensity leading to slope failure can be estimated using the following equation (Jibson, 2007) if a reasonable critical displacement can be specified:

$$\log D_N = 2.401 \log I_A - 3.481 \log a_c - 3.230 \pm 0.656, \qquad (8.6)$$

where D_N is the Newmark displacement (in cm), I_A is the Arias intensity (in m/s), a_c is the critical acceleration (in terms of g), and the last term is the model standard deviation.

8.6   Interpreting Results of Paleoseismic Landslide Studies

Once a landslide or group of landslides has been identified, dated, and linked to earthquake shaking, what can we learn about the magnitude and location of the triggering earthquake? Section 8.5 outlined a method for detailed geotechnical analysis to address this question, but in many cases such an analysis will be impossible owing to lack of data or the unsuitability of the landslide for detailed modeling. Several other approaches to this last level of paleoseismic interpretation are possible; in most cases, multiple lines of evidence will be required to make reasonable estimates of magnitude and location. Perhaps the most important aspect of such interpretation is a thorough understanding of the characteristics of landslides triggered by recent, well-documented earthquakes.

8.6.1   *Characteristics of Landslides Triggered by Earthquakes*

Keefer (1984, 2002) conducted by far the most comprehensive studies of landslides caused by historical earthquakes. He documented minimum earthquake magnitudes and intensities that have triggered landslides of various types, average and maximum areas affected by landslides as a function of magnitude, and maximum distances of landslides from earthquake sources as a function of magnitude. For these comparisons, he grouped different types of landslides into three categories: disrupted slides and falls (defined as falls, slides, and avalanches in rock and soil); coherent slides (defined as slumps and block slides in rock and soil and slow earth flows); and lateral spreads and flows (defined as lateral spreads and rapid flows in soil and subaqueous landslides).

Table 8.6: Minimum earthquake magnitude required to trigger landslides

Earthquake magnitude	Type of landslide
4.0	Rock falls, rock slides, soil falls, disrupted soil slides
4.5	Soil slumps, soil block slides
5.0	Rock slumps, rock block slides, slow earth flows, soil lateral spreads, rapid soil flows, subaqueous landslides
6.0	Rock avalanches
6.5	Soil avalanches

Data from Keefer (1984).

Table 8.7: Minimum modified Mercalli intensity required to trigger landslides

Landslide type	Lowest modified Mercalli intensity	Predominant modified Mercalli intensity
Disrupted slides and falls	IV	VI
Coherent slides	V	VII
Lateral spreads and flows	V	VII

Data from Keefer (1984).

8.6.1.1 Minimum Earthquake Magnitudes That Trigger Landslides

In a review of intensity reports from 300 earthquakes, Keefer (1984, 2002) found that the smallest earthquake reported to have caused landslides had a magnitude of 4.0. Landslides of various types have threshold magnitudes ranging from 4.0 to 6.5 (Table 8.6); disrupted landslides have lower threshold magnitudes than coherent slides. Although smaller earthquakes could conceivably trigger landslides, such triggering by very weak shaking probably would occur on slopes where failure was imminent before the earthquake.

8.6.1.2 Minimum Shaking Intensities That Trigger Landslides

Keefer (1984) also compared landslide initiation to MMI. Table 8.7 shows the lowest MMI values and the predominant minimum MMI values reported where the three categories of landslides occurred. Keefer's (1984) data show that landslides of various types are triggered one to five levels lower than indicated in the current language of the MMI scale. A subsequent study that used an expanded earthquake database confirmed Keefer's findings (Rodríguez *et al.*, 1999).

Solonenko (1977a,b) and Nikonov (1988a,b) correlated landslide initiation with threshold shaking levels using the Russian MSK intensity scale. Their observations indicated that small landslides are initiated at intensities IV–VII, large landslides at intensities VIII–IX, and "large landslides in basement rocks" at intensities of IX or greater.

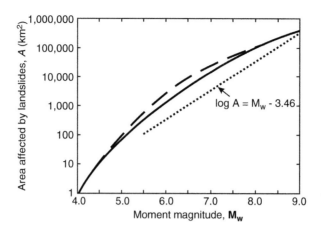

Figure 8.9: Area affected by seismically triggered landslides plotted as a function of earthquake magnitude. Solid line is upper bound of Keefer (1984); dashed line is from Rodriguez *et al.* (1999); dotted line is regression line from Keefer and Wilson (1989).

8.6.1.3   Areas Affected by Earthquake-Triggered Landslides

For 30 historical earthquakes, Keefer (1984) drew boundaries around all reported landslide locations and calculated the areas enclosed. His plot of area versus earthquake magnitude (Figure 8.9) shows a well-defined upper-bound curve representing the maximum area that can be affected for a given magnitude. Rodriguez *et al.* (1999) later expanded this data set to include 35 earthquakes, which slightly raised the upper-bound line (Figure 8.9). Keefer and Wilson (1989) fitted a regression line using data from 37 earthquakes (including the 30 from Keefer, 1984), to predict average area affected by landslides as a function of earthquake magnitude:

$$\log A = M - 3.46 \pm 0.47, \tag{8.7}$$

where A is the area affected by landslides (in km^2), and M is a composite magnitude term, which generally indicates surface-wave magnitudes below 7.5 and moment magnitudes above 7.5 (Keefer, 2002).

Keefer (1984) noted that the area affected by landslides is influenced, in part, by the geologic conditions that control the distribution of susceptible slopes. Also, he noted that earthquakes having focal depths greater than about 30-km plot on or near the upper bound (Figure 8.9), which indicates that deeper earthquakes can trigger landslides over larger areas. Surprisingly, he found no differences in the areas affected by landsliding that could be attributed to regional differences in seismic attenuation.

8.6.1.4   Maximum Distance of Landslides from Earthquake Sources

Keefer (1984) related earthquake magnitude to the maximum distance of the three categories of landslides from the earthquake epicenter and from the closest point on the fault-rupture surface (Figure 8.10). Again, upper-bound curves are well defined and are constrained to pass through the minimum threshold

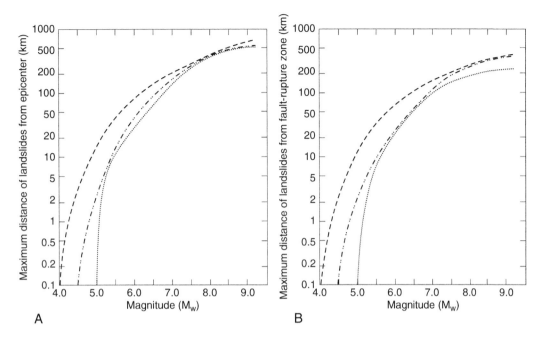

Figure 8.10: Maximum distance to landslides from (A) epicenter and (B) fault-rupture zone for earthquakes of different magnitudes. Dashed line is upper bound for disrupted slides and falls; dash-double-dot line is upper bound for coherent slides; and dotted line is upper bound for lateral spreads and flows. [Modified from Keefer (1984).]

magnitudes shown in Table 8.6 as distance approaches zero. Although the upper bounds shown have been exceeded a few times in subsequent earthquakes (Keefer, 2002), they remain fairly reliable indicators of the maximum possible distances at which the three classes of landslides could be triggered in earthquakes of various magnitudes.

Figure 8.10 indicates that disrupted slides and falls have the lowest shaking threshold and that lateral spreads and flows have the highest shaking threshold. As with area, earthquakes having focal depths greater than 30 km generally triggered landslides at greater distances than shallower earthquakes of similar magnitude.

8.6.2   Interpreting Earthquake Magnitude and Location

Keefer's (1984, 2002) results allow interpretation of earthquake magnitude and location in a variety of ways. If a single landslide is identified as being seismically triggered, then a minimum magnitude and MMI can be estimated based on the landslide type. For example, Schuster *et al.* (1992) used Keefer's (1984) magnitude of 6.5 as a lower-bound estimate for triggering of rock avalanches that formed dams. If several landslides in an area are identified as being seismically induced, then application of Keefer's (1984) magnitude–area and magnitude–distance relationships can yield minimum magnitude estimates. As the area in which landslides documented to have been triggered by the same earthquake increases, the estimated magnitude will increase toward the actual magnitude of the triggering earthquake. Therefore, documentation and analysis of landslides over a large area will produce more accurate magnitude estimates. If seismic source zones are well documented, then the distance from the closest source zone to the farthest landslide will yield a reasonable minimum magnitude estimate. The observation that greater source depth relates to greater areas affected and source distances for landslides of all types (Keefer, 1984) further complicates estimation of earthquake magnitude.

For a specific region, earthquake magnitude can be estimated based on comparison of paleoseismic landslide distribution with landslide distributions from recent, well-documented earthquakes in the region. This approach has been applied to landslide dams in New Zealand (Adams, 1981a,b) and to landslides in central Asia (Nikonov, 1988a,b).

Static and dynamic slope-stability analyses facilitate direct estimation of the minimum ground shaking, and hence magnitude, required to have caused failure of individual landslides (Jibson and Keefer, 1992, 1993), as described in detail in Section 8.5. If the critical acceleration of a landslide can be determined by stability analysis, and if a reasonable amount of displacement leading to catastrophic failure can be estimated, then Equation (8.5) can be used to estimate the minimum PGA, or Equation (8.6) can be used to estimate the minimum Arias intensity required to initiate failure. Appropriate attenuation equations can then be used to estimate possible magnitude–distance combinations.

Another approach for estimating earthquake magnitude from the results of slope-stability analyses was outlined by Crozier (1992) and is based on the work of Wilson and Keefer (1985), who defined a quantity referred to as $(A_c)_{10}$, which is the critical acceleration of a landslide that will yield 10 cm of displacement (the estimated critical displacement leading to catastrophic failure) in a given level of earthquake shaking. They selected 10 strong-motion records that spanned a range of Arias intensities and iteratively determined $(A_c)_{10}$ for each record. From these values, they developed a regression model relating Arias intensity to $(A_c)_{10}$:

$$\log (A_c)_{10} = 0.79 \log I_A - 1.095, \tag{8.8}$$

where $(A_c)_{10}$ is in g's and I_A is in meters per second. If the critical acceleration of a landslide can be determined, then this value can be used as the threshold value of $(A_c)_{10}$ in Equation (8.8), and the Arias intensity that would trigger the critical displacement of 10 cm can be calculated.

Stability analysis could also possibly be applied to speleothems, whose dynamic stability can be modeled, to estimate the ground shaking required to cause failure.

Earthquake locations generally are estimated based on the distribution of synchronous landslides attributed to a single seismic event. In a broad area of roughly similar susceptibility to landsliding, the earthquake epicenter probably will coincide fairly closely with the centroid of the landslide distribution. In areas of highly variable or asymmetrical landslide susceptibility, epicenter estimation is much more difficult and subject to error. In areas where seismic source zones are well defined, the epicentral location is best defined as the point in a known seismic source zone (or along a known seismogenic fault) closest to the centroid of the landslide distribution.

8.7   Final Comments

The use of landslides as paleoseismic indicators is a fairly recent development that is beginning to expand in scope and complexity. A few final comments on the advantages and limitations of paleoseismic landslide analysis are in order.

The primary limitation of paleoseismic analysis of landslides is the inherent uncertainty in interpreting a seismic origin. Unlike liquefaction, which can occur aseismically only in relatively rare conditions, landslides of all types form readily in the absence of earthquake shaking as a result of many different triggering mechanisms. In many cases, ruling out aseismic triggering will be impossible, and the level of confidence in any resulting paleoseismic interpretation will be limited. For this reason, paleoseismic landslide analysis should include, so far as possible, multiple lines of evidence to constrain a seismic origin. In this way, a strong case can be built for seismic triggering of one or more landslides, even if no single line of evidence is unequivocal. Where independent paleoseismic evidence from fault or liquefaction studies is available, paleoseismic landslide evidence can provide useful corroboration.

Detailed slope-stability analyses generally can be performed only on certain types of landslides. Failure conditions of falls, avalanches, and disrupted slides cannot easily be modeled using Newmark's (1965) method, and even static stability analyses of these types of slides can be very problematic. Also, the prelandslide geometry of slides in very steep terrain can be difficult or impossible to reconstruct. Thus, the analytical method described herein generally can be applied only to fairly coherent landslides where prelandslide geometry can be reconstructed with confidence, where groundwater conditions can be modeled reasonably, and where the geotechnical properties of the materials can be accurately measured.

Even allowing for these limitations, paleoseismic landslide studies have been extremely useful where applied successfully, and they hold great potential in the field of paleoseismology. Dating landslide deposits is, in many cases, easier than dating movement along faults because many different dating methods can be used on the same slide to produce redundant results. In addition, landslides have the potential for preserving large amounts of datable material in the various parts of the slide (scarp, body, toe, etc.). In areas containing multiple or poorly defined seismic sources, paleoseismic ground-failure analysis might be preferable to fault studies because landslides preserve a record of the shaking history of a site or region from all seismic sources. Knowing the frequency of strong shaking events could, in many cases, be more critical than knowing the behavior of any individual fault.

Paleoseismic landslide analysis could have greatest utility in assessing earthquake hazards in stable continental interiors, such as the eastern and central United States, where fault exposures are rare or absent but where earthquakes are known to have occurred. In such areas, analysis of earthquake-triggered ground failure, both landslides and liquefaction, might be one of the few paleoseismic tools available.

Another advantage of paleoseismic landslide analysis is that it gets directly at the effects of the earthquakes being studied. Ultimately, most paleoseismic studies are aimed at assessing earthquake hazards. Fault studies can be used to estimate slip rates, recurrence intervals, and, indirectly, magnitudes. From these findings, we extrapolate the effects of a possible earthquake on such a fault. In paleoseismic landslide studies, we observe the effects directly. Thus, if a seismic origin can be established, a landslide shows directly what the effects of some previous earthquake were. Even if magnitude and location are poorly constrained, at least we have a partial picture of the actual effects of seismic shaking in a locale or region. Thus, for example, a map of the distribution of landslides triggered by the 1811–1812 earthquakes in the New Madrid seismic zone (Jibson and Keefer, 1988) yields a very useful picture of the likely distribution of landslides in future earthquakes there.

In conclusion, paleoseismic landslide analysis can be applied in a variety of ways and can yield many different types of results. Although interpretations are limited by the certainty with which a seismic origin can be established, paleoseismic landslide studies can play a vital role in the paleoseismic interpretation of many areas, particularly those lacking fault exposures.

Subject Index

A

Accelerator mass spectrometry, 126
Active surface folding. *See also* Coseismic
    folding, paleoseismic evidence
    geomorphic evidence, 368–371
    stratigraphic evidence, 371–375
Aerial photography, surface deformation
    identification, 33
Airborne laser swath mapping, 295
Akademia Nauk volcano, eruption of, 310
Alaska
    earthquake
        coseismic deformation
            formation, 387
        Patton Bay fault, 332
        surface faulting, 383–384
    paleoseismic evidence
        coseismic subsidence, 411–413
        coseismic uplift, 403–406
    subduction earthquake, Montague Island
        formation by, 330
Alluvial fans, study of, 448. *See also*
    Paleoearthquakes, geomorphic
    evidence
Alpine Lakes, landslide and tsunami
    deposits, 163, 165
ALSM. *See* Airborne laser swath mapping
ALVIN observations, of San Clemente fault,
    139
AMRT. *See* Apparent mean residence time
AMS-radiocarbon dating,
    application of, 73
AMS. *See* Accelerator mass spectrometry
Angular unconformities
    displacements estimates from, 243–244
    in normal fault zones, 235–239
    and paleoseismic indicators, 479 (*see also*
        Paleoearthquakes, stratigraphic
        evidence of)
Anticline Ridge, earthquake, 369
Apparent mean residence time, 255
Archeoseismology, 112–117. *See also*
    Paleoseismology
Asal-Ghoubbet rift area, volcano-
    extensional structures, 277
Autonomous underwater vehicles
    (AUVs), 122

B

Back-tilting scarps, 193–194.
    See also Fault scarps
Bedrock fault planes and rock surfaces,
    184–186. *See also* Extensional
    tectonic environments, in
    paleoseismology
Bending-moment faults, characteristics of,
    319, 363. *See also* Seismogenic *vs.*
    nonseismogenic reverse faults
Benthic marine sampler (BMS), 126
Blind thrusts, seismic hazards assessment,
    375–379. *See also* Coseismic
    folding, paleoseismic evidence
Box corer, in seafloor mapping, 125
Bradley lake, landslide and tsunami
    deposits, 165, 167
British Columbia, rock avalanches, 575

C

Calderas
    definition, 278
    maximum-magnitude earthquakes, 307
Calibrated age methods, application of, 20
California–Oregon boundary, fluidization
    features, 546
Carrizo Plain segment, study of, 443
Cascadia Basin turbidite systems,
    investigation, 151–156. *See also*
    Sub-aqueous paleoseismology
Cascadia subduction zone, 407–412.
    See also Paleoseismic evidence
Catalina Ridge–San Clemente fault zone,
    investigation, 141, 143
CFS. *See* Coulomb failure stress
Channel Islands Thrust, 139
Charleston earthquake, magnitude
    estimation, 560
Chelungpu fault, Chi-Chi rupture, 350
Chi–Chi earthquake, surface deformation,
    367–368
Chikyu drilling vessel, 126
CHIRP. *See* Compressed high intensity
    radar pulse
Chuetsu offshore earthquake, 315
CIT. *See* Channel Islands Thrust

Clast-rich deposits fault zones,
    identification, 80–81
Coastal lakes, landslide and tsunami
    deposits, 165
Coastal South Carolina, liquefaction,
    510–512. *See also* Earthquake-
    induced liquefaction features
    craters, characteristics, 512–517
    prehistoric seismicity, 517–518
Coastal Washington State, liquefaction, 540.
    See also Earthquake-induced
    liquefaction features
    ancient marine-terrace features, 546
    Columbia river features, 541–545
    prehistoric shaking, strength of, 545–546
Coeval fault motion and fluid venting
    evidence, 169. *See also* Sub-aqueous
    paleoseismology
Collapse faults, characteristics, 368.
    See also Subsidence faults,
    characteristics
Collapse features and paleoseismic study,
    478–479. *See also* Paleoearthquakes,
    stratigraphic evidence
Colluvial wedge, 227–233
    displacement estimates, 242
    faulting age detemination by dating,
        255–259
    in fault-zone exposure, identification,
        230–231, 233
    and paleoseismic study, 479 (*see also*
        Paleoearthquakes, stratigraphic
        evidence)
Columbia river, characteristics of, 541–545.
    See also Coastal Washington State,
    liquefaction
Complex fault scarp, definition, 190
Compound fault scarp, definition, 215
Compressed high intensity radar pulse, 123
Compressional tectonic environments,
    315–317
    historic analog earthquakes, 323–327
    reverse faults, earthquake deformation
        cycle of, 322–323
    seismogenic structure in, 318–320
        (*see also* Compressional tectonic
        environments)

International Geophysics Series

EDITED BY

RENATA DMOWSKA
School of Engineering and Applied Sciences
Harvard University
Cambridge, Massachusetts

DENNIS HARTMANN
Department of Atmospheric Sciences
University of Washington
Seattle, Washington

H. THOMAS ROSSBY
Graduate School of Oceanography
University of Rhode Island, Narragansett
Rhode Island

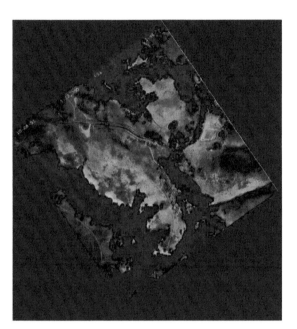

A. Traditional aerial photograph of the southeastern part of the study area.

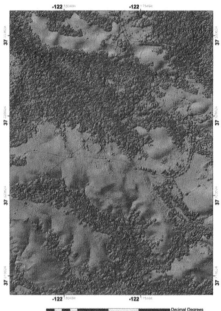

B. Unfiltered LiDAR return data show all surfaces imaged including vegetation. LiDAR images prepared by the National Center for Airborne Laser Mapping (NCALM)

Figure 2A.4: Comparison of fault visualization and mapping on aerial photographs and LiDAR images (A) Vertical aerial photograph of the San Andreas fault near southern end of the Mill Creek, California study area. Note heavy forest obscures details of fault geomorphology. **(B)** Hillshade image produced from the unfiltered DEM derived from first returns of LiDAR data showing tree canopy tops.

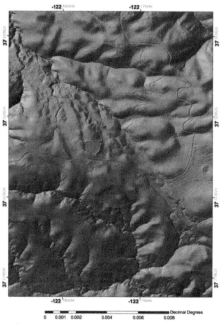

C. Filtered LiDAR data reveal 'bare Earth'
surface beneath vegetation

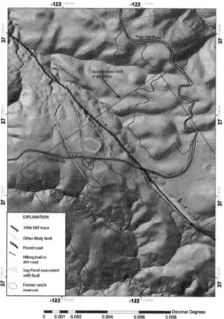

D. Fault trace and other features identified
and interpreted from LiDAR imagery.
Preliminary interpretation and air photo
provided by Dr. Carol Prentice, USGS.

Figure 2A.4 (Continued): (C) Hillshade produced from "bare-earth" DEM derived from filtered returns of LiDAR data showing ground surface. Note prominent fault features that are almost completely obscured by forest in aerial photographs. (D) Preliminary fault mapping on "bare-earth" LiDAR image. Mapping by Carol Prentice, U.S. Geological Survey, Menlo Park, CA; Rich Koehler and John Baldwin, William Lettis & Associates, Walnut Creek, CA. Source: U.S. Geological Survey, Menlo Park, CA.

Printed and bound by CPI Group (UK) Ltd, Croydon, CR0 4YY

03/10/2024

01040311-0006

Figure 2A.5: Three-dimensional perspective view of Karakax valley (northwestern Tibet). (A) from a DEM created by interferometry from two radar images (center at 36.1°N latitude, 79.2°E longitude). Scale varies in this perspective view, but the area is about 20 km (12 miles) wide in the middle of the image, and there is no vertical exaggeration. Elevations range from 4000 m (13,100 ft.) in the valley to over 6000 m (19,700 ft.) at the peaks of the glaciated Kun Lun mountains running from the front

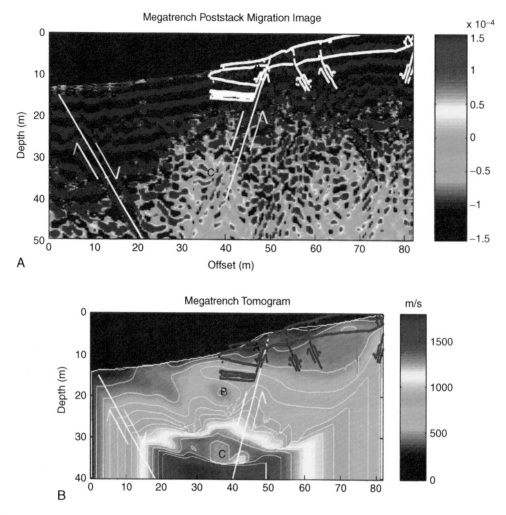

Figure 2A.11: Seismic refraction results from the Wasatch fault (Utah) megatrench of 1999 (McCalpin, 2002). (A) Poststack migration image and (B) refraction tomogram. White lines show faults interpreted by Sheley *et al.* (2003), black lines (red in color version) are their rendition of McCalpin's faults. The ground surface of McCalpin's trench does not exactly match the ground surface from the seismic profile because the survey line was about 20 m to the north and subparallel to the trench. From Sheley *et al.* (2003).

right toward the back. The active strand of the Altyn Tagh fault is visible as a sharp break in slope running diagonally up the valley side (between arrows). The original two radar images were acquired with spaceborne imaging radar C/X band (SIR-C/X-SAR) synthetic-aperture radar, aboard the space shuttle endeavor in October 1994. In this drape of C-band radar over the DEM, the L-band amplitude is assigned to red, L- and C-band (24 and 6 cm wavelengths) average to green, and C-band to blue (from NASA, http://southport.jpl.nasa.gov/pio/srl2/sirc/krkx.html); (B) from Google Earth; fault trace is between arrows.

West Mesa Fault 100 MHz GPR

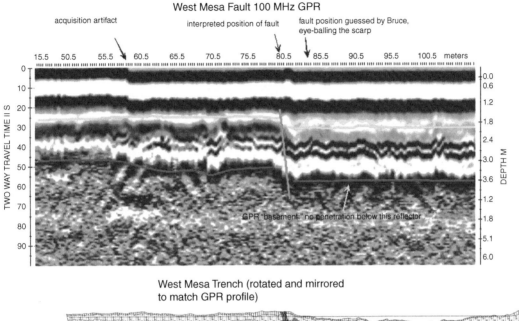

West Mesa Trench (rotated and mirrored
to match GPR profile)

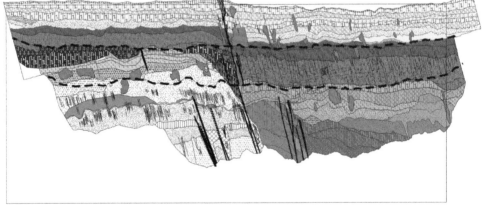

Figure 2A.15: 100 MHz GPR profile (top) and trench log (bottom) of the Calabacillas normal fault, Rio Grande rift, New Mexico, USA. Scale is in meters, total penetration depth is about 3.6 m, limited by a clay-rich paleosol. The fault position interpreted from GPR (green line at top) corresponds to the fault exposed in the trench walls. However, the stratigraphic contacts interpreted from GPR (yellow and blue lines on profile) did not represent the same stratigraphic unit on the footwall and hanging wall. Thus, the true vertical displacement was much greater than inferred from GPR, as can be appreciated from the trench log. From McCalpin and Harrison (2000).

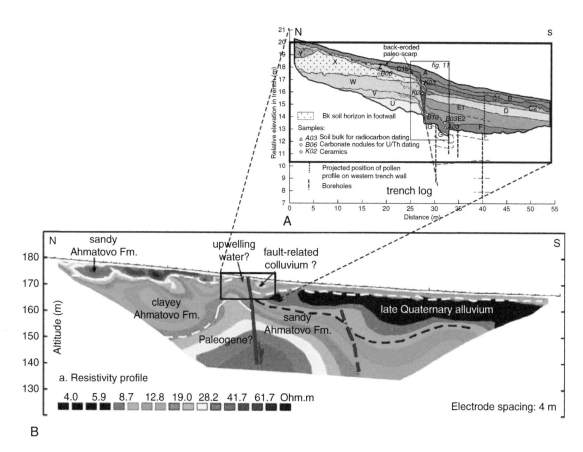

Figure 2A.16: ERT profile across the Chirpan normal fault, Bulgaria. This fault experienced 0.45 m displacement in the 1928 Chirpan earthquake. The trench log (top, ca. 4 m deep) shows three earlier colluvial wedges of middle to early Holocene age. The ERT profile (bottom) penetrates to about 30 m and shows that the major vertical displacement of pre-Quaternary deposits occurs on a buried fault farther south than the Holocene fault, something that is not obvious from the modern topography. From Vanneste *et al.* (2006).

Figure 2A.19: A large benched trench across a 23 m-high normal fault scarp on the Wasatch fault zone at Mapleton, Utah. Note the deep inner slot which is shored, and small vertical boards on the first bench, to assist in climbing up the trench. Preliminary results are described by Olig *et al.* (2005).

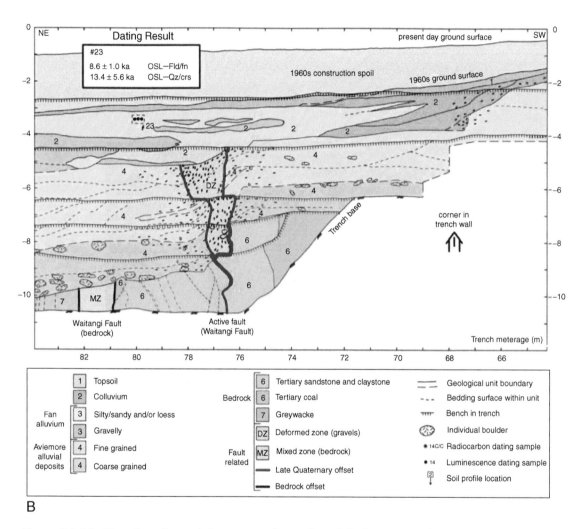

Figure 2A.25: Trench and trench log across the Waitangi fault, New Zealand. (B) Log of the left (south) wall of the trench, showing the currently active fault trace (thick black lines) and an older fault trace not active in the late Quaternary. Note how the active fault appears to shift to the left on the trench log as it crosses the benches, because the fault trends more easterly than a perpendicular to the trench walls. From Barrell *et al*. (2005).

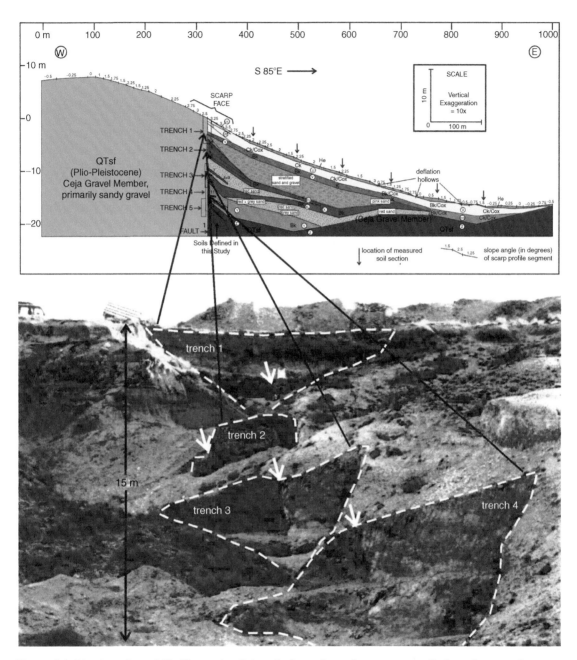

Figure 2A.31: A series of "half-trenches" (vertical cuts) used to expose a fault as it ascends up an erosional escarpment that trends roughly perpendicular to fault strike. (Top) Schematic cross section of the fault zone and colluvial wedge, showing the location of the half-trenches (rectangles along the fault plane). Note the high vertical exaggeration of this section; (Bottom) Telephoto view of the upper parts of trenches 1–4, outlined with white dashed lines. From McCalpin *et al.* (2006). Trench logs from this site are available on on-line Content.

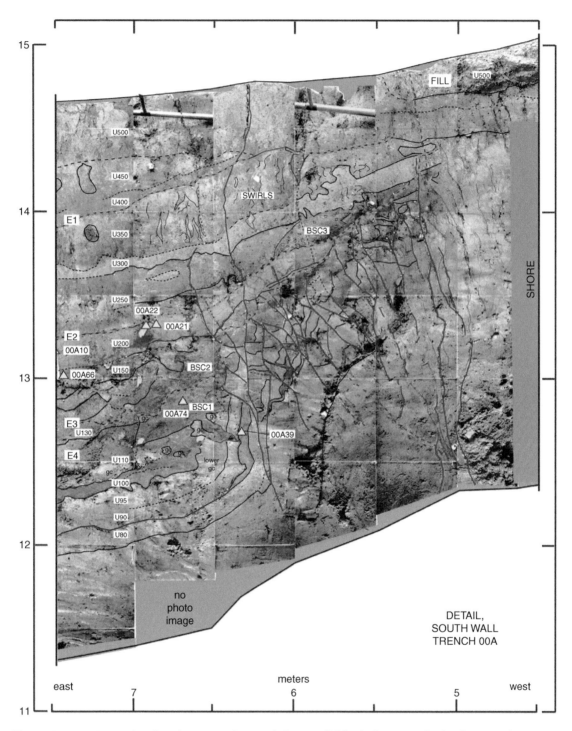

Figure 2A.43: Example of a photomosaic trench log. Individual photographs in the mosaic cover a 0.5 m by 0.5 m area. Trench units (labeled) are highlighted by thin lines, and faults (such as at right center) are highlighted by slightly thicker lines. Selected units can be further emphasized by semi-transparent color fills (see color version on Book's companion web site, Chapter 2A). Log is from the Hayward fault, a dextral fault east of the San Andreas fault in northern California. From Lienkaemper *et al.* (2002).

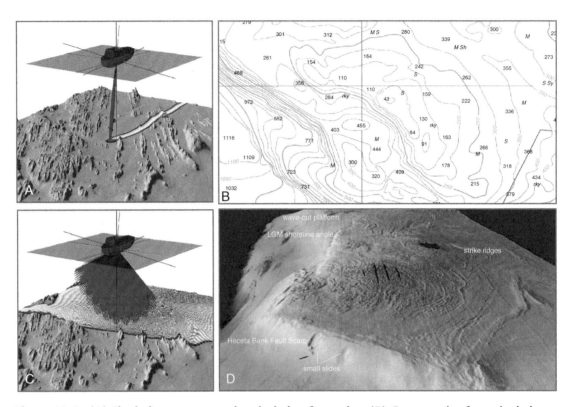

Figure 2B.2: (A) Single-beam sonar and typical ping foot print. (B) Contour plot from single-beam sounding data. (C) Typical multibeam sonar and swath footprint. Swath width varies by system, but ranges between two and seven time water depth. (D) Example of swath bathymetric survey of Heceta Bank, Oregon. Pixel size in this image is 10 m. Fault scarps, strike ridges, landslide scars, and a submerged last glacial maximum (LGM) submerged shoreline angle, and associated wave-cut platform are clearly visible. (A) and (C) courtesy John Hughes Clarke, University of New Brunswick. Image by the author, (D) Data courtesy R. W. Embley, NOAA Pacific Marine Environmental Laboratory, Newport Oregon.

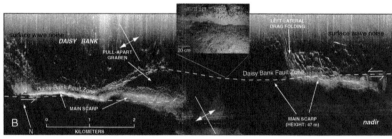

Figure 2B.3: (A) Sidescan towfish, typical of side scan sonar systems. Two transducer arrays transmit beams to port and starboard. This figure illustrates the port beam only. The beam is narrow in the horizontal plane and broad in the vertical plane, shown to the right in the figure. In a typical single-beam 100 kHz system this would be $\sim$1° in the horizontal and 40° in the vertical and for a 500 kHz system, $\sim$0.2° in the horizontal and 40° in the vertical. Radiation occurs also out the rear of the transducer, shown here as the pattern extending horizontally to the left of the figure (starboard for the towfish). Image courtesy L-3 Klein Associates, reprinted by permission. (B) SeaMARC 1A starboard swath image of the Daisy Bank Fault Zone, Cascadia margin. A left stepover in the left-lateral fault, associated pull-apart basin and anticline offset are shown. Offset anticlinal axis is highly reflective rubble. Drag folding and other fault details are shown. Inset shows active Holocene scarp from DELTA submersible photograph. Modified after Goldfinger *et al.* (1996).

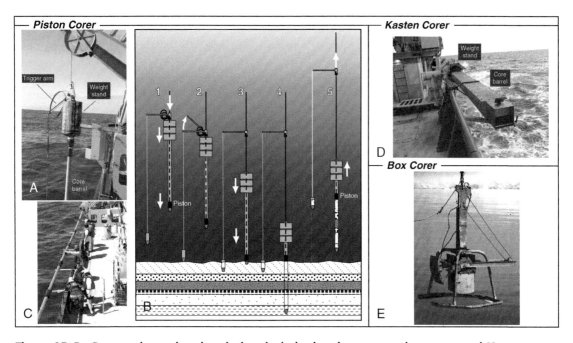

Figure 2B.5: Commonly used coring devices include the piston corer, box corer and Kasten corer. A piston corer is comprised of a piston positioned at the cutting head of a weighted, round core barrel of variable length (A). The piston corer is released to free-fall into the sediment when the trigger weight (or trigger core) hits the seafloor. The piston stays at the sediment surface as the core barrel punctures the sediment. The piston prevents the loss of sediment as the core is recovered (see sequence in (B)). A piston core ready for deployment is shown in (C). A Kasten corer is comprised of a weight stand on a rectangular core barrel (D), and does not require a trigger weight. Kasten cores are limited in length and are used where more volume of sediment is required. A box corer (E) is a weighted box with an open bottom, with a spring-loaded shovel that close over the bottom to prevent loss during recovery. Box cores penetrate a short distance into the seafloor and are typically used to capture a large amount of surface sediment. Photos (A), (C), and (D), OSU. Panel (B) by J. Patton and A. Morey. E. Courtesy Hannes Grobe.

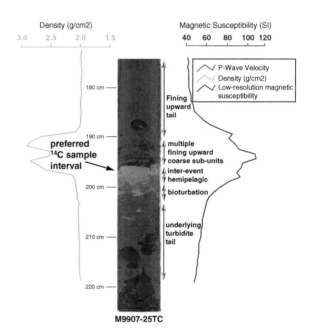

Figure 2B.7: (A) Detailed image of a Cascadia paleoseismic turbidite, its subunits and preferred ^{14}C sampling site. In this example, the turbidite tail/hemipelagic boundary is distinct visually, and variably disturbed by bioturbation. While turbidite bases can be erosive, dating is commonly done from planktonic foraminifera in the upper part of the underlying hemipelagic interval as the least problematic option. Typical sample location shown, with small "gap" above the sample.

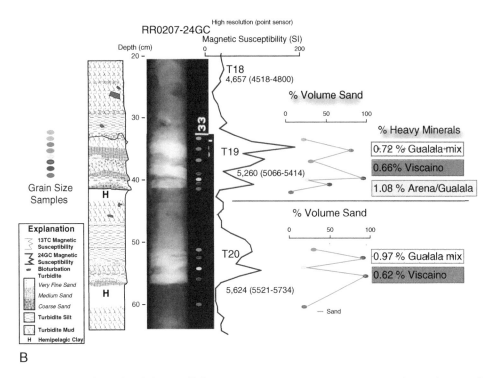

B

Figure 2B.7 (Continued): (B) Detail from Core RR0507-25TC event T4 along the Northern San Andreas margin. Example grain size analysis, magnetic susceptibility/density signatures and X-radiography in turbidites T19 and T20 in core 24GC below the Gualala–Noyo–Viscaino channel confluence. Light tones in the X-radiograph represent dense sand/silt intervals; darker gray tones represent clay/mud. Oval dots are grain size samples. Heavy trace is the magnetic susceptibility signature. Right plot is percent sand (obtained with Coulter laser counter method). The good correspondence between grain size, density, and magnetic susceptibility for the lithologies in both Cascadia and NSAF cores is established with selected analyses and permits the use of density and magnetics as mass/grain size proxies that show much greater resolution than possible with grain size analysis. These typical turbidites are composed of 1–3 fining upward sequences, each truncated by the overlying pulse. No hemipelagic clay exists between pulses, indicating the three pulses were deposited in a short time interval. Only the last pulse has a fine tail, indicating final waning of the turbidity current. We interpret these signatures as resulting from a single multipulse turbidity current. Number of coarse pulses commonly remains constant in multiple channel systems for a given event. Source provenance affinity for each sand pulse is shown to the right. Mineralogically distinct sandy units stacked vertically in order of arrival at a confluence near the core site. See Goldfinger et al. (2007) for further details and core locations. Modified after Goldfinger et al. (2007, 2008).

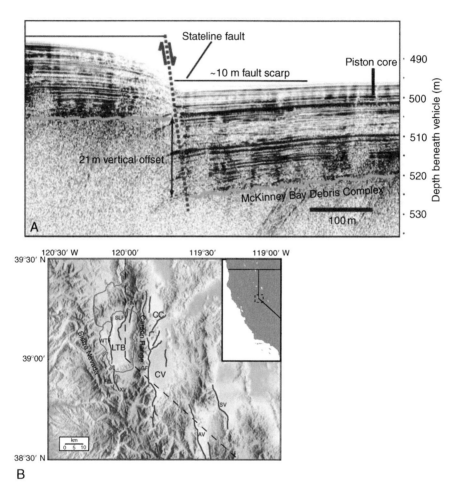

Figure 2B.10: Deep-towed Edgetech CHIRP reflection profile of across the Stateline Fault (SLF), Lake Tahoe California. Top of the McKinney Bay slide complex is offset 21 m, with possible small colluvial wedges visible against the fault plane. Location map shown in B. From Kent *et al.* (2005). Reprinted with permission from the Geological Society of America.

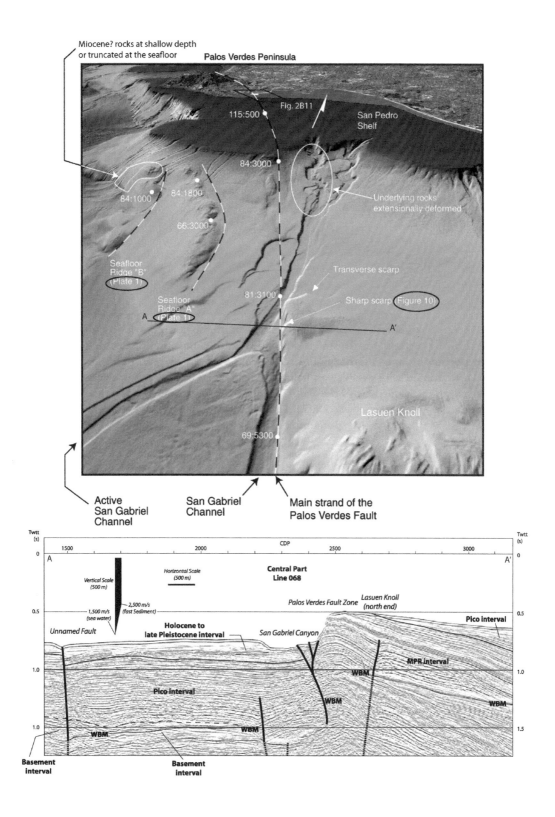

Miocene? rocks at shallow depth
or truncated at the seafloor

Palos Verdes Peninsula

115:500

Fig. 2B11

San Pedro
Shelf

84:3000

84:1000

84:1800

Underlying rocks
extensionally deformed

66:3000

Seafloor
Ridge "B"
(Plate 1)

Transverse scarp

Seafloor
Ridge "A"
(Plate 1)

31:3100

Sharp scarp (Figure 10)

A

A'

Lasuen Knoll

69:5300

Active
San Gabriel
Channel

San Gabriel
Channel

Main strand of the
Palos Verdes Fault

Twtt (s)

CDP

Twtt (s)

1500

2000

2500

3000

0

A

Vertical Scale
(500 m)

Horizontal Scale
(500 m)

Central Part
Line 068

A'

0

0.5

2,500 m/s
(fast Sediment)

Palos Verdes Fault Zone

Lasuen Knoll
(north end)

Pico interval

0.5

1,500 m/s
(sea water)

Unnamed Fault

Holocene to
late Pleistocene interval

San Gabriel Canyon

WBM

MPR interval

1.0

WBM

Pico interval

Pico interval

WBM

WBM

1.0

WBM

WBM

WBM

1.5

Basement
interval

Basement
interval

Figure 2B.12 (A) Shaded relief view of high-resolution multibeam bathymetry along the Palos Verdes Fault showing tectonic features of this strike-slip fault. Location of Figure 2B.11 shown on shelf at upper portion of image. Other figure callouts refer to the original publication. From Fisher *et al*. (2004). Reprinted with permission from the Geological Society of America. (B) Migrated multichannel reflection profile across the Palos Verdes Fault and Lasuen Knoll, a small restraining bend uplift along the fault (Legg *et al*., 2007). Line of section A–A' shown in A above. Figure from Bohannon *et al*. (2004).

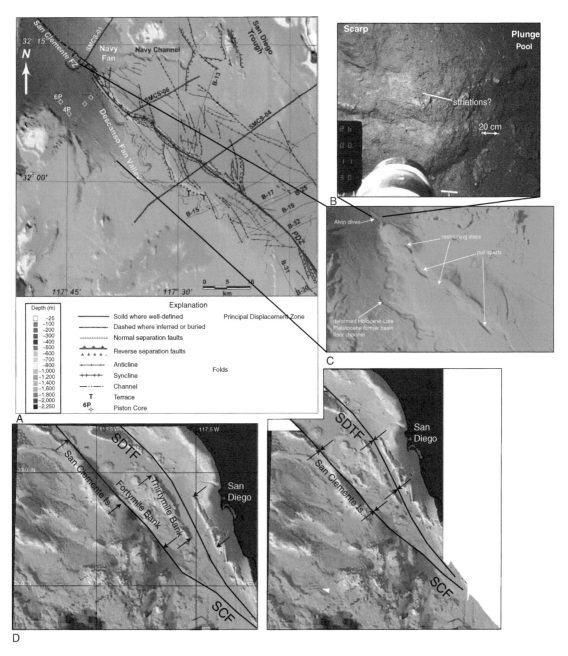

Figure 2B.13: (A) Bathymetric compilation map of the southern California Borderland showing the San Clemente Fault and the San Diego Trough Fault (SDTF). (B) Photograph from the DSV Alvin of fault scarp along the San Clemente Fault crossing the Navy fan (location in B). Sub-horizontal lineations may be slickensides consistent with strike slip motion. The scarp is composed of mud and layers of shells associated with ancient benthic communities at former cold seeps. Holocene scarp 0.3–1.5 m in height appears to be a single event scarp, indicated by the lack of multiple slope breaks and uniform "weathering" and bioturbation. The lightly bioturbated fresh scarp offsets Holocene and late Pleistocene strata, indicating a Holocene event that likely had a magnitude greater than 6 (Goldfinger *et al.*, 2000). Photo by C. Goldfinger, from Legg *et al.* (2007). (C) View of a restraining bend along the San Clemente Fault south of Navy Fan showing multiple pull apart and restraining bend features, superimposed on the larger uplift which itself is a restraining bend uplift due to a 5° strike change in the San Clemente Fault visible in (A). Location shown in (A). Channel at left is presently on the flank of the uplift, reflecting recent growth of this feature. (D) Retrodeformed San Clemente and SDTFs using morphologic and geologic piercing lines (Goldfinger *et al.*, 2000). San Clemente Fault (SCF) has a minimum horizontal separation of 50 km based on four piercing points (two are shown). The SDTF horizontal separation is 32 km, with 15 km extension as well (extension retrodeformation partially shown here to illustrate fit of offset features. (A and C) Reprinted with permission of the Royal Society of London.

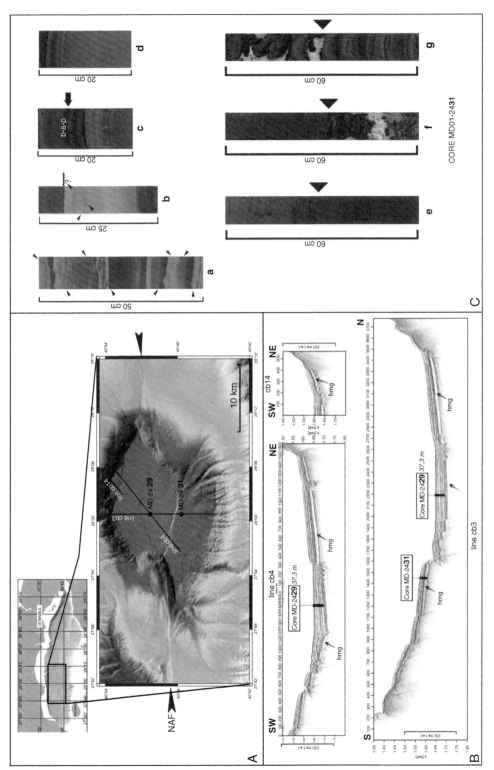

Figure 2B.14: (A) Detailed morpho-bathymetry of the Marmara Sea's Central Basin, location of giant piston cores and seismic profiles. (B) 3.5 kHz profiles across the Central Basin, showing active faults and evidence for a "homogenite." (C) Close-ups of selected portions of core MD01-2431. (a: conjugate microfractures; b: microfracturing with possible sealing by coeval turbidite arrival; c: possible *in situ* liquefaction, evidenced by ball-and-pillow –b-a-p– structure; d–g: details of the pre-Late Glacial event. In the continuous core section from 10.50 to 12.00 m X-ray scanning shows a constant orientation of microfractures). From Beck *et al.* (2007). Reprinted with permission from Elsevier.

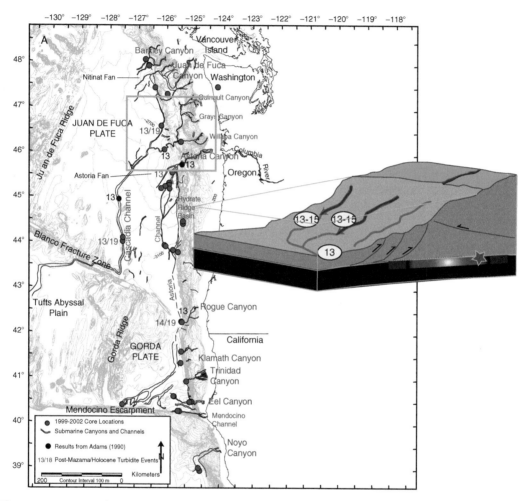

Figure 2B.18: Synchroneity test at a channel confluence as applied where Washington channels merge into the Cascadia Deep Sea Channel, indicated by box. The number of events downstream should be the sum of events in the tributaries, unless the turbidity currents were triggered simultaneously. Remarkable similarity of records in northern Cascadia supports the initial conclusion of Adams (1990) that these events are likely of earthquake origin. Modified after Goldfinger *et al*. (2009).

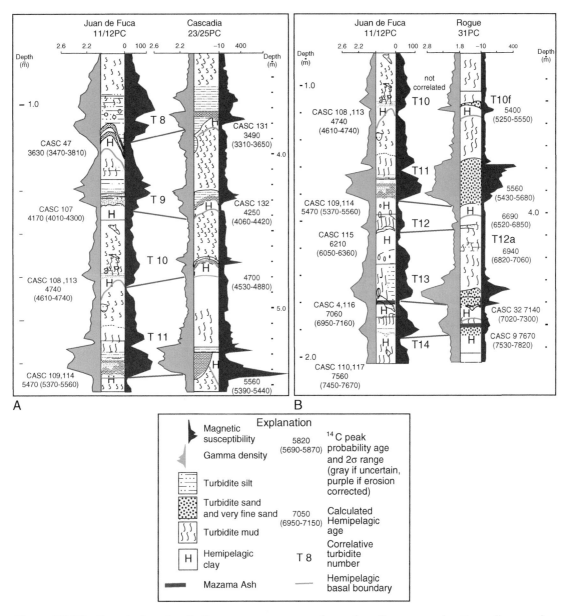

Figure 2B.19: Correlation details from two representative pairs of cores on the Cascadia margin.
(A) Events 8–11 in cores from Juan de Fuca Channel (left) and Cascadia Channel (right). Left traces
are raw gamma density, right traces are magnetic susceptibility. Lithologic logs are also shown.
Note correspondence of size, spacing, number of peaks, and trends of physical property traces
between these cores. (B) Similarly displays events T10–T14 in Juan de Fuca Channel (left) and
T10d–T14 in Rogue Channel (right). (A) Cores are part of the same channel system, distance along
channel = 475 km. (B) Cores are in channels that do not meet, separation distance = 500 km.
Note that correlation of longer sections and ¹⁴C data show that T10f and T10 do not correlate
in (B). Similarly, Mazama ash appears in T14, not T13 in Rogue apron, see text for discussion.
Modified after Goldfinger *et al.* (2008).

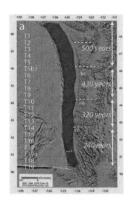

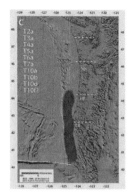
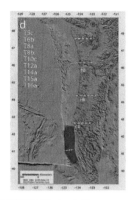

A

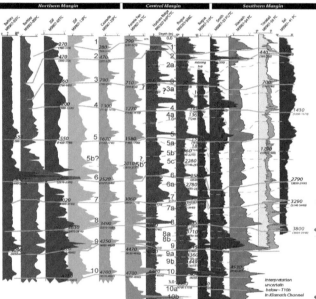
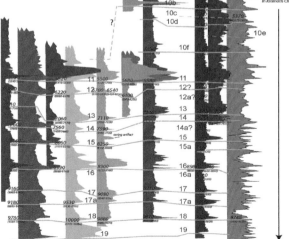

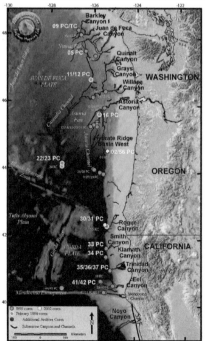

B

Figure 2B.20: (A) Holocene rupture lengths of Cascadia great earthquakes from marine and onshore paleoseismology. Four panels showing rupture modes inferred from turbidite correlation, supported by onshore radiocarbon data. (a) Full or nearly full rupture, represented at most sites by 20 turbidites, though with greater uncertainty in southern extent (we include Pleistocene T19 in the figure, but not in the statistics). (b) Mid-Southern rupture, represented by two (1?) events. (c) Southern rupture from central Oregon southward represented by 9 (10?) events. (d) Southern Oregon/northern California events, represented by eight events. Southern rupture limits vary with each event, and many events older than ~5000 years are limited by lack of core older data. Dashed white line offshore indicates reduced confidence in correlations south of Trinidad Canyon. Recurrence intervals for each segment shown in left panel. Each segment includes all full margin events, plus those exclusive to that segment. Rupture terminations are approximately located at three forearc structural uplifts, Nehalem Bank (NB), Heceta Bank (HB), and Coquille Bank (CB). Paleoseismic segmentation shown is also compatible with latitudinal boundaries of Episodic Tremor and Slip (ETS) events proposed for the downdip subduction interface (Brudzinski et al., 2007). These boundaries are shown by white-dashed lines. A northern segment proposed from ETS data at ~48N does not appear to have a paleoseismic equivalent. (B) Correlation plot of Holocene marine turbidite records and ^{14}C ages along the Cascadia margin from Barkley Channel to Eel Channel. All cores are vertically scaled to match Rogue core 31PC which is at true scale. Turbidite ages are shown using probability peaks and averaged where multiple ages at one site are available. Turbidites linked by stratigraphic correlations are shown by connecting lines. Full margin events correlated by using stratigraphy and ^{14}C are shown thicker, local southern Cascadia events are thinner and dashed. Modified after Goldfinger et al. (2009).

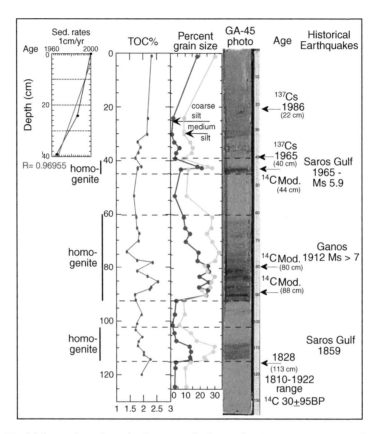

Figure 2B.21: Turbidite paleoseismologic example from the Marmara Sea. Grain size variability ranging from fine-sand to fine-silt and an increase in the total organic carbon (TOC %) of the sediments permit resolution of three homogenite (turbidite) deposits (40–44, 60–92, 100–112 cm). The homogenites are initiated by a sharp basal contact overlaid by multiple sand and silt-size laminae that fine upward to a thick wedge of homogenous fine-grained silt. The homogenites are separated by thin beds of clay (5–10 cm). Short-lived radioisotopes and radiocarbon chronology permit constructing an age model for correlation of the homogenites to the historical record of earthquakes: the large 1912 Ganos event $M_s > 7$ that lead to the deposition of a 30 cm thick homogenite and two smaller events that occurred in the Gulf of Saros. Sedimentation rates of 1 cm/yr were calculated for the upper 40 cm of the core that is apparently undisturbed. From McHugh *et al.* (2006). Reprinted by permission of Elsevier.

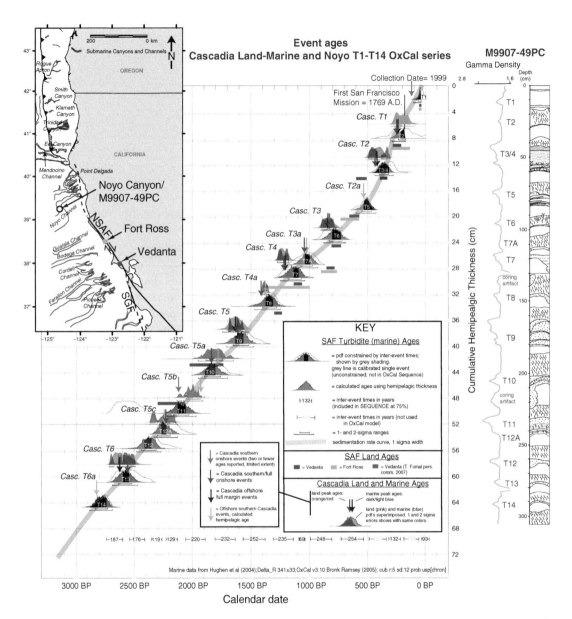

Figure 2B.22: OxCal age model for the youngest 15 events in the NSAF offshore system, and comparison to onshore NSAF ages. Cascadia OxCal PDFs are shown in blue, with lighter blue used where only Hemipelagic ages are available. Land ages from OxCal combines are shown in red. Cascadia mean event ages are also shown with blue arrows for well-dated turbidite events, Purple arrows for hemipelagic age estimates, and light red arrows for onshore paleoseismic events. See text for discussion and tables for data used and criteria, and discussion of temporal relationships. Inter-event times based on hemipelagic sediment thickness (represented by gray segments of NSAF PDFs) were used to constrain original ^{14}C calendar age distributions (gray traces) using the SEQUENCE option in OxCal. Inter-event times were estimated by converting hemipelagic sediment thickness between each pair of events to time using the sedimentation rate. Events dated more than once were combined in OxCal prior to calibration if results were in agreement; if not in agreement, the younger radiocarbon age was used in the final model. Five ages are calculated from sedimentation rates where not enough forams were present for ^{14}C dating. The resulting probability distributions (filled black, grey for undated events) are mostly in good agreement with land ages from Fort Ross except for T3–4 and T7a (green lines; Kelson *et al.*, 2006) Vedanta (red lines; Zhang *et al.*, 2006) Bolinas Lagoon and Bodega Bay (Purple lines, Knudsen *et al.*, 2002), and Point Arena (light blue lines, Prentice *et al.*, 2000). Additional Vedanta event is also shown (T. Fumal personal communication 2007). See inset for geographic locations. Figure from Goldfinger *et al.* (2008). Reprinted by permission of the Bulletin of the Seismological Society of America.

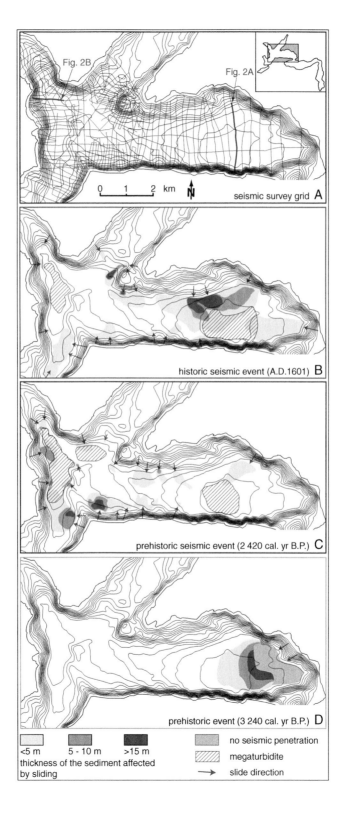

Fig. 2B

Fig. 2A

0   1   2   km

N

seismic survey grid A

historic seismic event (A.D.1601) B

prehistoric seismic event (2 420 cal. yr B.P.) C

prehistoric event (3 240 cal. yr B.P.) D

<5 m   5 - 10 m   >15 m

thickness of the sediment affected
by sliding

no seismic penetration

megaturbidite

slide direction

N

S

Quaternary surface

Quaternary surface

ROCK CREEK

Figure 3.9: Photograph of the trace of the Rock Creek fault west of Kemmerer, Wyoming (Class A fault, USGS ID No. 729). This scarp does not lie at the foot of the range front, but stays high on an erosional bedrock hillslope. Thus, the fault does not displace the Quaternary surfaces in the foreground, but only colluvium on steep slopes, except where the scarp crosses canyons mouths such as the one at center. This fault scarp occupies an anomalous topographic position (relative to most Basin and Range normal faults) over its entire 41 km length because it represents an immature fault formed by Neogene extension and tectonic inversion of a preexisting Mesozoic thrust fault. Photo by J. P. McCalpin (1991).

Figure 2B.23: Slide deposits in Lake Lucerne, Switzerland mapped with high-resolution reflection profiling. Slide deposits related to specific horizons. (A) Grid of 3.5 kHz seismic profiles acquired for this study. (B–D) Distribution and thickness of slide bodies corresponding to three event horizons identified in the reflection profiles. Hachured areas mark extent of megaturbidites directly overlying slide bodies. Bathymetric contour interval is 10 m. From Schnellmann *et al.* (2002) their Figure 3. Reprinted with permission of the Geological Society of America.

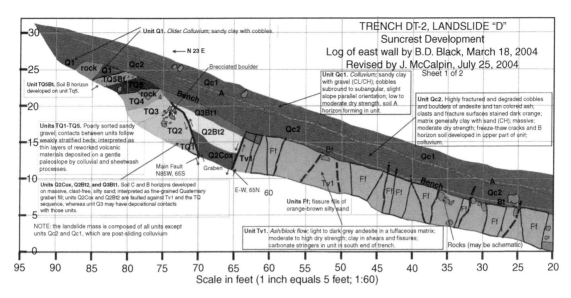

Figure 3.61: Log of a trench through a landslide headscarp, showing the similarities of landslide normal faults to coseismic normal faults. Note the graben at Sta. 65–70, the unusual back-tilted footwall, and extensive fracturing and fissuring of the hanging wall. Based on this exposure alone, it would be difficult to distinguish this fault zone from a seismogenic fault. From McCalpin (2005b).

Figure 4.3: (B) Small graben with 3-m scarp, formed by dike intrusion associated with the 1983–1990 eruptive episodes of Puu Oo (active vent in background), east rift zone of Kilauea volcano, Hawaii. From Smith *et al.* (1996); reprinted with permission of the American Geophysical Union.

Figure 4.6: Two contrasting styles of dike-induced faulting. (B) Southwest view of the Almannagja normal-fault scarp in Thingvellir National Park, Iceland, where it broke through the surface in basalt along the upper hinge of the monocline resulting in both vertical throw and horizontal opening. The limb of monocline (on left) occurs in the hanging wall of the fault. Sediments and basalt blocks fill the chasm along the fault. Hengill volcano is visible in the background. Reprinted with permission of Simon Kattenhorn.

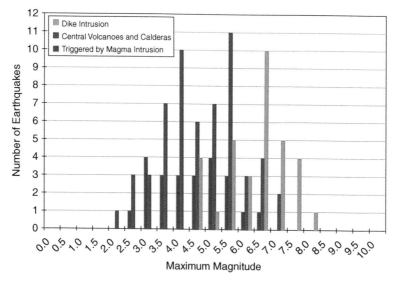

Figure 4.8: Histogram showing the distributions of maximum magnitudes of earthquakes associated with dike intrusion, occurring at central volcanoes and calderas, and triggered by magma intrusion (data are from Tables 4.2–4.4, respectively). The distributions show maximum magnitudes of earthquakes associated with dike intrusion are less than those at central volcanoes and calderas and triggered by magma intrusion.